ANALYTIC GEOMETRY
and
CALCULUS

By WILLIAM R. LONGLEY
Yale University

PERCEY F. SMITH
Yale University

and WALLACE A. WILSON
Late Professor of Mathematics, Yale University

BLAISDELL PUBLISHING COMPANY
A Division of Ginn and Company
NEW YORK • TORONTO • LONDON

PREFACE

The increased use of mathematics in science and engineering courses makes it desirable for the student to become acquainted with both the differential and integral calculus as early as possible. When considered as an aid to the study of these allied subjects, the introduction of the calculus cannot well be deferred until a course in analytic geometry has been completed. Fortunately, we have ample experience which shows that there is pedagogical value in a proper combination of the two subjects.

This text covers the material usually taught in introductory courses in plane and solid analytic geometry and the differential and integral calculus, including a chapter on differential equations. The elementary technique of integration, with significant applications, is introduced as early as experience has shown that this can be done profitably.

The presentation of the material, particularly in the early part of the book, is somewhat more rigorous than in earlier texts by these authors. Extensive trial by Professor Wilson has indicated that the rigor of a course in functions of a real variable is not entirely successful with college freshmen. However, some acquaintance with this type of reasoning has a distinct value and an attempt has been made to use such methods as far as possible without causing the subject to lose its meaning for a beginner.

All material in this text has, from time to time, been thoroughly tested in the classroom. While all three authors have contributed to this work, the final arrangement is the responsibility of the undersigned.

WILLIAM R. LONGLEY

CONTENTS

ANALYTIC GEOMETRY
AND CALCULUS

GREEK ALPHABET

Letters		Names	Letters		Names	Letters		Names
A	α	Alpha	I	ι	Iota	P	ρ	Rho
B	β	Beta	K	κ	Kappa	Σ	σ	Sigma
Γ	γ	Gamma	Λ	λ	Lambda	T	τ	Tau
Δ	δ	Delta	M	μ	Mu	Υ	υ	Upsilon
E	ϵ	Epsilon	N	ν	Nu	Φ	ϕ	Phi
Z	ζ	Zeta	Ξ	ξ	Xi	X	χ	Chi
H	η	Eta	O	o	Omicron	Ψ	ψ	Psi
Θ	θ	Theta	Π	π	Pi	Ω	ω	Omega

CHAPTER I

CARTESIAN COÖRDINATES. THE STRAIGHT LINE

1. Introduction. A survey of the problems which can be solved by elementary mathematics (algebra, geometry, and trigonometry) shows that although they are large in number they form a rather restricted class. For example, the problems of algebra consist mainly in finding one or more unknown quantities by the solution of equations; and in plane geometry and trigonometry we are confined to the study of figures bounded by straight lines and circles.

In order to solve more difficult problems we may proceed in two ways: (1) by using new processes of reasoning and calculation; (2) by combining our algebra and geometry so that we can use them together to greater advantage. The new processes above referred to, and to be described later, belong to the branch of mathematics known as calculus. We shall begin with the second way of proceeding, which, when carried out to its fullest extent, forms the subject of analytic geometry.

2. Real numbers. This book deals with real numbers. The only exception occurs when the roots of a quadratic equation may be imaginary.

Real numbers fall into two classes: *rational* and *irrational*. The rational numbers are the number 0, the integers, and the numbers which can be written as the quotients of two integers, as $\frac{2}{3}$, $-\frac{5}{2}$, etc. Other numbers are called irrational; for example, $\sqrt{2}$ and π are irrational numbers.

In performing the algebraic operations on real numbers it is to be emphasized that *division by zero is impossible* and that *there is no such number as infinity*. Such an expression as "$\tan 90° = \infty$" is merely a convenient abbreviation for the statement that "as x approaches $90°$, the value of $\tan x$ increases without bound."

An example of the confusion introduced by trying to use ∞ as a number is afforded by the identity $\tan (180° - x) = -\tan x$. Setting $x = 90°$ gives $\tan 90° = -\tan 90°$, whence $2 \tan 90° = 0$, which is absurd.

3. Inequalities. The notation $a < b$ means that a is less than b, while $a > b$ means that a is greater than b. If a is either less than or equal to b, we write $a \leqq b$. The following laws of inequalities from elementary algebra are given here for convenience.

3

LAW I.   If $a < b$ and $b < c$, then $a < c$.

LAW II.   If $a < b$, then $-a > -b$.

LAW III. If $a < b$, then $a + c < b + c$ and $a - c < b - c$.

LAW IV.   If $a < b$, then $c - a > c - b$.

LAW V.   If $a < b$ and c is positive, then $ac < bc$.

LAW VI.   If $a < b$ and both are positive, then $\dfrac{1}{a} > \dfrac{1}{b}$.

Other relations can be derived from these. For example, (III) shows that in inequalities we can transpose terms by changing the signs as in the case of equations. Thus if $x + y < z$, we obtain $x < z - y$ by subtracting y from each side of the inequality.

The notation $a < x < b$ means that each value of x is *both* greater than a and less than b; this relation is impossible unless $a < b$. The statement that "x is between a and b" means that $a < x < b$ if $a < b$, or that $a > x > b$ if $a > b$. The statement that "x is between a and b, inclusive" is sometimes used, with the meaning that $a \leqq x \leqq b$ or $a \geqq x \geqq b$. There is no briefer way of stating the fact that "either $x < a$ or $x > b$. Such a statement as $7 < x < -3$, intended to mean that the value of x is either greater than 7 or less than -3, is absurd, as this would require 7 to be less than -3.

4. Absolute values. The absolute value of a number a, as distinguished from its algebraic value, is denoted by $|a|$, and defined as follows.

$$\text{If } a \geqq 0, \text{ then } |a| = a;$$
$$\text{if } a < 0, \text{ then } |a| = -a.$$

Thus $|7| = 7$; $|-7| = -(-7) = 7$. If $x^2 = 4$, then $x = 2$ or $x = -2$, but $|x| = 2$, only.

The following important relations are readily verified.

$$|a + b| \leqq |a| + |b|; \quad |ab| = |a| \cdot |b|.$$

Another important fact is that the inequality $|a - b| < c$ is equivalent to each of the following.

$$-c < a - b < c, \qquad -c < b - a < c,$$
$$b - c < a < b + c, \qquad a - c < b < a + c.$$

These relations are proved by starting with the fact that $|a - b|$ equals $a - b$ if $a > b$ and equals $b - a$ if $a < b$.

5. Location of points on a line. The length of a segment of a straight line can be expressed in terms of a previously chosen unit by a real number, which is rational if the length of the segment and the unit are commensurable and irrational if they are not. For the sake of brevity we call the number itself the length of the segment. For example, if the unit is 1 inch and the length of the segment AB is 4 inches, we shall say indifferently that "the length of AB is 4" or that "the distance between A and B is 4" or simply that "$AB = 4$." Also, if A is a fixed point on a straight line and a is any real number, there are precisely two points B, one on each side of A, for which the distance between A and B is a.

If a fixed point O on a straight line is taken as a point of reference, the position of any other point A on the line is determined if we know the distance between O and A and the side of O on which A lies. The latter requirement is conveniently met by calling the distance from O to A positive if A is on one side of O and negative if it is on the other side.

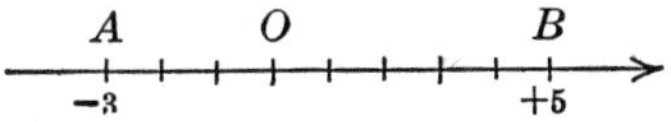

Fig. 5

In Fig. 5 distances to the right are taken as positive and those to the left as negative. Thus $OB = + 5$ and $OA = - 3$.

We have then a complete one-to-one correspondence between the points of the line and the real numbers. This means that to each point on the line there corresponds one and only one real number and to each real number there corresponds one and only one point on the line. This correspondence is so useful and so constantly employed that the points and the numbers locating them are identified with each other. Thus we may speak of A (**Fig. 5**) as "the point $- 3$" or speak of the "interval AB" when we really mean the set of numbers corresponding to points of the segment AB.

6. Rectangular coördinates. In order to describe the position of a point in a plane, two mutually perpendicular reference lines $X'X$ and $Y'Y$ are chosen. These are called *coördinate* axes. Their point of intersection O is called the *origin*. The line $X'X$ is called the *x-axis* and $Y'Y$ is called the *y-axis*.

The position of any point P in the plane is then described by giving its distance NP from the *y*-axis and its distance MP from the *x*-axis (Fig. 6.1, p. 6). In order to distinguish between points on opposite sides of the axes, the distance from the *y*-axis to a point P is regarded as positive if P is on the right of the *y*-axis and negative if it is on the left; the distance from the *x*-axis to P is regarded as positive if P is above the *x*-axis and negative if it is below. These distances or, to speak more precisely, the numbers representing them, are called *coördinates*.

DEFINITIONS. *The distance from the y-axis to a point is called the **abscissa**, or x-coördinate, of the point. The distance from the x-axis to a point is called the **ordinate**, or y-coördinate, of the point.*

Since parallel lines are everywhere equidistant, we can also say that the abscissa of a point P is the distance from the y-axis to the line which contains P and is parallel to the y-axis. Likewise, the ordinate of P is the distance from the x-axis to the line which contains P and is parallel to the x-axis.

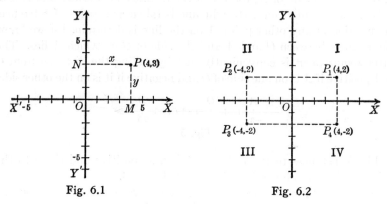

Fig. 6.1   Fig. 6.2

From the preceding discussion and the theorems of geometry, it is easy to show that, after a pair of axes and a unit of length have been chosen, every point in the plane determines uniquely a pair of real numbers and, conversely, every pair of real numbers determines one and only one point in the plane.

When a point is named by stating its coördinates, we write them in parentheses, putting the abscissa first. Thus the point which has the abscissa -4 and the ordinate 2 is designated by $(-4, 2)$. If this point is also designated by some letter (Fig. 6.2), as P_2, we say that $P_2 = (-4, 2)$ or speak of the point $P_2(-4, 2)$.

If the point is variable or unknown, the abscissa is usually denoted by x and the ordinate by y. Thus $P(x, y)$ would be a variable or unknown point. This notation is so common that the letters x and y have become almost synonymous with the words "abscissa" and "ordinate," respectively. If the position of a point is fixed, but arbitrary, subscripts are used. For example, if P_1P_2 denoted a fixed segment, we might denote the end points by $P_1(x_1, y_1)$ and $P_2(x_2, y_2)$ and let $P(x, y)$ be any point on this segment.

We have already used the notation AB to stand for the distance from A to B as well as for the segment AB itself. Thus we may say that the abscissa of the point P (Fig. 6.1) is NP and write $x = NP = 4$. Note that the abscissa of P should be read as NP or OM, but not as PN or MO, and the ordinate as MP or ON, but not as PM or NO. See Art. 9.

The four quadrants into which a pair of rectangular axes divides the plane are numbered as in trigonometry. If a point lies in the first quadrant, both coördinates are positive; if it lies in the second quadrant, its abscissa is negative and its ordinate is positive; if it lies in the third quadrant, both coördinates are negative; if it lies in the fourth quadrant, its abscissa is positive and its ordinate is negative.

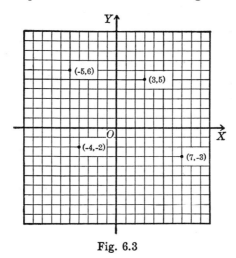

Fig. 6.3

7. Plotting. To "plot a point" given by rectangular coördinates is to mark it in its proper position, corresponding to the given coördinates. Thus, to plot the point (3, 5) means to mark the point 3 units to the right of the y-axis and 5 units above the x-axis. In practice we measure 3 units to the right along the x-axis and then 5 units upward from the x-axis. Coördinate, or plotting, paper is made by ruling off equidistant lines parallel to the axes so that distances may readily be measured Fig. 6.3 shows several points plotted on coördinate paper.

8. Symmetrical points. Two points, A and B, are said to be symmetrical with respect to a line l if this line is the perpendicular bisector of the segment AB. The line l is called the *axis of symmetry*. In Fig. 6.2 the points P_1 and P_2 are symmetrical with respect to the y-axis, and the points P_1 and P_4 are symmetrical with respect to the x-axis. It is evident that if l is the axis of symmetry of A and B, a rotation of the plane through 180° about l will interchange the points A and B.

Two points, A and B, are said to be symmetrical with respect to a point O, called the *center of symmetry*, if O is the mid-point of the segment AB. In the figure referred to, P_1 and P_3 (also P_2 and P_4) are symmetrical with respect to the origin.

PROBLEMS

1. State all values of x for which the following inequalities are true.

a. $x^2 - 6 > 3$.   **b.** $16 - x^2 > 0$.

c. $(x - 4)(x + 5) > 0$.   **d.** $x^2 + x > 12$.

e. $x^2 < x + 6$.   **f.** $(x - 2)^2(4 - x) < 0$.

2. Find the roots of the following equations.

a. $|x - 3| = 1$.   **b.** $|x|^2 - 5|x| - 6 = 0$.

c. $|x|^2 + 5|x| + 6 = 0$.   **d.** $|x|^2 - 5|x| + 6 = 0$.

3. For what values of x are the following inequalities true?

a. $|x - 3| < 1$.   **b.** $|x + 2| < 0.5$.

c. $|4 - x| < 2$.   **d.** $|a^2 - x^2| < b^2$.

4. Plot accurately the points $(5, 3)$, $(5, -3)$, $(-4, 2)$, $(-4, -2)$, $(6, 0)$, $(0, -7)$.

5. Plot as accurately as you can the points $(3.2, -4.6)$, $(-1.7, 5.3)$, $(\sqrt{3}, -\sqrt{3})$, $(-\sqrt{5}, \sqrt{10})$.

6. Let A be the point $(6, 8)$. What are the coördinates of the point symmetrical to A with respect to the x-axis? the y-axis? the origin?

7. Two vertices of a square are $(-3, 0)$ and $(5, 0)$, and the other vertices are above the x-axis. What are the coördinates of the other vertices? What is the length of a diagonal?

8. Three vertices of a rectangle are $(-1, -3)$, $(5, 5)$, and $(5, -3)$. What are the coördinates of the fourth vertex? What is the area of the rectangle? What is the length of a diagonal?

9. Draw the triangles which have the following points as vertices and find the area of each.

a. $(-3, 0)$, $(5, 0)$, $(4, 4)$.   **b.** $(0, 8)$, $(6, 2)$, $(0, -2)$.

c. $(0, -2)$, $(4, 3)$, $(6, -2)$.   **d.** $(-1, -1)$, $(5, 0)$, $(-1, 7)$.

10. Three vertices of a parallelogram are $(0, 0)$, $(a, 0)$, and (b, c). Find the coördinates of the fourth vertex. Prove your answer.

11. A square whose side has the length $2a$ has its center at the origin. What will be the coördinates of its vertices if **(a)** the sides are parallel to the axes? **(b)** the diagonals coincide with the axes?

12. An equilateral triangle whose side has the length a has its base on the x-axis. What are the coördinates of the vertices of the triangle if **(a)** the center of the base is at the origin? **(b)** one vertex is at the origin?

13. A regular hexagon whose side has the length a has its center at the origin and one diagonal along the x-axis. Find the coördinates of the vertices.

14. What can be said about the position of the point $P(x, y)$ under the following conditions?

a. $|x| < 2.$ **b.** $|y| > 3.$

c. $|x - 2| \leq 2.$ **d.** $|y - 4| \leq 1.$

e. $|x| < 4$ and $|y| < 4.$ **f.** $0 \leq x \leq 5$ and $0 \leq y \leq 6.$

g. $|3 - x| \leq 3$ and $|3 - y| \leq 3.$ **h.** $|x - 5| > 2$ and $|y| > 2.$

9. Directed lines. The great generality of analytic methods and for-mulas is due primarily to the use of *directed lines.* These are lines on which one direction is regarded as positive and the other as negative. The positive direction is often denoted by an arrowhead. If A and B are points on a directed line and the direction from A to B is positive, the distance from A to B and the length of the segment AB are positive; and in this case the distance from B to A and the length of the segment BA are negative. Thus, in Fig. 9.1, AB is positive and BA is negative.

If the direction of reading a segment of a directed line is changed, the sign must be changed; that is,

$$AB = -BA \quad \text{and} \quad BA = -AB.$$

Fig. 9.1

In adding segments of directed lines it is understood that the addi-tion is to be performed *algebraically.* For example, in Fig. 9.2,

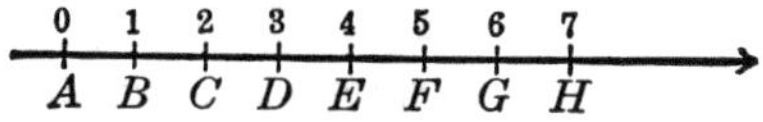

Fig. 9.2

$$AH = 7, \quad HA = -7, \quad DF = 2, \quad GD = -3, \text{ etc.,}$$
$$AC + CF = 2 + 3 = 5 = AF,$$
$$AC + CB = 2 + (-1) = 1 = AB,$$
$$GC + CE = -4 + 2 = -2 = GE.$$

The last three equations lead at once to the following theorem.

Theorem. *If C is any point on the directed line passing through A and B, then* $AC + CB = AB.$

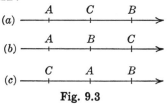

Fig. 9.3

Proof. When AB is positive there are three cases, as indicated by Fig. 9.3. These cases are as follows.

CASE I. If C is between A and B, the segments AC and CB are positive and we have at once

$$AB = AC + CB.$$

CASE II. If B is between A and C, the segments AC and BC are positive and so $AC = AB + BC$. Hence

$$AB = AC - BC = AC + CB.$$

CASE III. If A is between C and B, the segments CB and CA are positive and so $CB = CA + AB$. Hence

$$AB = CB - CA = AC + CB.$$

Exercise. Write out the proofs for the three cases when AB is negative.

This theorem illustrates the advantage of the idea of directed lines. If one goes from a point A to a point C and then from C to a point B, it requires no argument to show that the *net* result of the two motions is the same as going from A directly to B. The importance of the theorem lies in the fact that *one* equation (that is, $AC + CB = AB$) covers *all possible* cases. This equation is the same whether all the three motions are forward, or all backward, or one forward and two backward, etc.

Rule for Directed Lines in Analytic Geometry. *In using rectangular coördinates, unless the contrary is specifically stated, we shall assume that* (1) *the positive direction along the x-axis and along all lines parallel to the x-axis is from left to right;* (2) *the positive direction along all other lines is upward.*

The rules for the signs of coördinates given in Art. 6 agree with this, and hence an abscissa or an ordinate is simply a special case of length on a directed line.

10. Horizontal and vertical distances from one point to another. Let P_1 and P_2 be any two points, and let a line be drawn through P_1 parallel to the x-axis and meeting at Q the perpendicular from P_2 to the

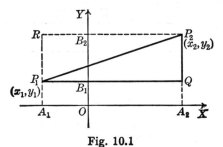

Fig. 10.1

x-axis. The directed line P_1Q is called the *horizontal distance* (or distance parallel to the x-axis) from P_1 to P_2. The directed line QP_2 is called the *vertical distance* (or distance parallel to the y-axis) from P_1 to P_2.

The horizontal distance from P_1 to P_2 is evidently the same as the projection of P_1P_2 on the x-axis or any line parallel to the x-axis. Thus, in the figure, $P_1Q = A_1A_2 = RP_2$. The vertical distance is evidently the same as the projection of P_1P_2 on the y-axis or any line parallel to the y-axis. Thus, in the figure, $QP_2 = B_1B_2 = P_1R$.

Theorem. *The horizontal distance from $P_1(x_1, y_1)$ to $P_2(x_2, y_2)$ is $x_2 - x_1$; the vertical distance from P_1 to P_2 is $y_2 - y_1$.*

Proof. With reference to Fig. 10.1, the horizontal distance is P_1Q, which is equal to A_1A_2. But the theorem of Art. 9, on adding segments, gives
$$A_1A_2 = A_1O + OA_2.$$

Now
$$x_1 = OA_1,$$

and
$$x_2 = OA_2;$$

hence, by substitution, $A_1A_2 = x_2 - x_1.$

In like manner we find that
$$QP_2 = B_1B_2 = y_2 - y_1.$$

The proofs for the special cases in which the line P_1P_2 is parallel to one of the coördinate axes follow in the same way.

Since each point can be in four different quadrants, it might seem that different proofs for the various cases would be necessary. This is not true. The proof as given can be used verbatim for all positions of P_1 and P_2. This happy result is due to the all-important theorem of Art. 9.

The truth of the theorem should be tested by marking several positions of the points P_1 and P_2. For example, in Fig. 10.2,

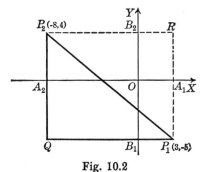

Fig. 10.2

$$x_1 = 3, \quad x_2 = -8, \quad y_1 = -5, \quad y_2 = 4.$$

Here $P_1Q = A_1A_2 = (-8) - (3) = -11,$

and $QP_2 = B_1B_2 = (4) - (-5) = 9.$

11. Distance between two points. Theorem. *The distance between two points $P_1(x_1, y_1)$ and $P_2(x_2, y_2)$ is given by the formula*

(I) $$d = \sqrt{(x_2 - x_1)^2 + (y_2 - y_1)^2}.$$

Two figures are drawn in connection with the proof, which may be used without change of wording for *any* positions of the points $P_1(x_1, y_1)$ and $P_2(x_2, y_2)$, unless P_1P_2 is parallel to an axis.

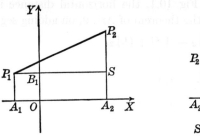 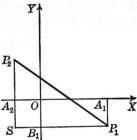

Fig. 11

Proof. Draw a line through P_1 parallel to OX, and one through P_2 parallel to OY; let these lines meet at S.

Now $$d = P_1P_2 = \sqrt{\overline{P_1S}^2 + \overline{SP_2}^2}.$$

But P_1S and SP_2 are the horizontal and vertical distances, respectively, from P_1 to P_2. Hence, by the theorem of Art. 10,

$$P_1S = x_2 - x_1 \quad \text{and} \quad SP_2 = y_2 - y_1.$$

Substituting these above, we have

$$d = \sqrt{(x_2 - x_1)^2 + (y_2 - y_1)^2}.$$

When the segment P_1P_2 is parallel to one of the coördinate axes, the formula remains valid. For example, if P_1P_2 is parallel to the x-axis, then $y_2 = y_1$ and the formula becomes

$$d = \sqrt{(x_2 - x_1)^2} = |\, x_2 - x_1 \,|.$$

12. Point dividing a segment in a given ratio. Let P_1 and P_2 be two fixed points on a directed line. Any third point P divides the segment

P_1P_2 into two segments P_1P and PP_2. The ratio of these two segments, denoted by r, is called the *ratio of division*. By definition,

$$r = \frac{P_1P}{PP_2}.$$

The ratio r is usually an integer or a common fraction, and its value is determined as in the following examples.

1. Let P be the mid-point of P_1P_2.

Then $\qquad\qquad\qquad P_1P = PP_2$

and $\qquad\qquad\qquad r = 1.$

2. Let P cut off one third of the segment P_1P_2.

Then $\qquad\qquad\qquad P_1P = \tfrac{1}{2} PP_2$

and $\qquad\qquad\qquad r = \tfrac{1}{2}.$

3. Let P be the point reached by extending P_1P_2 by half of its length.

Then $\qquad\qquad\qquad P_1P = - 3 PP_2$

and $\qquad\qquad\qquad r = - 3.$ $\qquad$ (Why is r negative?)

If the point of division P lies between P_1 and P_2, the two segments P_1P and PP_2 have the same sign, and r is positive. In this case P_1P_2 is said to be divided *internally*. If P does not lie between P_1 and P_2, the two segments have opposite signs, and r is negative. In this case P_1P_2 is said to be divided *externally*.

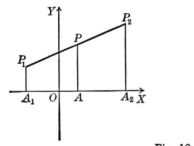 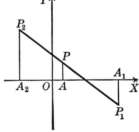

Fig. 12.1

Theorem. *If the segment P_1P_2 of the directed line passing through $P_1(x_1, y_1)$ and $P_2(x_2, y_2)$ is divided in the ratio r by the point $P(x, y)$, then*

(II) $\qquad\qquad x = \dfrac{x_1 + rx_2}{1 + r}, \qquad y = \dfrac{y_1 + ry_2}{1 + r}.$

Proof. Draw the ordinates A_1P_1, AP, and A_2P_2.

Since the segments of two transversals comprehended between parallel lines are proportional, we have

$$\frac{P_1P}{PP_2} = \frac{A_1A}{AA_2}.$$ (1)

But $\qquad A_1A = x - x_1 \quad$ and $\quad AA_2 = x_2 - x, \qquad$ by Art. 10

while $\qquad\qquad \dfrac{P_1P}{PP_2} = r. \qquad\qquad$ By definition

Hence (1) becomes, by substitution,

$$\frac{x - x_1}{x_2 - x} = r.$$

Solving for x, we have $\qquad x = \dfrac{x_1 + rx_2}{1 + r}.$

In a similar way it can be shown that

$$y = \frac{y_1 + ry_2}{1 + r}.$$

If $r = 1$, P is the mid-point of the segment P_1P_2. For this special case we have the following theorem.

Theorem. *The coördinates of the point bisecting the segment joining the points $P_1(x_1, y_1)$ and $P_2(x_2, y_2)$ are given by the formulas*

(II a) $\qquad\qquad x = \dfrac{x_1 + x_2}{2}, \qquad y = \dfrac{y_1 + y_2}{2}.$

Example 1. Given the triangle $A(6, 0)$, $B(2, 4)$, $C(1, -1)$. (a) Find the length of the median drawn from C. (b) Find the point of intersection of the medians.

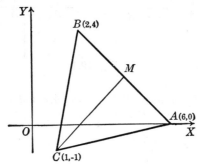

Fig. 12.2

Solution. (a) It is important that an accurate figure should be drawn, and that the coördinates of the points should be marked on the figure. As results are calculated they should, as far as possible, be marked on the figure, and all calculated results should be compared with the figure to see that they are reasonable.

The coördinates of M, the mid-point of AB, are given by the mid-point formulas

$$x = \tfrac{1}{2}(2 + 6) = 4, \qquad y = \tfrac{1}{2}(4 + 0) = 2.$$

The length CM is given by the distance formula

$$CM = \sqrt{(4 - 1)^2 + (2 + 1)^2} = \sqrt{18} = 4.24.$$

This result can be checked by measurement by using a strip cut from the coördinate paper.

(b) The medians intersect at a point E on the line CM such that $CE = \tfrac{2}{3} CM$. Hence E divides the line CM in the ratio

$$r = CE/EM = 2.$$

Applying (II), we have, for the coördinates of E,

$$x = \frac{1 + 2 \cdot 4}{1 + 2} = 3, \qquad y = \frac{-1 + 2 \cdot 2}{1 + 2} = 1.$$

The point of intersection of the medians is shown in mechanics to be the center of gravity of the triangle.

Example 2. Prove analytically that the lines joining the mid-points of adjacent sides of any rectangle form a rhombus.

Solution. Let $ABCD$ be the given rectangle, with $AB = a$ and $AD = b$. Letters a and b are chosen to represent the sides in order that the following proof may be valid for *any* rectangle. If numerical values were used for a and b, it would be shown that for *one particular* rectangle the lines joining the mid-points of adjacent sides form a rhombus. But this proves nothing about any other rectangle. Hence it is necessary to give a proof with letters which represent *any* (that is, *every possible*) rectangle.

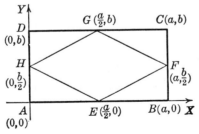

Fig. 12.3

To use analytic methods, coördinate axes must be introduced. Any set of axes could be used theoretically, but the most convenient will be those obtained by choosing the line AB for the x-axis and AD for the y-axis. The coördinates of the vertices will then be $A(0, 0)$, $B(a, 0)$, $C(a, b)$, and $D(0, b)$. The coördinates of the mid-points of the sides will be

$$E(a/2, 0), \quad F(a, b/2), \quad G(a/2, b), \quad H(0, b/2).$$

Application of the distance formula gives

$$EF = \sqrt{\left(a - \frac{a}{2}\right)^2 + \left(\frac{b}{2} - 0\right)^2} = \frac{1}{2}\sqrt{a^2 + b^2},$$

$$FG = \sqrt{\left(a - \frac{a}{2}\right)^2 + \left(\frac{b}{2} - b\right)^2} = \frac{1}{2}\sqrt{a^2 + b^2},$$

$$GH = \sqrt{\left(\frac{a}{2} - 0\right)^2 + \left(b - \frac{b}{2}\right)^2} = \frac{1}{2}\sqrt{a^2 + b^2},$$

$$HE = \sqrt{\left(\frac{a}{2} - 0\right)^2 + \left(0 - \frac{b}{2}\right)^2} = \frac{1}{2}\sqrt{a^2 + b^2}.$$

Since all four sides are equal, $EFGH$ is a rhombus.

PROBLEMS

1. Find the lengths of the sides of the following triangles.

   a. $(0, 1)$, $(5, 2)$, $(4, 6)$.   **b.** $(-3, 0)$, $(4, -2)$, $(2, 5)$.

   c. $(-3, -2)$, $(5, 1)$, $(-1, 4)$.   **d.** $(-4, -3)$, $(3, -3)$, $(-1, 5)$.

2. Show that the points $(-1, 6)$, $(2, 2)$, and $(3, 3)$ are the vertices of an isosceles triangle.

3. Show that the points $(2, -1)$, $(6, 1)$, and $(-2, 7)$ are the vertices of a right triangle. What is the area of the triangle?

4. (a) Show that the points $(0, 1)$, $(7, 0)$, $(5, 6)$, and $(12, 5)$ are the vertices of a parallelogram. **(b)** Show that one diagonal is twice as long as the other. **(c)** Show that the diagonals bisect each other.

5. Show that the points $(5, 0)$, $(8, 2)$, $(6, 5)$, and $(3, 3)$ are the vertices of a square. What is the area of the square?

6. Find the coördinates of the points which trisect the segment joining $A(-2, -4)$ and $B(4, 8)$.

7. (a) Show that the points $(4, 6)$, $(2, 2)$, and $(8, 4)$ are the vertices of an isosceles triangle. **(b)** Find the length of the perpendicular from the vertex $(4, 6)$ to the opposite side. **(c)** Find the area of the triangle.

8. Find the lengths of the medians and the center of gravity of the following triangles.

   a. $(0, 3)$, $(6, -3)$, $(12, 9)$.   **b.** $(6, -2)$, $(8, 6)$, $(3, 9)$.

9. Given the quadrilateral $(0, 1)$, $(3, 8)$, $(1, 10)$, $(-6, 6)$. Show that the lines joining the mid-points of opposite sides bisect each other.

10. If the mid-point of a segment is $(5, 3)$ and one end of the segment is $(9, -5)$, what are the coördinates of the other end?

11. If $A = (-3, -4)$ and $B = (2, 1)$, and AB is extended to C so that the length of AC is three times that of AB, find the coördinates of C.

12. Show analytically that the coördinates of the center of gravity of the triangle whose vertices are (x_1, y_1), (x_2, y_2), and (x_3, y_3) are $\frac{1}{3}(x_1 + x_2 + x_3)$ and $\frac{1}{3}(y_1 + y_2 + y_3)$.

13. Two vertices of a triangle are $(0, -4)$ and $(6, 0)$, and the medians intersect at $(2, 0)$. Find the third vertex of the triangle.

14. Prove analytically that the diagonals of any rectangle are equal.

15. Prove analytically that the middle point of the hypotenuse of any right triangle is equidistant from the three vertices.

16. Prove analytically that the diagonals of any parallelogram bisect each other.

17. Prove analytically that the area of any triangle is four times the area of the triangle formed by joining the mid-points of its sides.

HINT. Take two vertices on the x-axis and the third on the y-axis.

18. Prove analytically that the distance between the mid-points of the non-parallel sides of any trapezoid is equal to half the sum of the parallel sides.

19. Prove that the area of the triangle whose vertices are (x_1, y_1), (x_2, y_2), and (x_3, y_3) is given by the formula $\pm \frac{1}{2}(x_1y_2 + x_2y_3 + x_3y_1 - x_2y_1 - x_3y_2 - x_1y_3)$.

HINT. Circumscribe about the triangle a rectangle whose sides are parallel to the coördinate axes, and express the area of the given triangle as the difference between the area of the rectangle and three right triangles.

13. Angles, inclination, and slope. DEFINITION. *The angle between two directed lines is the angle whose sides extend from the vertex in the positive direction.*

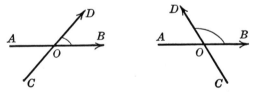

Fig. 13.1

In both figures the angle between the directed lines AB and CD is the angle BOD. The definition is merely a convention to enable us to distinguish which of the four angles formed by two intersecting lines is meant by the phrase "the angle between the lines."

DEFINITION. *The inclination of a line is the angle between the line and the x-axis.*

The inclination of a line parallel to the x-axis is zero. Since the positive direction along all lines not parallel to the x-axis is upward, the inclination is the angle between the given line, directed upward, and the x-axis (or a line parallel to the x-axis), directed toward the right. The inclination is taken always as a positive angle and is always less than 180°. It will be denoted by the Greek letter α.

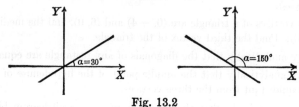

Fig. 13.2

DEFINITION. *The slope of a line is the tangent of its inclination.*

The slope will be denoted by m and the definition of slope may be written $m = \tan \alpha$. When α increases from 0° to 90°, $\tan \alpha$ increases from 0 to ∞, and when α increases from 90° to 180°, $\tan \alpha$ increases from $-\infty$ to 0. Hence the slope may be any real number, positive or negative. If α is less than 90°, the slope is positive; if α is greater than 90°, the slope is negative.

14. The slope formula. Theorem. *The slope of the line which passes through the points $P_1(x_1, y_1)$ and $P_2(x_2, y_2)$ is given by the formula*

(III) $$m = \frac{y_2 - y_1}{x_2 - x_1}.$$

Proof. We exclude from the proof the special cases when P_1P_2 is parallel to one of the coördinate axes. If P_1P_2 is parallel to the x-axis, the inclination is 0° and the slope $= \tan 0° = 0$. This agrees with the formula, since $y_2 = y_1$ and $x_2 \neq x_1$. If P_1P_2 is parallel to the y-axis, the inclination is 90° and the slope is infinite.

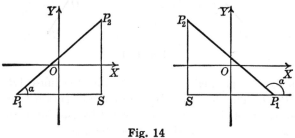

Fig. 14

If $y_1 < y_2$ (as in the figures), draw a line through P_1 parallel to the x-axis and one through P_2 parallel to the y-axis. Let these lines intersect at S. Two figures are possible, according as α is acute or obtuse. In both figures we have, by trigonometry,

$$m = \tan \alpha = \frac{SP_2}{P_1S}.$$

But $\qquad SP_2 = y_2 - y_1 \quad \text{and} \quad P_1S = x_2 - x_1.$ Art. 10

Hence $\qquad m = \dfrac{y_2 - y_1}{x_2 - x_1}.$

If in Fig. 14 we interchange P_1 and P_2, we obtain the proper figures for the case where $y_2 < y_1$. Reasoning similar to that above gives the alternative form

$$m = \frac{y_1 - y_2}{x_1 - x_2}.$$

The results are equal, since the value of a fraction is unchanged by changing the signs of both numerator and denominator. This shows that in problems it is immaterial which point is called P_1.

The slope is the most convenient way of giving the direction of a line. Any two points on the line will give the same value for the slope. The slope formula may be thought of as *the vertical distance from P_1 to P_2 divided by the horizontal distance from P_1 to P_2.*

15. To draw a line with a given slope. If the slope of a line is positive, we can see that from a given point on the line it extends upward and to the right (or downward and to the left). If the slope is negative the line extends, from a given point on it, upward and to the left (or downward and to the right).

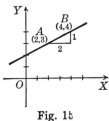

Fig. 15

The slope formula provides us with a simple method for constructing a line passing through a given point and having a given slope. For example, to construct a line passing through $A(2, 3)$ and having a slope $\frac{1}{2}$, we measure from the point A a distance 2 units to the right and 1 unit upward, which brings us to the point $B(4, 4)$. Then AB is the required line. This method of construction can be altered (1) by measuring any number of units to the right and then half as many units upward or (2) by measuring to the left and downward.

The essential idea is that the slope of a line is the *rate of change* of the ordinate with respect to the abscissa; that is, the change in the ordinate per unit change in the abscissa. In the example above, the change in the y-coördinate is $\frac{1}{2}$ unit per unit change in the x-coördinate.

16. Parallel and perpendicular lines. *Theorem. If two lines are parallel, their slopes are equal, and conversely.*

Proof. Let the lines l_1 and l_2 be parallel and let their respective inclinations and slopes be denoted by α_1, α_2, and m_1, m_2. Now α_1 and α_2 are exterior-interior angles of parallel lines cut by a transversal and so

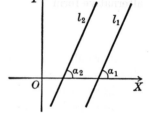

$$\alpha_1 = \alpha_2.$$

Then $$\tan \alpha_1 = \tan \alpha_2$$

or $$m_1 = m_2.$$

The converse is proved by reversing the order of the steps.

Fig. 16.1

Theorem. *If two lines are mutually perpendicular, the slope of one is the negative reciprocal of the slope of the other, and conversely.*

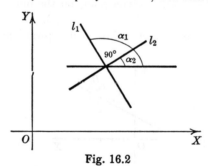

Fig. 16.2

Proof. Let the lines l_1 and l_2 be mutually perpendicular, where l_1 denotes the line of greater inclination. Let α_1, α_2 and m_1, m_2 denote the inclinations and slopes, respectively, of these lines. Through the point of intersection of the given lines draw a line parallel to the x-axis. Then α_1 and α_2 can be taken as angles between this line and the given lines.

Since $\alpha_1 > \alpha_2$ and the angle between them is $90°$,

$$\alpha_1 = \alpha_2 + 90°.$$

Then, by trigonometry,

$$\tan \alpha_1 = \tan (\alpha_2 + 90°)$$
$$= - \operatorname{ctn} \alpha_2$$
$$= - \frac{1}{\tan \alpha_2}.$$

Hence $\qquad\qquad m_1 = - \dfrac{1}{m_2} \quad$ or $\quad m_1 m_2 = - 1.$

The converse is proved by reversing the order of the steps.

For convenience in reference the theorems are restated as formulas.

Condition for parallelism:

(IV a) $\qquad\qquad\qquad m_1 = m_2.$

Condition for perpendicularity:

(IV b) $\qquad\qquad\qquad m_1 m_2 = - 1.$

17. The angle between two lines. *Theorem.* *If θ is the angle between the lines l_1 and l_2, and l_1 has the greater inclination,*

(V) $\qquad\qquad\qquad \theta = \alpha_1 - \alpha_2,$

where α_1 and α_2 denote the respective inclinations of the lines.

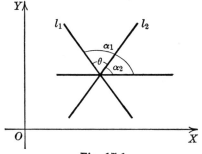

Fig. 17.1

Proof. Through the point of intersection of the given lines draw a line parallel to the x-axis. Then α_1 and α_2 can be taken as the angles between this line and the given lines. Since $\alpha_1 > \alpha_2$ by hypothesis, it follows that

$$\theta = \alpha_1 - \alpha_2.$$

Theorem. *If θ is the angle between the lines l_1 and l_2, and l_1 has the greater inclination,*

(V a) $$\tan \theta = \frac{m_1 - m_2}{1 + m_1 m_2},$$

where m_1 and m_2 denote the respective slopes of the lines.

Proof. By the preceding theorem

$$\theta = \alpha_1 - \alpha_2.$$

Hence, by trigonometry,

$$\tan \theta = \tan (\alpha_1 - \alpha_2) = \frac{\tan \alpha_1 - \tan \alpha_2}{1 + \tan \alpha_1 \tan \alpha_2} = \frac{m_1 - m_2}{1 + m_1 m_2}.$$

NOTATION. It will often be convenient to have a notation for the slope and inclination of a line which passes through two given points. Accordingly, let us agree to denote the slope and the inclination of the line AB by the abbreviations $m(AB)$ and $\alpha(AB)$, respectively.

Example 1. If the vertices of a triangle are $A(2, -1)$, $B(11, 1)$, $C(5, 8)$, prove that the median from B is perpendicular to AC.

Solution. By the mid-point formulas the coördinates of D, the mid-point of AC, are $(\frac{7}{2}, \frac{7}{2})$. Applying the slope formula, we obtain

$$m(AC) = \frac{8 - (-1)}{5 - 2} = 3,$$

and $$m(BD) = \frac{\frac{7}{2} - 1}{\frac{7}{2} - 11} = -\frac{1}{3}.$$

Since these slopes are negative reciprocals, the lines are perpendicular.

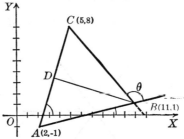

Fig. 17.2

Example 2. Find the angles of the triangle ABC above.

Solution. From the given data we have

$$m(AB) = \frac{1 + 1}{11 - 2} = \frac{2}{9} = 0.2222, \qquad \text{whence} \quad \alpha(AB) = 12° \, 32';$$

$$m(BC) = \frac{8 - 1}{5 - 11} = -\frac{7}{6} = -1.1667, \quad \text{whence} \quad \alpha(BC) = 130° \, 36';$$

$$m(AC) = \frac{8 + 1}{5 - 2} = 3, \qquad\qquad \text{whence} \quad \alpha(AC) = 71° \, 34'.$$

From Fig. 17.2, angle A is given by
$$A = \alpha(AC) - \alpha(AB) = 59°\ 2'.$$
Similarly,　　　　$$C = \alpha(BC) - \alpha(AC) = 59°\ 2'.$$

At B, however, the angle between the positive directions of AB and BC is the exterior angle θ. Hence
$$\theta = \alpha(BC) - \alpha(AB) = 118°\ 4',$$
and　　　　$$B = 180° - \theta = 61°\ 56'.$$

As a check on the work note that $A + B + C = 180°$.

The angles may be found also by applying (**V *a***) as follows.
$$\tan A = \frac{m(AC) - m(AB)}{1 + m(AC)m(AB)} = \frac{3 - \frac{2}{3}}{1 + \frac{2}{3}} = \frac{5}{3},$$
whence　　　　$$A = 59°\ 2'.$$
$$\tan C = \frac{m(BC) - m(AC)}{1 + m(BC)m(AC)} = \frac{-\frac{7}{6} - 3}{1 + (-\frac{7}{2})} = \frac{5}{3},$$
whence　　　　$$C = 59°\ 2'.$$
$$\tan \theta = \frac{m(BC) - m(AB)}{1 + m(BC)m(AB)} = \frac{-\frac{7}{6} - \frac{2}{3}}{1 + (-\frac{7}{27})} = -\frac{15}{8},$$
whence　　　　$$\theta = 118°\ 4'$$
and　　　　$$B = 180° - \theta = 61°\ 56'.$$

Example 3. In Fig. 17.3, $OBCA$ is a parallelogram. Find the coördinates of C.

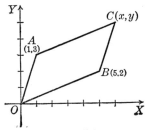

Fig. 17.3

Solution. Let the coördinates of C be (x, y). Since BC is parallel to OA, we have the slope of BC equal to the slope of OA, or
$$\frac{y - 2}{x - 5} = \frac{3 - 0}{1 - 0}. \tag{1}$$

Similarly, since AC is parallel to OB,
$$\frac{y - 3}{x - 1} = \frac{2 - 0}{5 - 0}. \tag{2}$$

Equations (1) and (2) are sufficient to determine x and y. We have, after simplification,

$$3x - y - 13 = 0,$$
$$2x - 5y + 13 = 0.$$

By solving these simultaneous equations for x and y, the result is found to be

$$x = 6, \quad y = 5.$$

Hence the coördinates of C are (6, 5).

PROBLEMS

1. Find the slope and the inclination of the lines which pass through the following pairs of points.

    a. (7, 5), (− 5, − 1).   **b.** (− 5, 6), (4, − 3).

    c. (6, 10), (2, 0).   **d.** (− 3, 5), (11, 9).

2. A point moves from left to right along each of the lines in Problem 1. How much does it rise per horizontal unit? How far does it rise in moving from the second point named to the point where $x = 15$?

3. Each of the following pairs of numbers gives the slopes of two lines. In each case find the angle between the lines.

    a. 3, 5.   **b.** − 2, − 3.   **c.** 2, − 3.

    d. $\frac{1}{2}, -\frac{2}{3}$.   **e.** $-\frac{1}{3}, -2$.   **f.** $-\frac{3}{7}, -\frac{2}{5}$.

4. The slope of one line is − 0.7743 and the inclination of another is 64° 38′. Find the acute angle between the lines.

5. The angle between two lines is 45° and the slope of one of the lines is 3. Find the slope of the other line exactly. If there are two possible solutions, find both.

6. Same as Problem 5 except that the given slope is

    a. $\frac{2}{5}$.   **b.** − 2.   **c.** − 3.

7. If the tangent of the angle between two lines is $\frac{3}{5}$ and the slope of one line is 2, find the slope of the other line.

8. Two lines have slopes $-\frac{7}{4}$ and $\frac{1}{5}$. Find exactly the slope of the line which bisects the angle between the given lines.

9. Prove by means of slopes that the points (− 3, − 1), (3, 2), and (7, 4) lie on the same straight line.

10. Are the points (6, − 3), (8, 7), and (2, 2) the vertices of a right triangle? Prove your answer.

11. Are the points (− 4, 4), (8, 16), (17, 6), and (− 6, − 6) the vertices of a rectangle? Prove your answer.

12. Find the angles of the following triangles.

    a. (7, 3), (4, 6), (1, 1).   **b.** (4, 8), (− 6, 4), (2, 2).

    c. (2, 3), (− 5, 5), (− 2, − 4).   **d.** (− 6, 4), (− 4, − 6), (2, − 8).

13. (a) Prove that the points $(-6, 0)$, $(0, -6)$, $(8, 6)$, and $(2, 12)$ are the vertices of a parallelogram. (b) Find the angles of the parallelogram. (c) Find the angle between the diagonals.

14. (a) Prove that the points $(-4, -2)$, $(-2, 4)$, $(4, 7)$, and $(10, 5)$ are the vertices of an isosceles trapezoid. (b) Find the angles of the trapezoid. (c) Find the angle between the diagonals.

15. Three vertices of a rectangle are $(0, -2)$, $(12, 10)$, and $(-10, 8)$. Find the fourth vertex.

16. The ends of the hypotenuse of an isosceles right triangle are $(8, -3)$ and $(-4, 6)$. Find the third vertex. How many solutions are there?

17. The ends of the base of an isosceles triangle are $(-4, 4)$ and $(8, -2)$ and the slope of one of the equal sides is -3. Find the third vertex.

18. If $(2, -1)$, $(-2, 4)$, and $(6, 6)$ are the mid-points of the sides of a triangle, what are the coördinates of the vertices of the triangle?

19. Let $A(a, 0)$ and $B(a + b, 0)$ denote any two points on the positive half of the x-axis, with A lying between B and the origin O. Two points C and D are taken in the first quadrant and a third point E is taken in the fourth quadrant so that OCA, ADB, and OEB are isosceles triangles which have their bases on the x-axis and their base angles equal to $30°$. Find the coördinates of C, D, and E.

HINT. $\tan 30° = 1/\sqrt{3}$.

20. Prove that the points C, D, and E in Problem 19 are the vertices of an equilateral triangle.

21. Prove analytically that the lines joining the middle points of the sides of any quadrilateral, taken in order, form a parallelogram.

22. Prove analytically that in any triangle the line joining the middle points of two sides is parallel to the third side and equal to one half of it.

23. Let $OABC$ be a parallelogram and let the points D and E trisect the diagonal OB. Prove analytically that the lines joining D and E to the vertices A and C form a parallelogram.

18. Equations of straight lines. We have seen that to every point in the plane there corresponds, in rectangular coördinates, a single pair of numbers, and, conversely, that to every pair of numbers representing coördinates there corresponds a single point. We now extend the connection between algebra and geometry by showing that to every straight line in the plane there corresponds an algebraic equation of the first degree in x and y, and that, conversely, to every algebraic equation of the first degree in x and y there corresponds a straight line in the plane.

DEFINITION. *The equation of a straight line is the equation satisfied by the coördinates of every point on the line and by those of no other point.*

Consider, for example, the straight line passing through the origin and making an angle of 45° with the x-axis. For any point on the line the abscissa is obviously equal to the ordinate and this is true for no other point, since the bisector of an angle is the locus of points equidistant from the sides of the angle. Hence $x = y$ is an equation of this line according to the definition. Similarly, $x = -y$ is an equation of the line which passes through the origin and has an inclination of 135°.

It is clear that instead of the equation $x = y$ we might use $5\,x = 5\,y$, $2\,x - y = x$, $(x - y)^2 = 0$, and many others. Since any two equations of the first degree which represent the line can be transformed into each other by algebraic operations not affecting the roots, it is natural to regard them as merely different forms of a single equation which expresses the relation between x and y. It has therefore become customary to use the phrase "*the* equation of the line" for any equation of the first degree which represents the line.

It is not necessary that both x and y appear in the equation. The equation $x = 1$ is satisfied by the coördinates of every point on the line which is parallel to the y-axis and one unit to the right of it, and it is satisfied by the coördinates of no other point, since no other point can have its abscissa 1. The equation of any line parallel to the y-axis is obviously $x = $ a constant; the equation of the y-axis is $x = 0$. The equation of the x-axis is $y = 0$, and the equation of any line parallel to the x-axis is $y = $ a constant.

The next three articles give standard forms for equations of lines not parallel to the coördinate axes.

19. The point-slope equation. *Theorem. The equation of the straight line which passes through the point (x_1, y_1) and has the slope m is*

(VI) $$y - y_1 = m(x - x_1).$$

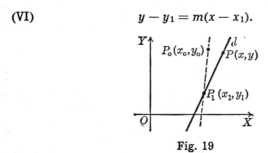

Fig. 19

Proof. We are given a line l which has the slope m and passes through the point $P_1(x_1, y_1)$. We must show (*a*) that the coördinates of every point on l satisfy the above equation and (*b*) that the coördinates of no other point satisfy the equation.

a. By inspection we see that the coördinates of P_1 satisfy the equation. Let $P_2(x_2, y_2)$ be any other point on l. By the slope formula

$$m = \frac{y_2 - y_1}{x_2 - x_1}.$$

Hence $y_2 - y_1 = m(x_2 - x_1).$

Thus the coördinates of P_2 satisfy the equation.

b. Let $P_0(x_0, y_0)$ be any point which is not on l. We must show that its coördinates cannot satisfy **(VI)**. The proof will be indirect. Let us assume that the coördinates of P_0 do satisfy the equation; that is, we suppose

$$y_0 - y_1 = m(x_0 - x_1). \tag{1}$$

If $x_0 = x_1$, then $y_0 = y_1$. This is false, since P_0 is not the same point as P_1.

If $x_0 \neq x_1$, we can divide both sides of (1) by $x_0 - x_1$. This gives

$$\frac{y_0 - y_1}{x_0 - x_1} = m. \tag{2}$$

The left-hand member of (2) is the slope of P_1P_0, whereas m is the slope of l. Thus (2) says that the slope of P_1P_0 equals the slope of l. This is false, since two intersecting lines cannot have the same slope.

Hence the assumption in (1) is false, and so the coördinates of P_0 do not satisfy **(VI)**.

The equation which we have derived above is known as the *point-slope form* of the straight-line equation. By means of it we can write down at once the equation of any straight line for which we can find the slope and the coördinates of one of its points.

Example 1. Find the equation of the line which has an inclination of $135°$ and which passes through $(1, -3)$.

Solution. Since $\alpha = 135°$, we have $m = -1$. Substituting in **(VI)**,

$$y + 3 = -1(x - 1),$$

and, simplifying, $x + y + 2 = 0.$

Example 2. Find the equation of the line through $(2, -3)$ which is perpendicular to the line joining $A(3, 1)$ and $B(4, -5)$.

Solution. The slope of the line AB is

$$\frac{1 + 5}{3 - 4} = -6.$$

Hence the slope of the required line is $\frac{1}{6}$. Substituting in **(VI)**,

$$y + 3 = \tfrac{1}{6}(x - 2);$$

and, simplifying, $x - 6y - 20 = 0.$

20. The slope-intercept form. The *intercepts* of a line are the distances from the origin to the points where the line cuts the coördinate axes. Thus the x-intercept is the abscissa of the point of intersection with the x-axis, and the y-intercept is the ordinate of the point of intersection with the y-axis.

The equation of the straight line which has the slope m and the y-intercept b is

(VII) $$y = mx + b.$$

This may be derived at once from the point-slope form, since we have given the slope and one point, $(0, b)$.

This form of the straight-line equation is called the *slope-intercept* form. Its most important use will be given in Art. 23.

21. Other forms of the straight-line equation. The point-slope form can be used for all lines except those parallel to the y-axis (where the equation is of the form $x = c$), but it is especially adapted to writing the equation of a line which is determined by a point and the direction of the line. Other special forms are convenient for other conditions determining a line. The slope-intercept form has already been found. Two others will be given here.

The two-point form. The equation of the straight line which passes through $P_1(x_1, y_1)$ and $P_2(x_2, y_2)$ is

(VIII) $$\frac{y - y_1}{x - x_1} = \frac{y_2 - y_1}{x_2 - x_1}.$$

This is derived from the point-slope form by substituting for m its value $\frac{y_2 - y_1}{x_2 - x_1}$. It may also be obtained directly from a figure.

The intercept form. The intercepts of a line were defined in the previous article. The equation of the straight line whose x-intercept is a and whose y-intercept is b is

(IX) $$\frac{x}{a} + \frac{y}{b} = 1.$$

Proof. Substituting $x_1 = a, y_1 = 0$ and $x_2 = 0, y_2 = b$ in the two-point form, we have

$$\frac{y - 0}{x - a} = \frac{b - 0}{0 - a}.$$

When simplified this becomes

$$- ay = bx - ba, \quad \text{or} \quad bx + ay = ab.$$

If each term is divided by ab, we obtain

$$\frac{x}{a} + \frac{y}{b} = 1.$$

Example 1. Find the equation of the line which passes through $(-1, 1)$ and $(2, 7)$.

Solution. Substituting in the two-point form,

$$\frac{y-1}{x+1} = \frac{7-1}{2+1},$$

and, simplifying, $\qquad 2x - y + 3 = 0.$

An absolute check on the correctness of the result is furnished by the ₋act that two points determine a straight line. Hence the equation is correct if, and only if, it is satisfied by the coördinates of both points.

Example 2. Find the equation of the line having the x-intercept 3 and the y-intercept -4.

Solution. Substituting in the intercept form,

$$\frac{x}{3} + \frac{y}{-4} = 1,$$

and, simplifying, $\qquad 4x - 3y - 12 = 0.$

PROBLEMS

Find the equations of the lines determined by the following conditions. **Draw** the lines and check the answers.

1. Passing through $(-3, 4)$ and $(6, 10)$.
2. Passing through $(5, 2)$ and $(-3, 5)$.
3. Passing through $(-3, 4)$ and having the slope $\frac{1}{3}$.
4. Passing through $(2, -3)$ and having the slope -2.
5. Passing through $(4, 0)$ and $(0, -2)$.
6. Passing through $(-3, 0)$ and having the inclination $45°$.
7. Passing through $(3, -6)$ and having the inclination $135°$.
8. Passing through $(6, 3)$ and parallel to the x-axis.
9. Passing through $(4, -2)$ and parallel to the y-axis.
10. Parallel to the y-axis and 5 units to the right of it.
11. Parallel to the x-axis and 4 units below it.
12. Having the x-intercept 3 and the y-intercept -6.
13. Having the x-intercept -5 and the slope -3.
14. Having the x-intercept 7 and the inclination $135°$.
15. Having the x-intercept -5 and passing through $(2, -1)$.
16. Having the y-intercept 3 and the slope $\frac{1}{3}$.
17. Having the y-intercept 5 and the inclination $45°$.

18. Having the y-intercept -4 and passing through $(5, 7)$.

19. Passing through $(4, -6)$ and rising $\frac{1}{2}$ unit per unit increase in x.

20. Passing through $(6, 3)$ and falling 2 units per unit increase in x.

22. The equation of the first degree. It has been seen that every straight line has an equation of the first degree. It will now be shown, conversely, that to every equation of the first degree in x and y there corresponds a straight line in the plane. This line is called the *locus* of the equation. Every equation of the first degree in x and y can be written in the form $Ax + By + C = 0$, where A, B, and C are any constants except that both A and B cannot be zero.

Theorem. *Every equation of the first degree in x and y is an equation of a definite straight line.*

Proof. Let the equation be $Ax + By + C = 0$. There are two cases.

CASE I. If $B = 0$ and $A \neq 0$, the equation can be solved for x and we get

$$x = -\frac{C}{A}.$$

We have seen that this is the equation of a line which is parallel to the y-axis and has its distance from the y-axis equal to $-C/A$.

CASE II. If $B \neq 0$, the equation may be solved for y, giving

$$y = -\frac{A}{B}x - \frac{C}{B}.$$

Comparison with the slope-intercept form $y = mx + b$ shows that this is the equation of the straight line that cuts the y-axis at $(0, -C/B)$ and has the slope $m = -A/B$.

NOTE. It should be noted that the values of x and y that satisfy an equation are not changed if every term of the equation is multiplied by the same constant factor. That is, $Ax + By + C = 0$ and $kAx + kBy + kC = 0$ represent the same straight line.

23. To construct a line when its equation is given. To draw a straight line when its equation is given, we find two points on the line. Any two points whose coördinates satisfy the equation will do, but the simplest to get are those where the line cuts the coördinate axes. Thus, to draw the line whose equation is $2x - 5y - 10 = 0$, we set $x = 0$, giving $y = -2$, and then set $y = 0$, giving $x = 5$. Hence two points on the line are $(0, -2)$ and $(5, 0)$. The straight line passing through these two points is the locus of the equation. If the line passes through the origin, another point not on the axes must be found. For accurate construction, this point should be as far from the origin as the plotting paper permits.

The slope of a line whose equation is given can be determined by inspection. It appears in the proof of the preceding theorem that the slope of the line whose equation is $Ax + By + C = 0$ is $m = -A/B$. This important fact can be stated also in the following form. *If the equation of a straight line is solved for y, the resulting coefficient of x is the slope.*

PROBLEMS

1. For each of the following equations find the intercepts of the corresponding line and draw the line. Also find the slope and inclination in each case.

a. $x - 2y + 8 = 0$. **b.** $3x + 5y - 15 = 0$.
c. $2x - y - 5 = 0$. **d.** $4x + 6y + 25 = 0$.

2. Write each equation in Problem 1 in the slope-intercept form, and in the intercept form.

3. Find the angle between each of the following pairs of lines taken from Problem 1.

i. *a* and *b*.   **ii.** *a* and *c*.   **iii.** *a* and *d*.
iv. *b* and *c*.   **v.** *b* and *d*.   **vi.** *c* and *d*.

4. Find the equations of the lines which pass through the origin and are parallel to the lines in Problem 1.

5. Find the equations of the lines which pass through the origin and are perpendicular to the lines in Problem 1.

6. Find the equations of the lines which pass through the point $(3, -2)$ and are parallel to the lines in Problem 1.

7. Find the equations of the lines which pass through the point $(-3, 2)$ and are perpendicular to the lines in Problem 1.

8. Find the equation of a line which passes through the point $(8, 6)$ and has its y-intercept equal to twice the x-intercept.

9. Find the equation of a line which passes through the point $(-4, 6)$ and has the sum of its intercepts equal to 4.

10. Find the equation of a line which passes through the point $(6, 0)$ and makes an angle of $45°$ with the line whose equation is $x - 2y + 8 = 0$.

11. Find the equation of a line which passes through the point $(5, 4)$ and intersects the line whose equation is $3x - 2y + 12 = 0$ at an angle whose tangent is $\frac{1}{2}$.

12. The vertices of a triangle are $A(3, -2)$, $B(-3, 4)$, and $C(6, 4)$. Find the equations of (**a**) the sides, (**b**) the medians, (**c**) the perpendicular bisectors of the sides, (**d**) the lines joining the mid-points of the sides, (**e**) the lines passing through the vertices and parallel to the opposite sides, (**f**) the lines passing through the vertices and perpendicular to the opposite sides.

13. The equations of the sides of a triangle are $2x + y = 8$, $2x - 3y = 24$, $x - 3y + 9 = 0$.

a. Draw the triangle.

b. Find the slope and inclination of each side.

c. Find the angles of the triangle.

14. The equations of two sides of a parallelogram are $3x - 5y + 15 = 0$ and $5x + 4y - 32 = 0$. Find the equations of the other two sides if one vertex is $(0, 5)$.

15. Find the equations of the sides of a square of which two opposite vertices are $(-4, 4)$ and $(6, -2)$.

16. A line of unknown slope m passes through the point $(2, 9)$ and forms with the coördinate axes a triangle of area 48 in the first quadrant. Find its slope and its equation.

HINT. Use the point-slope form, and find the intercepts in terms of m.

24. Points of intersection. Let the equations of two lines be

$$A_1x + B_1y + C_1 = 0,$$
$$A_2x + B_2y + C_2 = 0.$$

Every point which lies on the first line will have coördinates which satisfy the first equation, and every point which lies on the second line will have coördinates which satisfy the second equation. Consequently the coördinates of the point of intersection of the two lines will satisfy both equations. Conversely, if the coördinates of a point satisfy both equations, the point must lie on both lines. Hence the coördinates of the point of intersection of two lines can be found by solving their equations simultaneously.

Example 1. Find the point of intersection of the lines whose equations are $x + y - 6 = 0$ and $x - y + 1 = 0$.

Solution. By solving the equations simultaneously, we get $x = 2.5$ and $y = 3.5$.

Hence the point of intersection is $(2.5, 3.5)$. See Fig. 24.1.

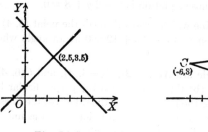

Fig. 24.1

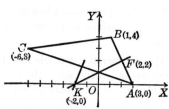

Fig. 24.2

Example 2. Find the center and the radius of the circumscribed circle of the triangle $A(3, 0)$, $B(1, 4)$, $C(-6, 3)$.

Solution. The center of the circumscribed circle is the point of intersection of the perpendicular bisectors of the sides (Fig. 24.2).

The mid-point of AB is $F(2, 2)$ and the slope of AB is -2. Consequently the slope of the perpendicular bisector is $\frac{1}{2}$ and its equation is

$$y - 2 = \tfrac{1}{2}(x - 2),$$

or

$$x - 2y + 2 = 0.$$

Similarly, the equation of the perpendicular bisector of AC is

$$y - \tfrac{3}{2} = 3(x + \tfrac{3}{2}),$$

or

$$3x - y + 6 = 0.$$

The coördinates of the center of the circumscribed circle are found by solving these equations simultaneously. The result is $x = -2$, $y = 0$.

The radius of the circumscribed circle is the distance from the center K to any vertex of the triangle. As a check on the correctness of the work we shall find the distance to each of the vertices.

$$KA = 5 \text{ (from the figure)},$$
$$KB = \sqrt{(-2 - 1)^2 + (0 - 4)^2} = 5,$$
$$KC = \sqrt{(-2 + 6)^2 + (0 - 3)^2} = 5.$$

Since these three distances are the same, we have proved that K is equidistant from the three vertices.

The center of the circumscribed circle is $(-2, 0)$, and its radius is 5.

It is important in every problem of this type to draw the figure accurately and to make sure that the calculated results agree with the figure.

PROBLEMS

1. For the following pairs of equations find the point of intersection of the corresponding lines and check by drawing a figure.

    a. $x - 2y - 2 = 0$ and $2x + y - 14 = 0$.
    b. $2x - 3y - 10 = 0$ and $x + y + 2 = 0$.
    c. $x - y + 6 = 0$ and $3x + 2y - 8 = 0$.
    d. $x - 3y + 9 = 0$ and $2x + y - 6 = 0$.

2. Draw the triangles of which the equations of the sides are as follows and find the coördinates of the vertices.

    a. $x + y = 7$, $x - y = 5$, $4x + y = 10$.
    b. $2x - 4y + 5 = 0$, $2x + 3y + 5 = 0$, $3x + y - 10 = 0$.

3. Each of the following sets of points are vertices of a triangle. Find the equations of the medians and prove analytically that they meet in a point.

    a. $A(3, 7)$, $B(-1, 5)$, $C(7, 3)$.
    b. $A(5, 5)$, $B(-3, 1)$, $C(6, -2)$.
    c. $A(2, -4)$, $B(8, 0)$, $C(14, 10)$.

4. For each of the triangles in Problem 3 find the equations of the perpendicular bisectors of the sides and prove analytically that they meet in a point.

5. For each of the triangles in Problem 3 find the center and radius of the circumscribed circle.

6. The vertices of a quadrilateral are $A(-4, -8)$, $B(12, 0)$, $C(2, 10)$, and $D(-6, 6)$. **(a)** Find the point of intersection of the diagonals. **(b)** Prove that the vertices lie on a circle.

7. Find the equation of the line which passes through the point $(7, 3)$ and is perpendicular to the line whose equation is $3x + y = 12$, the coördinates of the point of intersection of the two lines, and the length of the perpendicular from the given point to the given line.

8. By the method outlined in Problem 7 find the distance between the point $(0, 8)$ and the line whose equation is $x - 2y = 9$.

9. In each of the triangles of Problem 3 find the equation of the altitude from A to BC, the point where the altitude meets BC, the length of the altitude, and the area of the triangle.

10. The equations of two sides of a parallelogram are $3x - 5y + 15 = 0$ and $5x + 4y - 32 = 0$, and one vertex is the point $(0, -5)$. Find the other vertices.

11. Prove analytically that the medians of any triangle meet in a point.

HINT. For any triangle the vertices may be taken as $(a, 0)$, $(0, b)$, $(-c, 0)$.

12. Prove analytically that the perpendicular bisectors of the sides of any triangle meet in a point (see the hint under Problem 11).

13. Prove analytically that the altitudes of any triangle meet in a point (see the hint under Problem 11).

25. Systems of lines. In an equation which represents a line, the letters x and y are called the *variables* of the equation or the *running coördinates* of the line. They represent the coördinates of a point, and are different for different points of the same line. The other letters occurring in the equation represent fixed numerical values for any given fixed line, but the numerical values are different for different lines. For example, the equation of any line not parallel to the y-axis can be written in the form $y = mx + b$. If $m = 1$ and $b = 2$, the equation will represent the line which cuts the y-axis at $(0, 2)$ and has the slope 1.

It is important to note the correspondence between the *geometrical* fact that a straight line is determined by two geometrical conditions and the *algebraical* fact that its equation is determined by two constants, in the above case m and b. The apparent exception in the case of the equation $Ax + By + C = 0$ is disposed of by observing that this may be reduced to a form containing only two arbitrary constants if we divide through by A, B, or C.

Returning to the equation $y = mx + b$, suppose that $b = 2$ and that the value of m is not fixed. The equation $y = mx + 2$ will represent a line cutting the y-axis at $(0, 2)$, but its direction will depend on the value assigned to m. By assigning all possible values to m, the equation will represent all possible lines through the point $(0, 2)$ except the y-axis, the slope of which is infinite. These lines form a *system of lines*, and m is called an *arbitrary constant* or *parameter*. A system of straight lines, then, is a set of lines having one geometrical condition fixed, and it is represented by an equation of the first degree in x and y and containing an arbitrary constant. For example (see Fig. 25.1):

1. The equation $y = x + b$ represents the system of parallel lines making an angle of $45°$ with the x-axis.

2. The equation $\dfrac{x}{5} + \dfrac{y}{b} = 1$ represents the system of lines cutting the x-axis at $(5, 0)$.

3. The equation $y - 2 = m(x + 3)$ represents the system of lines passing through $(-3, 2)$.

4. If A, B, and C are fixed constants, the equation of the system of lines parallel to $Ax + By + C = 0$ is $Ax + By + k = 0$, where k is an arbitrary constant; and the equation of the system of lines perpendicular to the given line is $Bx - Ay + k = 0$.

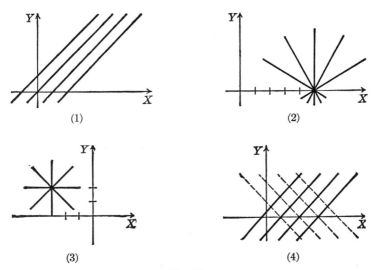

Fig. 25.1

The equation of a system of lines is useful in the solution of certain types of problems, as illustrated below.

Example 1. Find the equation of the line which passes through $(4, -2)$ and is parallel to the line $x + 3y - 10 = 0$.

Solution. The equation of the system of lines which are parallel to the given line is

$$x + 3y + k = 0. \tag{1}$$

If a line of this system passes through $(4, -2)$, its equation must be satisfied when $x = 4$, $y = -2$; that is,

$$4 - 6 + k = 0, \text{ whence } k = 2.$$

Substituting this value of k in (1), we get the final result

$$x + 3y + 2 = 0.$$

This method should be compared with the method of solving this problem without the use of the notion of a system of lines.

Example 2. Find the equation of the line which passes through $(6, 3)$ and cuts off from the first quadrant a triangle of area 48 square units.

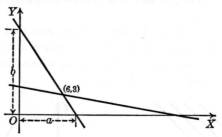

Fig. 25.2

Solution. The equation of the system of lines passing through $(6, 3)$ is

$$y - 3 = m(x - 6),$$

or

$$mx - y + 3 - 6m = 0. \tag{2}$$

The intercepts a and b of any one of these lines on the coördinate axes can readily be found in terms of m. We obtain

$$a = \frac{6m - 3}{m} \quad \text{and} \quad b = 3 - 6m.$$

The area of the triangle formed by the line and the coördinate axes is $ab/2$.

Hence

$$\frac{(3 - 6m)(6m - 3)}{2m} = 48,$$

which becomes, after simplification,

$$12m^2 + 20m + 3 = 0.$$

The two roots of this equation are

$$m = -\tfrac{3}{2} \quad \text{and} \quad m = -\tfrac{1}{6}.$$

Substituting in (2), we have the two results

$$3\,x + 2\,y - 24 = 0, \quad \text{and} \quad x + 6\,y - 24 = 0.$$

Both results are correct, as can be easily verified.

Example 3. Find the equation of the straight line which passes through the point (9, 6) and the point of intersection of the lines whose equations are $3\,x + 2\,y - 24 = 0$ and $4\,x - 3\,y + 12 = 0$.

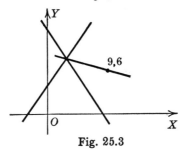

Fig. 25.3

Solution. This problem can be solved by first finding the point of intersection and then using the two-point form, but the following is an easier solution.

The equation

$$3\,x + 2\,y - 24 + k(4\,x - 3\,y + 12) = 0 \qquad (3)$$

is the equation of a system of lines which pass through the intersection of the given lines. For, in the first place, it is an equation of the first degree which contains one parameter (k) and so it is the equation of *some* system of lines. In the second place, if $(x_1,\ y_1)$ is the point of intersection of the given lines, these coördinates satisfy both of the given equations, that is,

$$3\,x_1 + 2\,y_1 - 24 = 0 \quad \text{and} \quad 4\,x_1 - 3\,y_1 + 12 = 0.$$

Consequently, x_1 and y_1 are values of x and y which satisfy (3) for all values of k. Hence every line of the system passes through $(x_1,\ y_1)$.

If a line of the system passes through (9, 6), equation (3) must be satisfied when $x = 9$ and $y = 6$. that is,

$$27 + 12 - 24 + k(36 - 18 + 12) = 0,$$

or $$15 + 30\,k = 0, \quad \text{whence} \quad k = -\tfrac{1}{2}.$$

Substituting this value of k in (3) and simplifying, we obtain the required equation
$$2\,x + 7\,y - 60 = 0.$$

Note. The general theorem which we have proved for a special case in the solution of this problem is as follows.

If $$A_1 x + B_1 y + C_1 = 0 \quad \text{and} \quad A_2 x + B_2 y + C_2 = 0$$

are the equations of two intersecting lines, then

$$A_1 x + B_1 y + C_1 + k(A_2 x + B_2 y + C_2) = 0,$$

where k is an arbitrary constant, is the equation of a system of lines which pass through the point of intersection of the given lines.

PROBLEMS

1. Write the equation of the system of lines determined by the following conditions.

    a. Having the slope $-\frac{2}{3}$.

    b. Having the inclination 135°.

    c. Passing through the point $(3, -5)$.

    d. Having the y-intercept equal to -4.

    e. Having the x-intercept equal to 2.

    f. Having the sum of the intercepts equal to 10.

    g. Having the x-intercept equal to twice the y-intercept.

    h. Forming with the positive halves of the coördinate axes a triangle whose area is 48 square units.

2. For each system in Problem 1 find the value of the parameter for which the corresponding line passes through the point $(6, 3)$, if this is possible.

3. Find the common geometrical property of each of the systems of lines having the following equations.

    a. $3x - 5y + k = 0$.   **b.** $kx - y = 4$.

    c. $kx + y - 2k = 0$.   **d.** $y + 6 = k(x - 2)$.

4. For each equation in Problem 3 find the value of the parameter for which the corresponding line **(a)** passes through the point $(9, 3)$, if this is possible; **(b)** has equal intercepts, if this is possible.

5. Write the equation of the system of lines which are parallel to each of the lines whose equations are

    a. $x + 2y - 6 = 0$.   **b.** $4x - 3y + 10 = 0$.

    c. $6x + 3y - 8 = 0$.   **d.** $3x - y - 7 = 0$.

6. Write the equation of the system of lines which are perpendicular to each of the lines in Problem 5.

7. Write the equation of the line which is parallel to each of the lines in Problem 5 and passes through the point $(5, 2)$.

8. Write the equation of the line which is perpendicular to each of the lines in Problem 5 and passes through the point $(5, 2)$.

9. Using the method of Example 3 above, write the equation of the system of lines which pass through the intersection of the lines whose equations are $2x + y - 12 = 0$ and $x - 3y + 8 = 0$. Then find the equations of the particular lines of the system which have the properties listed below.

    a. Passing through the point $(3, 6)$.

    b. Having the x-intercept equal to -4.

    c. Having the slope 3.

    d. Having equal intercepts.

    e. Perpendicular to the line whose equation is $x + 3y = 9$.

26. Distance from a line to a point. To find the distance from a line parallel to the y-axis to a point involves no difficulty. If the equation of the line is $x = c$, and the coördinates of the point are (x_1, y_1), the distance *from* the line *to* the point is obviously $x_1 - c$, whether the point is at the right or at the left of the line.

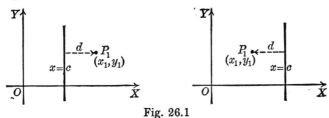

Fig. 26.1

For the more general case, where the line cuts the y-axis, we have the theorem below.

Lemma. *The distance from the line whose equation is $y = mx + b$ to the origin is given by the formula*

$$d = \frac{-b}{\sqrt{m^2 + 1}}.$$

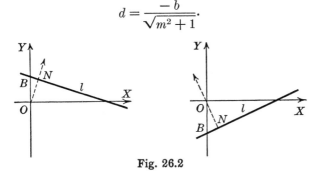

Fig. 26.2

Proof. In either figure let l be the given line whose equation is

$$y = mx + b. \tag{1}$$

From O draw the line which is perpendicular to l and denote the point of intersection of these two lines by N. Since ON is perpendicular to l and passes through the point $(0, 0)$, its equation is

$$y = -\frac{1}{m} x. \tag{2}$$

Solving (1) and (2) simultaneously, the coördinates of N are found to be

$$x = \frac{-bm}{m^2 + 1}, \quad y = \frac{b}{m^2 + 1}.$$

By (I) we find that the distance between O and N is

$$\sqrt{(x-0)^2+(y-0)^2} = \sqrt{\frac{b^2}{m^2+1}} = \frac{\pm b}{\sqrt{m^2+1}}.$$

The distance *from* the line l to the point O is the directed distance NO. The figures show that the sign of NO is opposite to the sign of $b\,(=OB)$.

Hence
$$NO = \frac{-b}{\sqrt{m^2+1}}.$$

Theorem. *The distance from the line whose equation is $y = mx + b$ to the point $P_1(x_1, y_1)$ is given by the formula*

(X)
$$d = \frac{y_1 - mx_1 - b}{\sqrt{m^2+1}}.$$

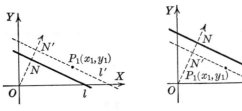

Fig. 26.3

Proof. In either figure let l be the given line whose equation is $y = mx + b$. Through P_1 draw a line l' parallel to l. From O draw the line which is perpendicular to l (and consequently perpendicular to l') and let this line meet l in the point N and l' in the point N'.

Since l' is parallel to l and passes through the point (x_1, y_1), its equation is found to be

$$y = mx + b', \text{ where } b' = y_1 - mx_1. \tag{3}$$

Now the distance d from the line l to the point P_1 is equal to NN'. This distance is positive if P_1 lies above l and negative if P_1 lies below l. Since O, N, and N' are points on a directed line, in all cases,

$$NN' = NO + ON' = NO - N'O. \tag{4}$$

From the preceding lemma,

$$NO = \frac{-b}{\sqrt{m^2+1}} \quad \text{and} \quad N'O = \frac{-b'}{\sqrt{m^2+1}}. \tag{5}$$

Substituting these values in (4) and using the value of b' from (3), we obtain the required formula.

Example 1. Find the distance to the point $(7, 2)$ from the line whose equation is $3x - 4y = 8$.

Solution. To apply **(X)**, first write the equation of the line in the slope-intercept form,
$$y = \tfrac{3}{4}x - 2$$
or
$$y - \tfrac{3}{4}x + 2 = 0.$$

Then the distance from the line to any point $P_1(x_1, y_1)$ is
$$d = \frac{y_1 - \tfrac{3}{4}x_1 + 2}{\sqrt{\tfrac{9}{16} + 1}}.$$

Substituting 7 for x_1 and 2 for y_1, we have
$$d = \frac{2 - \tfrac{3}{4} \cdot 7 + 2}{\sqrt{\tfrac{9}{16} + 1}}$$
$$= \frac{-\tfrac{5}{4}}{\tfrac{5}{4}} = -1.$$

To check the result draw the figure.

Example 2. Find the area of the triangle whose vertices are $A(-1, 1)$, $B(6, -2)$, $C(3, 6)$.

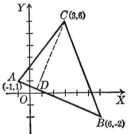

Fig. 26.4

Solution. The equation of AB is found to be
$$y = -\tfrac{3}{7}x + \tfrac{4}{7}.$$

The distance from AB to C is
$$d = \frac{6 - (-\tfrac{3}{7})(3) - \tfrac{4}{7}}{\sqrt{\tfrac{9}{49} + 1}} = \frac{47}{\sqrt{58}};$$

that is, the length of the altitude CD is $47/\sqrt{58}$.

The length of the base AB is given by the distance formula
$$AB = \sqrt{(-1 - 6)^2 + (1 + 2)^2} = \sqrt{58}.$$

Hence the area is $\quad \tfrac{1}{2} AB \cdot CD = \tfrac{1}{2}\sqrt{58} \cdot \frac{47}{\sqrt{58}} = 23.5.$

The above method should be compared with that suggested in Problem 9, Art. 24. It may be of interest to state that if the coördinates of the vertices are rational, the area will also be rational in every case.

Example 3. Find the equations of the bisectors of the angles between the lines whose equations are $y = x - 1$ and $y = 7x - 31$.

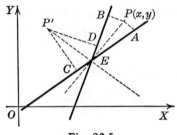

Fig. 26.5

Solution. We recall from plane geometry that any point on the bisector of an angle is equidistant from the sides. Thus (Fig. 26.5), if allowance is made for difference in sign,

$$AP = - BP.$$

By **(X)** this becomes

$$\frac{y - x + 1}{\sqrt{2}} = - \frac{y - 7x + 31}{\sqrt{50}},$$

which, upon simplification, gives for the equation of EP

$$2x - y - 6 = 0.$$

Similarly, $CP' = DP'$, whence the equation of EP' is found to be

$$x + 2y - 13 = 0.$$

The results show that the bisectors are mutually perpendicular.

PROBLEMS

1. Find the distance from the given line to the given point. Draw the figure in each case.

    a. $x + y = 8$, $(2, 3)$.   **b.** $2x - y = 4$, $(0, 6)$.
    c. $x + 2y = 8$, $(8, 4)$.   **d.** $3x + 4y + 5 = 0$, $(7, 0)$.
    e. $4x - 3y + 10 = 0$, $(-1, -3)$.   **f.** $5x + 12y - 15 = 0$, $(7, 8)$.

2. Find the distance between each pair of parallel lines.

        a. $x + y - 3 = 0$, $x + y + 5 = 0$.
        b. $3x - 4y + 2 = 0$, $3x - 4y - 12 = 0$.

Hint. The distance between two parallel lines is the distance from one of the lines to any point on the other.

3. In each of the following triangles find the length of the altitude, taking the first point as vertex and the line joining the other two as the base.

a. $(2, 4)$, $(-2, 0)$, $(4, -2)$.   **b.** $(3, 6)$, $(0, -1)$, $(5, 3)$.
c. $(0, 4)$, $(-4, 0)$, $(4, 2)$.   **d.** $(6, 4)$, $(-2, 6)$, $(2, -2)$.

4. Find the area of each of the triangles in Problem 3.

5. Find the area of the quadrilateral whose vertices are $(-2, -5)$, $(7, -6)$, $(9, 3)$, $(-6, 9)$.

6. Find the equations of the bisectors of the angles formed by the following pairs of lines.

a. $3x + 4y = 12$, $4x - 3y = 16$.
b. $5x - 12y = 60$, $3x + 4y = 8$.
c. $y = x + 1$, $y = 2x - 3$.

7. Find the equations of the bisectors of the interior angles of the triangle defined by the equations $3x - 4y - 12 = 0$, $4x + 3y - 8 = 0$, and $5x + 12y - 30 = 0$.

8. Find the equations of the bisectors of the exterior angles of the triangle in Problem 7.

9. Find the equations of the bisectors of the interior angles of the triangle defined by the equations $x = 4$, $y = 8$, and $x + y = 0$.

10. Find the equations of the bisectors of the exterior angles of the triangle in Problem 9.

11. Find the equations of the bisectors of the interior angles of the triangle whose vertices are $(10, 0)$, $(-2, 6)$, and $(-10, -10)$.

12. Find the equation of the locus of all points that are twice as far from the line whose equation is $4x - 3y = 12$ as from the x-axis.

13. Find the center and the radius of the circle inscribed in the triangle whose vertices are $(0, -2)$, $(5, 8)$, and $(-3, 4)$.

14. The equations of the sides of a certain triangle are $4x - 3y + 16 = 0$, $3x + 4y - 18 = 0$, and $5x - 12y - 2 = 0$. Find

a. The angles of the triangle.
b. The equations of the bisectors of the angles.
c. The center and the radius of the inscribed circle.
d. The equations of the lines passing through the center of the inscribed circle and perpendicular to the sides of the triangle.
e. The coördinates of the point of contact of the inscribed circle with each side of the triangle.

CHAPTER II

EQUATIONS OF CURVES

27. Equation of a curve, or locus. We have seen in the first chapter that the coördinates of every point on any given straight line satisfy an equation of the first degree and, conversely, that the locus of any given equation of the first degree in x and y is a definite straight line. In general it will be found that to every equation not of the first degree there corresponds a curve and that every curve has a definite equation. As a foundation for future work we have the following definitions.

The equation of a curve, or locus, is an equation which is satisfied by the coördinates of every point on the curve and by those of no other point.

Conversely, *the graph, or locus, of an equation in x and y is a figure which contains all points whose coördinates satisfy the equation, and no other points.*

The locus of an equation, if it exists, is unique, but a curve may have more than one equation. (See Art. 18.) If the equation is not linear, the locus is usually a curve, but in exceptional cases it may consist of isolated points or not exist at all.

These two definitions are really equivalent. From them we see that, in order to show that a particular point lies on a curve, we must substitute its coördinates in the equation of the curve and show that they verify the equation.

Example 1. It will be shown later that the graph of $x^2 + y^2 = 20\,x$ is a circle whose center is (10, 0) and whose radius is 10. In Fig. 27

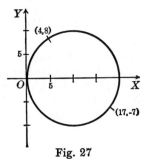

Fig. 27

the points (4, 8) and (17, -7) both appear to lie on this circle. If we substitute 4 for x and 8 for y in the equation, we have

$$16 + 64 - 80 = 0;$$

if we substitute 17 for x and -7 for y, we have

$$289 + 49 - 340 = -2 \neq 0.$$

Hence the first point is on the circle, but the second is not.

Example 2. Does the graph of $x^2 + 4xy + y^2 - 4x - y - 12 = 0$ pass through the origin? Where does it cut the x-axis?

Solution. Substituting $x = 0$ and $y = 0$ in the equation, we get $-12 = 0$, which is false. Hence the graph does not pass through the origin.

If it crosses the x-axis, $y = 0$ at the point of intersection. Substituting this in the equation, we get

$$x^2 - 4x - 12 = 0,$$
$$(x - 6)(x + 2) = 0;$$

whence $\qquad\qquad x = 6 \quad \text{or} \quad -2.$

Therefore the curve cuts the x-axis at the points $(6, 0)$ and $(-2, 0)$.

28. General problems related to a curve and its equation. As implied by the definitions in the preceding article, the study of curves by means of their equations involves two distinct, but closely related, problems. When the equation is known, the problem is to construct its graph and to study the properties of the curve by means of its equation. When a curve is known, or is defined by given properties, the problem is to find its equation.

The straight line has been treated in this manner in the first chapter, where it was shown that the graph of every equation of the first degree is a straight line, which can be drawn by finding two points on the line, and that the equation of any straight line can be found by appropriate methods.

The construction of the graphs of equations of degree higher than the first is explained in Arts. 29 and 30. The study of the properties of a curve by means of its equation is treated briefly in Arts. 31 and 32 and more fully in Chapter XI. The general method of finding the equation of a curve defined by given properties is illustrated in Art. 33.

29. Plotting graphs of equations. A few curves used in analysis are defined by geometric means, as is the circle in plane geometry, but usually a curve is defined by its equation. Since an equation in two unknowns is in general satisfied by an infinite number of pairs of values, we cannot plot all the points of a curve defined by an equation. But we can approximate the graph by finding a number of points whose coördinates satisfy the equation and by joining them with a smooth curve. This process is known as *plotting the graph of the equation.* In order to systematize the work it is customary to proceed in the following manner.

I. *Solve the equation for y in terms of x.* (Sometimes it is easier to solve for x in terms of y. If this is the case, interchange x and y in the following directions.)

II. 1. *Substitute convenient positive and negative values (usually integral) for x, and calculate the corresponding values of y.* Each pair of values of x and y satisfy the equation, and are therefore the coördinates of a point on the graph.

2. *Arrange these pairs of values in a table, with the values of x increasing algebraically.*

III. *Plot each point, and join the points in the order of the table by a smooth curve.*

Example. Plot the graph of the equation $x^2 - 2x - y = 0$.

x	y
− 2	8
− 1	3
0	0
1	− 1
2	0
3	3
4	8

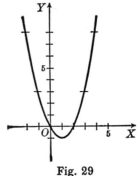

Fig. 29

Solution. Solving for y in terms of x, we have

$$y = x^2 - 2x.$$

Substituting $x = -2, -1, 0, 1, 2, 3, 4$, we get $y = 8, 3, 0, -1, 0, 3, 8$, respectively.

The points thus determined are tabulated at the left. If the table is extended by taking values of x greater than 4, it is apparent that the corresponding values of y will increase. The curve extends indefinitely to the right and upward. Similarly, by taking values of x less than -2, it can be seen that the curve extends indefinitely to the left and upward.

30. Further remarks on plotting. When the solution for y in terms of x involves a square root, both signs must be used before the radical and each value assigned to x will give two values of y and hence two points to be plotted.

The table of values should extend far enough to give a good idea of the shape of the graph.

The points where the curve crosses the axes, if there are any, should be found. This may be done as in the second example in Art. 27.

If the curve appears to change its direction abruptly at any point, look for an error in the table of values. If there is no such error, take intermediate fractional values of x (or y) and plot the corresponding points. This should also be done if the points plotted do not show clearly the form of the curve.

Applications of the above remarks are given in the solutions of the following illustrative examples.

Example 1. Plot the graph of $4\,x^2 + y^2 = 24$.

x	y
-3	imag.
-2	± 2.8
-1	± 4.5
0	± 4.9
1	± 4.5
2	± 2.8
3	imag.

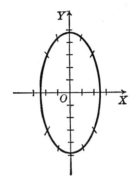

Fig. 30.1

Solution. Solving for y in terms of x,

$$y = \pm\sqrt{24 - 4\,x^2}$$
$$= \pm\,2\sqrt{6 - x^2}.$$

A table of square roots or a slide rule will be found convenient in calculating the various values of this radical. As y is imaginary for $x > 3$, but real for $x = 2$, the curve must cross the x-axis between $x = 2$ and $x = 3$. These points are readily found by setting $y = 0$, whence

$$4\,x^2 = 24, \quad\text{or}\quad x = \pm\sqrt{6} = \pm\,2.4^+.$$

Adding the points $(2.4, 0)$ and $(-2.4, 0)$ to the list, we can draw the entire curve.

Example 2. Plot the graph of $y = x^3 - x$.

Solution. (See next page.)

x	y
-3	-24
-2	-6
-1	0
0	0
1	0
2	6
3	24
$.5$	$-.375$
$-.5$	$+.375$

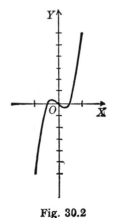

Fig. 30.2

There are no complications in finding the table of values. But when the points $(-1, 0)$, $(0, 0)$, and $(1, 0)$ are plotted, it is seen that the general rules for plotting appear to make the curve coincide with the x-axis in this region. By substituting $\pm \frac{1}{2}$ for x, we find two more points, which show that the curve rises above the x-axis between -1 and 0 and falls below it between 0 and 1.

Example 3. Plot the graph of $xy = -16$.

Solution. Solving for y, $y = -16/x$. Two things should be noticed. First, as y decreases slowly after x becomes greater than 5, it is not advisable to space these values of x closely. Second, there is no point for which $x = 0$. For if 0 is substituted for x in the equation, we have $0 \cdot y = 0$, not -16. Taking fractional values of x, we have the supplementary table. This indicates that the curve recedes indefinitely along the y-axis as well as along the x-axis. Since y increases indefinitely as x approaches 0, it is customary to say that for $x = 0$, $y = \pm \infty$. This abbreviation, however, should not be allowed to obscure the fact that for $x = 0$ there is *no* point on the curve.

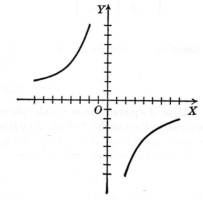

x	y	x	y
1	-16	-1	16
2	-8	-2	8
3	-5.3	-3	5.3
4	-4	-4	4
5	-3.2	-5	3.2
10	-1.6	-10	1.6
16	-1	-16	1
32	-0.5	-32	0.5
$\frac{3}{4}$	-21.3	$-\frac{3}{4}$	21.3
$\frac{1}{2}$	-32.0	$-\frac{3}{4}$	32.0

Fig. 30.3

PROBLEMS

1. Plot the graphs of the following equations.

a. $x = -3$.

b. $5x - 2y + 6 = 0$.

c. $y = x^2 - 6x + 8$.

d. $4y = 8 - 2x - x^2$.

e. $x^2 + y^2 = 32$.

f. $x^2 + y^2 = 36$.

g. $x^2 - y^2 = 4$.

h. $x^2 + 4y^2 = 36$.

i. $x^2 - 4y^2 = 16$.

j. $2x^2 + y^2 = 2$.

k. $y^2 = x^3$.

l. $y^3 = x^2$.

m. $y = x^3$.

n. $y^3 = x$.

o. $2xy = 9$.

p. $x^2y = 16$.

q. $xy^2 = 16$.

r. $x^2 + 4y = 0$.

s. $x^2 - y^2 + 16 = 0$.

t. $x^2y + 4y = 8$.

u. $x^2y + y = x$.

v. $x^2y - 4y = x$.

2. Find one of the curves in Problem 1 which passes through

      a. (0, 0).   **b.** (4, 0).

3. Plot the graphs of the following equations for the range of values indicated in each case.

Equation	Range of Values
a. $y^2 - 4y = x - 3$,	$y = -2$ to $y = 6$.
b. $x = y^2 + 5y + 4$,	$y = -7$ to $y = 2$.
c. $4x^2 - 9y^2 = 36$,	$x = -12$ to $x = 12$.
d. $x^2 - y^2 + 4x = 0$,	$x = -10$ to $x = 6$.
e. $y = x^3 - 4x$,	$x = -3$ to $x = 3$.
f. $2y = x^3 - 2x^2$,	$x = -2$ to $x = 4$.
g. $xy = 3y - 1$,	$x = -1$ to $x = 7$.
h. $xy - x^2 = 12$,	$x = -8$ to $x = 8$.
i. $5y = (x - 1)(x + 4)^2$,	$x = -6$ to $x = 3$.
j. $4y^2 = -x^3 + 4x^2$,	$x = -8$ to $x = 4$.
k. $y = x^3 - 9x - 5$,	$x = -3$ to $x = 4$.
l. $10y = x^4 - 9x^2$,	$x = -4$ to $x = 4$.
m. $y = (x - 1)(x + 1)(x - 2)$,	$x = -3$ to $x = 4$.

4. Find one of the curves in Problem 3 which passes through

      a. (0, 0).   **b.** (4, 0).

5. Choose convenient values (other than 0) for the arbitrary constants in each of the following equations, and plot the respective graphs.

Note. If there are two arbitrary constants, do not use the same value for both.

    a. $y = mx + b$.
    b. $x^2 + y^2 = a^2$.
    c. $b^2x^2 + a^2y^2 = a^2b^2$ (ellipse).
    d. $b^2x^2 - a^2y^2 = a^2b^2$ (hyperbola).
    e. $y^2 = 2px$ (parabola).
    f. $x^2 = 2py$ (parabola).
    g. $2xy = a^2$ (equilateral hyperbola).
    h. $y = ax^3$ (cubical parabola).
    i. $y^2 = ax^3$ (semicubical parabola).
    j. $y = a + bx + cx^2$ (parabola).

6. Find the points where the graph of Problem 3, m, crosses the x-axis. Write the equation of a curve crossing the x-axis at (4, 0), (0, 0), and (− 2, 0) and draw its graph.

7. Plot the graph of $y = 4x^2 + C$ for two values of C, using the same coordinate axes. What is the effect upon the graph of changing the value of C?

8. The point $P(x, y)$ lies on the graph of $y^2 = 8x$. Express in terms of x alone the distance between P and the point (2, 0).

9. The point $P(x, y)$ lies on the graph of $4y = x^2 - 4x$. Find the difference between the slopes of the lines which join P to the points (0, 0) and (4, 0).

31. Discussion of equations. An effective study of the properties of most curves requires a knowledge of differentiation, and so this topic will be reserved for a later chapter. A cursory examination of the equation, however, will often yield considerable information about the curve and facilitate the labor of plotting. The properties most readily found by this means are the intercepts and the symmetry of the curve with respect to the coördinate axes.

Intercepts. The intercepts are the distances from the origin to the points of intersection of the curve and the coördinate axes. That is, the x-intercepts are the abscissas of those points on the curve for which $y = 0$; similarly, the y-intercepts are the ordinates of those points on the curve for which $x = 0$. To find the intercepts on the x-axis, substitute 0 for y and solve the resulting equation for x. To find the intercepts on the y-axis, substitute 0 for x and solve the resulting equation for y. (See Art. 27.)

Symmetry. If there are only even powers of x in the equation, it is evident that negative values substituted for x will give the same values of y as the corresponding positive values of x. Thus the points of the curve may be arranged in pairs, each pair symmetrical with respect to the y-axis, since the ordinate is the same and the abscissas differ only in sign (see also Art. 8). Such a curve is said to be symmetrical with respect to the y-axis. In this case the values of y when x is negative may be obtained at once from the values of y when x is positive.

Similarly, if there are only even powers of y in the equation, the curve is symmetrical with respect to the x-axis.

If the equation of a curve is unchanged when $-x$ is substituted for x and $-y$ for y, the points of the curve may be arranged in pairs, each pair symmetrical with respect to the origin. Such a curve is symmetrical with respect to the origin.

Evidently this is also true if the substitutions mentioned change the sign of every term of the equation.

If a curve is symmetrical with respect to both axes, it is symmetrical also with respect to the origin; but a curve symmetrical with respect to the origin is not necessarily symmetrical with respect to both axes.

The tests for symmetry are summarized as follows.

The graph of an equation is symmetrical with respect to

1. *the x-axis, if there are only even powers of y;*
2. *the y-axis, if there are only even powers of x;*
3. *the origin, if the substitution of $-x$ for x and $-y$ for y leaves the equation unchanged.*

Example 1. Investigate the intercepts and the symmetry of the graph of $4x^2 + y^2 = 24$.

Solution. When $y = 0$, $x = \pm\sqrt{6}$; when $x = 0$, $y = \pm\sqrt{24}$. Hence the x-intercepts are $\pm\sqrt{6}$, and the y-intercepts are $\pm\sqrt{24}$, and four points of the curve are $(\pm\sqrt{6}, 0)$ and $(0, \pm\sqrt{24})$. As both x and y occur only with even exponents, the curve is symmetrical with respect to both axes and the origin, and a table of values for the first quadrant will suffice for the whole curve. This curve is plotted in the first illustrative example of Art. 30.

Example 2. Investigate the intercepts and the symmetry of the graph of $y = x^3 - x$.

Solution. The x-intercepts are 0 and ± 1; the y-intercept is 0. Hence the curve cuts the x-axis at three points: $(-1, 0)$, $(0, 0)$, and $(1, 0)$. The curve cuts the y-axis only at the origin. As both x and y occur with odd exponents, the curve is symmetrical with respect to neither axis. When $-x$ is substituted for x, and $-y$ for y, the equation becomes $-y = (-x)^3 - (-x)$. When the parentheses are removed, we have $-y = -x^3 + x$, which is the same as the original equation with the signs changed. Hence the curve is symmetrical with respect to the origin. This curve is plotted in the second illustrative example of Art. 30.

Variation and range of the coördinates. It is also frequently possible by inspection of the equation to determine whether y increases or decreases as x increases. Let us take as an illustration the equation discussed in Example 1. When we solve this for y in terms of x, in order to compute the table of values, we obtain $y = \pm 2\sqrt{6 - x^2}$. From this we see at once that as x increases numerically, y decreases numerically; also that x^2 can never be greater than 6, and hence that the largest possible value of $|x|$ is $\sqrt{6}$. Similarly, the largest possible value of $|y|$ is $2\sqrt{6}$.

32. Plotting by factoring. It sometimes happens that when we transpose all the terms of an equation to one side, we can factor this expression. If so, the locus of the equation is the group of lines or curves obtained by setting each factor equal to zero and by plotting the loci of the equations thus obtained on the same coördinate axes.

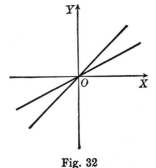

Fig. 32

Example. Plot the locus of
$$x^2 - 3xy = -2y^2.$$

Solution. Transposing $2y^2$ and factoring the expression, we have $(x - 2y)(x - y) = 0$. Setting each factor equal to zero, we have $x - 2y = 0$

and $x - y = 0$. The graphs of these equations are straight lines through the origin, with slopes $\frac{1}{2}$ and 1 respectively.

Thus, although the locus of an equation of the second degree is usually a true curve, it may sometimes be a pair of straight lines.

The validity of the process just described follows at once from the definition of the locus of an equation. For if a point lies on either of the lines thus obtained, its coördinates must satisfy one of the equations $x - 2y = 0$ or $x - y = 0$. But if either $x - 2y$ or $x - y$ is zero for a certain pair of values of x and y, their product, $x^2 - 3xy + 2y^2$, equals zero for the same values. Hence the point lies on the locus of the given equation. No other points have coördinates satisfying the given equation. No other points have coördinates satisfying the given equation; for $x^2 - 3xy + 2y^2$ cannot equal zero for any values of x and y unless one of its factors equals zero, in which case the point would be on one of the lines.

PROBLEMS

1. Investigate the intercepts and the symmetry of the graph of each of the following equations.

a. $y = 8 - x^2$.   **b.** $x^2 + 4y^2 = 16$.
c. $2x^2 + y^2 = 1$.   **d.** $x^2 - 9y^2 = 36$.
e. $y^2 + 8 = x$.   **f.** $y^2 = x^3/16$.
g. $x^2y = 36$.   **h.** $x^2y + 4y = 12$.
i. $y = 9x - x^3$.   **j.** $y = x^4 - 9x^2$.

2. Each of the following questions refers to the corresponding equation in Problem 1. Answer the question, giving the reason for the statement made.

    a. What is the largest value of y?
    b. What are the largest numerical values of x and y?
    c. What are the largest numerical values of x and y?
    d. What is the smallest numerical value of x?
    e. What is the smallest value of x?
    f. In which quadrants does this curve lie?
    g. How does y change as x increases?
    h. What is the largest value of y? Is y ever negative?
    i. For what values of x is y positive?
    j. For what values of x is y positive?

3. Using the results obtained in Problem 1, and such other information as the equation may yield, endeavor to sketch each of the curves without obtaining a table of values.

4. Which of the curves in Problem 1, Art. 30, have symmetry with respect to (**a**) the x-axis? (**b**) the y-axis? (**c**) both axes? (**d**) the origin only?

5. In the equation $4\,y = x^2 - 10\,x + 24$ substitute $-x$ for x and plot the graphs of both the new equation and the original equation on the same axes. In what way are the curves related?

6. Plot on the same axes the loci of $2\,y = x^2 - 16$ and $2\,y = 16 - x^2$. How are these loci related?

7. Plot on the same axes the loci of $y = x^3$ and $x = y^3$. How are these loci related?

8. For what value of p will the graph of $y^2 = 2\,px$ pass through the point $(16, 4)$? Plot the graph.

9. For what values of the constants a and b will the graph of $b^2x^2 + a^2y^2 = a^2b^2$ pass through $(1, 3)$ and $(4, -2)$? Plot the graph.

10. For what values of a, b, and c will the graph of $y = a + bx + cx^2$ pass through the points $(-1, 1)$, $(1, 7)$, and $(3, 5)$? Plot the graph.

11. What is the locus of

a. $x^2 + y^2 = 0$?   **b.** $(x - 2)^2 + (y + 3)^2 = 0$?

12. Show that $x^2 + y^2 + 6 = 0$ has no real locus.

13. Plot by factoring the loci of

        a. $xy = 0$.
        b. $y^2 + 5\,y - 6 = 0$.
        c. $4\,x^2 - y^2 = 0$.
        d. $4\,x^2 - y^2 - 8\,x + 4\,y = 0$.
        e. $x^2 - y^2 + 10\,x - 2\,y + 24 = 0$.
        f. $(3\,x + 2\,y)^2 + (3\,x + 2\,y) = 2$.
        g. $2\,x^2 - 3\,xy + 4\,x = 6\,y$.
        h. $(x^2 + y^2 - 25)(x^2 + y^2 - 16) = 0$.

14. What is the locus of $(x - k)(y - k) = 0$ if k is allowed to take on all positive and negative integral values?

15. Write a single equation whose locus is the pair of lines

a. Whose equations are $x - y = 0$ and $x = -y$.
b. Whose equations are $x - 2\,y = 6$ and $2\,x + y = 4$.
c. Which are parallel to the y-axis and which have the x-intercepts 2 and -2.
d. Which pass through the point $(1, 2)$ and have slopes -1 and $+1$.

16. Find the equations of the bisectors of the angles between the lines defined by $2\,x^2 - 5\,xy - 3\,y^2 + 12\,x + 6\,y = 0$.

33. Derivation of equations.

In order to find the equation of a given curve or of the locus of a point satisfying certain given conditions, it is customary to draw a figure showing the data of the problem, to take a typical point $P(x, y)$ which appears to satisfy the given conditions, and to express these conditions in the form of an equation containing x and y and no other variables. The following examples illustrate the method.

Example 1. Find the equation of the circle with the center at (10, 0) and with the radius 10.

Solution. Let $P(x, y)$ be any point on the circle, denote the center by C, and draw CP (Fig. 33.1). Then $CP = 10$, by the definition of a circle.
But $CP = \sqrt{(x - 10)^2 + y^2}$ by the distance formula.
Hence $\sqrt{(x - 10)^2 + y^2} = 10$.
After squaring and simplifying we obtain

$$x^2 + y^2 - 20\,x = 0,$$

which is the required equation of the circle.

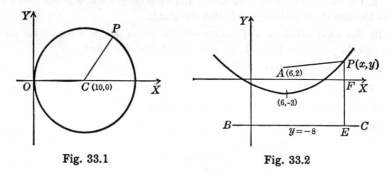

Fig. 33.1   Fig. 33.2

Example 2. Find the equation of the locus of a point equidistant from the line $y = -8$ and the point (6, 2).

NOTE. Here and in the future the phrase "equidistant from" has the ordinary meaning and does not refer to directed distances. In the *solution* the lines used are directed and must be read in the positive direction.

Solution. In Fig. 33.2 the given point is A and the given line is BC. Let $P(x, y)$ be a general point which seems to fulfill the condition of the problem. Since P is to be equidistant from A and BC, we draw AP and the line EP perpendicular to BC. By the statement of the problem these lines are equal in length; that is,

$$AP = EP.$$

We can replace AP by $\sqrt{(x - 6)^2 + (y - 2)^2}$, using the distance formula.
Since EP is the distance from E to P and the ordinate of E is -8, the vertical distance formula gives

$$EP = y - (-8) = y + 8.$$

Therefore $\qquad \sqrt{(x - 6)^2 + (y - 2)^2} = y + 8.$

Squaring and simplifying, we have

$$20\,y = x^2 - 12\,x - 24.$$

The graph should now be plotted on the original figure, so as to show its shape and relation to the given point and line.

Example 3. The ends of a line of variable length are on two fixed perpendicular lines. Find the locus of the mid-point of this line if the area of the triangle thus formed is 24.

Solution. The perpendicular lines should be taken as the coördinate axes. Let AB be one of the positions of the variable line and take $P(x, y)$ as its midpoint. The conditions of the problem are

$$AP = PB \quad \text{and} \quad \text{area of } AOB = 24.$$

The latter equation gives at once

$$\frac{OA \cdot OB}{2} = 24.$$

We must now express OA and OB in terms of x and y. Draw the coördinates of P, $EP = x$ and $FP = y$. The figure suggests the theorem of plane geometry

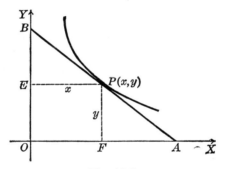

Fig. 33.3

that a straight line parallel to the base of a triangle and bisecting one side bisects the other.

Hence $\qquad OA = 2\,OF = 2\,x \quad \text{and} \quad OB = 2\,y.$

Substituting above, we get

$$\frac{(2\,x)(2\,y)}{2} = 24,$$

or $\qquad\qquad\qquad xy = 12,$

which is the equation of the locus. The curve can now be drawn in the usual manner.

NOTE. In none of the above examples is it shown that the equation obtained is satisfied by the coördinates of no point not on the locus, or, what amounts to the same thing, that every point whose coördinates satisfy the equation lies on the locus. This can be done by simply retracing the steps in the derivation of the equation in reverse order. This, however, is seldom necessary.

PROBLEMS

In each of the following problems the graph should be plotted after the equation has been found, and the data of the problem should be exhibited on the same figure.

1. Find the equation of the locus of points equidistant from

      a. $(-3, 6)$ and $(10, -2)$.   **b.** $(1, 4)$ and $(4, 1)$.

      c. $(9, 6)$ and $(15, -6)$.   **d.** $(0, 6)$ and $(4, 2)$.

2. Find the equation of the locus of points equidistant from

      a. The point $(-6, 0)$ and the line $x = 16$.

      b. The point $(0, 8)$ and the line $y = 5$.

      c. The point $(1, 4)$ and the line $y = 0$.

      d. The point $(8, 6)$ and the line $x = 12$.

3. Find the locus in each case of Problem 2 if the distance from the point is always two units greater than the distance from the line.

4. The base of an isosceles triangle is OB, where $O = (0, 0)$ and B is a variable point on the x-axis. Find the locus of the third vertex P if the area of OBP is 16.

5. Find the equation of the circle whose center and radius are as follows.

      a. Center $(4, -5)$, radius $= 10$.

      b. Center $(-6, 0)$, radius $= 6$.

      c. Center $(0, -5)$, radius $= 5$.

      d. Center $(0, 0)$, radius $= r$.

      e. Center (h, k), radius $= r$.

6. Find the locus of a point if its distance from the point $(-4, 0)$ is always twice its distance from the point $(2, 0)$.

7. Find the locus of a point if it is always three times as far from the origin as from the point $(4, 4)$.

8. The base of a triangle is AB, where A is $(-3, 0)$ and B is $(3, 0)$. Find the locus of the vertex P if the slope of BP is 1 unit greater than the slope of AP.

9. Find the locus of a point whose distance from the origin is a mean proportional between its distances from the points $(2, 0)$ and $(-2, 0)$.

10. The ends of the hypotenuse of a right triangle are $(2, 2)$ and $(4, -2)$. Find the equation of the locus of the vertex of the right angle.

11. The base of a triangle is AB, where A is $(-5, 0)$ and B is $(5, 0)$. Find the locus of the third vertex P if

      a. $(AP)^2 - (BP)^2$ is a constant, k.

      b. $(AP)^2 + (BP)^2$ is a constant, k.

      c. The slope of BP is k times the slope of AP.

      d. The sum of the slopes is a constant, k.

12. The base of a triangle is AB, where A is $(5, 0)$ and B is $(10, 0)$. Find the locus of the third vertex P if the median from A to BP is always 5 units long.

13. The base of a triangle is AB, where A is $(-4, 0)$ and B is $(4, 0)$. Find the locus of the third vertex P if

    a. One base angle is double the other.
    b. The sum of the base angles is 45°.
    c. The difference of the base angles is 45°.
    d. The difference of the other two sides is always 6.
    e. The sum of the other two sides is always 10.

14. Two vertices of a triangle are $A(-2, 3)$ and $B(6, 5)$. Find the locus of the third vertex C if

    a. The slope of AC is 2 less than the slope of BC.
    b. The slope of AC is one half the slope of BC.
    c. The sum of the slopes of AC and BC is 0.

15. Two vertices of a triangle are $(0, 0)$ and $(6, 0)$. Find the locus of the third vertex if

    a. The sum of the slopes of the sides of the triangle is 0.
    b. The slope of the median from $(0, 0)$ is 1 unit greater than the slope of the median from $(6, 0)$.
    c. The median from $(0, 0)$ is always perpendicular to the opposite side.

34. Intersection of curves. As in the case of straight lines, it follows from the definition of the equation of a curve that if two curves intersect, the coördinates of each common point must satisfy both equations. Hence, to find the points of intersection of two curves, solve the equations simultaneously.

If the results obtained by solving the equations simultaneously are imaginary, there is no point whose coördinates satisfy both equations, and the curves do not intersect.

Example. What are the points of intersection of the curves whose equations are $x^2 + y^2 = 36$ and $x^2 = 5\,y$?

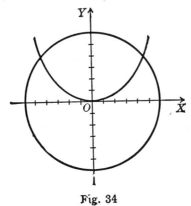

Fig. 34

Solution. The successive steps in solving the equations simultaneously are

$$5\,y + y^2 = 36;$$
$$y^2 + 5\,y - 36 = 0;$$
$$(y + 9)(y - 4) = 0;$$
$$y = -9 \quad \text{or} \quad 4.$$

Substituting these values in the equation $x^2 = 5\,y$, we have for $y = 4$,

$$x = \pm \sqrt{20} = \pm 4.47, \text{ approx.};$$

for

$$y = -9, x = \sqrt{-45}.$$

Hence the points of intersection are $(\sqrt{20},\ 4)$ and $(-\sqrt{20},\ 4)$. The graphs furnish a convenient check.

PROBLEMS

Find the points common to the loci of the following equations and check by plotting the loci. If any coördinates are fractional or irrational, approximations to the nearest tenth are accurate enough for checking.

1. $4\,x - y = 4,$
$y^2 - 8\,x = 0.$

2. $x^2 + y^2 = 41,$
$4\,x - 5\,y = 41.$

3. $4\,x^2 + 9\,y^2 = 36,$
$4\,x^2 + 3\,y = 30.$

4. $x^2 + y = 7,$
$y^2 - 81 = 0.$

5. $x^2 + y^2 = 64,$
$x + 2\,y = 2.$

6. $y = x^2 - 6\,x + 8,$
$y = 3\,x - 9.$

7. $x^2 + y^2 = 16,$
$xy = 7.$

8. $x^2 - y^2 = 64,$
$x - 2\,y = 0.$

9. $x^2 - y^2 = 64,$
$2\,x - y = 0.$

10. $x^2 = y,$
$y^2 = 2\,x.$

11. $y = \dfrac{8}{4 + x^2},$
$x^2 = 4\,y.$

12. $y = x^3,$
$y = x^2 + 2\,x.$

13. $x^2 y + y - x = 0,$
$4\,y = x.$

14. $3\,y = x^3 - 6\,x,$
$x + y = 0.$

15. Find the points on the line whose equation is $x - 7\,y = 34$ which are 5 units distant from the point $(2, -1)$.

HINT. First find the equation of the circle whose radius is 5 and whose center is $(2, -1)$.

16. Find the length of the chord of the circle $x^2 + y^2 = 64$ cut off on the line $x + y = 2$.

17. Find the distance, measured along the line $y = 2\,x$, from the origin to the ellipse $4\,x^2 + 9\,y^2 = 36$.

18. Find the coördinates of the points 4 units distant from $(1, 5)$ and 5 units distant from $(-3, 4)$.

19. For what values of m will the line $y = mx$ fail to meet the curve $y = x^2 + 5$?

CHAPTER III

DERIVATIVE OF A FUNCTION

35. Variables and constants. The symbols used in mathematics to represent numbers or numerical quantities are of two kinds, *variables* and *constants*. These terms are also applied to the quantities which the symbols represent.

A variable, as the name implies, does not need to have a fixed value and usually has an unlimited number of values. As simple examples of variable quantities we have the velocity v of a falling body and the abscissa x of a point tracing a curve.

A constant has a fixed value. There are two kinds of constants, namely, *absolute* constants, which have the same values in all discussions, as 2, -6, π, and *arbitrary* constants, which may have any values assigned, but have the same values throughout a particular discussion, as the letters representing the sides and angles of a triangle in a geometrical proof of the law of sines.

To avoid confusion we usually use the later letters of the alphabet for variables and the early ones for arbitrary constants. Thus in the equation of a straight line, $y = mx + b$, x and y are variables, being the coördinates of any point on the line, while the slope m and the y-intercept b are arbitrary constants, since these are fixed for a particular line, but are different for other lines.

It may happen that the values of a variable are restricted by the nature of the problem. Thus if the variable x denotes the sine of an angle, x has values only between -1 and $+1$, inclusive. This is often expressed by saying that "x ranges over the interval $(-1, +1)$." In general, if a and b are two constants and $a < b$, the set of all numbers between a and b inclusive is called the *interval* (a, b). To emphasize the fact that this includes the end values, we call it a *closed* interval. If the values of x are restricted to the interval (a, b), then $a \leq x \leq b$. If we wish to exclude the end values, we say that the interval is *open*, in which case $a < x < b$.

36. Functions. Two variables may be so related to each other that for each of a set of values assigned to one there are one or more definite values of the other. We then say that the variables are functionally related or that one is a function of the other. Such relations may be given by equations or by other means. The following definition covers all cases.

DEFINITION. *If one of two variables has one or more definite values corresponding to each one of a set of values assigned to the other variable, the first variable is said to be a function of the second.* *

The first variable is called the *dependent variable* or *function*, and the second variable is called the *independent variable* or *argument*. Thus the area of a circle is a function of the radius, the pressure of steam in a cylinder is a function of the temperature, the ordinate of a point $P(x, y)$ on a curve is a function of the abscissa of the point, etc. In the last example y is the dependent variable and x is the independent variable; but if we should regard the abscissa as a function of the ordinate, x would then be the dependent variable and y the independent variable.

If y is a function of x and to each value of x there corresponds just one value of y, we say that y is a *one-valued* function of x. In statements about functions in this book it will be understood, unless otherwise stated, that the functions are one-valued. If each value of x gives the same value of y, the function is constant.

37. Functional notation. To indicate that y is a function of x, we write $y = f(x)$. This is read "y equals f of x," or "y equals a function of x," and must not be confused with $y = f \cdot x$. Similarly, $y = f(x, t)$ denotes that y is a function of the two variables x and t.

Furthermore, the notation $y = f(x)$ may be used to represent the particular way in which y is related to x. For functional relations given by equations which are solved for y in terms of x, the symbol $f(x)$ may be regarded as an abbreviation for the right-hand member. For example, if $y = 6x - x^2$, we may represent this equation by $y = f(x)$, in which case $f(x)$ stands for the expression $6x - x^2$ throughout the discussion. In short, the symbol $f(x)$ may be used to represent any mathematical expression involving x as a variable, or any quantity which is a function of x. To denote several different functions of x, we use various letters, as $f(x)$, $g(x)$, $\phi(x)$, etc.

When a particular function $f(x)$ is under discussion, such symbols as $f(1)$ or $f(-2)$ mean the values of this function when 1 or -2, respectively, are substituted for x. Thus if $f(x) = 6x - x^2$,

$$f(1) = 5, \qquad f(-x) = -6x - x^2,$$
$$f(-2) = -16, \qquad f(3-x) = 6(3-x) - (3-x)^2 = 9 - x^2.$$

38. Inverse functions. The correspondence which gives y as a function of x also determines x as a function of y if we regard x as the dependent and y as the independent variable. Each of these functions is

*The word "function" by itself is often used in the sense of a particular kind of functional relation of the type described in this definition, with no thought of the variables involved. Thus we speak of "trigonometric functions," "algebraic functions," etc.

called the *inverse* of the other. Thus the functions $y = x^3$ and $x = \sqrt[3]{y}$ are inverse functions. (If we use the word "function" in the sense of a particular kind of functional relation, as mentioned in the footnote, Art. 36, we may even speak of x^3 and $\sqrt[3]{x}$ as inverse functions.) It should be noted that if $y = f(x)$ is one-valued, the inverse $x = g(y)$ may be multi-valued. If $y = f(x)$ is given by an equation solved for y, the inverse is found by solving this equation for x in terms of y. Thus if $y = 6x - x^2$, the inverse is $x = 3 \pm \sqrt{9 - y}$. Here $y = f(x)$ is one-valued, but $x = g(y)$ is two-valued for $y < 9$.

If $F(x, y)$ denotes a mathematical expression containing x and y, there may be pairs of values of x and y which satisfy the equation $F(x, y) = 0$. In this event the equation $F(x, y) = 0$ is said to define y as a function of x (or vice versa) *implicitly*. (This is merely a new way of expressing the functional relation, *not* a new kind of function.) If we can solve this for y in terms of x, we obtain $y = f(x)$ *explicitly*. The two functions at the end of the previous paragraph are both expressed implicitly by the equation $x^2 - 6x + y = 0$.

39. Domain of definition of a function. In defining a function of x it is always understood that there is a certain set of values of x for which the function is defined. This is called the *domain of definition* of the function. In careful work the domain of definition is stated when the function is defined. If the functional relation $y = f(x)$ is expressed by a formula and nothing further is stated, it is understood that the domain of definition comprises all values of x for which the formula yields real values of y. The set of all values of y thus obtained will be the domain of definition of the inverse function $x = g(y)$.

Sometimes the domain of definition includes every real number and often not. Thus the function $y = 6x - x^2$ is defined for every value of x, but the inverse function $x = 3 \pm \sqrt{9 - y}$ is not defined for $y > 9$ since imaginaries are excluded. In the former case we say that the domain is the interval $(-\infty, \infty)$ and in the latter case that it is the interval $(-\infty, 9)$.

Restrictions on the domain of definition of a function are sometimes due to the nature of the correspondence desired and sometimes to the nature of the problem. As an example consider the motion of a ball which is projected upward from the ground with a speed of 64 ft. per second and falls back to the ground where it stops. During the motion the height (h ft.) after t sec. is given by the function

$$h = 64t - 16t^2.$$

Setting $h = 0$ we find $t = 0$ or $t = 4$, which indicates that the ball was projected at a time $t = 0$ and fell back to the ground 4 sec. later. The equation has no physical meaning before the ball started to move or after the ball struck the ground. In this case the domain of definition of the function h is $0 \leqq t \leqq 4$.

PROBLEMS

1. Write each of the following equations in the forms $y = f(x)$ and $x = g(y)$.

    a. $16 x^2 - 64 x + y = 0$.   **b.** $x^2 + 4 y^2 = 16$.

    c. $x^2 - y^2 = 25$.   **d.** $y^2 + 2 x^2 - x^4 = 0$.

2. If $y = f(x)$ is the equation of a certain curve, what is the graph of each of the following equations?

    a. $y = 2 + f(x)$.   **b.** $y + f(x) = 0$.

    c. $y = f(-x)$.   **d.** $y + f(-x) = 0$.

3. Find

    a. $f(0)$ and $f(-1)$ if $f(x) = \sqrt{1 + x^2} - x$.

    b. $f(\sqrt{2})$ and $f(-6)$ if $f(t) = 6 - 3 t^2$.

    c. $g(0)$ and $g(\pi/2)$ if $g(\theta) = \sin 2 \theta$.

    d. $\phi(a)$ and $\phi(-x)$ if $\phi(x) = \sqrt{a^2 - x^2} - x$.

4. If $f(x) = \sin x$, prove that

    a. $f(\pi - x) = f(x)$.

    b. $f(2 x) = 2 f(x)\sqrt{1 - [f(x)]^2}$.

5. If $\phi(x) = a^x$, prove that $\phi(x) \cdot \phi(y) = \phi(x + y)$.

6. State the domain of definition of each of the following functions.

    a. $f(x) = \sqrt{3 x - 10}$.   **b.** $g(y) = \sqrt{8 - 3 y}$.

    c. $f(t) = \sqrt{25 - t^2}$.   **d.** $f(s) = \sqrt{4 s^2 - 25}$.

    e. $g(x) = \sqrt[3]{3 x + 2}$.   **f.** $\phi(r) = r + \sqrt{r + 3}$.

    g. $\psi(w) = \sqrt{4 w - w^2}$.   **h.** $f(t) = \sqrt{t^2 + 6 t}$.

40. Limit of a function. The idea of a variable approaching a constant as a limit occurs in elementary geometry. Suppose that a regular polygon of n sides is inscribed in a circle of unit radius. The area A of the polygon depends on the number n and, as n increases without bound, A approximates a fixed number denoted by π.

In order to make precise the idea of a limiting value of a variable, we give the following definition.

DEFINITION. *The variable v is said to have the constant l as a limit when the successive values of v are such that $|l - v|$ ultimately becomes and remains less than any preassigned positive number, however small.*

The relation defined is written $\lim v = l$. For convenience we shall use the notation $v \longrightarrow l$, read "v has l as a limit" or, more briefly, "v approaches l."

In applications of the preceding definition two variables are usually involved: (1) an independent variable which may be made arbitrarily to approach some constant as a limit, and (2) a dependent variable whose behavior is to be examined. If we call the independent variable x and the dependent variable v, which is a function of x, say $v = f(x)$, the question is to determine what constant value, if any, will be approached by v when x approaches some particular value a. If v approaches l when x approaches a, this fact is expressed by the notation

$$\lim_{x \to a} v = l,$$

read "The limit of v, as x approaches a, is l."

It should be observed that the statement above says nothing about the value of v when $x = a$. When $x = a$, the value of v may or may not be equal to l. The statement does imply that v is defined for every value of x such that $|a - x|$ is small enough, except the value $x = a$. It is not necessary that v be defined for $x = a$. The statement further implies that the value of $|l - v|$ can be made less than any preassigned positive number, however small, by taking $|a - x|$ less than some positive number which can be determined. The idea of the limit of a function is made precise as follows.

DEFINITION. *The number l is the limit of $f(x)$ as x approaches a if for every positive number ϵ, there is a positive number δ such that $|l - f(x)| < \epsilon$ if $0 < |a - x| < \delta$.*

The definition is equivalent to the statement that the value of $f(x)$ lies between $l - \epsilon$ and $l + \epsilon$ for every value of x between $a - \delta$ and $a + \delta$, except possibly a.

Example 1. If $f(x) = 6x - x^2$, prove by the formal definition that $\lim\limits_{x \to 3} f(x) = 9$.

Solution. Let ϵ be any positive number. In this example the supposed limit $l = 9$ and
$$|l - f(x)| = |9 - 6x + x^2| = (x - 3)^2.$$
Now $(x - 3)^2 < \epsilon$ if $|x - 3| < \sqrt{\epsilon}$.
Hence if we take $\delta \leqq \sqrt{\epsilon}$,
$$|9 - f(x)| < \epsilon \quad \text{if} \quad 0 < |x - 3| < \delta,$$
which is the requirement of the formal definition.

For this example we have found a rule which gives us a value of δ corresponding to any value of ϵ which may be specified. Thus if 0.0001 is proposed as a value of ϵ, we can use for δ any value less than or equal to $\sqrt{0.0001}$ or 0.01 and assert that $6x - x^2$ will differ from 9 by less than 0.0001 if x differs from 3 by less than 0.01. Or we may write

$$8.9999 < f(x) < 9.0001 \text{ if } 2.99 < x < 3.01.$$

Example 2. If $f(x) = \dfrac{x^2 - a^2}{x - a}$, show that

$$\lim_{x \to a} f(x) = 2a.$$

Solution. The given function is not defined for $x = a$ since division by zero is excluded. But for every other value of x, $f(x)$ is defined and $f(x) = x + a$. Hence

$$\lim_{x \to a} \frac{x^2 - a^2}{x - a} = \lim_{x \to a} (x + a) = 2a.$$

If $f(x) = 1/x^2$, then $\lim\limits_{x \to 0} f(x)$ does not exist. As $x \longrightarrow 0$, $f(x)$ increases without bound and approaches no fixed number. In such a case we say that $f(x)$ *becomes infinite* as $x \longrightarrow 0$ and write

$$\lim_{x \to 0} f(x) = \infty.$$

Since ∞ is not a number this symbol should *not* be read "$f(x)$ approaches infinity." If x increases without bound, then $f(x)$ decreases and approaches zero as a limit. In such a case we say that the limit of $f(x)$, as x becomes infinite, is zero and write

$$\lim_{x \to \infty} f(x) = 0.$$

41. Theorems on limits. It is apparent that the direct determination of a limit of a function, if the limit exists, is apt to be tedious and difficult except for the simplest functions. To assist in this determination the following theorems may be applied. Proofs* are omitted.

Let u and v be functions of x and suppose

$$\lim_{x \to a} u = l, \quad \lim_{x \to a} v = m.$$

Theorem I. $\lim\limits_{x \to a} (u \pm v) = \lim\limits_{x \to a} u \pm \lim\limits_{x \to a} v = l \pm m.$

Theorem II. $\lim\limits_{x \to a} uv = \left(\lim\limits_{x \to a} u\right)\left(\lim\limits_{x \to a} v\right) = lm.$

Theorem III. $\lim\limits_{x \to a} u^n = \left(\lim\limits_{x \to a} u\right)^n = l^n$, where n is any rational number.

Theorem IV. $\lim\limits_{x \to a} \dfrac{u}{v} = \dfrac{\lim\limits_{x \to a} u}{\lim\limits_{x \to a} v} = \dfrac{l}{m}$, provided $m \neq 0$.

*See Granville, Smith, Longley: *Elements of Calculus* (Ginn and Company, 1946), pages 17, 18.

If $m = 0$ and $l \neq 0$, the quotient has no limit, but becomes infinite. If both l and m are zero, the limit is indeterminate and can be evaluated only by some suitable transformation of the quotient. (See Example 2, Art. 40.)

In applying the four theorems above, some consequences of the definition of a limit should be noted, as listed below. The letter k denotes a constant.

1. If $f(x) = k$, then $\lim_{x \to a} f(x) = k$, for every number a.

2. If $f(x) = x$, then $\lim_{x \to a} f(x) = a$, for every number a.

3. If $\lim_{x \to a} u = 0$, then $\lim_{x \to a} \dfrac{k}{u} = \infty$.

4. If $\lim_{x \to a} u = \infty$, then $\lim_{x \to a} ku = \infty$.

5. If $\lim_{x \to a} u = \infty$, then $\lim_{x \to a} \dfrac{k}{u} = 0$.

Example 1. Find $\lim_{x \to 2} (5 x^2 - 3 x + 4)$.

Solution. $\lim_{x \to 2} (5 x^2 - 3 x + 4) = \lim_{x \to 2} 5 x^2 - \lim_{x \to 2} 3 x + \lim_{x \to 2} 4$,   By I

$$\lim_{x \to 2} 5 x^2 = \left(\lim_{x \to 2} 5\right)\left(\lim_{x \to 2} x^2\right) = 5(2)(2) = 20. \quad \text{By II, (1), (2), III}$$

$$\lim_{x \to 2} 3 x = 6. \qquad\qquad \text{By II, (1)}$$

$$\lim_{x \to 2} 4 = 4. \qquad\qquad \text{By (1)}$$

Hence $\lim_{x \to 2} (5 x^2 - 3 x + 4) = 20 - 6 + 4 = 18$.

Example 2. Find $\lim_{x \to \infty} \dfrac{3 x^3 - 4 x^2 + 8}{2 + 3 x - 5 x^3}$.

Solution. Divide numerator and denominator by x^3, the highest power of x in either. Then, since $x \neq 0$, we have

$$\lim_{x \to \infty} \frac{3 x^3 - 4 x^2 + 8}{2 + 3 x - 5 x^3} = \lim_{x \to \infty} \frac{3 - \dfrac{4}{x} + \dfrac{8}{x^3}}{\dfrac{2}{x^3} + \dfrac{3}{x^2} - 5}.$$

In the right-hand member the limit of each term containing x is zero, by (5). Hence, by I, (1), and IV, the limit of the quotient is $-\frac{3}{5}$.

Example 3. Find $\quad \lim\limits_{h \to 0} \dfrac{\sqrt{x+h} - \sqrt{x}}{h}$.

Solution. As $h \longrightarrow 0$ the limit of the denominator is zero and **IV** is not applicable. Since the limit of the numerator is also zero we transform the fraction by rationalizing the numerator.

$$\frac{\sqrt{x+h} - \sqrt{x}}{h} \cdot \frac{\sqrt{x+h} + \sqrt{x}}{\sqrt{x+h} + \sqrt{x}} = \frac{x+h-x}{h(\sqrt{x+h} + \sqrt{x})}.$$

If $h \neq 0$, a factor h can be cancelled from numerator and denominator of the right-hand member and

$$\lim_{h \to 0} \frac{\sqrt{x+h} - \sqrt{x}}{h} = \lim_{h \to 0} \frac{1}{\sqrt{x+h} + \sqrt{x}} = \frac{1}{2\sqrt{x}}. \quad \text{[By IV, I, III.}$$

PROBLEMS

Find the following limits.

1. $\lim\limits_{x \to 0} \dfrac{2x^3 + 4x^2 - 5}{3x^2 - 6x + 8}$.

2. $\lim\limits_{y \to 0} \dfrac{ay^3 + by^2 + c}{ey^4 + fy + g}$.

3. $\lim\limits_{x \to \infty} \dfrac{3x^2 + 5x - 7}{4x^2 - 3x - 8}$.

4. $\lim\limits_{t \to \infty} \dfrac{3t^4 + 5}{4t^5 + 6}$.

5. $\lim\limits_{x \to \infty} \dfrac{4x^2 + 2x - 3}{5x + 4}$.

6. $\lim\limits_{s \to a} \dfrac{s^4 - a^4}{s^2 - a^2}$.

7. $\lim\limits_{x \to 1} \dfrac{x^2 + x - 2}{x - 1}$.

8. $\lim\limits_{t \to 2} \dfrac{t^2 - 3t + 2}{t^2 - t - 2}$.

9. $\lim\limits_{x \to 0} \dfrac{a_0 x^n + a_1 x^{n-1} + \cdots + a_n}{b_0 x^n + b_1 x^{n-1} + \cdots + b_n}$.

10. $\lim\limits_{x \to \infty} \dfrac{a_0 x^n + a_1 x^{n-1} + \cdots + a_n}{b_0 x^n + b_1 x^{n-1} + \cdots + b_n}$.

11. In each of the following cases find

$$\lim_{h \to 0} \frac{f(x+h) - f(x)}{h}.$$

     a. $f(x) = x^2$.   **b.** $f(x) = x^3$.   **c.** $f(x) = 1/x$.

42. Continuity. In Example 1, Art. 41, it was shown that if
$$f(x) = 5x^2 - 3x + 4, \lim_{x \to 2} f(x) = 18.$$

Now the function is defined for $x = 2$ and $f(2) = 18$. The function is said to be continuous at the point where $x = 2$.

DEFINITION. *A function $f(x)$ is continuous at the point where $x = a$ if* $\lim\limits_{x \to a} f(x) = f(a)$.

This definition implies that a value has been assigned to the function for the value $x = a$ and that the limiting value of the function exists as $x \longrightarrow a$.

The function is said to be *discontinuous* at the point where $x = a$ if the conditions of the definition are not satisfied.

The function considered in Example 2, Art. 40, does not satisfy the definition of continuity at the point where $x = a$ because the function is not defined by the given expression. However, the function becomes continuous if we say, arbitrarily, that $f(a) = 2a$.

The function $f(x) = 1/x^2$ is discontinuous at the point where $x = 0$ because the function is not defined and also because the limit does not exist.

DEFINITION. *A function $f(x)$ is said to be continuous in an interval when it is continuous at every point in the interval.*

The theorems on limits, Art. 41, give corresponding theorems on continuity. Thus if $u(x)$ and $v(x)$ are continuous at the point where $x = a$, then $u(x) + v(x)$ is continuous at the point where $x = a$.

Since any explicit algebraic function is obtained by applying algebraic operations to x and constants, it follows from the theorems on limits that *an explicit algebraic function is continuous in any interval in which it is defined.*

It can be proved that the logarithmic, exponential, and trigonometric functions are continuous in any interval in which they are defined. Hence all functions considered in this book are continuous in any interval in which they are defined.

43. Increments. DEFINITION. *If a variable changes from one numerical value to another, the difference found by subtracting the first value from the second is called the increment of the variable.*

Thus, if x is first taken as 4 and then as 7, the increment of x is 3. More generally, if x_1 is the first or *initial* value of x and x_2 is the second or *final* value, the increment of x is $x_2 - x_1$. The term *increment* is used even when the variable decreases; thus, if x is first 4 and then 3, we say that the increment of x is $3 - 4 = -1$.

It is customary to denote an increment by the letter Δ (*delta*) prefixed to the letter designating the variable; thus, Δx denotes an increment of x, Δy an increment of y, $\Delta f(x)$ an increment of $f(x)$, etc. It should be noted that the symbol Δx is a composite symbol, not to be treated as Δ times x. For example, $x \cdot \Delta x$ means x times Δx and may not be written $\Delta \cdot x^2$; and $\Delta x \Delta y$ cannot be combined as $\Delta^2 xy$. Furthermore, if Δx is negative, it is not written $- \Delta x$, but is treated like any other algebraic variable.

If y is a function of x, and x takes on an increment Δx, then y will take on a *corresponding* increment Δy. In order to calculate the increment Δy we proceed as follows.

Suppose that $y = f(x)$. Let x_1 be the first value of x and x_2 be the second value. Then $\Delta x = x_2 - x_1$ and $x_2 = x_1 + \Delta x$. Let the corresponding values of y be y_1 and y_2. Then $y_1 = f(x_1)$ and $y_2 = f(x_2) = f(x_1 + \Delta x)$. Hence, by definition,

$$\Delta y = y_2 - y_1 = f(x_2) - f(x_1) = f(x_1 + \Delta x) - f(x_1).$$

Example 1. If $y = x^2 - 6x$, compute Δy when $x = 2$ and $\Delta x = 0.3$.

Solution. Let $x_1 = 2$. Then

$$y_1 = f(x_1) = 2^2 - 6 \cdot 2 = -8.$$

Since
$$\Delta x = 0.3,$$
$$x_2 = x_1 + \Delta x = 2.3,$$
and
$$y_2 = f(x_1 + \Delta x) = (2.3)^2 - 6(2.3) = -8.51.$$

Subtracting y_1 from y_2 we have

$$\Delta y = -8.51 - (-8) = -0.51.$$

Example 2. If $y = x^2 - 6x$, find Δy for any x and Δx.

Solution. Let x_1 be the initial value of x and y_1 the corresponding value of y. Then

$$y_1 = f(x_1) = x_1{}^2 - 6x_1. \tag{1}$$

When $x = x_1 + \Delta x$, we have

$$y_1 + \Delta y = f(x_1 + \Delta x) = (x_1 + \Delta x)^2 - 6(x_1 + \Delta x)$$
$$= x_1{}^2 + 2x_1\Delta x + (\Delta x)^2 - 6x_1 - 6\Delta x. \tag{2}$$

Subtracting equation (1) from equation (2), we have

$$\Delta y = 2x_1\Delta x + (\Delta x)^2 - 6\Delta x.$$

Since x_1 was *any* value of x, we may drop the subscript and write the desired result as follows.

$$\Delta y = (2x - 6)\Delta x + (\Delta x)^2.$$

NOTE 1. In these examples and the preceding paragraphs we have merely clothed a familiar problem in new language. To speak of the increment Δy corresponding to the increment Δx is another way of speaking of the change in y due to a change in x. Thus Example 1 can be worded in more familiar language as follows: If $y = x^2 - 6x$, how much will y change when x changes from 2 to 2.3?

By introducing the symbols Δx and Δy we are enabled to write in a compact form a *general formula* for the change in y in terms of x and the change in x, as was done in Example 2.

NOTE 2. The definition of continuity assumes the following form in the increment notation. *The function $y = f(x)$ is continuous at $x = x_1$ if and only if $\lim\limits_{\Delta x \to 0} \Delta y = 0$ for $x = x_1$.*

44. The derivative at a point. Let $y = f(x)$ be a one-valued function of x. Let x_1 be a fixed value of x and $y_1 = f(x_1)$. Consider the fraction

$$\frac{y - y_1}{x - x_1} = \frac{f(x) - f(x_1)}{x - x_1}.$$

This quantity is itself a function of x and is called the *difference quotient* of $f(x)$ at $x = x_1$. The difference quotient is defined for all values of x for which $f(x)$ is defined except x_1. When $x \longrightarrow x_1$ the difference quotient may or may not have a limit, but, if it does, this limit is called the *derivative of $f(x)$ with respect to x at $x = x_1$*. If we use the increment notation $\Delta x = x - x_1$, $\Delta y = f(x) - f(x_1)$, this statement may be made as follows.

DEFINITION. *If $y = f(x)$ and $\lim\limits_{\Delta x \to 0} \dfrac{\Delta y}{\Delta x}$ exists, this limit is the derivative of y with respect to x at $x = x_1$.*

The derivative is denoted by the symbol $f'(x_1)$ and the definition can be written briefly as follows.

$$f'(x_1) = \lim_{\Delta x \to 0} \frac{\Delta y}{\Delta x} \text{ at } x = x_1.$$

Example 1. If $f(x) = x^2 - 4x$, find $f'(1)$.

Solution. Set $y = f(x) = x^2 - 4x$. When $x = 1$, $y = -3$. When $x = 1 + \Delta x$, let $y = -3 + \Delta y$. Substitution in the equation gives

$$-3 + \Delta y = (1 + \Delta x)^2 - 4(1 + \Delta x)$$
$$= 1 + 2\Delta x + (\Delta x)^2 - 4 - 4\Delta x$$
$$= -3 - 2\Delta x + (\Delta x)^2.$$

Hence $\qquad\qquad \Delta y = -2\Delta x + (\Delta x)^2$

and $\qquad\qquad \Delta y/\Delta x = -2 + \Delta x.$

By the definition of a derivative and the theorems on limits (Art. 41),

$$f'(1) = \lim_{\Delta x \to 0} (-2 + \Delta x) = -2.$$

Example 2. If $s = 12/t$, find the derivative of s with respect to t when $t = 3$.

Solution. When $t = 3$, $s = 4$. When $t = 3 + \Delta t$, let $s = 4 + \Delta s$. Then

$$4 + \Delta s = \frac{12}{3 + \Delta t}.$$

Hence $\qquad\qquad \Delta s = \dfrac{12}{3 + \Delta t} - 4 = \dfrac{-4\Delta t}{3 + \Delta t},$

and $\qquad\qquad \dfrac{\Delta s}{\Delta t} = \dfrac{-4}{3 + \Delta t}.$

Then $\qquad\qquad s'(3) = \lim\limits_{\Delta t \to 0} \left(\dfrac{-4}{3 + \Delta t}\right) = -\dfrac{4}{3}.$

45. The derivative as a function. If the derivative of $f(x)$ exists for $x = x_1$, we say that $f(x)$ is *differentiable* for $x = x_1$. If $f(x)$ is differentiable for every value of x between a and b, inclusive, we say that $f(x)$ is *differentiable in the interval* (a, b).

If we associate with each value of x the value of the derivative of $f(x)$, if it exists, we obtain a new function of x, called simply *the derivative of $f(x)$ with respect to x*. This is denoted by $f'(x)$ and often has the same domain of definition as $f(x)$, but not always. If we can find a mathematical expression for $f'(x)$, we can then find the value of the derivative for a particular value of x by mere substitution.

Example. Find the derivative with respect to x of $f(x) = x^2 - 4x$.

Solution. Set $y = x^2 - 4x$. Let x_1 be a fixed value of x. For $x = x_1$ and $x = x_1 + \Delta x$ let the corresponding values of y be y_1 and $y_1 + \Delta y$. Then

$$y_1 = x_1{}^2 - 4x_1$$

and

$$y_1 + \Delta y = (x_1 + \Delta x)^2 - 4(x_1 + \Delta x)$$

$$= x_1{}^2 + 2x_1\Delta x + (\Delta x)^2 - 4x_1 - 4\Delta x.$$

Hence

$$\Delta y = 2x_1\Delta x - 4\Delta x + (\Delta x)^2$$

and

$$\Delta y/\Delta x = 2x_1 - 4 + \Delta x.$$

Then

$$f'(x_1) = \lim_{\Delta x' \to 0} (2x_1 - 4 + \Delta x) = 2x_1 - 4.$$

Since x_1 stood for *any* value of x, we may now drop the subscript and say that for all values of x

$$f'(x) = 2x - 4.$$

When $x = 1$, $f'(x) = -2$, as in Example 1, Art. 44.

Looking back over the work, we see that it is unnecessary to use the subscript if we regard the symbols x and y in the calculation of the derivative as *fixed values* of the *variables* x and y. For convenience this practice will be followed in the future.

NOTATION. If $y = f(x)$, the derivative of y with respect to x is denoted by various symbols, the most common of which are the following.

$\dfrac{dy}{dx}$, or dy/dx, read as "dy, dx";

y', read as "y prime";

$f'(x)$, read as "f prime of x";

$\dfrac{d}{dx} f(x)$, read as "d dx of $f(x)$";

$D_x y$, read as "D x of y."

Similarly, if $s = g(t)$, the derivative of s with respect to t is denoted by $\dfrac{ds}{dt}$, s', $g'(t)$, $\dfrac{d}{dt} g(t)$, or $D_t s$.

Using this notation we may write the definition of a derivative briefly as

$$y' = f'(x) = \frac{dy}{dx} = \lim_{\Delta x \to 0} \frac{\Delta y}{\Delta x} = \lim_{\Delta x \to 0} \frac{f(x + \Delta x) - f(x)}{\Delta x}.$$

The symbol dy/dx is not regarded as a fraction, but is to be used as a single symbol, just as Δx is a single symbol.

46. Calculation of derivatives. The process of finding the derivative is called *differentiation*. To differentiate one variable with respect to another we merely carry out the steps suggested by the definition.

I. *Let* $y = f(x)$. *Give* x *an increment* Δx *and, by substituting* $x + \Delta x$ *in the equation defining* y, *calculate* $y + \Delta y$, *the new value of* y.

II. *Subtract* y *from* $y + \Delta y$, *thus obtaining* Δy *in terms of* x *and* Δx.

III. *Divide* Δy *by* Δx, *thus obtaining the value of* $\Delta y/\Delta x$.

IV. *Find* $\lim\limits_{\Delta x \to 0} \dfrac{\Delta y}{\Delta x}$, *this being the value of* dy/dx.

This procedure is sometimes called the "Δ process." It should be clearly understood that during the process of differentiation x and y are regarded as fixed values of the variables x and y, and that Δx and Δy are variables.

Example 1. Find $\dfrac{dy}{dx}$, if $y = x^3 - 2x + 4$.

Solution. Substituting $x + \Delta x$ for x, we have

$$\begin{aligned}
y + \Delta y &= (x + \Delta x)^3 - 2(x + \Delta x) + 4 \\
&= x^3 + 3x^2 \Delta x + 3x(\Delta x)^2 + (\Delta x)^3 - 2x - 2\Delta x + 4.
\end{aligned}$$

Subtracting y from $y + \Delta y$,

$$\Delta y = 3x^2 \Delta x + 3x(\Delta x)^2 + (\Delta x)^3 - 2\Delta x.$$

Dividing by Δx,

$$\frac{\Delta y}{\Delta x} = 3x^2 + 3x\Delta x + (\Delta x)^2 - 2.$$

Let $\Delta x \longrightarrow 0$. The second and third terms of $\dfrac{\Delta y}{\Delta x}$ will evidently approach **0.**

Therefore $\qquad\qquad\qquad \dfrac{dy}{dx} = 3x^2 - 2$

Example 2. If $f(t) = 2/t$, find $f'(t)$.

Solution. $$f(t) + \Delta f(t) = f(t + \Delta t)$$

$$= \frac{2}{t + \Delta t}.$$

$$\Delta f(t) = \frac{2}{t + \Delta t} - \frac{2}{t}$$

$$= \frac{-2\,\Delta t}{t(t + \Delta t)}.$$

$$\frac{\Delta f(t)}{\Delta t} = \frac{-2}{t(t + \Delta t)}.$$

Finding the limit of this as $\Delta t \longrightarrow 0$ by means of the theorems on limits, we have

$$\frac{d}{dt} f(t), \text{ or } f'(t), = -\frac{2}{t^2}.$$

PROBLEMS

1. Find the increment of each of the following functions for the indicated values of the independent variable and of the increment of the independent variable.

    a. $y = 2 + 3x - x^2$; $x = 2$, $\Delta x = 0.1$.
    b. $y = \sqrt{25 - x^2}$; $x = 3$, $\Delta x = 0.2$.
    c. $p = 36/v$; $v = 4$, $\Delta v = -0.4$.
    d. $f(t) = 1 - 2t^2$; $t = -2$, $\Delta t = 0.4$.
    e. $y = x + \sqrt{x}$; $x = 4$, $\Delta x = 0.25$.

2. If $xy = -6$, $x = 3$, and $\Delta x = 0.2$, calculate the values of $y + \Delta y$ and Δy. Illustrate by a graph.

3. For the following equations find Δy for any x and Δx.

    a. $y = 4x - x^2$.   **b.** $y = x^3$.
    c. $y = 9 - x^2$.   **d.** $xy = 36$.
    e. $y = x^2 - 5x + 6$.   **f.** $y = ax^2 + bx + c$.

4. Find in terms of Δx the difference quotient of each of the following functions for the indicated initial value of the independent variable.

    a. $y = x^2$; $x_1 = 3$.   **b.** $y = 6x - x^2$; $x_1 = 2$.
    c. $y = 8/x$; $x_1 = 2$.   **d.** $y = \sqrt{x}$; $x_1 = 9$.

5. For each of the following functions find the value of the derivative at the given point.

    a. $f'(2)$ if $f(x) = x + \dfrac{8}{x}$.

    b. $f'(1)$ if $f(x) = 4 - x^2$.
    c. $g'(4)$ if $g(t) = 16t - t^2$.
    d. $g'(-2)$ if $g(t) = t^2 - 5t + 4$.

6. Find dy/dx for the following functions.

a. $y = 4 x^2$.

b. $y = 2 - x^2$.

c. $y = x^3 - x$.

d. $y = 8/x$.

e. $y = 2 x^3 - 3 x^2$.

f. $y = \dfrac{4}{x - 2}$.

g. $y = x + \dfrac{1}{x}$.

h. $y = 1/x^2$.

7. Show that $\dfrac{d}{dx} x^4 = 4 x^3$.

8. Find $\dfrac{ds}{dt}$ if $s = \dfrac{at}{t - a}$.

9. Find ds/dt if $s = c/t$.

10. Find s' if $s = at^3 - bt^2$.

11. Find $f'(x)$ if $f(x) = ax^2 + bx + c$.

12. Find $F'(z)$ if $F(z) = \dfrac{z - 1}{z + 1}$.

13. Find $\phi'(t)$ if $\phi(t) = \dfrac{1 + t^2}{t}$.

14. Find $\dfrac{dy}{d\theta}$ if $y = \dfrac{\theta}{\theta + 1}$.

15. Find $\dfrac{dy}{dz}$ if $y = \dfrac{a}{z^2 + a^2}$.

47. Tangents to a curve. If we have a simple continuous curve, our intuition and experience tell us that a point which moves along this curve constantly changes its direction of motion, but that at each point P of the curve there is a definite straight line which gives the direction of motion and closely approximates the curve near P. The line is called the tangent at P and may be approximated in drawing by turning a ruler about P until it appears to have the proper position. This method of approximation leads to the following definition.

DEFINITION.   *The tangent to the curve at the point P is the limit of a secant line through P and another point Q on the curve as Q approaches P.*

In Fig. 47.1, Q, Q', Q'', Q''' are several successive positions of the point Q as it approaches P. Evidently the secant PQ rotates about P and approaches a limiting position PT, which is the tangent.

The definition of *tangent* used in elementary geometry—namely, as "a line touching the circle at one

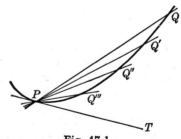

Fig. 47.1

and only one point"—cannot be used for other curves, as is shown by the following figures.

In figure (a) the line TPM is tangent to the curve at the point P. The tangent "touches" the curve at P, but also "cuts" it again at M.

In figure (b) the line TP is tangent to the curve at P. The tangent "cuts" the curve at the point of tangency instead of "touching" it. Such a tangent is called an *inflectional* tangent.

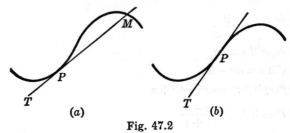

(a)   (b)

Fig. 47.2

In the case of the circle both definitions give the same line, which may be seen intuitively, and will be proved later.

48. The derivative as a slope. Let the curve (Fig. 48.1) be the graph of a one-valued function $y = f(x)$. Let the coördinates of P and Q be those marked in the figure and let PT be tangent to the curve at P. The slope of the secant PQ is $\Delta y/\Delta x$, and as Q approaches P it is clear that $\Delta x \longrightarrow 0$. Hence the slope of PT is given by $\lim\limits_{\Delta x \to 0} \dfrac{\Delta y}{\Delta x}$ for $x = x_1$. The truth of the following theorem is now intuitively evident and a formal proof will be omitted.

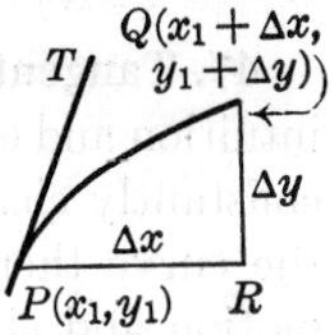

Fig. 48.1

Theorem. *If $y = f(x)$ is continuous and differentiable for $x = x_1$ and $P(x_1, y_1)$ lies on the graph of $y = f(x)$, the line through P which has its slope equal to $f'(x_1)$ is tangent to the graph at P.*

The slope of the tangent to a curve at a point is called the *slope of the curve* at that point.

The limitation in the discussion to one-valued functions causes no trouble in practice. If we desire to study the tangents to the graph of $y^2 = x + 2$, we merely consider the curve to be made up of two branches, one given by $y = +\sqrt{x + 2}$ and the other by $y = -\sqrt{x + 2}$. To each of these branches the theorem is applicable.

Example. Find the slope of the curve whose equation is $x^2 + 2y = 8$ at the point where $x = 2$; also its inclination at that point.

Solution. Solving the equation for y in terms of x,

$$y = 4 - \frac{x^2}{2}.$$

Differentiating with respect to x,

$$y + \Delta y = 4 - \frac{x^2 + 2x\,\Delta x + (\Delta x)^2}{2},$$

$$\Delta y = \frac{-2x\,\Delta x - (\Delta x)^2}{2},$$

$$\frac{\Delta y}{\Delta x} = \frac{-2x - \Delta x}{2},$$

$$\frac{dy}{dx} = -\frac{2x}{2} = -x.$$

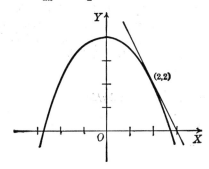

Fig. 48.2

Thus $-x$ is the slope at *any* point.

At the given point, $x = 2$; hence $m = -2$.

Since $m = \tan \alpha$, the inclination is the angle whose tangent is -2, or $116° 34'$. A carefully drawn figure verifies the result.

PROBLEMS

1. Find the slope of the tangent to the given curve at the given point for each of the following equations. Find also the slope of the secant line for $\Delta x = 0.5$. In each case draw the curve, the tangent, and the secant line.

a. $y = x^2$ at $(-2, 4)$.
c. $y = 2x - x^2$ at $(1, 1)$.

b. $y = x^2 - 3x$ at $(3, 0)$.
d. $y = 1/x$ at $(1, 1)$.

2. Find the slopes of the following curves at the points indicated, and in each case draw the curve and the tangent (or tangents) with the proper slope.

a. $y = x^3$; $x = 1, -1$.
c. $y = x^3 - 4x$; $x = 1$.
e. $y = 3 + 2x - x^2$; $x = 0$.

b. $y = x^2 - 5x + 4$; $y = 0$.
d. $xy = 12$; $x = -3$.
f. $y = (x - 2)^3$; $x = 3$.

$$\text{g. } y = \frac{8}{x + 4}; \; x = 0, \pm 1.$$

3. By setting $dy/dx = 0$, find the points at which the tangents to the following curves are parallel to the x-axis, and illustrate by a figure.

a. $4\,y = 16 + 12\,x - x^2$.

b. $6\,y = x^3 - 12\,x$.

c. $6\,y = 2\,x^3 - 3\,x^2 - 12\,x + 6$.

d. $xy - x^2 = 4$.

4. Write the equations of the tangents at the points given in Problem 2.

5. In Problem 2, c, find the slope of the secant through the points whose abscissas are 1 and 1.05, and compare your result with the one obtained by differentiating.

6. For the following equations tabulate the values of x, y, and m (the slope) corresponding to the values of x given; draw the tangents at the points indicated and *then* draw the curve.

a. $y = x^2 - 5\,x + 4$; $x = 0, 1, 2, 2.5, 3, 4, 5$.

b. $y = 3\,x + x^2$; $x = -4, -3, -2, -1.5, -1, 0, 1$.

c. $xy = -6$; $x = 0, \pm 1, \pm 2, \pm 3, \pm 4, \pm 6$.

d. $8\,y = x^3$; $x = 0, \pm 1, \pm 2, \pm 3$.

7. Differentiate $y = \sqrt{x}$.

HINT. Rationalize the numerator of $\Delta y/\Delta x$.

CHAPTER IV

DIFFERENTIATION OF ALGEBRAIC FUNCTIONS

49. Introduction. The general method used in the previous chapter for finding a derivative is applicable in theory to any function; but in practice the calculation of a derivative by this means involves a large amount of labor, unless the function is of the simplest type. Fortunately it is possible to obtain formulas which may be used for differentiating all the functions ordinarily met in elementary mathematics.

In this chapter we shall derive formulas for the differentiation of algebraic functions; that is, functions involving sums, products, and powers of the independent variable x.

50. Derivative of a power of x. Let $y = x^n$, where n is any positive integer. Applying the general rule, Art. 46, we have

I. $y + \Delta y = (x + \Delta x)^n$.

The second member may be expanded by the binomial theorem, giving

$$y + \Delta y = x^n + nx^{n-1}\,\Delta x + \frac{n(n-1)}{1 \cdot 2}\,x^{n-2}(\Delta x)^2 + \cdots + (\Delta x)^n.$$

II. $\Delta y = nx^{n-1}\,\Delta x + \dfrac{n(n-1)}{1 \cdot 2}\,x^{n-2}(\Delta x)^2 + \cdots + (\Delta x)^n.$

III. $\dfrac{\Delta y}{\Delta x} = nx^{n-1} + \dfrac{n(n-1)}{1 \cdot 2}\,x^{n-2}(\Delta x) + \cdots + (\Delta x)^{n-1}.$

When Δx approaches zero as a limit, the limit of the first term on the right is nx^{n-1}, and the limit of each of the other terms is zero. Hence, by Theorem I, Art. 41,

IV. $\dfrac{dy}{dx} = nx^{n-1}.$

Hence

(I) $$\frac{d}{dx}\,x^n = nx^{n-1}.$$

Corollary I. If $n = 1$, we have

(I *a*) $$\frac{d}{dx}\,x = 1.$$

Corollary II. If c is a constant,

(I *b*) $$\frac{d}{dx}\,cx^n = cnx^{n-1}.$$

The first of these corollaries is obtained by reproducing the proof of (I) for the special case when $n = 1$. The second is proved by multiplying the right-hand member of each equation in the proof of (I) by c.

Formula (I) has been proved only for the case when n is a positive integer. However, it is valid for all values of x and n such that x^n is a real number, except for the case when $x = 0$ and $n < 1$, and we shall assume the truth of this without proof for the present. Thus,

$$\frac{d}{dx} x^5 = 5\,x^4, \text{ for all values of } x;$$

$$\frac{d}{dx} x^{\frac{5}{2}} = \tfrac{5}{2} x^{\frac{3}{2}}, \text{ for } x \geqq 0;$$

$$\frac{d}{dx} x^{\frac{1}{3}} = \tfrac{1}{3} x^{-\frac{2}{3}} = \frac{1}{3\,x^{\frac{2}{3}}}, \text{ for all values of } x \text{ except } x = 0.$$

The function $x^{\frac{1}{3}}$ is not differentiable for $x = 0$. We may sometimes say that the derivative becomes infinite for $x = 0$.

51. Derivative of a Constant. Suppose that $y = c$, where c is a constant. Applying the general rule, we get

I. $\qquad\qquad y + \Delta y = c.$

II. $\qquad\qquad \Delta y = 0.$

III. $\qquad\qquad \dfrac{\Delta y}{\Delta x} = 0.$

IV. $\qquad\qquad \dfrac{dy}{dx} = 0.$

Hence

(II) $\qquad\qquad \dfrac{dc}{dx} = 0.$

In words, *the derivative of a constant with respect to any variable is zero.*

52. Derivative of a sum of functions. Theorem. *If u and v are functions of x which are differentiable in the interval (a, b), their sum is differentiable in (a, b) and*

(III) $\qquad\qquad \dfrac{d}{dx}(u + v) = \dfrac{du}{dx} + \dfrac{dv}{dx}.$

Proof. Let $y = u + v$. Let Δy, Δu, and Δv be the increments of y, u, and v, respectively, corresponding to the increment Δx of x. Then

$$y + \Delta y = u + \Delta u + v + \Delta v.$$
$$\Delta y = \Delta u + \Delta v.$$
$$\frac{\Delta y}{\Delta x} = \frac{\Delta u}{\Delta x} + \frac{\Delta v}{\Delta x}.$$

Since u and v are differentiable functions of x,

$$\lim_{\Delta x \to 0} \frac{\Delta u}{\Delta x} = \frac{du}{dx} \quad \text{and} \quad \lim_{\Delta x \to 0} \frac{\Delta v}{\Delta x} = \frac{dv}{dx}.$$

Hence by the theorem on the limit of a sum

$$\frac{dy}{dx} = \lim_{\Delta x \to 0} \frac{\Delta y}{\Delta x} = \frac{du}{dx} + \frac{dv}{dx}.$$

It is obvious that we have a similar theorem for the difference of two differentiable functions and that the theorem can be extended to any algebraic sum of a finite number of differentiable functions. In words, *the derivative of an algebraic sum of differentiable functions is the same algebraic sum of the derivatives of the functions.*

The reason for introducing the restriction to an interval (a, b) in the theorem lies in cases like the following. Consider the functions $x^{\frac{3}{2}}$ and $(-x)^{\frac{3}{2}}$. Both are differentiable wherever defined. The first is defined only in the interval $(0, +\infty)$, and the second in the interval $(-\infty, 0)$. Hence the sum is defined only at $x = 0$ and consequently has no derivative.

53. Summary and examples. The formulas obtained in Arts. 50–52 enable us to differentiate any polynomial at sight. We can also differentiate products of polynomials and fractions whose denominators consist of a single term by first reducing them to the form of a polynomial.

Example 1. If $y = 5\,x^4 - 3\,x^2 + 6 + 2\,x^{-1}$, find $\dfrac{dy}{dx}$.

Solution. $\dfrac{dy}{dx} = \dfrac{d}{dx}\,(5\,x^4) - \dfrac{d}{dx}\,(3\,x^2) + \dfrac{d}{dx}\,(6) + \dfrac{d}{dx}\,(2\,x^{-1})$

$\qquad\quad = 20\,x^3 - 6\,x - 2\,x^{-2}.$

The first line of the solution is found by **(III)** extended. The derivatives of the first, second, and fourth terms are found by means of **(I b)**. The derivative of 6 is zero, by **(II)**.

Example 2. If $y = (2\,x + 3)(x^2 - 2)$, find dy/dx.

Solution. Multiplying together the two polynomials, we have

$$y = 2\,x^3 + 3\,x^2 - 4\,x - 6.$$

Differentiating, as in Example 1, we find that

$$\frac{dy}{dx} = 6\,x^2 + 6\,x - 4.$$

In a later section we shall obtain a method of differentiating a product without multiplying the factors.

Example 3. Find the slope of the tangents to the curve whose equation is $xy = x^2 + 2$ at the points $(2, 3)$ and $(-2, -3)$. Draw a figure, showing the curve and the tangents.

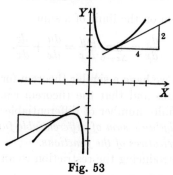

Fig. 53

Solution. Solving the equation for y in terms of x, we get $y = x + \dfrac{2}{x}$. This can be written in the form

$$y = x + 2\,x^{-1}.$$

The slopes of the tangents are given by the values of dy/dx when $x = 2$ and $x = -2$.

Differentiating, we have $\dfrac{dy}{dx} = 1 - 2\,x^{-2} = 1 - \dfrac{2}{x^2}$.

Substituting $x = 2$ in the derivative, we find that the slope at $(2, 3)$ is $1 - \frac{2}{4} = \frac{1}{2}$. Similarly, the slope at $(-2, -3)$ is also $\frac{1}{2}$.

In drawing the tangent the method of Art. 15 should be used. Since the slope is $\frac{1}{2}$, this means that the tangent rises at the rate of $\frac{1}{2}$ vertical unit per horizontal unit, or 2 vertical units for each 4 horizontal units. Hence a second point on the tangent at $(2, 3)$ is $(6, 5)$.

PROBLEMS

1. Write the following functions in a column, and opposite each one write its derivative with respect to the independent variable: $5\,x^4$, $2\,x^{-3}$, $4\sqrt{x}$, $6\,x^{-\frac{1}{3}}$, $-7\,t$, πr^2, x^{-n}, $\frac{4}{3}\pi r^3$, $3\,z^{\frac{7}{3}}$.

2. Find dy/dx for each of the following functions.

a. $y = 4\,x^5 - 2\,x^3 - 12$.

b. $y = 5\,x^7 - 3\,x^2 + 6\,x^{-1}$.

c. $y = 3\,x^2 - 5\,x$.

d. $y = 6 - \frac{2}{3}\,x^3$.

e. $y = 3\,x^{11} - 7\,x^9 + 11$.

f. $y = ax^2 + bx + c$.

3. Find

a. $\dfrac{d}{dx}\,(3\,x^{-2} - 2\,x^{-5})$.

b. $\dfrac{dz}{dx}$, if $z = x^3 - \dfrac{1}{x}$.

c. $f'(t)$, if $f(t) = 8\,t^5 + \dfrac{7}{t^3} - \dfrac{3}{t^4}$.

d. y', if $y = 1 + \dfrac{3}{t} - \dfrac{6}{t^2}$.

e. $\phi'(x)$, if $\phi(x) = -4\,x^{-5} + 6\sqrt{x}$.

4. Differentiate each of the following functions (in each case first reduce to the form of a polynomial by multiplication or division).

a. $y = \dfrac{x^3 - 1}{x^2}.$

b. $y = \dfrac{1 + x}{\sqrt{x}}.$

c. $f(t) = (t^2 - 1)^2.$

d. $\phi(t) = (t - 1)^3.$

e. $y = x(a + x)(a - x).$

f. $z = (u + 2)(2\,u^2 - 1).$

$$\text{g. } z = \frac{6\,u^2 - 4\,u^3}{3\,u^2}.$$

5. Differentiate the following functions, and in each case compute the value of the derivative for the given value of x.

a. $y = \dfrac{1}{x} + \dfrac{1}{x^2} - \dfrac{1}{x^3},\ x = \dfrac{1}{2}.$

b. $y = \sqrt{x} + \dfrac{1}{\sqrt{x}},\ x = 4.$

c. $y = x^{\frac{3}{2}} + x^{-\frac{3}{2}},\ x = 9.$

d. $y = x^2 - 2 + \dfrac{1}{x^2},\ x = 3.$

6. Find dy/dx from each of the following equations.

HINT. First solve for y.

a. $x^2 + 2\,xy = 6.$

b. $xy = 18.$

c. $xy - 2\,x = 5.$

d. $\sqrt{x} + \sqrt{y} = 1.$

e. $x^2 y = -10.$

f. $xy^2 = 4.$

7. Find the slope of each of the following curves at the point indicated and draw an illustrative figure.

a. $y = \dfrac{x^2}{4} - 3\,x;\ x = 4.$

b. $3\,y = 25\,x - x^2;\ x = 10.$

c. $y = x^3 - x^2 - 2\,x;\ x = 2.$

d. $x^2 + 2\,xy = 12;\ x = 2.$

e. $x^2 y = 16;\ x = 3.$

f. $8\,y = x^4 - 4\,x^2;\ x = 3$

g. $xy = 12;\ x = 2, 3, 4, 6.$

h. $y^2 = 8\,x;\ x = 8.$

8. Differentiate $y = x^{-4} = 1/x^4$ by the Δ-process. By the same method show that the derivative of x^{-n} is $\dfrac{-nx^{n-1}}{x^{2n}} = -nx^{-n-1}.$

9. Find a point on the graph of $y^2 = 8\,x$ where the inclination of the tangent is $45°$.

10. Sketch the graph of $y = x^3$. Then find the slope of this curve at the origin and correct your graph accordingly.

54. A property of differentiable functions. In order to prove later formulas we need the following important property of differentiable functions.

Theorem. If $y = f(x)$ is differentiable for $x = x_1$, then $\lim\limits_{\Delta x \to 0} \Delta y = 0$ for $x = x_1$.

Proof. Since $\Delta x \neq 0$, we can set

$$\Delta y = \frac{\Delta y}{\Delta x} \cdot \Delta x.$$

Now $\lim\limits_{\Delta x \to 0} \frac{\Delta y}{\Delta x} = f'(x_1)$ by the definition of a derivative and $\lim\limits_{\Delta x \to 0} \Delta x = 0$.
Hence by the theorem on the limit of a product

$$\lim_{\Delta x \to 0} \Delta y = f'(x_1) \cdot 0 = 0.$$

The truth of this theorem is intuitively clear from the definition of a derivative. For, since $\lim\limits_{\Delta x \to 0} \Delta x = 0$, the limit of the fraction $\Delta y/\Delta x$ cannot very well be finite unless the limit of the numerator Δy is also zero.

55. Derivative of the product of two functions. *Theorem. If u and v are functions of x which are differentiable in the interval (a, b), the product is differentiable in (a, b) and*

(IV) $$\frac{d}{dx}(uv) = u\frac{dv}{dx} + v\frac{du}{dx}.$$

Proof. Let $y = uv$. Let Δy, Δu, and Δv be the increments of y, u, and v, respectively, corresponding to the increment Δx of x. Then

$$y + \Delta y = (u + \Delta u)(v + \Delta v)$$
$$= uv + u\,\Delta v + v\,\Delta u + \Delta u\,\Delta v.$$
$$\Delta y = u\,\Delta v + v\,\Delta u + \Delta u\,\Delta v.$$
$$\frac{\Delta y}{\Delta x} = u\frac{\Delta v}{\Delta x} + v\frac{\Delta u}{\Delta x} + \Delta u\frac{\Delta v}{\Delta x}. \tag{1}$$

Let us recall once more that in the calculation of $\lim\limits_{\Delta x \to 0} \frac{\Delta y}{\Delta x}$ the value
of x is fixed. Consequently the functions u, v, and y have corresponding definite values. The variables are the increments Δx, Δu, Δv, and Δy.

Consider separately the three terms of the second member of (1). Since v is differentiable,

$$\lim_{\Delta x \to 0} \frac{\Delta v}{\Delta x} = \frac{dv}{dx}.$$

Since u is constant,

$$\lim_{\Delta x \to 0} u = u.$$

Hence by the theorem on the limit of a product,

$$\lim_{\Delta x \to 0} u\frac{\Delta v}{\Delta x} = u\frac{dv}{dx}.$$

Similarly,

$$\lim_{\Delta x \to 0} v\frac{\Delta u}{\Delta x} = v\frac{du}{dx}.$$

In the third term the first factor is Δu. Since u is differentiable with respect to x, we know from Art. 54 that

$$\lim_{\Delta x \to 0} \Delta u = 0.$$

The theorem on the limit of a product gives

$$\lim_{\Delta x \to 0} \Delta u \frac{\Delta v}{\Delta x} = 0 \cdot \frac{dv}{dx} = 0.$$

The definition of a derivative and the theorem on the limit of a sum now show that

$$\lim_{\Delta x \to 0} \frac{\Delta y}{\Delta x} = \frac{dy}{dx} = u \frac{dv}{dx} + v \frac{du}{dx}.$$

In words, *the derivative of the product of two functions is equal to the first function times the derivative of the second plus the second function times the derivative of the first.*

Example. Differentiate $y = (2x + 3)(x^2 - 2)$.

Solution. Here $u = 2x + 3$ and $v = x^2 - 2$. Hence $du/dx = 2$ and $dv/dx = 2x$. Applying **(IV)**, we have

$$dy/dx = (2x + 3)(2x) + (x^2 - 2)(2)$$
$$= 4x^2 + 6x + 2x^2 - 4$$
$$= 6x^2 + 6x - 4.$$

Compare this result with that obtained in Example 2, Art. 53.

An important special case of **(IV)** occurs when one factor is a constant. If $u = c$, $du/dx = 0$, by **(II)**, and **(IV)** becomes

(IV a)
$$\frac{d}{dx}(cv) = c \frac{dv}{dx}.$$

In words this may be stated thus: *The derivative of a constant times a function is the constant times the derivative of the function.*

56. Derivative of the product of several functions. If y is the product of more than two functions, the derivative may be obtained by repeated application of **(IV)**. Thus, if $y = uvw$, we may write

$$y = u(vw).$$

Hence
$$\frac{dy}{dx} = u \frac{d}{dx}(vw) + vw \frac{du}{dx}$$

$$= u \left[v \frac{dw}{dx} + w \frac{dv}{dx} \right] + vw \frac{du}{dx}.$$

Hence

(IV b)
$$\frac{d}{dx}(uvw) = uv \frac{dw}{dx} + vw \frac{du}{dx} + wu \frac{dv}{dx}.$$

This result can be extended to the product of any finite number of functions and can be stated as follows.

The derivative of the product of any number of differentiable functions is the sum of the terms obtained by multiplying the derivative of each function by the product of the other functions.

57. Derivative of the quotient of two functions. *Theorem.* *If u and v are functions of x which are differentiable in the interval* (a, b), *then u/v is differentiable in* (a, b) *with respect to x except for values of x such that v = 0 and the derivative is given by the formula*

$$\textbf{(V)} \qquad \frac{d}{dx}\left(\frac{u}{v}\right) = \frac{v\dfrac{du}{dx} - u\dfrac{dv}{dx}}{v^2}.$$

Proof. Let $y = u/v$. Proceeding as in the case of the product, we have

$$y + \Delta y = \frac{u + \Delta u}{v + \Delta v}.$$

Then
$$\Delta y = \frac{u + \Delta u}{v + \Delta v} - \frac{u}{v}$$
$$= \frac{v\,\Delta u - u\,\Delta v}{v^2 + v\,\Delta v}.$$

and
$$\frac{\Delta y}{\Delta x} = \frac{v\dfrac{\Delta u}{\Delta x} - u\dfrac{\Delta v}{\Delta x}}{v^2 + v\,\Delta v}. \qquad (1)$$

Since u and v are differentiable, the theorems on limits give

$$\lim_{\Delta x \to 0}\left[v\frac{\Delta u}{\Delta x} - u\frac{\Delta v}{\Delta x}\right] = v\frac{du}{dx} - u\frac{dv}{dx}.$$

By Art. 54,
$$\lim_{\Delta x \to 0} \Delta v = 0$$

and so
$$\lim_{\Delta x \to 0} (v^2 + v\,\Delta v) = v^2.$$

When $v \neq 0$, the theorem on the limit of a quotient is applicable and shows that $\lim\limits_{\Delta x \to 0}\dfrac{\Delta y}{\Delta x}$ exists. Hence **(V)** follows from (1).

In words, *the derivative of the quotient of two functions is equal to the denominator times the derivative of the numerator minus the numerator times the derivative of the denominator, all divided by the square of the denominator.*

SPECIAL CASE: *the Denominator a Constant.* If $v = k$, a constant, **(V)** becomes

$$\frac{d}{dx}\left(\frac{u}{k}\right) = \frac{1}{k}\left(\frac{du}{dx}\right),$$

which is the same as **(IV a)**, where c is replaced by $\dfrac{1}{k}$. Hence, to differentiate $\dfrac{u}{k}$, where k is a constant, write $\dfrac{u}{k}$ in the form $\dfrac{1}{k}(u)$ and use **(IV a)**.

Example 1. Differentiate $y = \dfrac{x^2 + 1}{x^3 - 3\,x}$.

Solution. $\dfrac{dy}{dx} = \dfrac{(x^3 - 3\,x)\dfrac{d}{dx}(x^2 + 1) - (x^2 + 1)\dfrac{d}{dx}(x^3 - 3\,x)}{(x^3 - 3\,x)^2}$,   by **(V)**

$$= \frac{(x^3 - 3\,x)(2\,x) - (x^2 + 1)(3\,x^2 - 3)}{(x^3 - 3\,x)^2}$$

$$= \frac{-x^4 - 6\,x^2 + 3}{(x^3 - 3\,x)^2}.$$

Example 2. Differentiate $y = \dfrac{4\,x^3 - 7\,x^2}{5}$.

Solution. This may be written as

$$y = \tfrac{1}{5}(4\,x^3 - 7\,x^2).$$

Hence, by **(IV a)**, we have

$$\frac{dy}{dx} = \frac{1}{5}\frac{d}{dx}(4\,x^3 - 7\,x^2)$$
$$= \tfrac{1}{5}(12\,x^2 - 14\,x)$$
$$= \frac{12\,x^2 - 14\,x}{5}.$$

Example 3. Differentiate $y = \dfrac{5}{4\,x^3 - 7\,x^2}$.

Solution. $\dfrac{dy}{dx} = \dfrac{(4\,x^3 - 7\,x^2)(0) - 5(12\,x^2 - 14\,x)}{(4\,x^3 - 7\,x^2)^2}$,   by **(V)**

$$= \frac{-10(6\,x - 7)}{x^3(4\,x - 7)^2}.$$

PROBLEMS

1. Differentiate the following functions by means of **(IV)**, and check your answers by obtaining the derivative according to the method of Example 2, Art. 53.

a. $y = (2\,x - 1)(3\,x + 2)$.

b. $y = (x^3 - 4)(2\,x^4 - 3)$.

c. $s = (t^2 - 3\,t)(t^3 + 4\,t^2)$.

d. $z = (u^3 - 2\,u)\left(1 + \dfrac{2}{u}\right)$.

e. $f(x) = x(1 - x)(1 + x)$.

f. $y = (x^2 - x + 1)(x^2 + x + 1)$.

2. Differentiate the following functions.

a. $y = \left(x^6 - \dfrac{1}{x}\right)\left(\dfrac{3}{x^2} + 2x\right).$

b. $y = \dfrac{x}{3 - 5x}.$

c. $y = \dfrac{x^2 - 1}{x^2 - 2x}.$

d. $y = (3x^{\frac{1}{3}} + 1)(2x^{\frac{1}{2}} + 1).$

e. $f(x) = (x^2 + 2x)\left(\dfrac{1}{x^2} + \dfrac{2}{x}\right).$

f. $s = \dfrac{1 - t + t^2}{1 + t + t^2}.$

g. $u = (t - 1)(t + 1)(t + 2).$

h. $u = \dfrac{z^3}{3 - z^3}.$

i. $y = \dfrac{(1 - x)(1 + 2x)}{1 - 3x}.$

j. $g(x) = \dfrac{x - \sqrt{x}}{x + \sqrt{x}}.$

k. $y = \dfrac{2 - x^{\frac{1}{3}}}{2 + x^{\frac{1}{3}}}.$

l. $s = \dfrac{a^2 - t^2}{a^2 + t^2}.$

m. $u = \dfrac{z^2}{z^2 + a^2}.$

n. $y = \dfrac{x}{(x^2 + 4)^2}.$

o. $f(x) = \dfrac{7 - \sqrt{x}}{7 + \sqrt{x}}.$

p. $y = \dfrac{8a^3}{x^2 + 4a^2}.$

q. $y = \dfrac{-8}{1 - x^4}.$

r. $y = \dfrac{x}{x^2 + a^2}.$

3. Differentiate each of the following functions, and compute the value of the derivative for the indicated value of the independent variable.

a. $y = (x + 4)\left(3 - \dfrac{1}{\sqrt{x}}\right),\ x = 4.$

b. $s = \dfrac{t - 4}{t + 2},\ t = 2.$

c. $u = \dfrac{z}{\sqrt{z} + 2},\ z = 9.$

d. $y = \dfrac{5}{5 - x^2},\ x = 3.$

e. $y = \dfrac{(x - 1)(x - 4)}{2x - 5},\ x = 4.$

f. $y = \dfrac{\sqrt{x} + 1}{\sqrt{x} - 1},\ x = 7.$

4. Find dy/dx from the following equations.

a. $x + y - 2xy = 0.$

b. $xy = y + 2.$

c. $x^2y + a^2y = ax^2.$

d. $(x + 2)(y - 3) = 6.$

5. Find the slope of the curve whose equation is $x^2 + 2xy = 6$ at each point where it crosses the x-axis.

6. Find the slope of the curve whose equation is $y = \dfrac{8}{x^2 + 4}$ at the point whose abscissa is 2. Also write the equation of the tangent at this point.

7. Show that the slope of the curve whose equation is $y = \dfrac{4}{x - 1}$ at the point where $x = 3$ is -1. At what other point is the slope also -1?

8. What is the inclination of the curve whose equation is $y = \dfrac{x}{x^2 + 1}$ at the origin?

9. Find the slope of the curve whose equation is $y = \dfrac{x^2}{x^2 + 4}$ at the points where $x = 0$ and $x = \pm 2$. What is the limit of the slope as $x \longrightarrow \infty$?

10. Find the slope and also the inclination of the curve whose equation is $y = (x - 1)(x - 2)(x + 1)$ at the points where it crosses the x-axis.

11. Find the slope of the curve whose equation is $y = \dfrac{x + 3}{x^3}$ at the point where $x = -3$.

12. Find the points where the tangent to the curve whose equation is $y(1 + x^2) = 6\,x$ is parallel to the x-axis.

13. Find the slope of the curve whose equation is

$$y = \frac{(x - a)(x - b)}{2\,x - (a + b)}$$

at each of the points where the curve cuts the x-axis. Illustrate graphically for the case when $a = 2$ and $b = 8$.

58. Derivative of a function of a function. Suppose that y is a function of u and that u is a function of x. For example, we might have $y = \sqrt[3]{u}$, where $u = 4 + 3\,x - x^2$. In such cases y is said to be a *function of a function*. The two correspondences may give a correspondence between values of y and some or all of the values of x for which u is defined. If so, y is a function of x. In this example we can find dy/du and du/dx by the rules already given and the question of the differentiability of y with respect to x is to be considered.

Theorem. *Let y be a function of u and u be a function of x such that these relations define y as a function of x in some interval (a, b). Let u be differentiable with respect to x in (a, b) and let y be differentiable with respect to u for the corresponding values of u. Then y is differentiable with respect to x in (a, b) and*

(VI)
$$\frac{dy}{dx} = \frac{dy}{du}\frac{du}{dx}.$$

Proof. For all functions considered in this book $\Delta u \neq 0$ if $\Delta x \neq 0$ and $|\Delta x|$ is sufficiently small. Under these circumstances we may write

$$\frac{\Delta y}{\Delta x} = \frac{\Delta y}{\Delta u} \cdot \frac{\Delta u}{\Delta x}$$

and
$$\lim_{\Delta x \to 0} \frac{\Delta y}{\Delta x} = \left(\lim_{\Delta x \to 0} \frac{\Delta y}{\Delta u}\right)\left(\lim_{\Delta x \to 0} \frac{\Delta u}{\Delta x}\right). \tag{1}$$

Since u is differentiable with respect to x,

$$\lim_{\Delta x \to 0} \frac{\Delta u}{\Delta x} = \frac{du}{dx}$$

and
$$\lim_{\Delta x \to 0} \Delta u = 0. \qquad\qquad \text{By Art. 54}$$

Then $$\lim_{\Delta x \to 0} \frac{\Delta y}{\Delta u} = \lim_{\Delta u \to 0} \frac{\Delta y}{\Delta u} = \frac{dy}{du}.$$

Substituting in the right-hand side of (1), we have

$$\lim_{\Delta x \to 0} \frac{\Delta y}{\Delta x} = \frac{dy}{du}\frac{du}{dx},$$

which proves the theorem.

Example. Let $y = u^{\frac{1}{3}}$ and $u = 4 + 3x - x^2$. These equations are seen to define y as a function of x for every value of x. Now

$$\frac{du}{dx} = 3 - 2x \text{ for every value of } x,$$

and $$\frac{dy}{du} = \frac{1}{3}u^{-\frac{2}{3}} \text{ for every value of } u \neq 0.$$

Hence by the theorem y is differentiable with respect to x except for those values of x which make $u = 0$ and, by **(VI)**,

$$\frac{dy}{dx} = \frac{3 - 2x}{3u^{\frac{2}{3}}}.$$

A result in terms of x only is found by substituting for u its value in terms of x. The values of x for which this result is not valid are found by setting $u = 4 + 3x - x^2 = 0$ and solving for x. These values are $x = 4$ and $x = -1$.

59. Derivative of a power of a function. An important special case of **(VI)** occurs when $y = u^n$, where n is a constant.

Theorem. *If u is a differentiable function of x and n is a positive integer, u^n is differentiable with respect to x and*

(VII) $$\frac{d}{dx}u^n = nu^{n-1}\frac{du}{dx}.$$

The proof follows immediately by the use of **(I)** and **(VI)**.

This formula will be used constantly, and care must be taken to avoid the common error of forgetting the factor du/dx.

Formula **(VII)** may also be used when n is any constant different from zero, provided that the conditions of the theorem in Art. 58 are fulfilled.

Example 1. Differentiate $y = \sqrt{8x}$.

Solution. If this is written as $y = (8x)^{\frac{1}{2}}$, we see that y is of the form u^n, where

$$u = 8x, \quad n = \frac{1}{2}, \quad \text{and} \quad \frac{du}{dx} = 8.$$

Hence, by **(VII)**,

$$\frac{dy}{dx} = \frac{1}{2}(8\,x)^{-\frac{1}{2}}(8)$$

$$= \frac{4}{\sqrt{8\,x}} = \frac{4}{2\sqrt{2}\,x} = \frac{\sqrt{2}}{\sqrt{x}}.$$

If we write the given expression as $y = 8^{\frac{1}{2}}\,x^{\frac{1}{2}}$, we can differentiate by means of **(I)** and **(IV a)**, and we obtain

$$\frac{dy}{dx} = (\sqrt{8})\frac{1}{2}x^{-\frac{1}{2}} = \frac{\sqrt{8}}{2\sqrt{x}} = \frac{2\sqrt{2}}{2\sqrt{x}} = \frac{\sqrt{2}}{\sqrt{x}},$$

as above. Note that neither method gives as a result $\frac{1}{2}(8\,x)^{-\frac{1}{2}}$, the result of the error alluded to above.

Example 2. Differentiate $y = \sqrt{a^2 - x^2}$.

Solution. Here $u = a^2 - x^2$, $n = \frac{1}{2}$, and $du/dx = -2\,x$. Hence, by **(VII)**,

$$\frac{dy}{dx} = \frac{1}{2}(a^2 - x^2)^{-\frac{1}{2}}(-2\,x)$$

$$= \frac{-x}{\sqrt{a^2 - x^2}}.$$

Example 3. Differentiate $y = \dfrac{7}{(2\,x^2 - 3)^2}$.

Solution. This can be differentiated by the rule for a fraction; but when the denominator is a power and the numerator is a constant, it is better to use the power formula, as follows.

$$y = 7(2\,x^2 - 3)^{-2}.$$

$$\frac{dy}{dx} = -14(2\,x^2 - 3)^{-3}(4\,x)$$

$$= \frac{-56\,x}{(2\,x^2 - 3)^3}.$$

60. The derivative as an aid in curve-sketching. The fact that the slope of a curve at any point is given by the value of the derivative at that point is of considerable assistance in quickly obtaining a fair sketch of a curve. Usually, if the points where the tangents are parallel to the x-axis are found, a very good idea of the general shape of the curve can be obtained by finding two or three more points. This is because the points where the tangents are horizontal are usually (not always) the highest or lowest points of the curve in that neighborhood. The slope at points of intersection with the x-axis is sometimes a help.

Example 1. Find the points where the tangents to the curve whose equation is $y = \dfrac{x^3}{3} - \dfrac{x^2}{2} - 2x + 2$ are parallel to the x-axis, and sketch the curve.

Solution. Since the tangents are to be parallel to the x-axis, we must find the points where the slope is zero. Differentiating, we have

$$\frac{dy}{dx} = x^2 - x - 2.$$

Setting this equal to 0, we obtain

$$x^2 - x - 2 = 0;$$

whence

$$(x - 2)(x + 1) = 0,$$

or

$$x = -1 \text{ and } x = 2.$$

Substituting these values in the given equation, $y = \frac{19}{6}$ and $y = -\frac{4}{3}$. Hence the required points are $(-1, \frac{19}{6})$ and $(2, -\frac{4}{3})$. Three other points easily obtained are $(0, 2)$, $(3, \frac{1}{2})$, and $(-3, -\frac{11}{2})$. Since there are but two points where the tangent is horizontal, the curve must rise continuously at the right of the point $(3, \frac{1}{2})$ and fall at the left of the point $(-3, -\frac{11}{2})$. See Fig. 60.1.

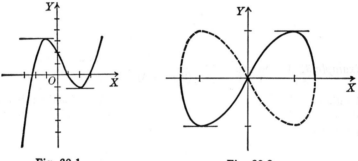

Fig. 60.1   Fig. 60.2

Example 2. Find the points where the tangents to the curve whose equation is $y^2 = x^2(2 - x^2)$ are parallel to the x-axis, and sketch the curve.

Solution. Solving for y, we obtain $y = \pm x\sqrt{2 - x^2}$. Thus y is a two-valued function of x which is defined for $|x| \leq \sqrt{2}$. Let us take first the case when the sign is positive. Writing the equation in the form $y = x(2 - x^2)^{\frac{1}{2}}$ and differentiating, we have

$$\frac{dy}{dx} = \frac{x}{2}(2 - x^2)^{-\frac{1}{2}}(-2x) + (2 - x^2)^{\frac{1}{2}}(1)$$

$$= \frac{2(1 - x^2)}{(2 - x^2)^{\frac{1}{2}}}.$$

Setting the derivative equal to 0, we obtain $x = -1$ and $+1$. The corresponding values of y are -1 and $+1$, and so the required points are $(-1, -1)$ and $(1, 1)$.

It is unnecessary to repeat the work for $y = -x\sqrt{2-x^2}$, since only the sign of dy/dx is changed and so we have the same values of x as before. But now the corresponding values of y are 1 and -1, and we have two more points, $(-1, 1)$ and $(1, -1)$, where the tangents are parallel to the x-axis.

From the given equation we find that the x-intercepts are 0 and $\pm\sqrt{2}$. For the latter values of x, dy/dx becomes infinite. Hence at the corresponding points the slope becomes infinite and the tangents are perpendicular to the x-axis. In Fig. 60.2 the heavy line is the graph of $y = +x\sqrt{2-x^2}$ and the dotted line is the graph of $y = -x\sqrt{2-x^2}$.

PROBLEMS

1. Differentiate the following functions.

a. $y = \sqrt{4+3x}$.

b. $y = (5-2x)^3$.

c. $y = \left(x + \dfrac{1}{x}\right)^5$.

d. $y = \sqrt{7-2x+x^2}$.

e. $y = \dfrac{1}{\sqrt{a^2-x^2}}$.

f. $y = (2-3x)^{\frac{3}{2}}$.

g. $y = \sqrt{1+x^3}$.

h. $y = \dfrac{1}{\sqrt{x^2+4x+10}}$.

i. $y = \sqrt{3x} - \dfrac{1}{\sqrt{3x}}$.

j. $f(t) = \sqrt{at^2 - t^3}$.

k. $z = (4 - u^2)^3$.

l. $g(z) = \left(\dfrac{z}{1-z}\right)^2$.

2. Differentiate the following functions in two ways and show that the answers thus obtained are identical.

a. $y = (3+4x)^3$.

b. $y = \sqrt{2t} + (2t)^{\frac{3}{2}}$.

3. Find dy/dx in each of the following equations.

a. $x^2 + y^2 = 9$.

b. $b^2x^2 + a^2y^2 = a^2b^2$.

c. $y^2 = 2px$.

d. $x^{\frac{2}{3}} + y^{\frac{2}{3}} = a^{\frac{2}{3}}$.

4. In the following problems state the domain of definition of y as a function of x, find dy/dx by **(VI)**, and state for what values of x the result is valid.

a. $y = u^8$, $u = 2 + \sqrt{x}$.

b. $y = \sqrt{u} - u$, $u = x^2 + 2x$.

c. $y = u^4 - u^2$, $u = 2x^3 - x + 1$.

d. $y = \dfrac{u}{u-2}$, $u = \dfrac{2x-3}{2x+3}$.

5. If $y = u^{\frac{3}{2}}$ and $u = x^6 - 4x^4$, find dy/dx by **(VI)** and state for what values of x the result is valid.

6. Find the points where the following curves have horizontal tangents, and sketch the curve in each case.

a. $y = 3x - x^3$.

b. $y = x^2(x-2)^2$.

c. $y = \dfrac{x^4}{4} - 2x^2 - 1$.

d. $y = (x-3)^3$.

e. $y^2 = x^3$.

f. $y = (x-1)(x+1)(x-2)$.

7. For the following equations find the points where the tangents to the graphs are horizontal and vertical and then plot the curves.

a. $y^2 = x(x-4)^2$.   **b.** $y = x^2(x-2)^{\frac{2}{3}}$.

8. Show that the graph of any equation which has the form $y = a + bx + cx^2$ has one point where the tangent is parallel to the x-axis.

9. Find the inclination of the tangent to the circle whose equation is $x^2 + y^2 = 20$ at the point $(2, 4)$.

10. Find the slope of the curve whose equation is $x^2 - 4y^2 = 20$ at the point $(6, 2)$ and write the equation of the tangent.

61. Derivative of an inverse function. Suppose that $y = f(x)$ and $x = g(y)$ are inverse functions (Art. 38). In this case we have the following formulas.

(VIII)
$$\frac{dy}{dx} = \frac{1}{\dfrac{dx}{dy}}.$$

Also

(VIII a)
$$\frac{dx}{dy} = \frac{1}{\dfrac{dy}{dx}}.$$

To establish **(VIII)** we suppose, as usual, that $y = f(x)$ and $x = g(y)$ are one-valued, continuous functions. Suppose that dx/dy exists and is not zero. Then $\Delta x/\Delta y \neq 0$ for $|\Delta y|$ sufficiently small and we can write

$$\frac{\Delta y}{\Delta x} = \frac{1}{\dfrac{\Delta x}{\Delta y}}$$

and
$$\lim_{\Delta y \to 0} \frac{\Delta y}{\Delta x} = \lim_{\Delta y \to 0} \left(\frac{1}{\dfrac{\Delta x}{\Delta y}}\right) = \frac{1}{\lim\limits_{\Delta y \to 0} \dfrac{\Delta x}{\Delta y}} = \frac{1}{\dfrac{dx}{dy}}. \qquad (1)$$

Now
$$\lim_{\Delta y \to 0} \Delta x = 0 \qquad\qquad \text{by Art. 54}$$

and
$$\lim_{\Delta y \to 0} \frac{\Delta y}{\Delta x} = \lim_{\Delta x \to 0} \frac{\Delta y}{\Delta x} = \frac{dy}{dx}.$$

Substitution in (1) gives the result.

62. Formulas for the differentiation of algebraic functions. With the formulas derived in the previous articles any explicit algebraic function can be differentiated. The most important ones are collected here for convenience.

(I) $$\frac{d}{dx} x^n = nx^{n-1}.$$

(I a) $$\frac{d}{dx} x = 1.$$

(II) $$\frac{d}{dx} c = 0.$$

(III) $$\frac{d}{dx} (u + v) = \frac{du}{dx} + \frac{dv}{dx}.$$

(IV) $$\frac{d}{dx} (uv) = u \frac{dv}{dx} + v \frac{du}{dx}.$$

(IV a) $$\frac{d}{dx} (cv) = c \frac{dv}{dx}.$$

(V) $$\frac{d}{dx} \left(\frac{u}{v}\right) = \frac{v \dfrac{du}{dx} - u \dfrac{dv}{dx}}{v^2}.$$

(VI) $$\frac{dy}{dx} = \frac{dy}{du} \frac{du}{dx}.$$

(VII) $$\frac{d}{dx} u^n = nu^{n-1} \frac{du}{dx}.$$

(VIII) $$\frac{dy}{dx} = \frac{1}{\dfrac{dx}{dy}}.$$

The following examples illustrate the differentiation of several types of functions relatively more complicated than those given before.

Example 1. Find the derivative of $y = \sqrt{3 + 4\sqrt{x}}$.

Solution. This may be written in the form

$$y = (3 + 4\sqrt{x})^{\frac{1}{2}}.$$

$$\frac{dy}{dx} = \frac{1}{2} (3 + 4\sqrt{x})^{-\frac{1}{2}} \frac{d}{dx} (4\sqrt{x}) \qquad \text{by (VII)}$$

$$= \frac{1}{2} (3 + 4\sqrt{x})^{-\frac{1}{2}} (2 x^{-\frac{1}{2}}) \qquad \text{by (I) and (IV a)}$$

$$= \frac{1}{(3 + 4\sqrt{x})^{\frac{1}{2}} x^{\frac{1}{2}}} = \frac{1}{\sqrt{3 x + 4 x^{\frac{3}{2}}}}.$$

Example 2. Differentiate $y = x\sqrt{2 - 3\,x}$.

Solution.
$$y = x(2 - 3\,x)^{\frac{1}{2}}.$$

$$\frac{dy}{dx} = x\frac{d}{dx}(2 - 3\,x)^{\frac{1}{2}} + (2 - 3\,x)^{\frac{1}{2}}\frac{d}{dx}x \qquad \text{by (IV)}$$

$$= x\frac{1}{2}(2 - 3\,x)^{-\frac{1}{2}}(-3) + (2 - 3\,x)^{\frac{1}{2}} \quad \text{by (VII) and (IV a)}$$

$$= \frac{-3\,x}{2\sqrt{2 - 3\,x}} + \sqrt{2 - 3\,x}$$

$$= \frac{4 - 9\,x}{2\sqrt{2 - 3\,x}}. \qquad \text{By simplification}$$

Example 3. Differentiate $y = \dfrac{x^2}{\sqrt{x^2 + a^2}}$.

Solution. $\dfrac{dy}{dx} = \dfrac{\sqrt{x^2 + a^2}\,\dfrac{d}{dx}x^2 - x^2\dfrac{d}{dx}\sqrt{x^2 + a^2}}{x^2 + a^2}$ \qquad by **(V)**

$$= \frac{\sqrt{x^2 + a^2}\cdot 2\,x - x^2\dfrac{1}{2}(x^2 + a^2)^{-\frac{1}{2}}\dfrac{d}{dx}(x^2 + a^2)}{x^2 + a^2}$$

by **(I)** and **(VII)**

$$= \frac{2\,x\sqrt{x^2 + a^2} - \dfrac{x^2\cdot 2\,x}{2\sqrt{x^2 + a^2}}}{x^2 + a^2} \qquad \text{by (I)}$$

$$= \frac{x^3 + 2\,a^2x}{(x^2 + a^2)^{\frac{3}{2}}}. \qquad \text{By simplification}$$

PROBLEMS

1. Differentiate the following functions.

a. $y = \dfrac{(x - 1)^3}{(x + 1)^2}$.

b. $s = \dfrac{a + t^2}{a - t^2}$.

c. $y = x^2\sqrt{1 - 2\,x}$.

d. $v = u\sqrt{2 + u^2}$.

e. $f(x) = x\sqrt[3]{2 + 3\,x}$.

f. $s = \dfrac{t}{\sqrt{2 - t}}$.

g. $y = \sqrt{\dfrac{a + x}{a - x}}$.

h. $s = \sqrt{\dfrac{a - t^2}{a + t^2}}$.

i. $y = \dfrac{a^2 - 2\,x^2}{\sqrt{a^2 - x^2}}$.

j. $y = (2 - x^2)^{\frac{1}{2}}(3 + x^3)^{\frac{1}{3}}$.

k. $f(x) = \sqrt{(a + x)(a - x)^3}$.

2. In the following problems find the value of dy/dx for the given value of x.

a. $y = x^2 + \dfrac{4}{x^2}$, $x = 2$.

b. $y = \dfrac{x}{3 - x}$, $x = 2$.

c. $y = \sqrt{7\,x} + 7\sqrt{x} + \dfrac{1}{\sqrt{7\,x}} + \dfrac{7}{\sqrt{x}}$, $x = 7$.

d. $y = \dfrac{x^2}{a^2 - x^2}$, $x = 0$.

e. $y = x\sqrt{4\,x - x^2}$, $x = 2$.

3. In the following equations find dy/dx.

a. $x = \sqrt{y} + \sqrt[3]{y^2}$.   b. $x = \sqrt{4 + y^2} - y^4$.

c. $x = 3\sqrt{4 - y^2}$.

4. Draw the graphs in a, b, d, and e of Problem 2, showing the tangent at the point indicated. In each case check the graph by finding the slope at two other points.

5. Show that the slope at any point of the curve whose equation is $y^2 = 2\,px$ is p/y. Illustrate by a graph.

6. Show that the slope at any point of the curve whose equation is $x = ay^3$ is $\dfrac{1}{3\,ay^2}$. Illustrate by a graph.

7. Find the point at which the tangent to the curve whose equation is $x = a + by + cy^2$ is perpendicular to the x-axis. Take $a = b = c = 1$ and draw a figure.

63. Implicit functions. Let a relation between x and y be expressed by the equation

$$x^2 + y^2 - a^2 = 0. \qquad (1)$$

This equation may be solved for y, in which case $y = \sqrt{a^2 - x^2}$ is called an *explicit* function of x, or it may be solved for x, in which case $x = \sqrt{a^2 - y^2}$ is called an explicit function of y. If the equation is not solved but stands in the original form (1), it is said that y is an *implicit* function of x, or that x is an implicit function of y.

A derivative may be calculated from either the implicit or explicit form of the function. Suppose that, in the example above, it is desired to find the derivative of y with respect to x.

If the explicit form is used, $y = \sqrt{a^2 - x^2}$,

whence $$\dfrac{dy}{dx} = \dfrac{-x}{\sqrt{a^2 - x^2}}. \qquad (2)$$

If the implicit form is used, it is necessary only to differentiate the terms of (1) as they stand *with respect to x.* The result is

$$2x + 2y\frac{dy}{dx} = 0;$$

whence, by solving, $$\frac{dy}{dx} = -\frac{x}{y}. \tag{3}$$

The results in (2) and (3) are easily seen to be identical.

In the example above there is little choice between the explicit and implicit forms of the function. In some cases the implicit form is more convenient.

Consider, for example,

$$x^2 + 3xy + y^2 - 4x + 8y - 12 = 0.$$

It is possible to solve this equation and to get an explicit expression for *y* as a function of *x*. It is far more convenient, however, to calculate *dy/dx* from the implicit form. Thus, differentiating the terms as they stand, we have

$$2x + \left(3x\frac{dy}{dx} + 3y\right) + 2y\frac{dy}{dx} - 4 + 8\frac{dy}{dx} = 0.$$

Solving this equation for the derivative gives

$$\frac{dy}{dx} = \frac{4 - 2x - 3y}{8 + 3x + 2y}.$$

In still other cases it is impossible to solve the equation for either *y* or *x*, and the implicit form must be used to calculate the derivative. For example,

$$x^6 + xy + y^6 - 3 = 0$$

cannot be solved algebraically. The derivative of *y* with respect to *x* can be calculated, however, by differentiating the terms as they stand.

Thus, $$6x^5 + \left(x\frac{dy}{dx} + y\right) + 6y^5\frac{dy}{dx} = 0;$$

whence, solving for the derivative,

$$\frac{dy}{dx} = -\frac{6x^5 + y}{x + 6y^5}.$$

Note that when the derivative is found by the implicit method, it is usually a function of *y* as well as of *x*.

Example. Find the slope of the tangent to the ellipse whose equation is $x^2 + 2y^2 = 24$ at the points where $x = 4$.

Solution. Differentiating implicitly, we have

$$2\,x + 4\,y\,\frac{dy}{dx} = 0,$$

whence

$$\frac{dy}{dx} = -\frac{x}{2\,y}.$$

Since the derivative involves both x and y, we must find the value of y from the given equation. When $x = 4$, $y = \pm 2$. Substituting in the expression for dy/dx, we find that the slope at $(4, 2)$ is -1 and the slope at $(4, -2)$ is $+1$.

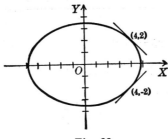

Fig. 63

64. Successive differentiation. If y is a differentiable function of x, the derivative dy/dx is also a function of x. That is, we have given $y = f(x)$ and obtain by differentiation the function $y' = f'(x)$. If this new function can be differentiated with respect to x, the result dy'/dx is called the *second derivative* of y with respect to x. Continuing this process, we get the *third* derivative, *fourth* derivative, etc.

NOTATION. If $y = f(x)$, the following notations are used for the derivatives.

First derivative, $\qquad \dfrac{dy}{dx} = y' = f'(x).$

Second derivative, $\qquad \dfrac{d^2y}{dx^2} = y'' = f''(x).$

Third derivative, $\qquad \dfrac{d^3y}{dx^3} = y''' = f'''(x).$ Etc.

Example 1. Find the successive derivatives of

$$f(x) = x^3 - 2\,x^2 + 6.$$

Solution. $\qquad\qquad f'(x) = 3\,x^2 - 4\,x.$

$\qquad\qquad\qquad\quad f''(x) = 6\,x - 4.$

$\qquad\qquad\qquad\quad f'''(x) = 6.$

$\qquad\qquad\qquad\quad f^{iv}(x) = 0.$

Evidently all the derivatives from this point on are zero.

Example 2. If $x^2 + y^2 = a^2$, find d^2y/dx^2.

First solution. Solving for y, we have $y = \pm \sqrt{a^2 - x^2}$. Taking the positive sign with the radical, we have

$$\frac{dy}{dx} = \frac{1}{2} (a^2 - x^2)^{-\frac{1}{2}}(-2x) = \frac{-x}{\sqrt{a^2 - x^2}}.$$

Differentiating again, we have

$$\frac{d^2y}{dx^2} = \frac{\sqrt{a^2 - x^2}(-1) + x \dfrac{-x}{\sqrt{a^2 - x^2}}}{a^2 - x^2} = \frac{-a^2}{(a^2 - x^2)^{\frac{3}{2}}}.$$

Second solution. Differentiating with respect to x implicitly, we have

$$2x + 2y \frac{dy}{dx} = 0, \quad \text{or} \quad x + yy' = 0.$$

Solving for y', the result is

$$y' = -\frac{x}{y}.$$

Differentiating again, we have

$$y'' = -\frac{y - xy'}{y^2} = -\frac{y + \dfrac{x^2}{y}}{y^2} = -\frac{x^2 + y^2}{y^3} = -\frac{a^2}{y^3}.$$

The results of the two solutions are equivalent.

PROBLEMS

1. Find $\dfrac{dy}{dx}$ in the following equations.

a. $x^2 + xy - y^5 = 0$.

b. $x^2y - xy^2 + y^3 - 9 = 0$.

c. $y^3 + xy + x + y = 0$.

d. $x^2 + 3xy + y^2 = 31$.

e. $x^2 + \sqrt{xy} + y^2 = 1$.

f. $x + \sqrt{xy} + y = 6$.

g. $(2x)^{\frac{2}{3}} + (9y)^{\frac{2}{3}} = 13$.

h. $x^3 + y^3 - 3axy = 0$.

2. Show that the slope at any point of the circle whose equation is $x^2 + y^2 = a^2$ is $-x/y$.

3. Show that the slope at any point of the ellipse whose equation is $b^2x^2 + a^2y^2 = a^2b^2$ is $-b^2x/a^2y$.

4. Show that the slope at any point of the hyperbola whose equation is $\dfrac{x^2}{a^2} - \dfrac{y^2}{b^2} = 1$ is $\dfrac{b^2x}{a^2y}$.

5. The locus of the equation $\sqrt{x} + \sqrt{y} = \sqrt{a}$ is an arc of a parabola whose axis of symmetry is the line whose equation is $y = x$. Show that the slope at any point is $-\sqrt{y}/x$. Plot the locus for $a = 25$, taking values of x which are perfect squares. Check the graph by noting the slope at each point plotted.

6. The locus of the equation $x^{\frac{2}{3}} + y^{\frac{2}{3}} = a^{\frac{2}{3}}$ is called a hypocycloid of four cusps. Show that its slope at any point is $-\sqrt[3]{y/x}$. Sketch the graph by finding the intercepts and the points where $x = \pm y$, and the slope at each of these points.

7. Find the intercepts of the curve whose equation is $\left(\dfrac{x}{a}\right)^2 + \left(\dfrac{y}{b}\right)^{\frac{2}{3}} = 1$. By differentiating implicitly, show that the slope at each of the corresponding points is 0, and sketch the curve.

8. The curve whose equation is $x^3 + y^3 - 3\,axy = 0$ is called the folium of Descartes. Find the slope of the tangent at the point where $x = y$. Draw the tangent at this point and a small arc of the curve.

9. Find the second derivative of the following functions.

a. $y = 2\,x - 4\,x^2$.

b. $y = 5\,x^5 - 3\,x^3 + 8$.

c. $y = x(a + x)(a - x)$.

d. $v = (u + 1)(2\,u^2 - 3)$.

e. $f(x) = 3\,x^{-2} - 2\,x^{-3}$.

f. $s = t^2 - 2\,t + \dfrac{1}{t^2}$.

g. $y = \dfrac{x + a}{x - a}$.

h. $f(t) = (t^2 - 1)^2$.

i. $y = x\sqrt{4 - x}$.

j. $y = \dfrac{2\,x^2}{x^2 + 4}$.

k. $y = \sqrt{\dfrac{x - 1}{x + 1}}$.

l. $s = (a + t)\sqrt{a^2 - t^2}$.

m. $f(s) = \sqrt{\dfrac{s^2 - 1}{s^2 + 1}}$.

10. Find d^2y/dx^2 from the following equations.

a. $y^2 = 2\,px$.

b. $x = y + \dfrac{1}{y}$.

c. $x = 4\,y^2 - y^4$.

d. $y^2 + 2\,xy = 9$.

e. $x - 2 = y + 2\sqrt{y}$.

f. $b^2x^2 + a^2y^2 = a^2b^2$.

g. $\sqrt{x} + \sqrt{y} = \sqrt{a}$.

h. $x^3 + y^3 = a^3$.

i. $x^{\frac{2}{3}} + y^{\frac{2}{3}} = a^{\frac{2}{3}}$.

j. $x^3 + y^3 - 3\,axy = 0$.

65. Applications of the derivative in geometry. It has been shown that the slope of a curve at any point can be found by getting the value of the derivative dy/dx at that point. Other problems which are based upon this theorem are the following.

1. To find the points on a curve where the tangent has a given slope.
2. To find the point of contact of a tangent which passes through a given point which is not on the curve.
3. To find the angle of intersection between two curves. *By the angle of intersection between two curves is meant the angle between their tangents at the point of intersection.*

4. To find the equations of the tangent and the normal to a curve at a given point. *The normal to a curve at a given point is the perpendicular to the tangent at that point.*

Example 1. Find on the circle whose equation is $x^2 + y^2 = 20$ the points where the tangents are parallel to the line whose equation is $2x + y = 0$.

Solution. From the equation of the circle we get by implicit differentiation $dy/dx = -x/y$. Since the slope of the given line is -2, we now have

$$-x/y = -2,$$

or
$$x = 2y, \tag{1}$$

if x and y are the coordinates of the point of contact. Since this point lies on the circle, we have

$$x^2 + y^2 = 20. \tag{2}$$

Solving (1) and (2) simultaneously, we get $y = \pm 2$ and $x = \pm 4$. The proper pairing of these values is found from (1) to be (4, 2) and $(-4, -2)$. (Fig. 65.1)

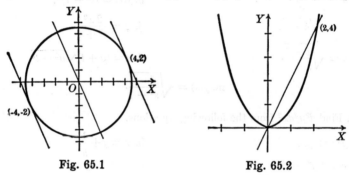

Fig. 65.1   Fig. 65.2

Example 2. Find the angles of intersection between the curves whose equations are $y = x^2$ and $y = 2x$.

Solution. Solving the equations simultaneously, the points of intersection are found to be (0, 0) and (2, 4). The slope of the first curve at any point is given by $dy/dx = 2x$. The second curve is a straight line, and its slope at every point is 2.

Angle at (0, 0). The inclination of the straight line is 63° 26′. The inclination of the curve at (0, 0) is 0. Hence, by subtraction, the angle of intersection at the origin is 63° 26′.

Angle at (2, 4). For $x = 2$ the slope of the curve is 4 and its inclination is 75° 58′. The inclination of the straight line is 63° 26′. Hence, by subtraction, the angle of intersection at (2, 4) is 12° 32′. (Fig. 65.2)

Example 3. Find the equations of the tangent and of the normal to the ellipse whose equation is $x^2 + 4y^2 = 8$ at the point (2, 1).

Solution. We first make sure that the point $(2, 1)$ lies on the curve. Differentiating the equation of the curve, we have

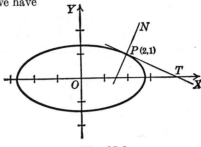

$$2\,x + 8\,y\,\frac{dy}{dx} = 0,$$

whence $\quad\dfrac{dy}{dx} = -\dfrac{x}{4\,y}.$

The value of the derivative at $(2, 1)$ is $-\frac{1}{2}$. Hence the slope of the tangent is $-\frac{1}{2}$ and the slope of the normal is 2. Using the point-slope form, the equation of the tangent PT is

Fig. 65.3

$$y - 1 = -\tfrac{1}{2}(x - 2), \quad \text{or} \quad x + 2\,y - 4 = 0,$$

and the equation of the normal PN is

$$y - 1 = 2(x - 2), \quad \text{or} \quad 2\,x - y - 3 = 0.$$

If the equation of a tangent with a given slope is required, we first determine the point or points of tangency, as in Example 1.

Example 4. Find the points of contact of the tangents to the ellipse in Example 3 which pass through the external point $(6, -1)$.

Solution. If $P(x, y)$ is a point of contact, the slope of the tangent at this point is $\dfrac{y + 1}{x - 6}$, by the slope formula. But, as in Example 3, the slope is $dy/dx = -x/4\,y$. Hence

$$-\frac{x}{4\,y} = \frac{y + 1}{x - 6},$$

or $\qquad\qquad x^2 - 6\,x + 4\,y^2 + 4\,y = 0. \qquad\qquad (1)$

But P is on the curve, and so

$$x^2 + 4\,y^2 = 8. \qquad\qquad (2)$$

Solving (1) and (2) simultaneously gives $x = 2$ or $\frac{2}{5}$ and $y = 1$ or $-\frac{7}{5}$. Hence there are two tangents and the points of contact are $(2, 1)$ and $(\frac{2}{5}, -\frac{7}{5})$.

PROBLEMS

1. Find the points on the following curves where the tangent is horizontal. Draw the figure in each case.

    a. $y = 3\,x + x^2.$ **b.** $y = 4 - x - x^2.$
    c. $y = 3\,x - x^3.$ **d.** $y = 2\,x^2 - x^4.$
    e. $y = x^3 + 1.$ **f.** $y = x^3 + x.$

2. Find the points on the curve $y^2 = x$ where the tangent is parallel to the line $x - 4\,y - 4 = 0.$

3. Find the points on the curve $4 y = x^2$ where the inclination is 45°; 135°; 60°.

4. Find the points on the curve $10 y = 6 x^2 - x^3$ where the tangent is perpendicular to the line $2 x - 3 y = 16$.

5. Find the equations of the tangent and of the normal to the following curves at the points indicated.

a. $y = x^2 - 2 x$, at $(0, 0)$.   **b.** $y = x^2 - 4$, at $(1, -3)$.
c. $x^2 + y^2 = 10$, at $(-1, 3)$.   **d.** $x^2 + 4 y^2 = 20$, at $(2, -2)$.
e. $x^2 - y^2 = 7$, at $(4, 3)$.   **f.** $x^2 - 2 y^2 + 4 = 0$, at $(-2, 2)$.

6. For each curve in Problem 5 find the points of contact of the tangents which have the indicated slope.

     a. 2.   **b.** -2.   **c.** $-\frac{1}{3}$.
     d. $\frac{1}{4}$.   **e.** $\frac{4}{3}$.   **f.** $-\frac{1}{2}$.

7. For each curve in Problem 5 find the points of contact of the tangents from the indicated external point and write the equations of the tangents.

     a. $(3, -6)$.   **b.** $(4, 3)$.   **c.** $(11, 1)$.
     d. $(10, 0)$.   **e.** $(0, -3)$.   **f.** $(-6, 4)$.

8. Find the length of the tangent to the curve $y = x^2 - 6 x$ at $(2, -8)$ from the point of tangency to the point where it cuts the x-axis.

9. In Problem 8 find the length of the normal between the curve and the y-axis.

10. In Problem 5, d, find the length of the segment of the tangent intercepted by the coördinate axes.

11. Find the angles of intersection of the following pairs of curves.

     a. $y^2 = 2 x$, $x - y - 4 = 0$.
     b. $2 y = x^2 - 4$, $8 y = x^2 - 4$.
     c. $x^2 + y^2 = 17$, $3 x - 5 y + 17 = 0$.
     d. $x^2 + 4 y^2 = 32$, $x = 2 y$.
     e. $x^2 + y^2 = 36$, $x^2 = 5 y$.
     f. $x^2 = 4 y$, $y = \dfrac{8}{4 + x^2}$.
     g. $x^2 + y^2 = 25$, $xy = 12$.

12. Show that the curves whose equations are $xy = k$ and $x^2 - y^2 = k$ intersect at right angles.

13. Show that the curve whose equation is $y = Ax^3 + Bx^2 + Cx + D$ has not more than two points at which its tangent is horizontal. Show that the tangent is horizontal at no point if $B^2 < 3 AC$.

14. Show that the tangents to the parabola $y^2 = 2 px$ at the points where $x = p/2$ are perpendicular to each other.

CHAPTER V

RATES AND DIFFERENTIALS

66. Introduction. When one variable is expressed as a function of a second variable, the relation may be used to study the effect upon the function of a variation in the independent variable. We may draw the graph, taking the values of the independent variable as abscissas and those of the function as ordinates, and apply the results of previous chapters. These show that the function is increasing when its derivative is positive, and is decreasing when its derivative is negative. Compare Art. 110. We also note from the graph that at points where the curve is steep the function is changing *rapidly* with respect to the independent variable, and vice versa; in other words, the rate of change of the function is measured by the slope of its graph. This is the intuitive idea upon which will be based the further study of the variation of a function corresponding to a given variation of the independent variable.

67. Average and constant rates of change. DEFINITION. *If y is a function of x, x_1 and x_2 are two values of x, and y_1 and y_2 are the corresponding values of y, the difference quotient $\dfrac{y_2 - y_1}{x_2 - x_1}$ is called the average rate of change of y with respect to x in the interval (x_1, x_2).*

The average rate of change merely expresses in another form the fact that y changes by the amount $y_2 - y_1$ when x changes from x_1 to x_2 and, if it is not constant or nearly constant, it is of little value except as a means of comparing the rapidity of change of y in two different intervals.

If, however, the average rate of change is the same for all possible intervals (x_1, x_2), we say that y is changing with respect to x at a constant rate, and the value of the difference quotient is called the rate of change of y with respect to x. In this case the rate of change is merely the amount of change in y for each unit increase in x.

For example, suppose that oil is being pumped at a constant rate into a tank containing 10 gal. at 10.02 A.M. and 50 gal. at 10.12 A.M. We see easily that the contents are increasing at the rate of 40 gal. in 10 min., or 4 gal. per minute, and conclude that in the next 5 min. $5 \times 4 = 20$ more gallons will be added, in the next 10 min. 40 more, etc. This is expressed more formally in the language of the definition as follows. The volume (V) of oil in the tank is a function of the time (t), which it is convenient to measure from 10 A.M. The given values of t are $t_1 = 2$ and $t_2 = 12$, and the corresponding values of V are $V_1 = 10$

and $V_2 = 50$. By definition, the average rate of change of V with respect to t in the interval $(2, 12)$ is

$$\frac{V_2 - V_1}{t_2 - t_1} = \frac{50 - 10}{12 - 2} = 4 \text{ gal. per min.}$$

Since we are told that the rate of change is constant, this means that the quantity of oil in the tank will increase by 4 gal. each minute that the pump continues to work at this constant rate.

The foregoing is an example of a time rate, but many other kinds occur. As a second example, if a man walks along a railroad track with a 1 per cent grade, his altitude above sea level is a function of the distance walked, and is increasing at the rate of 52.8 ft. per mile.

68. Formula for a variable changing at a constant rate. If a variable y changes with respect to another variable x at a known constant rate, and if we know a pair of corresponding values of x and y, we can at once write down the formula for y in terms of x. This is a consequence of the following general theorem.

Theorem. *If y is a linear function of x, the rate of change of y with respect to x is constant, and conversely.*

Proof. If y is a linear function of x, the equation which gives the relation between x and y is of the first degree and can be written in the form $y = b + mx$, where b and m are constants. Let x_1 and x_2 be any two values of x and y_1 and y_2 be the corresponding values of y. Then

$$y_2 = b + mx_2,$$
$$y_1 = b + mx_1,$$

whence
$$\frac{y_2 - y_1}{x_2 - x_1} = m,$$

which shows that the rate of change of y with respect to x is constant, by the definition.

Conversely, let the rate of change of y with respect to x be m, let x_1 and y_1 be a fixed pair of corresponding values, and x and y be any other pair of values. By definition of a constant rate,

$$\frac{y - y_1}{x - x_1} = m,$$

or
$$y - y_1 = m(x - x_1).$$

This equation is of the first degree and so y is a linear function of x.

Corollary. If y is a linear function of x, the rate of change of y with respect to x is the slope of the graph of the function.*

The last equation in the proof of the theorem is the point-slope form of the equation of a line. Hence, to find the relation between x and y, if the rate of change of y with respect to x is m and x_1 and y_1 are two corresponding values, we merely have to write the equation of the straight line which passes through the point (x_1, y_1) and has the slope m. If two pairs of values are given, we can use the two-point form of the equation of a line.

For example, in the oil-tank problem of Art. 67, $V = 10$ when $t = 2$ and the rate of change of V with respect to t is 4. Hence to find the relation between t and V we have to write the equation of a straight line which passes through the point $(2, 10)$ and has the slope 4. This is

$$V - 10 = 4(t - 2),$$
$$V = 2 + 4\,t.\dagger$$

The same result could have been obtained by using the point $(12, 50)$, or by using the two points $(2, 10)$ and $(12, 50)$ in the two-point form of the equation of a line.

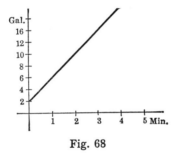

Fig. 68

NOTATION. For the sake of compactness the rate unit is usually written as a fraction. Thus a rate of 4 gal. per minute is written as 4 gal./min. or as $4\ \dfrac{\text{gal.}}{\text{min.}}$. Similarly, a speed of 44 ft. per second is written as 44 ft./sec. or as $44\ \dfrac{\text{ft.}}{\text{sec.}}$. In all problems involving rates the units should be clearly stated.

---

*In a strict sense this is true only when the same scales are used on both coördinate axes, but it has become customary to use the name "slope" for the difference quotient in all cases.

†This formula has no meaning when $t < -\frac{1}{2}$, since then V would be negative.

PROBLEMS

1. In experiments on the temperatures at various depths in an artesian well the temperature ($T°$) was found to be connected with the depth (d ft.) by the equation $T = 52 + 0.017\,d$. What was the rate of change in the temperature with respect to the depth? Draw the graph of the equation.

2. In experiments with a pulley block the pull p (lb.) required to lift a weight w (lb.) was found to be $p = 0.03\,w + 0.5$. How much does the pull increase per hundredweight increase in w?

3. A metal bar increases in length at a constant rate if heated. Suppose that at 20° C. its length is 1000 cm. and that at 60° C. its length is 1000.76 cm. Find its increase in length per degree increase in temperature. Express the length as a function of the temperature.

4. The extension of the spring of a balance is proportional to the weight attached to it. The length of the spring is 4.25 in. under a weight of 2 lb., and 5 in. under a weight of 8 lb. Express the length of the spring as a function of the weight. What is the natural length of the spring?

5. Fahrenheit temperature (F) is connected with centigrade temperature (C) by the relation $F = \frac{9}{5}\,C + 32$. Draw the graph of this function on a large enough scale to make it possible to read off corresponding values of F and C correct to one degree.

6. At a height of 110 m. above sea level the barometer reads 750 mm.; at a height of 770 m. it reads 695 mm. If it is assumed that the barometric pressure decreases constantly as the height increases, find the rate at which it decreases with respect to the height. What will be the pressure at a height of 850 m.? Express the pressure as a function of the height.

7. The boiling-point of water decreases as the altitude above sea level increases. At an altitude of 500 ft. the boiling-point is 211° and at 2500 ft. it is 207°. Find the rate of change, assuming that it is constant. What will be the boiling-point at an altitude of 12,000 ft.?

8. Water is flowing out of a cylindrical tank of radius 5 ft. at the constant rate of 20 gal./sec. How fast is the surface falling? If the depth was 15 ft. when $t = 0$, express the depth as a function of t and draw its graph. (1 cu. ft. = 7.5 gal.)

9. Oil is being pumped into a cylindrical tank of diameter 4 ft. at the constant rate of 5 gal./min. How fast is the surface of the oil in the tank rising?

10. If $3\,x + 5\,y = 8$, at what rate does y change with respect to x? Illustrate by a graph.

11. Show that if $y = 4 - 8\,x - x^2$, y does not change with respect to x at a constant rate.

12. The velocity of a moving point is given by the formula $v = 50 - 6\,t$, where v is measured in feet per second and t is measured in seconds. At what rate is v changing?

13. Show that the circumference of a circle changes at a constant rate with respect to the radius.

14. If a variable point P moves from $(2, 8)$ to $(6, -2)$ along a straight line, at what rate is the ordinate changing with respect to the abscissa?

69. Average rates in general. As stated in Art. 67 the average rate of change is of little value by itself unless it is constant or nearly so. It is, however, useful for the purpose of developing another concept, to be explained in Art. 70. For this purpose it is convenient to express the definition in the increment notation, setting $\Delta y = y_2 - y_1$ and $\Delta x = x_2 - x_1$.

DEFINITION. *If y is a function of x, the average rate of change of y with respect to x in the interval $(x_1, x_1 + \Delta x)$ is the value of $\Delta y / \Delta x$ for $x = x_1$.*

Example. What is the average rate at which area is added to a square if the length of its side increases from 5 to 9 in.?

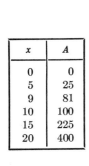

x	A
0	0
5	25
9	81
10	100
15	225
20	400

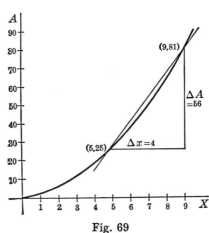

Fig. 69

Solution. Let A be the area and x the length of the side of the square. From geometry we have

$$A = x^2.$$

Give x an increment Δx, and we have

$$A + \Delta A = x^2 + 2\,x\,\Delta x + (\Delta x)^2.$$

Hence
$$\Delta A = 2\,x\,\Delta x + (\Delta x)^2,$$

and the average rate of increase for any values of x and Δx is

$$\frac{\Delta A}{\Delta x} = 2\,x + \Delta x.$$

For the case given, $\qquad x = 5$

and $\qquad\qquad\qquad \Delta x = 9 - 5 = 4.$

$$\frac{\Delta A}{\Delta x} = 10 + 4 = 14.$$

Hence the required average rate is 14 sq. in. per inch increase in the length of the side. Note that the average rate is a function of both x and Δx.

In this case the average rate appears in the graph as the slope of the secant (if allowance is made for the different scales used for A and x) joining the points $(5, 25)$ and $(9, 81)$.

Consequently, if we try to compute successive values of A by multiplying the average rate 14 by the increments of x and adding the results to the value of A for $x = 5$, as can be done when the rate is constant, we merely get the ordinates of points on the secant.

70. Instantaneous rates. In the example of Art. 69 it was found that as x increases from 5 to 9 the area increases at the average rate of 14 sq. in. per inch increase in the length of the side. As x increases from 5 to 8 we find, by substituting $x = 5$ and $\Delta x = 3$ in the expression for $\Delta A / \Delta x$, that the average rate of increase in the area is 13 sq. in./in. Similarly, as x increases from 5 to 7 ($\Delta x = 2$), the average rate of increase of the area is 12 sq. in./in. As Δx is taken smaller and smaller and made to approach zero as a limit, the average rate of increase of the area approaches the limiting value 10 sq. in./in., which is the *instantaneous rate of change* of the area with respect to the side at the instant when the latter is 5 in.

In general we have the following definition.

DEFINITION. *If y is a function of x, the instantaneous rate of change of y with respect to x when $x = x_1$ is the limit (if it exists) of the average rate of change in the interval $(x_1, x_1 + \Delta x)$ as Δx approximates zero.*

The following important theorem is an immediate consequence of the definition.

Theorem. *If $y = f(x)$, the instantaneous rate of change of y with respect to x for $x = a$ is the value of dy/dx for $x = a$.*

Proof. For any values of x and for $\Delta x \neq 0$ the average rate of change of y is $\dfrac{\Delta y}{\Delta x}$.

The instantaneous rate is the limit of the average rate as $\Delta x \longrightarrow 0$, by definition. Hence

$$\text{Instantaneous rate} = \lim_{\Delta x \to 0} \frac{\Delta y}{\Delta x} = \frac{dy}{dx}.$$

Corollary. The instantaneous rate of change of any function with respect to its variable is equal to the slope of the graph of the function at the given point.

Fig. 70.1 represents the graph of the function $y = f(x)$ for the interval $(x_1, x_1 + \Delta x)$. By the definition of slope, $MT = f'(x_1)\Delta x$ is the increment of the ordinate tracing the tangent PT when x changes from x_1 to $x_1 + \Delta x$. Since the tangent closely approximates the curve in the vicinity of the point P, it is clear that $\Delta y = MQ$ is nearly equal to MT when Δx is small. Con-

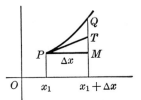

Fig. 70.1

sequently, if we assume that when x changes from x_1 to $x_1 + \Delta x$, y changes at a constant rate equal to the instantaneous rate $f'(x_1)$, a good approximation to the value of Δy is given by $f'(x_1)\Delta x$ if Δx is small enough.

Example 1. In the example of Art. 69 find the instantaneous rate of change of A with respect to x when $x = 5$ and use this result to approximate A when $x = 5.2$.

Solution. Since $A = x^2$, the instantaneous rate for any value of x is $dA/dx = 2x$. When $x = 5$, this is equal to 10 sq. in./in. When x changes from 5 to 5.2, $\Delta x = 0.2$ and so ΔA is approximately equal to $10 \times 0.2 = 2$ sq. in. Since $A = 25$ sq. in. when $x = 5$ in., A is approximately equal to $25 + 2$, or 27 sq. in., when $x = 5.2$ in. The exact value of A when $x = 5.2$ is $(5.2)^2$, or 27.04 sq. in., and the exact value of ΔA is 2.04 sq. in.

Example 2. At noon a ship bound north is 60 mi. south of another ship, which is bound east. If the first ship sails at the rate of 15 mi./hr., and the second ship at the rate of 10 mi./hr., how fast will the distance between them be changing at 2 o'clock? at 3 o'clock?

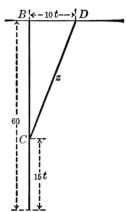

Fig. 70.2

Solution. Let A and B be the first positions of the ships and C and D their positions after t hours. Then $BD = 10t$ and $CB = 60 - 15t$. Let z be the distance between them. Then we have at once

$$z = \sqrt{CB^2 + BD^2}$$
$$= \sqrt{(60 - 15t)^2 + 100t^2}$$
$$= \sqrt{3600 - 1800t + 325t^2}.$$

To find the rate at which z is changing, differentiate:

$$\frac{dz}{dt} = \frac{325t - 900}{\sqrt{3600 - 1800t + 325t^2}}.$$

At 2 o'clock $t = 2$ and

$$\frac{dz}{dt} = \frac{-250}{\sqrt{1300}} = -6.9;$$

that is, the ships will be approaching each other at the rate of 6.9 mi./hr. When $t = 3$, $\frac{dz}{dt} = \sqrt{5} = 2.2$; that is, the ships will be separating at the rate of 2.2 mi./hr.

71. Rectilinear velocity and acceleration. Several kinds of rates occur so often that names have been given to them. Among these are *velocity* and *acceleration*.

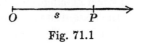

Fig. 71.1

The velocity of a point moving along a straight line is merely the rate at which its distance from some fixed point on the line is changing with respect to the time. Thus, if s represents the distance of P from the fixed point O at any time t, the velocity of P is given by the formula

$$v = \frac{ds}{dt}.$$

Since a function is increasing if its derivative is positive, and decreasing if its derivative is negative, it follows that the point is moving in the positive direction along the line if the velocity is positive and that it is moving in a negative direction if the velocity is negative. The *speed* is the numerical value of the velocity.

The acceleration of a point moving along a straight line is the rate at which its velocity is changing with respect to the time. Hence, if a denotes acceleration and v denotes velocity, we have at once

$$a = \frac{dv}{dt}.$$

Since v itself is the derivative of s with respect to t, this makes a the second derivative (see Art. 64) of s with respect to t; that is,

$$a = \frac{dv}{dt} = \frac{d}{dt}\left(\frac{ds}{dt}\right) = \frac{d^2s}{dt^2}.$$

Example. A point moves in a straight line directed vertically upward according to the law $s = 96\,t - 16\,t^2$. Find **(a)** its velocity and acceleration after 4 sec., **(b)** how high it will rise, **(c)** how far it will move in the fifth second.

Solution. At any time,

$$v = \frac{ds}{dt} = 96 - 32\,t$$

and

$$a = \frac{dv}{dt} = -32.$$

(a) When $t = 4$, $v = 96 - 32\,t = 96 - 128 = -32$ ft./sec., indicating that the point is coming *down* at the rate of 32 ft./sec. Since the acceleration is negative, the velocity is always decreasing at the rate of 32 ft. per second per second. The speed is decreasing when the point is moving upward and is increasing when the point is moving downward.

(b) The body will cease to rise when $v = 0$; therefore

$$96 - 32\,t = 0, \quad \text{or} \quad t = 3 \text{ sec.}$$

Its height will be the value of s for $t = 3$; that is,

$$s = 96 \cdot 3 - 16 \cdot 3^2 = 288 - 144 = 144 \text{ ft.}$$

(c) To find the distance moved in the fifth second, find the values of s for $t = 4$ and $t = 5$; by substitution in the given formula these are seen to be 128 and 80 respectively. As the latter height is less than the former, it is seen that during the fifth second the body came down $128 - 80 = 48$ ft. Compare this result with the velocity at the beginning of the fifth second, which was found in (a) to be -32 ft./sec.

Fig. 71.2

NOTE. The acceleration of a body falling freely in a vacuum is called the acceleration of gravity and is denoted by g. This number varies slightly in different parts of the earth, but is approximately 32 ft. per second per second (sometimes written 32 ft./sec.²). That is, if the positive direction is taken as downward, the velocity of a body moving under the given conditions increases by 32 ft./sec. each second. If the positive direction is taken as upward, the effect is to decrease the velocity, and the acceleration is $-g$, or -32. The same acceleration is frequently used in studying the motion of bodies falling in the air, and fairly good approximations may be obtained provided the velocities are not excessive and the falling body is heavy in proportion to its volume.

PROBLEMS

1. If $y = x^2 + \dfrac{4}{x^2}$, find (a) the average rate of change of y with respect to x as x increases from 2 to 4, (b) the instantaneous rate when $x = 2$, (c) the actual change in y when x changes from 2 to 2.5.

2. If $y = x^3 - 4\,x^2 + 6$, find the rate at which y is changing with respect to x when $x = 1$. Using this rate, approximate the change in y when x changes from 1 to 1.2. Why is your result not exact?

3. If $xy^2 = 36$, what is the rate of change of y with respect to x when $y = 3$? Illustrate your result by a graph.

4. If $s = a + bt$, where a and b are constants, show that the acceleration is 0.

5. If $s = a + bt + ct^2$, where a, b, and c are constants, show that the acceleration is constant.

6. A particle falls according to the law $s = \frac{1}{2} gt^2$. Find its average velocity during the first 5 sec. What is the speed at the end of 4 sec.? How far does it fall during the next 0.1 sec.?

7. If the distance s is measured in feet and t in seconds, find the velocity at the end of 2 sec. when (**a**) $s = (1 - t^2)^{-1}$, (**b**) $s = \sqrt{t} + 1$.

8. In Problem 7 find the acceleration at the end of 2 sec.

9. How fast is the reciprocal of x changing with respect to x when $x = 0.1$? when $x = 10$?

10. Find the rate of increase in the volume of a sphere with respect to the radius when the radius is 14 in. From this result approximate the change in the volume if the radius increases from 14 to 14.3 in.

11. In the following problems a point moves in a straight line according to the law given. Find (1) the values of t for which the velocity is positive and for which it is negative; (2) the values of t for which the acceleration is positive and for which it is negative. Describe the motion.

> **a.** $s = 6\, t^2 - 2\, t^3$, $(0 \leqq t \leqq 3)$.
> **b.** $s = t^3 - 12\, t^2$, $(0 \leqq t \leqq 12)$.
> **c.** $s = t^3 - 12\, t^2 + 36\, t$, $(0 \leqq t \leqq 6)$.
> **d.** $s = 16\, t^2 - t^4$, $(0 \leqq t \leqq 4)$.

12. At a certain instant a ship bound north is 6 mi. west of another ship, which is bound east. If the first ship is sailing at the rate of 15 mi./hr. and the second ship at the rate of 12 mi./hr., how fast are they separating at the end of one hour?

13. Two railroad tracks intersect at right angles. At noon there is a train on each track approaching the crossing at 40 mi./hr., one being 100 mi. and the other 200 mi. distant. Find (**a**) how fast they are approaching each other at 1 o'clock, (**b**) when they will be nearest together, (**c**) what will be their minimum distance apart.

14. As a man walks across a bridge at the rate of 4 ft./sec. a boat passes directly beneath him going downstream at 10 ft./sec. If the bridge is 30 ft. above the water, how fast are the man and the boat separating 3 sec. later?

15. It is found by experiment that the volume of water which at 4° C. has unit volume is given by the equation $V = 1 + a(t - 4)^2$, where t denotes the temperature and $a = 0.00000838$. Find the rate at which the volume is changing when $t = 0°$ and when $t = 20°$.

16. The time (t seconds) of a complete oscillation of a pendulum of length l inches is given by the formula $t = 0.324\sqrt{l}$. Find the rate of change of the time with respect to the length of the pendulum when $l = 9$ in. By means of this result approximate the change in t due to a change in l from 9 in. to 9.2 in.

17. If there is no transfer of heat, the pressure and volume of compressed air are connected by the relation $pv^{1.41} = C$, where C is a constant. If $p = 20$ lb./sq. in. when $v = 800$ cu. in., find the rate of change in p with respect to v when $v = 800$ cu. in., and approximate the resulting change in p if v is decreased 10 cu. in.

72. Related rates. Frequently the time rate of change of one variable is known, and it is desired to find the time rate of change of a second variable which is related to the first. Such problems are easily solved by differentiating the equation connecting the variables implicitly with respect to the time and substituting the given values of the variables.

Example 1. A barge whose deck is 5 ft. below the level of a dock is drawn up to it by means of a cable running over a pulley at the edge of the dock. When the barge is 12 ft. away from the dock, the cable is being hauled in at the rate of 4 ft./min. At what rate is the barge moving at this time?

Solution. Let z denote the length of the cable from the pulley to the barge and let x denote the distance of the barge from the dock *at any time*. It is given that $dz/dt = -4$ when $x = 12$ and it is required to find the value of dx/dt at this time.

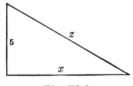

Fig. 72.1

The equation connecting x and z is seen from Fig. 72.1 to be

$$z^2 = x^2 + 25. \tag{1}$$

Differentiating with respect to t,

$$2z\frac{dz}{dt} = 2x\frac{dx}{dt}. \tag{2}$$

We are given the values of $\dfrac{dz}{dt}$ and x; we find by substitution in (1) that $z \approx 13$. Substituting these values in (2), we obtain

$$2(13)(-4) = 2(12)\frac{dx}{dt},$$

or

$$\frac{dx}{dt} = -\frac{13}{3} = -4\tfrac{1}{3}.$$

That is, the barge is moving toward the dock at the rate of $4\tfrac{1}{3}$ ft./min. Note that a positive result would be erroneous, for that would indicate that x is increasing, not decreasing.

The above problem is typical, and in general the same procedure should be followed. The various steps are

I. *Determine the variables involved. These always include those for which the rates of change are given or desired.*

II. *Set up the equation connecting the variables.*

III. *Differentiate the equation implicitly with respect to t.*

IV. *Substitute the given values of the variables in the two equations and solve for the unknowns.*

Example 2. The radius of the base of a right circular cone is decreasing at the rate of 4 in./min., and the height is increasing at the rate of 6 in./min. At what rate is the volume changing when the height is 12 in. and the radius is 6 in.?

Solution. Let V, r, and h denote the volume, radius, and height, respectively, of the cone. We are given $dr/dt = -4$ and $dh/dt = 6$; we are to find dV/dt when $h = 12$ and $r = 6$.

The formula for the volume of a cone is $V = \dfrac{\pi r^2 h}{3}$.

Differentiating this with respect to t, we have

$$\frac{dV}{dt} = \frac{1}{3}\pi r^2 \frac{dh}{dt} + \frac{2}{3}\pi r h \frac{dr}{dt}.$$

Substituting the given values of the variables, we obtain

$$\frac{dV}{dt} = \frac{\pi}{3}(36)(6) + \frac{2}{3}\pi(6)(12)(-4)$$
$$= 72\,\pi - 192\,\pi = -120\,\pi = -377,$$

indicating that the volume is decreasing at the rate of about 377 cu. in./min.

Example 3. A point $P(x, y)$ moves along the line whose equation is $x - 2y + 4 = 0$ in such a way that y increases at the rate of 3 units/sec. The point $A(0, 6)$ is joined to P and the segment AP is prolonged to meet the x-axis in a point Q. Find how fast the distance from the origin to Q is changing when P reaches the point $(4, 4)$.

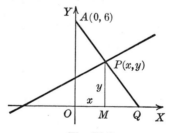

Fig. 72.2

Solution. The rate of change of y is given and it is desired to find the rate of change of OQ, which we denote by z. If MP is perpendicular to the x-axis, $MP = y$ and $OM = x$. The triangles OAQ and MPQ are similar, whence

$$\frac{z}{6} = \frac{z - x}{y}.$$

This gives

$$yz = 6\,z - 6\,x, \tag{1}$$

or

$$z = \frac{6\,x}{6 - y}.$$

Substituting the value of x from the equation of the given line, we have

$$z = \frac{12(y - 2)}{6 - y}$$

and

$$\frac{dz}{dt} = \frac{48}{(6 - y)^2}\frac{dy}{dt}.$$

Setting $y = 4$ and $\frac{dy}{dt} = 3$, we obtain

$$\frac{dz}{dt} = 36;$$

that is, z is increasing at the rate of 36 units/sec.

Another method is to differentiate implicitly in (1). This gives

$$y\frac{dz}{dt} + z\frac{dy}{dt} = 6\frac{dz}{dt} - 6\frac{dx}{dt}. \tag{2}$$

We know that $dy/dt = 3$. From the equation $x - 2\,y + 4 = 0$ we find that $\frac{dx}{dt} = 2\frac{dy}{dt} = 6$. From (1) we find that $z = 12$ when $x = 4$ and $y = 4$. Substitution in (2) gives $dz/dt = 36$, as before.

PROBLEMS

1. A ladder 24 ft. long leans against a vertical wall. If the lower end is being moved away from the wall at the rate of 3 ft./sec., how fast is the top descending when the lower end is 8 ft. from the wall?

2. In Problem 1 find when the lower and the upper ends are moving with the same speed.

3. A conical funnel is 14 in. across the top and 12 in. deep. A liquid is flowing in at the rate of 60 cu. in./sec., and flowing out at the rate of 40 cu. in./sec. Find how fast the surface of the liquid is rising when it is 6 in. deep.

4. A man 6 ft. tall walks away from an arc light 15 ft. high at the rate of 3 mi./hr. **(a)** How fast is the farther end of his shadow moving? **(b)** How fast is his shadow lengthening?

5. A kite is 80 ft. high, with 100 ft. of cord out. If the kite starts moving away horizontally at the rate of 5 mi./hr., how fast is the cord being paid out?

6. A boat is fastened to a rope which is wound about a windlass 20 ft. above the level at which the rope is attached to the boat. The boat is drifting away at the rate of 8 ft./sec. How fast is it unwinding the rope when 30 ft. from the point directly under the windlass?

7. The volume of a sphere is increasing at the rate of 16 cu. in./sec. How fast is the radius increasing when it is 6 in.?

8. Find how fast the surface is increasing in Problem 7.

9. Sand is being poured on the ground from an elevated pipe and forms a pile which has always the shape of a right circular cone whose height is equal to the radius of the base. If the sand falls at the rate of 6 cu. ft./min., how fast is the height of the pile increasing when the height is 5 ft.?

10. The radius of a cone is decreasing at the rate of 2 in./min., and the altitude is increasing at the rate of 3 in./min. When the radius is 18 in. and the altitude is 20 in., find **(a)** the rate at which the volume is changing, **(b)** the rate at which the curved surface is changing.

11. The diameter of a hemispherical bowl is 18 in. If the depth of the water in it is increasing at the rate of $\frac{1}{8}$ in./sec. when it is 8 in. deep, how fast is the water flowing in? (The volume of a segment of a sphere of radius r is $\pi h^2\left(r - \frac{h}{3}\right)$, where h is the height of the segment.)

12. If $y = x^2$, and x is increasing at the rate of $\frac{1}{2}$ unit/min. when $x = 2$, find **(a)** how fast y is changing, **(b)** how fast the slope of the graph is changing.

13. The path of a moving point is the curve $x^2 - 4y^2 = 36$. If x increases steadily at the rate of 2 units/min., find how fast y is changing at the point $(10, -4)$.

14. The velocity of a jet of liquid issuing from an orifice is given by the formula $v^2 = 2gh$, where h is the height of the liquid surface above the orifice. If h is decreasing at the rate of 3 in./min., find how fast the velocity of flow is changing when $h = 100$ ft.

15. If $y^2 = 2x$, and x is decreasing steadily at the rate of 0.25 units/min., find how fast the slope of the graph is changing at the point $(8, -4)$.

16. If $100y = 400 - x^2$, and y is increasing at the rate of 0.1 unit/min. when $x = 20$, find **(a)** how fast x is changing, **(b)** how fast the slope of the graph is changing.

17. A point moves along the parabola whose equation is $4y = x^2$ so that the abscissa increases at the constant rate of 2 units per second. At what rate is the distance between this point and the point $(0, 4)$ changing when $x = 2$ and when $x = \sqrt{8}$?

18. The velocity of a point moving along a straight line is given by $v^2 = a + \frac{2b}{s}$, where a and b are constants. Show that the acceleration is $-b/s^2$.

73. Differentials defined. *The differential of a function of a variable is the product of its derivative with respect to the variable by the increment of the variable.*

The differential of a function is denoted by the letter d prefixed to the symbol denoting the function. The above definition may then be stated thus: If $y = f(x)$, $dy = df(x) = f'(x)\Delta x = \dfrac{dy}{dx}\Delta x$. Evidently dy is a function both of x and of Δx.

If $y = f(x) = x$, $f'(x) = 1$ and so $dy = 1 \cdot \Delta x = \Delta x$ in this case. On this account the differential of the *independent* variable is defined as equal to the increment and it is customary to write the relation $dy = f'(x)\Delta x$ in the form $dy = f'(x)dx$. This practice will be followed in the future.

The student must not, however, make the mistake of thinking that the differential and the increment of the dependent variable are identical. The difference is shown geometrically in the following discussion.

In Fig. 73 let P be any point on the graph of $y = f(x)$, let PS be the tangent at P, and let Q be a neighboring point on the curve.

Evidently $PM = \Delta x = dx$, and $MQ = \Delta y$. Moreover, angle $SPM = \alpha$.

Hence $\qquad\qquad\qquad m = \tan \alpha = MS/dx;$

or, since $m = f'(x)$, $\qquad MS = f'(x)dx.$

But, by definition, $\qquad dy = f'(x)dx.$

Therefore $\qquad\qquad\quad dy = MS.$

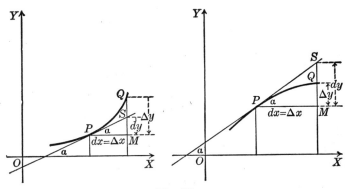

Fig. 73

Thus we have shown that *if $P(x, y)$ is a point on the curve $y = f(x)$, then, for a particular value of x and an arbitrarily chosen value of the increment dx, Δy is the corresponding increment of the ordinate drawn to the curve, and dy is the corresponding increment of the ordinate drawn to the tangent at P.*

74. Ratio of the differential to the increment of the function. If we recall that $f'(x)$ is the instantaneous rate of change of $y = f(x)$ with respect to x and compare the discussion just concluded with the one following the theorem in Art. 70, it is evident that *dy is the approximation to the increment of y corresponding to Δx, which we obtain by assuming that Δx is small enough to permit us to regard y as changing at a constant rate.*

In Art. 70 it was asserted that this approximation is a good one when Δx is small. The following theorem gives the justification for this assertion.

Theorem. *If for a given value of x, $y = f(x)$ is differentiable and $f'(x)$ $\neq 0$, then* $\lim\limits_{\Delta x \to 0} \dfrac{\Delta y}{dy} = 1$.

Proof. By the definition of a differential,

$$\frac{\Delta y}{dy} = \frac{\Delta y}{f'(x)\Delta x} = \frac{\dfrac{\Delta y}{\Delta x}}{f'(x)}. \tag{1}$$

By hypothesis and the definition of a derivative,

$$\lim_{\Delta x \to 0} \frac{\Delta y}{\Delta x} = \frac{dy}{dx} = f'(x) \neq 0.$$

Using this in relation (1), we have

$$\lim_{\Delta x \to 0} \frac{\Delta y}{dy} = \frac{f'(x)}{f'(x)} = 1.$$

Corollary. Under the hypothesis of the theorem

$$\lim_{\Delta x \to 0} \frac{\Delta y - dy}{dy} = 0.$$

Thus the corollary shows that the difference between Δy and dy is not only small when Δx is small, but is also small *in comparison to the value of dy.* For this reason dy is often called the *principal part* of the increment Δy.

For example, if $y = x^2$, $dy = 2\,x\,dx$ and

$$\Delta y = 2\,x\,\Delta x + (\Delta x)^2 = dy + (\Delta x)^2.$$

Thus the difference between Δy and dy is $(\Delta x)^2$, a quantity which is comparatively negligible if Δx is small.

75. How to find differentials. The definition indicates the method of finding the differential of a given function, namely, to find the derivative, and to multiply it by the differential of the variable. Therefore, to every formula giving a derivative there corresponds one giving the differential. For example, $dc = 0$; $d(u^n) = nu^{n-1} du$; etc.

Note that, since $dy = f'(x)dx$, the symbol dy/dx, which we have used hitherto as a single symbol for the derivative of y with respect to x, may now be regarded as the quotient of corresponding differentials dy and dx, if we so desire. There is no inconsistency in doing this. For in regard to Fig. 73 we may think of dy/dx as the derivative of y with respect to x, in which case it is the slope of the tangent PS; or we may think of dy/dx as the quotient of the corresponding differentials $dy = MS$ and $dx = PM$.

Example 1. Find the differential of $x(2 - 3\,x)^2$.

Solution.
$$d[x(2 - 3\,x)^2] = \frac{d}{dx}\,[x(2 - 3\,x)^2]dx$$
$$= (2 - 3\,x)(2 - 9\,x)dx.$$

When y is given implicitly as a function of x by an equation connecting these variables, it is usually simplest to differentiate implicitly, using the differential notation, and then solve the resulting equation for dy in terms of x, y, and dx.

Example 2. Find dy, if $xy = x^2 - 1$.

Solution.
$$d(xy) = d(x^2) - d(1).$$
$$x\,dy + y\,dx = 2\,x\,dx - 0.$$
$$dy = \frac{2\,x\,dx - y\,dx}{x} = \frac{2\,x - y}{x}\,dx.$$

PROBLEMS

1. Find the differentials of the following functions.

a. $x(1 - x^2)^3$.   b. $\sqrt{1 + x}/4\,x$.   c. $1/x$.

d. $\dfrac{y - 1}{y + 1}$.   e. $t\sqrt{1 + t}$.   f. $x\sqrt{a - bx}$.

g. $1/t^2$.   h. $\frac{4}{3}\,\pi r^3$.

2. Find dy in terms of x, y, and dx if

a. $\sqrt{x} + \sqrt{y} = \sqrt{a}$.   b. $x^3 + y^3 = 3\,axy$.
c. $b^2x^2 - a^2y^2 = a^2b^2$.   d. $xy = 6$.
e. $x^2y^3 + 4\,y = 4$.   f. $xy + y^2 + 4\,x = 0$.

3. Find algebraically dy and Δy if $y = x^3$. Also find their difference if $x = 1$ and $dx = 0.1$.

4. Find algebraically ds and Δs if $s = 16\,t - 8\,t^2$. Also find their difference if $t = 3$ and $\Delta t = 0.5$.

5. Show geometrically that, if y is a linear function of x, dy and Δy are equal.

6. In each of the following equations find dy for the values of x and dx given. In each case draw a small arc of the curve near the point named and mark dx, dy, and Δy.

$$\textbf{a. } y = 7 - 6\,x + x^2,\ x = 2,\ dx = 0.3.$$
$$\textbf{b. } y = \sqrt{x},\ x = 1,\ dx = 0.4.$$
$$\textbf{c. } y = x^3 - 4\,x^2,\ x = 3,\ dx = 0.2.$$
$$\textbf{d. } x^2 + y^2 = 25,\ x = 4,\ y = -3,\ dx = 0.3.$$

7. Show that $(x + dx)^2$ is approximately equal to $x^2 + 2\,x\,dx$ if dx is small. Use this result to approximate $(15.2)^2$.

8. Show that $(x + dx)^3$ is approximately equal to $x^3 + 3\,x^2\,dx$ if dx is small. Use this result to approximate $(10.2)^3$.

9. Show that $f(x + \Delta x) = f(x) + df(x)$, approximately.

76. Approximation of magnitudes by means of differentials. The fact that for small values of dx the corresponding increment of $y = f(x)$ is closely approximated by dy is of considerable utility in approximating quantities which can be expressed as the increments of functions.

Example 1. Estimate the amount of material needed to make a hemispherical shell of inner radius 10 in. and of thickness $\frac{1}{8}$ in.

Solution. The exact amount required is the difference of two hemispheres of radii 10 in. and 10.125 in. respectively. If we let V be the volume of a hemisphere of radius R, $V = \frac{2}{3}\pi R^3$, and the desired result is the value of ΔV when $R = 10$ and $\Delta R = 0.125$. To calculate this is troublesome, and, if an approximation only is desired, the more easily obtained value of dV will suffice, since ΔR is small. Hence the amount required is approximately

$$dV = 2\,\pi R^2 dR = 2\,\pi(100)(0.125) = 25\,\pi = 78.5 \text{ cu. in.}$$

Note that $dV = 2\,\pi R^2 dR$ is merely the product of the inner surface of the shell by the thickness.

Good judgment must be employed in using this method. If, in the example given above, dR were too large, the result would be worthless. In this particular case the exact value of ΔV is nearly 79.5 cu. in.

In other cases we may know the value of $y = f(x)$ for some value of x and the value of y is required for another value of x, which may be designated by $x + \Delta x$. If Δx is small and it is easy to calculate dy, the quantity $y + dy$ is a convenient approximation to the true value $y + \Delta y$.

Example 2. Use differentials to approximate $\sqrt{67}$ and $\sqrt{61}$.

Solution. Since the number 67 differs but little from the perfect square 64, we can get a good approximation by finding the value of the differential of $y = \sqrt{x}$ for $x = 64$ and $dx = 3$. Differentiating,

$$dy = \frac{dx}{2\sqrt{x}} = \frac{3}{16}.$$

Hence $\sqrt{67} = y + dy = 8\frac{3}{16}$, approximately.

To find $\sqrt{61}$, observe that $dx = -3$ and so $dy = -\frac{3}{16}$. Therefore we have $\sqrt{61} = 8 - \frac{3}{16} = 7\frac{13}{16}$, approximately.

If we generalize this problem, we obtain the useful approximate formula

$$\sqrt{x + dx} = \sqrt{x} + \frac{dx}{2\sqrt{x}}.$$

If we take for x the nearest perfect square to the number whose square root is desired, we get fairly accurate results and the formula has the remarkable property of becoming more accurate as x increases. For example, to find $\sqrt{73}$ take $x = 81$ (since 73 is nearer to 81 than to 64) and $dx = -8$; then $\sqrt{73} = 9 - \frac{8}{18} = 8\frac{5}{9}$. The error in this is less than 0.012. Similarly, to find $\sqrt{240}$ take $x = 225$; then $\sqrt{x}$ and dx are both 15 and $\sqrt{240} = 15 + \frac{15}{30} = 15.5$. The error here is about 0.008.

77. Approximation of errors. When the value of a function is obtained by calculation, it may happen that the value of the independent variable is not exactly known, as is always the case when it is obtained by measurement. It is then desirable to know approximately the error in the function due to the possible small error in the variable. This is equivalent to approximating the increment in the function corresponding to a small increment in the variable. Hence the differential of the function is the desired error, approximately.

Example 1. The radius of a sphere is measured as 3 in., with a possible error of 0.02 in. Find, approximately, (a) the greatest possible error in the calculated volume; (b) how accurately the radius must be measured in order that the error in the calculated volume shall not exceed 1 cu. in.

Solution. (a) Let r be the radius and V the volume of the sphere. The actual error in r may be anything between 0.02 in. and -0.02 in., but in the solution of problems of this kind it is usually best to assume that the error actually is the largest possible positive value. In this case then we are given that $r = 3$ and $dr = 0.02$, and we are required to find the value of dV. Differentiating $V = \frac{4}{3}\pi r^3$, we obtain

$$dV = 4\pi r^2 dr. \tag{1}$$

Substituting the values of r and dr we get

$$dV = (4\pi)(9)(0.02) = 0.72\pi = 2.26 \text{ cu. in., approximately.}$$

Thus the error in the calculated value of the volume may be anything between + 2.26 cu. in. and − 2.26 cu. in.

(b) For this part of the problem we are given that $r = 3$ and $dV = 1$. Substitution in (1) gives

$$1 = (4\,\pi)(9)dr,$$

whence
$$dr = \frac{1}{36\,\pi} = 0.009 \text{ in., approximately.}$$

Thus if the calculated value of the volume is to be accurate to within 1 cu. in., the error in measuring the radius must not exceed 0.009 in.

Relative error. In judging the accuracy of measured quantities or of others dependent upon them we are often less interested in the actual error than what is called *relative error*. The relative error is the ratio of the actual error to the quantity under consideration. For example, if a mile is measured with a possible error of 1 ft., the relative error is $\frac{1}{5280} = 0.00019$, and this measurement is relatively as accurate as that of a foot with a possible error of $0.00019 \times 12 = 0.0023$ in.

The relative error in a quantity is sometimes expressed as a percentage, and is known as the *percentage error*. In the example just considered, the percentage error was 0.019 per cent,—less than two hundredths of 1 per cent.

By definition the true value of the relative error in a value of y may be expressed as $\Delta y/y$; but unless the measurements are grossly inaccurate, Δy is nearly equal to dy, and the relative error is therefore closely approximated by the value of dy/y, except when y is small in comparison with Δy and dy.

Example 2. Find the greatest possible relative error in the calculated volume of the sphere in Example 1.

Solution. There are two ways to proceed. The first is to calculate the value of V for $r = 3$, which is $36\,\pi$ cu. in., and to divide this into the value of dV. The relative error is then

$$\frac{dV}{V} = \frac{0.72\,\pi}{36\,\pi} = 0.02 = 2 \text{ per cent.}$$

It is usually better, however, to find dV/V in terms of r and dr before substituting. By formula,

$$V = \tfrac{4}{3}\,\pi r^3 \quad \text{and} \quad dV = 4\,\pi r^2 dr.$$

Dividing dV by V, we find

$$\frac{dV}{V} = \frac{4\,\pi r^2 dr}{\dfrac{4\,\pi r^3}{3}} = \frac{3\,dr}{r} = \frac{3(0.02)}{3} = 0.02 = 2 \text{ per cent.}$$

Note that this last equation shows that the relative error in the volume is three times the relative error in the radius for all values of r.

Example 3. Show that the relative error in the product of two numbers may be as great as the sum of the relative errors of the numbers.

Solution. Let the numbers be u and v and let y be their product; that is, $y = uv$. We are required to investigate dy/y.

Differentiating, we have

$$dy = u\,dv + v\,du.$$

The relative error in y is obtained by dividing by $y = uv$:

$$\frac{dy}{y} = \frac{u\,dv + v\,du}{uv} = \frac{dv}{v} + \frac{du}{u}.$$

But dv/v and du/u are the relative errors in v and u respectively. Hence we have proved that the relative error in the product is the sum of the relative errors of the factors if du and dv are both positive or both negative.

The result just obtained is of some importance in calculation. It shows that the product may be less accurate than either of the factors forming it. For example, if u is 2 per cent too large and v is 3 per cent too large (that is, if $du/u = 0.02$ and $dv/v = 0.03$), then y will be about 5 per cent too large. Only in the case where one number is too large and the other too small will the errors tend to compensate each other.

PROBLEMS

1. Find an approximate formula for the area of a circular ring of radius r and width dr. What is the exact formula?

2. If A is the area of a square of side x, find dA. Draw squares of sides x and $x + dx$, one within the other, and mark on the figure the areas equivalent to dA and ΔA.

3. Find by means of differentials an approximate formula for the volume of a thin cylindrical shell with open ends if the radius is r, the length h, and the thickness t.

4. The acceleration (a) due to gravity varies inversely as the square of the distance (s) from the center of the earth and is 32.2 at the surface of the earth (that is, $a = 32.2$ when s equals the radius of the earth, which is about 21,000,000 ft.). Find the formula for a in terms of s and the change in a produced by going up 21,000 ft. in a balloon.

5. A box is to be constructed in the form of a cube to hold 8 cu. ft. How accurately must the inner edge be made so that the volume will be correct to within 10 cu. in.?

6. The time of one vibration of a pendulum is given by the formula

$$t^2 = \frac{\pi^2 l}{g},$$

where t is measured in seconds, $g = 32.2$, and l, the length of the pendulum, is measured in feet. Find **(a)** the length of a pendulum vibrating once a second, **(b)** the change in t if the pendulum in **(a)** is lengthened 0.01 ft., **(c)** how much a clock with this error would lose or gain in a day.

7. If $y = x^{\frac{3}{2}}$ and the possible error in measuring x is 0.2 when $x = 16$, what is the possible error in the value of y? Use this result to obtain approximate values of $(16.2)^{\frac{3}{2}}$ and $(15.8)^{\frac{3}{2}}$.

8. Use differentials to find an approximate value of $1/\sqrt{25.5}$.

9. Use differentials to find approximate values of $\sqrt[3]{127}$ and $\sqrt[3]{123}$.

10. Show by means of differentials that

$$\frac{1}{x + dx} = \frac{1}{x} - \frac{dx}{x^2} \text{ (approximately).}$$

11. The diameter and altitude of a cylinder are found by measurement to be 16 in. and 12 in., respectively. If there is a possible error of 0.05 in. in each measurement, what is the greatest possible error in the calculated volume?

12. The horsepower of a certain type of engine with m cylinders of diameter d in. is given by the formula $P = 0.4 \, md^2$. Estimate the amount added to the horsepower of a four-cylinder engine by increasing the diameter of each cylinder from $3\frac{1}{4}$ to $3\frac{3}{8}$ in.

13. The volume of a sphere is increasing at the rate of 16 cu. in./sec. Approximate the increment in the radius in the next $\frac{1}{2}$ sec. after the radius becomes 6 in.

14. How exactly must the diameter of a circle be measured in order that the area shall be correct to within 1 per cent?

15. Show that the relative error in the nth power of a number is n times the relative error in the number.

16. When a cubical block of metal is heated, each edge increases $\frac{1}{10}$ per cent per degree increase in temperature. Show that the surface increases $\frac{2}{10}$ per cent per degree, and that the volume increases $\frac{3}{10}$ per cent per degree.

17. Show that the relative error in the quotient of two numbers may be as great as the sum of the relative errors in the numbers.

18. The value of g may be found by timing the vibration of a pendulum. Find the relative error in g due to a relative error of 1 per cent in measuring the time of vibration of a pendulum.

‒ HINT. First solve the formula given in Problem 6 for g in terms of t and l.

19. The value of g is found by timing the vibrations of a pendulum whose length was measured as 7.34 ft., with an uncertainty of 0.005 ft. The time of each vibration was 1.5 sec., which was assumed to be exact. Find the value of g, the greatest possible error in this value, and the greatest possible percentage error. (See hint in Problem 18.)

20. The boiling-point of water at altitude H (ft.) above sea level is given by $H = 517(212° - T) - (212° - T)^2$, T being the boiling temperature in degrees Fahrenheit. Find the uncertainty in the calculated value of H if the uncertainty in the measured value of T is 1° and T is measured as 200°.

CHAPTER VI

INDEFINITE INTEGRALS—CONSTANT OF INTEGRATION

78. Integration. The process of differentiation enables us to find the differential (or derivative) of a given function. Integration is the inverse process of finding the function when the differential (or derivative) is given.

For example, if the given function is x^3, the differential is found to be $3\,x^2\,dx$. Inversely, if the differential $3\,x^2\,dx$ is given, it is clear that a function having this differential is x^3. But this is not the only function having the given differential. Obviously $3\,x^2\,dx$ is the differential of $x^3 - 7$, of $x^3 + 10$, and, in general, of $x^3 + C$, where C is any constant.

NOTATION. In the example above, $x^3 + C$ is called the *integral* of $3\,x^2\,dx$, and this is indicated by the notation

$$\int 3\,x^2\,dx = x^3 + C.$$

The function $3\,x^2$ is called the *integrand*.

In general the notation

$$\int f(x)dx = F(x) + C$$

means that $F(x) + C$ is the integral of $f(x)dx$; $f(x)$ is the integrand and, by definition,

$$dF = f(x)dx, \quad \text{or} \quad \frac{dF}{dx} = f(x).$$

In words, the integral may be defined as follows: *The indefinite integral of a given differential expression is the function whose differential is the given expression.* The integral is called *indefinite* because the constant C can have any value. In order to determine a value for C, additional data must be given (see Art. 82).

It will appear later that if $f(x)$ is continuous in an interval, there is always a function $F(x)$ such that

$$\frac{d}{dx}F(x) = f(x),$$

and so the indefinite integral of a continuous function always exists. Also the indefinite integral is unique, as shown in the following theorem.

Theorem. *If two functions have the same derivative in an interval, their difference is a constant.*

125

Proof. Let $F(x)$ and $G(x)$ be differentiable in the interval (a, b) and $F'(x) = G'(x)$. Let

$$H(x) = F(x) - G(x).$$

Then, by hypothesis,

$$H'(x) = F'(x) - G'(x) = 0.$$

That is, the rate of change of $H(x)$ with respect to x is zero for every value of x in the interval. Hence $H(x)$ is a constant.

79. Formulas for integration. Since integration is an inverse process, the formulas are obtained by inverting the formulas for differentiation. In fact, every solution of a problem in differentiation yields a formula for integration. For practical use, tables of integrals have been compiled containing many forms. For present purposes we give only four formulas for integration.

(I) $$\int du = u + C.$$

(II) $$\int [f(x) + g(x)]dx = \int f(x)dx + \int g(x)dx.$$

(III) $$\int a\, du = a \int du.$$

(IV) $$\int u^n\, du = \frac{u^{n+1}}{n+1} + C, \text{ if } n \neq -1.$$

In these formulas u denotes a function of x (or some other independent variable).

Formula **(I)** merely states the definition of an integral.

Formula **(II)** states that the integral of a sum of differentials is equal to the sum of the integrals of the differentials.

In Formula **(III)**, a is any constant, and this formula shows that a constant can be moved from one side of the integral sign to the other. The student is warned that a *variable* cannot be moved from one side of the integral sign to the other.

Formula **(IV)** is the *power formula*, in which n is any constant except -1. If the exponent of u in the integral is -1, the integral involves a logarithm, which will be considered later (Chapter XII).

Proofs of the formulas. It follows from the definition of an integral that each formula may be proved by showing that the differential of the right-hand side is the expression under the integral sign on the left. We give the proof for Formula **(II)**; the others are proved in a similar manner.

$$d\left[\int f(x)dx + \int g(x)dx\right] = d\int f(x)dx + d\int g(x)dx$$
$$= f(x)dx + g(x)dx \text{ by definition of integral}$$
$$= [f(x) + g(x)]dx.$$

80. Use of the formulas. Any function consisting of algebraic expressions can be differentiated by direct application of the formulas already given. But comparatively few expressions can be integrated in terms of known functions, and, for the present, only such expressions as can be reduced, by a proper transformation, to the exact form of one of the formulas of the preceding paragraph. Since the integrand is the derivative of an integral, the correctness of the work in any problem of integration can be tested by showing that the derivative of the result is equal to the integrand. The method of using the formulas will be illustrated by some examples.

Example 1. Find $\int 5\,x^2\,dx$.

Solution.
$$\int 5\,x^2\,dx = 5\int x^2\,dx, \qquad\qquad \text{by (III)}$$
$$= \frac{5\,x^3}{3} + C. \qquad\qquad \text{By (IV)}$$

Example 2. Find $\int (x^3 + 2\,x^2 + 3)dx$.

Solution. $\int (x^3 + 2\,x^2 + 3)dx = \int x^3\,dx + \int 2\,x^2\,dx + \int 3\,dx, \qquad \text{by (II)}$

$$= \int x^3\,dx + 2\int x^2\,dx + 3\int dx, \qquad \text{by (III)}$$
$$= \frac{x^4}{4} + \frac{2\,x^3}{3} + 3\,x + C. \qquad \text{By (IV) and (I)}$$

It should be noticed that a separate constant might be added with each of the three integrals of the second step of the solution. But this is unnecessary, since one arbitrary constant is equivalent to the sum of any number of constants.

Example 3. Find $\int \dfrac{x^2 + 2}{\sqrt{x}}\,dx$.

Solution. Since the denominator is a monomial, we can divide term by term and apply the power formula to each term of the result. This gives

$$\int \frac{x^2 + 2}{\sqrt{x}}\,dx = \int \left(x^{\frac{3}{2}} + 2\,x^{-\frac{1}{2}}\right)dx$$
$$= \int x^{\frac{3}{2}}\,dx + 2\int x^{-\frac{1}{2}}\,dx$$
$$= \tfrac{2}{5}\,x^{\frac{5}{2}} + 4\,x^{\frac{1}{2}} + C.$$

Example 4. Find $\int \sqrt{x+1}\,dx$.

Solution. This integral can be found by the power formula. Comparison with **(IV)** shows that $u = x + 1$, $du = dx$, and $n = \frac{1}{2}$. Hence

$$\int \sqrt{x+1}\,dx = \frac{(x+1)^{\frac{3}{2}}}{\frac{3}{2}} + C = \frac{2}{3}\,(x+1)^{\frac{3}{2}} + C.$$

Example 5. Find $\int \sqrt{8\,x}\,dx$.

Solution. This can be solved in two ways. The first is to factor out the constant, which gives

$$\int \sqrt{8\,x}\,dx = \sqrt{8} \int x^{\frac{1}{2}}\,dx = \sqrt{8}\,\frac{x^{\frac{3}{2}}}{\frac{3}{2}} + C = \frac{2\,x}{3}\sqrt{8\,x} + C.$$

A second solution is as follows: Let $u = 8\,x$; then $du = 8\,dx$. Since by **(III)** a constant factor may be moved from one side of the integral sign to the other, we can multiply the integrand by 8 without changing the value of the given expression, provided that we place the compensating factor $\frac{1}{8}$ before the integral sign. We then have

$$\int \sqrt{8\,x}\,dx = \tfrac{1}{8}\int \sqrt{8\,x}\;8\,dx = \tfrac{1}{8}\int (8\,x)^{\frac{1}{2}}8\,dx.$$

The integral is now in the form of the power formula with $n = \frac{1}{2}$. Hence

$$\int \sqrt{8\,x}\,dx = \frac{1}{8}\cdot\frac{(8\,x)^{\frac{3}{2}}}{\frac{3}{2}} + C = \frac{(8\,x)^{\frac{3}{2}}}{12} + C.$$

This is easily seen to be equal to the result obtained above. (Compare Example 1, Art. 59.)

Example 6. Find $\int \dfrac{x\,dx}{\sqrt{a^2 - x^2}}$.

Solution. If this is written in the form

$$(a^2 - x^2)^{-\frac{1}{2}}x\,dx,$$

the power formula is suggested, with $n = -\frac{1}{2}$ and $u = a^2 - x^2$. But then $du = -2\,x\,dx$ instead of $x\,dx$. However, by **(III)**, a constant factor may be moved from one side of the integral sign to the other. Hence we may multiply $x\,dx$ by -2 if we place the compensating factor before the integral sign. This gives

$$\int \frac{x\,dx}{\sqrt{a^2 - x^2}} = -\frac{1}{2}\int (a^2 - x^2)^{-\frac{1}{2}}(-2\,x\,dx)$$

$$= -\frac{1}{2}\frac{(a^2 - x^2)^{\frac{1}{2}}}{\frac{1}{2}} + C, \qquad\qquad \text{by (IV)}$$

$$= -\sqrt{a^2 - x^2} + C.$$

Example 7. Find $\int \dfrac{dx}{(3\,x - 2)^2}$.

Solution. Using the procedure employed in Example 6, we have

$$\int \frac{dx}{(3\,x - 2)^2} = \frac{1}{3}\int (3\,x - 2)^{-2}(3\,dx)$$

$$= \frac{1}{3}\frac{(3\,x - 2)^{-1}}{-1} + C$$

$$= \frac{-1}{3(3\,x - 2)} + C.$$

PROBLEMS

Integrate the following.

1. $\int 6 \, dx.$

2. $\int \sqrt[3]{t^2} \, dt.$

3. $\int \frac{dx}{x^3}.$

4. $\int x^{-n} \, dx.$

5. $\int \frac{dy}{\sqrt{y}}.$

6. $\int (2 - x) dx.$

7. $\int (x^4 - 3x^2 + x - 2) dx.$

8. $\int (a^2 - t^2) dt.$

9. $\int \left(x^2 - 3\sqrt{x} + \frac{6}{x^2} \right) dx.$

10. $\int (a + bx + cx^2) dx.$

11. $\int (t^2 - t^{-2}) dt.$

12. $\int (1 - x)(2 + x^2) dx.$

13. $\int x^2 (2 - 3x) dx.$

14. $\int \left(\frac{4}{x^3} + \frac{9}{x^4} \right) dx.$

15. $\int (2 - 3x)^2 x \, dx.$

16. $\int (2 - x) \sqrt[3]{x} \, dx.$

17. $\int \frac{4 - x^2}{x^4} \, dx.$

18. $\int (\sqrt{a} - \sqrt{x})^2 \, dx.$

19. $\int \left(a^{\frac{2}{3}} - x^{\frac{2}{3}} \right)^3 \, dx.$

20. $\int \left(\frac{3}{\sqrt{x}} - \frac{2}{\sqrt[3]{x}} - \frac{1}{x^2} \right) dx.$

21. $\int \sqrt{1 - x} \, dx.$

22. $\int \left(a^{\frac{2}{3}} - x^{\frac{2}{3}} \right)^2 \frac{dx}{x^{\frac{1}{3}}}.$

23. $\int x(a + bx^2)^{\frac{5}{4}} dx.$

24. $\int \left(\frac{y}{8} - 3 \right)^2 dy.$

25. $\int \sqrt[3]{(2ax + x^2)} \, (a + x) dx.$

26. $\int \frac{(x^2 - 2)^3 dx}{x^2}.$

27. $\int \frac{6s \, ds}{\sqrt[3]{1 - 2s^2}}.$

28. $\int \sqrt{5x} \, dx.$

29. $\int \frac{dx}{\sqrt{3x}}.$

30. $\int \sqrt{x - 2} \, dx.$

31. $\int \frac{dx}{\sqrt{5 - x}}.$

32. $\int (3x - 2)^2 \, dx.$

33. $\int (5 - x)^3 \, dx.$

34. $\int (4 - 3x)^2 \, dx.$

35. $\int x(1 + \sqrt{x}) dx.$

36. $\int \left(\frac{x^2}{2} + \frac{2}{x^2} \right) dx.$

37. $\int \left(7\sqrt{x} + \frac{7}{\sqrt{x}} \right) dx.$

38. $\int \frac{(1 + \sqrt{x})}{x^2} \, dx.$

39. $\int \dfrac{5\,x\,dx}{(x^2-3)^2}.$

40. $\int x\sqrt{3\,x}\;dx.$

41. $\int x\sqrt[3]{2\,x}\;dx.$

42. $\int \dfrac{3\,x\,dx}{\sqrt{5\,x}}.$

43. $\int x\sqrt{a^2-x^2}\;dx.$

44. $\int s\sqrt{2\,s^2-3}\;ds.$

45. $\int \dfrac{7\,t\,dt}{\sqrt{6-t^2}}.$

46. $\int 2\,y^3\sqrt{y^4+a^4}\;dy.$

47. $\int \dfrac{3\,v\,dv}{(9-4\,v^2)^3}.$

48. $\int z^2\sqrt{9+z^3}\;dz.$

81. Integration by a change of variable. Many integrals can be reduced to those of the standard formulas by making a substitution which introduces a new independent variable. The following substitution of this type is frequently useful: when the integrand contains a radical of the form $\sqrt[n]{a+bx}$, where n is an integer, set $\sqrt[n]{a+bx}=v$; then $a+bx=v^n$ and $b\,dx=nv^{n-1}\,dv$.

Example. Find $\displaystyle\int \dfrac{x\,dx}{\sqrt[3]{1+2\,x}}.$

Solution. Let $\sqrt[3]{1+2\,x}=v$; then $1+2\,x=v^3$, $2\,dx=3\,v^2\,dv$, and $x=\dfrac{v^3-1}{2}.$

Now

$$\int \frac{x\,dx}{\sqrt[3]{1+2\,x}}=\int\left(\frac{v^3-1}{2}\right)\left(\frac{1}{v}\right)\left(\frac{3\,v^2\,dv}{2}\right)$$

$$=\frac{3}{4}\int(v^4-v)dv=\frac{3}{4}\left[\frac{v^5}{5}-\frac{v^2}{2}\right]+C.$$

Substitution of the value of v gives

$$\int \frac{x\,dx}{\sqrt[3]{1+2\,x}}=\frac{3}{4}\left[\frac{(1+2\,x)^{\frac{5}{3}}}{5}-\frac{(1+2\,x)^{\frac{2}{3}}}{2}\right]+C.$$

A more compact form of the answer which saves labor when numerical values are to be computed is obtained by writing

$$\frac{3}{4}\left[\frac{v^5}{5}-\frac{v^2}{2}\right]=\frac{3}{40}\left[2\,v^5-5\,v^2\right]=\frac{3\,v^2}{40}\,(2\,v^3-5).$$

Now substituting the value of v we get

$$\frac{3(4\,x-3)(1+2\,x)^{\frac{2}{3}}}{40}+C.$$

The method of substitution should not be used if the integration can be performed by the power formula.

PROBLEMS

Work out the following integrals.

1. $\int x\sqrt{a-x}\,dx.$

2. $\int \dfrac{x\,dx}{\sqrt{1+x}}.$

3. $\int \dfrac{x\,dx}{\sqrt{1+x^2}}.$

4. $\int x^2\sqrt{1+x}\,dx.$

5. $\int x\sqrt{1+x^2}\,dx.$

6. $\int \dfrac{x^2\,dx}{\sqrt{1-x}}.$

7. $\int \dfrac{x^2\,dx}{\sqrt{1-x^3}}.$

8. $\int \dfrac{t\,dt}{\sqrt{a+bt}}.$

9. $\int y\sqrt{a+by}\,dy.$

10. $\int y\sqrt{a+by^2}\,dy.$

11. $\int \dfrac{(x-2)^2\,dx}{\sqrt{x}}.$

12. $\int t^3\sqrt{2-t}\,dt.$

13. $\int x^2\sqrt{4\,x+3}\,dx.$

14. $\int \dfrac{x\,dx}{\sqrt[3]{x+1}}.$

15. $\int \dfrac{x\,dx}{\sqrt[4]{3+4\,x}}.$

16. $\int x\sqrt[3]{x-1}\,dx.$

17. $\int x\sqrt[3]{x^2-1}\,dx.$

18. $\int x\sqrt[4]{5+2\,x}\,dx.$

82. Initial conditions. It has appeared that the process of integration introduces an arbitrary constant, and, consequently, a function is not completely determined when its differential is known. For the complete determination of a function by integration it is necessary to know the numerical value of the function corresponding to some numerical value of the independent variable. The particular values of the independent variable and of the function which serve to determine the constant of integration are known as *initial conditions*. The use of initial conditions to determine the constant of integration will be illustrated by an example.

Example. Find the function $y = f(x)$ whose differential is $x^2\,dx$ and which is equal to 5 when x is equal to 3.

Solution. Given

$$dy = x^2\,dx.$$

By integration

$$y = \int x^2\,dx,$$

or

$$y = \frac{x^3}{3} + C.$$

Since $y = 5$ when $x = 3$, we have, by substitution,

$$5 = \tfrac{27}{3} + C;$$

whence

$$C = -4.$$

Substituting this value of C above, the final result is

$$y = \frac{x^3}{3} - 4.$$

PROBLEMS

In the following problems find the function whose differential is given and which satisfies the given initial conditions.

1. $dy = dx/x^2$, $y = 0$ when $x = 1$.

2. $dy = \sqrt{x}\, dx$, $y = 2$ when $x = 4$.

3. $dy = x\sqrt{x^2 + 16}\, dx$, $y = 5$ when $x = 3$.

4. $dy = \dfrac{dx}{\sqrt{5 - 4\,x}}$, $y = 4$ when $x = 1$.

5. $dy = \left(x^{\frac{1}{3}} + \dfrac{1}{x^{\frac{1}{3}}}\right) dx$, $y = 0$ when $x = 8$.

6. $ds = \sqrt{7\,t}\, dt$, $s = 0$ when $t = 0$.

7. $ds = (32\,t - 10)dt$, $s = 20$ when $t = 0$.

8. $ds = (t^2 - 2\sqrt{t})dt$, $s = 7$ when $t = 0$.

9. $dy = x\sqrt{9 + x}\, dx$, $y = 0$ when $x = 0$.

10. $dy = \dfrac{x\, dx}{\sqrt{5 + x}}$, $y = 1$ when $x = 4$.

11. $dy = \dfrac{x^2\, dx}{\sqrt{1 + x^3}}$, $y = 0$ when $x = 0$.

12. $dy = x\sqrt{25 - x^2}\, dx$, $y = 0$ when $x = 0$.

13. $ds = \dfrac{dx}{\sqrt{5 - x}}$, $s = -2$ when $x = -4$.

14. $dy = \sqrt{a^2 - x^2}\, x\, dx$, $y = 0$ when $x = a$.

15. $dy = \dfrac{x\, dx}{\sqrt{1 + x}}$, $y = \dfrac{10}{3}$ when $x = 3$.

16. $dz = (2 - x)\sqrt[3]{x}\, dx$, $z = -1$ when $x = 8$.

17. $dy = \dfrac{6 - x^2}{x^4}\, dx$, $y = 2$ when $x = 1$. Find the value of y when $x = -1$.

18. $ds = \sqrt{1 - t}\, dt$, $s = \frac{1}{3}$ when $t = 0$. Find the value of s when $t = -1$.

19. $dA = \sqrt{2\, px}\, dx$, $A = p^2/3$ when $x = p/2$. Find the value of A when $x = 2\,p$.

20. $dz = (2 - y^2)^3\, dy$, $z = -6$ when $y = 0$. Find the value of z when $y = 2$.

21. $dy = \sqrt{x + 2}\, dx$, $y = \frac{1}{3}$ when $x = 2$. Find the value of y when $x = 7$.

22. $dz = dx/\sqrt{x}$, $z = -4$ when $x = 0$. Find the value of z when $x = 4$.

23. $dy = x\sqrt{9 + 4\,x^2}\, dx$, $y = 0$ when $x = 0$. Find the value of y when $x = 2$.

24. $dy = x\sqrt{2\,x + 1}\, dx$, $y = 0$ when $x = 0$. Find the value of y when $x = 4$.

83. Curves with given slope. Suppose that x and y represent rectangular coördinates of a point on a curve and that the derivative of y with respect to x is known in terms of x and y. Thus, let

$$\frac{dy}{dx} = f(x, y).$$

This equation is called a *differential equation*, because it involves a derivative, and it gives the slope of the curve at every point for which $f(x, y)$ has a real value. In order to determine the equation of the curve it is obvious that an integration must be performed. Since the process of integration introduces an arbitrary constant, the differential equation represents a *system of curves*. A particular curve of the system can be determined by requiring that it shall pass through a given point. The method of finding the equations of curves having given slopes will be illustrated by two examples.

Example 1. Find the equation of the curve whose slope at every point is equal to its abscissa and which passes through the point $(0, -4)$.

Solution. By the conditions of the problem,

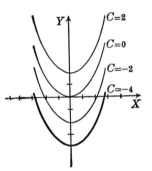

$$\frac{dy}{dx} = x, \qquad (1)$$

whence $dy = x\,dx$

and $y = \int x\,dx.$

Hence $y = \dfrac{x^2}{2} + C.$ (2)

The value of C is determined by substituting $x = 0$, $y = -4$, in (2):

$$-4 = 0 + C.$$

Fig. 83.1

Hence $C = -4$, and, substituting this value of C in equation (2), the final result is $y = \dfrac{x^2}{2} - 4$.

Equation (2), which is the equivalent of the differential equation (1), represents a system of parabolas, as shown in the figure. The heavy curve of the system is the one which satisfies the given initial conditions.

Example 2. Find the equation of the curve which passes through $(3, 2)$ and whose slope at every point is given by

$$\frac{dy}{dx} = -\frac{x}{y}.$$

Solution. In order to integrate the given equation it is necessary first to *separate the variables*; that is, to write the given equation in a form such that each term contains only x or only y. Thus, clearing of fractions, we have

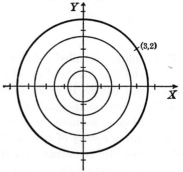

$$y \, dy = - \, x \, dx.$$

Each term of this equation may be integrated, giving

$$\frac{y^2}{2} = - \frac{x^2}{2} + C,$$

or $\qquad\qquad x^2 + y^2 = 2\,C. \qquad\qquad (1)$

The value of C is determined by substituting $x = 3$, $y = 2$, in (1):

$$9 + 4 = 2\,C.$$

Fig. 83.2

Hence $2\,C = 13$, and, on substitution of this value in equation (1), the final result becomes

$$x^2 + y^2 = 13.$$

Equation (1) represents the system of circles with center at the origin.

PROBLEMS

1. In the following problems find the equation of the curve which has the given slope at every point and which passes through the given point.

a. $\dfrac{dy}{dx} = \sqrt{x}$, $(0, 0)$.

b. $\dfrac{dy}{dx} = \sqrt{y}$, $(0, 0)$.

c. $\dfrac{dy}{dx} = \sqrt{4 - x}$, $(0, 0)$.

d. $\dfrac{dy}{dx} = \dfrac{x}{y}$, $(5, 0)$.

e. $\dfrac{dy}{dx} = \sqrt{xy}$, $(1, 1)$.

f. $\dfrac{dy}{dx} = y^2$, $(1, -1)$.

g. $\dfrac{dy}{dx} = -\dfrac{4\,x}{9\,y}$, $(0, 6)$.

h. $\dfrac{dy}{dx} = \dfrac{x}{\sqrt{y}}$, $(1, 0)$.

i. $\dfrac{dy}{dx} = -3\,x$, $(0, 2)$.

j. $\dfrac{dy}{dx} = \dfrac{-bx}{a\sqrt{a^2 - x^2}}$, $(0, b)$.

k. $\dfrac{dy}{dx} = \dfrac{x}{\sqrt{x^2 + 4}}$, $(0, 2)$.

l. $\dfrac{dy}{dx} = \dfrac{2}{3\sqrt[3]{x}}$, $(0, 0)$.

m. $\dfrac{dy}{dx} = -\sqrt{\dfrac{y}{x}}$, $(a, 0)$.

n. $\dfrac{dy}{dx} = -\sqrt[3]{\dfrac{y}{x}}$, $(a, 0)$.

2. In the following problems find the system of curves defined by the given equation. Draw three curves of each system.

a. $\dfrac{dy}{dx} = \dfrac{1}{2}.$

b. $\dfrac{dy}{dx} = \dfrac{4\,x}{y}.$

c. $\dfrac{dy}{dx} = -\dfrac{4\,x}{y}.$

d. $\dfrac{dy}{dx} = \dfrac{x}{2}.$

e. $\dfrac{dy}{dx} = \dfrac{1}{2\,y}.$

f. $\dfrac{dy}{dx} = x^2 - 1.$

g. $\dfrac{dy}{dx} = 3\,x^2.$

h. $\dfrac{dy}{dx} = \dfrac{4 - x}{2 + y}.$

i. $\dfrac{dy}{dx} = \dfrac{y^2}{x^2}.$

j. $\dfrac{dy}{dx} = \dfrac{x^2}{y^2}.$

k. $\dfrac{dy}{dx} = \dfrac{x}{y^2}.$

l. $\dfrac{dy}{dx} = \dfrac{x^2}{y}.$

3. At every point of a certain curve $y'' = -3$. Find the equation of this curve if it passes through the origin and has the slope 1 at that point.

4. At every point of a certain curve $y'' = 24/x^4$. Find the equation of the curve if it passes through (2, 1) and has the slope -1 at that point.

5. At every point of a curve $y'' = 6\,x$, and it is tangent to the line whose equation is $2\,x - 3\,y = 6$ at the point $(0, -2)$. Find its equation.

6. Find the equation of the curve at every point of which $y'' = 15\sqrt{x}$ and which passes through the point (4, 0) with an inclination of 45°.

7. Find the equation of the curve at every point of which $y'' = 6/x^3$ and which passes through the point (1, 1) with an inclination of 135°.

8. What is the system of curves determined by the equation $dy/dx = k$, where k is a constant?

84. Straight-line motion. If the distance s of a point moving along a straight line from a fixed point O taken as an origin is known as a function of the time t, the velocity and acceleration, as we have shown in Chapter V, are obtained by differentiation. Thus

$$v = \frac{ds}{dt}$$

and

$$a = \frac{dv}{dt}.$$

It follows from these formulas that if the acceleration is known as an explicit function of the time, the velocity is determined by integration. Thus

$$v = \int a\,dt.$$

And if the velocity is known as an explicit function of the time, the distance from the origin O is determined by integration. Thus

$$s = \int v\,dt.$$

Of course, each integration introduces an arbitrary constant, which can be determined if proper initial conditions are known.

Example 1. A particle moves with an acceleration of $\dfrac{1+t}{10}$ ft./sec.^{2}, its initial speed being 6 ft./sec. Find its speed and the distance moved after t seconds.

Solution. By the statement of the problem, the initial conditions are that when $t = 0$, $v = 6$; and we may choose the starting-point as the origin, so that $s = 0$.

Since
$$a = \frac{1+t}{10},$$
$$v = \int \frac{1+t}{10}\,dt;$$
whence
$$v = \frac{t}{10} + \frac{t^2}{20} + C. \tag{1}$$

If we substitute $t = 0$ and $v = 6$ in equation (1), we have $C = 6$, and
$$v = 6 + \frac{t}{10} + \frac{t^2}{20}. \tag{2}$$

But
$$s = \int v\,dt, \quad \text{or} \quad s = \int \left(6 + \frac{t}{10} + \frac{t^2}{20}\right)dt.$$

Integrating, we have
$$s = 6\,t + \frac{t^2}{20} + \frac{t^3}{60} + C'. \tag{3}$$

The substitution of $t = 0$ and $s = 0$ in equation (3) gives $C' = 0$, whence
$$s = 6\,t + \frac{t^2}{20} + \frac{t^3}{60}. \tag{4}$$

Equations (2) and (4) are the required results.

Example 2. A stone is projected straight upward from the edge of the roof of a building 120 ft. high with a speed of 20 ft./sec. Neglecting air resistance, find the speed with which the stone will strike the ground.

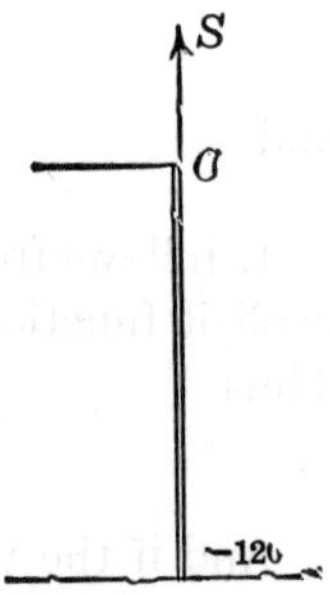

Solution. In all problems of this kind the following preliminary steps are essential. (1) Choose the origin and the positive direction on the line of motion. (2) Determine the initial conditions, that is, the given values of the variables. (3) Determine what values are to be found.

In this problem let us take the origin at the top of the building and the positive direction upward, as indicated in Fig. 84. Let t be measured from the instant when the stone is projected. The initial conditions are: when $t = 0$, $v = 20$ and $s = 0$. It is required to find the value of v when $s = -120$.

Fig. 84

Since air resistance is neglected and the acceleration of gravity is downward, $a = -32$ ft./sec.^{2} Hence
$$v = \int -32\,dt = -32\,t + C. \tag{1}$$

Substituting $t = 0$, $v = 20$ in (1), we find $C = 20$. Hence

$$v = -32\,t + 20, \tag{2}$$

and $$s = \int (-32\,t + 20)dt = -16\,t^2 + 20\,t + C'. \tag{3}$$

Since $s = 0$ when $t = 0$, (3) shows that $C' = 0$. Hence

$$s = -16\,t^2 + 20\,t. \tag{4}$$

The function s is defined for $0 \leqq t \leqq T$, where T is the time when the stone strikes the ground. To find T, set $t = T$ and $s = -120$ in (4). The result is

$$16\,T^2 - 20\,T - 120 = 0,$$

from which $$T = \frac{5 \pm \sqrt{505}}{8}.$$

Since T is positive, the $+$ sign must be taken with the radical. Substitution of the value of T for t in (2) gives

$$v = -4\sqrt{505} = -89.9 \text{ ft./sec., approximately.}$$

The sign of v is negative, as it should be, since the motion is downward. The stone actually moves more than 120 ft., as it goes up until $v = 0$, that is, until $t = \frac{5}{8}$ sec., when $s = \frac{25}{4}$ ft. But since s is a directed distance, it is unnecessary to consider the upward and downward motions separately.

PROBLEMS

1. With what velocity will a stone strike the ground if dropped from the top of a building 120 ft. high?

2. With what velocity will the stone of Problem 1 strike the ground if thrown downward with a speed of 20 ft./sec.?

3. A ball projected upward from the ground reaches a height of 384 ft. in 4 sec. Find how high the ball will go.

4. A stone dropped from a balloon, which was rising at the rate of 15 ft./sec., reached the ground in 8 sec. How high was the balloon when the stone was dropped?

5. In Problem 4, if the balloon had been falling at the rate of 15 ft./sec., how long would the stone have taken to reach the ground?

6. A train leaving a railroad station has an acceleration of $0.5 + 0.02\,t$ feet per second per second. Find how far it will have moved in 20 sec.

7. A car makes a trip in 10 min., and its velocity is given by $v = 500\,t - 5\,t^3$, where t is measured in minutes and v in feet per minute. (**a**) How far does the car go? (**b**) What is its maximum speed? (**c**) How far has the car moved when its maximum speed is reached?

8. If $v = 2$ when $t = 3$, find the relation between v and t, provided that the acceleration is

a. $2t - t^2$.

b. $\dfrac{1}{t^2} - t$.

c. $3t^{-\frac{3}{2}}$.

d. $4 - t^2$.

e. $\sqrt{t} + 2$.

f. $6 - \dfrac{1}{t^2}$.

9. Find the relation between s and t if $s = 2$ when $t = 1$ and if

a. $v = \sqrt{t - 1}$.

b. $v = t^2 + \dfrac{1}{t^2}$.

c. $v = a + bt$.

d. $v = t\sqrt{t^2 - 1}$.

10. A particle starts with an initial velocity u and is subject to a constant acceleration a. Show that the velocity and distance after time t are given by the formulas $v = u + at$ and $s = ut + \frac{1}{2}at^2$.

11. If the acceleration of a particle moving with a variable velocity v is $- kv^2$, where k is a constant, and if u is the initial velocity, show that $\dfrac{1}{v} = \dfrac{1}{u} + kt$.

12. A particle sliding on a certain inclined plane 48 ft. long is subject to an acceleration downward of 6 ft./sec.^{2} (**a**) If the particle is started upward from the bottom of the plane with a velocity of 12 ft./sec., how far will it go before stopping? (**b**) With what initial speed must the particle be started from the bottom so that it may just reach the top?

CHAPTER VII

DEFINITE INTEGRALS —
THE FUNDAMENTAL THEOREM

85. The definite integral. Let $f(x)$ be continuous in the interval (a, b) and suppose that

$$\int f(x)dx = \phi(x) + C.$$

In this indefinite integral, $\phi(x) + C$, put first $x = b$, then $x = a$, and subtract. Thus

$$\phi(b) + C - [\phi(a) + C] = \phi(b) - \phi(a).$$

This result, from which the constant of integration has disappeared, is known as a definite integral. It is indicated by the notation

$$\int_a^b f(x)dx,$$

which is read "the integral from a to b of $f(x)dx$." The number a is the lower limit, and b is the upper limit, of the integral.

Since the constant of integration does not appear in the final result, it is unnecessary to add it when finding the value of a definite integral. The operations involved in evaluating a definite integral as explained above are shown by the following notation.

(I) $$\int_a^b (x)dx = [\phi(x)]_a^b = \phi(b) - \phi(a).$$

Example 1. Find the value of $\int_{-1}^2 x^2\, dx$.

Solution. $$\int_{-1}^2 x^2\, dx = \left[\frac{x^3}{3}\right]_{-1}^2 = \left[\frac{8}{3}\right] - \left[-\frac{1}{3}\right] = 3.$$

Example 2. Find the value of $\int_{-2}^5 \frac{5\, x\, dx}{\sqrt{x^2 + 3}}$.

Solution. $$\int_{-2}^5 \frac{5\, x\, dx}{\sqrt{x^2 + 3}} = 5\int_{-2}^5 (x^2 + 3)^{-\frac{1}{2}}\, x\, dx.$$

$$= \frac{5}{2}\int_{-2}^5 (x^2 + 3)^{-\frac{1}{2}}\, 2\, x\, dx$$

$$= \left[5\sqrt{x^2 + 3}\,\right]_{-2}^5$$

$$= 5\sqrt{28} - 5\sqrt{7} = 5\sqrt{7} = 13.23.$$

86. Change of limits in integration by substitution. In Chapter VI, Art. 81, we learned that certain expressions which cannot be integrated by the power formula can be integrated by making a substitution of the type $\sqrt{a+bx}=v$. For example, to find $\int x\sqrt{9-x}\,dx$ we substitute $\sqrt{9-x}=v$, obtaining

$$\int x\sqrt{9-x}\,dx = \int (2\,v^4 - 18\,v^2)dv$$
$$= \tfrac{2}{5}v^5 - 6\,v^3 + C$$
$$= \tfrac{2}{5}(9-x)^{\frac{5}{2}} - 6(9-x)^{\frac{3}{2}} + C.$$

If the *definite integral* $\int_0^5 x\sqrt{9-x}\,dx$ is required, the work is shortened by changing the limits for x when substituting $\sqrt{9-x}=v$ to the *corresponding* limits for v. Thus, when $x=0$, $v=\sqrt{9-0}=3$; when $x=5$, $v=\sqrt{9-5}=2$.

Hence
$$\int_0^5 x\sqrt{9-x}\,dx = \int_3^2 (2\,v^4 - 18\,v^2)dv$$
$$= \left[\tfrac{2}{5}v^5 - 6\,v^3\right]_3^2$$
$$= [\tfrac{64}{5} - 48] - [\tfrac{486}{5} - 162] = \tfrac{148}{5}.$$

This method enables us to avoid the troublesome substitution in terms of the original variable, and should always be employed.

PROBLEMS

Find the value of each of the following definite integrals.

1. $\int_1^4 \dfrac{dx}{\sqrt{x}}.$

2. $\int_2^6 \sqrt{x-2}\,dx.$

3. $\int_0^4 x\sqrt{x^2+9}\,dx.$

4. $\int_1^3 (x^2+x)dx.$

5. $\int_0^3 x\sqrt{x^2+5}\,dx.$

6. $\int_1^8 \left(x^{\frac{1}{2}}+x^{\frac{1}{3}}\right)dx.$

7. $\int_0^5 \sqrt{5\,x}\,dx.$

8. $\int_0^2 \dfrac{x^2\,dx}{(10-x^3)^2}.$

9. $\int_0^3 \sqrt{25-3\,x}\,dx.$

10. $\int_1^3 x^2(2-3\,x)dx.$

11. $\int_a^{8a} \left(a^{\frac{2}{3}} - x^{\frac{2}{3}}\right)\dfrac{dx}{x^{\frac{1}{3}}}.$

12. $\int_1^4 \dfrac{dt}{\sqrt{5-t}}.$

13. $\int_0^a x\sqrt{a^2-x^2}\,dx.$

14. $\int_{-3}^{-1} (s^2 - s^{-2})ds.$

15. $\int_1^4 \dfrac{4 - x^2}{x^4}\, dx.$

16. $\int_9^{16} x(1 + \sqrt{x})\, dx.$

17. $\int_2^3 s\sqrt{2\, s^2 - 5}\, ds.$

18. $\int_0^4 (4 - x)^3\, dx.$

19. $\int_0^a (\sqrt{a} - \sqrt{x})^2\, dx.$

20. $\int_1^2 z^2\sqrt{1 + z^3}\, dz.$

21. $\int_{-1}^0 (1 - t)(2 + t^2)\, dt.$

22. $\int_1^4 \dfrac{dx}{x^3}.$

23. $\int_0^a \sqrt{a - y}\, dy.$

24. $\int_1^4 \dfrac{1 + \sqrt{u}}{u^2}\, du.$

25. $\int_2^3 4\, dt.$

26. $\int_0^a \dfrac{x\, dx}{(x^2 + a^2)^2}.$

27. $\int_{-2}^1 x\sqrt{x + 3}\, dx.$

28. $\int_{-2}^2 \dfrac{x\, dx}{\sqrt{2\, x + 5}}.$

29. $\int_2^5 \dfrac{(x + 1)\, dx}{\sqrt{x - 1}}.$

30. $\int_4^1 \dfrac{4 - x^2}{x^4}\, dx.$

31. $\int_0^a x\sqrt{a - x}\, dx.$

32. $\int_0^3 \dfrac{x^2\, dx}{\sqrt{x + 1}}.$

33. $\int_2^7 \dfrac{x\, dx}{\sqrt{x + 2}}.$

34. $\int_4^8 x(x - 4)^{\frac{3}{2}}\, dx.$

35. $\int_{-7}^1 x\sqrt[3]{1 - x}\, dx.$

36. $\int_0^a \dfrac{x\, dx}{\sqrt{x + a}}.$

37. Prove that $\displaystyle\int_a^b f(x)\, dx = -\int_b^a f(x)\, dx.$

38. A body moves with the speed $v = 4 - 6\, t$. Show that the distance moved between the times $t = 3$ and $t = 6$ is given by $\displaystyle\int_3^6 (4 - 6\, t)\, dt.$

87. Derivative of an area. Let $y = f(x)$ be continuous and nowhere negative in some interval (a, b). Let A be the area of the figure bounded by the x-axis, the graph of $y = f(x)$, the ordinate at the point where $x = a$, and a variable ordinate at any point x in (a, b). For the sake of brevity we shall call A the "area under the graph of $y = f(x)$ from a to x." Clearly A is a function of x in (a, b); in fact we can imagine that the area is generated by the variable ordinate moving from a to the right, just as a point generates a curve. If the ordinate moves to the left, the area generated will be negative. We have as yet no expression for A as a function of x, but with the aid of the figure on the following page we can get the derivative of A with respect to x.

Let x_1 and $x_1 + h$, where $h > 0$, be two values of x within (a, b). Assuming that $f(x)$ has a finite number of maxima and minima in (a, b), we see that h can be taken small enough so that $f(x)$ either (I) does not decrease or (II) does not increase in the interval $(x_1, x_1 + h)$. We consider first Case I as illustrated in Fig. 87.

Let Δx be an increment such that $0 < \Delta x < h$. The corresponding increment ΔA is the area under the graph from x_1 to $x_1 + \Delta x$ (the area $DEKG$ in the figure) and the corresponding increment Δy is HK in the figure.*

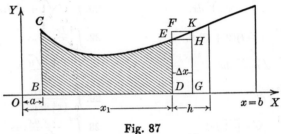

Fig. 87

Completing the rectangles $DEHG$ and $DFKG$, we see from the figure that

$$DEHG < \Delta A < DFKG. \tag{1}$$

But $DEHG = y\,\Delta x$ and $DFKG = (y + \Delta y)\Delta x;$

hence $y\,\Delta x < \Delta A < (y + \Delta y)\Delta x.$

Dividing this by Δx, we have

$$y < \frac{\Delta A}{\Delta x} < y + \Delta y.$$

When Δx approaches zero as a limit, it is evident that $y + \Delta y$ approaches y as a limit, and hence $\Delta A/\Delta x$ must also approach y as a limit, since its value is always between y and $y + \Delta y$; that is,

$$\lim_{\Delta x \to 0} \frac{\Delta A}{\Delta x} = y = f(x). \tag{2}$$

In Case II the inequality signs in (1) are reversed and (2) follows.

For $\Delta x < 0$ we consider an interval $(x_1 - h, x_1)$. Then $\Delta A < 0$ and we are led to (2) as before.

But, by definition of a derivative,

$$\lim_{\Delta x \to 0} \frac{\Delta A}{\Delta x} = \frac{dA}{dx}.$$

Hence, from (2),

(II) $$\frac{dA}{dx} = f(x).$$

*Assuming that $f(x)$ is not constant, $\Delta y > 0$ in Case (I) and $\Delta y < 0$ in Case (II).

In words, *the derivative of the area bounded by the curve $y = f(x)$, the x-axis, a fixed ordinate, and a variable ordinate is equal to $f(x)$.*

If the differential notation is used, **(II)** may be reduced at once to

(II a) $$dA = f(x)dx = y\ dx.$$

This may be remembered easily if it is noted that the differential $dA = y\ dx$ is merely the area of the rectangle $DEHG$ in Fig. 87. This is a close approximation to ΔA, which is the area of $DEKG$. (Compare Art. 76.)

REMARK. In the above proof we took x_1 *between* a and b. If $x_1 = a$, we need consider only positive values for Δx and obtain a right-hand derivative. If $x_1 = b$, Δx can only be negative and we get a left-hand derivative.

88. Area under a curve as a definite integral. *Theorem.* *Let $y = f(x)$ be continuous and nowhere negative in the interval (a, b). Then the area A under the graph of $y = f(x)$ from a to b is given by*

(III) $$A = \int_a^b f(x)dx = \int_a^b y\ dx.$$

Proof. Let x be any point in (a, b) and u be the area under the graph from a to x. By **(II)** u is differentiable with respect to x and $du = f(x)dx$. Hence if $\phi\ (x)$ is any indefinite integral of $f(x)\ dx$, we may write

$$u = \int f(x)dx = \phi(x) + C.$$

When $x = a$, it is clear that $u = 0$. Therefore,

$$0 = \phi(a) + C \quad \text{or} \quad C = -\phi(a).$$

Hence for any value of x

$$u(x) = \phi(x) - \phi(a).$$

When $x = b$, $$u(b) = \phi(b) - \phi(a)$$

or $$A = \int_a^b f(x)dx. \qquad \text{By definition}$$

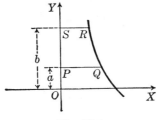

Fig. 88.1

If we interchange the variables x and y, an argument similar to the preceding gives the following statement. (Fig. 88.1)

If $x = g(y)$ is continuous and nowhere negative in the interval (a, b), the area bounded by the graph, the y-axis, and the lines $y = a$ and $y = b$ is given by

$$\int_a^b g(y)dy.$$

If $y = f(x)$ is continuous and nowhere positive in the interval (a, b), the area bounded by the graph, the x-axis, and the ordinates at a and b lies below the x-axis. Rotation of the figure through $180°$ about the x-axis gives a new figure having the same area and the equation of the new curve is $y = g(x)$, where $g(x) = -f(x)$. Hence, by the theorem,

$$A = \int_a^b g(x)dx = -\int_a^b f(x)dx.$$

Granting the existence of the area considered in the preceding proof, the theorem implies the existence of an indefinite integral of any continuous function. The statement often made that a certain function is "not integrable" usually means that the indefinite integral is not one of the elementary functions and so cannot be found by the rules for integration given in books on the calculus. Thus $\sqrt{1 - x^4}$ is continuous in the interval $(-1, 1)$ and so has an indefinite integral, but it would be called "not integrable" in the sense of the previous sentence.

Formula **(III)** enables us to find an area if the indefinite integral is known. Conversely, it enables us to find the value of the definite integral if the area is known. Suppose it is desired to find

$$\int_0^a \sqrt{a^2 - x^2}\, dx.$$

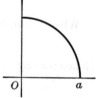

Fig. 88.2

As yet we have no means of performing the integration required. However, the function $\sqrt{a^2 - x^2}$ is continuous and non-negative in the interval $(0, a)$. Consequently the required definite integral is the area under the graph of the function $y = \sqrt{a^2 - x^2}$ from 0 to a. The graph (Fig. 88.2) is a quadrant of a circle of radius a. Hence

$$\int_0^a \sqrt{a^2 - x^2}\, dx = \frac{\pi a^2}{4}.$$

The fact that **(III)** can be utilized to identify an integral of a non-negative continuous function with the area under the graph is one of the most important in integral calculus. In many practical applications the functions obtained are not integrable and it is necessary to obtain approximations to the definite integrals involved by various devices. Two of the simplest of these (Arts. 89, 90) are obtained by approximating the area under the graph. The fact also serves as a means of obtaining a simple proof of the "fundamental theorem" (Art. 92), which is the principal tool used in finding magnitudes of all kinds.

Example 1. Find the area bounded by the line $2y = x + 8$, the x-axis, and the ordinates $x = -2$ and $x = 4$. (Fig. 88.3)

Solution. The required area is given by

$$\int_{-2}^{4} y\, dx = \int_{-2}^{4} \left(\frac{x}{2} + 4\right) dx$$

$$= \left[\frac{x^2}{4} + 4x\right]_{-2}^{4} = 20 - (-7) = 27.$$

Hence the area is 27 square units.

This result can be verified by geometry, since $PQRS$ is a trapezoid in which one base is $PQ = 3$, the other base is $SR = 6$, and the altitude is $PS = 6$. Hence the area is

$$\tfrac{1}{2}(3 + 6) \times 6 = 27.$$

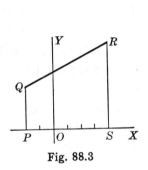

Fig. 88.3

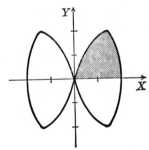

Fig. 88.4

Example 2. Find the entire area bounded by the curve whose equation is $y^2 = x^2(4 - x^2)$. (Fig. 88.4)

Solution. Since the curve has symmetry with respect to both axes, it will be sufficient to find the area of the portion in the first quadrant and multiply this result by 4. To get the limits of integration, find the x-intercepts. These are 0 and 2.

Hence we have, by (III),

$$\frac{A}{4} = \int_{0}^{2} y\, dx$$

$$= \int_{0}^{2} x\sqrt{4 - x^2}\, dx.$$

(Continued on next page)

Integrating this, we find

$$\frac{A}{4} = \left[-\frac{1}{3}(4 - x^2)^{\frac{3}{2}} \right]^2$$

$$= 0 + \tfrac{8}{3} = \tfrac{8}{3}.$$

Therefore $A = \tfrac{32}{3} = 10.67.$

PROBLEMS

In each of the following problems the result should be checked approximately by drawing the given curve carefully on coördinate paper and counting the number of squares in the required area.

1. Find the area bounded by the given curve, the x-axis, and the given ordinates.

a. $y = \dfrac{x^2}{4};\ x = 0,\ x = 4.$ **b.** $y^2 = 2\,x;\ x = 2,\ x = 8.$

c. $y = 9\,x - x^3;\ x = 0,\ x = 3.$ **d.** $y = 8\,x - x^2;\ x = 2,\ x = 6.$

e. $y^2 = 4 - x;\ x = 0,\ x = 3.$ **f.** $y = \dfrac{10}{\sqrt{x + 9}};\ x = 0,\ x = 16.$

g. $y = x + \dfrac{4}{x^2};\ x = 2,\ x = 4.$ **h.** $y = \dfrac{4\,x}{(x^2 + 1)^2};\ x = 0,\ x = 3.$

i. $y^2 = 16 - 5\,x;\ x = -3,\ x = 0.$ **j.** $xy^2 = 12;\ x = 1,\ x = 64.$

k. $x^2 y - x^2 + 9 = 0;\ x = 3,\ x = 12.$ **l.** $y^2 = x^3 + 4\,x^2;\ x = -4,\ x = 0.$

2. Find the area bounded by the given curve, the y-axis, and the given lines.

a. $x + y = 10;\ y = 2,\ y = 8.$ **b.** $y^2 = 2\,x;\ y = 1,\ y = 3.$

c. $y = x^2/4;\ y = 1,\ y = 4.$ **d.** $y^3 = x;\ y = 1,\ y = 2.$

3. Find the area bounded by the following curves and the x-axis.

a. $y = 4 - x^2.$ **b.** $y = 8 + 2\,x - x^2.$

c. $y = x - x^3.$ **d.** $y = x - \sqrt[3]{x}.$

e. $4\,y = x^4 - 8\,x^2.$ **f.** $x^2 = (y + 4)^3.$

4. Find the area bounded by the parabola $y^2 = 4\,x$ and the straight line $y = x$.

5. Find the area bounded by the two parabolas $y^2 = 8\,x$ and $x^2 = 8\,y$.

6. Find the area in the first quadrant bounded by the curve whose equation is $y = x^3$ and by the line whose equation is $y = 2\,x$.

7. Find the area bounded by the parabola whose equation is $\sqrt{x} + \sqrt{y} = \sqrt{a}$ and by the x-axis and y-axis.

8. Find the area of the segment of the parabola whose equation is $y = 6 + x - x^2$ cut off by the chord joining the points $(-1, 4)$ and $(3, 0)$.

9. Find in two ways the area bounded by the curve whose equation is $y^2 = 8 - 4\,x$ and the y-axis.

10. The velocity of a moving point is given by the formula $v = 8\,t - \sqrt{t}$. By determining the constant of integration, as in the previous chapter, show that the distance moved from time $t = 2$ to $t = 8$ is given by the definite integral

$$\int_2^8 (8\,t - \sqrt{t})dt,$$

and evaluate the integral.

11. Generalize Problem 10 by showing that if the velocity is $v = f(t)$, the distance moved from $t = a$ to $t = b$ is $\int_a^b v\,dt$. What area is equal to this distance?

12. Find the area inclosed by the loop of the curve whose equation is $4\,y^2 = x^2(4 - x)$.

13. Find the area bounded by the curve whose equation is $y^2 = x^2(x^2 - 1)$ and by the line whose equation is $x = 2$.

14. Find the area inclosed by the loop of the curve whose equation is $y^2 = x^2(9 - x)$.

15. Find the area bounded by the curve whose equation is $y^2 = x^3 - x^2$ and by the line whose equation is $x = 2$.

16. Find the area inclosed by the loop of the curve whose equation is $y^2 = x(x - 2)^2$.

17. Find the area inclosed by the loop of the curve whose equation is $4\,y^2 = x^4(4 - x)$.

18. Write the integral giving the area bounded by the curve whose equation is $xy = 12$ and by the lines whose equations are $x = 1$ and $x = 5$. Can you evaluate this integral?

89. The trapezoidal rule. This rule for approximating the definite integral of a continuous function is derived by first assuming that the function is nowhere negative and then approximating the area under the graph by means of trapezoids (Fig. 89.1). The interval (a, b) is divided into n equal sub-intervals, each of length Δx, and ordinates are erected at the ends of the sub-intervals. The upper ends of consecutive ordinates are then joined by straight lines forming trapezoids. The abscissas of the points of division are $x_0(= a)$, x_1, x_2, $\cdots$, $x_n(= b)$. The corresponding values of $f(x)$ are $y_0 = f(x_0)$, $y_1 = f(x_1)$, $y_2 = f(x_2)$, $\cdots$, $y_n = f(x_n)$. The areas of the trapezoids are given by the following expressions.

$$0.5(y_0 + y_1)\Delta x,$$
$$0.5(y_1 + y_2)\Delta x,$$
$$0.5(y_2 + y_3)\Delta x,$$

$$\cdot \quad \cdot \quad \cdot \quad \cdot \quad \cdot \quad \cdot$$

$$0.5(y_{n-1} + y_n)\Delta x.$$

The sum of the areas of the n trapezoids, which is obviously an approximation to the area under the curve, will be

$$(0.5\, y_0 + y_1 + y_2 + \cdots + y_{n-1} + 0.5\, y_n)\Delta x.$$

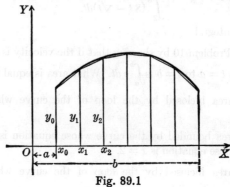

Fig. 89.1

The method of procedure is formulated into a working rule as follows.

To evaluate approximately the integral $I = \int_a^b f(x)dx$, the interval $b - a$ is divided into n parts, each being equal to Δx. The abscissas of the points of division are $x_0(= a)$, x_1, x_2, $\cdots$, $x_n(= b)$. The corresponding values of $f(x)$ are $y_0 = f(x_0)$, $y_1 = f(x_1)$, $\cdots$, $y_n = f(x_n)$. Then

(T) $I = (0.5\, y_0 + y_1 + y_2 + \cdots + y_{n-1} + 0.5\, y_n)\Delta x.$

Example. Using $n = 4$, compute the approximate value of

$$\int_0^2 \sqrt{4 + x^3}\, dx.$$

Solution. The table of values for x and y is first computed.

$x_0 = 0.0;\ y_0 = \sqrt{4 + 0}$
$\qquad = 2.000.$

$x_1 = 0.5;\ y_1 = \sqrt{4 + 0.125}$
$\qquad = 2.031.$

$x_2 = 1.0;\ y_2 = \sqrt{4 + 1}$
$\qquad = 2.236.$

$x_3 = 1.5;\ y_3 = \sqrt{4 + 3.375}$
$\qquad = 2.716.$

$x_4 = 2.0;\ y_4 = \sqrt{4 + 8}$
$\qquad = 3.464.$

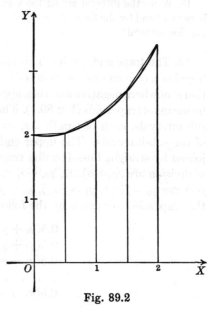

Fig. 89.2

Applying **(T)**,

$$I = (1.000 + 2.031 + 2.236 + 2.716 + 1.732) \times 0.5 = 4.858.$$

(If we take $n = 10$, we obtain the more accurate value 4.826.)

PROBLEMS

1. Approximate the value of $\displaystyle\int_0^4 \frac{dx}{\sqrt{16 + x^2}}$ by constructing a graph and estimating the area under it.

2. Compute the approximate values of the following integrals by the trapezoidal rule, using the values of n indicated.

a. $\displaystyle\int_0^6 \sqrt[3]{10 + x^2}\, dx,\ n = 3.$

b. $\displaystyle\int_0^4 \frac{dx}{\sqrt{1 + x^3}},\ n = 4.$

c. $\displaystyle\int_0^2 \sqrt{10 - x^3}\, dx,\ n = 4.$

d. $\displaystyle\int_0^5 \sqrt{125 - x^3}\, dx,\ n = 5.$

e. $\displaystyle\int_4^{10} \sqrt[3]{x^2 - 16}\, dx,\ n = 3.$

f. $\displaystyle\int_0^{10} \sqrt{25 + 0.01\, x^4}\, dx,\ n = 5.$

3. Compute the approximate values of the following integrals by the trapezoidal rule, using the values of n indicated. Check your results by performing the integrations.

a. $\displaystyle\int_0^3 x\sqrt{9 - x^2}\, dx,\ n = 6.$

b. $\displaystyle\int_1^4 \frac{x^2\, dx}{\sqrt{1 + x^3}},\ n = 3.$

90. Simpson's (parabolic) rule. As in the case of the trapezoidal rule we first assume that the continuous function $f(x)$ is nowhere negative in the interval (a, b) and approximate $\displaystyle\int_a^b f(x)dx$ by approximating the area under the graph. For this rule, however, it is necessary that n, the number of parts into which the interval $b - a$ is divided, shall be an *even* number. After the ordinates have been constructed as before, they are

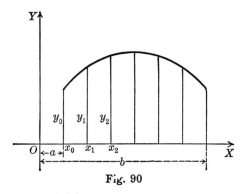

Fig. 90

arranged in groups of three, the first group being y_0, y_1, y_2, the second y_2, y_3, y_4, etc. Through the extremities of the first group is passed an arc of a parabola having its axis parallel to the y-axis. The area under this parabola between x_0 and x_2 is given by the formula (the proof of which is omitted)

$$\frac{\Delta x}{3}(y_0 + 4\,y_1 + y_2).^*$$

Similarly, if a parabola is passed through the upper ends of the next group of three ordinates, the area between x_2 and x_4 is

$$\frac{\Delta x}{3}(y_2 + 4\,y_3 + y_4).$$

The area under the parabola passed through the upper ends of the last group of ordinates is

$$\frac{\Delta x}{3}(y_{n-2} + 4\,y_{n-1} + y_n).$$

By addition, the sum of the areas of the strips, which form an approximation to the area under the original curve, is

$$\frac{\Delta x}{3}(y_0 + 4\,y_1 + 2\,y_2 + 4\,y_3 + 2\,y_4 + \cdots + 4\,y_{n-1} + y_n).$$

The method of procedure is formulated into a working rule as follows.

To evaluate approximately the integral $I = \int_a^b f(x)dx$, the interval $(b - a)$ is divided into n (an even number of) parts, each being equal to Δx. The abscissas of the points of division are $x_0\ (= a)$, x_1, x_2, $\cdots$, $x_n\ (= b)$. The corresponding values of $f(x)$ are $y_0 = f(x_0)$, $y_1 = f(x_1)$, $\cdots$, $y_n = f(x_n)$. Then

(S) $\quad I = (y_0 + 4\,y_1 + 2\,y_2 + 4\,y_3 + 2\,y_4 + \ldots + 4\,y_{n-1} + y_n)\dfrac{\Delta x}{3}.$

Example. Using $n = 4$, compute the approximate value of

$$\int_0^2 \sqrt{4 + x^3}\; dx.$$

---

*The proof of this formula is simple, although the algebraic work is tedious. The equation of a parabola with its axis parallel to the y-axis is $y = a + bx + cx^2$. Let $x_0 = h$, $x_1 = h + \Delta x$, and $x_2 = h + 2\,\Delta x$. The area beneath the parabola between the ordinates $x = h$ and $x = h + 2\,\Delta x$ can be found by integration. Calculating the values of y_0, y_1, and y_2 in terms of h and Δx, and substituting in the formula in the text, we find the same result as by the integration.

Solution. The values of y were computed in Art. 89. The first two columns in the table below give values of x and y, the third column the multipliers used in Simpson's formula, and the last column the values of y_0, $4\,y_1$, $2\,y_2$, etc.

Applying Simpson's formula we get

$$I = (28.924)\,\frac{0.5}{3} = 4.821.$$

x	y		
0.0	2.000	1	2.000
0.5	2.031	4	8.124
1.0	2.236	2	4.472
1.5	2.716	4	10.864
2.0	3.464	1	3.464
			28.924

PROBLEMS

1. Compute the approximate values of the following integrals by Simpson's rule, using the value of n indicated.

a. $\displaystyle\int_0^4 \sqrt[3]{10 + x^2}\, dx$, $n = 4$.

b. $\displaystyle\int_0^4 \sqrt{100 - x^3}\, dx$, $n = 4$.

c. $\displaystyle\int_0^2 \sqrt{9 + x^4}\, dx$, $n = 4$.

d. $\displaystyle\int_4^{10} \sqrt[3]{x^2 - 16}\, dx$, $n = 6$.

2. In the following problems compute the approximate values of the integrals by both the trapezoidal and Simpson's rules. If the indefinite integral can be found, calculate also the exact value of the integral.

a. $\displaystyle\int_0^6 \sqrt{36 - x^2}\, dx$, $n = 6$.

b. $\displaystyle\int_1^5 \frac{dx}{x}$, $n = 8$.

c. $\displaystyle\int_0^4 (16 - x^2)dx$, $n = 8$.

d. $\displaystyle\int_0^6 x\sqrt{36 - x^2}\, dx$, $n = 6$.

e. $\displaystyle\int_1^6 \frac{dx}{x^2}$, $n = 8$.

f. $\displaystyle\int_2^6 \frac{dx}{\sqrt{x^2 - 1}}$, $n = 4$.

g. $\displaystyle\int_2^6 \frac{x\, dx}{\sqrt{x^2 - 1}}$, $n = 4$.

3. A field is bounded by two parallel straight lines and two irregular curves. The length between the straight sides is 50 rd. Starting at one end, the width parallel to the straight sides is measured at 5-rod intervals with the results 35, 37, 35, 32, 30, 31, 30, 28, 25, 21, and 12 rd. Find the area in acres by Simpson's rule. (1 acre $= 160$ sq. rd.)

91. Area under a curve as the limit of a sum. In Art. 88 it was shown that if $y = f(x)$ is a non-negative continuous function, the value of $\displaystyle\int_a^b f(x)dx$ is identical with the area under the graph of $y = f(x)$ from a to b. Let us now take a small positive number p, divide the interval (a, b) into subintervals Δx_1, Δx_2, $\cdots$, Δx_n, each of length less than p, and on each subinterval Δx_i as a base construct a rectangle of height $f(x_i)$, where x_i denotes any value of x in the subinterval Δx_i. The construction is illustrated in Fig. 91. It is intuitively evident that the sum

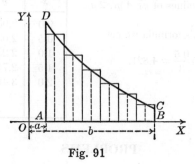

Fig. 91

of the areas of the rectangles is approximately equal to the area under the graph (and consequently to the value of $\int_a^b f(x)dx$) and, moreover, this approximation becomes closer when p is made smaller. Now the areas of the various rectangles are

$$f(x_1)\Delta x_1, f(x_2)\Delta x_2, \cdots, f(x_n)\Delta x_n.$$

Thus it appears that for this type of function the definite integral $\int_a^b f(x)dx$ may be regarded as the limit of a sum of products of the form $f(x_i)\Delta x_i$.

To indicate the sum of a number of terms of like character we can write the first three or four followed by a series of dots and finally the last term as was done in the case of the formula for the trapezoidal rule. Since this notation is long and sometimes ambiguous, the so-called sigma notation is preferred. This is best explained by an example. Instead of indicating the sum of the squares of the first n integers by

$$1^2 + 2^2 + 3^2 + \cdots + n^2,$$

we write

$$\sum_{i=1}^{n} i^2,$$

which is read as "the sum, from $i=1$ to $i=n$, of i^2." The letter $\sum$ (sigma) is the first letter of the Greek word for "sum," the letter i is called the index, and the notation $\sum_{i=1}^{n}$ means that we have the sum of n terms, obtained by successively replacing the index i in the type term, which follows the letter $\sum$, by the integers 1, 2, 3, $\cdots$, n.

In the preceding discussion the sum of the products approximating the integral could have been written concisely as

$$\sum_{i=1}^{n} f(x_i)\Delta x_i.$$

This notation will be employed in the future.

92. The fundamental theorem. This theorem is suggested by the discussion of Art. 91. Every definite integral can be expressed geometrically as the area under a curve by simply plotting the curve $y = f(x)$, where $f(x)$ is the integrand. But in most applications the primary meaning of the definite integral is something quite different from an area, and it is convenient to state the result reached in the previous section as a theorem in analysis without reference to its possible geometric representation. This statement is so important that it is usually referred to as the "Fundamental Theorem of the Integral Calculus."

The Fundamental Theorem. Let $y = f(x)$ *be continuous in the interval* (a, b).

Let $p > 0$ *and let the interval* (a, b) *be divided into n subintervals* Δx_1, $\Delta x_2, \cdots, \Delta x_n$, *each of length less than* p.

Let x_i *be any value of* x *in the interval* Δx_i.

Then

$$\lim_{p \to 0} \sum_{i=1}^{n} f(x_i)\Delta x_i = \int_{a}^{b} f(x)dx.$$

A rigorous proof is omitted. The truth of the theorem is intuitively evident if $f(x) \geqq 0$ for every value of x in (a, b). If $f(x) < 0$ for some values of x in (a, b), we may proceed as follows.

Let m be the least value of $f(x)$ in (a, b) and set $g(x) = f(x) - m$. Then $g(x)$ is continuous and nowhere negative in (a, b). Hence

$$\lim_{p \to 0} \sum_{i=1}^{n} g(x_i)\Delta x_i = \int_{a}^{b} g(x)dx.$$

$$\lim_{p \to 0} \sum_{i=1}^{n} [f(x_i) - m]\Delta x_i = \int_{a}^{b} [f(x) - m]dx.$$

$$\lim_{p \to 0} \sum_{i=1}^{n} f(x_i)\Delta x_i - m \lim_{p \to 0} \sum_{i=1}^{n} \Delta x_i = \int_{a}^{b} f(x)dx - m \int_{a}^{b} dx.$$

But
$$\sum_{i=1}^{n} \Delta x_i = b - a \quad \text{and} \quad \int_{a}^{b} dx = b - a.$$

Hence
$$\lim_{p \to 0} \sum_{i=1}^{n} f(x_i)\Delta x_i = \int_{a}^{b} f(x)dx.$$

The use of this theorem in applications will be illustrated in the next chapter.

CHAPTER VIII

APPLICATIONS OF THE FUNDAMENTAL THEOREM

93. Introduction. The importance of the fundamental theorem is due to the great variety of problems which can be solved by its use. The theorem says that any magnitude whatever which can be expressed as the limit of a sum of the form $\sum_{i=1}^{n} f(x_i)\Delta x_i$, where $f(x)$ is continuous and the sum is formed as directed in the theorem, is equal to the definite integral $\int_{a}^{b} f(x)dx$. Although the previous discussion used the idea of plane areas, the theorem itself is not limited to this particular magnitude. It is a theorem in analysis and there are proofs which make no use of areas.*

In applying the theorem there are three steps. First, we satisfy ourselves that the magnitude in question is the limit of a sum of the specified kind. Second, applying the fundamental theorem, we turn this limit into a definite integral. Third, performing the integration and substituting the limits, we get the desired numerical results.

Example. It is known that a point moves along a straight line with the speed $v = 2 + 3\sqrt{t}$, where t is measured in seconds and v in ft./sec. It is desired to obtain the distance moved in the first 9 sec., that is, between the times $t = 0$ and $t = 9$.

We first observe that $v = f(t)$ is a continuous function of t in the interval $(0, 9)$. Now take a small positive number p and divide the interval $(0, 9)$ into subintervals $\Delta t_1, \Delta t_2, \cdots, \Delta t_n$, each of length less than p. Now let t_i be any value of t in the interval Δt_i $(i = 1, 2, \cdots, n)$, and $v_i = f(t_i) = 2 + 3\sqrt{t_i}$ be the corresponding speed at this instant. Since Δt_i is small, the speed at any time in this subinterval will be approximately v_i and consequently the distance moved in the time interval Δt_i will be approximately $v_i \cdot \Delta t_i = (2 + 3\sqrt{t_i})\Delta t_i$. Hence we conclude that the distance moved in the whole time interval $(0, 9)$ will be *approximately* equal to

$$\sum_{i=1}^{n} (2 + 3\sqrt{t_i})\Delta t_i$$

and will be *exactly* equal to

$$\lim_{p \to 0} \sum_{i=1}^{n} (2 + 3\sqrt{t_i})\Delta t_i.$$

*For example, see *Elements of Calculus* by Granville, Smith, and Longley (Ginn and Company, 1946), page 244.

154

Applying the fundamental theorem, we see that the distance moved in the first 9 sec. is

$$\int_0^9 (2 + 3\sqrt{t})dt = \left[2\,t + 2\,t^{\frac{3}{2}}\right]_0^9 = 72 \text{ ft.}$$

This result may be checked by the methods used in Art. 84. The remainder of this chapter consists of various other applications of the fundamental theorem.

94. Plane areas. In the problems at the end of Art. 88 we used the fact that the area under the graph of a non-negative continuous function $f(x)$ from a to b is equal to $\int_a^b f(x)dx$ to find such areas and also to find other areas which could be expressed as the sum or difference of such areas. For the most general cases it is usually convenient to make use of the fundamental theorem as illustrated in the following examples. In all problems a correct figure is absolutely necessary.

Example 1. Find the area bounded by the parabola $y^2 = 2\,x$ and the straight line $2\,y = x$.

First solution. The curves intersect at the origin and at (8, 4). The area to be found is shown in Fig. 94.1. An approximate value of the area may be ob-

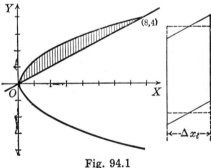

Fig. 94.1

tained by dividing it into n narrow strips, by lines parallel to the y-axis, and then replacing each strip by a rectangle. The figure represents an enlargement of strip number i, of width Δx_i. The area of this strip is approximated by that of a rectangle having the same width Δx_i and a length equal to the difference between the ordinate of the parabola and that of the straight line at some point x_i within the interval. At this point the ordinate of the parabola is $\sqrt{2\,x_i}$, and the ordinate of the straight line is $x_i/2$. Hence the length of the rectangle is $\sqrt{2\,x_i} - \frac{x_i}{2}$, and the area of the rectangle is

$$\left(\sqrt{2\,x_i} - \frac{x_i}{2}\right)\Delta x_i.$$

Now the important point is that this formula applies to *every* strip into which the required area has been divided; for the lower end of *every* strip is on the straight line, and the upper end is on the parabola. Hence an approximate value of the area will be

$$\sum_{i=1}^{n} \left(\sqrt{2\,x_i} - \frac{x_i}{2}\right)\Delta x_i,$$

and this approximation is evidently improved as the width of the strips is diminished.

Consequently, if every Δx_i is less than some previously chosen positive number p, the exact value of the area will be

$$A = \lim_{p \to 0} \sum_{i=1}^{n} \left(\sqrt{2\,x_i} - \frac{x_i}{2}\right)\Delta x_i.$$

Applying the fundamental theorem, we have

$$A = \int_0^8 \left(\sqrt{2\,x} - \frac{x}{2}\right)dx,$$

the limits being assigned to include all the strips. Integrating,

$$A = \left[\frac{1}{3}(2\,x)^{\frac{3}{2}} - \frac{x^2}{4}\right]_0^8 = \frac{64}{3} - \frac{64}{4} = \frac{16}{3}.$$

The detailed reasoning given in the solution above must be clearly understood in order to avoid mistakes. But not all the details need be reproduced in every solution. The area of one of the small rectangles is called an *element of area* and is denoted by dA. In writing this element of area we may omit the subscript i and replace Δx_i by the differential dx, which gives at once the expression to be integrated. The essential details will be illustrated by a second solution.

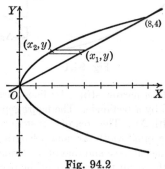

Fig. 94.2

Second solution. Let the area be divided into strips by lines parallel to the x-axis. One of these strips is shown in Fig. 94.2. Every such strip will have its left-hand end on the parabola and its right-hand end on the straight line. The width of a strip is dy, and its length is $x_1 - x_2$, where x_1 is the abscissa of a point on the line (since this is the larger one) and x_2 is the abscissa of a point on the parabola. Hence the element of area is

$$dA = (x_1 - x_2)dy = \left(2\,y - \frac{y^2}{2}\right)dy,$$

by substitution from the equations of the curves. When all the strips are included, y varies from 0 to 4. Hence

$$A = \int_0^4 \left(2\,y - \frac{y^2}{2}\right)dy = \left[y^2 - \frac{y^3}{6}\right]_0^4 = 16 - \frac{32}{3} = \frac{16}{3}.$$

The result may be checked approximately by drawing the figure on squared paper and counting the squares within the area.

It has appeared in the preceding example that the elementary strips may be taken parallel to either the x-axis or the y-axis. This freedom of choice does not exist in all problems, as shown in the next example.

When the strips are taken parallel to the y-axis, the element of area is expressed in terms of x and dx. In this case it is said that we *integrate with respect to* x. When the strips are taken parallel to the x-axis, we *integrate with respect to* y. The elements of area are respectively equal to

$$dA = (y_1 - y_2)dx,$$

and
$$dA = (x_1 - x_2)dy.$$

Example 2. Find the area bounded by the parabola $y^2 = 2\,x$ and the straight line $x - y = 4$.

Solution. The curves intersect at $(2, -2)$ and $(8, 4)$.

If the area is divided into strips parallel to the x-axis, every strip will have its left-hand end on the parabola and its right-hand end on the straight line. It is possible then to write *one* formula which will give the area of *every* strip, and thus to apply the fundamental theorem.

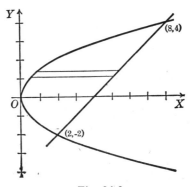

Fig. 94.3

If the area is divided into strips parallel to the y-axis, the strips between $x = 0$ and $x = 2$ will extend from the lower to the upper branch of the parabola, while those between $x = 2$ and $x = 8$ will extend from the straight line to the parabola. In this case it is not possible to write one formula which will give the area of every strip, and the area cannot be expressed by a single integral.

In order to express the area by a single integral, it is necessary to integrate with respect to y. The element of area is

$$dA = (x_1 - x_2)dy,$$

where x_1 is the abscissa of a point on the line and where x_2 is the abscissa of a point on the curve. Substitution from the equations of the line and the curve gives

$$dA = \left(y + 4 - \frac{y^2}{2}\right)dy.$$

When all the strips are included, y varies from -2 to 4. Hence

$$A = \int_{-2}^{4}\left(y + 4 - \frac{y^2}{2}\right) dy = \left[\frac{y^2}{2} + 4y - \frac{y^3}{6}\right]_{-2}^{4}$$
$$= (8 + 16 - \tfrac{32}{3}) - (2 - 8 + \tfrac{4}{3}) = 18.$$

PROBLEMS

1. Find the areas bounded by the following curves. In each case draw the figure, showing the element of area.

a. $y = x^3$, $y = 4x$.

b. $y = 6x - x^2$, $y = x$.

c. $x = 4y - y^2$, $y = x$.

d. $y = 4x - x^2$, $y = 2x - 3$.

e. $y^2 = 4x$, $2x - y = 4$.

f. $y = x^3 - 3x$, $y = x$.

g. $y = x^2$, $2x - y + 3 = 0$.

h. $y = x^2 - 2x - 3$, $y = 6x - x^2 - 3$.

i. $y^2 = 4x$, $x = 12 + 2y - y^2$.

j. $y = x^3 - x$, $y = x - x^2$.

2. Find the area included between the two parabolas $y^2 = 2px$ and $x^2 = 2py$.

3. Find the area included between the two parabolas $y^2 = ax$ and $x^2 = by$.

4. Find the entire area of the curve $y^2 = 9x^2 - x^4$.

5. Find the area bounded by the curve $\sqrt{x} + \sqrt{y} = \sqrt{a}$ and the coördinate axes.

6. Find the area bounded by the curve whose equation is $y = 6 - x^2$ and the line whose equation is $y = -3$.

7. Find the area bounded by the curve whose equation is $y = 6 + 4x - x^2$ and the chord joining $(-2, -6)$ and $(4, 6)$.

8. Find the area bounded by the curve whose equation is $y^3 = x^2$ and the chord joining $(-1, 1)$ and $(8, 4)$.

9. Find the area bounded by the curves whose equations are $y^2 = x$, $y = x$, and $2y = x$.

10. Find the area under the hyperbola $xy = 24$ from $x = 4$ to $x = 8$ (use Simpson's rule, taking $n = 8$).

11. Write the integral giving the area of a segment of a circle of radius 10 cut off by a chord distant 6 from the center. Approximate the integral by Simpson's rule, taking $n = 4$.

12. Find the area bounded by the curve whose equation is $y = x(1 \pm \sqrt{x})$ and the line whose equation is $x = 4$.

13. Find the entire area of the loop of the curve whose equation is $y^2 = 4x^2 - x^3$.

14. Find the area bounded by the curve whose equation is $x^2y = x^2 - 1$ and the lines whose equations are $y = 1$, $x = 1$, and $x = 4$.

15. Find the area bounded by the y-axis, the curve whose equation is $y = x^3 - 9x^2 + 24x - 7$, and the line whose equation is $y = 29$.

16. Find the area of a circle of radius 6 by dividing it into elements of area which are concentric rings of width Δr.

17. Plot on the same coördinate axes the circle whose equation is $x^2 + y^2 = a^2$ and the ellipse whose equation is $b^2x^2 + a^2y^2 = a^2b^2$. Set up the integrals giving the areas within the curves and, without integrating, show that the area of the ellipse is b/a times the area of the circle, and therefore is πab.

95. Length of a curve. We consider only curves or arcs of curves which are the graphs of one-valued functions continuous in a finite interval (a, b). In order to compute the length of an arc when the equation of the curve is given we must formulate a definition of length which can be used mathematically and which gives reasonable results. For example, if the arc represented a crooked wire, the result obtained by straightening it out and measuring it with a ruler should agree with the result given by our mathematical definition. To obtain such a definition we proceed as follows.

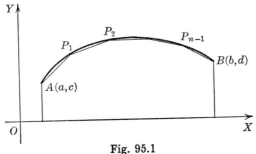

Fig. 95.1

Let $p > 0$ and let $P_1, P_2, \cdots, P_{n-1}$ be points on the arc AB (Fig. 95.1) such that the length of each of the chords $AP_1, P_1P_2, \cdots, P_{n-1}B$ is less than p. These chords form a broken line and we shall call this a "broken line inscribed in AB." Let L_p denote the length of the broken line, that is, the sum of the lengths of the chords. It seems reasonable that when p is small, L_p is a good approximation to the length AB and this leads to the following definition.

DEFINITION. *Let AB be an arc of the graph of a continuous function. Let $p > 0$ and let L_p be the length of a broken line inscribed in AB, each of whose chords has a length less than p. Then the length of AB is* $\lim\limits_{p \to 0} L_p$.

For all curves considered in this book $\lim\limits_{p \to 0} L_p$ exists, that is, is finite, and the curves are called *rectifiable*. The process of finding the length of a curve is sometimes referred to as the "rectification of the curve."

Theorem. *Let $y = f(x)$ be continuous and have a continuous first derivative in (a, b). Let A and B be the points on the graph corresponding to a and b and let s denote the length of AB. Then*

(I) $$s = \int_a^b \sqrt{1 + \left(\frac{dy}{dx}\right)^2}\, dx.$$

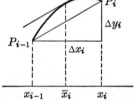

Fig. 95.2

Proof. Let $p > 0$ and inscribe in AB a broken line $P_0 P_1 P_2 \cdots P_n$, where $P_0 = A$, $P_n = B$ and the length of each chord is less than p. Let the abscissa of each point P_i be x_i, where $x_0 = a$ and $x_n = b$. With each chord $P_{i-1}P_i$ as a hypotenuse draw a right triangle whose legs are parallel to the coördinate axes and denote these legs by Δx_i and Δy_i. One of these triangles is illustrated on an enlarged scale in Fig. 95.2. It seems clear that

$$|P_{i-1}P_i| = \sqrt{(\Delta x_i)^2 + (\Delta y_i)^2} = \sqrt{1 + \left(\frac{\Delta y_i}{\Delta x_i}\right)^2}\, \Delta x_i. \tag{1}$$

The length of the broken line is

$$L_p = \sum_{i=1}^n \sqrt{1 + \left(\frac{\Delta y_i}{\Delta x_i}\right)^2}\, \Delta x_i.$$

Now the slope of the chord $P_{i-1}P_i$ is $\Delta y_i / \Delta x_i$ and it is shown in Art. 190 that for a certain value $\bar{x}_i$ in the interval Δx_i the tangent to the curve at the corresponding point is parallel to the chord. That is,

$$\frac{\Delta y_i}{\Delta x_i} = \left(\frac{dy}{dx}\right)_{x = \bar{x}_i}$$

the notation signifying the value of dy/dx at $x = \bar{x}_i$. Hence

$$L_p = \sum_{i=1}^{n} \sqrt{1 + \left(\frac{dy}{dx}\right)_{x=\bar{x}_i}^2} \, \Delta x_i.$$

Now every $\Delta x_i < p$ and $\sqrt{1 + \left(\frac{dy}{dx}\right)^2}$ is a continuous function of x, since dy/dx is continuous by the hypothesis. Hence the fundamental theorem is applicable and, by definition,

$$s = \lim_{p \to 0} L_p = \int_a^b \sqrt{1 + \left(\frac{dy}{dx}\right)^2} \, dx.$$

If the solution of the equation for x in terms of y yields a one-valued, continuous function of y whose derivative with respect to y is continuous in the interval (c, d), where c is the ordinate of A and d is the ordinate of B, we obtain the formula

(II) $$s = \int_c^d \sqrt{1 + \left(\frac{dx}{dy}\right)^2} \, dy.$$

This is proved by factoring Δy_i out of the radical in (1).

An easy way of remembering both forms is to regard the element of arc ds as a chord which is the hypotenuse of a small right triangle whose sides are dx and dy. Then

$$ds = \sqrt{(dx)^2 + (dy)^2}. \tag{2}$$

Factoring out dx, we obtain **(I)**; factoring out dy, we obtain **(II)**.

Example. Find the length of the curve $9\,y^2 = 4(1 + x^2)^3$ from the point where $x = 0$ to the point where $x = 3$.

Solution. From the equation of the curve,

$$y = \tfrac{2}{3}(1 + x^2)^{\frac{3}{2}}.$$

Differentiating, $\dfrac{dy}{dx} = 2\,x\sqrt{1 + x^2}.$

Using **(I)**, $\quad ds = \sqrt{1 + 4\,x^2(1 + x^2)}\,dx = (1 + 2\,x^2)dx.$

Hence $\quad s = \int_0^3 (1 + 2\,x^2)dx = \left[x + \frac{2\,x^3}{3}\right]_0^3 = 21.$

Gross errors may usually be detected by comparing the result with the length of the chord joining the end points of the curve. In this problem the end points are $(0, 0.67)$ and $(3, 21.1)$; hence the length of the chord is 20.6.

PROBLEMS

1. Find the length of the curve whose equation is $y^2 = x^3$ from the point where $x = 0$ to the point where $x = 5$.

2. Find the length of the curve whose equation is $9\,y^2 = (2 + x^2)^3$ from the point where $x = 0$ to the point where $x = 2$.

3. Find the length of the curve whose equation is $y^3 = ax^2$ between the points $(0, 0)$ and (a, a).

4. Find the length of that part of the curve whose equation is $(y - 8)^2 = x^3$ which is intercepted between the coördinate axes.

5. Find the entire length of the hypocycloid whose equation is $x^{\frac{2}{3}} + y^{\frac{2}{3}} = a^{\frac{2}{3}}$.

6. Find the length of the curve whose equation is $y^3 = x^2$ between the points $(0, 0)$ and $(8, 4)$.

7. Find the length of the curve whose equation is $y = \dfrac{x^3}{6} + \dfrac{1}{2\,x}$ from the point where $x = 1$ to the point where $x = 3$.

8. The points $A(-1, 1)$, $O(0, 0)$, and $B(8, 4)$ lie on the graph of $y^3 = x^2$. Find the length of arcs AO, OB, and AB. Check by finding the length of the chord AB.

In the following problems the indefinite integrals cannot be found by the rules previously given. The values of the definite integrals must be approximated.

9. Approximate the length of the parabola whose equation is $y = x^2$ between $(0, 0)$ and $(2, 4)$. Use Formula I for ds, and evaluate the integral by Simpson's rule, taking $n = 4$.

10. Approximate the length of the parabola whose equation is $y^2 = 2\,x$ between $(0, 0)$ and $(8, 4)$. Use Formula II for ds, and evaluate the integral by Simpson's rule, taking $n = 4$.

11. Approximate the length of the curve whose equation is $y = x^3$ between $(0, 0)$ and $(2, 8)$. Use Formula I for ds, and evaluate the integral by Simpson's rule, taking $n = 4$.

12. Approximate the length of the hyperbola whose equation is $x^2 - y^2 = 9$ from $(3, 0)$ to $(5, 4)$. Use Formula II for ds, and evaluate the integral by the trapezoidal rule, taking $n = 4$.

13. Approximate the length of the arch of the parabola whose equation is $y = 4\,x - x^2$ which lies above the x-axis.

96. Volumes of solids of revolution.

A solid of revolution is a solid having an axis of symmetry such that every section of the solid made by a plane perpendicular to this axis is a circle with its center on the axis. Such a solid may be regarded as *generated* by revolving about the axis an area bounded by a segment AB of the axis, two segments AC and BD

perpendicular to the axis at A and B (either of which may be of zero length), and a curve which joins C and D and does not cut AB. Thus, a right circular cone is generated by revolving a right triangle about one of its legs, and a sphere is generated by revolving a semicircular area about its diameter.

Let the area under the graph of the continuous function $y = f(x)$ from a to b be revolved about the x-axis, forming a solid of revolution (Fig. 96.1). To find the volume of such a solid we proceed much as in Art. 94. Let the segment (a, b) of the x-axis be divided into subintervals Δx_1, Δx_2, $\cdots$, Δx_n, each of length less than some positive number p, and let planes perpendicular to the x-axis be passed through each point of division. These section planes divide the solid into n circular plates, or discs.

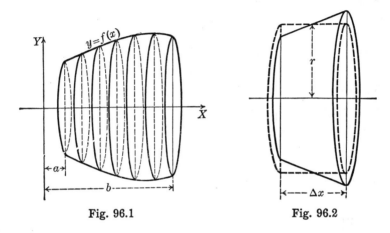

Fig. 96.1   Fig. 96.2

The volume of any one of these discs will be approximately equal to that of a cylinder having the same thickness Δx and a radius, r, equal to the radius of the disc at some point within the corresponding subinterval Δx.* The cylinder would be generated by the revolution of a rectangle, as shown in Fig. 96.2. The volume of such a cylinder is $\pi r^2 \Delta x$. An approximation to the volume of the solid is then given by the sum of the volumes of the small cylinders; that is, by $\sum \pi r^2 \Delta x$, the summation extending over all the subintervals Δx of the interval (a, b). We then agree that the limit of this sum as $p \longrightarrow 0$ is the exact volume of the solid.

In the present problem the radius r is equal to an ordinate to the curve $y = f(x)$; that is, $r = y = f(x)$. Since $f(x)$ is continuous, r^2 is a continu-

---

*In this and the following applications the subscripts hitherto used will be omitted for the sake of simplifying the notation.

ous function of x and so the fundamental theorem is applicable. Hence the exact volume of the solid is given by

$$\lim \sum \pi [f(x)]^2 \Delta x = \int_a^b \pi [f(x)]^2 dx.$$

This formula is valid only for the particular type of problem described; that is, when the area under the curve $y = f(x)$ is revolved about the x-axis. The general method can be applied to any case of a solid of revolution, and should always be used without attempting to memorize formulas. The essential steps in the general method are the following.

I. *Draw the figure showing an element of volume, a right cylinder. Denote the thickness of the element by dx if the axis of revolution is parallel to the x-axis (or by dy if the axis of revolution is parallel to the y-axis).*

II. *Let r denote the radius of the element of volume. With the aid of the figure express r in terms of x (or y) from the equation of the given curve.*

III. *Form the element of volume*

$$dV = \pi r^2 \, dx \text{ (or } \pi r^2 \, dy).$$

IV. *Determine the limits of integration from the figure.*

The following examples illustrate the application of the general method to the different types of problems. These examples are based on Fig. 96.3, in which the equation of the curve OA is $y^2 = 2x$. The letters x and y as used in the problems below always represent the coördinates of a point on the curve OA.

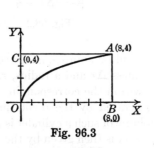

Fig. 96.3

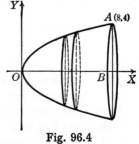

Fig. 96.4

Example 1. Find the volume generated by revolving the area OAB about the x-axis.

Solution. An element of volume is shown in Fig. 96.4. The thickness is dx, and the radius is an ordinate of the curve.

Hence　　　　$r = y = \sqrt{2x}$

and　　　　$dV = \pi r^2 \, dx = \pi \, 2x \, dx.$

$$V = \int_0^8 2\pi x \, dx = \pi [x^2]_0^8 = 64\pi = 201.1.$$

Gross errors may often be detected by comparing the result obtained with the volume of an inscribed or circumscribed cone or cylinder. In this example the volume of the inscribed cone is $\frac{128}{3}\pi$ and that of the circumscribed cylinder is 128π.

Example 2. Find the volume generated by revolving the area OAB about the line AB.

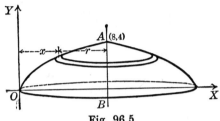

Fig. 96.5

Solution. An element of volume is shown in Fig. 96.5. The thickness is dy, and the radius is $8 - x$.

Hence
$$r = 8 - x = 8 - \frac{y^2}{2}$$

and
$$dV = \pi r^2\, dy = \pi\left(64 - 8\,y^2 + \frac{y^4}{4}\right)dy.$$

$$V = \int_0^4 \pi\left(64 - 8\,y^2 + \frac{y^4}{4}\right)dy = \pi\left[64\,y - \frac{8\,y^3}{3} + \frac{y^5}{20}\right]_0^4$$

$$= \frac{2048\,\pi}{15} = 428.9.$$

Example 3. Find the volume generated by revolving the area OAC about the y-axis.

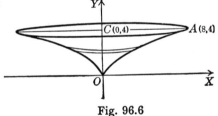

Fig. 96.6

Solution. An element of volume is shown in Fig. 96.6. The thickness is dy and the radius is an abscissa of the curve.

Hence
$$r = x = \frac{y^2}{2}$$

and
$$dV = \pi r^2\, dy = \frac{\pi}{4}\,y^4\, dy.$$

$$V = \int_0^4 \frac{\pi}{4}\,y^4\, dy = \frac{\pi}{4}\left[\frac{y^5}{5}\right]_0^4$$

$$= \frac{256\,\pi}{5} = 160.8.$$

Example 4. Find the volume generated by revolving the area OAC about the line CA.

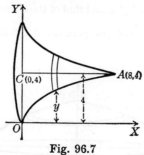

Fig. 96.7

Solution. An element of volume is shown in Fig. 96.7. The thickness is dx and the radius is $4 - y$.

Hence

$$r = 4 - y = 4 - \sqrt{2\,x}$$

and

$$dV = \pi r^2\, dx$$

$$= \pi(16 - 8\sqrt{2\,x} + 2\,x)dx.$$

$$V = \int_0^8 \pi(16 - 8\sqrt{2\,x} + 2\,x)dx$$

$$= \pi\Big[16\,x - \tfrac{8}{3}(2\,x)^{\frac{3}{2}} + x^2\Big]_0^8$$

$$= \frac{64\,\pi}{3} = 67.02.$$

97. Simplification by a change of variable. In Example 4, above, the desired volume is given by the integral

$$V = \pi \int_0^8 (4 - y)^2\, dx.$$

Instead of substituting for y its value in terms of x, we may change the variable of integration by substituting for dx in terms of y and dy.

Thus, since $y^2 = 2\,x$, $2\,y\,dy = 2\,dx$, or $dx = y\,dy$. When $x = 0$, $y = 0$; when $x = 8$, $y = 4$. Substituting and changing the limits, we have

$$V = \pi \int_0^8 (4 - y)^2\, dx = \pi \int_0^4 (4 - y)^2 y\, dy$$

$$= \pi\Big[8\,y^2 - \frac{8}{3}\,y^3 + \frac{y^4}{4}\Big]_0^4$$

$$= \frac{64\,\pi}{3} \text{ (as before).}$$

This process is merely integration by an appropriate substitution, similar to that discussed in Arts. 81 and 86. It frequently enables one to avoid troublesome fractional exponents.

PROBLEMS

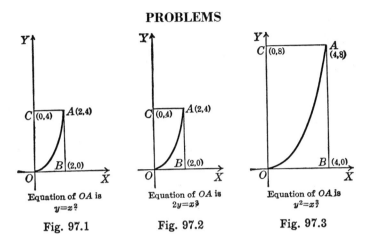

Equation of *OA* is
$y=x^2$

Fig. 97.1

Equation of *OA* is
$2y=x^3$

Fig. 97.2

Equation of *OA* is
$y^2=x^3$

Fig. 97.3

1. In Fig. 97.1 above find the volume generated when the area

a. *OAB* is revolved about *OX*. **b.** *OAB* is revolved about *AB*.
c. *OAB* is revolved about *CA*. **d.** *OAB* is revolved about *OY*.
e. *OAC* is revolved about *OY*. **f.** *OAC* is revolved about *CA*.
g. *OAC* is revolved about *AB*. **h.** *OAC* is revolved about *OX*.

2. The same as Problem 1 for Fig. 97.2.

3. The same as Problem 1 for Fig. 97.3.

4. Find by integration the volume of the cone generated by revolving about the x-axis the triangle whose vertices are $(0, 0)$, $(a, 0)$, (a, b).

5. Find by integration the volume of the cone generated by revolving about the y-axis the triangle whose vertices are $(0, 0)$, (a, b), $(0, b)$.

6. Find the volume of the paraboloid of revolution generated by revolving about the x-axis the area bounded by $y^2 = 2\,px$ and $x = h$. Show that the result is one half the volume of the cylinder having the same base and altitude.

7. Find the volume generated by revolving about the y-axis the area bounded by $y^2 = 2\,px$, $y = b$, and the y-axis. Show that the result is one fifth the volume of the cylinder having the same base and altitude.

8. Find the volume of the oblate spheroid generated by revolving the area bounded by the ellipse $\dfrac{x^2}{a^2} + \dfrac{y^2}{b^2} = 1$ about the y-axis.

9. Find the volume of the prolate spheroid generated by revolving the area bounded by the ellipse $\dfrac{x^2}{a^2} + \dfrac{y^2}{b^2} = 1$ about the x-axis.

10. A hyperboloid of revolution is generated by revolving the hyperbola $x^2 - y^2 = a^2$ about the x-axis. Find the volume of a segment of one base of this solid of thickness a.

11. Find the volume generated by revolving about the x-axis the area of the hypocycloid $x^{\frac{2}{3}} + y^{\frac{2}{3}} = a^{\frac{2}{3}}$.

12. Find the volume generated by revolving about the x-axis the area bounded by the coördinate axes and the parabola $x^{\frac{1}{2}} + y^{\frac{1}{2}} = a^{\frac{1}{2}}$.

13. Find the volume of a sphere by integration.

14. The plane surface of a spherical segment of one base is a circle of radius 8 in., and the greatest thickness of the segment is 4 in. Find its volume by integration.

HINT. What is the radius of the sphere from which the segment was cut?

15. A segment of one base of thickness h is cut from a sphere of radius r. Show by integration that its volume is $\left(\pi \dfrac{h^2}{3} \right)(3\,r - h)$.

16. Find the volume generated by revolving about the x-axis the area bounded by one of the ovals whose equation is $x^2 y^2 = (x^2 - 25)(4 - x^2)$.

17. Find the volume generated by revolving about the x-axis the entire area of the curve whose equation is $y^2 = x^2(4 - x^2)$.

18. Find the volume generated by revolving the area bounded by the loop of the curve whose equation is $y^2 = x(x^2 - 4)$ about the x-axis.

19. The smaller segment of the circle whose equation is $x^2 + y^2 = 25$ cut off by the line whose equation is $x = 3$ is revolved about this line, generating a spindle-shaped solid. Set up the integral giving its volume and evaluate it by Simpson's rule, taking $n = 4$.

20. The area bounded by the parabola whose equation is $y = 4 + 6\,x - 2\,x^2$ and the line whose equation is $y = -4$ is revolved about the line. Find the volume generated.

98. Volume of known cross section. Volumes of certain solids may be found by the same general method as is used for volumes of revolution. We shall first develop in general terms the principles used.

Let us consider a solid which satisfies the following requirements.

1. For some line which we may take as an x-axis (or a y-axis) the solid lies between two planes which are perpendicular to the axis at the points $x = a$ and $x = b$.

2. A plane perpendicular to the axis at any point in the interval (a, b) cuts the solid in a cross section of area A, where A is a continuous function of x in (a, b). We shall therefore use the notation $A(x)$.

3. The change in shape of the cross section as x changes from a to b is continuous in a geometric sense, so that we can conceive of the solid as generated by the motion of a variable cross section, just as we regarded the area under a curve as generated by a moving variable ordinate. A solid of revolution is a simple solid of this kind and another example is illustrated in Fig. 98.1.

Using the figure, we find the volume V of the solid as follows. Let p be a small positive number and divide the segment (a, b) of the axis into subintervals $\Delta x_1, \Delta x_2, \cdots, \Delta x_n$, each of length less than p. Through the points of division pass planes perpendicular to the x-axis. These planes divide the solid into thin slices, one of which is marked in the figure. Since the slice is thin and the area $A(x)$ of the cross section is a continuous function of x, we see that the volume of any slice is nearly equal to the product of its thickness Δx_i and the value of $A(x_i)$, where x_i is any point in the interval Δx_i. If the volumes of all the slices from $x = a$ to $x = b$ are summed, we get an approximation to the volume of the solid which gets closer as the slices get thinner. Hence we have

$$V = \lim_{p \to 0} \sum_{i=1}^{n} A(x_i) \Delta x_i.$$

Since $A(x)$ is continuous in (a, b), we can apply the fundamental theorem and this relation becomes

$$V = \int_a^b A(x)dx.$$

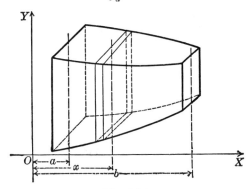

Fig. 98.1

In order to use this result with facility we must have solids of such a character that the area of any cross section perpendicular to the x-axis (or y-axis) can be represented by a formula in terms of x (or y). The element of volume is a cylindrical solid whose thickness is dx (or dy) and whose base has the area $A(x)$ (or $A(y)$). Hence

$$dV = A(x)dx \quad \text{or} \quad dV = A(y)dy.$$

We now illustrate the method by examples.

Example 1. Derive the formula for the volume of a pyramid.

Solution. Let $OP(=h)$ be the altitude of the pyramid $O\text{-}CDE*$ and let B denote the area of the base. Let OP be the x-axis with the origin at O. Then x varies from 0 to h. For every value of x the cross section of the pyramid is a polygon (in the figure a triangle) whose area may be denoted by $A(x)$. The pyramid may be regarded as generated by the motion of this variable polygon from the vertex O to the base DCE. It is clear that the solid satisfies the three requirements given above.

The element of volume dV is a prism whose altitude or thickness is dx and whose base is a polygon of area $A(x)$. It is known from solid geometry that

$$\frac{A(x)}{B} = \frac{x^2}{h^2}, \quad \text{or} \quad A(x) = \frac{Bx^2}{h^2}.$$

Hence

$$dV = \frac{Bx^2}{h^2}\,dx$$

and

$$V = \int_0^h \frac{Bx^2\,dx}{h^2} = \frac{B}{h^2}\left[\frac{x^3}{3}\right]_0^h = \frac{Bh}{3}.$$

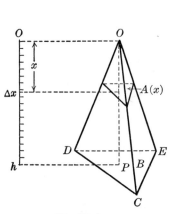

Fig. 98.2

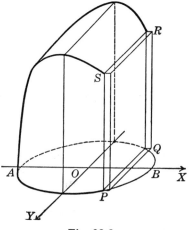

Fig. 98.3

Example 2. A solid has a circular base of radius 10 in. The line AB is the diameter of its base, and every section of the solid made by a plane perpendicular to AB is a square. Find the volume.

Solution. In Fig. 98.3 the right-hand end of the solid is cut away to show an element of volume whose base is $PQRS$. Let AB be taken as the x-axis, with the origin at O, the center of the circular base. With this notation the thickness of the element of volume is dx. Since every section perpendicular to AB is a square, the element of volume is a rectangular parallelepiped with square base

*Fig. 98.2 is drawn with a triangular base, but the reasoning applies to any base.

of edge PQ. In order to express the length PQ in terms of x, we take the y-axis in the plane of the circular base. It is then evident that PQ is a double ordinate of the circle. That is,

$$dV = \overline{PQ}^2 \, dx = 4 \, y^2 \, dx.$$

But the equation of the circle is

$$x^2 + y^2 = 100.$$

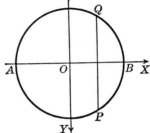

Hence $dV = 4(100 - x^2)dx.$

The symmetry of the figure shows that half the volume will be obtained by integrating over the segment OB.

Fig. 98.4

$$\frac{V}{2} = \int_0^{10} 4(100 - x^2)dx = 4\left[100 \, x - \frac{x^3}{3} \right]_0^{10} = \frac{8000}{3}.$$

Hence the volume of the solid is

$$V = \tfrac{16000}{3} = 5333\tfrac{1}{3} \text{ cu. in.}$$

Example 3. Find the volume generated by revolving about the y-axis the area bounded by the parabola $4 \, y = x^2$, the x-axis, and the line $4 \, x - y = 12$.

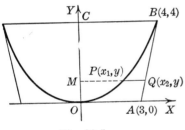

Fig. 98.5

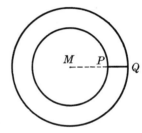

Fig. 98.6

Solution. The solid is that part of the frustum of a cone generated by the revolution of $OABC$ (Fig. 98.5) which remains after removing the paraboloid generated by OBC. A plane perpendicular to the y-axis at any point M between O and C cuts the solid in a cross section which is a circular ring with inner radius MP and outer radius MQ. The element of volume (Fig. 98.6) will be a cylindrical solid shaped like a washer, which has this ring as a base and a thickness equal to dy.

If we set $x_1 = MP$ and $x_2 = MQ$, the equations of the parabola and the line give $x_1 = \sqrt{4 \, y}$ and $x_2 = \tfrac{1}{4}(y + 12)$. Then the area of the ring is

$$A(y) = \pi x_2{}^2 - \pi x_1{}^2 = \pi\left[\tfrac{1}{16}(y + 12)^2 - 4 \, y\right]$$

$$= \frac{\pi}{16} (y^2 - 40 \, y + 144),$$

and $$dV = A(y)dy = \frac{\pi}{16}(y^2 - 40\,y + 144)dy.$$

Hence $$V = \frac{\pi}{16}\int_0^4 (y^2 - 40\,y + 144)dy = \frac{52\,\pi}{3}.$$

99. The shell method. Suppose that it is desired to find the volume of the solid generated by revolving about the y-axis the area bounded by the parabola $y = 4\,x - x^2$ and the x-axis.

We can solve this problem in three ways. First, referring to Fig. 99.1, we can find the volume of the solids of revolution generated by OBC and $OABC$ and subtract the former from the latter. Second, we can proceed as in Example 3, Art. 98, using as elements of volume "washers" whose thickness is dy and whose bases are circular rings cut out by planes perpendicular to the y-axis. The inner and outer radii of a typical ring are $x_1 = MP$ and $x_2 = MQ$, respectively, where x_1 and x_2 are found by solving the equation $y = 4\,x - x^2$ for x in terms of y. This gives $x_1 = 2 - \sqrt{4 - y}$ and $x_2 = 2 + \sqrt{4 - y}$. Completion of the solution leads to the result $V = \frac{128}{3}\,\pi$.

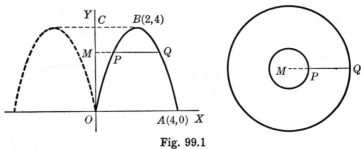

Fig. 99.1

A third method of solving this problem is provided by the "shell method," which follows. Let us take a small positive number p and divide the interval $(0, 4)$ on the x-axis into subintervals $\Delta x_1, \Delta x_2, \cdots,$ Δx_n, each of length less than p. On each Δx_i as a base construct a rectangle whose height is the value of y for some value x_i of x in Δx_i. As the figure revolves about OY each of these rectangles generates a hollow right cylindrical shell. The volume of the whole solid is approximated by the sum of the volumes of these shells and is equal to the limit of this sum as $p \longrightarrow 0$. In short, the solid is regarded as made up of a set of very thin shells, of varying heights and radii, "nested" within each other.

The volume of a thin cylindrical shell is approximately equal to the product of the thickness by the lateral area. The lateral area of a cylinder is $2\,\pi rh$. If the shell is thin, we can take for r either the inner radius, the outer radius, or any value between. In this example the thickness is Δx_i, the height is $y_i = 4\,x_i - x_i^2$, and the radius may be taken as x_i; hence the volume of any shell is approximately $2\,\pi x_i (4\,x_i - x_i^2)\Delta x_i$.

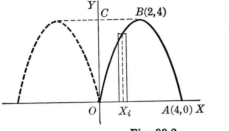

Fig. 99.2

Application of the fundamental theorem leads to the result

$$V = 2\,\pi \int_0^4 x(4\,x - x^2)dx = \frac{128\,\pi}{3}.$$

The shell method is useful when the axis of revolution is perpendicular to the x-axis (or y-axis) and it is possible to express r and h as continuous functions of x (or y). The element of volume may be written as

$$dV = 2\,\pi r h\,dx \quad \text{or} \quad dV = 2\,\pi r h\,dy,$$

according as the axis of revolution is perpendicular to the x-axis or the y-axis, respectively.

Which of these three methods should be used depends upon the problem. The second method requires the solution of the equation for x in terms of y, which may be inconvenient or impossible. If the equation above were replaced by $y = 4\,x - x^3$, a somewhat similar solid would be obtained. In this case only the shell method is available, since we cannot solve this equation for x in terms of y.

PROBLEMS

1. A solid has a circular base of radius 10 in. The line AB is a diameter of the base. Find the volume of the solid if every section perpendicular to AB is

   a. An equilateral triangle.
   b. An isosceles right triangle with its hypotenuse in the plane of the base.
   c. An isosceles right triangle with one leg in the plane of the base.
  *d. An isosceles triangle with its altitude equal to 10 in.
   e. An isosceles triangle with its altitude equal to its base.

*In these problems approximate the definite integrals by Simpson's rule, or use the result in connection with Fig. 88.2.

2. A solid has a base in the form of an ellipse whose equation is $x^2 + 4y^2 = 100$, where the unit of length is 1 inch. Find the volume of the solid if every section perpendicular to the x-axis is

> **a.** A square.
> **b.** An equilateral triangle.
> ***c.** An isosceles triangle with altitude 5 in.

3. The base of a solid is a segment of a parabola whose equation is $y^2 = 8x$ cut off by the line whose equation is $x = 8$. The unit of length is 1 inch. Find the volume of the solid if every section perpendicular to the axis of the base is

> **a.** A square.
> **b.** An equilateral triangle.
> **c.** An isosceles triangle with altitude 10 in.

4. A plane section passed through two opposite seams of a football is an ellipse whose equation is $x^2 + 4y^2 = 49$. Find the volume (**a**) if the leather is so stiff that every section perpendicular to the x-axis is a square; (**b**) if every such section is a circle.

5. Two cylinders of equal radius r have their axes meeting at right angles. Find the volume of the common part.

HINT. What is the shape of a section made by a plane parallel to the two axes?

*6. The circle whose equation is $x^2 + (y - 6)^2 = 9$ is revolved about the x-axis. The solid generated is called a *torus*. Describe it. Show that the volume is equal to $48\pi \int_0^3 \sqrt{9 - x^2}\, dx$. Evaluate the integral.

7. A wedge is cut from a cylinder of radius 5 in. by two planes, one perpendicular to the axis of the cylinder and the other passing through the diameter of the section made by the first plane and inclined to this plane at an angle of 45°. Find the volume of the wedge.

8. Find the volume of the solid generated by revolving about the x-axis the area bounded by the curves whose equations are

> **a.** $y = 6x - x^2,\ y = x$.　　　　**b.** $y = 4x - x^2,\ 2y = 4x - x^2$.
> **c.** $y = x^3 - 3x,\ y = -2x$.　　　**d.** $y^2 = 4x,\ 2x - y = 4$.

9. Find the volume of the solid generated by revolving about the y-axis the area bounded by the curves whose equations are

> **a.** $x = 4y - y^2,\ y = x$.　　　　**b.** $y = x^3,\ y = 4x$.
> **c.** $y^2 = 2px,\ x^2 = 2py$.　　　　**d.** $y = x^2,\ 2x - y + 3 = 0$.

10. Find the volume of the solid generated by revolving about the y-axis the area in the first quadrant bounded by the curve $y = 16x - x^2$ and the x-axis.

11. Find the volume of the solid generated by revolving about the x-axis the area bounded by the curve $x = 8y - y^2 - 7$ and the y-axis.

*See footnote, page 173.

100. Area of surfaces of revolution. A surface of revolution is the lateral surface of a solid of revolution. Such a surface may be generated by the revolution of an arc of a plane curve about an axis in its plane.

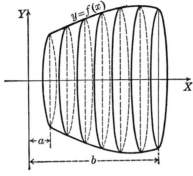

Fig. 100.1

Suppose that a surface of revolution is generated by the revolution about the x-axis of the arc of the curve $y = f(x)$ from $x = a$ to $x = b$.

Also let $f(x)$ be differentiable and let $f'(x)$ be continuous in (a, b). Now let the interval (a, b) be divided into subintervals $\Delta x_1, \Delta x_2, \cdots,$ Δx_n, each of length less than some small positive number p, and let planes perpendicular to the axis be passed through each point of division. These planes divide the surface into narrow strips, and we wish to get an expression for an approximate value of the area of each strip so that the surface may be found by the limiting process of the fundamental theorem.

For this purpose consider any subinterval of length Δx. At the mid-point, x, of this interval erect the ordinate y to the curve, and at the extremity of this ordinate draw the tangent to the curve. If α is the inclination of this tangent, then

$$\tan \alpha = \frac{dy}{dx}$$

and $$\sec \alpha = \sqrt{1 + \left(\frac{dy}{dx}\right)^2},$$

where dy/dx is the value of the derivative for this mid-point. The length of the tangent between the ordinates at the ends of the sub-interval is

$$l = \Delta x \sec \alpha = \sqrt{1 + \left(\frac{dy}{dx}\right)^2} \, \Delta x.$$

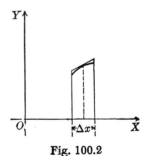

Fig. 100.2

When the preceding construction has been carried out for each of the subintervals, and when the curve together with the tangent lines is revolved about the x-axis, each elementary strip of surface will be circumscribed by a frustum of a cone. Since y is the radius of the mid-section, and l is the slant height of the frustum, its lateral surface will be

$$2\,\pi y \sqrt{1 + \left(\frac{dy}{dx}\right)^2}\,\Delta x, \tag{1}$$

by a formula from geometry.

The area of the solid of revolution is the limit of the sum of the lateral areas of the frusta as $p \longrightarrow 0$. Since y and dy/dx are continuous functions of x by hypothesis, the function (1) is continuous and the fundamental theorem is applicable. Hence the surface S is given by

$$S = \lim \sum 2\,\pi y \sqrt{1 + \left(\frac{dy}{dx}\right)^2}\,\Delta x = \int_a^b 2\,\pi y \sqrt{1 + \left(\frac{dy}{dx}\right)^2}\,dx.$$

Since the element of arc on the curve is $ds = \sqrt{1 + \left(\frac{dy}{dx}\right)^2}\,dx$, the element of surface, dS, may be written

$$dS = 2\,\pi y\,ds.$$

Either of the two forms for ds may be used in applications.

If the arc is revolved about the y-axis, the element of surface is

$$dS = 2\,\pi x\,ds.$$

Example. Derive the formula for the area of the surface of a sphere.

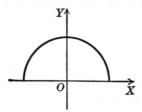

Fig. 100.3

Solution. A spherical surface is generated by the revolution of a semicircle about a diameter. Let the x-axis be the axis of revolution and let the semicircle have its center at the origin. The equation of the curve is

$$x^2 + y^2 = r^2;$$

whence

$$\frac{dy}{dx} = -\frac{x}{y}.$$

Then

$$ds = \sqrt{1 + \frac{x^2}{y^2}}\,dx = \sqrt{\frac{y^2 + x^2}{y^2}}\,dx = \frac{r}{y}\,dx$$

and

$$dS = 2\,\pi y\,ds = 2\,\pi r\,dx.$$

The limits of integration are from $-r$ to $+r$, or, because of the symmetry, we may integrate from 0 to r and multiply the result by 2.

Hence
$$S = 2 \int_0^r 2\pi r \, dx = 4\pi r [x]_0^r = 4\pi r^2.$$

PROBLEMS

1. Find the area of the surface generated by revolving about the y-axis the arc of the parabola $y = x^2$ from $y = 0$ to $y = 2$.

2. Find the area of the surface generated by revolving about the x-axis the arc of the parabola $y^2 = 2\,px$ from $x = 0$ to $x = 4\,p$.

3. Find the area of the surface generated by revolving about the x-axis the arc of $y = x^3$ from $x = 0$ to $x = 2$.

4. Find the area of the surface generated by revolving about the x-axis the arc of the parabola $y^2 = 4 - x$ which lies in the first quadrant.

5. Find the area of the surface generated by revolving about the x-axis the hypocycloid $x^{\frac{2}{3}} + y^{\frac{2}{3}} = a^{\frac{2}{3}}$.

*6. Find the area of the surface generated by revolving about the x-axis the arc of the parabola $y = x^2$ from $(0, 0)$ to $(2, 4)$.

*7. Find the area of the surface generated by revolving about the y-axis the arc of $y = x^3$ from $(0, 0)$ to $(2, 8)$.

*8. Find the area of the surface generated by revolving about the x-axis the arc of $y^2 = x^3$ from $(0, 0)$ to $(4, 8)$.

9. The slope of the tractrix at any point of the curve in the first quadrant is given by $\dfrac{dy}{dx} = \dfrac{-y}{\sqrt{c^2 - y^2}}$. Show that the surface generated by revolving about the x-axis the arc joining the points (x_1, y_1) and (x_2, y_2) on the tractrix is $2\pi c(y_1 - y_2)$.

10. The area in the first quadrant bounded by the curves whose equations are $y = x^3$ and $y = 4\,x$ is revolved about the x-axis. Find the total surface of the solid generated.

11. The area bounded by the y-axis and the curves whose equations are $x^2 = 4\,y$ and $x - 2\,y + 4 = 0$ is revolved about the y-axis. Find the total surface of the solid generated.

12. A *zone* is cut from the surface of the sphere generated by revolving the circle whose equation is $x^2 + y^2 = r^2$ about the x-axis by planes perpendicular to the x-axis at $x = a$ and $x = a + h$. Show that the area of the zone is $2\pi rh$.

13. The arc of the curve whose equation is $y^2 = x^3$ extending from the origin to the point $(4, 8)$ is revolved about the y-axis. Find the surface generated.

*Use Simpson's rule for approximating the definite integral.

14. Find the surface generated by revolving about the x-axis the arc of the curve whose equation is $y = \dfrac{x^3}{6} + \dfrac{1}{2\,x}$ from $x = 1$ to $x = 3$.

15. Find the area of the surface generated by revolving about the x-axis the loop of the curve whose equation is $9\,y^2 = x(3 - x)^2$.

16. Find the area of the surface generated by revolving about the y-axis the loop of the curve whose equation is $9\,y^2 = x(3 - x)^2$.

HINT. Integrate with respect to x.

101. Centroids of plane areas. If a piece of stiff cardboard is cut in any shape, there is one point on which it can be balanced. This point is its center of gravity, or, in mathematical terms, it is the centroid of the area represented by the cardboard. Experience teaches us that the centroid of a square, of a circle, or of any regular polygon is its geometrical center. In general, if the boundary of any plane area has a center of symmetry, this center will be the centroid of the area. The centroid of a rectangle is its geometrical center (the point of intersection of its diagonals), and it can be proved that the centroid of a triangle is the point of intersection of the medians.

In order to calculate the position of the centroid of a general plane area, we must first define the terms *moment arm* and *moment of area*. The *moment arm* of an area with respect to an axis is the distance from the axis to the centroid of the area. The *moment of an area* with respect to an axis is the product of the area and its moment arm with respect to that axis.

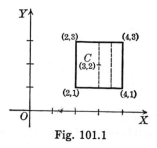

Fig. 101.1

For example, consider the square whose vertices are $(2, 1)$, $(4, 1)$, $(4, 3)$, $(2, 3)$. The area is 4 sq. in. if the unit of length is 1 in. The centroid is at $(3, 2)$. The moment arm with respect to the y-axis is 3 in., and the moment of area with respect to the y-axis is $4 \times 3 = 12$.* The

*Since the moment of area is defined as the product of an area by a length, it is a quantity involving the third power of the unit of length. In engineering applications the unit of length is usually one inch, and the unit of moment of area is written in^3.

moment arm with respect to the x-axis is 2 in., and the moment of area with respect to the x-axis is 8.

Let us now divide the above square into three rectangles by the lines $x = 3$ and $x = 3.5$. The abscissas of the centroids of these rectangles are easily seen to be 2.5, 3.25, and 3.75. Their areas are 2, 1, and 1. Hence the moments with respect to the y-axis are 5, 3.25, and 3.75, and the sum of the moments is 12, which was the moment of the whole square. If the square is divided in any other way, the same result will be obtained. From this example we infer the following principle, which we state without proof.

If a plane area is divided into a finite number of parts, the moment of the area about any axis is the sum of the moments of the different parts about the same axis.

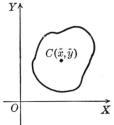

Fig. 101.2

Now let A denote the number of square units in any given area, M_x the moment of this area with respect to the x-axis, M_y the moment of the area with respect to the y-axis, and $(\bar{x}, \bar{y})$ the coördinates of the centroid of the area. Then, from the definition of moment,

$$M_x = A\bar{y}, \qquad M_y = A\bar{x}.$$

From these equations we get immediately

$$\bar{x} = M_y/A, \qquad \bar{y} = M_x/A,$$

from which the coördinates of the centroid may be calculated when the area and its two moments are known.

In order to calculate M_x and M_y for any area, the area is divided into strips, as in Art. 94, and the area of each strip is approximated by means of a rectangle. The moment of each element of area may be written down from the figure. The limiting value of the sum of the moments of all the elements is the moment of the whole area, which will then be represented by a definite integral.

For the simple case of the area under a curve the process would be as follows. To find M_x and M_y for the area bounded by $y = f(x)$, the

x-axis, and the ordinates $x = a$ and $x = b$, the area is divided into strips by lines parallel to the y-axis. A typical strip of width Δx is shown in Fig. 101.3. Let x be the abscissa of the mid-point of this subinterval

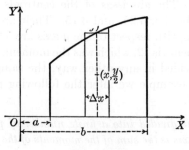

Fig. 101.3

Δx and let y be the ordinate of the corresponding point on the curve. On Δx as a base construct a rectangle of height y. The coördinates of the centroid of this element of area are $(x, y/2)$, and its area is $y\,\Delta x$. Hence the moment of this element with respect to the x-axis is

$$\frac{y}{2}\,(y\,\Delta x) = \frac{1}{2}\,y^2\,\Delta x,$$

and, by the fundamental theorem,

$$M_x = \int_a^b \tfrac{1}{2}\,y^2\,dx.$$

Furthermore, the moment of this area with respect to the y-axis is $xy\,\Delta x$, and

$$M_y = \int_a^b xy\,dx.$$

If not otherwise known, the area is given by

$$A = \int_a^b y\,dx,$$

and the formula on page 179 may be used to calculate the coördinates of the centroid.

In general, if dM_x and dM_y denote the moments of the element dA about the x-axis and the y-axis respectively, and the coördinates of the centroid of dA are (x', y'), then

$$dM_x = y'\,dA$$

and

$$dM_y = x'\,dA.$$

It is obvious that an area having an axis of symmetry will have its centroid on this axis. Hence if an area is symmetrical with respect to the x-axis, $\bar{y} = 0$; and if it is symmetrical with respect to the y-axis, $\bar{x} = 0$.

Example 1. Find the centroid of the area of a quadrant of a circle.

Solution. Let the axes be chosen so that they bound the quadrant of the circle, as shown in Fig. 101.4. Then the equation of the circular arc is

$$x^2 + y^2 = r^2.$$

Let the area be divided into elements parallel to the y-axis, one of which is shown. Then, reasoning as above, we have

$$dM_x = \frac{y}{2}\,dA = \frac{y^2}{2}\,dx = \frac{1}{2}(r^2 - x^2)\,dx$$

and $$dM_y = x\,dA = xy\,dx = x\sqrt{r^2 - x^2}\,dx.$$

Hence $$M_x = \frac{1}{2}\int_0^r (r^2 - x^2)\,dx = \frac{1}{2}\left[r^2 x - \frac{x^3}{3}\right]_0^r = \frac{r^3}{3},$$

$$M_y = \int_0^r x\sqrt{r^2 - x^2}\,dx = -\frac{1}{3}\left[(r^2 - x^2)^{\frac{3}{2}}\right]_0^r = \frac{r^3}{3}.$$

The area of a quadrant of a circle is $\frac{1}{4}\pi r^2$. Hence we have

$$\bar{x} = \frac{r^3}{3} \div \frac{\pi r^2}{4} = \frac{4\,r}{3\,\pi}, \qquad \bar{y} = \frac{r^3}{3} \div \frac{\pi r^2}{4} = \frac{4\,r}{3\,\pi}.$$

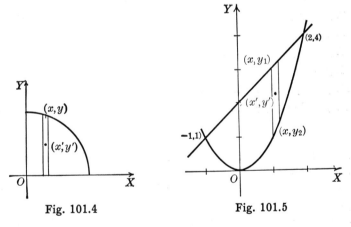

Fig. 101.4   Fig. 101.5

Example 2. Find the centroid of the area bounded by the curves whose equations are $y = x^2$ and $y = x + 2$.

Solution. Inspection of Fig. 101.5 shows that the simplest way to take the elementary strips is parallel to the y-axis. Let x denote the abscissa of the midpoint of the typical subinterval Δx, and y_1 and y_2 denote the ordinates of the corresponding points on the line and the parabola respectively. Then the area of the element is $(y_1 - y_2)\Delta x$, and the centroid of the element will have the coördinates

$$x' = x, \quad y' = \tfrac{1}{2}(y_1 + y_2) = \tfrac{1}{2}(x + 2 + x^2).$$

Also
$$dA = (y_1 - y_2)dx = (x + 2 - x^2)dx.$$

Hence
$$dM_x = y'dA = \frac{x^2 + 4x + 4 - x^4}{2}\,dx;$$

$$dM_y = x\,dA$$
$$= x(x + 2 - x^2)dx$$
$$= (x^2 + 2x - x^3)dx.$$

Therefore

$$A = \int_{-1}^{2} (x + 2 - x^2)dx = \left[\frac{x^2}{2} + 2x - \frac{x^3}{3}\right]_{-1}^{2} = \frac{9}{2};$$

$$M_x = \frac{1}{2}\int_{-1}^{2} (x^2 + 4x + 4 - x^4)dx = \frac{1}{2}\left[\frac{x^3}{3} + 2x^2 + 4x - \frac{x^5}{5}\right]_{-1}^{2} = \frac{36}{5};$$

$$M_y = \int_{-1}^{2} (x^2 + 2x - x^3)dx = \left[\frac{x^3}{3} + x^2 - \frac{x^4}{4}\right]_{-1}^{2} = \frac{9}{4}.$$

Hence the coördinates of the centroid are

$$\bar{x} = \tfrac{9}{4} \div \tfrac{9}{2} = \tfrac{1}{2}, \quad \bar{y} = \tfrac{36}{5} \div \tfrac{9}{2} = \tfrac{8}{5}.$$

PROBLEMS

Find the centroids of the areas bounded by the following curves.

1. $y^2 = x, x = 4.$

2. $y^2 = 2\,px, x = h.$

3. $y^2 = x, y = 2, x = 0.$

4. $y = x^3, x = 2, y = 0.$

5. $y = x^3, y = 8, x = 0.$

6. $y^2 = x^3, x = 4, y = 0.$

7. $y^2 = x^3, y = 8, x = 0.$

8. $y = x^3, y = 4x.$

9. $y = 6x - x^2, y = x.$

10. $x = 4y - y^2, y = x.$

11. $y = 4x - x^2, y = 2x - 3.$

12. $y^2 = 4x, 2x - y = 4.$

13. $y = x^3 - 3x, y = x.$

14. $y = x^2, 2x - y + 3 = 0.$

15. $y = x^2 - 2x - 3,$
$y = 6x - x^2 - 3.$

16. Find the centroid of the area of the triangle whose vertices are $(0, 0)$, $(a, 0)$, and $(0, b)$.

17. Find the centroid of the area bounded by the parabola whose equation is $\sqrt{x} + \sqrt{y} = \sqrt{a}$ and the coördinate axes.

18. Find the centroid of the area bounded by the loop of the curve whose equation is $y^2 = 4x^2 - x^3$.

19. Find the centroid of the portion in the first quadrant of the ellipse whose equation is $\dfrac{x^2}{a^2} + \dfrac{y^2}{b^2} = 1$.

HINT. The area of the ellipse is πab.

102. Centroids of other figures. The definitions and methods of the previous section may be easily extended to the finding of centroids of volumes, surfaces of revolution, and arcs of curves. The chief difference is that the moment of volume with respect to an axis is the product of the *volume* by the moment arm, etc. A thorough consideration of the subject belongs to more advanced courses, but it may be of interest to work a few simple problems.

Suppose, for example, that we wish to find the centroid of the solid generated by revolving about the x-axis the segment of the parabola $y^2 = 2\,px$ cut off by the line $x = h$.

The centroid is on the x-axis. Cutting the solid by planes perpendicular to the x-axis, we have elements of volume

$$dV = \pi y^2\,dx.$$

The moment of each of these with respect to the y-axis is

$$dM_y = x\,dV = \pi x y^2\,dx.$$

Hence the moment of the solid is

$$M_y = \pi \int_0^h xy^2\,dx = \pi \int_0^h 2\,px^2\,dx = \frac{2\,\pi p h^3}{3}.$$

On the other hand, the volume of the solid is

$$V = \pi \int_0^h y^2\,dx = \pi p h^2.$$

Hence
$$\bar{x} = M_y \div V = \frac{2\,h}{3}.$$

PROBLEMS

1. Find the distance of the centroid of a solid hemisphere of radius r from the center.

2. Find the distance of the centroid of the arc of a semicircle of radius r from the center.

3. Find the distance of the centroid of the area of the curved surface of a hemisphere of radius r from the center.

103. Fluid pressure. If a right cylindrical tank filled with water (or other fluid) has a horizontal base, the base is subjected to a force equal to the weight of the water. The force on the base would remain the same as before if we altered the shape of the tank in any way not affecting the area of the base or the depth of the water. The force per unit area is called the *pressure* and in English units this is expressed in lb./sq. ft. or lb./sq. in.

The water also exerts a force against the vertical surface of the tank and the pressure on the surface is different at different depths below the surface of the water. In order to determine the pressure at a given depth, we make use of the physical principle that pressure at a point within a fluid is the same in all directions. Hence the pressure against a vertical surface at a given point is equal to the pressure downward at that point, and the pressure will be the same at all points which are the same distance below the surface. The method of applying these general principles will be illustrated by some examples.

Example 1. One side of a rectangular water tank is 10 ft. long and 4 ft. deep. Find the force of pressure on this side when the tank is full of water.

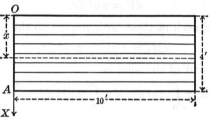

Fig. 103.1

Solution. Let the vertical edge OA of the side of the tank be taken as the x-axis, and let the segment OA be divided into n parts. Horizontal lines through the points of division will divide the side of the tank into elementary strips. Let Δx be the width of any one of these strips and let x be the depth to the center of this strip. Imagine this vertical strip to be turned about its center line until its surface is horizontal. The force of pressure on it will then be equal to the weight of a column of water having this strip as a base and a height equal to the depth. The volume of this column is the product of the area $10\,\Delta x$ by the depth x. If w is the weight of a cubic foot of water, the weight of the column is $w \cdot 10\,x \cdot \Delta x$ and this is the force of pressure on the strip.

Now, since the pressure is the same in all directions and since the strip is very narrow, the pressure is approximately the same on the strip in its vertical position as in its horizontal position. Hence $w \cdot 10\,x \cdot \Delta x$ may be taken as representing approximately the force of pressure on the typical strip, and an approximation to the total force will be

$$\sum w\,10\,x\,\Delta x.$$

As the width of each strip approaches zero as a limit this approximate value approaches the exact value of the force as a limit. Hence, by the fundamental theorem, the force P on the side of the tank is given by

$$P = \lim \sum w\,10\,x\,\Delta x = \int_0^4 w\,10\,x\,dx = 10\,w\left[\frac{x^2}{2}\right]_0^4 = 80\,w.$$

The weight of a cubic foot of water is about 62.5 lb. Using this value for w gives as the final result 5000 lb.

The essential part of the above reasoning is that the force of pressure on an elementary horizontal strip is equal to the product of the area of the strip (dA), the depth of the strip (h), and the weight of a cubic unit of the fluid (w). That is,

$$dP = wh\, dA.$$

Example 2. The vertical end of a water trough is in the form of a right triangle with dimensions as shown in Fig. 103.2. Calculate the force on this end when the trough is full of water.

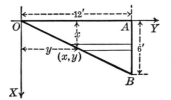

Fig. 103.2

Solution. Imagine that the triangular area is divided into elementary strips, one of which is shown in the figure, and that coördinate axes are introduced as shown. The equation of the line OB, which joins the origin to the point B (6, 12), is $y = 2\,x$. With (x, y) representing the coördinates of a point on OB, it may be seen from the figure that

1. The width of the elementary strip is dx and its length is $12 - y$. Hence $dA = (12 - y)dx$.

2. The depth of the elementary strip is x.

3. The limits of integration are from $x = 0$ to $x = 6$. Hence, reasoning as in Example 1, the element of force dP is

$$dP = wx(12 - y)dx = 62.5\, x(12 - 2\,x)dx.$$

Hence the total force is

$$P = 62.5 \int_0^6 x(12 - 2\,x)dx = 62.5[6\,x^2 - \tfrac{2}{3}\,x^3]_0^6 = 4500 \text{ lb.}$$

NOTE. The purpose of the particular choice of axes in the above example was to avoid negative coördinates. This is usually the best way to draw the axes in pressure problems if the origin is taken at the top of the surface; if the origin is taken at the bottom (as in the next example), the axes may be conveniently drawn in the usual way.

Example 3. A gate in a dam is in the form of an isosceles trapezoid, as shown in Fig. 103.3. Calculate the force of pressure on the gate when the surface of the water is 8 ft. above the top of the gate.

Solution. Choosing rectangular axes as shown, we see from Fig. 103.3 that

1. $$dA = 2\,x\,dy.$$

2. The depth of the elementary strip below the surface is $12 - y$.

3. The limits of integration are $y = 0$ to $y = 4$.

4. The equation of AB is $y = 2\,x - 8$. Hence

$$dP = 62.5(12 - y)2\,x\,dy$$
$$= 62.5(y + 8)(12 - y)dy,$$

and the total force is

$$P = 62.5 \int_0^4 (y + 8)(12 - y)dy$$

$$= 62.5 \int_0^4 (96 + 4\,y - y^2)dy$$

$$= 62.5 \left[96\,y + 2\,y^2 - \frac{y^3}{3} \right]_0^4 = 24{,}667 \text{ lb.}$$

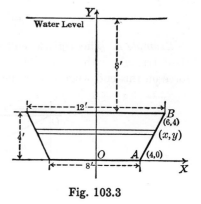

Fig. 103.3

PROBLEMS

1. A horizontal cylindrical tank of diameter 8 ft. is half full of oil weighing 60 lb./cu. ft. Calculate the force on one end.

2. The vertical end of a water trough is an isosceles triangle with dimensions as shown in Fig. 103.4. Calculate the force on the end when the trough is full of water.

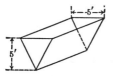

Fig. 103.4

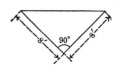

Fig. 103.5

3. The vertical end of a water trough is an isosceles right triangle of which each leg is 8 ft. (Fig. 103.5). Calculate the force on the end when the trough is full of water.

4. The vertical end of a trough is an isosceles trapezoid with dimensions as shown in Fig. 103.6. Calculate the force on the end when the trough is filled with a liquid weighing 50 lb./cu. ft.

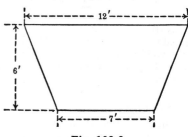

Fig. 103.6

5. The vertical end of a tank is an isosceles trapezoid with dimensions as shown in Fig. 103.7. Calculate the force on the end when the tank is filled with a liquid weighing 50 lb./cu. ft.

Fig. 103.7

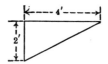

Fig. 103.8

6. A gate in a dam is in the form of a right triangle with dimensions as shown in Fig. 103.8. Calculate the force on the gate when the surface of the water is 6 ft. above the top of the gate.

7. A gate in a dam is in the form of a right triangle as shown in Fig. 103.9. Calculate the force on the gate when the surface of the water is 4 ft. above the top of the gate.

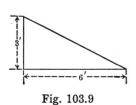

Fig. 103.9

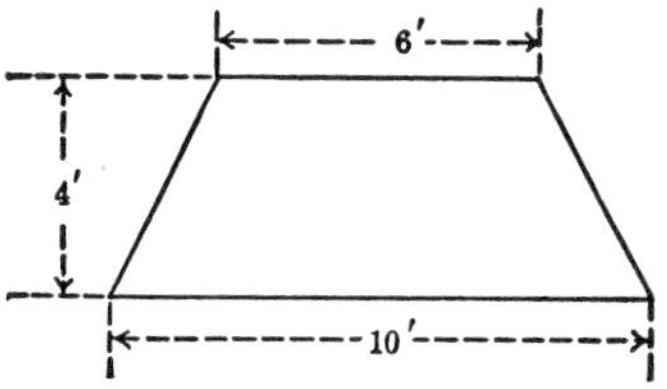

Fig. 103.10

8. A gate in a dam is in the form of an isosceles trapezoid as shown in Fig. 103.10. Calculate the force on the gate when the surface of the water is 8 ft. above the top of the gate.

9. A rectangular gate in a vertical dam is 10 ft. wide and 6 ft. deep. Find (**a**) the force when the level of the water is 8 ft. above the top of the gate; (**b**) how much higher the water must rise to double the pressure found in (**a**).

10. Show that the force on any vertical surface is the product of the weight of a cubic unit of the liquid, the area of the surface, and the depth of the centroid.

11. A vertical cylindrical tank, of diameter 30 ft. and height 50 ft., is full of water. Find the force on the curved surface.

12. A horizontal cylindrical tank is 10 ft. long and has a diameter of 6 ft. It is half full of a liquid weighing 60 lb./cu. ft. Find (**a**) the weight of the liquid; (**b**) the force of pressure on the curved surface.

104. Work. *Work* is a technical term used in mechanics. The work done in overcoming a resistance is defined as the product of the distance through which a body is moved and of the force opposed to the motion. The definition is applicable if the force is constant and all parts of the body move the same distance. If either of these conditions is not fulfilled, it is necessary to frame a new definition and to use the calculus, as in the case of other magnitudes. In the applications of this section the only force considered is the force of gravity, and the work considered is that done in lifting a body. In this case the work done is the product of the weight of the body by the vertical distance through which it is lifted. The usual unit of work is the foot-pound (ft.-lb.). Thus, if a weight of 100 lb. is lifted upward 2 ft., the amount of work done is 200 ft.-lb.

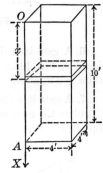

Fig. 104.1

The use of the calculus in calculating work is illustrated by the following simple example. Suppose that a rectangular tank 10 ft. deep, with a base 4 ft. square, is filled with water. How much work must be done to pump the water to the top of the tank? The definition given above cannot be used immediately, because the water near the bottom of the tank must be lifted much farther than the water near the top. To answer the question, we must apply the Fundamental Theorem of the Integral Calculus.

Let the edge OA of the tank be taken as the x-axis and let the segment OA be divided into subintervals each of length less than some positive number p. Imagine horizontal planes passed through each point of division so that consecutive planes divide the water into horizontal layers. Let us fix our attention on a typical layer of thickness Δx and let x be the depth to some point in this layer. Assuming that Δx is small, each drop of water in the layer must be lifted a distance nearly equal to x, and hence an approximate value of the work done in lifting this layer to the top will be x times the weight of the water in the layer. If w denotes the weight of a cubic foot of water, the weight of the layer is $w\,16\,\Delta x$, and an approximate value of the work done is $w\,16\,x\,\Delta x$.

An approximation to the total work done will be

$$\sum w\,16\,x\,\Delta x,$$

the summation extending over all the layers from $x = 0$ to $x = 10$.

As the thickness of each layer approaches zero as a limit this approximate value approaches the exact value of the work as a limit. Hence, by the fundamental theorem, the work W is given by

$$W = \lim \sum w\,16\,x\,\Delta x = \int_0^{10} 16\,wx\,dx$$

$$= 16(62.5)\left[\frac{x^2}{2}\right]_0^{10} = 50{,}000 \text{ ft.-lb.}$$

The essential principle of the reasoning in this example is that the element of work (dW) done in lifting an elementary layer of water (dV) is equal to the weight of the layer multiplied by the height (h) it is lifted.

That is, $$dW = hw\, dV.$$

Example. A conical cistern is 20 ft. across the top and 15 ft. deep. If the surface of the water is 5 ft. from the top, calculate the work necessary to pump the water to the top of the cistern.

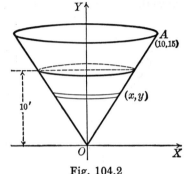

Solution. Let a system of rectangular axes be introduced with the origin at the vertex of the cone and with the x-axis horizontal. The element of the cone lying in the xy-plane joins $(0, 0)$ and $(10, 15)$; hence its equation is $2\,y = 3\,x$. An elementary horizontal layer of water is shown in Fig. 104.2; as usual this is regarded as a cylinder. If (x, y) is a point on OA, the following facts appear from the figure.

Fig. 104.2

1. The thickness of the layer of water is dy.
2. The radius of the layer of water is x.
3. The height through which this layer must be raised is $15 - y$.
4. The limits of integration are from $y = 0$ to $y = 10$.

The volume of the elementary layer is
$$dV = \pi x^2\, dy = \tfrac{4}{9}\,\pi y^2\, dy.$$
Hence
$$dW = (15 - y)w\, dV$$
$$= (15 - y)(62.5)(\tfrac{4}{9}\,\pi y^2\, dy)$$
and
$$W = \frac{4\,\pi}{9}\,(62.5)\int_0^{10}(15\,y^2 - y^3)dy$$
$$= \frac{4\,\pi}{9}\,(62.5)\left[5\,y^3 - \frac{y^4}{4}\right]_0^{10} = \frac{625000\,\pi}{9}$$
$$= 218{,}167 \text{ ft.-lb.}$$

PROBLEMS

1. A vertical cylindrical cistern of diameter 16 ft. and depth 20 ft. is full of water. Calculate the work necessary to pump the water to the top of the cistern.

2. A vertical cylindrical cistern of diameter 10 ft. and depth 12 ft. is half full of water. Calculate the work necessary to pump the water to the top of the cistern.

3. A vertical cylindrical cistern of diameter 12 ft. and depth 12 ft. is full of water. Calculate the work necessary to pump the water to a height of 10 ft. above the top of the cistern.

4. A conical cistern 20 ft. across the top and 20 ft. deep is full of water. Calculate the work necessary to pump the water to a height of 15 ft. above the top of the cistern.

5. A hemispherical tank of diameter 10 ft. is full of oil weighing 60 lb./cu. ft. Calculate the work necessary to pump the oil to the top of the tank.

6. A vat is built in the shape of a regular pyramid 12 ft. square at the top and 10 ft. deep. If the vat is full of a liquid weighing 65 lb./cu. ft., calculate the work necessary to pump the liquid to the top of the vat. (Fig. 104.3)

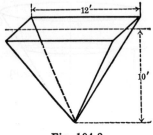
Fig. 104.3

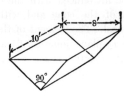
Fig. 104.4

7. Calculate the work necessary to pump the liquid in the vat of Problem 6 to a height of 8 ft. above the top of the vat.

8. The ends of a horizontal trough are vertical isosceles right triangles. The width across the top is 8 ft., and the length is 10 ft. If the trough is full of water, calculate the work necessary to pump the water to the top of the trough. (Fig. 104.4)

9. Each end of a trough is a vertical right triangle. The dimensions are shown in Fig. 104.5. If the trough is full of water, calculate the work necessary to pump the water to the top of the trough.

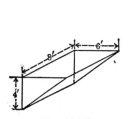
Fig. 104.5

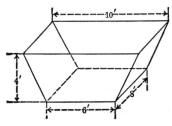
Fig. 104.6

10. Each end of a trough is a vertical isosceles trapezoid. The dimensions are shown in Fig. 104.6. If the trough is full of water, calculate the work necessary to pump the water to the top of the trough.

11. A bucket of weight M is to be lifted from the bottom of a shaft h feet deep. The weight of the rope used is m pounds per foot. Find the work done.

105. Mean value of a function. The arithmetic mean (or average value or mean value) of n numbers $y_1, y_2, \cdots, y_n$ is

$$MV = \frac{y_1 + y_2 + \cdots + y_n}{n}. \tag{1}$$

We proceed to establish the formula for the mean value of a function.

Let $y = f(x)$ be continuous in the interval (a, b). The graph is shown in Fig. 105. The mean value of the ordinates of the arc DC with respect

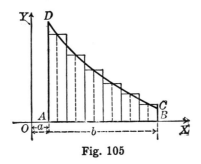

Fig. 105

to equal divisions of the segment AB is to be defined. Let the interval (a, b) be divided into n subintervals *each* equal to Δx and let x_1 be the abscissa of the mid-point of the first subinterval, x_2 be the mid-point of the second subinterval, etc. Let $y_1, y_2, \cdots, y_n$ be the corresponding ordinates. The mean value of these ordinates is given by (1). Multiply numerator and denominator by Δx and note that $n\,\Delta x = b - a$. Equation (1) becomes

$$MV = \frac{\sum_{i=1}^{n} y_i\,\Delta x}{b-a} = \frac{\sum_{i=1}^{n} f(x_i)\,\Delta x}{b-a}. \tag{2}$$

The mean value of the function over the interval (a, b) is defined as the limit of the mean value (2) as $n \longrightarrow \infty$, that is, as $\Delta x \longrightarrow 0$. Denoting the mean value of the function by M and applying the fundamental theorem, (2) gives the formula

$$M = \frac{\int_a^b f(x)\,dx}{b-a}.$$

If $f(x)$ is nowhere negative in (a, b), the numerator gives the area under the graph of $y = f(x)$ from a to b and the denominator is the length of the base AB. Hence the mean value is the altitude of a rectangle with the base AB and area equal to the area under the graph.

Example. If a body falls from rest and air resistance is neglected, the velocity after t sec. is given by

$$v = gt. \tag{3}$$

The velocity after falling s ft. is given by

$$v = \sqrt{2\,gs}. \tag{4}$$

Taking $g = 32$, the body will fall 64 ft. in 2 sec. Find the average value of the velocity (a) with respect to the time; (b) with respect to the distance.

Solution. a. Using (3), the formula gives

$$M(t) = \frac{\int_0^2 32\,t\,dt}{2 - 0} = \frac{64}{2} = 32 \text{ ft./sec.}$$

b. Using (4), the formula gives

$$M(s) = \frac{\int_0^{64} 8\sqrt{s}\,ds}{64 - 0} = \frac{\left[\frac{16}{3}s^{\frac{3}{2}}\right]_0^{64}}{64} = 42\tfrac{2}{3} \text{ ft./sec.}$$

Thus the mean value of the function depends on the independent variable with respect to which the function is averaged.

PROBLEMS

1. Show that the average value of the slope of the graph of $y = f(x)$ from $x = a$ to $x = b$ is equal to the slope of the chord joining the end points of the arc.

2. If a particle falls from rest in a vacuum for t sec., show that the average value of the velocity during this time is one half the velocity at the end of the fall.

3. If a particle falls from rest in a vacuum through a distance h, show that the average value of the velocity over this distance is two thirds the velocity at the end of the fall.

4. Find the average value of the ordinates of $y^2 = 4\,x$ from $(0, 0)$ to $(4, 4)$ taken uniformly along the x-axis.

5. Find the average value of the abscissas of $y^2 = 4\,x$ from $(0, 0)$ to $(4, 4)$ when uniformly distributed along the y-axis.

6. If air resistance is neglected and a body is thrown downward with an initial velocity of v_0 ft./sec., the velocity after t sec. is given by

$$v = v_0 + gt.$$

The velocity after dropping s ft. is given by

$$v = \sqrt{v_0^2 + 2\,qs}.$$

If $g = 32$ and $v_0 = 60$, the body will drop 100 ft. in $1\frac{1}{4}$ sec. Find the average value of the velocity (a) with respect to the time; (b) with respect to the distance.

7. Given the circle whose equation is $x^2 + y^2 = r^2$. Find the mean value of the ordinates in the first quadrant when y is expressed as a function of the abscissa x.

8. Rectangles with sides parallel to the coördinate axes are inscribed in the circle $x^2 + y^2 = a^2$. Find the average value of their areas if the vertical sides are equally spaced along the x-axis.

9. A point is taken at random on a straight line of length a. Prove (a) that the mean area of the rectangle whose sides are the two segments is $a^2/6$; (b) that the mean value of the sum of the squares on the two segments is $2\,a^2/3$.

10. A quantity of steam expands so that it follows the law $pv^{0.8} = 1000$, where p is measured in lb./sq. in. Find the mean pressure as v increases from 2 to 5 cu. in.

CHAPTER IX

EXTREME VALUES

106. Introduction. The work of the first four chapters was concerned mainly with problems of geometry. The variables were interpreted as the rectangular coördinates of a point in a plane and a functional relation between the variables as a curve. The introduction of derivatives has made it possible to solve problems which are not solvable by the methods of elementary geometry (Art. 65). We now turn to the study of certain problems which furnish a wider scope for the employment of the methods developed in Chapters I–IV, but which are not primarily geometrical.

107. Functions in general. Many of the problems of mathematics involve the study of quantities of varying magnitude. We shall confine ourselves for the present to the case where there are two variables of which one is a function of the other.

In Chapter III the term *function* was defined as follows.

If one of two variables has one or more definite values corresponding to each one of a set of values assigned to the other variable, the first variable is said to be a function of the second.

We there used the notation "$y = f(x)$" to represent an equation connecting x and y; and when x and y had concrete meanings, they were coördinates of a point tracing the graph of the equation $y = f(x)$.

It is clear, however, that the definition is capable of much wider applications. The variables may represent any quantities, and we may know that one variable is a function of another (in the sense of the definition) without thinking of the equation connecting them, or, indeed, without even knowing any such equation. For example, the velocity of a falling body is a function of the distance through which it has fallen; the premium paid on a \$1000 life-insurance policy is a function of the age of the insured; the normal rate of gasoline consumption of an automobile is a function of the speed.

In examples like those just stated the functional law is sometimes given merely by a table of values of the two variables. For example, it is found by experiment that the temperature $\theta°$ of a vessel of cooling water t minutes after the beginning of the observation is as given in the adjoining

t	θ
0	92.0
1	85.3
2	79.5
3	74.5
5	67.0
7	60.5
10	53.5
15	45.0
20	39.5

table. Evidently θ is a function of t, and decreases when t increases. The limitations of this method of stating a functional relation are equally evident: for values of t not given in the table we can only approximate values of θ by interpolation if t is less than 20, and we cannot do even this if t is greater than 20.

108. Derivation of equations of functions. For mathematical purposes it is desirable, when possible, to express the functional relation by an equation or formula connecting the variables. From such an equation we can deduce many facts concerning the behavior of the function. To find the equation connecting the variables, we must know the way in which they are related and translate this relation into algebraic language.

Example 1. Experiments in an artesian well showed that the temperature increased with the depth at the rate of $4°$ C per hundred meters. If the average temperature at the surface was $12°$ C, express the temperature below the surface as a function of the depth.

Solution. Let T be the temperature in degrees and d the depth in meters. Since T increased at the rate of $4°$ per hundred meters, it increased at the rate of $0.04°$ per meter. Hence at a depth of d meters T had increased $0.04\,d$ degrees. But at the surface, $T = 12$. Thus we have the formula

$$T = 12 + 0.04\,d.$$

Example 2. The parcel-post regulations prescribe that the sum of the length and the girth of a package must not exceed 100 inches. Rectangular boxes with two square sides are to be constructed so that they come just within the rule. Express the volume of such a box as a function of the edge of the square side.

Solution. Let V cu. in. be the volume of the box and let the edges be x in., x in., and y in. By geometry,

$$V = x^2y. \tag{1}$$

CASE I. If $y > x$, the length of the box is y in. and the girth is $4\,x$ in. The regulations prescribe that $y + 4\,x = 100$, or

$$y = 100 - 4\,x. \tag{2}$$

Substituting in (1), the result is

$$V = 100\,x^2 - 4\,x^3. \tag{3}$$

CASE II. If $y < x$, the length of the box is x in. and the girth is $(2\,x + 2\,y)$ in. Hence, by the regulations, $3\,x + 2\,y = 100$, or

$$y = 50 - 1.5\,x. \tag{4}$$

Substituting in (1), the result is

$$V = 50\,x^2 - 1.5\,x^3. \tag{5}$$

When $y = x$, the box is a cube and either (2) or (4) gives $x = 20$. Hence the volume is given by (3) if $x \leqq 20$ and by (5) if $x \geqq 20$.

These equations enable us to calculate a value for V for any value of x. However, the results have no physical meaning unless the values of x and V are both positive. From (3) $V = 0$ when $x = 0$ and from (5) $V = 0$ when $x = \frac{100}{3}$ and $V < 0$ when $x > \frac{100}{3}$. Hence the domain of definition of the function V is $0 < x < \frac{100}{3}$ and the function is given by

$$V = 100\, x^2 - 4\, x^3 \quad \text{for} \quad 0 < x \leqq 20;$$
$$V = 50\, x^2 - 1.5\, x^3 \quad \text{for} \quad 20 \leqq x < \frac{100}{3}.$$

109. Graphs of functions. Some of the properties of a function can be studied by means of a graph. It is customary to take the horizontal axis as the axis of the independent variable and the vertical axis as that of the dependent variable. If the values of one variable are much larger than those of the other, it may be necessary to use different scales on the coördinate axes in order to construct a graph of reasonable size. In every case the scales used must be indicated on the axes.

A functional relation may be given in three ways: (*a*) by an equation, (*b*) by a table of values, (*c*) by a graph.

a. If a functional relation is given by an equation, a table of values can be calculated and the graph can be drawn by the methods explained in Chapter II.

b. If a functional relation is given by a table of values, the corresponding points can be plotted. If the function is not defined for values of the independent variable other than those given in the table, nothing more can be done. If it is known, or is assumed, that the function is defined and is continuous for all values of the independent variable in the interval considered, the points may be joined by a smooth curve or by segments of straight lines and corresponding values of the variables, other than those occurring in the table, can be estimated from the graph.

c. Functional relations are sometimes shown by a graph which has been constructed by a machine. The simplest example is that of a recording thermometer. From such a graph certain properties of the function can be observed. For example, let $y = f(x)$ be a continuous function defined in the interval (a, b) and let Fig. 109.1 be its graph.

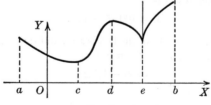

Fig. 109.1

We see from the figure that when $x = a$ the value of $f(x)$ is larger than for other values of x near a. As x increases from a to c, $f(x)$ steadily decreases, the slope of the graph is negative, and hence $f'(x) < 0$. When $x = c$, $f(x)$ is smaller than for values of x near c, and $f'(c) = 0$. As x increases from c to d, $f(x)$ steadily increases and $f'(x) > 0$. When $x = d$, $f(x)$ is larger than for values of x near d, and $f'(d) = 0$. When $x = e$, $f(x)$ is smaller than for values of x near e and $f'(x)$ becomes infinite as $x \longrightarrow e$.

When a function is defined by an equation, it may have properties similar, in some respects, to those described above. It is the purpose of this chapter to study such properties.

Example. Draw the graph of the function obtained in Example 2, Art. 108. Estimate from the graph the values of x for which $V = 5000$.

Solution. The function is defined as follows.

$$V = f(x) = 100\, x^2 - 4\, x^3 \quad \text{for} \quad 0 < x \leq 20, \tag{1}$$

$$V = g(x) = 50\, x^2 - 1.5\, x^3 \quad \text{for} \quad 20 \leq x < \tfrac{100}{3}. \tag{2}$$

From these equations the following table of values is calculated and the graph is plotted. The large values of V make it necessary to use different scales for x and V in order to get the curve on the paper. The first and last entries in the table do not belong to the domain of definition and obviously have no meaning in the physical problem. When $V = 5000$, the graph shows that $x = 9$ or $x = 29$, approximately.

x	V	x	V
0	0	18	9072
2	368	20	8000
4	1344	22	8228
6	2736	24	8064
8	4352	26	7436
10	6000	28	6272
12	7488	30	4500
14	8624	32	2048
16	9216	$33\tfrac{1}{3}$	0

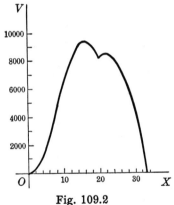

Fig. 109.2

Inspection of the table in the above solution reveals several things. As x increases, V increases until $x = 16$, then decreases until $x = 20$, then increases until $x = 22$, then decreases to the end of the interval. The graph suggests that there is some value of x near 16, say $x = x_1$, for which V is larger than for other values of x near x_1; that there is some value of x near 20, say $x = x_2$, for which V is smaller than for other values of x near x_2; and that there is some value of x near 22, say $x = x_3$, for which V is larger than for other values of x near x_3. It will be shown later that $x_1 = 16\frac{2}{3}$ and, by similar methods, $x_2 = 20$, and $x_3 = 22\frac{2}{9}$.

From (1) and (2) it is apparent that V is differentiable for every value of x for which V is defined except, perhaps, $x = 20$. To investigate the situation at $x = 20$, we proceed according to the definition (Art. 44) and set $x = 20 + \Delta x$. After performing the necessary calculations, we find that, for $\Delta x < 0$,

$$\frac{\Delta V}{\Delta x} = \frac{\Delta f}{\Delta x} = -800 - 140 \, \Delta x - 4(\Delta x)^2, \tag{3}$$

and that for $\Delta x > 0$,

$$\frac{\Delta V}{\Delta x} = \frac{\Delta g}{\Delta x} = 200 - 4 \, \Delta x - \tfrac{3}{2}(\Delta x)^2. \tag{4}$$

Equation (3) gives

$$\lim_{\Delta x \to 0} \frac{\Delta V}{\Delta x} = -800,$$

and (4) gives

$$\lim_{\Delta x \to 0} \frac{\Delta V}{\Delta x} = 200.$$

In this case we say that, at $x = 20$, V has a *left-hand* derivative, the value of which is -800, and a *right-hand* derivative, the value of which is 200. The graph of the function is said to have a "corner point" at $x = 20$ and the derivative is said to have a "finite discontinuity."

In general, we say that dV/dx exists for $x = x_1$ if and only if both the left-hand and right-hand derivatives exist and are equal. An exception to this statement occurs at the end points of the domain of definition. Referring to the function represented by Fig. 109.1, we see that a right-hand derivative only is possible for $x = a$ and a left-hand derivative only is possible for $x = b$. In such cases we frequently use the notation $f'(a)$ to mean the value of the right-hand derivative for $x = a$ and $f'(b)$ to mean the value of the left-hand derivative for $x = b$.

It is not necessary to use the Δ-process to investigate the derivative at a point where the function is continuous and where the derivative may have a finite discontinuity. Consider the function defined as follows.

$$y = f(x) \quad \text{for} \quad a \leqq x \leqq c,$$
$$y = g(x) \quad \text{for} \quad c \leqq x \leqq b,$$

where it is supposed that $f(c) = g(c)$ and that $f(x)$ is differentiable for $a \leq x \leq c$ and that $g(x)$ is differentiable for $c \leq x \leq b$. Then

$$\frac{dy}{dx} = f'(x) \quad \text{for} \quad a \leq x < c,$$

$$\frac{dy}{dx} = g'(x) \quad \text{for} \quad c < x \leq b.$$

Then the left-hand derivative of y for $x = c$ is $f'(c)$ and the right-hand derivative is $g'(c)$. If $f'(c) = g'(c)$, then y is differentiable for $x = c$; otherwise dy/dx does not exist for $x = c$. In geometric language, if at the point where $x = c$ the tangent to the curve $y = f(x)$ coincides with the tangent to the curve $y = g(x)$, the graph of the function is "smooth" and has a tangent. If, at the point where $x = c$, the tangents to the two curves are distinct, the graph of the function has a corner point.

PROBLEMS

1. A body is projected vertically up from the ground with an initial velocity of 100 ft. per sec. and falls back to the ground where it stops. Its distance from the ground after t sec. is s ft., where $s = 100\,t - 16\,t^2$. What is the domain of definition of the function s? Plot the graph of this function and estimate from the graph when the body reaches its greatest height.

2. A rectangle is inscribed in a circle of radius 10. Express the area of the rectangle as a function of the length of one side. Plot the graph of the function and estimate the greatest value of the area.

3. In a circle of radius 10 express the length of a chord as a function of its distance from the center. Tabulate the various lengths of the chord for integral values of its distance from the center.

4. The cost of setting the type for a pamphlet is $500, and the charge for paper and printing is 50¢ per copy. Express the cost per copy (C) as a function of the number (n) of copies printed.

5. A closed cylindrical can is to hold 58 cu. in. Express the amount of material (M) required to make it as a function of the radius (r). Calculate this amount for radii 1.5 in., 2 in., and 2.5 in.

6. A rectangle with sides parallel to the axes is inscribed in the curve whose equation is $4\,x^2 + 9\,y^2 = 36$. Express the area of this rectangle as a function of x.

NOTE. If $y = f(x)$, we say that y varies *with* x. The expression y *varies as* x means that y is proportional to x, that is, $y = kx$, where k is a constant. If $y = k/x$, we say that y varies *inversely* as x.

7. Above the surface of the earth the weight of a body varies inversely as the square of its distance from the center of the earth. If an object weighs 100 lb. at the surface of the earth, express its weight as a function of its distance from the center of the earth, and find its weight when it is 100 mi. above the surface of the earth. (Take the radius of the earth as 4000 mi.)

8. The time required for one swing of a pendulum varies as the square root of its length. If the time required for one swing of a pendulum whose length is 3.25 ft. is 1 sec., express the time as a function of the length. Also find the length of a pendulum which makes one swing in 0.5 sec.

9. A rectangle has an area of 320 sq. rd. Express the perimeter as a function of the length of one side and plot the graph. From the graph estimate the smallest value of the perimeter.

10. The radius of a circle is 6, and a tangent is drawn to it from a point x units distant from the center of the circle. Express the length of the tangent as a function of x and plot this function. From the graph estimate the values of x for which the length of the tangent is 12; 18.

11. It is desired to construct an open-topped rectangular box, with a square base, which will hold just 6 cu. ft. Express the amount of material required as a function of the side of the base. Plot the graph of this function and estimate from the graph the dimensions which will require the least material.

12. The dimensions of a rectangular block are 2 in., 4 in., and 8 in. If each dimension is decreased by x inches, what is the volume remaining? Plot the graph of this function and determine from the graph what value of x will make the remaining volume just one half the given volume.

13. A box is to be made from a piece of cardboard 10 in. by 15 in. by cutting a square of side x from each corner and turning up the sides. Express the volume in terms of x. Plot the graph of this function and from the graph find the values of x which correspond to $V = 60, 70, 80, 90, 100, 110, 120,$ and 130 cu. in.

14. Express the area of a rectangle inscribed in a semicircle of radius 8 as a function of the base. Plot the graph of this function and find from the graph for what value of the base the area is greatest.

15. Express the perimeter of the rectangle of Problem 14 as a function of the base and determine from the graph of the function the value of the base which makes the perimeter a maximum.

16. A function is defined for the interval $(0, 2)$ as follows.

$$f(x) = x^2 \qquad \text{for} \quad 0 \leqq x \leqq 1,$$
$$= 2x - 1 \quad \text{for} \quad 1 \leqq x \leqq 2.$$

Draw the graph and show that $f'(1) = 2$. Does $f''(x)$ exist for $x = 1$?

17. A function is defined for the interval $(0, 2)$ as follows.

$$f(x) = x^3 - x \quad \text{for} \quad 0 \leqq x \leqq 1,$$
$$= x - 1 \qquad \text{for} \quad 1 \leqq x \leqq 2.$$

Draw the graph. Is $f(x)$ differentiable for $x = 1$?

18. Given $f(x) = x - |x|$. Draw the graph and show that $f'(0)$ does not exist.

19. Given $f(x) = x^2 - |x|$. Draw the graph and show that $f'(0)$ does not exist.

20. Given $f(x) = \frac{1}{4}(x - |x|)^2$. Draw the graph and show that $f'(0) = 0$.

21. Given $f(x) = \dfrac{x}{1 + |x|}$. Draw the graph and show that $f'(0) = 1$.

110. Functions increasing or decreasing at a point. Let $f(x)$ be defined for all values of x such that $a - h \leqq x \leqq a + h$, where h is some conveniently chosen positive constant.

DEFINITION 1. *A function $f(x)$ is increasing at the point $x = a$ if, for all values of x in the interval $(a - h, a + h)$, the following statements are true. When $x < a$, $f(x) < f(a)$ and when $x > a$, $f(x) > f(a)$.*

DEFINITION 2. *A function $f(x)$ is decreasing at the point $x = a$ if, for all values of x in the interval $(a - h, a + h)$, the following statements are true. When $x < a$, $f(x) > f(a)$, and when $x > a$, $f(x) < f(a)$.*

NOTE. If a is a left-hand (right-hand) end point of the domain of definition of $f(x)$, we consider only values of x such that $a \leqq x \leqq a + h$ ($a - h \leqq x \leqq a$).

It may happen that the requirements of neither definition are satisfied and in that case the function is neither increasing nor decreasing when $x = a$.

Figures 110.1 and 110.3 illustrate functions which are increasing when $x = a$ and Fig. 110.2 illustrates a function which is decreasing when $x = a$.

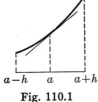 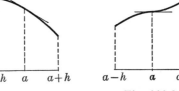

$a - h$ $\quad a$ $\quad a + h$	$a - h$ $\quad a$ $\quad a + h$	$a - h$ $\quad a$ $\quad a + h$
Fig. 110.1	**Fig. 110.2**	**Fig. 110.3**

The following theorems are usually more convenient to use than the definitions. The truth of the theorems is apparent from Figures 110.1 and 110.2. Formal proofs are omitted.

Theorem I. *If $f'(a)$ exists and is positive, $f(x)$ is increasing when $x = a$.*

Theorem II. *If $f'(a)$ exists and is negative, $f(x)$ is decreasing when $x = a$.*

Example 1. Determine whether the function $f(x) = x(x - 2)^3$ is increasing or decreasing when $x = 1$ and when $x = 2$.

Solution. Differentiating, we obtain

$$f'(x) = 3\,x(x - 2)^2 + (x - 2)^3 = (x - 2)^2(4\,x - 2).$$

Then $f'(1) = +2$ and, by Theorem I, $f(x)$ is increasing when $x = 1$.

Since $f'(2) = 0$, neither theorem applies and we must have recourse to the definition. Now $f(2) = 0$ and for all values of x such that $0 < x < 2$, $f(x) < 0$. For all values of x such that $x > 2$, $f(x) > 0$. Hence, by Definition 1, $f(x)$ is increasing when $x = 2$. Cf. Fig. 110.3.

Example 2. For what values of x is the function $f(x) = 3\,x^2 - 2\,x^3$ increasing and for what values is it decreasing?

Solution. Differentiating, we obtain

$$f'(x) = 6\,x - 6\,x^2 = 6\,x(1 - x).$$

The derivative is written in factored form so that its sign may be determined by considering the sign of each factor. The derivative is zero when $x = 0$ and when $x = 1$. We now test in succession values less than 0, between 0 and 1, and greater than 1.

When $x < 0$, $1 - x$ is positive and $f'(x) < 0$. Hence $f(x)$ is decreasing.

When $0 < x < 1$, $1 - x$ is positive and $f'(x) > 0$. Hence $f(x)$ is increasing.

When $x > 1$, $1 - x$ is negative and $f'(x) < 0$. Hence $f(x)$ is decreasing. See Fig. 110.4.

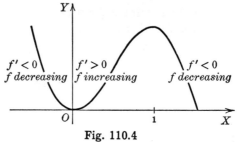

Fig. 110.4

Example 3. For what values of x is the function $f(x) = (x - 2)^2\sqrt{x}$ increasing, and for what values is it decreasing?

Solution. The solution, which is similar to that in Example 2, may be condensed as follows.

$$f'(x) = \frac{(x - 2)^2}{2\sqrt{x}} + 2(x - 2)\sqrt{x}$$

$$= \frac{(x - 2)(5\,x - 2)}{2\sqrt{x}}.$$

The function is defined for $x \geqq 0$, but it is not differentiable when $x = 0$. $f'(x) = 0$ when $x = \frac{2}{5}$ and when $x = 2$. Testing the sign of $f'(x)$, we have the following results.

When $0 < x < \frac{2}{5}$, $\qquad f'(x) = \frac{(-)(-)}{+} = +.$

When $\frac{2}{5} < x < 2$, $\qquad f'(x) = \frac{(-)(+)}{+} = -.$

When $x > 2$, $\qquad f'(x) = \frac{(+)(+)}{+} = +.$

Hence the function is increasing for all values of x such that $0 < x < \frac{2}{5}$ and for all values such that $x > 2$. The function is decreasing for all values such that $\frac{2}{5} < x < 2$.

PROBLEMS

1. Is the given function increasing or decreasing for the given values of x?

a. $y = \dfrac{x}{1 + x^2}$, $x = 0, 2$.

b. $y = x\sqrt{x^2 + 1}$, $x = 0, 2$.

c. $y = x^2 + \dfrac{1}{x}$, $x = -1, +1$.

d. $y = \sqrt[3]{x^2 + 2}$, $x = -1, 0, +1$.

2. For what values of x are the following functions increasing? In each case illustrate your answer by a sketch of the graph.

a. $y = x^3 + 3x$.

b. $y = x - x^2$.

c. $y = x + \dfrac{4}{x}$.

d. $y = 2x^3 - 3x^2 - 12x$.

3. Is the function $y = x^5 + 1$ increasing or decreasing when $x = 0$? Explain your answer.

4. In each of the following equations determine the values of the independent variable for which the function is increasing and those for which it is decreasing.

a. $s = 100t - 16t^2$.

b. $s = \sqrt{30(l - 80)}$.

c. $V = 6r^2 - r^3$.

d. $A = x^2 + \dfrac{16}{x}$.

e. $F = 4.21 + 0.043w$.

f. $V = 4x^3 - 50x^2 + 150x$.

5. From the following equations find the values of x for which y is increasing.

a. $xy = -24$.

b. $y = (x - 7)^2 x^{\frac{1}{3}}$.

6. Is the function $f(x) = 4x - x^2$ increasing or decreasing **(a)** when $x < 2$; **(b)** when $x > 2$? What can you say about this function when $x = 2$? Justify your last answer.

7. For what values of x is the function $f(x) = x(x - 4)^3$ increasing, and for what values is it decreasing? What can you say about this function when $x = 1$? Justify your last answer.

111. Functions increasing or decreasing in an interval. The theorems of Art. 110 give definite information concerning the behavior of a function at the point where $x = a$, provided $f'(a) > 0$ or $f'(a) < 0$. If $f'(a) = 0$, the behavior of the function is not determined without additional information. Consider the following examples.

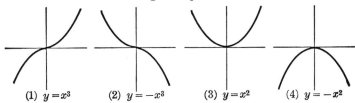

(1) $y = x^3$ (2) $y = -x^3$ (3) $y = x^2$ (4) $y = -x^2$

Fig. 111.1

In each case $dy/dx = 0$ when $x = 0$. It appears from the graphs, however, that in (1) y is increasing at $x = 0$; in (2) y is decreasing; in (3) y has a minimum value; and in (4) y has a maximum value. To investigate the situation further, we make the following definitions. Let h be a conveniently chosen positive constant and denote the interval $(c - h, c + h)$ by I.

DEFINITIONS. *If $f(c) > f(x)$ for all values of x in I, except $x = c$, $f(c)$ is called a maximum value of $f(x)$. If $f(c) < f(x)$ for all values of x in I, except $x = c$, $f(c)$ is called a minimum value of $f(x)$.*

Maximum and minimum values are called collectively *extreme* values. Note that a maximum is not necessarily the greatest value of $f(x)$ but is merely the greatest value in some vicinity of c, which may be very small, just as the top of a hill is a point of maximum elevation in the neighborhood but not necessarily the highest point in a country.

Theorem III. *Let $f(x)$ be defined in (a, b), let c lie within (a, b), and let $f(c)$ be an extreme value. If $f'(c)$ exists, $f'(c) = 0$.*

Proof. Take first the case when $f(c)$ is a maximum value. We give an indirect proof.

If $f'(c) > 0$, $f(x)$ is increasing when $x = c$. Then for $x > c$ and near enough to c, $f(x) > f(c)$, which is a contradiction, since $f(c)$ is a maximum value.

If $f'(c) < 0$, $f(x)$ is decreasing when $x = c$. Then for $x < c$ and near enough to c, $f(x) > f(c)$, which is a contradiction.

Hence the only possibility is that $f'(c) = 0$.

A similar argument is applicable to the case when $f(c)$ is a minimum.

The truth of the following theorem is intuitively apparent from a graph and from the discussion of Art. 110. We will accept the theorem without a rigorous proof, which can be made by use of the Mean Value Theorem to be discussed later (Art. 190).

Theorem IV. *Let $f(x)$ be continuous in the interval (a, b) and be differentiable within (a, b). If $f'(x) > 0$ for every value of x between a and b, $f(b) > f(a)$. If $f'(x) < 0$ for every value of x between a and b, $f(b) < f(a)$.*

We also note the following property of continuous functions.

If $f(x)$ is continuous in the closed interval (a, b), it has a greatest value and a least value in that interval.

Theorem III tells us that *if $f(x)$ is differentiable in some interval (a, b) and if there are values of x between a and b which yield extreme* values of $f(x)$ then these values will be found among the roots of $f'(x) = 0$. These are called *critical values* of x. Further investigation will be needed to determine which, if any, of these critical values actually do give extreme values of $f(x)$. Thus, if $f(x) = x^3$, $f'(x) = 0$ when $x = 0$, but $f(x)$ has neither a maximum nor a minimum value.

Consider the function $f(x) = (x - 2)^2\sqrt{x}$ of Example 3, Art. 110. Since this is a continuous function, it has a greatest and a least value in the closed interval $(0, 2)$. Since $f(0) = 0$ and $f(2) = 0$ and $f(x) > 0$ for $0 < x < 2$, the function has a maximum value for some value of x between 0 and 2. By Theorem III this value occurs at the point where $f'(x) = 0$, that is by the example above, when $x = \frac{2}{5}$. The corresponding value of the function is found to be 1.62, approximately.

Note that if $f(x)$ is defined only in a closed interval (a, b), extreme values at the ends of the interval may not be revealed by setting $f'(x) = 0$, since the theorem is applicable only to points *between* a and b. As an example, see Fig. 109.1. It is evident that $f(a)$ and $f(b)$ are maximum values, but $f'(a) < 0$ and $f'(b) > 0$.

As another example of the same kind, let $f(x)$ be the distance between the point $(-3, 0)$ and any point (x, y) on the circle whose equation is $x^2 + y^2 = 25$. By means of the distance formula it is found that $f(x) = \sqrt{34 + 6\,x}$, whence

$$f'(x) = \frac{3}{\sqrt{34 + 6\,x}}.$$

Now $f'(x)$ is positive for every value of x for which it is defined and so Theorem III reveals no extreme values; in fact $f(x)$ is everywhere an increasing function of x. But on this circle x varies only from -5 to $+5$. Hence $f(-5) = 2$ is a minimum value and $f(5) = 8$ is a maximum value.

There may be extreme values at points where $f(x)$ is not differentiable. An example of this kind is $f(x) = x^{\frac{2}{3}}$ which clearly has a minimum value at $x = 0$. But $f'(0)$ does not exist. (Fig. 111.2.)

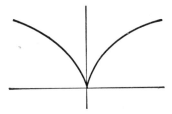

Fig. 111.2

112. Extreme values. The discussion of Art. 111 shows that for extreme values of a function $f(x)$ whose domain of definition is an interval (a, b), there are only the following possibilities.

1. There may be extreme values at the ends of the interval, that is, $f(a)$ and $f(b)$ may be extreme values.

2. A value of x for which $f(x)$ is not differentiable may give an extreme value.

3. Values of x between a and b for which $f'(x) = 0$ may give extreme values.

Values of x belonging to any of the three classes just named will be called *critical values*. All three kinds, if they exist, must be investigated if we are to find all extreme values of a given function. If $f(x)$ is defined for all values of x, there will be no critical values of the first class. The functions usually employed are everywhere differentiable and critical values of the second class will not often occur. In most practical problems we encounter only the third class of critical values, given by Theorem III, and the major emphasis in our discussion will be on this class.

113. Two tests for maxima and minima. Three tests are available for the determination of the behavior of a function at a critical point of the third class, that is, a point $x = c$ which is *within* the domain of definition and such that $f'(c) = 0$. These tests are called the direct test, the first derivative test, and the second derivative test. The direct test is embodied in the following theorem.

Theorem V. *Let $x_1 < c$ and $x_2 > c$, and let both $f(x_1)$ and $f(x_2)$ be less than $f(c)$. Then $f(c)$ is a maximum value provided that (1) $f(x)$ is continuous in the interval (x_1, x_2) and (2) c is the only critical value of x within this interval. If the word "less" is replaced by "greater," $f(c)$ is a minimum.*

Proof. Let $f(x_1)$ and $f(x_2)$ be less than $f(c)$. By the property of continuous functions (Art. 111), $f(x)$ has a greatest value M in (x_1, x_2). As M is a maximum value and c is the only critical point between x_1 and x_2, $M = f(c)$, or $f(c)$ is a maximum value.

Example. Consider the parcel-post problem of Example 2, Art. 108, and suppose $y \geqq x$.

Then
$$V = 100\, x^2 - 4\, x^3$$

and the domain of definition is $(0, 20)$.

$$\frac{dV}{dx} = 4\, x(50 - 3\, x).$$

The only critical point within the domain of definition is $x = \frac{50}{3}$. For convenience of calculation let $x_1 = 10$ and $x_2 = 20$. Corresponding values of x and V are shown in the table. The conditions of Theorem V are satisfied and the maximum value of V is $9259\frac{7}{27}$ which occurs when $x = 16\frac{2}{3}$.

x	V
10	6000
$16\frac{2}{3}$	$9259\frac{7}{27}$
20	8000

The first derivative test is embodied in the following theorem.

Theorem VI. Let $x_1 < c$ and $x_2 > c$, and let $f(x)$ be continuous in the interval (x_1, x_2). If $f'(x)$ exists and is positive for all values of $x < c$ in this interval and $f'(x)$ exists and is negative for all values of $x > c$ in this interval, $f(c)$ is a maximum value. If $f'(c)$ is negative for $x < c$ and positive for $x > c$, $f(c)$ is a minimum value.

Proof. The following is a proof of the first half of the theorem. The second half can be proved in a similar manner.

By hypothesis $f(x)$ is continuous in the interval (x_1, c) and $f'(x) > 0$ between x_1 and c. Hence, by Theorem IV,

$$f(x_1) < f(c).$$

Likewise, $f'(x) < 0$ between c and x_2 and, by Theorem IV,

$$f(c) > f(x_2).$$

Since both $f(x_1)$ and $f(x_2)$ are less than $f(c)$ and c is the only critical value between x_1 and x_2, $f(c)$ is a maximum value by Theorem V.

Observe that Theorem VI says nothing about the existence of $f'(c)$. The theorem is illustrated graphically by Figs. 113.1 and 113.2.

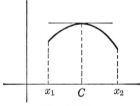

Fig. 113.1

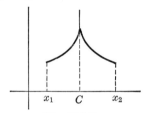

Fig. 113.2

The function whose graph is shown in Fig. 113.1 has a maximum value $f(c)$ where $x = c$ is a critical value of the third class. The function whose graph is shown in Fig. 113.2 has a maximum value $f(c)$ where $x = c$ is a critical value of the second class and $f'(c)$ does not exist.

In the example above, the conditions of Theorem VI are satisfied. It is readily verified that $V'(x) > 0$ for $10 < x < \frac{50}{3}$ and $V'(x) < 0$ for $\frac{50}{3} < x < 20$. Hence, by Theorem VI, $V(\frac{50}{3})$ is a maximum value.

In applying this test we do not have to find values for x_1 and x_2. We merely have to know that there are values such that the conditions stated in Theorem VI are true. It is not necessary to compute any numerical values of $f'(x)$, but we must be able to state the sign for *all* values of x under consideration. This is a simple matter when $f'(x)$ is written in factor form; otherwise it may be difficult.

114. The second derivative test. *Theorem VII. Let $f(x)$ be differentiable in an interval (a, b). For some value c between a and b let $f'(c) = 0$ and $f''(c)$ be negative. Then $f(c)$ is a maximum value. If $f''(c)$ is positive, $f(c)$ is a minimum value.*

Proof. Since $f''(c)$ is the derivative of $f'(x)$ when $x = c$ and $f''(c) < 0$, it follows that the function $f'(x)$ is decreasing when $x = c$. Hence, by the definition of a decreasing function, for all values of x in some conveniently chosen interval $(c - h, c + h)$ which is contained in (a, b),

$$f'(x) > f'(c) \text{ if } x < c, \text{ (1)} \qquad\qquad f'(x) < f'(c) \text{ if } x > c. \text{ (2)}$$

As $f'(c) = 0$, relations (1) and (2) show that $f'(x) > 0$ if $x < c$ and $f'(x) < 0$ if $x > c$. Hence the first derivative test is applicable and $f(c)$ is a maximum value.

Example. Apply the second derivative test to the Example of Art. 113.

Solution. The second derivative is

$$\frac{d^2V}{dx^2} = 200 - 24\,x.$$

For the critical value $x = \frac{50}{3}$ we find $d^2V/dx^2 = -200$. Hence $x = \frac{50}{3}$ gives a maximum value for V.

115. Comparison of the tests. The second derivative test is the most convenient to use if the labor of finding $f''(x)$ is not excessive and if $f''(c) \neq 0$. Observe that it is unnecessary to find a numerical value for $f''(c)$. Only the sign of $f''(c)$ is required.

The direct test is usually the most convenient if the differentiation is complicated. The extreme value of the function is often required and the computation of the other two values of the function is not difficult if the values of x on either side of the critical value are judiciously chosen.

The first derivative test is convenient if it is possible to write $f'(x)$ as the product (or quotient) of linear factors or factors which have the same sign for all values of x. Observe that the first derivative test is available when $f'(c)$ and $f''(c)$ do not exist, provided the conditions of Theorem VI are fulfilled.

116. Extreme values at the ends of an interval. If the critical value to be investigated is an end point of the domain of definition of the function, the three tests given can still be used if they are properly modified. The principal difference is that in the direct and the first derivative tests we need consider only points on one side of the critical value, instead of on both sides. Moreover, as remarked in Art. 111, we can have an extreme value at an end point if $f'(x) \neq 0$ at this point. There are, in all, four sets of criteria, one of which is as follows.

Let $f(x)$ be defined only in the interval (a, b). Then $f(a)$ is a minimum value if any one of the following conditions is satisfied.

1. $f'(a) > 0$.

2. For some value $x_1 > a$, $f(x)$ is continuous in the interval (a, x_1), $f(a) < f(x_1)$, and there is no critical value between a and x_1.

3. $f(x)$ is continuous in some interval $(a, a + h)$ and $f'(x)$ exists and is positive for all values of x *within* this interval.

4. $f(x)$ is differentiable in some interval $(a, a + h)$, $f'(a) = 0$, and $f''(a) > 0$.

EXERCISES

1. Prove the criteria above.

State and prove similar criteria for

2. $f(a)$ to be a maximum value.

3. $f(b)$ to be a maximum value.

4. $f(b)$ to be a minimum value.

117. Directions for solving problems. Directions for a complete investigation of the extreme values of a function $f(x)$ are summarized here, although in applications a complete investigation is often unnecessary.

I. Note the domain of definition of $f(x)$. If there are any end points, these are critical values. Usually it is possible to tell by inspection whether an end point yields a maximum or a minimum value. In the contrary event the criteria of Art. 116 are available.

II. Find $f'(x)$ in terms of x and simplify the result as much as possible. In particular, if any fractions are involved, write $f'(x)$ as a single fraction, factor the numerator and denominator if possible, and cancel out any common factors.

III. Set $f'(x) = 0$ and solve the equation for x. If $f'(x)$ is a fraction and the denominator contains x, set the denominator equal to zero and solve the equation for x. If c is a root of the first equation, it is a critical value where $f'(c) = 0$. If c is a root of the second equation, it is a critical value where $f'(c)$ does not exist. For algebraic functions, which are the only ones under consideration at present, this process gives all critical values except the end points.

IV. Examine each critical value by one of the tests.

Direct test. Choose values x_1 and x_2 such that $x_1 < c < x_2$, $f(x)$ is defined in (x_1, x_2), and c is the only critical value between x_1 and x_2. Calculate the values of $f(x_1)$, $f(c)$, and $f(x_2)$. If $f(c)$ is larger (smaller) than $f(x_1)$ and $f(x_2)$, $f(c)$ is a maximum (minimum) value.

First derivative test. Write $f'(x)$ in factor form. Find the signs of $f'(x)$ for values of x near c. If $f'(x)$ is positive (negative) for all values of x a little less than c and negative (positive) for all values of x a little greater than c, then $f(c)$ is a maximum (minimum) value.

Second derivative test. This is applicable if $f'(c) = 0$. If $f''(c) < 0$, $f(c)$ is a maximum value. If $f''(c) > 0$, $f(c)$ is a minimum value. If $f''(c) = 0$, the test fails to answer the question.

Example 1. Examine $f(x) = x(x-2)^3$ for extreme values.

Solution. This function is defined for all values of x and it is everywhere differentiable. Hence the only critical values are those for which $f'(x) = 0$. Differentiating, we obtain

$$f'(x) = (5x - 2)(x - 2)^2.$$

Setting $f'(x) = 0$ and solving for x, gives $x = \frac{2}{5}$ and $x = 2$. We apply the first derivative test to the critical value $x = \frac{2}{5}$. For all values of x, except $x = 2$, the factor $(x - 2)^2$ is positive. For $0 < x < \frac{2}{5}$, the factor $5x - 2$ is negative and for $\frac{2}{5} < x < 2$, this factor is positive. Hence

$$\text{for } 0 < x < \tfrac{2}{5}, \qquad f'(x) = (-)(+) = -,$$
$$\text{for } \tfrac{2}{5} < x < 2, \qquad f'(x) = (+)(+) = +.$$

By Theorem VI, $f(\frac{2}{5}) = -1.6384$ is a minimum value.

If $x > \frac{2}{5}$, the factor $5x - 2 > 0$ and the factor $(x - 2)^2 > 0$ both for $x < 2$ and $x > 2$. Hence the test indicates that $f(2)$ is not an extreme value. In fact, inspection of the equation which defines $f(x)$ shows that $f(x) < 0$ when $0 < x < 2$; $f(x) = 0$ when $x = 2$; and $f(x) > 0$ when $x > 2$. By definition $f(x)$ is increasing when $x = 2$

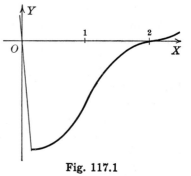

Fig. 117.1

Example 2. Examine $f(x) = x + \dfrac{1}{x}$ for extreme values.

Solution. Differentiating, $f'(x) = 1 - \dfrac{1}{x^2}.$

Setting $f'(x) = 0$ and solving for x, we get $x = \pm 1$. Applying the direct test, we calculate the values shown in the tables. The first table shows that $f(-1) = -2$ is a maximum value and the second table shows that $f(1) = 2$ is a minimum value. For $x = 0$, $f(x)$ is not defined.

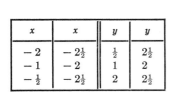

x	x	y	y
-2	$-2\frac{1}{2}$	$\frac{1}{2}$	$2\frac{1}{2}$
-1	-2	1	2
$-\frac{1}{2}$	$-2\frac{1}{2}$	2	$2\frac{1}{2}$

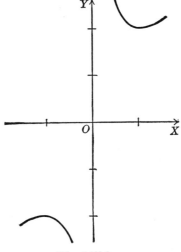

Fig. 117.2

If the first derivative test is preferred, the derivative should be written as a fraction and then factored.

$$f'(x) = \frac{x^2 - 1}{x^2} = \frac{(x + 1)(x - 1)}{x^2}.$$

Hence   for $0 < x < 1$, $f'(x) = \dfrac{(+)(-)}{+} = -,$

        for $1 < x$, $f'(x) = \dfrac{(+)(+)}{+} = +.$

By Theorem VI, $f(1)$ is a minimum value.

Example 3. A boat is anchored at a point A six miles from the shore, which is assumed to be a straight line. The nearest point on the shore is B and a point C is eight miles from B along the shore. A man can row 3 mi. per hour and walk 5 mi. per hour. Where should he land in order to go from A to C in the least time?

Solution. Let D be the landing point and x be the distance between B and D. Then $AD = \sqrt{x^2 + 36}$ and $DC = 8 - x$. If T is the time (in hours) for the trip between A and C at the given speeds, we have

$$T = \frac{\sqrt{x^2 + 36}}{3} + \frac{8 - x}{5}.$$

The domain of definition of $T(x)$ is $(0, 8)$.

Differentiation gives

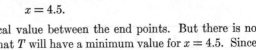

$$T'(x) = \frac{x}{3\sqrt{x^2 + 36}} - \frac{1}{5} = \frac{5x - 3\sqrt{x^2 + 36}}{15\sqrt{x^2 + 36}}.$$

Fig. 117.3

Setting $T'(x) = 0$, we have
$$5x = 3\sqrt{x^2 + 36},$$
$$25 x^2 = 9 x^2 + 324,$$
whence
$$x = 4.5.$$

Thus 4.5 is the only critical value between the end points. But there is no a priori reason for believing that T will have a minimum value for $x = 4.5$. Since there is no easy way of telling the signs of $T'(x)$ for values of x near 4.5, we use the direct test and find that $T(0) = 3.6$; $T(4.5) = 3.2$; $T(8) = 3\frac{1}{3}$. Hence T has a minimum value when $x = 4.5$.

Exercise. Discuss the solution of the preceding example if $BC = 4$ mi.

Example 4. Find the altitude of the largest right circular cone which can be inscribed in a sphere of radius r.

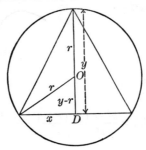

Fig. 117.4

Solution. Fig. 117.4 represents a vertical section. Let x be the radius of the cone and let y be its height. The volume V is to be a maximum.

By geometry,
$$V = \tfrac{1}{3} \pi x^2 y.$$

But V must be expressed as a function of *one* variable only. From the figure we see that
$$x^2 = r^2 - OD^2 = r^2 - (y - r)^2 = 2ry - y^2.$$

Substituting this above, we have

$$V = \frac{\pi}{3} (2 ry^2 - y^3).$$

Differentiating, and setting the derivative equal to 0,

$$\frac{dV}{dy} = \frac{\pi}{3} (4\,ry - 3\,y^2) = 0.$$

Solving, $y = 0$ or $\frac{4\,r}{3}$.

It is obvious from the conditions of the problem that there must be a maximum value of V and that the value $y = 0$ is meaningless. Hence we have the maximum cone when $y = \frac{4}{3}\,r$.

Example 5. A manufacturing plant has a capacity of 25 articles per week. Experience has shown that n articles per week can be sold at a price of p dollars each, where $p = 110 - 2\,n$, and the cost of producing n articles is $(600 + 10\,n + n^2)$ dollars. How many articles should be made each week to give the largest profit?

Solution. The profit (P dollars) on the sale of n articles is

$$P = np - (600 + 10\,n + n^2),\ \text{that is,}$$
$$P = 100\,n - 600 - 3\,n^2.$$

In this problem n must be an integer, and, since it is not a continuous variable, it is impossible to differentiate P with respect to n. The formula shows that P is negative if n is less than 8. By direct calculation we may construct the table below, which shows that the largest profit is obtained when 17 articles per week are manufactured.

n	P	n	P	n	P
8	8	14	212	20	200
9	57	15	225	21	177
10	100	16	232	22	148
11	137	17	233	23	113
12	168	18	228	24	72
13	193	19	217	25	25

To avoid excessive computation in such problems, we may proceed as follows. Consider the function

$$y = 100\,x - 600 - 3\,x^2,$$

in which x varies continuously from 0 to 25. The graph of this function is a continuous curve, and for integral values of x the ordinates correspond to the values of P in the table.

Differentiating, $\frac{dy}{dx} = 100 - 6\,x,$

whence $x = 16\frac{2}{3}$ and $y = 233\frac{1}{3}$, the maximum ordinate on the curve. It is now apparent that the largest ordinate corresponding to an integral value of x will occur for $x = 16$ or $x = 17$, and calculation shows that $x = 17$ is the correct value.

PROBLEMS

1. Examine the following functions for maxima and minima. Draw the graph in each case.

a. $y = x^2 - 5x + 3$.

b. $y = 1 + 7x - 2x^2$.

c. $y = \frac{1}{3}x^3 - \frac{1}{2}x^2 - 2x + 2$.

d. $y + x^3 + 12x^2 + 45x + 52 = 0$.

e. $6y = 2x^3 - 3x^2 - 36x$.

f. $y = 2x^3 - 3x^2 + 6x - 2$.

g. $y = x^4 - 2x^2$.

h. $15y = 3x^5 - 25x^3 + 60x$.

i. $y = x^4 - 4x^3 + 4x^2 - 4$.

j. $y = 6x^2 - x^4$.

k. $y = x^2 + \dfrac{16}{x}$.

l. $y = x^2 + \dfrac{1}{x^2}$.

m. $y = x^3 + \dfrac{3}{x}$.

n. $y = \dfrac{6x}{x^2 + 1}$.

2. Find two positive numbers whose sum is 20, such that

a. The sum of their squares is a minimum.

b. The sum of their cubes is a minimum.

c. Their product is a maximum.

d. The difference between one and the reciprocal of the other is a maximum.

3. A rectangular field to contain 40 A. is to be fenced off along the bank of a straight river. If no fence is needed along the river, what must be the dimensions requiring the least amount of fencing? (1 A. = 160 sq. rd.)

4. Show that of all isosceles triangles inscribed in a circle of radius a the equilateral triangle has the greatest area.

5. The legs of an isosceles triangle are each 20 in. long. Find the length of the base if the area is a maximum.

6. A trough is to be made of a long rectangular piece of tin by bending up two edges so as to give a rectangular cross section. If the width of the piece is 14 in., how deep should the trough be made in order that its carrying capacity may be a maximum?

7. Two upright poles, AB and CD, are 40 ft. apart. AB is 30 ft. high and CD is 20 ft. high. Find the distance AE if the length of the rope BED is a minimum.

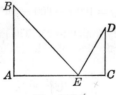

Fig. 117.5

8. A rectangular box is to be made from a sheet of tin 16 in. by 20 in. by cutting a square from each corner and turning up the sides. Find the edge of this square which makes the volume a maximum.

9. A rectangular box with a square base and a cover is to be built to contain 800 cu. ft. If the cost per square foot for the bottom is 15¢, for the top 20¢, and for the sides 10¢, what are the dimensions for a minimum cost?

10. A sheet of paper for a poster is to contain 16 sq. ft. The margins at the top and the bottom are to be 6 in., and those on the sides 4 in. What are the dimensions if the printed area is to be a maximum?

11. A rectangular box with a square base and an open top is to be made. Find the volume of the largest box that can be made from 320 sq. ft. of material.

12. The strength of a rectangular beam varies as the product of the breadth and the square of the depth. Find the dimensions of the strongest beam that can be cut from a cylindrical log whose diameter is a.

13. The stiffness of a rectangular beam varies as the product of the breadth and the cube of the depth. Find the dimensions of the stiffest beam that can be cut from a cylindrical log whose radius is a.

14. Two vertices of a rectangle are on the diameter of a semicircle of radius a, and the other two vertices are on the arc. Find the dimensions of the rectangle if its area is a maximum.

15. Two roads intersect at right angles, and a spring is located in an adjoining field 10 rd. from one road and 5 rd. from the other. How should a straight path just passing the spring be laid out from one road to the other so as to cut off the least amount of land? How much land is cut off?

16. Rectangles are inscribed in a circle of radius a. Find the dimensions of the rectangle whose perimeter has an extreme value, and show whether it is a maximum or a minimum.

17. One base of an isosceles trapezoid is the diameter of a circle of radius a, and the ends of the other base lie on the circumference of the circle. Find the length of the other base if the area is a maximum.

18. A frame for a cylindrically shaped lamp shade is made from a piece of wire 16 ft. long. The frame consists of two equal circles, two diametral wires in the upper circle, and four equal wires from the upper to the lower circle. For what radius will the volume of the cylinder be a maximum?

19. What should be the diameter of a tin can holding 1 qt. (58 cu. in.) and requiring the least amount of tin **(a)** if the can is open at the top, **(b)** if the can has a cover?

20. A vertical cylindrical water tank, open at the top, is to contain 15,000 gal. Find the diameter if the material used is a minimum. (1 cu. ft. = 7.5 gal.)

21. Find the volume of the largest cylinder which can be cut from a given right circular cone whose height is h and whose base has the radius r.

22. The slant height of a right circular cone is a given constant a. Find the altitude if the volume is a maximum.

23. Find the dimensions of the right circular cylinder of maximum volume which can be cut from a solid wooden sphere of diameter 16 in.

24. Find the dimensions of the largest inscribed rectangular parallelepiped with a square base which can be cut from a solid sphere of radius r.

25. An oil can is made in the shape of a cylinder surmounted by a cone. If the radius of the cone is three fourths of its height find the most economical proportions.

26. If the cost per hour for fuel required to run a given steamer varies as the cube of its speed and is $40 per hour for a speed of 10 mi. per hour, and if other expenses amount to $200 per hour, find the most economical rate to run it a distance of 500 mi.

27. A railroad company agreed to run a special train for 50 passengers at a uniform fare of $10 each. In order to secure more passengers, the company agreed to deduct 10¢ from this uniform fare for each passenger in excess of the 50 (that is, if there were 60 passengers, the fare would be $9 each). What number of passengers would give the company maximum gross receipts?

28. Find the area of the largest rectangle which can be inscribed in the ellipse whose equation is $x^2 + 4y^2 = 16$.

29. Find the dimensions of the largest rectangle which can be inscribed in the ellipse whose equation is $\dfrac{x^2}{a^2} + \dfrac{y^2}{b^2} = 1$.

30. Find the area of the largest rectangle which can be drawn with its base on the x-axis and with two vertices on the witch whose equation is $y = \dfrac{8a^3}{x^2 + 4a^2}$.

31. On the circle whose equation is $x^2 + y^2 = 100$ find the coördinates of the point which is nearest to the point $(12, 16)$. Is there any other way of solving this problem?

32. What point on the curve $4y = x^2$ is nearest to the point $(0, 4)$?

33. The x-axis and y-axis are joined by lines which are tangent to the circle whose equation is $x^2 + y^2 = r^2$, where r is a constant. Find the length of the shortest of these lines.

34. A ditch is to be dug to connect the points A and B of Fig. 117.6. The earth at the left of the line AD is soft, and the cost of digging the portion AC is $10 per foot. The earth at the right of AD is hard, and the cost of digging the portion CB is $20 per foot. Where should the turn C be made for a minimum cost?

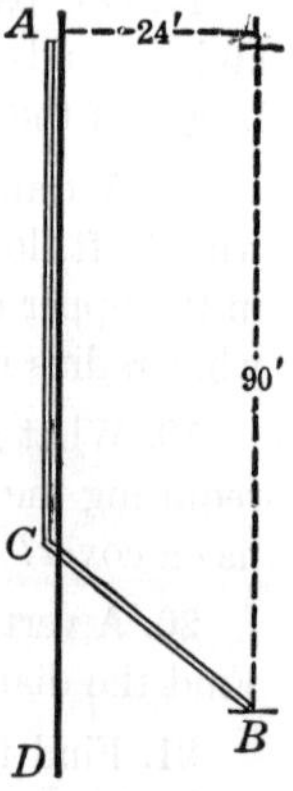

35. A brick conduit, designed to accommodate underground cables, is to be built with a cross section in the form of a rectangle surmounted by a semicircle. Its carrying capacity (that is, the area of the cross section) is to be 24 sq. ft. If the cost of construction is assumed to be proportional to the perimeter of the cross section, find the width which will involve the least cost.

Fig. 117.6

36. A radio manufacturer finds that he can sell x instruments per week at p dollars each, where $5x = 375 - 3p$. The cost of production is $(500 + 15x + \frac{1}{5}x^2)$ dollars. Show that the largest profit is obtained when the production is about 30 instruments per week.

37. In Problem 36 suppose the relation between x and p is
$$x = 100 - 20 \sqrt{p/5}.$$
Show that the manufacturer should produce only about 25 instruments per week for maximum profit.

38. In Problem 36 suppose the relation between x and p is
$$x^2 = 2500 - 20 p.$$
How many instruments should be produced each week for maximum profit?

39. In Problem 36 suppose a tax of t dollars per instrument is imposed by the government. The manufacturer adds the tax to his cost and determines the output and price under the new conditions.

a. Show that the price increases by a little less than half the tax.

b. Express the receipts from the tax in terms of t and determine the tax for maximum return.

c. When the tax determined in **b** is imposed, show that the price is increased by about 33 per cent.

40. A steel plant is capable of producing x tons per day of a low-grade steel and y tons per day of a high-grade steel, where $y = \dfrac{40 - 5x}{10 - x}$. If the fixed market price of low-grade steel is half that of high-grade steel, show that about $5\frac{1}{2}$ tons of low-grade steel are produced per day for maximum receipts.

41. Two towns are situated at distances of 3 mi. and 9 mi., respectively, from the shore of a lake, which is assumed to be a straight line. If the points on the shore nearest the towns are 5 mi. apart, at what point on the shore should a pumping-station, designed to supply both towns, be located so as to require the least amount of water mains? How many miles of mains are required?

42. Let $P(a, b)$ be a point in the first quadrant of a set of rectangular axes. Draw a line through P cutting the positive ends of the axes at A and B. Calculate the intercepts of this line on OX and OY in the following cases.

a. When the area OAB is a minimum.

b. When the length AB is a minimum.

c. When the sum of the intercepts is a minimum.

d. When the perpendicular distance from O to AB is a maximum.

43. An electric current flows about a coil of radius r, and exerts a force F on a small magnet the axis of which is on a line drawn through the center of the coil and perpendicular to its plane. This force is given by $F = \dfrac{x}{(r^2 + x^2)^{\frac{5}{2}}}$, where x is the distance to the magnet from the center of the coil. Show that F is a maximum for $x = r/2$.

CHAPTER X

THE CIRCLE, PARABOLA, ELLIPSE, AND HYPERBOLA

THE CIRCLE

118. The standard equation of the circle. In the first chapter it was shown that the equation of every straight line is of the first degree and, conversely, that the locus of every equation of the first degree is a straight line. In a similar way we shall now find the form of the equation of every circle.

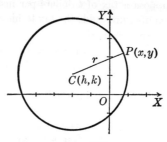

Fig. 118

Let $P(x, y)$ be any point on a circle whose radius is r and whose center is $C(h, k)$. Since, by the definition of a circle, the radius is constant, we have at once, for all positions of P,

$$CP = r;$$

whence

$$\sqrt{(x - h)^2 + (y - k)^2} = r,$$

or

(I) $$(x - h)^2 + (y - k)^2 = r^2.$$

Equation **(I)** is the *standard equation* of a circle, and may be used to write the equation of any circle when its center and radius are known. In Fig. 118 the center is $(-2, 1)$ and the radius is 3. Hence the equation is

$$(x + 2)^2 + (y - 1)^2 = 9.$$

The position and size of the circle depend upon the three arbitrary constants h, k, and r. Since h and k are merely the coördinates of the center, they may be either positive or negative, while r is necessarily positive.

If the center is at the origin, $h = 0$ and $k = 0$, and form (I) reduces to

(I a) $x^2 + y^2 = r^2$.

This is usually employed when it is desired to study some property of the circle which does not depend upon its position.

119. The general form of the equation of the circle. If we expand the parentheses in (I) and collect the terms, we get

$$x^2 + y^2 - 2\,hx - 2\,ky + (h^2 + k^2 - r^2) = 0.$$

Since $-2\,h$, $-2\,k$, and $h^2 + k^2 - r^2$ are constants, this may be written in the form

(II) $x^2 + y^2 + Dx + Ey + F = 0,$

which may be called the *general form*. If the equation of a circle is given in this form, it can be reduced back to form (I) by merely completing the squares in x and y.

The most general equation of the second degree in x and y is

$$Ax^2 + Bxy + Cy^2 + Dx + Ey + F = 0.$$

This equation can be reduced to the general form of the equation of a circle if, and only if, $B = 0$ and $C = A$. When $B = 0$ and $C = A \neq 1$, the equation can be reduced to form (II) by merely dividing through by the common value of A and C. Thus we have proved the following theorem.

Theorem. *The locus of an equation of the second degree in x and y is a circle if, and only if, the coefficients of x^2 and y^2 are equal and there is no term in xy, unless the equation has no real locus.* See Example 2, Art. 120, for the exceptional case.

120. To find the center and radius of a circle. As stated in the preceding article, if the equation of a circle is given in the general form it can be reduced to the standard form by completing the squares in x and y. The center and the radius can then be identified at once.

Example 1. Find the center and the radius of the circle whose equation is

$$2\,x^2 + 2\,y^2 + 8\,x - 6\,y + 7 = 0.$$

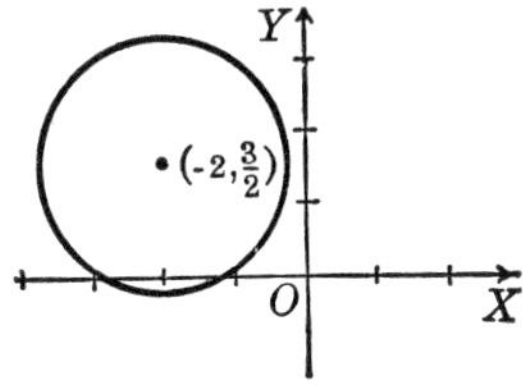

Fig. 120

Solution. Dividing through by 2, we have the general form

$$x^2 + y^2 + 4x - 3y + \tfrac{7}{2} = 0.$$

To find the center and the radius, reduce this to form **(I)** by transposing $\tfrac{7}{2}$ and completing the squares.

$$x^2 + 4x + 4 + y^2 - 3y + \tfrac{9}{4} = -\tfrac{7}{2} + 4 + \tfrac{9}{4} = \tfrac{11}{4},$$

$$(x + 2)^2 + \left(y - \frac{3}{2}\right)^2 = \frac{11}{4} = \left(\frac{\sqrt{11}}{2}\right)^2.$$

Comparing this with the standard form, we see that $h = -2$, $k = \tfrac{3}{2}$, and $r = \tfrac{1}{2}\sqrt{11} = 1.66$, approximately. Hence the center is $(-2, \tfrac{3}{2})$, and the radius is about 1.66.

The algebraic work is so simple that a check is scarcely necessary; but if one is desired, the simplest is to find one or more points from the given equation and compare them with the figure. Here the x-intercepts are

$$-2 \pm \tfrac{1}{2}\sqrt{2} = -2 \pm 0.71 = -2.71 \quad \text{or} \quad -1.29.$$

Example 2. What is the locus of $x^2 + y^2 - 8x + 4y + 21 = 0$?

Solution. According to the theorem of Art. 119, the locus is a circle if the equation has a real locus. Completing the squares, we have

$$(x - 4)^2 + (y + 2)^2 = -21 + 16 + 4 = -1.$$

But here $r^2 = -1$, making r imaginary. An inspection of the equation shows that the locus is imaginary, as $(x - 4)^2$ and $(y + 2)^2$ are squares and therefore never less than zero; hence their sum cannot equal -1.

To avoid having exceptions we say that the locus is an *imaginary circle*. When $r = 0$, making the locus a point, we call it a *point-circle*. Such forms are called *degenerate forms*.

121. The slope of the tangent. As in the case of other curves, we find the slope of the tangent to a circle by finding $\dfrac{dy}{dx}$ from the equation.

It is also easy to show that the tangent is perpendicular to the radius drawn to the point of contact. For from the standard equation

$$(x - h)^2 + (y - k)^2 = r^2,$$

we have $$2(x - h) + 2(y - k)\frac{dy}{dx} = 0;$$

whence $$m = \frac{dy}{dx} = -\frac{x - h}{y - k}.$$

But the slope m' of the radius joining the center $C(h, k)$ to the point $P(x, y)$ on the circumference is

$$m' = \frac{y - k}{x - h}.$$

Hence $mm' = -1$, which is the criterion for perpendicularity.*

In problems involving the tangent to a circle the slope may be found either by differentiating or by using the negative reciprocal of the slope of the radius. The former method is more convenient when the center is not given.

Example. Find the equation of the tangent to the circle whose equation is $x^2 + y^2 - 4x - 2y = 20$ at the point $(5, 5)$.

Solution. Differentiating the equation, we find

$$m = \frac{dy}{dx} = \frac{2 - x}{y - 1}.$$

Substitution of the coördinates $(5, 5)$ gives $m = -\frac{3}{4}$. Hence the equation of the tangent is

$$y - 5 = -\tfrac{3}{4}(x - 5), \quad \text{or} \quad 3x + 4y = 35.$$

PROBLEMS

1. In the following cases write the equation of the circle and reduce it to the general form.

a. Center $(6, 4)$, radius 6.   **b.** Center $(4, 3)$, radius 5.
c. Center $(5, -12)$, radius 13.   **d.** Center $(a, 0)$, radius a.
e. Center $(0, a)$, radius a.   **f.** Center (a, a), radius $a\sqrt{2}$.

2. Find the center and the radius and draw the following circles.

a. $(x + 2)^2 + (y - 3)^2 = 25$.   **b.** $(x - 1)^2 + (y + 4)^2 = 0$.
c. $x^2 + y^2 - 8x + 6y + 24 = 0$.   **d.** $x^2 + y^2 - 13x = 0$.
e. $2x^2 + 2y^2 + 15y = 0$.   **f.** $4x^2 + 4y^2 - 4x - 4y + 1 = 0$.
g. $4x^2 + 4y^2 - 12x - 81 = 0$.   **h.** $3x^2 + 3y^2 + 36x - 14y = 0$.
i. $x^2 + y^2 - 20x + 40y + 379 = 0$.   **j.** $6x^2 + 6y^2 = 25y$.

3. Determine by inspection which of the circles in Problem 2

> **a.** Have their centers on the x-axis.
> **b.** Have their centers on the y-axis.
> **c.** Pass through the origin.

4. Find the equations of the circles satisfying the following conditions and draw the figure in each case.

a. Center at $(1, 2)$ and passing through $(-2, 3)$.
b. Having the line joining $(2, -4)$ and $(4, 6)$ for a diameter.
c. Center at $(-3, 5)$ and tangent to the x-axis.

---

*We have now reconciled the definition of tangent given in Chapter III with that given in plane geometry. For the tangent as defined in plane geometry is there shown to be perpendicular to the radius at the point of contact; and as there can be but one perpendicular to a line at a given point, both definitions give the same line.

d. Center at $(4, -2)$ and tangent to the y-axis.

e. Center on the line $y = 7$, radius 2, and tangent to the y-axis.

f. Center on the line $x + 2 = 0$, radius 3, and tangent to the x-axis.

g. Radius 5 and tangent to both axes.

h. Center on the line $x = 5$, radius 13, and passing through the origin.

i. Center at the origin and tangent to the line $x + y = 6$.

j. Center at $(1, 2)$ and tangent to the line $x - y - 4 = 0$.

5. Draw the system of circles defined by $(x + 4)^2 + (y - 3)^2 = k$ for the following values of k: $k = 25, 16, 9, 4, 1, 0, -1$.

6. a. The point $(\frac{7}{2}, \frac{1}{2})$ bisects a chord of the circle $x^2 + y^2 = 25$. Find the equation of the chord, and its length.

b. Solve the same problem for the point $(4, 7)$ and the circle whose equation is $x^2 + y^2 - 4x - 6y = 12$.

7. Find the coördinates of the points on the line whose equation is $x - y = 9$ which are 10 units distant from the point $(0, 5)$.

8. Find the locus of the vertex of a right triangle which has the ends of its hypotenuse at $(0, -3)$ and $(6, 5)$.

9. Find the locus of a point the sum of the squares of whose distances from the points $(\pm c, 0)$ is k. For what values of k does the locus fail to exist?

10. The ends of the base of a triangle are the points $(\pm c, 0)$. Find the equation of the locus of the vertex if the median to one of the sides has a constant length k.

11. Find the equation of the locus of a point whose distance from the origin is always twice its distance from $(6, 0)$. Draw the locus.

12. Find the equation of the locus of a point whose distance from $(-2, 2)$ is always three times its distance from $(2, -2)$. Draw the locus.

13. Show that the locus of a point whose distance from $(c, 0)$ is k times its distance from $(-c, 0)$ is a circle. Find its center and radius.

14. The base of a triangle is AB, where $A = (-c, 0)$ and $B = (c, 0)$. The third vertex C is above the x-axis and the angle $ACB = 45°$. Find the locus of C.

15. The point $P_1(x_1, y_1)$ lies without the circle whose center is (h, k) and whose radius is r. If t is the length of a tangent from P to the circle, show that

$$t^2 = (x_1 - h)^2 + (y_1 - k)^2 - r^2.$$

16. Find the equations of the tangent and of the normal to each of the following circles at the point indicated.

a. $(x + 2)^2 + (y - 3)^2 = 25$, $(2, 0)$.

b. $x^2 + y^2 - 8x + 6y + 24 = 0$, $x = 4$.

c. $x^2 + y^2 - 13x = 0$, $x = 4$.

d. $4x^2 + 4y^2 - 4x - 4y + 1 = 0$, $(0.8, 0.9)$.

e. $4x^2 + 4y^2 - 12x - 81 = 0$, $x = 6$.

f. $3x^2 + 3y^2 + 36x - 14y = 0$, $(0, 0)$.

g. $6x^2 + 6y^2 = 25y$, $(-\frac{5}{3}, \frac{10}{3})$.

17. Find the angles of intersection of each of the following circles and straight lines.

  a. $(x + 2)^2 + (y - 3)^2 = 25$, $7x - y - 8 = 0$.
  b. $x^2 + y^2 - 8x + 6y + 24 = 0$, $3x + 4y = 0$.
  c. $x^2 + y^2 - 13x = 0$, $y = x$.
  d. $4x^2 + 4y^2 - 4x - 4y + 1 = 0$, $2x - 14y + 11 = 0$.
  e. $4x^2 + 4y^2 - 12x - 81 = 0$, $4x - 2y - 21 = 0$.
  f. $3x^2 + 3y^2 + 36x - 14y = 0$, $x + y = 0$.
  g. $6x^2 + 6y^2 = 25y$, $y = 3x$.

18. Find the angles of intersection of each of the following pairs of circles.

  a. $x^2 + y^2 + 4y = 5$, $x^2 + y^2 - 4x = 1$.
  b. $x^2 + y^2 - 8y - 1 = 0$, $x^2 + y^2 - 10x - 2y + 9 = 0$.

122. Note on the analytic method of solving problems. In the problems of this chapter the methods developed previously should be freely used. These have been mainly analytical in character, but in solving the easier problems there is danger of forgetting the fundamental principle involved; namely, that *the coördinates of every point on the locus must satisfy the equation.* In certain problems a keen realization of this principle is essential. This will be illustrated in the two following examples.

Example 1. Find the dimensions of the largest cylinder that can be cut from a spherical segment which is cut off from a sphere of radius 5 in. by a plane 2 in. distant from the center of the sphere.

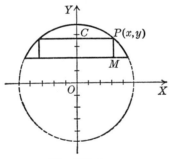

Fig. 122.1

Solution. Fig. 122.1 represents a section made by a plane passing through the axis of the cylinder, which we here take as the y-axis. The section of the cylinder is a rectangle of which one vertex $P(x, y)$ lies on the circle. We see that the radius of the cylinder is

$$CP = x$$

and that the altitude is $MP = y - 2$.

Hence the volume which is to be a maximum is given by the formula

$$V = \pi x^2(y - 2).$$

Since P lies on the circle, its coördinates must satisfy the equation of the circle, which is

$$x^2 + y^2 = 25.$$

Using this relation, we find that

$$V = \pi(25 - y^2)(y - 2)$$
$$= \pi(25\,y - 50 - y^3 + 2\,y^2),$$

which satisfies the requirement of expressing V in terms of one variable.

Differentiating, we have

$$\frac{dV}{dy} = \pi(25 - 3\,y^2 + 4\,y).$$

Setting this equal to zero and solving the resulting equation, we get

$$y = \frac{4 \pm \sqrt{316}}{6}.$$

The negative result obviously has no meaning; the positive sign gives $y = 3.63$, approximately. Hence the altitude of the cylinder is $y - 2 = 1.63$, and the radius is $x = \sqrt{25 - y^2} = 3.44$.

Example 2. Find the points of contact of the tangents to the circle whose equation is $x^2 + y^2 - 4\,x - 2\,y - 20 = 0$ from the external point $(3, 8)$.

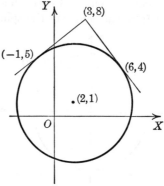

Fig. 122.2

Solution. By differentiation we find that the slope of the tangent at any point (x, y) on the circle is

$$m = \frac{dy}{dx} = \frac{2 - x}{y - 1}.$$

But if (x, y) is one of the points of contact of the required tangents, the slope of the tangent is given by the slope formula as

$$\frac{y - 8}{x - 3}.$$

Hence the coördinates of the points of contact satisfy the equation

$$\frac{2-x}{y-1} = \frac{y-8}{x-3}. \tag{1}$$

But they must also satisfy the equation of the circle. Therefore they may be found by solving the two equations simultaneously. The algebraic work follows.

Clearing (1) of fractions, we obtain

$$x^2 + y^2 - 5x - 9y + 14 = 0. \tag{2}$$

Subtracting this from the equation of the circle, we get

$$x = 34 - 7y. \tag{3}$$

Substituting this in (2), we get

$$y^2 - 9y + 20 = 0.$$

Thus
$$y = 4 \quad \text{or} \quad 5,$$

and
$$x = 6 \quad \text{or} \quad -1.$$

Hence the points of contact of the two tangents are $(6, 4)$ and $(-1, 5)$.

123. The circle determined by any three conditions. In Chapter I we noticed that a straight line is determined by two points upon it and that its standard equations each involve two independent arbitrary constants. Since the equation of the circle contains three arbitrary constants, we should be able to find its equation if we are given three points upon it or any other three geometrical conditions determining it.

The simplest case is where three points are given; but the method of procedure is the same in all cases. Express the geometrical conditions in the form of three equations having as unknowns h, k, and r (or D, E, and F), and solve these equations simultaneously for the unknowns. Then write down the equation of the circle.

Example 1. Find the equation of the circle passing through $(7, 1)$, $(6, 8)$, and $(-1, 7)$.

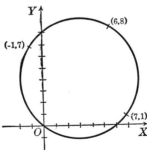

Fig. 123.1

Solution. Each of these pairs of coördinates must satisfy the equation of the circle. Therefore, using the standard form of the equation,

$$(7 - h)^2 + (1 - k)^2 = r^2,$$
$$(6 - h)^2 + (8 - k)^2 = r^2,$$
$$(-1 - h)^2 + (7 - k)^2 = r^2.$$

These equations are easy to solve simultaneously; for if we expand them and subtract the first successively from the second and third, we eliminate all terms of the second degree, having left

$$2\,h - 14\,k = -50,$$
$$16\,h - 12\,k = 0.$$

These give $k = 4$, $h = 3$, and $r = 5$. Therefore the desired equation is

$$(x - 3)^2 + (y - 4)^2 = 5^2,$$

which may be reduced to $x^2 + y^2 - 6\,x - 8\,y = 0.$

If we use the general form of the circle equation, we obtain

$$49 + 1 + 7\,D + E + F = 0,$$
$$36 + 64 + 6\,D + 8\,E + F = 0,$$
$$1 + 49 - D + 7\,E + F = 0.$$

Solving, $D = -6, \quad E = -8, \quad \text{and} \quad F = 0.$

In problems of this kind the circle should be drawn with the correct radius and center. As a check upon the work the coördinates of the given points should be substituted in the final equation. When three points are given, this check is absolute.

Example 2. Find the equation of the circle which is tangent to the line $3\,x - 4\,y = 2$ at the point $(2, 1)$ and which passes through the origin.

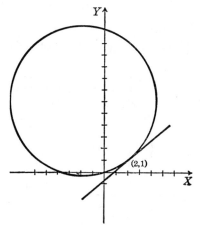

Fig. 123.2

Solution. There are given two points on the circle, $(2, 1)$ and $(0, 0)$. Substituting these coördinates in the standard form of the equation of a circle, we have

$$(2 - h)^2 + (1 - k)^2 = r^2, \tag{1}$$
$$h^2 + k^2 = r^2. \tag{2}$$

We need a third equation. This is provided by the fact that the given line is tangent to the circle at $(2, 1)$ which makes the slope of the tangent at this point equal to $\frac{3}{4}$.

Since the center is the point $C(h, k)$, the slope of the radius to the point $(2, 1)$ is $\dfrac{k - 1}{h - 2}$. Consequently the slope of the tangent at $(2, 1)$ is $\dfrac{2 - h}{k - 1}$, and the third equation is

$$\frac{2 - h}{k - 1} = \frac{3}{4}. \tag{3}$$

Solving (1), (2), and (3) simultaneously, we obtain $h = -\frac{7}{4}$, $k = 6$, and $r = \frac{25}{4}$. Hence the equation of the circle is

$$(x + \tfrac{7}{4})^2 + (y - 6)^2 = \tfrac{625}{16},$$

or

$$2x^2 + 2y^2 + 7x - 24y = 0.$$

PROBLEMS

1. Find the equation of the circle determined by the following points.

a. $(4, 2)$, $(2, 4)$, $(-4, 2)$.   **b.** $(1, 3)$, $(5, 1)$, $(3, -3)$.

c. $(8, 0)$, $(0, 12)$, $(7, 5)$.   **d.** $(1, -8)$, $(9, -4)$, $(10, -5)$.

e. $(2, 1)$, $(2, 9)$, $(10, 5)$.   **f.** $(6, 3)$, $(0, 6)$, $(-6, -6)$.

2. Find the equation of a circle which

a. Has the center $(4, 1)$ and passes through $(-2, 5)$.

b. Passes through $(6, -3)$ and $(8, -1)$ and has the radius 10.

c. Passes through $(0, 4)$ and $(6, 8)$ and has its center on the x-axis.

d. Passes through $(2, 5)$ and $(10, 1)$ and has its center on the line $x - y - 4 = 0$.

e. Passes through $(16, 12)$ and $(2, -2)$ and is tangent to the y-axis.

f. Passes through the point $(4, 4)$ and is tangent to the line $x - y - 4 = 0$ at $(4, 0)$.

g. Passes through $(18, -25)$ and is tangent to both axes.

h. Has its center at $(-2, 4)$ and is tangent to the line $x - y - 6 = 0$.

i. Passes through $(1, -1)$ and is tangent both to the y-axis and the line whose equation is $y = 7$.

3. Prove that the four points $(0, 6)$, $(4, 8)$, $(12, 0)$, and $(4, -6)$ lie on the same circle.

4. Find the equations of the lines which are tangent to the given circle and parallel to the given line.

a. $x^2 + y^2 - 4x - 2y - 20 = 0$, $4x + 3y = 12$.
b. $x^2 + y^2 - 8x + 6y - 20 = 0$, $x + 2y = 10$.

5. Find the coördinates of the points of contact of the tangents drawn from the given point to the given circle.

a. $x^2 + y^2 = 13$, $(5, -1)$.   **b.** $x^2 + y^2 - 10\,x = 0$, $(0, 10)$.
c. $x^2 + y^2 + 10\,y = 0$, $(7, -4)$.   **d.** $x^2 + y^2 + 6\,x - 8\,y = 0$, $(4, 3)$.

6. Find the dimensions of the largest rectangle that can be cut from a semicircle whose radius is a.

7. Find the length of the upper base of the largest trapezoid that can be cut from a semicircle whose radius is a.

8. Find the dimensions of the largest rectangle that can be cut from the segment bounded by the circle whose equation is $x^2 + y^2 = 25$ and the line whose equation is $x = 2$.

9. Show that the equation of the tangent to the circle $x^2 + y^2 = r^2$ at any point (x_1, y_1) is $x_1 x + y_1 y = r^2$.

10. In Problem 9 show that the equation of the normal is $x_1 y - y_1 x = 0$.

11. A square is inscribed in a circle, and another circle is drawn with a vertex of the square as a center and a side of the square as a radius. Find the angle at which it meets the first circle.

12. The equation of a certain circle is $x^2 + y^2 + 6\,x + 4\,y = 87$. Find the equations of circles tangent to this circle and satisfying the following conditions.

a. Center at $(2, -2)$.
b. Point of tangency $(-13, -2)$, radius 6.

THE PARABOLA

124. Definitions. *The locus of a point equidistant from a given fixed point and a given fixed line is called a parabola.*

The fixed point is called the *focus*; and the given fixed line is called the *directrix*. It is understood that the focus is never on the directrix. The distance between the directrix and the focus is denoted by p. The line perpendicular to the directrix and passing through the focus is called the *axis* of the parabola.

By definition of the parabola, the point on the axis halfway between the focus and the directrix lies upon the locus; it is called the *vertex*.

125. Construction of the parabola. There is no simple instrument for drawing a parabola, but, when the focus and directrix are given, as many points as desired may be constructed with ruler and compasses. Let DD' be the directrix and F the focus (Fig. 125). The line AFB, perpendicular to DD', is the axis, and V, midway between A and F, is the vertex. Through any point C on the axis draw a line CC' parallel to the directrix. With AC as radius and F as center describe a circular arc cutting CC' in P and P'. The points P and P' are equidistant from F and from DD' and hence lie on the parabola.

This construction is facilitated by the use of coördinate paper on which the lines CC' are drawn at regular intervals.

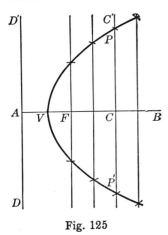

Fig. 125   Fig. 126

126. Equations of the parabola. Any line may be taken as the directrix, and any point not upon the directrix may be taken as the focus. In general, however, the equations thus obtained are very complicated. We shall therefore direct our attention for the present to the equations obtained by taking the vertex at the origin and the axis of the parabola as the x-axis or y-axis.

Let us first take the axis of the parabola as the x-axis. The focus F will then lie upon the x-axis, and the directrix AB will be parallel to the y-axis. Since the vertex bisects the segment of the axis between the focus and the directrix, the coördinates of F are $(\frac{1}{2} p, 0)$, and the equation of AB is $x = -\frac{1}{2} p$ (Fig. 126).

Now let $P(x, y)$ be any point on the locus. Join F and P and draw PM perpendicular to AB. By definition of the parabola,

$$FP = MP.$$

By inspection and by use of the distance formula this becomes at once

$$\sqrt{\left(x - \frac{p}{2}\right)^2 + y^2} = x + \frac{p}{2}.$$

Squaring and simplifying, we have the standard equation

(III) $y^2 = 2\,px.$

If the focus is taken at the point $(0, \frac{1}{2} p)$, and the directrix as the line $y = -\frac{1}{2} p$, we obtain in the same way the form

(III a) $x^2 = 2 py.$

Taking the focus at the left or below the directrix yields two other standard forms, as follows.

Focus $\left(-\frac{p}{2}, 0\right)$, directrix $x = \frac{p}{2}.$

(III b) $y^2 = -2 px.$

Focus $\left(0, -\frac{p}{2}\right)$, directrix $y = \frac{p}{2}.$

(III c) $x^2 = -2 py.$

Focal radius. The distance between the focus of a parabola and any point on the parabola is called the focal radius of that point. In the derivation of **(III)** it is shown that the focal radius $FP = MP = \frac{1}{2} p + x$. Likewise, corresponding to **(III a)**, **(III b)**, and **(III c)**, the formulas for the focal radius are $\frac{1}{2} p + y$, $\frac{1}{2} p - x$, and $\frac{1}{2} p - y$, respectively.

127. Discussion of the equations. We shall discuss only one of the standard equations; the others are treated in a similar manner. Let us take

$$y^2 = 2 px.$$

The form of the equation shows that the intercepts are both 0. From this we deduce that the parabola crosses its axis at the vertex alone.

Since the only term involving y is y^2, there is symmetry with respect to the x-axis; that is, a parabola is symmetrical with respect to its axis.

Since $y^2 \geqq 0$, x is never negative; hence the nearest point of a parabola to its directrix is the vertex. As x increases, $|y|$ also increases. Hence the parabola recedes indefinitely from both its axis and its directrix.

The discussion makes it easy to distinguish between the various standard forms. For a form involving y^2 the axis of symmetry is the x-axis, and the focus is to the left or right of the vertex according to the sign before $2 px$. Similar remarks apply to the forms containing x^2.

128. The latus rectum. The chord through the focus of a parabola perpendicular to its axis is called the *latus rectum*. In Fig. 128, F is the focus, ACB the directrix, and $P'P$ the latus rectum. Draw PG perpendicular to AB.

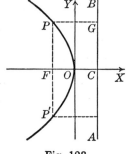

Fig. 128

By definition of the parabola, the distance of P from the focus is the same as that from the directrix. Hence

$$FP = PG = FC = p,$$

and the length of the whole latus rectum is $2\,p$.

129. Drawing the parabola. When the equation of a parabola is given, the curve may be plotted by using as many points as necessary for the degree of accuracy required. If it is desired merely to sketch the curve to show its position, size, and general shape, the method of the following example is usually sufficient.

Example. Sketch the parabola $x^2 = -\,12\,y$.

Solution. This is of the form $x^2 = -\,2\,py$; hence $2\,p = 12$, or $p = 6$. Since the term in x is of the second degree and the sign before $2\,p$ is minus, the axis of the parabola is the y-axis and the curve lies wholly below the x-axis. There-

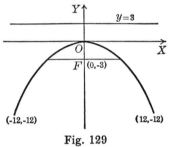

fore the focus is $(0,\,-\frac{1}{2}\,p)$, or $(0,\,-3)$; and the equation of the directrix is $y = 3$. Measuring off $p = 6$ to the right and the left of the focus gives the ends of the latus rectum. Two more points easy to find are those for which $y = -\,2\,p$, or $y = -\,12$. They are $(\pm\,2\,p,\,-\,2\,p)$, or $(\pm\,12,\,-\,12)$. These points, together with the vertex O, make five points on the curve, which are ample for a sketch.

Fig. 129

PROBLEMS

1. With coördinate paper and compasses construct a parabola having (**a**) its focus 10 units from its directrix, (**b**) its latus rectum 16 units long.

2. Find the coördinates of the focus, the equation of the directrix, and the length of the latus rectum of each of the following parabolas. Sketch the curve.

a. $y^2 = 16\,x$.   **b.** $x^2 + 8\,y = 0$.   **c.** $2\,x^2 - 9\,y = 0$.
d. $3\,y = 4\,x^2$.   **e.** $y^2 = 6\,ax$.   **f.** $15\,x + 2\,y^2 = 0$.

3. Write the equations of the parabolas satisfying the given conditions and draw the figure in each case.

a. Directrix $y = 4$, focus $(0,\,-\,4)$.
b. Directrix $x = -\,8$, focus $(8,\,0)$.
c. Directrix $y = -\,2$, vertex $(0,\,0)$.
d. Latus rectum $= 8$, vertex $(0,\,0)$, focus on x-axis.
e. Vertex $(0,\,0)$, focus on x-axis, and passing through $(-\,2,\,6)$.
f. Vertex $(0,\,0)$, and passing through $(4,\,6)$.

4. One end of a chord through the focus of a parabola which has the x-axis as the principal axis and the origin as the vertex is the point $(4, -4)$. Find the coördinates of the other end.

5. Find the equation of the circle passing through the vertex and the ends of the latus rectum of the parabola $y^2 = 8x$.

6. In Problem 5 find the angle between the circle and the parabola at each point of intersection.

7. Find the slope of the parabola $y^2 = 2px$ in terms of x and show that as x approaches infinity the slope approaches 0.

8. Find the equations of the tangent and the normal to each of the following parabolas at the point indicated.

a. $2y^2 = 3x$, $(6, 3)$.   b. $x^2 = 16y$, $(8, 4)$.
c. $y^2 + 5x = 0$, $(-5, 5)$.   d. $y = 6x^2$, $(1, 6)$.

9. Find the points at which the tangent to the corresponding curve in Problem 8 has the indicated slope.

   a. $\frac{3}{4}$.   b. $\frac{1}{2}$.   c. $\frac{5}{6}$.   d. 6.

10. Show that the line from the focus of the parabola $y^2 = 24x$ to the point where the tangent to the parabola at the point $(24, 24)$ cuts the y-axis is perpendicular to the tangent.

11. Find the angles of intersection of the graphs of the following pairs of equations.

   a. $4y = x^2$, $2x = y$.   b. $y = x^2$, $x^2 + y^2 = 12$.

$$c.\ x^2 = 4y,\ y = \frac{8}{x^2 + 4}.$$

12. Find the tangent of the angle of intersection of the parabolas $y^2 = 2px$ and $x^2 = 2py$ at their point of intersection in the first quadrant.

13. What are the dimensions of the largest rectangle which can be inscribed in the segment of the parabola $x^2 = 24y$ cut off by the line $y = 6$?

14. Show that for any point on the parabola $y^2 = 2px$ the segment of the tangent between the point of tangency and the x-axis has a projection upon the x-axis of $2x$. From this deduce a geometric method of drawing a tangent to a parabola at any point.

15. Show that for any point on the parabola $y^2 = 2px$ the segment of the normal between the point of tangency and the x-axis has a projection upon the x-axis of length p.

16. Prove that the tangents to a parabola at the ends of the latus rectum
   a. Are perpendicular.   b. Intersect on the directrix.

17. Show that the equation of the tangent to the parabola $y^2 = 2px$ at the point (x_1, y_1) is $y_1y = p(x + x_1)$.

18. Generalize Problem 13 to find the dimensions of the largest rectangle which can be inscribed in the segment of a parabola cut off by the latus rectum.

19. Let a circle whose center is the focus F of a parabola cut the parabola at a point P and the axis in points A and B. Show that one of the lines AP and BP is a tangent to the parabola and the other is a normal. (See Problem 14.)

20. Generalize Problem 10 to show that the line from the focus of a parabola whose equation is $y^2 = 2\,px$ to the point where the tangent to the parabola at the point $(x_1,\, y_1)$ cuts the y-axis is perpendicular to the tangent.

130. Applications of the parabola: the parabolic reflector. The parabola is frequently met with in the applications of mathematics to the sciences. The following are examples: the paths of some comets seem to be parabolic; arches are sometimes made in this shape; the first approximation to the path of a projectile is the parabola; the cable of a suspension bridge has the form of a parabola.

Among the most interesting of the applications is the parabolic reflector. The inner surface of such a reflector is that generated by rotating a parabola about its axis; the lamp is placed at the focus. The reason for its use is that every ray of light from the lamp at the focus is reflected along a line parallel to the axis of the parabola. We will now prove this fact.

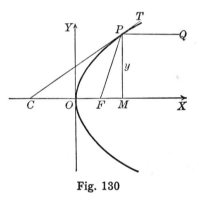

Fig. 130 represents a sectional view of the reflector. Let FP be any ray of light from the focus, which is reflected at P along the line PQ. Let CPT be tangent to the parabola at P. We have to prove that PQ is parallel to the axis CX for any position of P.

Fig. 130

From the equation $y^2 = 2\,px$ we find that the slope of the parabola at P is $p/y = \tan PCX$. On the other hand, the figure shows that $\tan PCX = y/CM$. Hence

$$\frac{y}{CM} = \frac{p}{y}, \quad \text{or} \quad CM = \frac{y^2}{p} = \frac{2\,px}{p} = 2\,x.$$

But $$OM = x,$$
and therefore $$CO = x;$$
whence $$CF = x + \frac{p}{2}.$$

Now FP is a focal radius and, by Art. 126,
$$FP = \tfrac{1}{2}\,p + x.$$

Hence $FP = CF$ and the triangle CFP is isosceles. Therefore
$$\text{angle } PCX = \text{angle } CPF.$$

By a principle of physics the ray FP and the reflected ray PQ make equal angles with the tangent CT; that is,

$$\text{angle } CPF = \text{angle } TPQ.$$

Hence
$$\text{angle } PCX = \text{angle } TPQ,$$

and the line PQ is parallel to CX by a well-known theorem of plane geometry. Thus all reflected rays have the direction of the axis of the parabola.

131. The parabolic arch. In practical problems involving the construction of a parabolic arch the dimensions given are the span ($AB = 2a$, in Fig. 131) and the height ($CO = h$, in the figure).

The curve is then constructed in the following way. The rectangle $OCBD$ is drawn and OD is divided into a certain number (four in the figure) of equal parts by the points of division K_1, K_2, K_3. The side DB is divided into the same number of equal parts by the points L_1, L_2, L_3. Lines $K_1K'_1$, $K_2K'_2$, $K_3K'_3$ are drawn parallel to OC, and OL_1, OL_2, OL_3 are drawn. The intersections of OL_1 and $K_1K'_1$, OL_2 and $K_2K'_2$, etc. are points on the curve.

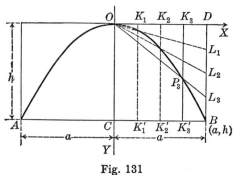

Fig. 131

To prove that this process gives points on the required parabola it is convenient to take the origin at the top of the arch and the positive direction of the y-axis as downward. Then the coördinates of the point B are (a, h). Let the coördinates of any one of the constructed points, say P_3, be (x, y); then $x = OK_3$ and $y = K_3P_3$.

Since the triangles OK_3P_3 and ODL_3 are similar,

$$\frac{x}{a} = \frac{y}{DL_3}.$$

Since OD and DB were divided into the same number of equal parts,

$$\frac{x}{a} = \frac{DL_3}{h}.$$

Multiplying these two equations together, we get

$$\frac{x^2}{a^2} = \frac{y}{h},$$

or
$$hx^2 = a^2 y.$$

This is evidently the equation of a parabola which is symmetrical with respect to the y-axis and it is satisfied by the coördinates of O, A, and B.

132. Other equations of the parabola. The equations of the parabola so far given are valid for a very restricted case, namely, that in which the vertex is at the origin and the directrix is parallel to one of the coordinate axes. If these conditions are not fulfilled, the equation is more complicated. In more extensive books on analytic geometry it is shown that this equation is always of the second degree, and transformations are given which enable us to find the vertex, the focus, etc.

These we shall not consider; for in problems where the focus and directrix of the parabola are known, it is usually possible to choose the origin and the axes in such a manner that we can use the simpler forms already derived.

On the other hand, it is frequently convenient to recognize the locus of the equation $y = a + bx + cx^2$ as a parabola with its axis perpendicular to the x-axis. Similarly, $x = a + by + cy^2$ is the equation of a parabola with its axis parallel to the x-axis.

If the vertex is not at the origin, but the axis is identical with or parallel to one of the coördinate axes, special equations can be found by using the definitions in Art. 124. The results are as follows. Proofs are left as exercises.

If the vertex is $V(h, k)$ and the focus is $F(h + \frac{1}{2} p, k)$, the equation of the parabola is

(III d) $(y - k)^2 = 2\,p(x - h).$

If the vertex is $V(h, k)$ and the focus is $F(h - \frac{1}{2} p, k)$, the equation of the parabola is

(III e) $(y - k)^2 = -\,2\,p(x - h).$

If the vertex is $V(h, k)$ and the focus is $F(h, k + \frac{1}{2} p)$, the equation of the parabola is

(III f) $(x - h)^2 = 2\,p(y - k).$

If the vertex is $V(h, k)$ and the focus is $F(h, k - \frac{1}{2} p)$, the equation of the parabola is

(III g) $(x - h)^2 = -\,2\,p(y - k).$

When solved for x, **(III d)** and **(III e)** take the form $x = a + by + cy^2$. When solved for y, **(III f)** and **(III g)** take the form $y = a + bx + cx^2$.

133. Discussion of the equation $y = a + bx + cx^2$. For very large values of x the value of y will depend chiefly on the term cx^2. Hence if c is positive, y will be positive for large values of x, which means that the branches of the parabola extend upward. If c is negative, y will be negative for very large values of x, which means that the branches extend downward.

Differentiating, we find
$$y' = b + 2\,cx.$$

Setting $y' = 0$, we find that $x = -b/2\,c$. Since $y'' = 2\,c$, y will have a maximum value if $c < 0$, and a minimum value if $c > 0$. Hence the point where $x = -b/2\,c$ is the vertex and the equation of the axis is $x = -b/2\,c$.

No attempt should be made to memorize these results, but each equation of the type given should be treated in the above manner.

Example 1. Find the vertex and sketch the parabola whose equation is $4\,y = 4 + 8\,x - x^2$.

Solution. Since the coefficient of x^2 is negative, the branches of the parabola extend downward.

Differentiating, $y' = \dfrac{8 - 2\,x}{4}$. Setting $y' = 0$, $x = 4$. The substitution of this value in the original equation gives $y = 5$. Hence the vertex is $(4, 5)$ and the axis is $x = 4$ (Fig. 133.1).

A few more points are easily computed from the given equation. For example, $x = 0$ and $x = 8$ give the symmetrical points $(0, 1)$ and $(8, 1)$.

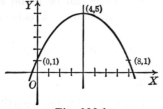

Fig. 133.1

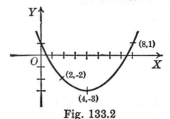

Fig. 133.2

Example 2. Find the equation of a parabola with a vertical axis and passing through the points $(8, 1)$, $(4, -3)$, and $(2, -2)$ (Fig. 133.2).

Solution. Substituting these coördinates in the equation
$$y = a + bx + cx^2,$$
we obtain
$$1 = a + 8\,b + 64\,c,$$
$$-3 = a + 4\,b + 16\,c,$$
$$-2 = a + 2\,b + 4\,c.$$

Solving these equations simultaneously, we have $a = 1$, $b = -2$, and $c = \tfrac{1}{4}$. The desired equation is then
$$y = 1 - 2\,x + \tfrac{1}{4}\,x^2,$$
which reduces to
$$4\,y = 4 - 8\,x + x^2.$$

PROBLEMS

1. Find the vertex of the following parabolas and draw the graph in each case.

a. $y = 4x - x^2$.

b. $2y = 8x - x^2$.

c. $4y = 8 + 16x - x^2$.

d. $3y = 4x^2 - 3$.

e. $5y = x^2 + 4x - 6$.

f. $4x = 9y^2 - 18y - 2$.

2. Find the equation of a parabola with a vertical axis through each of the following sets of points.

a. $(2, 5)$, $(3, 0)$, $(-7, 0)$.

b. $(5, -8)$, $(2, -6)$, $(-5, -3)$.

c. $(3, 11)$, $(5, 3)$, $(4, 5)$.

d. $(-2, 5)$, $(-8, -4)$, $(2, 4)$.

e. $(-1, 2)$, $(-2, -3)$, $(3, -1)$.

f. $(1, 6)$, $(2, 4)$, $(4, -2)$.

3. Find the equation of a parabola with its axis parallel to the x-axis and passing through each of the sets of points in Problem 2.

4. Using the definition of a parabola, find the equations of the parabolas satisfying the following data and draw their graphs.

a. Directrix $y = 7$, focus $(2, -1)$.

b. Directrix $y = -2$, focus $(4, 4)$.

c. Directrix $x = 6$, focus $(-2, 3)$.

d. Directrix $x = -4$, focus $(6, 2)$.

5. Plot carefully the graph of the parabola $\sqrt{x} + \sqrt{y} = \sqrt{a}$. What is its axis of symmetry? Where is its vertex?

6. At what point on the curve will the tangent to the parabola $y = 4 + 2x - x^2$ be parallel to the line $2x + y - 6 = 0$?

7. At what point on the curve will the tangent to the parabola $y = x^2 - 7x + 3$ be perpendicular to the line $x + 3y = 3$?

8. The path of a projectile from a mortar cannon lies on the parabola $y = 2x - x^2$; the unit is 1 mi., OX being horizontal and OY vertical, and the origin is the point of projection.

a. Find the direction of motion of the projectile at the instant of projection.

b. Find the direction of motion of the projectile when it strikes a vertical cliff $1\frac{1}{2}$ mi. distant.

c. Where will the path have an inclination of $45°$ to the horizontal?

d. What will be the highest point reached by the projectile?

9. Find the angle at which the parabolas $y = 4 - x^2$ and $y = x^2 - 14$ intersect.

10. Find the equation of the tangent to the parabola whose equation is $y = 4x - x^2$ which is parallel to the line joining the points $(1, 3)$ and $(4, 0)$.

11. In Problem 10 find the angles of intersection of the line and the parabola.

12. Find the points on the parabola whose equation is $y = 8x - x^2$ which are nearest to the point $(4, 0)$.

13. A parabola which has its vertex at the origin and the y-axis as its axis is tangent to the line whose equation is $x - 2y = 4$. Find the equation of the parabola and the coördinates of the point of contact.

14. A parabola whose axis is parallel to the y-axis passes through the origin and is tangent to the line whose equation is $x - 2y = 8$ at the point $(20, 6)$. What is the equation of the parabola and where is its vertex?

15. A point $P(x, y)$ moves along the parabola whose equation is $2y = 16 - x^2$. When P is at the point $(-2, 6)$, x is increasing at the rate of 2 units/sec. How fast is the distance between P and the vertex changing?

16. The cable of a suspension bridge assumes the shape of a parabola if the weight of the suspended roadbed (together with that of the cable) is uniformly distributed horizontally. Suppose that the towers of a bridge are 240 ft. apart and 60 ft. high and that the lowest point of the cable is 20 ft. above the roadway; find the vertical distance from the roadway to the cables at intervals of 20 ft.

THE ELLIPSE

134. Definitions. *The locus of a point the sum of whose distances from two fixed points is constant is an ellipse.*

The fixed points are called the *foci*, and the distance between the foci is denoted by $2c$.

The sum of the distances of any point of the ellipse from the foci is denoted by $2a$. Obviously $2a > 2c$, or $a > c$.

135. Construction of an ellipse. The definition suggests at once a simple mechanical construction of an ellipse. Let pins be placed at the foci F and F', and a loop of string $F'PFF'$ of length $F'F + 2a$ be placed over the pins. If a pencil is placed at P and moved so as to keep the string taut, it will describe an ellipse. For then $F'P + FP$ will constantly equal $2a$.

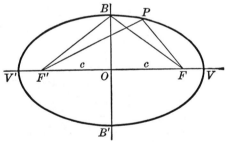

Fig. 135

136. Some properties of the ellipse. The preceding construction shows intuitively (as will be proved later analytically) that the curve is symmetrical with respect to the indefinite line through F and F', and also that it is symmetrical with respect to the line BB', which is the perpendicular bisector of FF'. The indefinite line through the foci is called the *principal axis* of the ellipse. The points V and V', where the principal axis cuts the curve, are the *vertices* of the ellipse.

Since the curve is symmetrical with respect to each of two perpendicular axes, it is symmetrical with respect to their point of intersection. Consequently the point O midway between the foci is called the *center* of the ellipse.

The distance OV from the center to one of the vertices is a. For, by definition, $FV + F'V = 2\,a$. But, by symmetry, $V'F' = FV$. Hence $V'V = V'F' + F'V = 2\,a$ and so $OV = a$.

The point B is equidistant from F and F'. Hence $FB = a$. Let $OB = b$. Then, from the right triangle OBF, $b^2 = a^2 - c^2$. Hence $b < a$; that is, $OB < OV$ and $B'B < V'V$. For this reason the axis $V'V$ $(= 2\,a)$ is called the *major axis* of the ellipse, and the axis $B'B$ $(= 2\,b)$ is called the *minor axis*.

137. An equation of the ellipse. Take the principal axis as the x-axis and the point midway between the foci as the origin. Then the coördinates of the foci are $(c, 0)$ and $(-c, 0)$.

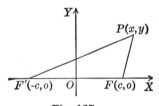

Fig. 137

If $P(x, y)$ is any point of the ellipse, the definition requires that $F'P + FP = 2\,a$. Hence, by the distance formula,

$$\sqrt{(x + c)^2 + y^2} + \sqrt{(x - c)^2 + y^2} = 2\,a.$$

Solving for the first of the radicals, squaring, and collecting terms, we have

$$4\,cx - 4\,a^2 = -\,4\,a\sqrt{(x - c)^2 + y^2}.$$

Dividing by 4 and squaring again, we get

$$c^2x^2 - 2\,a^2cx + a^4 = a^2x^2 - 2\,a^2cx + a^2c^2 + a^2y^2,$$

or $\qquad (a^2 - c^2)x^2 + a^2y^2 = a^4 - a^2c^2 = a^2(a^2 - c^2).$

But $\qquad\qquad a^2 - c^2 = b^2.$

This gives $\qquad b^2x^2 + a^2y^2 = a^2b^2.$

Dividing both sides by a^2b^2, we obtain the standard equation

(IV) $$\frac{x^2}{c^2} + \frac{y^2}{b^2} = 1.$$

138. Discussion of the ellipse. The discussion of **(IV)** verifies the properties of the ellipse stated in Art. 136 and gives some additional ones.

Intercepts and symmetry. It appears from the equation that the x-intercepts are $\pm a$ and the y-intercepts are $\pm b$. All three tests for symmetry are satisfied. Hence the ellipse is symmetrical with respect to its principal axis and the perpendicular bisector of the segment joining the foci.

Extent. The terms x^2/a^2 and y^2/b^2 are both squares and hence never negative. Since their sum is constantly equal to 1, two facts are at once apparent: (1) As x increases numerically y decreases numerically, and vice versa. (2) Neither term can be greater than 1; hence x is never greater numerically than a, and y is never greater numerically than b.

Thus the ellipse lies wholly within a rectangle whose sides are parallel to the axes of symmetry and pass through the vertices and the ends of the minor axis.

The circle as a limiting form of the ellipse. If we take $b = a$, the standard equation of the ellipse reduces to $x^2 + y^2 = a^2$, which is the equation of a circle of radius a. Thus the circle is a special form of the ellipse. It is instructive to see just how this happens.

If in the equation $b^2 = a^2 - c^2$ we take $b = a$, c becomes 0. Hence a circle is an ellipse whose foci coincide. This is also obvious from a consideration of the construction explained in Art. 135.

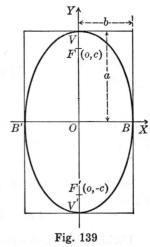

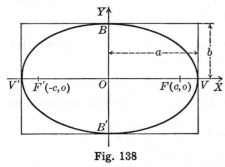

Fig. 138   Fig. 139

139. Equation of ellipse with foci on the y-axis. If the foci are on the y-axis and the center is at the origin, the coördinates of the foci are $(0, c)$ and $(0, -c)$. Taking $F'P + FP = 2\,a$, and proceeding as in Art. 137, we obtain the equation

(IV a)
$$\frac{y^2}{a^2} + \frac{x^2}{b^2} = 1.$$

In this case the coördinates of the vertices are $(0, \pm a)$, and those of the ends of the minor axis are $(\pm b, 0)$.

140. Sketching the Ellipse. The standard equations **(IV)** and **(IV *a*)** show that the locus of any equation of the form $Ax^2 + Cy^2 = F$, where A, C, and F are positive, is an ellipse with its center at the origin and with its foci on one of the coördinate axes. For such an equation can be reduced to one of the two standard forms by dividing through by F and writing the equation in the form

$$\frac{x^2}{\frac{F}{A}} + \frac{y^2}{\frac{F}{C}} = 1.$$

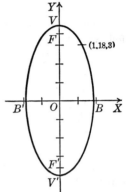

To sketch the locus of such an equation we first find the intercepts. The larger intercept is half the major axis (semi-major axis), and the smaller intercept is half the minor axis (semi-minor axis). To avoid the tendency to make the ellipse too pointed, compute the coördinates of an additional pair of points near each vertex.

Example. Sketch the ellipse whose equation is $5\,x^2 + y^2 = 16$. Also find its foci.

Fig. 140

Solution. The intercepts on the x-axis are $\pm\sqrt{\frac{16}{5}} = \pm 1.79$. The intercepts on the y-axis are ± 4. The major axis lies along the y-axis. Hence $a = 4$ and $b = \sqrt{\frac{16}{5}} = 1.79$. From the relation $c^2 = a^2 - b^2$ we find that $c = \sqrt{\frac{64}{5}} = 3.58$. The foci are then $(0, \pm 3.58)$. Four additional points are obtained by taking $y = \pm 3$, giving $x = \pm\sqrt{\frac{7}{5}} = \pm 1.18$.

141. Generalized standard forms of the equation of the ellipse. If the center is not at the origin, but the axes are identical with or parallel to the coördinate axes, special equations can be found by using the methods of Arts. 137 and 139. The results are as follows. Proofs are left as exercises.

If the center of an ellipse is the point (h, k) and the principal axis is parallel to the x-axis, the equation is

(IV *b*)
$$\frac{(x - h)^2}{a^2} + \frac{(y - k)^2}{b^2} = 1.$$

If the center of an ellipse is the point (h, k) and the principal axis is parallel to the y-axis, the equation is

(IV *c*)
$$\frac{(y - k)^2}{a^2} + \frac{(x - h)^2}{b^2} = 1.$$

Both equations can be reduced to the form

$$Ax^2 + Cy^2 + Dx + Ey + F = 0,$$

where A and C have like signs. Conversely an equation in this form can be reduced to one of the forms **(IV b)** and **(IV c)** unless the locus is imaginary or a point, and the various facts regarding the ellipse can then be stated by inspection.

PROBLEMS

1. Find the major and minor axes and the coördinates of the foci of the following ellipses, and sketch the curves.

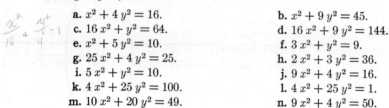

<table>
<tr><td>a. $x^2 + 4\,y^2 = 16.$</td><td>b. $x^2 + 9\,y^2 = 45.$</td></tr>
<tr><td>c. $16\,x^2 + y^2 = 64.$</td><td>d. $16\,x^2 + 9\,y^2 = 144.$</td></tr>
<tr><td>e. $x^2 + 5\,y^2 = 10.$</td><td>f. $3\,x^2 + y^2 = 9.$</td></tr>
<tr><td>g. $25\,x^2 + 4\,y^2 = 25.$</td><td>h. $2\,x^2 + 3\,y^2 = 36.$</td></tr>
<tr><td>i. $5\,x^2 + y^2 = 10.$</td><td>j. $9\,x^2 + 4\,y^2 = 16.$</td></tr>
<tr><td>k. $4\,x^2 + 25\,y^2 = 100.$</td><td>l. $4\,x^2 + 25\,y^2 = 1.$</td></tr>
<tr><td>m. $10\,x^2 + 20\,y^2 = 49.$</td><td>n. $9\,x^2 + 4\,y^2 = 50.$</td></tr>
</table>

2. Find the equation of the ellipse whose center is at the origin and which satisfies the following conditions.

    a. x-intercepts $\pm\, 14$, y-intercepts $\pm\, 7$.
    b. Major axis 20, minor axis 12, vertices on the x-axis.
    c. Major axis 20, minor axis 12, vertices on the y-axis.
    d. Major axis 16, minor axis 8, vertices on the x-axis.
    e. Major axis 16, minor axis 8, vertices on the y-axis.

3. Find the equation of the ellipse whose center is at the origin and which satisfies the following conditions.

    a. One vertex $(5, 0)$, one focus $(3, 0)$.
    b. One focus $(4, 0)$, minor axis 8.
    c. One vertex $(0, 5)$, minor axis 6.
    d. One vertex $(0, 6)$, one focus $(0, 4)$.

4. Find the equation of the ellipse which satisfies the following conditions. The center is at the origin, the principal axis is one of the coördinate axes, and two of its points are

    a. $(4, 3)$ and $(6, 2)$.   **b.** $(-\,1, 6)$ and $(2, 3)$.

5. Are the following points on, inside, or outside the ellipse $x^2 + 4\,y^2 = 16$?

    a. $(2, \tfrac{9}{5})$.   **b.** $(\tfrac{17}{5}, -\,1)$.   **c.** $(\tfrac{3}{2}, \tfrac{7}{4})$.

6. The chord of an ellipse through the focus perpendicular to the principal axis is called the *latus rectum*. Prove that its length is $\dfrac{2\,b^2}{a}$.

7. The quotient c/a is called the *eccentricity* of the ellipse and is denoted by e. Show that for a circle $e = 0$ and that for all ellipses $e < 1$. Prove that if two ellipses have the same eccentricity, their major and minor axes are proportional.

8. From points on the circumference of the circle whose equation is $x^2 + y^2 = 64$, perpendiculars are drawn to the x-axis. Find the equation of the locus of a point which moves so as to bisect each perpendicular.

9. Find the equation of the locus of the vertex of a triangle whose base is the line joining $(a, 0)$ and $(- a, 0)$ and in which the product of the tangents of the base angles is b^2/a^2.

10. Show that the slope of the line which is tangent to the ellipse whose equation is $b^2x^2 + a^2y^2 = a^2b^2$ at any point is $- b^2x/a^2y$.

11. Find the equations of the tangent and normal to the ellipse $x^2 + 3\,y^2 = 21$ at the point in the fourth quadrant where $x = 3$.

12. Find the equations of the two tangents to the ellipse $x^2 + 4\,y^2 = 18$ which pass through the point $(2, 2)$.

13. A circle is circumscribed about the ellipse whose equation is $4\,x^2 + y^2 = 16$, and a tangent is drawn to each curve in the first quadrant at the point whose ordinate is 2. Find the point of intersection of these tangents.

14. Find the angles of intersection of the following curves.

> **a.** $x^2 + 4\,y^2 = 8$, $x^2 = 2\,y + 2$.
> **b.** $9\,x^2 + 4\,y^2 = 36$, $4\,x^2 + 9\,y^2 = 36$.
> **c.** $2\,x^2 + y^2 = 24$, $x + 2\,y = 10$.
> **d.** $4\,x^2 + y^2 = 8$, $y^2 = 4\,x$.
> **e.** $x^2 + 4\,y^2 = 65$, $y = 4\,x$.

15. At what point on the ellipse $16\,x^2 + 9\,y^2 = 400$ does y decrease at the same rate that x increases?

16. The lower base of an isosceles trapezoid is the major axis of an ellipse; the ends of the upper base are points on the ellipse. Show that the maximum trapezoid of this type has the length of its upper base half that of the lower.

17. Find the area of the largest rectangle that can be inscribed in an ellipse.

18. An isosceles triangle is to be inscribed in the ellipse $b^2x^2 + a^2y^2 = a^2b^2$, the vertex being taken at $(0, b)$. Find the equation of the base if the triangle is a maximum.

19. Using the result of Problem 10 and the formula for the angle between two lines, show that the acute angle between a tangent at any point and the line joining this point to either focus is given by the formula $\tan \theta = b^2/cy$.

20. Prove that the equation of the tangent to the ellipse $b^2x^2 + a^2y^2 = a^2b^2$ at the point (x_1, y_1), which lies on the ellipse, is $b^2x_1x + a^2y_1y = a^2b^2$.

21. Using the generalized standard forms write the equations of the ellipses which satisfy the following conditions.

> **a.** Major axis 20, minor axis 12, center $(- 3, 2)$, principal axis $x = - 3$.
> **b.** Foci $(3, - 2)$ and $(9, - 2)$, minor axis 8.
> **c.** Vertices $(\pm 10, 6)$, latus rectum 10. See Problem 6.
> **d.** Vertices $(0, 0)$ and $(10, 0)$, foci $(1, 0)$ and $(9, 0)$.

22. Reduce the following equations of ellipses to the generalized standard forms and sketch the graphs.

a. $x^2 - 8x + 4y^2 = 0$.　　　　　　**b.** $16x^2 + y^2 - 32x + 4y - 44 = 0$.
c. $4x^2 + 25y^2 + 100y = 0$.　　　　**d.** $9x^2 + 4y^2 + 36x - 24y - 252 = 0$.

THE HYPERBOLA

142. Definitions. *The locus of a point the difference of whose distances from two fixed points is constant is a hyperbola.*

The fixed points are called the *foci*, and the distance between the foci is denoted by $2c$.

The difference of the distances of any point of the hyperbola from the foci is denoted by $2a$. By a theorem in plane geometry, $2a < 2c$, or $a < c$.

143. Construction of a hyperbola. Let F and F' be the foci, and suppose one end of a string, indefinite in length, is passed through a hole

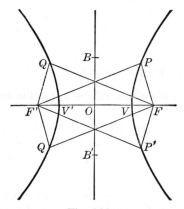

Fig. 143

in the paper at F, while the other end is passed through a hole at F'. When the string is drawn tight, suppose a pencil is knotted in it at V so that $F'V - FV = 2a$. Then let the string be drawn out through each hole *at the same rate* while the pencil is moved so as to keep the two parts taut. Then, for any position P of the pencil, $F'P - FP = 2a$, and, by definition, the pencil will describe a hyperbola. The definition does not prescribe the order in which the difference of the distances shall be taken. If the pencil is knotted at V' so that $FV' - F'V' = 2a$, the same process will generate another curve, $QV'Q$. These two curves together constitute the hyperbola, each curve being called a *branch* of the hyperbola.

144. Some properties of the hyperbola. The preceding construction shows intuitively (as will be proved later analytically) that the hyperbola is symmetrical with respect to the indefinite line through F and F' and also that it is symmetrical with respect to the line BB', which is the perpendicular bisector of FF'. The indefinite line through the foci is called the *principal axis* of the hyperbola. The points V and V', where the principal axis cuts the curve, are the vertices of the hyperbola.

Since the curve is symmetrical with respect to the perpendicular lines FF' and BB', it is symmetrical with respect to their point of intersection. The point O midway between the foci is called the *center* of the hyperbola.

The distance OV from the center to one of the vertices is a. The proof is similar to that in Art. 134.

The segment of the principal axis VV' ($= 2\,a$) is called the *transverse* axis of the hyperbola.

The points B and B' are each at a distance b from O, where $b = \sqrt{c^2 - a^2}$. The segment BB' ($= 2\,b$) is called the *conjugate* axis of the hyperbola. Note that since, by definition, $b^2 = c^2 - a^2$, whence $c^2 = a^2 + b^2$, b may be either larger or smaller than a.

145. An equation of the hyperbola. Take the principal axis as the x-axis and the point midway between the foci as the origin. Then the coördinates of the foci are $(c, 0)$ and $(-c, 0)$.

If $P(x, y)$ is any point on the hyperbola, the definition requires that

$$F'P - FP = \pm 2\,a.$$

Hence, by the distance formula,

$$\sqrt{(x + c)^2 + y^2} - \sqrt{(x - c)^2 + y^2} = \pm 2\,a.$$

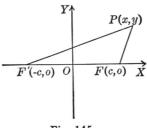

Fig. 145

Solving for the first of the radicals, squaring and collecting terms, we have

$$4\,cx - 4\,a^2 = \pm 4\,a\sqrt{(x - c)^2 + y^2}.$$

If this equation is divided by 4 and squared again, it may be reduced to

$$(c^2 - a^2)x^2 - a^2y^2 = a^2(c^2 - a^2).$$

Set $b^2 = c^2 - a^2$. This gives

$$b^2x^2 - a^2y^2 = a^2b^2.$$

Dividing by a^2b^2, we obtain the standard equation

(V) $$\frac{x^2}{a^2} - \frac{y^2}{b^2} = 1.$$

The student should compare the above carefully with Art. 137.

146. Discussion of the hyperbola. A discussion of (V) verifies the properties stated in Art. 144 and gives some additional ones.

Intercepts and symmetry. The x-intercepts are $\pm a$. The y-intercepts are imaginary. Hence the curve does not cross the y-axis and there are two *branches*. All three tests for symmetry are satisfied. Hence the hyperbola is symmetrical to its principal axis and to the perpendicular bisector of the segment joining the foci.

Extent. If we write (V) in the form $\frac{x^2}{a^2} = 1 + \frac{y^2}{b^2}$, we see at once that $\frac{x^2}{a^2}$ is never less than 1, and hence that x is always numerically greater than or equal to a. Also, as y increases, x increases, numerically. On the other hand, y may have any value.

Therefore the two branches of the hyperbola lie outside the lines $x = \pm a$ and recede to infinity from both axes in all four quadrants.

147. Asymptotes. If the equation is solved for y in terms of x, we obtain $y = \pm \frac{b}{a}\sqrt{x^2 - a^2}$, a result which indicates that, for values of x large in comparison with that of a, y is approximated by $\pm bx/a$. This suggests that the hyperbola approaches the lines whose equations are $y = \pm bx/a$ as its tracing point recedes to infinity. Lines approached by a curve in this manner are called *asymptotes*.

To prove the above statements, take a point $P(x, y)$ on the hyperbola in the first quadrant and let $P_1(x, y_1)$ be the point on the line $y = bx/a$ which has the same abscissa. We must show that $\lim\limits_{x \to \infty} (y_1 - y) = 0$.

From the equations of the line and of the hyperbola, we have

$$y_1 - y = \frac{bx}{a} - \frac{b\sqrt{x^2 - a^2}}{a}$$

$$= \frac{b(x - \sqrt{x^2 - a^2})}{a}.$$

Multiplying numerator and denominator by $x + \sqrt{x^2 - a^2}$, we have

$$y_1 - y = \frac{ab}{x + \sqrt{x^2 - a^2}}.$$

Both terms in the denominator of this fraction are positive and increase indefinitely with x. Hence the limit of the denominator as $x \longrightarrow +\infty$ is $+\infty$. Since the numerator is a constant, we conclude that the limit of the fraction is zero; that is

$$\lim_{x \to \infty} (y_1 - y) = 0.$$

The other four quadrants are treated in the same way.

We have thus shown that as the point tracing the hyperbola recedes to infinity it approaches asymptotically lines through the origin with slopes $\pm b/a$. These are easily seen to be the diagonals of the rectangle whose sides are the lines $x = \pm a$ and $y = \pm b$. Another use for this rectangle is in locating the foci; for half the length of a diagonal is c and so a circle which circumscribes the rectangle will cut the principal axes at the foci.

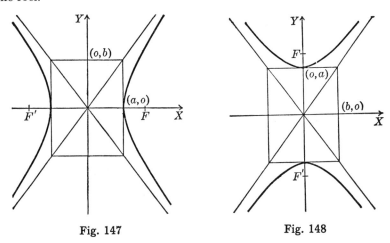

Fig. 147   Fig. 148

148. **Equation of hyperbola with foci on the y-axis.** Let us take the foci at $(0, c)$ and $(0, -c)$. Proceeding exactly as in Art. 145, we obtain a second standard form,

(V a) $$\frac{y^2}{a^2} - \frac{x^2}{b^2} = 1.$$

For this form the vertices are $(0, \pm a)$. No value of y is numerically less than a, and x may have any value.

Solving for y,

$$y = \pm \frac{a}{b} \sqrt{x^2 + b^2}.$$

From this it may be shown, as in Art. 147, that for this form the asymptotes are the lines

$$y = \pm \frac{ax}{b}.$$

These asymptotes are the diagonals of the rectangle whose sides are the lines $x = \pm b$, $y = \pm a$.

149. Sketching the hyperbola. Inspection of the standard equations **(V)** and **(V a)** shows that the locus of any equation of the form $Ax^2 - Cy^2 = \pm F$, where A, C, and F are positive, is a hyperbola with its center at the origin and with the foci on one of the coördinate axes. For such an equation can be reduced to one of the two standard forms by dividing through by F and writing the equation in the form

$$\frac{x^2}{\dfrac{F}{A}} - \frac{y^2}{\dfrac{F}{C}} = \pm 1.$$

To sketch the locus of such an equation we first find the intercepts, which will be real on one of the coördinate axes and imaginary on the other. The real intercept is half the transverse axis (*semi-transverse axis*). The foci and vertices lie upon the coördinate axis on which the intercepts are real. A sketch of the curve may be made by proceeding as in the example below.

Example. Find the vertices and foci of the hyperbola $4x^2 - 9y^2 = 36$, and draw the curve.

Solution. The real intercepts are $x = \pm 3$, and the imaginary intercepts are $y = \pm 2\sqrt{-1}$. Thus $a = 3$, $b = 2$, and $c = \sqrt{13} = 3.61$. Hence the vertices are $(\pm 3, 0)$ and the foci are $(\pm 3.61, 0)$.

Four other points are obtained by taking $x = \pm 4$, whence $y = \pm 1.76$. The slopes of the asymptotes are $\pm \dfrac{b}{a} = \pm \dfrac{2}{3}$, and they may be drawn as the diagonals of the rectangle whose sides are $x = \pm 3$ and $y = \pm 2$. Using the asymptotes as guiding lines, we draw each branch of the curve through the three points plotted.

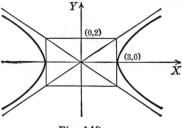

Fig. 149

Conjugate hyperbolas. If we change the signs of the left-hand member of the given equation we get $9\,y^2 - 4\,x^2 = 36$, which is the equation of a hyperbola having the same asymptotes as the given one but with its vertices and foci on the y-axis. The values of a and b are interchanged but c remains the same, and so the circle circumscribing the rectangle passes through the foci of the new hyperbola. Such pairs of hyperbolas are called *conjugate* to each other.

PROBLEMS

1. Sketch the following hyperbolas with their asymptotes. Find the foci in each case.

a. $x^2 - y^2 - 4 = 0$.   **b.** $x^2 - y^2 + 4 = 0$.

c. $x^2 - 4\,y^2 - 16 = 0$.   **d.** $4\,x^2 - y^2 - 16 = 0$.

e. $4\,x^2 - y^2 + 25 = 0$.   **f.** $9\,x^2 - 16\,y^2 + 144 = 0$.

g. $2\,x^2 - y^2 - 8 = 0$.   **h.** $x^2 - 2\,y^2 + 1 = 0$.

i. $5\,x^2 - 7\,y^2 - 35 = 0$.   **j.** $3\,x^2 - y^2 + 9 = 0$.

k. $6\,x^2 - 2\,y^2 + 25 = 0$.   **l.** $3\,x^2 - 8\,y^2 - 36 = 0$.

2. Find the equation of the hyperbola with center at the origin which satisfies each of the following conditions.

a. $a = 10$, $b = 5$, foci on the x-axis.   **b.** $a = 10$, $b = 5$, foci on the y-axis.

c. $a = 4$, $b = 8$, foci on the x-axis.   **d.** Vertex $(3, 0)$, focus $(4, 0)$.

e. Vertex $(0, 2)$, focus $(0, 5)$.   **f.** Vertex $(4, 0)$, focus $(6, 0)$.

g. Focus $(2, 0)$, conjugate axis 2.   **h.** Focus $(0, 5)$, conjugate axis 6.

               i. Vertex $(0, 4)$, conjugate axis 8.

3. Find the equation of the hyperbola having its center at the origin and its foci on the x-axis, and passing through

a. $(6, 2)$ and $(2\sqrt{6}, 1)$.   **b.** $(2, 2)$ and $(4, 5)$.

4. Find the equation of the locus of a point which moves so that the difference of its distances from the points $(0, \pm 13)$ is 10.

5. The *latus rectum* of a hyperbola is a chord through the focus perpendicular to the principal axis. From the equation show that the length of the latus rectum is $2\,b^2/a$.

6. What are the equations of the hyperbolas conjugate to those given in Problems 1 **c** and 1 **d**? What is the graph of $4\,x^2 - y^2 = 0$? How is this graph related to the hyperbolas of Problems 1 **d** and 1 **e**?

7. Find the equations of the tangent and the normal to each of the following hyperbolas at the point indicated.

a. $x^2 - y^2 = 9$, $(5, 4)$.   **b.** $4\,x^2 - y^2 - 39 = 0$, $(4, -5)$.

c. $x^2 - y^2 + 9 = 0$, $(-4, 5)$.   **d.** $x^2 - 4\,y^2 = 9$, $(-5, 2)$.

8. Find the equations of the tangents to the hyperbola $2\,x^2 - y^2 = 9$ which are parallel to the line $2\,x - y + 7 = 0$.

9. Find the equations of the tangents to the hyperbola $x^2 - y^2 + 16 = 0$ which are perpendicular to the line $5x + 3y - 15 = 0$.

10. Find the equations of the tangents to the hyperbola $x^2 - 8y^2 = 1$ which pass through the point $(-5, -2)$.

11. Find the angles of intersection of the line $2x - 3y = 0$ and the hyperbola $x^2 - y^2 = 5$.

12. A certain ellipse and a certain hyperbola have the same foci and latera recta. The foci are $(\pm 5, 0)$ and the length of each latus rectum is 4.5. Find the equations of the curves and show that the curves are perpendicular to each other at their points of intersection.

13. Show that the equation of the tangent to the hyperbola $b^2x^2 - a^2y^2 = a^2b^2$ at the point (x_1, y_1), which lies on the curve, is $b^2x_1x - a^2y_1y = a^2b^2$.

14. A point moves along the hyperbola $x^2 - 4y^2 = 20$ in such a way that x is increasing at the rate of 3 units/sec. Find the rate at which y is changing when the point passes through $(6, -2)$.

15. A segment is bounded by the line $x = 8$ and the hyperbola $x^2 - y^2 = 16$. Find the dimensions of the largest rectangle which can be inscribed in this segment.

16. Find the point on the hyperbola $x^2 - y^2 - 16 = 0$ which is nearest to the point $(0, 6)$.

150. The equilateral hyperbola. If $a = b$, the standard forms of the equation reduce to

$$x^2 - y^2 = a^2$$

and $\qquad y^2 - x^2 = a^2.$

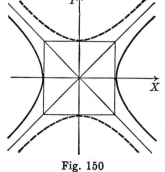

Fig. 150

These are called *equilateral* or *rectangular* hyperbolas. The latter name is due to the fact that the asymptotes have their slopes ± 1 and are therefore at right angles to each other.

It is shown in more extensive books on analytic geometry that the locus of the equation $xy = C$ is an equilateral hyperbola. The asymptotes are easily seen to be the x-axis and the y-axis. If C is negative the two branches are in the second and fourth quadrants. The vertices are on the line $y = x$ or $y = -x$.

A slightly more general form is $xy + ax + by + c = 0$. It may be shown that the asymptotes are $x = -b$ and $y = -a$.

The importance of these forms is due chiefly to the fact that many functional relations have equations of these types.

151. Generalized standard forms of the equation of the hyperbola.
If the center is not at the origin, but the axes are identical with or
parallel to the coördinate axes, special equations can be found by the
methods of Arts. 145 and 148. The results are as follows. Proofs are
left as exercises.

If the center of a hyperbola is the point (h, k) and the principal axis
is parallel to the x-axis, the equation is

(V b) $$\frac{(x - h)^2}{a^2} - \frac{(y - k)^2}{b^2} = 1.$$

If the center of a hyperbola is the point (h, k) and the principal axis
is parallel to the y-axis, the equation is

(V c) $$\frac{(y - k)^2}{a^2} - \frac{(x - h)^2}{b^2} = 1.$$

Both equations can be reduced to the form

$$Ax^2 + Cy^2 + Dx + Ey + F = 0,$$

where A and C have unlike signs. Conversely an equation in this form
can be reduced to one of the forms **(V b)** and **(V c)**, unless the locus is a
pair of lines, and the various facts can then be stated by inspection.

152. The conics. The equations of the circle, the parabola, the
ellipse, and the hyperbola, discussed in this chapter, are all special
forms of the general quadratic equation

$$Ax^2 + Bxy + Cy^2 + Dx + Ey + F = 0.$$

It is shown in more extensive books on analytic geometry that the
locus of this equation is always one of the curves mentioned, exception
being made of certain degenerate forms. Furthermore, all these curves
may be obtained as the intersection of a plane and a right circular conical surface. For this reason they are called *conic sections* or simply *conics*.
The ellipse and hyperbola are called *central conics*.

PROBLEMS

1. Name and sketch the following curves.

a. $xy + 8 = 0$.   **b.** $x^2 + y^2 - 4 = 0$.

c. $x^2 - y^2 + 4 = 0$.   **d.** $x^2 + y - 4 = 0$.

e. $4 x^2 + y^2 - 64 = 0$.   **f.** $x + 4 y^2 - 64 = 0$.

g. $4 x^2 - y^2 - 64 = 0$.   **h.** $4 x^2 + y = 0$.

i. $4 x^2 - y^2 = 0$.   **j.** $4 x^2 + y^2 = 0$.

k. $xy - 6 x + 7 y - 42 = 0$ **ı.** $xy + 4 x - 2 y + 8 = 0$.

2. Two points are 2000 ft. apart. At one of these points the report of a cannon is heard one second later than at the other. By means of the definition of a hyperbola show that the cannon is somewhere on a certain hyperbola, and write its equation after making a suitable choice of axes. (The velocity of sound is 1090 ft./sec.)

3. Find the equation of the hyperbola through the point (1, 1), with asymptotes $y = \pm\, 2\, x$.

4. Find the points of the hyperbola $xy = 48$ where the slope is $-\,\frac{3}{4}$.

5. Find the equations of the tangent and normal to each of the following curves at the point indicated.

a. $y^2 + 16\, x = 0$, $(-\, 4,\, 8)$.   **b.** $xy + 24 = 0$, $(4,\, -\, 6)$.
c. $4\, x^2 + y^2 = 100$, $(-\, 3,\, 8)$.   **d.** $x^2 - 4\, y^2 + 64 = 0$, $(6,\, 5)$.

6. Show that the hyperbola $x^2 - y^2 = 5$ and the ellipse $4\, x^2 + 9\, y^2 = 72$ intersect at right angles.

7. Show that the hyperbolas $x^2 - y^2 = a^2$ and $xy = a^2\sqrt{2}$ intersect at right angles.

8. Find the angles of intersection of the following pairs of curves.

a. $y = x^2$, $x^2 - y^2 + 12 = 0$.   **b.** $x^2 + y^2 = 104$, $x^2 - 9\, y^2 = 64$.
c. $xy = 16$, $x^2 + 4\, y^2 = 80$.   **d.** $xy = -\, 48$, $x^2 + y^2 = 100$.

9. Show that the tangent to the hyperbola $xy = C$ at any point forms with the coördinate axes a right triangle whose area is $2\, xy = 2\, C$.

10. Show that the equation of the tangent to the hyperbola $xy = C$ at the point $(x_1,\, y_1)$, which lies on the curve, is $x_1y + y_1x = 2\, C$.

11. For what value of k will the line $y = x + k$ be tangent to the hyperbola $x^2 - 9\, y^2 = 72$?

12. Prove that the triangle formed by a tangent to the curve $2\, xy = a^2$, the x-axis, and the line joining the point of contact to the origin is isosceles.

HINT. Compare the slopes of the sides.

13. If $P(h,\, k)$ is any point on the equilateral hyperbola $y^2 - x^2 = c$, prove that a circle which has P as its center and passes through the origin cuts the x-axis in a point A such that AP is normal to the hyperbola.

14. Write the equations of hyperbolas which satisfy the following data.

a. Vertices $(3,\, \pm\, 6)$, foci $(3,\, \pm\, 10)$.
b. Vertices $(-\, 2,\, 3)$ and $(6,\, 3)$, one focus $(7,\, 3)$.
c. Center $(-\, 3,\, -\, 2)$, axis $y = -\, 2$, $a = 4$, $b = 6$.

15. By completing squares reduce the following equations to a standard form if possible and draw the locus.

a. $x^2 - 4\, y^2 + 4\, x + 24\, y - 48 = 0$.   **b.** $16\, x^2 - 9\, y^2 - 96\, x = 0$.
c. $16\, x^2 - 9\, y^2 + 36\, y + 108 = 0$.   **d.** $4\, x^2 - y^2 - 6\, y = 0$.
e. $4\, x^2 - y^2 - 24\, x + 2\, y + 35 = 0$.

CHAPTER XI

CURVE–TRACING

153. Introduction. In previous chapters the student has learned how to plot curves by points and how to find their simpler properties— symmetry, slope, and maximum and minimum points. He has also studied the conics in some detail. It is now in order to consider curves of a more general nature and to study the methods applicable to their discussion.

The graph of any equation in x and y can be constructed by assuming values for x and computing the corresponding values of y, or vice versa. But this method alone is often laborious and ineffective. It is laborious because all facts about the graph are discovered by trial and because there is no advance information concerning the most important values of x (or y) to be chosen. It is ineffective because the most painstaking and extensive calculations may fail to show certain important properties of the curve.

In Chapter II we utilized the notions of intercepts and symmetry, for assistance in making tables of values. It was there stated that for a thorough discussion of a curve the methods of the calculus are necessary. Some of the methods have already been given; these will be amplified and others will be added in this chapter.

154. Tracing a curve. To "trace a curve" is to draw the graph after having determined its most important properties. This always involves the computation of values of one variable corresponding to certain values of the other; but these values are carefully chosen so as to include those that show the nature of the curve and to avoid the necessity of computing many values that have no special significance.

The "discussion of an equation" is the determination of the properties of the curve represented by the equation. The points to be considered are taken up in the following paragraphs.

155. Intercepts. The method of finding the intercepts of a curve was given in Art. 31. The points determined by the intercepts are usually the easiest points to find from the equation and the curve can cross the coördinate axes at no other points. Thus, if there are no x-intercepts and the curve is symmetrical with respect to the x-axis, the curve is discontinuous and has at least two branches.

156. Symmetry. The principle of symmetry was explained in Art. 31. The rules for symmetry with respect to the coördinate axes and the origin are repeated for convenience.

1. A curve is symmetrical with respect to the x-axis if its equation contains only even powers of y, and conversely.

2. A curve is symmetrical with respect to the y-axis if its equation contains only even powers of x, and conversely.

3. A curve is symmetrical with respect to the origin if its equation remains unchanged after substituting $-x$ for x and $-y$ for y, and conversely.

The following additional tests for symmetry are sometimes useful. If the solution of the equation for y has the form $y = b \pm f(x)$, the line $y = b$ is an axis of symmetry. If the solution for x has the form $x = a \pm f(y)$, the line $x = a$ is an axis of symmetry. If the interchange of x and y in the equation leaves the equation unchanged, the line $y = x$ is an axis of symmetry.

157. Excluded values. If the solution of the equation for y in terms of x involves a radical of even index, values of x which make the expression under the radical negative must be excluded; for the corresponding values of y are imaginary and there are no corresponding points on the curve. If the expression under the radical can be resolved into linear factors,* these excluded values of x can be determined by inspection. Similarly, to determine what values of y must be excluded the equation must be solved for x in terms of y.

The determination of the values to be excluded shows the *extent* of the curve. For example, if it appears that all negative values of x must be excluded, there is no part of the curve to the left of the y-axis, and if no other values of x need to be excluded, the curve extends indefinitely to the right.

158. Examples. In applying the previous and the following sections to specific problems, it is important to keep in mind the two purposes of a discussion: first, it facilitates the labor of computing the table of values; second, it serves as a check upon the table of values, and vice versa. It should be unnecessary to state that the curve as drawn must have the properties given in the discussion. If some of the points given by the computed table of values do not agree with the discussion, there is an error in either the computation or the discussion.

---

*If this is not possible, the determination of the excluded values may be quite arduous and in general will not be attempted. It will be helpful to recall that a quadratic factor of the form $ax^2 + bx + c$ always has the same sign if the roots of the equation $ax^2 + bx + c = 0$ are imaginary and that this is true if $b^2 - 4ac < 0$. If the roots r_1 and r_2 of $ax^2 + bx + c = 0$ are real and unequal, $ax^2 + bx + c = a(x - r_1)(x - r_2)$.

Example 1. Given the curve whose equation is $y^2 = x + 4$.

a. Find the intercepts.   **b.** Discuss the symmetry.
c. Find the extent.   **d.** Draw the curve.

Solution. a. Setting $y = 0$, the x-intercept is found to be -4. Setting $x = 0$, the y-intercepts are found to be $+2$ and -2.

b. The curve is symmetrical with respect to the x-axis (see Art. 156). It is not symmetrical with respect to the y-axis or the origin.

c. Solving for y, $y = \pm \sqrt{x + 4}$.

All values of x such that $x + 4$ is negative must be excluded. Hence $x < -4$ must be excluded.

Solving for x, $x = y^2 - 4$.

No values of y need be excluded.

Hence the curve does not extend to the left of the line $x = -4$. It extends indefinitely to the right and indefinitely upward and downward. (Fig. 158.1)

d. The curve is drawn by plotting a few points in addition to the intercepts already found. The shaded area indicates the region of excluded values.

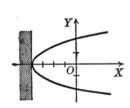

Fig. 158.1

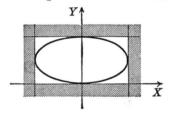

Fig. 158.2

Example 2. The equation of a curve is $x^2 + 4y^2 - 16y = 0$.

a. Find the intercepts.   **b.** Discuss the symmetry.
c. Find the extent.   **d.** Draw the curve.

Solution. a. When $y = 0$, $x = 0$; when $x = 0$, $y = 0$ or 4.

b. The tests for symmetry show symmetry with respect to the y-axis, but not with respect to the x-axis or the origin.

c. Solving the equation for y, we obtain

$$y = 2 \pm \frac{\sqrt{16 - x^2}}{2}.$$

Hence x^2 cannot be greater than 16; that is, x cannot be numerically greater than 4. This fact is expressed by stating that $|x| > 4$ is excluded. The form of the equation shows that $y = 2$ is an axis of symmetry.

Solving the equation for x, we obtain

$$x = \pm 2\sqrt{4y - y^2}$$
$$= \pm 2\sqrt{y(4 - y)}.$$

Hence we must exclude $y > 4$ and $y < 0$.

The excluded values show that the curve does not extend outside the rectangle bounded by the lines $x = 4$, $x = -4$, $y = 0$, and $y = 4$.

d. The curve is shown in Fig. 158.2, the shaded area indicating the region of excluded values.

PROBLEMS

(a) Find the intercepts, (b) discuss the symmetry, (c) find the extent, and (d) draw the graph of the following equations.

1. $x^2 + 4y^2 - 6x = 91$.

2. $x^2 + y = 2$.

3. $x^3 + y^3 = 1$.

4. $y = x^3$ (*cubical parabola*).

5. $y^2 = x^3$ (*semicubical parabola*).

6. $y = x^3 - 4x$.

7. $y^2 = x^3 - 4x$.

8. $x^2 y = 4$.

9. $x^2 y^2 = 4$.

10. $y^3 = x^2$ (*semicubical parabola*).

11. $x^4 + y^4 = a^4$.

12. $x^2 + y^2 + 6x + 8y = 0$.

13. $9x^2 + 25y^2 - 54x - 144 = 0$.

14. $x^2 - y^2 + 4x + 3 = 0$.

15. $y = \dfrac{8a^3}{x^2 + 4a^2}$ (*witch*).

16. $x^{\frac{2}{3}} + y^{\frac{2}{3}} = a^{\frac{2}{3}}$ (*hypocycloid of four cusps*).

17. $y = 2^{-x^2}$.

18. $y^2(2a - x) = x^3$ (*cissoid*).

19. $\sqrt{x} + \sqrt{y} = \sqrt{a}$ (*parabola*).

20. $y^2 = \dfrac{x^2(a + x)}{(a - x)}$ (*strophoid*).

21. $(y - x^2)^2 = x^5$.

22. $y = x^4 - 4x^2$.

23. $y^2 = 4x^2 - x^4$.

24. $y = \dfrac{x}{x^2 + 1}$.

159. Asymptotes. An asymptote is sometimes defined as a line which is approached by a curve as it recedes to infinity. This somewhat loose statement is intended to express the idea that as a point P traces a branch of the unbounded curve in question and moves off to infinity the limit of the distance from the asymptote to P is zero.

We shall here consider only vertical and horizontal asymptotes of the simpler algebraic curves, that is, asymptotes parallel to the coördinate axes. These are usually easy to find and are of much help in sketching curves. There are two methods.

First method. The definition gives at once the following method of finding vertical and horizontal asymptotes if the equations can be solved for x and y.

The line $x = a$ is a vertical asymptote of the curve if

$$\lim_{y \to +\infty} x = a \quad \text{or} \quad \lim_{y \to -\infty} x = a.$$

The line $y = b$ is a horizontal asymptote if

$$\lim_{x \to +\infty} y = b \quad \text{or} \quad \lim_{x \to -\infty} y = b.$$

Consider the equation $xy - y = 1$. Solving for y, we have

$$y = \frac{1}{x-1}. \tag{1}$$

Clearly $\lim\limits_{x \to +\infty} y = 0$ and $\lim\limits_{x \to -\infty} y = 0$. Hence the curve approaches the x-axis asymptotically in both the positive and negative directions.

Solving for x, we have

$$x = \frac{y+1}{y} \quad \text{or} \quad x = 1 + \frac{1}{y}. \tag{2}$$

Hence $\qquad \lim\limits_{y \to +\infty} x = 1 \quad \text{and} \quad \lim\limits_{y \to -\infty} x = 1.$

Thus the line $x = 1$ is a vertical asymptote approached by the curve in both directions.

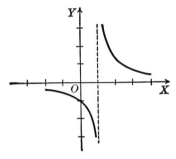

Fig. 159.1

Second method. Since the denominator in (1) is zero when $x = 1$, it is clear that $|y|$ increases indefinitely as x approaches 1 from either side. Hence we conclude that $x = 1$ is a vertical asymptote. When $x > 1$, y is positive and the curve approaches the positive half of this asymptote from the right. When $x < 1$, y is negative and the curve approaches the negative half of this asymptote from the left.

Similarly from (2) we see that the line $y = 0$ is a horizontal asymptote. When $y > 0$, x is positive and the curve approaches the positive half of this asymptote from above. When $y < 0$, $x < 1$ and the curve approaches the negative half of this asymptote from below.

The procedure employed above can be summarized in the following working rule.

To find the vertical asymptotes of a curve solve its equation for y. If the result is a fraction, set the denominator equal to zero and solve for x. The real roots will give the vertical asymptotes.* If there are no real roots or if the denominator is constant, the curve has no vertical asymptote.

To find the horizontal asymptotes solve the equation for x. If the result is a fraction, set the denominator equal to zero and solve for y. The real roots will give the horizontal asymptotes.

In the preceding discussion it has been tacitly assumed that x and y are one-valued continuous functions of each other. If they are multi-valued, the curve must be considered as made up of two or more branches for each of which the functions are one-valued.

Example. Find the asymptotes of the curve whose equation is

$$x^2y - y - 2\,x^2 + x = 0$$

and plot the curve.

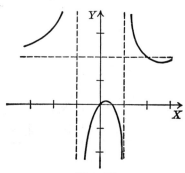

Fig. 159.2

Solution. Solving the equation for y, we obtain

$$y = \frac{2\,x^2 - x}{x^2 - 1}.$$

Setting the denominator equal to zero and solving gives $x = \pm 1$. Hence there are two vertical asymptotes: $x = -1$ and $x = +1$.

To solve the equation for x write it in the form

$$(y - 2)x^2 + x - y = 0$$

and use the quadratic formula. We obtain

$$x = \frac{-1 \pm \sqrt{1 - 8\,y + 4\,y^2}}{2(y - 2)}.$$

Hence the horizontal asymptote is $y = 2$.

---

*Note exceptions of this nature. If a is a root of the denominator and y is imaginary for all values of x near a, the line $x = a$ will not be an asymptote. For example, if $y = \dfrac{1}{\sqrt{x^4 - 4\,x^2}}$ the line $x = 0$ is not an asymptote since y is imaginary when $|\,x\,| < 2$.

In plotting the curve consideration should be given to the intercepts, symmetry, and extent. Furthermore, the table of values should be extensive enough to show how the curve approaches its asymptotes. In this example the x-intercepts are 0 and $\frac{1}{2}$, there is no symmetry, and the solution for y shows that no value of x is excluded except ± 1. The values of y for $x = -2, -\frac{1}{2}, \frac{3}{4}$, and $\frac{3}{2}$ show the directions in which the curve approaches the vertical asymptotes and the values for $x = \pm 10$ perform a similar service for the horizontal asymptote. Evidently y has a maximum value for some value of x between 0 and $\frac{1}{2}$, and a minimum value for some value of x between 2 and 10.

In this example it is easier to find the horizontal asymptote by the first method. To do this we must find

$$\lim_{x \to \infty} y = \lim_{x \to \infty} \frac{2x^2 - x}{x^2 - 1}.$$

x	y	x	y
-10	2.12	$\frac{3}{4}$	-0.86
-2	3.33	1	$\mp\infty$
-1	$\pm\infty$	$\frac{3}{2}$	2.40
$-\frac{1}{2}$	-1.33	2	2
0	0	3	1.88
$\frac{1}{4}$	0.13	10	1.92
$\frac{1}{2}$	0		

This can be found by dividing numerator and denominator by the highest power of x, namely x^2, and then finding the limit of each term in the resulting fraction. Or we can reason as follows. For very large values of x, positive or negative, the terms of highest degree in the numerator and denominator are relatively much larger than the other terms and hence for x numerically large the value of the fraction is approximately equal to $2x^2/x^2$, or 2, and so the limit of the fraction as $x \longrightarrow +\infty$ or $x \longrightarrow -\infty$ is 2. Hence the line $y = 2$ is a horizontal asymptote. This procedure is especially useful when it is easy to solve for one variable and difficult or impossible to solve for the other.

PROBLEMS

In the following problems find the intercepts and the horizontal and vertical asymptotes and draw the curves.

1. $4xy + 2y - 2x + 1 = 0$.
2. $xy - x^2 + 4 = 0$.
3. $xy^2 - 4x + 4 = 0$.
4. $x^2y - 2y - x^2 = 0$.
5. $xy - 2x - y + 3 = 0$.
6. $xy^2 - y^2 - 4x + 1 = 0$.
7. $x^2y + x^2 - 4y - 1 = 0$.
8. $x^2y(y^2 - 4) = 1$.
9. $xy(x - 1)(x - 2) = 1$.
10. $y^3 + 3y^2 + x^2y - 3x^2 = 0$.

160. Direction of bending. It has been seen that any function is increasing when its derivative is positive, and decreasing when its derivative is negative, and conversely. Let $y = f(x)$, and consider the derived function $y' = f'(x)$. As a consequence of the fact stated above, y' is increasing if *its* derivative, y'', is positive, and y' is decreasing if y'' is negative, and conversely. The geometric significance of this statement is important.

Let Fig. 160.1 represent the graph of $y = f(x)$. Then $y' = f'(x)$ will give the value of the slope of the graph at any point. At A the slope is positive and equal approximately to 3 (as estimated from the figure). Proceeding to the right along the curve (that is, as x increases), the graph shows that the slope decreases until the maximum point B is reached, where $y' = 0$. From B to C the slope continues to decrease from zero to a value approximately -3. As x increases from a to c the slope y' decreases; hence the derivative of y', which is y'', is negative. From A to C the curve is bending downward, or is *concave downward*.

Proceeding from C to D along the curve, the slope increases from -3 (approximately) to zero, and from D to E the slope continues to increase from zero to a value approximately 1. As x increases from c to e the slope y' increases; hence y'' is positive. From C to E the curve is bending upward, or is *concave upward*.

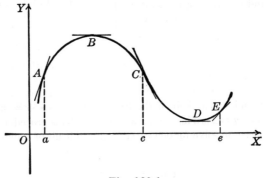

Fig. 160.1

The preceding discussion suggests the following theorem. A formal proof is omitted.

Theorem. Let $y = f(x)$ be continuous in the interval (a, b) and have first and second derivatives within this interval. For all values of x between a and b let $f''(x) > 0$. Then the corresponding arc of the graph is concave upward. If $f''(x) < 0$, the arc is concave downward.

Example. Given the equation $y = \dfrac{x^3}{3} - \dfrac{3 x^2}{2} + 3$. Find the maximum and minimum points of its graph. Find also the values of x for which the graph is concave upward and those for which it is concave downward.

Solution. Finding the first and second derivatives, we have

$$y' = x^2 - 3 x,$$
$$y'' = 2 x - 3.$$

Setting $y' = 0$, we obtain $x = 0$ or 3, the critical values. When $x = 0$, $y'' = -3$; hence $x = 0$ gives a maximum point $(0, 3)$. When $x = 3$, $y'' = +3$. Hence $x = 3$ gives a minimum point $(3, -\frac{3}{2})$.

The curve is concave downward when $y'' < 0$ or when $x < \frac{3}{2}$; it is concave upward when $y'' > 0$ or when $x > \frac{3}{2}$. (Fig. 160.2)

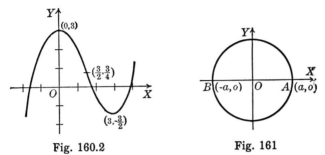

Fig. 160.2   Fig. 161

161. Points of inflection.

A point of inflection is a point at which a curve changes its direction of bending.* That is, as the tracing point moves from left to right the curve changes from concave downward to concave upward, or vice versa, at a point of inflection. In Fig. 160.1, C is a point of inflection. In the preceding example the inflectional point is $(\frac{3}{2}, \frac{3}{4})$.

The discussion of Art. 160 shows that $f''(x)$ changes sign when the tracing point of the graph passes through a point of inflection. Hence in general† $f''(c) = 0$ if $x = c$ gives a point of inflection. To find the points of inflection, we have the following working rule.

Find the roots of the equation $f''(x) = 0$. Let $x = c$ be one of the roots and suppose it is the only root in some interval (x_1, x_2), where $x_1 < c < x_2$. If $f''(x_1)$ and $f''(x_2)$ have opposite signs, the point on the graph of $y = f(x)$ where $x = c$ is a point of inflection.

It should be noted that a root of $f''(x) = 0$ does not necessarily give a point of inflection. For example, if $y = x^4$, then $f''(x) = 12 x^2$ and $f''(0) = 0$. Also $f''(x) > 0$ if $x < 0$ and $f''(x) > 0$ if $x > 0$. The origin is a minimum point on the graph and there is no point of inflection.

*In connection with this definition it should be remembered that we are considering only functions which are single-valued and continuous, and which have continuous first and second derivatives. Consider the circle $x^2 + y^2 = a^2$. The points $A(a, 0)$ and $B(-a, 0)$ separate an arc of the curve which is concave upward from an arc which is concave downward. But A and B are not points of inflection. The function of x is $y = \pm \sqrt{a^2 - x^2}$, which is not single-valued, and dy/dx becomes infinite at A and B.

†This statement is always true if y'' varies continuously. In certain curves, which will not be considered in this book, points of inflection exist at which y'' is not zero; that is, y'' changes sign by becoming infinite instead of zero.

The tangent to the curve at a point of inflection is called an *inflectional tangent*. It should be observed that an inflectional tangent cuts through the curve at the point of tangency, and that it also approximates the curve very closely for a considerable distance on either side of the point of tangency. For this reason any inflectional tangent which may exist should be drawn with its proper slope and used as a guiding line in drawing the curve.

162. Information given by the derivatives. For convenience we summarize here certain results which have been obtained earlier and which bear on the tracing of a curve.

1. If y' is positive (negative), the curve is rising (falling).

2. If y'' is positive (negative), the curve is concave upward (downward).

3. If $y' = 0$ and y'' is positive (negative), the curve has a minimum (maximum) point.

4. If $y'' = 0$, the curve has a point of inflection.

PROBLEMS

In the following problems (**a**) find the maximum, minimum, and inflectional points; (**b**) find the slope of each inflectional tangent and draw it; (**c**) draw the curve.

1. $y = \dfrac{x^3}{3} - x^2 + 2.$ **2.** $6\,y = 6 + 30\,x - 2\,x^3.$

3. $6\,y = x^3 - 12\,x - 6.$ **4.** $2\,y = 2\,x^3 + 3\,x^2.$

5. $6\,y = x^3 - 6\,x^2 + 12\,x.$ **6.** $y = x(x-1)(x-2).$

7. $y = \dfrac{x^4}{4} - 2\,x^2 + 2.$ **8.** $20\,y = x^4 - 24\,x^2 - 12.$

9. $6\,y = 6 + 18\,x^2 + 8\,x^3 - 3\,x^4.$ **10.** $y = (x^2-1)(x^2-4).$

11. $20\,y = x^5 - 10\,x^2.$ **12.** $15\,y = 3\,x^5 - 25\,x^3 + 60\,x.$

13. $y = x + \dfrac{4}{x}.$ **14.** $y = x^2 + \dfrac{2}{x}.$

15. $y = x + \dfrac{4}{x^2}.$ **16.** $y = x^3 + \dfrac{1}{x^2}.$

17. Show that the following curves have no point of inflection: (**a**) the ellipse, (**b**) the hyperbola, (**c**) the parabola.

18. Find the points of inflection of the curve whose equation is $y = \dfrac{x}{x^2+1}$ and show that they lie on a straight line.

19. Show that the curve whose equation is $x^3 + y^3 = 1$ has a point of inflection where it crosses the y-axis.

163. Discussion of an equation. The discussion of an equation is resorted to for the purpose of getting all possible advance information about the properties of the curve before making a table of values and drawing the curve. The five points to be considered are those which have been mentioned in Arts. 155–162; namely,

1. Intercepts on the axes.   2. Symmetry.
3. Extent.   4. Horizontal and vertical asymptotes.
5. Maximum, minimum, and inflectional points.

It is not to be expected that complete information on all five points will be obtained from every equation. For example, to determine the extent of a curve, it is necessary to solve the equation algebraically for both x and y. This is usually not practicable if the equation is of higher degree than the second. But partial information about the extent may be obtained if the equation is solved for one variable only. Again, the determination of intercepts may involve the solution of a numerical equation of higher degree than the second. This will not be undertaken unless the equation can be readily factored, because the results do not justify the labor.

But in every case of curve-tracing the five questions of the discussion should be answered as far as practicable. As each fact concerning the curve appears in the discussion it should be indicated, if possible, on the plotting-paper. Thus when the first question has been answered, the intercepts should be marked on the coördinate axes. If certain values are to be excluded, this fact should be indicated as in Art. 158. If the answer to the fourth question shows that the curve has horizontal or vertical asymptotes, these should be drawn on the plotting-paper to serve as guide lines in tracing the curve. The maximum, minimum, and inflectional points should be carefully plotted and a short tangent line with proper slope drawn through each. The knowledge thus gained frequently makes it possible to sketch the curve completely. If the information obtained by the discussion is not sufficient for drawing the curve completely, point-by-point plotting must be used.

Example 1. Discuss the equation $x^2y - x^2 + 4 = 0$ and trace its graph.

Solution. We are given $\qquad x^2y - x^2 + 4 = 0.$ \hfill (1)

Solving for y, $\qquad\qquad y = 1 - \dfrac{4}{x^2}.$ \hfill (2)

Solving for x, $\qquad\qquad x = \pm \sqrt{\dfrac{4}{1-y}}.$ \hfill (3)

Differentiating (2), $\qquad\quad y' = \dfrac{8}{x^3},$ \hfill (4)

$$y'' = -\frac{24}{x^4}.$$ \hfill (5)

Discussion. 1. *Intercepts.* Inspection of equation (1) shows that the x-intercepts are ± 2 and that there are no y-intercepts. Hence the curve cuts the x-axis at $(\pm 2, 0)$ and does not cross the y-axis.

2. *Symmetry.* The curve is symmetrical with respect to the y-axis. It is not symmetrical with respect to the x-axis or the origin.

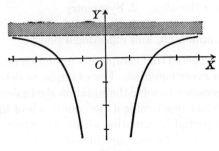

Fig. 163.1

3. *Extent.* Equation (2) shows that no values of x (except 0) need be excluded. Hence the curve extends indefinitely to the right and to the left. Equation (3) shows that values of y greater than 1 must be excluded. Hence the curve has no part above the line $y = 1$, but extends indefinitely downward.

4. *Asymptotes.* Equation (2) shows that the y-axis is a vertical asymptote and (3) shows that the line $y = 1$ is a horizontal asymptote.

5. *Maximum, minimum, and inflectional points.* Since y' cannot equal zero, there are no extreme points. Since y'' is always negative, the curve is everywhere concave downward and there is no point of inflection. Since there are no extreme or inflectional points, it is necessary to calculate the coördinates of a few points besides the two already found in order to make an accurate drawing, although the general shape is obvious from the discussion.

Example 2. Discuss the equation $x^2y + a^2y - 4a^2x = 0$ and trace its graph.

Solution. We are given $x^2y + a^2y - 4a^2x = 0$. $\qquad$ (1)

Solving for y, $\qquad y = \dfrac{4a^2x}{x^2 + a^2}.$ $\qquad$ (2)

Solving for x, $\qquad x = \dfrac{2a^2 \pm a\sqrt{4a^2 - y^2}}{y}.$ $\qquad$ (3)

Differentiating (2), $\qquad y' = \dfrac{4a^2(a^2 - x^2)}{(x^2 + a^2)^2},$ $\qquad$ (4)

$$y'' = \frac{8a^2x(x^2 - 3a^2)}{(x^2 + a^2)^3}.$$ $\qquad$ (5)

Discussion. 1. *Intercepts.* If $y = 0$, $x = 0$, and if $x = 0$, $y = 0$. Hence the curve cuts the coördinate axes only at the origin.

2. *Symmetry.* Equation (1) contains an odd power of y which shows that the curve is not symmetrical with respect to the x-axis. Since the equation contains an odd power of x, the curve is not symmetrical with respect to the y-axis. Since the equation remains unchanged after substituting $-x$ for x and $-y$ for y, the curve is symmetrical with respect to the origin.

3. *Extent.* Equation (2) shows that no values of x need be excluded. Hence the curve extends indefinitely to the right and to the left. Equation (3) shows that y^2 cannot be greater than $4\,a^2$. Hence the curve lies entirely in the horizontal strip bounded by the lines $y = 2\,a$ and $y = -2\,a$.

4. *Asymptotes.* Equation (2) shows that there is no vertical asymptote. This is also apparent from the fact that y cannot exceed $2\,a$ in numerical value. Equation (3) shows that $y = 0$ is a horizontal asymptote.

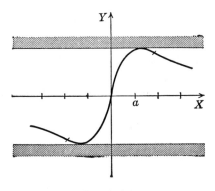

Fig. 163.2

5. *Maximum, minimum, and inflectional points.* Setting $y' = 0$ gives $x = +a$ and $-a$. For $x = +a$, $y = 2\,a$ and y'' is negative. Hence $(a, 2\,a)$ is a maximum point. For $x = -a$, $y = -2\,a$ and y'' is positive. Hence $(-a, -2\,a)$ is a minimum point.

Setting $y'' = 0$ gives $x = 0$, $a\sqrt{3}$, and $-a\sqrt{3}$. These values divide the x-axis into four intervals. Testing the sign of y'' for x within each of these intervals, we find that when $x < -a\sqrt{3}$, $y'' < 0$; when $-a\sqrt{3} < x < 0$, $y'' > 0$; when $0 < x < a\sqrt{3}$, $y'' < 0$; and when $a\sqrt{3} < x$, $y'' > 0$. Hence the curve has three inflectional points: $(-a\sqrt{3}, -a\sqrt{3})$, with slope $-\frac{1}{2}$; $(0, 0)$, with slope 4; $(a\sqrt{3}, a\sqrt{3})$, with slope $-\frac{1}{2}$.

The curve is concave upward between $(-a\sqrt{3}, -a\sqrt{3})$ and $(0, 0)$ and to the right of $(a\sqrt{3}, a\sqrt{3})$. It is concave downward between $(0, 0)$ and $(a\sqrt{3}, a\sqrt{3})$ and to the left of $(-a\sqrt{3}, -a\sqrt{3})$.

Choosing any convenient numerical value for a, the curve can be sketched from the information given by the discussion.

PROBLEMS

Discuss the following equations and trace their graphs.

1. $x^3 + y - 1 = 0.$

2. $12\,y = 24 + 24\,x^2 - x^4.$

3. $xy + y - x = 0.$

4. $x^2 y + y - x^2 = 0.$

5. $y = x^2(x^2 - 4).$

6. $y = x^2(4 - x^2).$

7. $y = x + \dfrac{4}{x}.$

8. $y = x + \dfrac{4}{x^2}.$

9. $8\,x^2 y - x^5 + 32 = 0.$

10. $y = x^2 + \dfrac{4}{x}.$

11. $x^3 y - 8\,y - x^3 = 0.$

12. $y = \dfrac{8\,a^3}{x^2 + 4\,a^2}.$

13. $y = \dfrac{2\,x^2}{x^2 + 4}.$

14. $y = x^5 - \dfrac{3\,x^3}{5}.$

15. $y = x^5 - x.$

16. $y = x^5 - x^2.$

164. More complicated curves. When the solution of the equation for y involves radicals, the derivatives are frequently so complicated that the labor of finding all maximum, minimum, and inflectional points is not justified for the purpose of sketching the curve. In these cases it is best to discuss the equation for intercepts, symmetry, extent, and asymptotes, and then to make a sketch from a table of values. When it is necessary to use the derivatives, care must be exercised in handling the double signs which occur when the solution for y involves an even root. It is best to operate with the positive sign alone. The results for the negative sign can usually be inferred from those obtained for the positive sign.

Example 1. Discuss the equation $(y - x)^2 = x^3$ and trace its graph.

Solution. Solving for y, we have

$$y = x \pm x^{\frac{3}{2}}. \tag{1}$$

Discussion. 1. *Intercepts.* The given equation shows that the curve contains the points $(0, 0)$ and $(1, 0)$ on the coördinate axes.

2. *Symmetry.* There is no symmetry with respect to the axes or the origin.

3. *Extent.* Equation (1) shows that we must exclude $x < 0$. It also shows that for each value of x (except $x = 0$) there are two values of y, and that the curve consists of two branches which start at the origin and lie on opposite sides of the line $y = x$. For the upper branch $\lim\limits_{x \to +\infty} y = +\infty$ and for the lower branch $\lim\limits_{x \to +\infty} y = -\infty$. Hence the curve extends indefinitely up and down.

4. *Asymptotes.* Equation (1) shows that there is no vertical asymptote and the discussion of the extent in y shows that there is no horizontal asymptote.

5. *Maximum, minimum, and inflectional points.* Taking the positive sign in (1) and differentiating, we get

$$y' = 1 + \tfrac{3}{2} x^{\frac{1}{2}},$$
$$y'' = \tfrac{3}{4} x^{-\frac{1}{2}}.$$

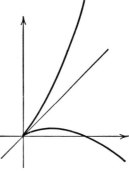

Since $y' \neq 0$, the upper branch has no extreme point except the origin, which is a minimum point since it is a left-hand end point and the slope there is $+1$. Since $y'' > 0$ for $x > 0$, the upper branch is concave upward and has no point of inflection.

For the lower branch we have

$$y' = 1 - \tfrac{3}{2} x^{\frac{1}{2}},$$
$$y'' = - \tfrac{3}{4} x^{-\frac{1}{2}}.$$

Fig. 164.1

Here again the origin is a minimum point as in the case of the upper branch. Setting $y' = 0$, we get $x = \tfrac{4}{9}$ and $y = \tfrac{4}{27}$. This point is a maximum point since $y'' < 0$ for $x > 0$. The last statement also tells us that the lower branch is concave downward and has no point of inflection.

The two branches of this curve are tangent to the same line at the same point, forming a "sharp" point on the curve. Such a point is called a *cusp*.

Example 2. Discuss the equation $(a - x)y^2 - (a + x)x^2 = 0$ and trace its graph.

Solution. Solving for y, we have

$$y = \pm x \sqrt{\frac{a + x}{a - x}}. \tag{2}$$

The solution for x is not practicable.

Discussion. We will assume that a is positive.

1. *Intercepts.* The y-intercept is 0; the x-intercepts are $-a$ and 0.

2. *Symmetry.* The curve is symmetrical with respect to the x-axis only.

3. *Extent.* Equation (2) shows that $x > a$, and $x < -a$ must be excluded. Hence the curve lies entirely within the vertical strip bounded by the lines $x = a$ and $x = -a$.

4. *Asymptotes.* The line $x = a$ is a vertical asymptote. There is no horizontal asymptote, because x cannot exceed a numerically.

Taking the positive sign with the radical, equation (2) shows that y has the same sign as x. Hence the curve lies below the x-axis between $-a$ and 0 and above the x-axis between 0 and a, approaching the positive end of the asymptote

$x = a$. Obviously the curve must have a minimum point between $-a$ and 0. The other half of the curve, obtained by taking the minus sign in equation (2), can be drawn in by symmetry.

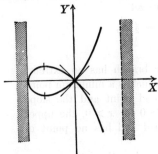

Fig. 164.2

When a numerical value of a is given, the shape of the curve is readily obtained by plotting a few points. If it is desired to locate accurately the minimum point between $-a$ and 0, we find, from equation (2),

$$y' = \frac{a^2 + ax - x^2}{(a - x)\sqrt{a^2 - x^2}}.$$

Setting $y' = 0$ gives $\qquad x = \frac{a}{2}\left(1 \pm \sqrt{5}\right).$

The positive sign must be rejected, since x cannot be greater than a. Hence the minimum point is

$$x = \frac{a}{2}\left(1 - \sqrt{5}\right) = -0.618\,a, \qquad y = -0.300\,a.$$

By symmetry the point $(-0.618\,a, +0.300\,a)$ is a maximum point for the other branch.

The form of the derivative y' shows that both branches of the curve are tangent to the line $x = -a$ at the point $(-a, 0)$. This point is a maximum point for one branch and a minimum point for the other.

PROBLEMS

Discuss the following equations and trace their graphs.

1. $x^3 + y^2 - 1 = 0.$
2. $y^2 + x^3 - x = 0.$
3. $xy^2 + y^2 - x = 0.$
4. $y^2 = x^2(x^2 - 16).$
5. $x^2y^2 + y^2 - x^2 = 0.$
6. $y^2 = x^2(16 - x^2).$
7. $x^2 - 2xy + y^2 - x - y = 0.$
8. $x^2 + xy + y^2 = 4.$
9. $x^2 + 3xy + y^2 = 4.$
10. $y^2(2\,a - x) = x^3.$
11. $\sqrt{x} + \sqrt{y} = \sqrt{a}.$
12. $(y - x^2)^2 = x^5.$
13. $y^2(a - x) = ax(a + x).$
14. $y^2(x^2 - 4x + 3) = x.$
15. $y^2 = (4 - x)(3 - x)(2 - x).$
16. $x^4 + y^2 = 16$

CHAPTER XII

THE EXPONENTIAL AND
LOGARITHMIC FUNCTIONS

165. The exponential function. The functions so far used in this book have been algebraic, that is, combinations of the variables and constants involving sums, products, quotients, and constant powers and roots. All other functions are called *transcendental*. The more common transcendental functions are the exponential, logarithmic, trigonometric, and inverse trigonometric functions.

The quantity $y = a^x$, in which a is a constant, is known as an *exponential function*. The number a is called the base. While any positive number except unity may serve as the base of an exponential function, the most important base is a certain irrational number denoted by e, whose value is approximately 2.71828. The quantity $y = e^x$ is usually called *the* exponential function.

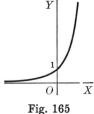

Fig. 165

The graph of $y = e^x$ is readily plotted from a table of values of the exponential function. The value of y is always positive, the graph crosses the y-axis at $(0, 1)$, and approaches the negative x-axis as an asymptote. The exponential function is an increasing function; that is, y always increases when x increases.

In higher algebra the definition of the number e is

$$e = \lim_{t \to 0} (1 + t)^{\frac{1}{t}}.$$

It is proved that this limit exists, that it is irrational, and that it is approximately equal to 2.71828. This approximation may be roughly checked by calculating $(1 + t)^{\frac{1}{t}}$ for a few small values of t, using a table of logarithms when convenient. Thus

when $t = \frac{1}{2}$, $(1 + t)^{\frac{1}{t}} = (\frac{3}{2})^2 = 2.25$;

when $t = \frac{1}{3}$, $(1 + t)^{\frac{1}{t}} = (\frac{4}{3})^3 = 2.370$;

when $t = 0.1$, $(1 + t)^{\frac{1}{t}} = (1.1)^{10} = 2.594$;

when $t = 0.01$, $(1 + t)^{\frac{1}{t}} = (1.01)^{100} = 2.705$; etc.

For numerical values of the exponential function see Table 9, Chapter XXV.

269

166. The logarithmic function. Since $y = a^x$ is a function of x, x is also a function of y. This inverse function is the logarithm of y to the base a by virtue of the definition of logarithms, which is as follows.

The logarithm of a number to any base is the exponent of the power to which the base must be raised to equal the number.

In other words, if y denotes the number, a denotes the base, and x denotes the logarithm of y to the base a, the above definition may be stated algebraically as follows.

$$x = \log_a y \quad \text{if and only if} \quad y = a^x.$$

The equivalence of the two equations, $y = a^x$ and $x = \log_a y$, is perhaps the most important single fact in connection with logarithms.

Interchanging variables, so as to make y the dependent variable, we have $y = \log_a x$. This, by definition, is equivalent to $x = a^y$. Hence the graph of $y = \log_a x$ is the same as that of $y = a^x$ with the axes interchanged. If $a = e$, the graph of $y = \log_e x$ is obtained from the graph of Art. 165 by interchanging the axes. It could, of course, be plotted directly from a table of logarithms using the base e. Properties of the graph to be noted are the following.

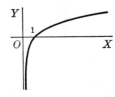

Fig. 166

The intercept on the x-axis is 1, since $\log_e 1 = 0$.

The y-axis is an asymptote. The numerical relation associated with this statement is expressed by

$$\lim_{x \to 0} \log_e x = -\infty.$$

This is sometimes written more briefly $\log_e 0 = -\infty$.

167. The two bases. Any positive number except 1 can be used as the base of a system of logarithms, but for practical purposes we are restricted to two bases.

Logarithms to the base 10 are called *common*, or *Briggsian*, logarithms (see Table 2, Chapter XXV). Since our number system is decimal, common logarithms are most convenient in numerical computation such, for example, as is necessary in trigonometry.

Logarithms to the base e are called *natural*, or *Napierian*, logarithms (see Table 7, Chapter XXV). They are most convenient in all problems involving differentiation and integration. To distinguish between natural logarithms and common logarithms when the base is not explicitly stated, the following notation will be used.

Natural logarithm of v (base e) $= \ln v$.
Common logarithm of v (base 10) $= \log v$.

For convenience the formulas governing operations with logarithms are assembled here.

(I) $$\log_a xy = \log_a x + \log_a y.$$

(II) $$\log_a \left(\frac{x}{y}\right) = \log_a x - \log_a y.$$

(III) $$\log_a x^n = n \log_a x.$$

(IV) $$\log_a \sqrt[n]{x} = \frac{1}{n} \log_a x.$$

(V) $$\log_b x = \frac{\log_a x}{\log_a b}.$$

These formulas are all proved by reference to the definition of logarithms. As (V) is not commonly used in elementary work, the proof will be given here.

By definition, if we set $u = \log_b x$, then $b^u = x$.

Taking the logarithms of both sides by (III),

$$u \log_a b = \log_a x.$$

Hence
$$u = \frac{\log_a x}{\log_a b};$$

or
$$\log_b x = \frac{\log_a x}{\log_a b}.$$

This last formula is used for transferring logarithms from one base to another. For example, the Napierian logarithm of 446 may be found as follows from a table of common logarithms.

$$\log_{10} 446 = 2.6493 \text{ and } \log_{10} e = \log_{10} 2.71828 = 0.4343.$$

Hence
$$\log_e 446 = \frac{2.6493}{0.4343} = 6.1002.$$

If in (V) we set $x = a$, we get

$$\log_b a = \frac{\log_a a}{\log_a b} = \frac{1}{\log_a b}.$$

This relationship enables us to write (V) in the form

(V a) $$\log_b x = \log_a x \cdot \log_b a.$$

When $b = 10$ and $a = e$, the quantity $\log_b a$, or $\log_{10} e$, is known as the *modulus* of the common system of logarithms; its value is 0.43429, the reciprocal of $\log_e 10 = 2.30259$.

Example. Discuss and sketch the graph of $y = \ln \sqrt{25 - x^2}$.

Solution. We first simplify the equation by means of **(IV)**, obtaining

$$y = \tfrac{1}{2} \ln (25 - x^2).$$

1. *Intercepts.* If $x = 0$, then $y = \tfrac{1}{2} \ln 25 = \ln 5 = 1.6$. If $y = 0$, then $25 - x^2 = 1$ (since $\ln 1 = 0$) and $x = \pm \sqrt{24} = \pm 4.9$.

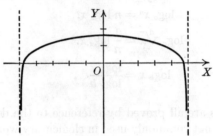

Fig. 167.1

2. *Symmetry.* The graph is symmetric to the y-axis only.

3. *Extent.* Since negative numbers have no real logarithms, x^2 cannot exceed 25. Hence the graph does not extend to the right of $x = 5$ nor to the left of $x = -5$.

4. *Asymptotes.* By Art. 166 $y \longrightarrow -\infty$ when $(25 - x^2) \longrightarrow 0$, and the graph has two vertical asymptotes, $x = \pm 5$.

PROBLEMS

1. Solve each of the following equations for x.

   a. $y = \log x$.   **b.** $y = \ln \sqrt{5} \, x$.   **c.** $y = 6^x$.   **d.** $y = e^{-\frac{1}{2}x}$.

2. Prove that if $y = x^{\frac{1}{\ln x}}$, then $y = e$.

3. Sketch the following curves on the same axes.

   a. $y = e^x$.   **b.** $y = e^{-x}$.   **c.** $y = -e^x$.   **d.** $y = -e^{-x}$.

4. Sketch the following curves on the same axes.

   a. $y = \ln x$.   **b.** $y = \ln \dfrac{1}{x}$.   **c.** $y = \ln (-x)$.   **d.** $y = \ln \left(-\dfrac{1}{x}\right)$.

5. Sketch the graphs of $y = \ln x$ and $y = \log x$ on the same axes.

6. Plot the graph of the probability curve $y = e^{-x^2}$ (see Fig. 167.2).

7. Plot the graph of the catenary $y = \tfrac{1}{2} a\left(e^{\frac{x}{a}} + e^{-\frac{x}{a}}\right)$ (see Fig. 167.3). (For $a = 1$ this function is called the hyperbolic cosine of x and is denoted by $y = \cosh x$.)

8. Plot the graph of the hyperbolic sine curve $y = \tfrac{1}{2}(e^x - e^{-x})$, denoted by $y = \sinh x$.

9. Plot the graphs of the following functions.

a. $y = e^{x+2}$. **b.** $2y = e^{4-x}$. **c.** $y = e^x - x$. **d.** $10y = xe^x$. **e.** $y = e^x/x$.

10. Sketch the following curves on the same axes.

a. $y = \ln x$. **b.** $y = \ln x^2$. **c.** $y = \ln \sqrt{x}$.

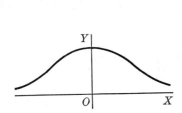

Fig. 167.2

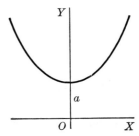

Fig. 167.3. Catenary

11. Discuss and sketch the graphs of the following equations.

a. $y = \ln (x + 3)$.

b. $y = \ln (3 - x)$.

c. $y = \ln \left(\dfrac{2}{x+3}\right)$.

d. $y = \ln (10 - x^2)$.

e. $y = \ln \left(\dfrac{1}{25 - x^2}\right)$.

f. $y = \ln (10 x - x^2)$.

12. Discuss and plot the graph of $y = \ln (36 - x^2)$. Using this graph and **(II)** and **(IV)**, sketch the graphs of $y = \ln \sqrt{36 - x^2}$, $y = \ln \left(\dfrac{1}{36 - x^2}\right)$, and $y = \ln \left(\dfrac{1}{\sqrt{36 - x^2}}\right)$.

13. Solve each of the following equations for x in terms of y.

a. $y = \frac{1}{2}(e^x + e^{-x})$.

b. $y = \frac{1}{2}(e^x - e^{-x})$.

168. Differentiation of logarithmic functions. Suppose that u is a function of x which is differentiable and greater than zero for every value of x in the interval (a, b). Let $y = \ln u$. We first find dy/du by the general rule used in Chapter III. Following the four steps of this rule we have

$$\text{I.} \quad y + \Delta y = \ln (u + \Delta u).$$

$$\text{II.} \quad \Delta y = \ln (u + \Delta u) - \ln u$$

$$= \ln \left(1 + \frac{\Delta u}{u}\right). \qquad \text{By (II), Art. 167}$$

$$\text{III.} \quad \frac{\Delta y}{\Delta u} = \frac{1}{\Delta u} \ln \left(1 + \frac{\Delta u}{u}\right).$$

In order to evaluate the limit when $\Delta u \longrightarrow 0$, the second member of this equation is multiplied by $\dfrac{u}{u}$ and written in the form

$$\frac{\Delta y}{\Delta u} = \frac{1}{u}\frac{u}{\Delta u}\ln\left(1+\frac{\Delta u}{u}\right)$$

$$= \frac{1}{u}\ln\left(1+\frac{\Delta u}{u}\right)^{\frac{u}{\Delta u}}. \qquad \text{By (III), Art. 167}$$

Then

IV. $$\frac{dy}{du} = \lim_{\Delta u \to 0}\frac{1}{u}\ln\left(1+\frac{\Delta u}{u}\right)^{\frac{u}{\Delta u}}.$$

In order to find the value of this limit set $\Delta u/u = t$. Since $u > 0$, we see that $t \longrightarrow 0$ when $\Delta u \longrightarrow 0$. Hence

$$\lim_{\Delta u \to 0}\left(1+\frac{\Delta u}{u}\right)^{\frac{u}{\Delta u}} = \lim_{t \to 0}(1+t)^{\frac{1}{t}} = e. \qquad \text{By definition}$$

Since the logarithmic function is continuous,

$$\lim_{\Delta u \to 0}\ln\left(1+\frac{\Delta u}{u}\right)^{\frac{u}{\Delta u}} = \ln e = 1.$$

Consequently, from IV, we have

$$\frac{dy}{du} = \frac{1}{u}.$$

But $$\frac{dy}{dx} = \frac{dy}{du}\frac{du}{dx},$$

by the theorem on differentiation of a function of a function. We therefore have the formula

(VI) $$\frac{d}{dx}\ln u = \frac{1}{u}\frac{du}{dx}.$$

When $u = x$ this formula reduces to

(VI a) $$\frac{d}{dx}\ln x = \frac{1}{x}.$$

If some number a, different from e, is used as the base of the logarithms, the derivative may be found as follows:

Suppose $y = \log_a u$. Then, by (V a), Art. 167,

$$y = \log_a u = \ln u \log_a e.$$

Since $\log_a e$ is a constant, this function may be differentiated by **(VI)**, giving

$$\frac{dy}{dx} = \log_a e \frac{d}{dx} \ln u = \frac{1}{u} \frac{du}{dx} \log_a e.$$

Hence

(VI b)
$$\frac{d}{dx} \log_a u = \frac{1}{u} \frac{du}{dx} \log_a e.$$

In the case of common logarithms, $\log_{10} e = 0.434$, to three decimals, and formula **(VI b)** becomes

(VI c)
$$\frac{d}{dx} \log_{10} u = \frac{0.434}{u} \frac{du}{dx}.$$

Example 1. Differentiate $y = \ln\left(\dfrac{x^2}{\sqrt{3 - 4x}}\right)$.

Solution. It is convenient first to simplify the expression for y as much as possible before differentiating. Thus

$$y = \ln x^2 - \ln \sqrt{3 - 4x} \qquad\qquad \text{by (II), Art. 167}$$
$$= 2 \ln x - \tfrac{1}{2} \ln (3 - 4x). \qquad\qquad \text{By (III), Art. 167}$$

Hence, by **(VI)** and **(VI a)**,

$$\frac{dy}{dx} = \frac{2}{x} - \frac{1}{2} \frac{(-4)}{3 - 4x}$$

$$= \frac{2}{x} + \frac{2}{3 - 4x} = \frac{6 - 6x}{x(3 - 4x)}.$$

Example 2. Differentiate $y = x^2 \ln (1 + 2x)$.

Solution. In differentiating this function we first apply the product formula

$$\frac{dy}{dx} = x^2 \frac{d}{dx} \ln (1 + 2x) + \ln (1 + 2x) \frac{d}{dx} x^2$$

$$= \frac{2x^2}{1 + 2x} + 2x \ln (1 + 2x).$$

PROBLEMS

Differentiate the following functions.

1. $y = \ln (ax + b).$

2. $y = \ln \left(\dfrac{x^2 - 4}{x^2 + 4}\right).$

3. $y = \ln (x^2 + 2x).$

4. $y = \log (2x - x^4).$

5. $y = x^2 \ln x.$

6. $f(x) = \ln x^3.$

7. $f(x) = \ln^3 x.*$

8. $f(x) = \ln \left(\dfrac{a - x}{a + x}\right).$

*Note that $\ln^3 x$ means $(\ln x)^3$ and is not the same as $\ln x^3 = 3 \ln x$.

9. $f(x) = \ln (x + \sqrt{1 + x^2})$.

10. $z = \ln \left(\dfrac{1 - 2\,y}{1 + 2\,y}\right)$.

11. $s = t \ln \sqrt{t}$.

12. $s = \ln \left(\dfrac{t^2}{\sqrt{3 - 2\,t}}\right)$.

13. $u = \ln \sqrt{\dfrac{1 + y}{1 - y}}$.

14. $y = \ln \sqrt{(2\,x - 1)(2\,x^2 - 1)}$.

15. $y = x^n \ln x$.

16. $f(x) = \ln \left(\dfrac{3\,x + 4}{4\,x + 3}\right)$.

17. $f(x) = \ln \left(\dfrac{1 - x}{\sqrt{1 + x}}\right)$.

18. $y = \dfrac{\ln x}{x}$.

19. $y = \ln (x^4 \sqrt{1 + x^2})$.

20. $y = \log (x^2 + 5\,x)^{\frac{1}{3}}$.

21. Find y and dy/dx for the indicated value of x.

a. $y = \ln (x \sqrt{7 - x})$, $x = 4$.

b. $y = \ln \left(\dfrac{x^2}{6 - x^2}\right)$, $x = 2$.

c. $y = x^2 \ln \sqrt{x}$, $x = 3$.

d. $y = \dfrac{\ln x}{x^2}$, $x = 2$.

e. $y = \log (x^3 + 3)$, $x = 5$.

f. $y = \log (x \sqrt{20 - 7\,x})$, $x = 2$.

22. Sketch the following curves and find the slope at each point where the curve crosses the axes of coördinates.

a. $y = \ln (x + 2)$.

b. $y = \ln (9 - x^2)$.

c. $y = \ln (x + 3)^2$.

d. $y = \log x$.

e. $y = \ln (4 - x)$.

f. $y = \ln \sqrt{4 - x^2}$.

g. $y = \ln \sqrt{3 - x}$

h. $y = \log (16 - x^2)$.

23. Find the point on the curve $y = 2 \ln x$ where the tangent is (a) perpendicular to the line $2\,x + y - 4 = 0$; (b) parallel to the line $x - y + 2 = 0$.

24. Find the maximum and minimum and inflectional points of the following curves and draw their graphs.

a. $y = \ln (1 + x^2)$.

b. $y = x/\ln x$.

c. $y = x \ln x$.

d. $y = \ln (8\,x - x^2)$.

25. If $\ln 10 = 2.303$, approximate $\ln 10.2$ by means of $d(\ln x)$.

169. Logarithmic differentiation. When $y = f(x)$ is a product or quotient, it may often be differentiated more easily by taking logarithms, simplifying, and using the rule for differentiation of logarithms. This process is called differentiating logarithmically.

Example. Differentiate $y = \dfrac{\sqrt{x^2 + 1}}{(x - 2)(3 - x^2)}$.

Solution. Taking Napierian logarithms of both sides and simplifying (Art. 167), we have

$$\ln y = \tfrac{1}{2} \ln (x^2 + 1) - \ln (x - 2) - \ln (3 - x^2).$$

Differentiation with respect to x gives

$$\frac{1}{y}\frac{dy}{dx} = \frac{x}{x^2+1} - \frac{1}{x-2} + \frac{2x}{3-x^2}.$$

Multiplying both sides by y, we have

$$\frac{dy}{dx} = \left[\frac{x}{x^2+1} - \frac{1}{x-2} + \frac{2x}{3-x^2}\right]\frac{\sqrt{x^2+1}}{(x-2)(3-x^2)}$$
$$= \frac{2x^4 - 2x^3 + 3x^2 - 10x - 3}{\sqrt{x^2+1}(x-2)^2(3-x^2)^2}.$$

PROBLEMS

Differentiate logarithmically

1. $y = (x+1)(x+2)^2(x+3)^3.$ **2.** $y = x\sqrt{(x+3)(2-x^2)}.$

3. $y = \dfrac{x(1+x^2)}{\sqrt{1-x^2}}.$ **4.** $y = \dfrac{\sqrt{x^3+2}}{x\sqrt{3-2x}}.$

Find y and dy/dx for the indicated value of x.

5. $y = x\sqrt{(x+1)(x+2)}$, $x=3$.   **6.** $y = \sqrt{x(2x-1)(3x-2)}$, $x=2$.

7. $y = \dfrac{(x+2)(2x-1)}{x^2}$, $x=2$.   **8.** $y = x^2\sqrt{\dfrac{x+5}{x-5}}$, $x=7$.

170. Differentiation of the exponential functions. Let $y = a^u$, where u is a function of x which is differentiable in some interval. Taking the Napierian logarithms of both sides,

$$\ln y = u \ln a,$$

whence $$u = \frac{\ln y}{\ln a}.$$

Differentiating with respect to y, since $\ln a$ is a constant,

$$\frac{du}{dy} = \frac{1}{y \ln a},$$

whence $$\frac{dy}{du} = y \ln a = a^u \ln a.$$

Applying the rule for differentiating a function of a function, we now have

(VII) $$\frac{d}{dx} a^u = a^u \ln a \frac{du}{dx}.$$

Special cases. When $u = x$,

(VII a) $$\frac{d}{dx} a^x = a^x \ln a.$$

When $a = e$, $\ln a = 1$ and

(VII b)
$$\frac{d}{dx} e^u = e^u \frac{du}{dx}.$$

When $a = e$ and $u = x$,

· **(VII c)**
$$\frac{d}{dx} e^x = e^x.$$

The formulas for the exponential functions must not be confused with that for the power function. In the former case the base is a constant and the exponent a variable, while in the latter the base is variable and the exponent is constant. In the unusual case where both base and exponent are variable, logarithmic differentiation is usually used (see Example 3 below).

Example 1. Differentiate $y = e^{\sqrt{x}}$.

Solution. By **(VII b)** we have

$$\frac{dy}{dx} = e^{\sqrt{x}} \frac{d\sqrt{x}}{dx} = \frac{e^{\sqrt{x}}}{2\sqrt{x}}.$$

Example 2. Differentiate $y = \dfrac{e^x}{x}$.

Solution. Applying the formula for the derivative of a quotient, we have

$$\frac{dy}{dx} = \frac{x \dfrac{de^x}{dx} - e^x \dfrac{dx}{dx}}{x^2} = \frac{e^x(x - 1)}{x^2}.$$

Example 3. Differentiate $y = x^x$.

Solution. Taking the Napierian logarithms of both sides, we have
$$\ln y = x \ln x.$$

Differentiating with respect to x,

$$\frac{1}{y}\frac{dy}{dx} = x \frac{d}{dx} \ln x + \ln x \frac{dx}{dx}$$
$$= 1 + \ln x,$$

whence
$$\frac{dy}{dx} = (1 + \ln x)x^x.$$

PROBLEMS

Differentiate the following functions.

1. $y = e^{3x - 2}$.

2. $y = e^{a^2 - x^2}$.

3. $y = 10^{x^2}$.

4. $y = (1 - x)e^x$.

5. $y = x^2 e^{ax}$.

6. $s = \dfrac{e^t - 1}{e^t + 1}$.

7. $f(x) = \dfrac{\ln x}{e^x}$.

8. $y = \ln\left(\dfrac{e^x}{1 + e^x}\right)$.

9. $y = \dfrac{e^x - e^{-x}}{e^x + e^{-x}}$.

10. $s = t^{\frac{1}{t}}$.

11. $y = x^{\frac{1}{\ln x}}$.

12. $s = \left(\dfrac{a}{t}\right)^t$.

13. Show that if $y = \frac{1}{2} a\left(e^{\frac{x}{a}} + e^{-\frac{x}{a}}\right)$, then $y'' = \dfrac{y}{a^2}$.

14. What is the minimum value of $y = ae^{kx} + be^{-kx}$?

15. Find the maximum point and the points of inflection of the graph of $y = e^{-x^2}$ and draw the curve.

16. Show that the rectangle of maximum area and with base on the x-axis which can be inscribed under the curve $y = e^{-x^2}$ has two of its vertices at the points of inflection.

17. Find the minimum point and the point of inflection of the curve whose equation is $y = xe^x$. Draw the curve.

18. Given $e^2 = 7.39$. Using differentials, find approximately the value of $e^{2.1}$.

171. The integrals of du/u and $e^u\, du$. The formulas for differentiation derived above give formulas for integration as follows.

From **(VI a)** we obtain

(VIII) $$\int \frac{du}{u} = \ln u + C.$$

This is the exceptional case, $n = -1$, of the general power formula

$$\int u^n\, du = \frac{u^{n+1}}{n+1} + C,$$

which holds good for all values of n except -1.

From **(VII c)** we obtain

(IX) $$\int e^u\, du = e^u + C.$$

From **(VII a)** we obtain

(X) $$\int a^u\, du = \frac{a^u}{\ln a} + C.$$

Example 1. Find the value of $\displaystyle\int_0^2 \frac{x\, dx}{x^2 + 3}$.

Solution. Let $u = x^2 + 3$; then $du = 2 x\, dx$. Hence, substituting in the given integral

$$x^2 + 3 = u \quad \text{and} \quad x\, dx = \tfrac{1}{2} du,$$

and finding the limits for u, we have

$$\int_0^2 \frac{x\, dx}{x^2 + 3} = \frac{1}{2}\int_3^7 \frac{du}{u} = \frac{1}{2}\Big[\ln u\Big]_3^7$$
$$= \tfrac{1}{2}(\ln 7 - \ln 3) = \tfrac{1}{2}(1.946 - 1.099) = 0.424.$$

Example 2. Integrate $\int \dfrac{x^3\,dx}{x-2}$.

Solution. Since the degree of the denominator is not higher than the degree of the numerator, we must divide the numerator by the denominator until the remainder is of lower degree than the denominator. This gives

$$\frac{x^3}{x-2} = x^2 + 2\,x + 4 + \frac{8}{x-2}.$$

Hence $\quad \int \dfrac{x^3\,dx}{x-2} = \int x^2\,dx + \int 2\,x\,dx + \int 4\,dx + \int \dfrac{8\,dx}{x-2}$

$$= \tfrac{1}{3}\,x^3 + x^2 + 4\,x + 8\ln(x-2) + C.$$

Example 3. Find the value of $\displaystyle\int_0^2 e^{\frac{1}{2}x}\,dx$.

Solution. Here $u = \tfrac{1}{2}\,x$ and $du = \tfrac{1}{2}\,dx$. Hence, substituting in the given integral

$$\tfrac{1}{2}\,x = u \quad \text{and} \quad dx = 2\,du,$$

and changing the limits to the values of u corresponding to $x = 0$ and $x = 2$, we have

$$\int_0^2 e^{\frac{1}{2}x}\,dx = 2\int_0^1 e^u\,du = 2\Big[\,e^u\,\Big]_0^1 = 2[e-1] = 3.436.$$

PROBLEMS

Integrate the following.

1. $\int e^{ax}\,dx$.

2. $\int e^{-x}\,dx$.

3. $\int e^{2s}\,ds$.

4. $\int \dfrac{3\,dx}{e^x}$.

5. $\int a^{2x}\,dx$.

6. $\int xa^{x^2}\,dx$.

7. $\int \dfrac{x\,dx}{x^2+1}$.

8. $\int \dfrac{x^2\,dx}{x+1}$.

9. $\int \dfrac{(x-1)dx}{x^2-2\,x-5}$.

10. $\int \dfrac{5\,x^2\,dx}{10\,x^3+6}$.

11. $\int \dfrac{(y^2-2)^3\,dy}{y^5}$.

12. $\int \dfrac{5\,bx\,dx}{8\,a-6\,bx^2}$.

13. $\int \dfrac{(\ln x)^3\,dx}{x}$.

14. $\int \dfrac{2\,e^x\,dx}{e^x+1}$.

15. $\int \left(e^{\frac{x}{a}} + e^{-\frac{x}{a}}\right)dx$.

16. $\int (e^y + e^{-y})^2\,dy$.

17. $\int (e^{-2t} - 2\,t)dt$.

18. $\int \dfrac{(x-3)dx}{6\,x-x^2}$.

Evaluate the following definite integrals.

19. $\int_0^1 e^{2x}\,dx$.

20. $\int_2^5 \dfrac{dx}{x+2}$.

21. $\int_1^e \dfrac{dx}{x}$.

22. $\int_1^2 xe^{x^2}\,dx$.

23. $\int_0^3 \dfrac{x\,dx}{x^2+1}$.

24. $\int_2^3 \dfrac{t\,dt}{t^2+1}$.

25. $\int_0^2 \dfrac{x^3\,dx}{x+1}$.

26. $\int_0^1 \dfrac{dx}{e^{3x}}$.

27. Find the area bounded by the equilateral hyperbola $xy = 12$, the x-axis, and the lines $x = 2$ and $x = 4$.

28. Find the area bounded by the equilateral hyperbola $xy = 12$, the y-axis, and the lines $y = 3$ and $y = 6$.

29. P and Q are any two points on an equilateral hyperbola $xy = k$. Show that the area bounded by the arc PQ, the ordinates of P and Q, and the x-axis is equal to the area bounded by PQ, the abscissas of P and Q, and the y-axis.

30. Find the area in the first quadrant bounded by the line $y = x/4$ and the curve $y(1 + x^2) = x$.

31. Find the area bounded by the curve $y = e^x$, the y-axis, the x-axis, and any ordinate.

32. Find the area bounded by the catenary $y = \dfrac{a}{2}\left(e^{\frac{x}{a}} + e^{-\frac{x}{a}}\right)$, the x-axis, and the lines $x = a$ and $x = -a$.

33. Find the area bounded by the curve $(x^2 + 1)y = x$, the x-axis, and the lines $x = 1$ and $x = 4$.

34. Find the area bounded by the curve $y = e^x$, the y-axis, and the line $y = 6$.

35. Find the length of an arc of the catenary $y = \dfrac{a}{2}\left(e^{\frac{x}{a}} + e^{-\frac{x}{a}}\right)$ extending from $x = 0$ to $x = a$.

36. Find the length of the arc of the curve whose equation is $y = \dfrac{x^2}{4} - \dfrac{1}{2}\ln x$ extending from $x = 1$ to $x = 4$.

37. Find the volume generated by revolving about the x-axis the area bounded by $y^2(6 - x) = x$, $y = 0$, and $x = 4$.

38. Find the volume generated by revolving about the x-axis the area bounded by $y^2(2a - x) = x^3$, $x = a$, and $y = 0$.

39. Find the volume generated by revolving about the x-axis the area bounded by $y = e^x$, $x = 0$, $x = 1$, and $y = 0$.

40. Find the volume generated by revolving about the x-axis the loop of the curve $(x - 4)y^2 = x(x - 3)$.

41. Find the volume generated by revolving about the x-axis the area found in Problem 32.

42. Find the volume generated by revolving about the x-axis the area bounded by the curve $y = e^{-x}$, the lines $x = 0$ and $x = 10$, and the x-axis.

172. Compound-interest law. Numerous processes in nature proceed according to the compound-interest law (C. I. Law), namely, a magnitude varies so that its rate of change is always proportional to the magnitude itself. The connection between the name of the law and the statement of it is seen as follows.

Let　　　$y = $ a sum of money in dollars accumulating at compound interest;

$i = $ interest on one dollar for a year;

$\Delta t = $ an interval of time measured in years;

$\Delta y = $ the interest on y dollars for the interval of time Δt.

Then　　　　　　　　　　$\Delta y = iy\, \Delta t.$

Therefore　　　　　　　　$\dfrac{\Delta y}{\Delta t} = iy.$　　　　　　　　　(1)

Equation (1) states that the average rate of change of y for the period of time Δt is proportional to y itself. In business, interest is added to the principal at stated times only—yearly, quarterly, etc. In other words, y changes discontinuously with t. But in nature, changes proceed on the whole in a continuous manner; so that to adapt (1) to natural phenomena we must imagine the sum y to accumulate continuously, that is, assume the interval of time Δt to decrease and approach zero as a limit. Then (1) becomes

$$\frac{dy}{dt} = iy,\qquad\qquad(2)$$

and the rate of change of y is proportional to y, agreeing with the definition of the C. I. Law.

Let $y_0 = $ the sum of money placed at interest, that is, the value of y when $t = 0$. To integrate (2), we multiply both sides by dt and divide both members by y. Then we get

$$\frac{dy}{y} = i\, dt.$$

Integrating, the result is

$$\ln y = it + C.\qquad\qquad(3)$$

To determine the constant C, we have given that $y = y_0$ when $t = 0$.

Hence　　　　　　　　　　$C = \ln y_0.$

Substituting this value of C in (3), transposing, and using **(II)**, Art. 167, we get

$$\ln \frac{y}{y_0} = it.\qquad\qquad(4)$$

By Art. 166 this is the same as

$$\frac{y}{y_0} = e^{it},\quad\text{or}\quad y = y_0 e^{it}.\qquad\qquad(5)$$

The amount after t years, when interest is added annually, is given by the formula

$$y = y_0(1 + i)^t.\qquad\qquad(6)$$

Example 1. The sum of \$100 is placed at compound interest at 6 per cent, and interest accumulates (**a**) annually, (**b**) continuously. Compare the amounts in (**a**) and (**b**) after 10 yr.

Solution. (**a**) By (6), $y = 100(1.06)^{10} = 179.08.$

(**b**) By (5), $y = 100\,e^{0.6} = 182.20.$

If a, b, c are constants, the equation

$$\frac{d}{dx}(y+a) = b(y+a), \quad \text{or} \quad \frac{dy}{dx} = by + c, \text{ where } c = ab, \tag{7}$$

states that $(y+a)$ follows the C. I. Law. Examples of this sort are very common in nature. (See Problem 5 below.)

Example 2. Washing down a solution. Water is run into a tank containing a saline (or acid) solution with the purpose of reducing its strength. The volume v of the mixture in the tank is kept constant. If $s =$ quantity of salt (or acid) in the tank at any time, and $x =$ amount of water which has run through, show that the rate of decrease of s with respect to x varies as s, and, in fact, that $\dfrac{ds}{dx} = -\dfrac{s}{v}.$

Solution. Since $s =$ quantity of salt in the mixture of total volume v, the quantity of salt in any other volume u of the mixture is $\dfrac{s}{v}\,u.$

Suppose a volume Δx of the mixture is dipped out of the tank. The amount of salt thus dipped out will be $\dfrac{s}{v}\,\Delta x$, and hence the change in the amount of salt in the tank is given by

$$\Delta s = -\frac{s}{v}\,\Delta x. \tag{8}$$

Suppose now that a volume of water Δx is added to fill the tank to its original volume v. Then from (8) the ratio of the amount of salt removed to the volume of water added is given by

$$\frac{\Delta s}{\Delta x} = -\frac{s}{v}.$$

When $\Delta x \longrightarrow 0$ we obtain the instantaneous rate of change of s with respect to x; namely,

$$\frac{ds}{dx} = -\frac{s}{v}.$$

PROBLEMS

1. In Example 2 above, if $v = 10{,}000$ gal., how much water must be run through to wash down 50 per cent of the salt?

2. *Newton's law of cooling.* If the excess temperature of a body above the temperature of the surrounding air is x degrees, the time-rate of decrease of x is proportional to x. If this excess temperature was at first 80 degrees, and after 1 min. is 70 degrees, what will it be after 2 min.? In how many minutes will it decrease 20 degrees?

3. Atmospheric pressure p at points above the earth's surface as a function of the altitude h above sea level follows the C. I. Law. Assuming $p = 15$ lb. per square inch at sea level, and 10 lb. per square inch at an altitude of 10,000 ft., find p (**a**) when $h = 5000$ ft.; (**b**) when $h = 15,000$ ft.

4. In the inversion of raw sugar, the time-rate of change varies as the amount of raw sugar remaining. If, after 10 hr., 1000 lb. of raw sugar has been reduced to 800 lb., how much raw sugar will remain at the expiration of 24 hr.?

5. *Building up a saline (or acid) solution* by adding salt (or acid), maintaining constant volume, leads to the equation

$$\frac{dy}{dx} = \frac{1}{v}(v - y),$$

where $v =$ the constant volume, $y =$ salt (or acid) in the tank at any moment, $x =$ salt (or acid) added from the beginning. Compare this result with Example 2, above.

6. In Problem 5, if $v = 10,000$ gal., how much acid must be run in to build up to a 50 per cent solution?

7. In an electric circuit with given voltage E and current i (amperes), the voltage E is consumed in overcoming (1) the resistance R (ohms) of the circuit, and (2) the inductance L, the equation being

$$E = Ri + L\frac{di}{dt}, \quad \text{or} \quad \frac{di}{dt} = \frac{1}{L}(E - Ri).$$

This process therefore comes under (7) above, E, R, L being constants. Given $L = 640$, $R = 250$, $E = 500$, and $i = 0$ when $t = 0$, show that the current will approach 2 amperes as t increases. Also find in how many seconds i will reach 90 per cent of its maximum value.

8. In a condenser discharging electricity, the rate of change of the voltage e is proportional to e, and e decreases with the time. Hence $de/dt = -e/k$. Given $k = 40$, in how many seconds will e decrease to 10 per cent of its original value?

CHAPTER XIII

THE TRIGONOMETRIC FUNCTIONS

173. Measurement of angles. In the sexagesimal system of angular measurement a right angle is divided into ninety degrees, each degree into sixty minutes, and each minute into sixty seconds. For purposes of plotting the graphs of trigonometric functions and measuring slopes, rates, areas, volumes, etc., it is necessary to use *radian* measure. A *radian* is a central angle whose intercepted arc equals the radius of the circle. From this definition it follows that

$$\pi \ (= 3.1416) \ \text{radians} = 180°.$$
$$1 \ \text{radian} = 57° \ 18', \quad 1° = 0.01745 \ \text{radian}.$$

Tables of equivalents for radians and degrees are given in Chapter XXV (see Tables 3, 4, 5). These equivalents are all approximate, owing to the fact that the circumference and radius of a circle are incommensurable.

By definition a central angle of one radian intercepts an arc of length equal to the radius. Then a central angle of θ radians will intercept an arc s given by

$$s = \theta r,$$

where the radius r and the arc s are measured in the same linear units. This equation follows from the theorem in elementary geometry to the effect that the angular measures of two central angles in a circle are in the same ratio as their intercepted arcs.

In the analytic use of the trigonometric functions a thorough familiarity with the usual formulas connecting these functions is necessary, and this knowledge is assumed in what follows. A collection of the most important formulas is given in Chapter XXV.

x (degrees)	x (radians)	y (sin x)	y (cos x)
0	0.00	0.00	1.00
15	$\frac{1}{12}\pi = 0.26$	0.26	0.97
30	$\frac{1}{6}\pi = 0.52$	0.50	0.87
45	$\frac{1}{4}\pi = 0.79$	0.71	0.71
60	$\frac{1}{3}\pi = 1.05$	0.87	0.50
75	$\frac{5}{12}\pi = 1.31$	0.97	0.26
90	$\frac{1}{2}\pi = 1.57$	1.00	0.00
180	$\pi = 3.14$	0.00	-1.00
270	$\frac{3}{2}\pi = 4.71$	-1.00	0.00
360	$2\pi = 6.28$	0.00	1.00
etc.	etc.	etc.	etc.

174. Graphs of $y = \sin x$ and $y = \cos x$. In plotting graphs of the trigonometric functions the independent variable x is always measured in radians. If tables are available which give the values of sine and cosine for angles measured in radians (see Table 6, Chapter XXV), the corresponding values of x and y may be read off directly from the tables. If the ordinary tables for angles measured in degrees are used, it is best to make three columns—the first giving the values of x in degrees, the second the corresponding values in radians taken from the conversion table, and the third the values of $\sin x$ or $\cos x$. This is done in the table given on page 285.

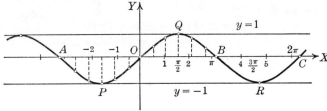

Fig. 174.1

Since $\sin(-x) = -\sin x$ and $\sin(x + 2\pi) = \sin x$, it is unnecessary to continue the table for the sine curve beyond the range 0 to 360°, or 0 to 2π radians. The complete curve consists simply of an indefinite repetition of the portion $OQBRC$ in the figure on both sides of the y-axis. In any problem involving the geometric interpretations of the calculus the same scale is always used for x and y.

The graph of $y = \cos x$ may be plotted from the table of values, but this is not necessary. For the formula $\sin(x + \frac{1}{2}\pi) = \cos x$ shows that the graph of $y = \sin x$ becomes that of $y = \cos x$ if the origin is moved to the point $(\frac{1}{2}\pi, 0)$. That is, the figure for $y = \cos x$ may be obtained from the figure for $y = \sin x$ by drawing a new y-axis through the first maximum point to the right. The curves $y = \sin x$ and $y = \cos x$ (see Fig. 174.2) are, therefore, identical curves, differing only in the position of the y-axis.

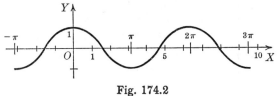

Fig. 174.2

These graphs should be known well enough so that they can be reproduced from memory. They are of considerable aid in recalling many details regarding the functions $\sin x$ and $\cos x$. For $\sin x$, note that it is 0 when $x = \pm n\pi$ ($n =$ any integer), that it has its maximum value ($+1$) when $x = \frac{1}{2}\pi \pm 2\pi n$; that it has its minimum value (-1) when $x = \frac{3}{2}\pi \pm 2\pi n$. For $\cos x$, note that it is 0 when $x = \frac{1}{2}\pi \pm 2\pi n$; that it has its maximum value ($+1$) when $x = 2\pi n$; and that it has its minimum value (-1) when $x = \pi \pm 2\pi n$.

175. Periodic functions. A function whose values are repeated after the independent variable has passed through a certain range of values is called *periodic*, and may be formally defined as follows.

The function $f(x)$ is periodic if there is a positive constant p such that $f(x + p) = f(x)$ for all values of x.

The constant p is called the *period*. From the previous article we see that $\sin x$ and $\cos x$ have the period 2π.

NOTE. Obviously, if p is a period of $f(x)$, $2p$, $3p$, etc. are periods. In this and all that follows it is assumed that p is the smallest period.

Let us now investigate the effect upon the period of multiplying the independent variable by a constant. We have the following theorem.

Theorem. *If $f(x)$ has the period p, the period of $f(bx)$ is $\frac{p}{b}$.*

Proof. We must show that $f(bx)$ is unchanged if x is replaced by $x + \frac{p}{b}$.

Substituting, we have

$$f\left[b\left(x + \frac{p}{b}\right)\right] = f(bx + p) = f(bx),$$

since p is a period of f.

Thus, if the independent variable is multiplied by a constant b, the period is divided by the same constant. For example, the period of $y = \sin 3x$ is $\frac{2}{3}\pi$.

Since the maximum value of the sine (or cosine) is 1, the maximum value of $a \sin bx$ (or $a \cos bx$) is a. The number a is called the *amplitude*.

Any curve whose equation is of the type $y = a \sin bx$ or $y = a \cos bx$ can be sketched by the aid of the general properties mentioned above and the knowledge of the graphs of $y = \sin x$ and $y = \cos x$. The method of sketching such a graph is as follows.

1. Determine the period.

2. Plot the points on the curve for each successive quarter period. These points determine the intercepts and the maximum and minimum points, by means of which the curve may be drawn in.

Example 1. Sketch the graph of $y = \frac{3}{2} \sin \frac{1}{2} x$.

Solution. The period is $\dfrac{2\pi}{\frac{1}{2}} = 4\pi$.

The values of y at each quarter period are given in the table.

x	y
0	0
π	$\frac{3}{2}$
2π	0
3π	$-\frac{3}{2}$
4π	0

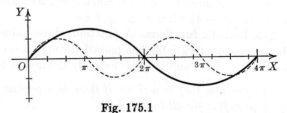

Fig. 175.1

The graph is drawn in Fig. 175.1, where the graph of $y = \sin x$ is also shown by a dotted curve for the purpose of comparison.

Example 2. Sketch the graph of $y + 2\cos \frac{1}{2}\pi x = 0$.

Solution. The period is $\dfrac{2\pi}{\frac{1}{2}\pi} = 4$.

The values of y at each quarter period are given in the table.

x	y
0	-2
1	0
2	2
3	0
4	-2

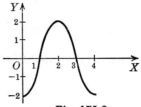

Fig. 175.2

The locus of the equation

$$y = \tfrac{3}{2} \sin \left(\tfrac{1}{2} x + 1\right) \tag{1}$$

is also a sine curve. For, by taking the coefficient of x, namely, $\frac{1}{2}$, outside the parenthesis, (1) becomes

$$y = \tfrac{3}{2} \sin \tfrac{1}{2}(x + 2). \tag{2}$$

Substituting x' for $(x + 2)$ in (2), we have

$$y = \tfrac{3}{2} \sin \tfrac{1}{2} x'. \tag{3}$$

Equation (3) is derived from (2) by the transformation

$$x + 2 = x', \quad \text{or} \quad x = x' - 2, \tag{4}$$

the value of y being unchanged. But this transformation merely translates the axes of coördinates in the locus of (1) to the new origin $(-2, 0)$. The locus of (3) is shown above in Example 1. The origin in this figure is the point $(-2, 0)$ for the locus of (1). Hence to obtain the figure for

(1) from the figure for (3) we must translate the y-axis two units to the right.

Observe that the period of (1) is determined, as before, by the coefficient of x. The added term $(+1)$ merely affects the intercepts on the x-axis.

In the same way, the locus of

$$y = a \sin (bx + c). \qquad\qquad (5)$$

can be obtained from the locus of

$$y = a \sin bx' \qquad\qquad (6)$$

by translating the axes. The figures for (5) and (6) differ only in the position of the y-axis.

176. Graphs of the tangent, cotangent, secant, and cosecant functions. The graphs of these functions are plotted in much the same way as the

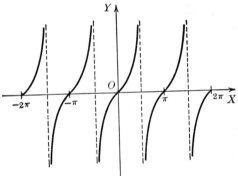

Fig. 176.1. Graph of $y = \tan x$

sine and cosine curves. Since all become infinite for certain finite values of x and all are periodic, they will have indefinitely many vertical asymptotes and will have indefinitely many branches.

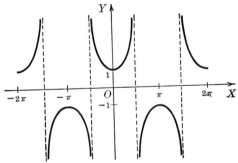

Fig. 176.2. Graph of $y = \sec x$

The sketching of the graphs is much simplified by recalling the relations between these functions and the sine and cosine. Thus $y = \tan x = \sin x / \cos x$ shows that $\tan x = 0$ when $\sin x = 0$ and $\tan x = \infty$ when $\cos x = 0$. Also $y = \sec x = 1/\cos x$ shows that $\sec x = \infty$ when $\cos x = 0$. Also, since $\cos x$ is never greater than unity, $\sec x$ is never less than unity. The following table gives the intercepts and asymptotes, and periods.

Function	Period	Intercepts	Asymptotes	Range of Values
$\tan x$	π	$x = \pm n\pi$	$x = \frac{1}{2}\pi \pm n\pi$	$-\infty$ to $+\infty$
$\operatorname{ctn} x$	π	$x = \frac{1}{2}\pi \pm n\pi$	$x = \pm n\pi$	$-\infty$ to $+\infty$
$\sec x$	2π	None	$x = \frac{1}{2}\pi \pm n\pi$	$-\infty$ to -1 and $+1$ to $+\infty$
$\csc x$	2π	None	$x = \pm n\pi$	$-\infty$ to -1 and $+1$ to $+\infty$

The values of these four functions for values of x in radians are given in Table 6, Chapter XXV. The graphs of $y = \tan x$ and $y = \sec x$ are given; those of $y = \operatorname{ctn} x$ and $y = \csc x$ should be plotted from a table of values and kept for reference.

PROBLEMS

1. Sketch the graph of each of the following equations.

a. $y = \sin 2x$.

b. $y = 2\cos 2x$.

c. $y = 3\sin \frac{1}{2}\pi x$.

d. $y = \cos \frac{1}{3}\pi x$.

e. $y + 2\sin x = 0$.

f. $y + 2\cos \pi x = 0$.

g. $y = \frac{1}{2}\tan x$.

h. $y = \operatorname{ctn} \frac{1}{2}x$.

i. $y = \sec \frac{1}{2}x$.

j. $y = \csc \frac{1}{2}x$.

k. $y = \tan \frac{1}{4}\pi x$.

l. $y = \frac{1}{2}\operatorname{ctn} \frac{1}{4}\pi x$.

2. Make a table of values for the range indicated and plot the graph of each of the following equations.

a. $y = \frac{1}{2}x + \sin x$; $(-2\pi \text{ to } 2\pi)$.

b. $y = x - \cos \frac{1}{2}\pi x$; $(-4 \text{ to } 4)$.

c. $y = 2\cos x - x$; $(-2\pi \text{ to } 2\pi)$.

d. $y = \sin \frac{1}{3}\pi x - \frac{1}{2}x$; $(-6 \text{ to } 6)$.

e. $y = x\sin x$; $(-\pi \text{ to } 2\pi)$.

f. $y = \frac{1}{2}x\cos x$; $(-\pi \text{ to } 2\pi)$.

3. Sketch the graph of each of the following equations.

a. $y = \sin (x - \frac{1}{2})$.

b. $y = \cos (x + \frac{1}{4}\pi)$.

c. $y = 2\sin (\frac{1}{4}\pi x - \frac{1}{2}\pi)$.

177. The limit of $\dfrac{\sin \theta}{\theta}$ **as** θ **approaches 0.** In order to differentiate the sine function, we must first evaluate this limit. In Fig. 177 the angle 2θ (AOB) is drawn with its vertex at the center of a circle of unit radius. Let AC and BC be lines tangent to the circle at A and B respectively. It is known from geometry that the line OC bisects the angle AOB and the arc AB, and also that chord $AB <$ arc $ATB < AC + BC$. Since the radius of the circle is 1,

$$BS = AS = \sin \theta,$$

and $$BC = AC = \tan \theta.$$

But if the angle 2θ is measured in radians, arc $ATB = 2\theta$.

Substituting these values in the inequality above, we have

$$2 \sin \theta < 2\theta < 2 \tan \theta.$$

Dividing by $2 \sin \theta$, we get

$$1 < \frac{\theta}{\sin \theta} < \frac{1}{\cos \theta}.$$

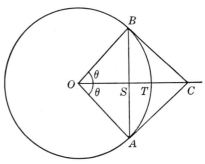

Fig. 177

Inverting the terms of this inequality, we get

$$1 > \frac{\sin \theta}{\theta} > \cos \theta.$$

Now as θ approaches 0 as a limit, $\cos \theta$ approaches 1. Since $\dfrac{\sin \theta}{\theta}$ is, in value, between 1 and $\cos \theta$, it must also approach 1 as a limit. Hence

$$\lim_{\theta \to 0} \frac{\sin \theta}{\theta} = 1.$$

Note that this relation holds only if θ is measured in radians. As stated before, it is essential for the next article; but it has also an importance of its own, since it shows that, for small values of θ, $\sin \theta$ is nearly equal to θ itself.

178. Differentiation of sin u. Suppose that u is a function of x which is differentiable in the interval (a, b) and let $y = \sin u$. We first differentiate with respect to u. The General Rule gives at once the following equations.

$$y = \sin u, \tag{1}$$
$$y + \Delta y = \sin (u + \Delta u),$$
$$\Delta y = \sin (u + \Delta u) - \sin u. \tag{2}$$

We now prove that

$$\Delta y = 2 \cos (u + \tfrac{1}{2} \Delta u) \sin \tfrac{1}{2} \Delta u. \tag{3}$$

The last equality is given by the trigonometric formula

$$\sin A - \sin B = 2 \cos \tfrac{1}{2}(A + B) \sin \tfrac{1}{2}(A - B),$$

writing $A = u + \Delta u$, $B = u$, and hence (by solving)

$$\tfrac{1}{2}(A + B) = u + \tfrac{1}{2}\,\Delta u, \quad \tfrac{1}{2}(A - B) = \tfrac{1}{2}\,\Delta u.$$

Dividing (3) by Δu, we obtain

$$\frac{\Delta y}{\Delta u} = \frac{2}{\Delta u}\cos\left(u + \tfrac{1}{2}\,\Delta u\right)\sin\tfrac{1}{2}\,\Delta u \tag{4}$$

$$= \cos\left(u + \tfrac{1}{2}\,\Delta u\right)\left(\frac{\sin\tfrac{1}{2}\,\Delta u}{\tfrac{1}{2}\,\Delta u}\right).$$

As $\Delta u \longrightarrow 0$, $\lim \cos\left(u + \tfrac{1}{2}\,\Delta u\right) = \cos u$. The fraction in the second parenthesis is in the form considered in the previous article; hence its limit is 1. We therefore have, for all values of u,

$$\frac{dy}{du} = \cos u.$$

But

$$\frac{dy}{dx} = \frac{dy}{du}\cdot\frac{du}{dx}.$$

Hence

(I) $$\frac{d}{dx}\sin u = \cos u\,\frac{du}{dx}.$$

179. Differentiation of cos u. Since $\cos u = \sin\left(u + \tfrac{1}{2}\,\pi\right)$ we can use the formula of the previous article.

$$\frac{d}{dx}\cos u = \frac{d}{dx}\sin\left(u + \tfrac{1}{2}\,\pi\right)$$

$$= \cos\left(u + \tfrac{1}{2}\,\pi\right)\frac{d}{dx}\left(u + \tfrac{1}{2}\,\pi\right)$$

$$= -\sin u\,\frac{du}{dx}.$$

The formula is, then,

(II) $$\frac{d}{dx}\cos u = -\sin u\,\frac{du}{dx}.$$

180. Differentiation of tan u, ctn u, sec u, csc u. Since

$$\tan u = \frac{\sin u}{\cos u},$$

we have, by the formula for differentiating a quotient,

$$\frac{d}{dx}\tan u = \frac{\cos u\,\dfrac{d}{dx}\sin u - \sin u\,\dfrac{d}{dx}\cos u}{\cos^2 u}$$

$$= \frac{\cos^2 u + \sin^2 u}{\cos^2 u}\frac{du}{dx}$$

$$= \frac{1}{\cos^2 u}\frac{du}{dx} = \sec^2 u\,\frac{du}{dx}.$$

Hence

(III)
$$\frac{d}{dx} \tan u = \sec^2 u \frac{du}{dx}.$$

Similarly,

(IV)
$$\frac{d}{dx} \operatorname{ctn} u = - \csc^2 u \frac{du}{dx}.$$

Setting $\sec u = \dfrac{1}{\cos u}$, we find

(V)
$$\frac{d}{dx} \sec u = \sec u \tan u \frac{du}{dx}.$$

Similarly,

(VI)
$$\frac{d}{dx} \csc u = - \csc u \operatorname{ctn} u \frac{du}{dx}.$$

The formulas above are valid in any interval (a, b) in which u is differentiable with respect to x, except for values of x for which $\tan u$, $\operatorname{ctn} u$, etc. are not defined.

Example 1. Differentiate $y = \cos^2 3\,x$.

Solution. Applying first the formula for differentiating u^n, we have, taking $u = \cos 3\,x$ and $n = 2$,

$$\frac{dy}{dx} = 2 \cos 3\,x \frac{d}{dx} \cos 3\,x$$
$$= 2 \cos 3\,x \left[- \sin 3\,x \frac{d}{dx}\,(3\,x) \right]$$
$$= - 6 \cos 3\,x \sin 3\,x = - 3 \sin 6\,x.$$

Example 2. Differentiate $r = \theta \tan \theta$.

Solution. Applying first the formula for differentiating a product, we have

$$\frac{dr}{d\theta} = \theta \frac{d}{d\theta} \tan \theta + \tan \theta \frac{d\theta}{d\theta} = \theta \sec^2 \theta + \tan \theta.$$

Example 3. A number of iron brackets in the form of a capital **Y** are to be constructed. The height of the **Y** is to be 12 in. and the width across the top is to be 10 in. What shape requires the least material?

Solution. The total length of the stem and the two branches must evidently be a minimum. Let $2\,x$ be the angle between the branches. Then $BC = 5 \csc x$, $BD = 5 \operatorname{ctn} x$, and $BA = 12 - 5 \operatorname{ctn} x$. Hence the length required for any shape is

$$L = 12 - 5 \operatorname{ctn} x + 10 \csc x.$$

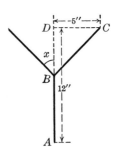

Fig. 180

Differentiating, we have

$$\frac{dL}{dx} = 5\csc^2 x - 10\csc x \operatorname{ctn} x.$$

Setting this equal to 0 and dividing by $5\csc x$ (which cannot equal 0), the result is

$$\csc x - 2\operatorname{ctn} x = 0.$$

That is, $\csc x = 2\operatorname{ctn} x$,   or $\dfrac{1}{\sin x} = \dfrac{2\cos x}{\sin x}$,

whence $\cos x = \frac{1}{2}.$

Therefore $x = 60°$ gives either a maximum or a minimum length.

That the length is a minimum can be proved by getting d^2L/dx^2. This problem can be solved without recourse to trigonometric functions, but their use avoids the necessity of troublesome radicals.

PROBLEMS

1. Differentiate the following functions.

a. $y = 3\sin\frac{1}{2}x.$
c. $r = \tan\frac{1}{2}\theta.$
e. $y = 3\sec 2x.$
g. $y = x\sin x.$
i. $y = \ln\sin x.$
k. $s = \cos\dfrac{a}{t}.$
m. $y = e^x\ln\sin x.$

o. $f(x) = \sin(x+a)\cos(x-a).$

q. $y = \cos\sqrt{5}\,x.$
s. $r = e^{\sin\theta}.$
u. $s = e^t\tan t.$
w. $y = x^{\sin x}.$
y. $y = (\cos x)^x.$

b. $y = \cos(ax+b).$
d. $s = a\operatorname{ctn} bt.$
f. $y = 4\csc\frac{1}{2}x.$
h. $s = e^{-2t}\cos t.$
j. $y = \dfrac{\cos x}{x}.$
l. $r = \sin^3\frac{1}{3}\theta.$
n. $y = \ln\sqrt{\dfrac{1+\sin x}{1-\sin x}}.$

p. $r = \dfrac{1+\cos\theta}{1-\cos\theta}.$

r. $xy = \cos x.$
t. $y = \ln\sec^2 x.$
v. $s = e^{-\frac{1}{5}t}\sin 5t.$
x. $y = (\sin x)^x.$
z. $s = e^{-\frac{1}{4}t}\cos 2t.$

2. For the following functions find the value of dy/dx for the given value of x.

a. $y = 2\sin x;\; x = 1.$
c. $y = \tan\frac{1}{4}\pi x;\; x = 1.$
e. $y = x\operatorname{ctn} x;\; x = \frac{1}{4}\pi.$
g. $y = \ln\sin x;\; x = \frac{1}{4}\pi.$
i. $y = x\sin x;\; x = \frac{1}{3}\pi.$
k. $y = e^{\sin x};\; x = \frac{1}{2}\pi.$
m. $y = e^{-x}\cos\pi x;\; x = \frac{1}{2}$

b. $y = \cos 4x;\; x = 0.5.$
d. $y = x\sin\frac{1}{2}x;\; x = 2.$
f. $y = x\tan x;\; x = 1.$
h. $y = e^{-\frac{1}{2}x}\cos x;\; x = 2.$
j. $y = x\cos 2x;\; x = \frac{1}{4}\pi.$
l. $y = e^x\sin 2x;\; x = 0.5.$
n. $y = e^{-\frac{1}{10}x}\sin\frac{1}{3}\pi x;\; x = 6.$

3. Find the second derivatives of the following functions.

a. $\rho = \sin 4\,\theta.$

b. $s = \cos \frac{1}{2}\,\pi t.$

c. $y = \tan x.$

d. $\rho = \ln \sin \theta.$

e. $\rho = \theta \cos \theta.$

f. $s = e^t \sin 2\,t.$

g. $y = \sec 2\,x.$

h. $y = x \sin x.$

i. $s = e^{-t} \cos t.$

j. $s = e^{-\frac{1}{3}t} \sin \pi t.$

k. $y = e^{\cos x}.$

l. $\rho = \sin^2 \frac{1}{2}\,\theta.$

4. Find the slopes of $y = \tan x$ and $y = x^3$ at the origin. Sketch both curves on the same axes.

5. Find the angle which the curves $y = \sin x$ and $y = \cos x$ make with each other at their points of intersection.

6. Find the angle between the curves $y = \tan x$ and $y = \operatorname{ctn} x$ at their points of intersection.

7. Find the angles of intersection of the graphs of $y = x$ and $y = x - \sin 2\,x$ at the points where $x = 0$ and $x = \frac{1}{2}\,\pi.$

8. Find the angles between the curves $y = \cos x$ and $y = \sin 2\,x$ at their points of intersection.

9. The equation of motion of a point that moves along a straight line is $s = 4 \sin \frac{1}{2}\,\pi t.$ Find the position, velocity, and acceleration when $t = 0, 1, 2, 3, 4.$

10. The distance of the head of a piston from the center of the driving shaft is given by the formula

$$y = k + r \cos \theta + \sqrt{a^2 - r^2 \sin^2 \theta},$$

where a is the length of the connecting rod, k is another constant, r is the length of the crank in feet, and θ is the angle the crank makes with the horizontal. If the crank rotates at a constant angular velocity of 600 revolutions per minute, find the rate at which the piston is moving when $\theta = 90°$; when $\theta = 0°.$

11. An angle is increasing at a constant rate. Show that the tangent and the sine are increasing at the same rate when the angle is zero, and that the tangent increases eight times as fast as the sine when the angle is 60°.

12. A revolving light sending out a bundle of parallel rays is at a distance of $\frac{1}{2}$ mi. from the shore and makes 1 revolution per minute. Find how fast the light is traveling along the straight beach when at a distance of 1 mi. from the nearest point of the shore.

13. A tapestry 7 ft. in height is hung on a wall so that its lower edge is 9 ft. above an observer's eye. At what distance from the wall should he stand in order to obtain the most favorable view, that is, so that the vertical angle subtended by the tapestry at his eye is a maximum?

14. A body of weight W is dragged along a horizontal plane by means of a force P whose line of action makes an angle x with the plane. The magnitude of the force is given by the equation

$$P = \frac{mW}{m \sin x + \cos x},$$

in which m denotes the coefficient of friction. Show that the pull is least when $\tan x = m$.

15. A steel girder 30 ft. long is carried along a passage 10 ft. wide and into a corridor at right angles to the passage. The thickness of the girder being neglected, how wide must the corridor be in order that the girder may go round the corner?

16. Given $\sin 60° = 0.86603$ and $\cos 60° = 0.5$; use differentials to compute the values of the following functions to four decimal places.

    a. $\sin 62°$.   **b.** $\cos 61°$.   **c.** $\sin 59°$.   **d.** $\cos 58°$.

17. Two sides of a triangle are 3 ft. and 4 ft. respectively, and the included angle θ is 60°. Express the third side y as a function of θ, find its value for $\theta = 60°$, and by means of differentials find the error in y caused by an error of $3'$ in measuring θ.

181. Addition of ordinates. When the equation of a curve has the form

$$y = \text{the algebraic sum of two expressions,}$$

as, for example,

$$y = \sin x + \cos x, \qquad y = \tfrac{1}{2} x + \sin^2 x,$$

the principle known as addition of ordinates may be employed to advantage. Characteristic features of the locus are more easily discovered by this principle.

For example, to construct the locus of

$$y = 2 \sin \tfrac{1}{4} \pi x + \tfrac{1}{2} x, \tag{1}$$

we employ the auxiliary curves

$$y_1 = 2 \sin \tfrac{1}{4} \pi x, \tag{2}$$
$$y_2 = \tfrac{1}{2} x. \tag{3}$$

Plot these curves one below the other, keeping the y-axes in a straight line. The same scales must be used in both figures. The locus of (2) is the sine curve of Fig. 181.1. The locus of (3) is the straight line in Fig. 181.2.

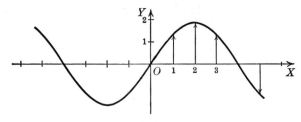

Fig. 181.1

The ordinates of Fig. 181.1 are now added to the corresponding ones in Fig. 181.2, attention being given to the algebraic signs. That is,

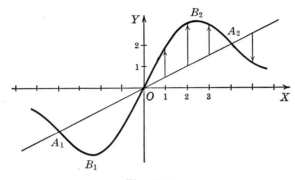

Fig. 181.2

positive ordinates in Fig. 181.1 are laid off *above* the straight line in Fig. 181.2. *Negative* ordinates in Fig. 181.1 are laid off *below* the straight line in Fig. 181.2. The derived curve $A_1B_1OB_2A_2$ has the equation

$$y = y_1 + y_2 = 2 \sin \tfrac{1}{4} \pi x + \tfrac{1}{2} x, \qquad (4)$$

as required. It is now seen that the locus winds back and forth across the line $y = \tfrac{1}{2} x$, crossing it at $x = 0, \pm 4, \pm 8$, etc., that is, directly under or over the points where the sine curve in Fig. 181.1 crosses the x-axis.

Example. Find the maximum and minimum points and the points of inflection of $y = x - \sin 2x$ between 0 and π, and plot the graph.

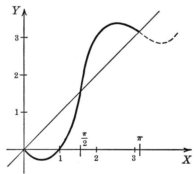

Fig. 181.3

Solution. Differentiating, we have

$$y' = 1 - 2 \cos 2x,$$
$$y'' = 4 \sin 2x.$$

Setting the first derivative equal to zero gives $\cos 2x = \frac{1}{2}$, or $2x = 60°$ or $300°$. Hence two of the critical values are $x = 30° = \frac{1}{6}\pi$ and $x = 150° = \frac{5}{6}\pi$. Testing these by substitution in the second derivative, we find that $x = \frac{1}{6}\pi$ gives a minimum value,

$$y = \tfrac{1}{6}\pi - \sin \tfrac{1}{3}\pi = 0.5236 - 0.8660 = -0.3424;$$

and $x = \frac{5}{6}\pi$ gives a maximum value,

$$y = \tfrac{5}{6}\pi - \sin \tfrac{5}{3}\pi = 2.6180 + 0.8660 = 3.4840.$$

Hence the minimum point is $(0.5236, -0.3424)$, and the maximum point is $(2.6180, 3.4840)$. Setting the second derivative equal to zero gives $\sin 2x = 0$; whence $x = 0°$, $90°$, or $180°$. Calculating the corresponding values of y, we find that the points of inflection are $(0, 0)$, $(1.5708, 1.5708)$, $(3.1416, 3.1416)$.

In the accompanying table are assembled the results found (abbreviated); these are used in checking Fig. 181.3, which is drawn by addition of ordinates.

x	y	y'	y''	
0	0	-1	0	Inflectional point
0.52	-0.34	0	$+$	Minimum point
1.57	1.57	3	0	Inflectional point
2.62	3.48	0	$-$	Maximum point
3.14	3.14	-1	0	Inflectional point

PROBLEMS

In each of the following problems find the maximum, minimum, and inflectional points in the range indicated, and draw the graph.

1. $y = x + 2\sin x$; (0 to 2π).
2. $y = x + \sin 2x$; (0 to π).
3. $y = \frac{1}{2}x - \sin x$; (0 to 2π).
4. $y = \sin x + \cos x$; (0 to 2π).
5. $y = 2x - \tan x$; (0 to π).
6. $y = \tan x - 4x$; (0 to π).
7. $y = 3\sin x - 4\cos x$; (0 to 2π).
8. $y = x + \cos 2x$; (0 to π).
9. $y = \sin \pi x - \cos \pi x$; (0 to 2).
10. $y = \frac{1}{2}x + \sin 2x$; (0 to π).
11. $y = x - 2\cos 2x$; (0 to π).
12. $y = \frac{1}{2}\pi x + \sin \pi x$; (0 to 2).
13. Show that the maximum value of $y = a\sin x + b\cos x$ is $\sqrt{a^2 + b^2}$.

182. Addition of ordinates of sine curves with equal periods. In the equation of the sine curve

$$y = 2\sin \left(\tfrac{1}{3}\pi x + \tfrac{1}{6}\pi\right), \qquad (1)$$

the right-hand member may be expanded by the formula for $\sin (x + y)$. Then

$$y = 2\sin \tfrac{1}{3}\pi x \cos \tfrac{1}{6}\pi + 2\cos \tfrac{1}{3}\pi x \sin \tfrac{1}{6}\pi. \qquad (2)$$

Fig. 182

Substituting $\cos \frac{1}{6} \pi = \frac{1}{2}\sqrt{3}$, $\sin \frac{1}{6} \pi = \frac{1}{2}$, the result is

$$y = \sqrt{3} \sin \tfrac{1}{3} \pi x + \cos \tfrac{1}{3} \pi x. \tag{3}$$

Hence the ordinates in (1) may now be obtained by addition of the ordinates of two sine curves with equal periods ($= 6$). Conversely, when the ordinates of sine curves with equal periods are added, the final curve is a simple sine curve with the *same* period.

For example, given,

$$y = A \sin kx + B \cos kx. \tag{4}$$

The period is $2\pi/k$ for both terms. Draw the right triangle with sides A and B, the amplitudes of the sine curves. Denote the angle opposite the side B by γ. Then the hypotenuse is $C = \sqrt{A^2 + B^2}$, and $B = C \sin \gamma$, $A = C \cos \gamma$. Substitute these values of A and B in (4). Then (4) becomes

$$y = C \sin kx \cos \gamma + C \cos kx \sin \gamma = C \sin (kx + \gamma). \tag{5}$$

But this is a simple sine curve with period $2\pi/k$ and amplitude $C = \sqrt{A^2 + B^2}$. Note that $\tan \gamma = B/A$, from which γ is found in radians. The locus of (4) is now more readily constructed from (5), as in Art. 175, than by addition of ordinates.

PROBLEMS

Change the following functions to the form (5) and draw their graphs. Compute y from the original equation for the value of x given, and check in the figure.

1. $y = \sin x + \cos x$; $x = \frac{1}{2}\pi$.
2. $y = 3 \sin x + 4 \cos x$; $x = 1$.
3. $y = 2 \sin x - \cos x$; $x = 1$.
4. $y = \sin 2x - 2 \cos 2x$; $x = \frac{1}{2}$.
5. $y = 3 \sin \frac{1}{2} \pi x + 2 \cos \frac{1}{2} \pi x$; $x = 1$.
6. $y = 5 \sin \frac{1}{4} \pi x - 3 \cos \frac{1}{4} \pi x$; $x = 2$.

183. Multiplication of ordinates; boundary curves. In plotting the locus of an equation of the form

$$y = \text{product of two factors}, \tag{1}$$

one of which is a sine or cosine, as, for example,

$$y = e^x \sin x \quad \text{or} \quad s = t^2 \cos \tfrac{1}{4} \pi t,$$

the following considerations are of much value.

For example, let us construct the locus of

$$y = e^{-\frac{1}{4}x} \sin \frac{1}{2} \pi x. \tag{2}$$

We make the following observations.

1. Since the numerical value of the sine never exceeds unity, the value of y in (2) for any value of x will not exceed in numerical value the value of the first factor, $e^{-\frac{1}{4}x}$.

Consequently, if the curves

$$y_1 = e^{-\frac{1}{4}x} \quad \text{and} \quad y_1 = -e^{-\frac{1}{4}x} \tag{3}$$

are drawn, the locus of (2) will lie entirely between these curves. They are accordingly called *boundary curves*.

Draw these curves as in Fig. 183. The second is obviously symmetric to the first with respect to the x-axis. To plot, find three points on the first curve, as in the table.

x	y_1
0	1
2	$e^{-\frac{1}{2}} = .61$
4	$e^{-1} = .37$

2. When $\sin \frac{1}{2} \pi x = 0$, then in (2) $y = 0$, since the first factor is always finite. Hence the locus of (2) meets the x-axis in the same points as the auxiliary sine curve,

$$y_2 = \sin \frac{1}{2} \pi x. \tag{4}$$

3. The required curve touches (see below for proof) the boundary curves when the second factor, $\sin \frac{1}{2} \pi x$, is $+1$ or -1; that is, when the ordinates of the auxiliary curve (4) have a maximum or minimum value.

Hence draw the sine curve (4). The period is 4 and the amplitude is 1. This curve is the dotted line of the figure.

The discussion shows these facts: The locus of (2) crosses the x-axis at $x = 0, \pm 2, \pm 4, \pm 6$, etc., and touches the boundary curves (3) at $x = \pm 1, \pm 3, \pm 5$, etc.

We may then readily sketch the curve, as in the figure; that is, the winding curve between the boundary curves (3).

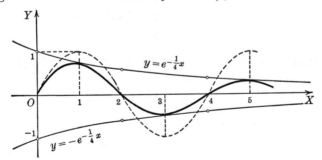

Fig. 183

4. For a check remember that the ordinate of (2) is the product of the ordinates of the boundary curve $y_1 = e^{-\frac{1}{4}x}$ and the sine curve (4). In the figure, for example, the required curve lies above the x-axis between $x = 0$ and $x = 2$, for the ordinates of $y_1 = e^{-\frac{1}{4}x}$ and of the sine curve are now all positive. But between $x = 2$ and $x = 4$ the required curve lies below the x-axis, for the ordinates of $y_1 = e^{-\frac{1}{4}x}$ and the sine curve now have unlike signs.

To prove the statement made in 3 above, differentiate (2). We find

$$y' = e^{-\frac{1}{4}x}(-\tfrac{1}{4}\sin \tfrac{1}{2}\pi x + \tfrac{1}{2}\pi \cos \tfrac{1}{2}\pi x). \tag{5}$$

When $\sin \tfrac{1}{2}\pi x = 1$, $y' = -\tfrac{1}{4}e^{-\frac{1}{4}x}$, the same as y' for the upper boundary curve.

From (5) we see that $y' = 0$ when

$$-\tfrac{1}{4}\sin \tfrac{1}{2}\pi x + \tfrac{1}{2}\pi \cos \tfrac{1}{2}\pi x = 0. \tag{6}$$

From this equation we find $\tan \tfrac{1}{2}\pi x = 2\pi$. Hence $x = 0.90 + 2n$, n any integer. That is, the maximum and minimum points are slightly to the *left* of the points of contact with the boundary curves.

PROBLEMS

Sketch the graphs of the following equations, using the intervals indicated.

1. $y = 5 e^{-\frac{1}{4}x} \cos \pi x$; $(0, 4)$.　　　　2. $y = 4 e^{-\frac{1}{5}x} \sin \pi x$; $(0, 3)$.

3. $y = \tfrac{1}{4} x \sin x$; $(0, 3\pi)$.　　　　4. $y = \tfrac{1}{2} x \cos 2x$; $(0, 2\pi)$.

5. $y = \dfrac{\sin x}{x}$; $(0, 2\pi)$.

Sketch the graphs of the following equations for the interval indicated and find the coördinates of the maximum, minimum, and inflectional points.

6. $y = 2 e^{-\frac{1}{2}x} \sin 2x$; $(0, \pi)$.　　　　7. $y = 3 e^{-\frac{1}{4}x} \cos x$; $(0, 2\pi)$.

8. $y = 10 e^{-x} \sin \tfrac{1}{2}\pi x$; $(0, 4)$.　　　　9. $y = 5 e^{-\frac{1}{5}x} \cos \tfrac{1}{3}\pi x$; $(0, 6)$.

184. Distance-time diagrams in mechanics. In studying motion on a straight line the graph of the equation of motion,

$$s = f(t), \tag{1}$$

where s and t denote distance and time, respectively, is useful. Values of t are plotted as abscissas, values of s as ordinates. The path of the moving point is on the axis of ordinates OS.

ILLUSTRATION. Fig. 184.1 shows a distance-time diagram. Let P be a point on the graph, and draw the ordinate MP and the abscissa RP.

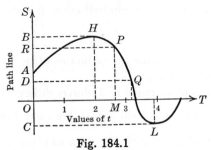

Fig. 184.1

Then $s = OR$ when $t = OM$. At this instant the point is at R and is moving *downward*. By Art. 71,

$$\text{Velocity } v = \frac{ds}{dt}, \quad \text{Acceleration } a = \frac{dv}{dt} = \frac{d^2s}{dt^2}. \tag{2}$$

The value of s ($= OB$) at H is a *maximum*; at L, s ($= OC$) is a *minimum*. In both cases $v = 0$. Also, Q is a point of inflection and $a = 0$ (Art. 161). Then the following statements based on the considerations of Chapter V may be made: When $t = 0$, the point is at A, moving upward. It moves upward until $s = OB$, and then $v = 0$. It then moves downward from B to C (when v is again zero), and thereafter moves upward. The speed is zero at B and at C and is a maximum at D. The values of t for these positions are to be read from the diagram.

When v has a maximum or minimum value, the acceleration is zero. That is, the positions of the moving point on OS which correspond to the points of inflection on the distance-time diagram are those at which the velocity is greatest or least.

Example. The equation of a straight-line motion is

$$s = \tfrac{1}{3} t + \cos \tfrac{1}{2}\, \pi t. \tag{3}$$

a. Find s, v, and a when $t = 0$.
b. Draw the distance-time diagram for $t = 0, \cdots, 4$.
c. Find t and s when $v = 0$, and mark the positions on the path.
d. Find t, s, and v when $a = 0$.
e. Find s, v, and a when $t = 4$.
f. Tabulate all results of (a), (c), (d), and (e).

Solution. (**a**) Differentiating (3), we find

$$v = \tfrac{1}{3} - \tfrac{1}{2}\, \pi \sin \tfrac{1}{2}\, \pi t, \tag{4}$$
$$a = -\tfrac{1}{4}\, \pi^2 \cos \tfrac{1}{2}\, \pi t. \tag{5}$$

When $t = 0$, $s = 1$,
$$v = \tfrac{1}{3},$$
$$a = -\tfrac{1}{4}\pi^2 = -2.47.$$

Hence when $t = 0$, the point is at a unit distance from $s = 0$ and is moving up. The speed is decreasing.

   (b) The graph of (3) is drawn by addition of ordinates as in Art. 181. The result is a winding curve crossing the line $s_1 = \tfrac{1}{3}t$ (OL in Fig. 184.2) at $t = 1, 3$, etc. A table of values of t and s to be used in drawing the curve is appended.

t	0	0.5	1	1.5	2	2.5	3	3.5	4
s	1	0.87	0.33	-0.21	-0.33	0.13	1	1.87	2.33

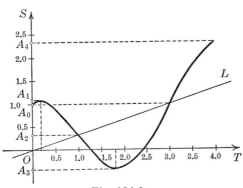

Fig. 184.2

From the results in (a) we know that the slope of the curve in the figure at $A_0(0, 1)$ is $\tfrac{1}{3}$ and that the curve is concave downward (since a is negative).

   (c) From (4) the values of t when $v = 0$ satisfy

$$\sin \tfrac{1}{2}\pi t = \frac{2}{3\pi} = 0.212.$$

Hence $\tfrac{1}{2}\pi t = 0.21$ and 2.93.   (Table 6, Chap. XXV.)
Then $t = 0.13$ and 1.86,
and $s = 1.02$ and -0.36.

   The points $(0.13, 1.02)$ and $(1.86, -0.36)$ on the diagram are the points of contact of horizontal tangents meeting the path line OS at A_1 and A_3 respectively. At A_1 and A_3, therefore, $v = 0$.

   (d) From (5) the values of t when $a = 0$ satisfy

$$\cos \tfrac{1}{2}\pi t = 0.$$

Hence $t = 1$ and 3,
$$s = \tfrac{1}{3} \text{ and } 1,$$
and $v = -1.24$ and 1.90.

The points $(1, \frac{1}{3})$ and $(3, 1)$ are points of inflection on the distance-time diagram. They lie on the line OL. The horizontal line through $(1, \frac{1}{3})$ meets the path line OS at A_2. When the moving point descends from A_1 to A_3, its acceleration at A_2 is zero and the speed has a maximum value equal to 1.24 units of velocity. When the point moves upward from A_3, its acceleration at A_0 is zero and the speed has a maximum value equal to 1.90 units of velocity.

(e) When $t = 4$, we find
$$s = 2\tfrac{1}{3},$$
$$v = \tfrac{1}{3},$$
$$a = -2.47.$$

That is, the point is moving upward with decreasing speed.

(f) A table summarizing the results and giving the significant successive positions of the point is shown below.

t	v	s	Point at
0	$\frac{1}{3}$	1	A_0 moving up
0.13	0	1.02	A_1 at rest
1	-1.24	$\frac{1}{3}$	A_2 moving down with the speed a maximum
1.86	0	-0.36	A_3 at rest
3	1.90	1	A_0 moving up with the speed a maximum
4	$\frac{1}{3}$	2.33	A_4 moving up

From this table the motion may be described for the given interval of time.

185. Types of motion on a straight line. Some interesting types of rectilinear motion have for distance-time diagrams curves studied in this chapter. Consider the following cases.

Simple harmonic vibration. A typical equation of motion is

$$s = a \cos kt. \tag{1}$$

The distance-time diagram here is a simple sine curve with period $2\pi/k$ and amplitude a (Art. 175). The motion is a simple oscillation between successive fixed extreme positions $(0, a)$ and $(0, -a)$ with elapsed time π/k.

From (1), differentiating, we obtain

$$v = \frac{ds}{dt} = -ak \sin kt, \tag{2}$$

$$a = \frac{dv}{dt} = -ak^2 \cos kt = -k^2 s. \tag{3}$$

The last equation shows the property which is characteristic of simple harmonic vibration; namely, the acceleration is directly proportional to the distance and differs from it in sign.

Damped vibration. A typical equation of motion is

$$s = ae^{-bt} \cos kt. \qquad (b > 0) \qquad (4)$$

This equation, which differs from (1) in having the exponential factor e^{-bt}, has, as graph, the type of curve discussed in Art. 183. The exponential factor is called the *damping factor.* Bearing in mind the properties of the graph of (4), we see that the point oscillates between extreme positions whose distances from the center of motion ($s = 0$) constantly decrease. (In fact, these distances form a decreasing geometric series.) The time between consecutive extreme positions is, however, the same as in the motion (1), namely, a half-period π/k. To see this, differentiate (4) to find v. Then

$$v = ae^{-bt}(- b \cos kt - k \sin kt). \qquad (5)$$

Now $v = 0$ when $b \cos kt + k \sin kt = 0$, that is, for all values of t satisfying

$$\tan kt = -\frac{b}{k}. \qquad (6)$$

Since the period of $\tan kt = \pi/k$, successive values of t satisfying (6) differ by π/k. Hence the elapsed time between two successive extreme positions is π/k.

Forced vibration. A typical equation of motion is

$$s = at \cos kt. \qquad (7)$$

This equation differs from (1) by the presence of the factor t. The graph is readily drawn by the methods explained in Art. 183. (See also Problems 3, 4, page 301.) The vibration is now between extreme positions whose distances from the center ($s = 0$) increase. The periodic character has disappeared. In fact, differentiating (7), we obtain

$$v = a(\cos kt - kt \sin kt). \qquad (8)$$

From this equation we see that the values of t when $v = 0$ satisfy

$$\cos kt - kt \sin kt = 0, \quad \text{or} \quad \operatorname{ctn} kt = kt. \qquad (9)$$

Successive values of t which satisfy the equation (9) do not differ by a constant.

PROBLEMS

1. Show that the compound harmonic vibration

$$s = 3 \sin t + 4 \cos t$$

is a simple harmonic vibration with amplitude 5 and period 2π (see page 299, Problem 2).

2. Given the equation of a damped vibration,

$$s = 3 \, e^{-\frac{1}{2}t} \sin 2 \, t.$$

a. Draw the distance-time diagram for $t = 0, \cdots, 4$.

b. Find t and s when $v = 0$. Mark the positions of the point on the s-axis.

c. Find t and s when $a = 0$. Mark the positions on the path.

d. Make a table summarizing the results. (See the table of the example, Art. 184.)

3. a. Draw the distance-time diagram for the straight-line motion $s = t(t - 3)^2$ for $t = 0, \cdots, 4$.

b. Find t and s when $v = 0$.

c. When does a change sign? Find s and v when $a = 0$.

d. Find s, v, and a when $t = 4$. Make a table summarizing the results in (b) and (c).

4. In Problem 2, show that the values of s for successive extreme positions form a decreasing geometric progression whose ratio is $- e^{-\frac{1}{4}\pi}$.

5. Given the equation of a forced vibration, $s = \frac{1}{2} t \sin t$.

a. Draw the distance-time diagram for $t = 0, \cdots, 4$.

b. Find s, v, and a when $t = 0, 1, 2$. Mark these positions on the s-axis.

c. Mark the position (approximately) of the point on the path line when $v = 0$ for the first time (not $t = 0$), and measure it.

6. Draw the distance-time diagram for the straight-line motion

$$s = t^3 - 6 \, t.$$

Find the values of t, s, v, and a missing from the accompanying table. Describe the motion.

t	s	v	a
0			
		0	
			0
	12		

7. In each of the following problems the given equation represents a straight-line motion for a limited time, specified in each case. Sketch the distance-time diagram. Find **(a)** the values of t and s when $v = 0$; **(b)** the values of t and s when $a = 0$; **(c)** the greatest speed. Describe the motion.

     a. $s = t(t - 5)^2$; $(t = 0 \text{ to } t = 5)$.

     b. $s = 250 \, t^2 - \frac{5}{4} t^4$; $(t = 0 \text{ to } t = 10)$.

     c. $s = 2 - 3 \cos 2 \, t$; $(t = 0 \text{ to } t = 4)$.

     d. $s = \frac{1}{2} t - 2 \sin \pi t$; $(t = 0 \text{ to } t = 3)$.

     e. $s = 10 \, e^{-\frac{1}{5}t} \sin \frac{1}{2} \pi t$; $(t = 0 \text{ to } t = 4)$.

     f. $s = 5 \, e^{-\frac{1}{4}t} \cos 2 \, t$; $(t = 0 \text{ to } t = \pi)$.

8. In the simple harmonic vibration $s = a \cos kt$ find the mean value for one period of v^2 with respect to the time. Show that it equals half the square of the maximum velocity.

9. In the simple harmonic vibration $s = a \cos kt$ show that the mean value of v^2 with respect to s between $s = - a$ and $s = + a$ equals two thirds of the maximum value of v^2. Compare with the preceding problem.

186. The inverse trigonometric functions. The meaning of the equation $y = \text{arc sin } x$* is that y is the angle (arc) whose sine is x. Thus $y = \text{arc sin } x$ and $x = \sin y$ are equivalent. Hence the graph of the function $y = \text{arc sin } x$ can be constructed by interchanging the axes in the graph of $y = \sin x$. Similar statements hold for the other trigonometric functions. Thus

$$y = \text{arc cos } x \text{ means } x = \cos y,$$
$$y = \text{arc tan } x \text{ means } x = \tan y,$$
$$y = \text{arc ctn } x \text{ means } x = \text{ctn } y,$$
$$y = \text{arc sec } x \text{ means } x = \sec y,$$
$$y = \text{arc csc } x \text{ means } x = \csc y.$$

It should be noted that the functions $y = \text{arc sin } x$ and $y = \text{arc cos } x$ are defined only for values of x between -1 and $+1$, inclusive; the functions $y = \text{arc tan } x$ and $y = \text{arc ctn } x$ are defined for all values of x; the functions $y = \text{arc sec } x$ and $y = \text{arc csc } x$ are defined for all values of x except those between -1 and $+1$.

The inverse trigonometric functions are multiple-valued functions. Thus, if $y = \text{arc sin } x$ and if $x = \frac{1}{2}$, we may have $y = \frac{1}{6} \pi$ radians (30°), or $y = \frac{5}{6} \pi$ radians (150°), or $y = \frac{13}{6} \pi$ radians (390°), or $y = -\frac{7}{6} \pi$ radians ($-210°$), etc. In practical applications the nature of the problem usually determines the value or values to be chosen. In purely formal calculations any one of the infinite number of values might be used. To avoid ambiguity when only *one* value is wanted we shall agree to choose values according to the table below. With the choice indicated, we may always take the positive sign with each square root in the formulas for differentiation which follow. To see the truth of this statement note the sign of the derivative in each of the formulas in Art. 188 and compare with the slope of the corresponding graph.

Choice of values of the inverse trigonometric functions. 1. If the value of x in any one of the inverse trigonometric functions is positive, we shall choose the value of the angle y between 0 and $\frac{1}{2} \pi$.

2. If the value of x is negative and

$y = \text{arc sin } x$, we shall choose y between 0 and $-\frac{1}{2} \pi$;
$y = \text{arc cos } x$, we shall choose y between $\frac{1}{2} \pi$ and π;
$y = \text{arc tan } x$, we shall choose y between 0 and $-\frac{1}{2} \pi$;
$y = \text{arc ctn } x$, we shall choose y between $\frac{1}{2} \pi$ and π;
$y = \text{arc sec } x$, we shall choose y between $-\frac{1}{2} \pi$ and $-\pi$;
$y = \text{arc csc } x$, we shall choose y between $-\frac{1}{2} \pi$ and $-\pi$.

*The notation $y = \sin^{-1} x$ is also used.

ILLUSTRATION. By the above rule, we have

arc sin $\frac{1}{2} = \frac{1}{6} \pi = 0.524$ arc sin $(-\frac{1}{2}) = -0.524$.

arc cos $\frac{1}{2} = \frac{1}{3} \pi = 1.047$ arc cos $(-\frac{1}{2}) = \frac{2}{3} \pi = 2.094$.

arc tan $1 = \frac{1}{4} \pi = 0.785$ arc tan $(-1) = -0.785$.

arc ctn $1 = \frac{1}{4} \pi = 0.785$ arc ctn $(-1) = \frac{3}{4} \pi = 2.356$.

arc sec $2 = \frac{1}{3} \pi = 1.047$ arc sec $(-2) = -\frac{2}{3} \pi = -2.094$.

arc csc $2 = \frac{1}{6} \pi = 0.524$ arc csc $(-2) = -\frac{5}{6} \pi = -2.618$.

Since the inverse circular functions must be evaluated in radians, Table 6, Chapter XXV, should be used.

Figures are given on the opposite page for the graphs of each of the inverse circular functions. The portion in each graph drawn in a solid heavy line corresponds to the statement made above.

187. Differentiation of arc sin u. Suppose that u is a function of x, which is differentiable and such that $|u(x)| < 1$ for every value of x in the interval (a, b).

Let $y = $ arc sin u.

Then $u = \sin y,$

and, differentiating with respect to y, we have

$$\frac{du}{dy} = \cos y,$$

whence $$\frac{dy}{du} = \frac{1}{\cos y}.$$

Applying the formula for differentiating a function of a function, we have

$$\frac{dy}{dx} = \frac{dy}{du}\frac{du}{dx} = \frac{1}{\cos y}\frac{du}{dx}.$$

But $$\cos y = \sqrt{1 - \sin^2 y} = \sqrt{1 - u^2}.$$

Hence $$\frac{d}{dx} \text{ arc sin } u = \frac{\frac{du}{dx}}{\sqrt{1 - u^2}}.$$

Either sign may be taken with the radical. The choice depends upon the value of y. The positive sign is taken if y is chosen as indicated in Art. 186. However, the conditions of a problem may make this choice impossible. For example, it may be necessary to choose y in the second quadrant, that is, $\pi/2 < y < \pi$. In this case $\cos y < 0$ and the negative sign must be taken with the radical. It must be remembered that the double sign is possible with each radical in the formulas of Art. 188. For example, the negative sign must be taken with the radical in **(XI)** if $-\pi/2 <$ arc sec $u < 0$.

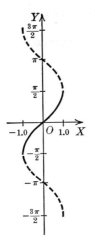

Fig. 186.1. $y = \text{arc sin } x$

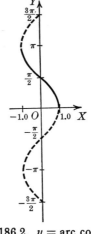

Fig. 186.2. $y = \text{arc cos } x$

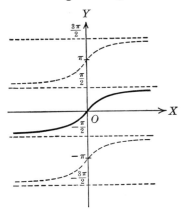

Fig. 186.3. $y = \text{arc tan } x$

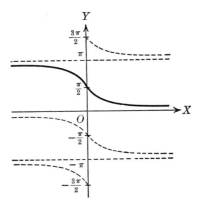

Fig. 186.4. $y = \text{arc ctn } x$

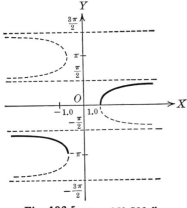

Fig. 186.5. $y = \text{arc sec } x$

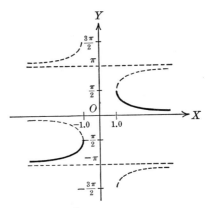

Fig. 186.6. $y = \text{arc csc } x$

309

188. Formulas for differentiating the inverse trigonometric functions.
Following the method of the last section, we may derive formulas for
differentiating the other inverse trigonometric functions. The complete
list follows. Proofs of **(VIII)–(XII)** are left as exercises.

(VII) $$\frac{d}{dx} \text{ arc sin } u = \frac{\dfrac{du}{dx}}{\sqrt{1 - u^2}}.$$

(VIII) $$\frac{d}{dx} \text{ arc cos } u = - \frac{\dfrac{du}{dx}}{\sqrt{1 - u^2}}.$$

(IX) $$\frac{d}{dx} \text{ arc tan } u = \frac{\dfrac{du}{dx}}{1 + u^2}.$$

(X) $$\frac{d}{dx} \text{ arc ctn } u = - \frac{\dfrac{du}{dx}}{1 + u^2}.$$

(XI) $$\frac{d}{dx} \text{ arc sec } u = \frac{\dfrac{du}{dx}}{u\sqrt{u^2 - 1}}.$$

(XII) $$\frac{d}{dx} \text{ arc csc } u = - \frac{\dfrac{du}{dx}}{u\sqrt{u^2 - 1}}.$$

Example 1. Find the value of y and also of y' when $x = -1$, if
$y = 2 \text{ arc sin } \frac{1}{2} x$.

Solution.

$$\frac{dy}{dx} = 2\left(\frac{\frac{d}{dx}\left(\frac{x}{2}\right)}{\sqrt{1 - \frac{1}{4}x^2}}\right) = 2\left(\frac{\frac{1}{2}}{\frac{1}{2}\sqrt{4 - x^2}}\right) = \frac{2}{\sqrt{4 - x^2}}.$$

When $x = -1$, $y = 2 \text{ arc sin } (-\frac{1}{2}) = 2(-\frac{1}{6}\pi) = -\frac{1}{3}\pi = -1.047$.

$$y' = \frac{2}{\sqrt{4 - 1}} = 1.155.$$

Example 2. Differentiate $y = x \text{ arc tan } x$.

Solution. $\dfrac{dy}{dx} = x \dfrac{d}{dx} \text{ arc tan } x + \text{ arc tan } x \cdot \dfrac{dx}{dx} = \dfrac{x}{1 + x^2} + \text{ arc tan } x$.

PROBLEMS

1. Solve the following equations for x in terms of y.

a. $y = \text{arc sin } \dfrac{x}{a}$.

b. $y = \text{arc tan } \dfrac{a}{x}$.

c. $y = \text{arc cos } (x - a)$.

d. $y = \text{arc ctn } (2\,x + a)$.

2. Differentiate the following functions.

a. $y = \text{arc sin } \dfrac{x}{a}$.

b. $y = \text{arc cos } \tfrac{1}{2}\, x$.

c. $y = \text{arc tan } \dfrac{x}{a}$.

d. $y = \text{arc sin } \dfrac{a}{x}$.

e. $y = \text{arc ctn } (x^2 - 3)$.

f. $y = \text{arc tan } \dfrac{2\,x}{1 - x^2}$.

g. $y = \text{arc csc } \dfrac{3}{2\,x}$.

h. $y = x\sqrt{a^2 - x^2} + a^2 \text{ arc sin } \dfrac{x}{a}$.

i. $\theta = \text{arc sin } (3\,r - 1)$.

j. $\theta = \text{arc tan } \dfrac{r + a}{1 - ar}$.

k. $\theta = \text{arc sec } \dfrac{1}{\sqrt{1 - r^2}}$.

l. $y = \text{arc tan } \dfrac{e^x - e^{-x}}{2}$.

m. $s = \text{arc cos } \dfrac{e^t - e^{-t}}{e^t + e^{-t}}$.

3. For each of the following functions find the values of y and y' for the given value of x.

a. $y = \text{arc tan } 2\,x;\ x = \tfrac{1}{2}$.

b. $y = \text{arc cos } \tfrac{1}{2}\, x;\ x = 1$.

c. $y = \text{arc sec } x;\ x = \sqrt{2}$.

d. $y = \text{arc ctn } x;\ x = -1$.

e. $y = \text{arc sin } 2\,x;\ x = -\tfrac{1}{4}$.

f. $y = \text{arc csc } \tfrac{1}{2}\, x;\ x = -4$.

g. $y = \text{arc ctn } (x + 1);\ x = -2$.

h. $y = x \text{ arc sin } \tfrac{1}{4}\, x;\ x = 1$.

4. Sketch each of the following curves. Find the values of y and y' for the given values of x, and draw the tangents at the corresponding points on the curve.

a. $y = \text{arc sin } \tfrac{1}{2}\, x;\ x = 1,\ -\tfrac{1}{2}$.

b. $y = \dfrac{1}{\pi} \text{ arc cos } x;\ x = 0.5,\ -0.6$.

c. $y = \text{arc tan } 2\,x;\ x = 1,\ -\tfrac{1}{2}$.

d. $y = \dfrac{2}{\pi} \text{ arc ctn } x;\ x = 1,\ -2$.

e. $y = 2 \text{ arc sec } x;\ x = 2,\ -2$.

f. $y = \dfrac{4}{\pi} \text{ arc csc } x;\ x = 2,\ -2$.

CHAPTER XIV

THEOREM OF MEAN VALUE. INDETERMINATE FORMS

189. Rolle's theorem. This theorem lies at the foundation of the theoretical development of the calculus and may be stated as follows.

Theorem I. *Let $f(x)$ be continuous in the interval (a, b) and suppose that $f(a) = 0$, $f(b) = 0$. Let $f'(x)$ exist for every value of x such that $a < x < b$. Then there is at least one value $x = x_1$, where $a < x_1 < b$, such that $f'(x_1) = 0$.*

Proof. (*i*) If $f(x)$ is constant, then, since it is continuous, $f(x) = 0$ and $f'(x) = 0$ for every x such that $a < x < b$ and the theorem is obviously true.

(*ii*) If $f(x)$ is not constant, then $f(x)$ is positive (or negative) in some part of the interval (a, b). Then there is a value $x = x_1$ such that $f(x_1)$ is a maximum (or minimum) value and, consequently, $f'(x_1) = 0$. See Art. 111.

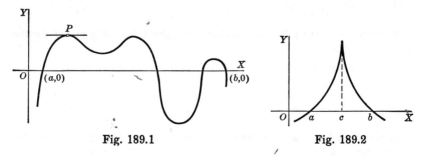

Fig. 189.1   Fig. 189.2

Fig. 189.2 illustrates a case in which Rolle's Theorem does not apply; $f(x)$ is continuous throughout the interval (a, b). $f'(x)$, however, does not exist for $x = c$, but becomes infinite. At no point of the graph is the tangent parallel to the x-axis.

190. Theorem of Mean Value. This theorem is also called the Law of the Mean for derivatives and may be stated in various forms.

Theorem II. *Let $f(x)$ and $F(x)$ be continuous and differentiable in the interval (a, b) and suppose $F'(x) \neq 0$ for $a < x < b$. Then*

(A)
$$\frac{f(b) - f(a)}{F(b) - F(a)} = \frac{f'(x_1)}{F'(x_1)}.$$
$$(a < x_1 < b)$$

Proof. Form the function

$$\phi(x) \equiv \frac{f(b) - f(a)}{F(b) - F(a)} [F(x) - F(a)] - [f(x) - f(a)]. \tag{1}$$

Evidently $\phi(a) = \phi(b) = 0$, and Rolle's Theorem may be applied. Differentiating,

$$\phi'(x) = \frac{f(b) - f(a)}{F(b) - F(a)} F'(x) - f'(x). \tag{2}$$

This must vanish for a value $x = x_1$ between a and b.

$$\therefore \frac{f(b) - f(a)}{F(b) - F(a)} F'(x_1) - f'(x_1) = 0. \tag{3}$$

Dividing through by $F'(x_1)$ (remembering that $F'(x_1)$ does not vanish), and transposing, the result is **(A)**.

If $F(x) = x$, **(A)** becomes

(B) $$\frac{f(b) - f(a)}{b - a} = f'(x_1). \qquad (a < x_1 < b)$$

In this form the theorem has a simple geometric interpretation. In Fig. 190 the curve is the graph of $f(x)$. Also,

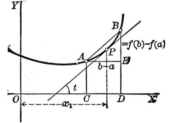

$$OC = a, \quad CA = f(a),$$
$$OD = b, \quad DB = f(b).$$

Hence

$$\frac{f(b) - f(a)}{b - a} = \text{slope of chord } AB.$$

Fig. 190

Now $f'(x_1)$ in **(B)** is the slope of the curve at a point on the arc AB, and **(B)** states that the slope at this point equals the slope of AB. Hence *there is at least one point on the arc AB at which the tangent line is parallel to the chord AB.*

Clearing **(B)** of fractions, we may also write the theorem in the form

(C) $$f(b) = f(a) + (b - a)f'(x_1).$$

PROBLEMS

1. Verify Rolle's Theorem by finding the values of x for which $f(x)$ and $f'(x)$ vanish in each of the following cases.

a. $f(x) = x^3 - 3x$.　　　　　　　　**b.** $f(x) = 6x^2 - x^3$.
c. $f(x) = a + bx + cx^2$.　　　　　　**d.** $f(x) = \sin x$.
e. $f(x) = \sin \pi x - \cos \pi x$.　　　　**f.** $f(x) = x \ln x$.
g. $f(x) = xe^x$.

191. Indeterminate forms. When, for a particular value of the independent variable, a function takes on one of the forms

$$\frac{0}{0}, \quad \frac{\infty}{\infty}, \quad 0 \times \infty, \quad \infty - \infty, \quad 0^0, \quad \infty^0, \quad 1^\infty,$$

it is said to be *indeterminate*, and the function is *not* defined for that value of the independent variable by the given analytical expression. For example, suppose we have

$$y = \frac{f(x)}{F(x)},$$

where for some value of the variable, as $x = a$,

$$f(a) = 0, \quad F(a) = 0.$$

For this value of x the function is *not* defined and we may therefore assign to it any value we please. It is evident from what has gone before that it is desirable to assign to the function a value that will make it continuous when $x = a$ whenever it is possible to do so.

192. Evaluation of a function taking on an indeterminate form. If the function $f(x)$ assumes an indeterminate form when $x = a$, then if

$$\lim_{x \to a} f(x)$$

exists and is finite, we assign this value to the function for $x = a$, which now becomes continuous for $x = a$.

The limiting value can sometimes be found after simple transformations, as the following examples show.

Example 1. Given $f(x) = \dfrac{x^2 - 4}{x - 2}$. Find $\lim_{x \to 2} f(x)$.

Solution. Direct substitution gives $f(2) = 0/0$, which is indeterminate. If $x \neq 2$, the numerator may be divided by the denominator, giving

$$f(x) = x + 2, \quad x \neq 2.$$

Then $$\lim_{x \to 2} f(x) = \lim_{x \to 2} (x + 2) = 4.$$

Example 2. Given $f(x) = \sec x - \tan x$. Find $\lim_{x \to \frac{1}{2}\pi} f(x)$.

Solution. Direct substitution gives $f(\frac{1}{2}\pi) = \infty - \infty$, which is indeterminate. If $\cos x \neq 0$, the expression may be transformed as follows.

$$\sec x - \tan x = \frac{1 - \sin x}{\cos x} = \frac{1 - \sin x}{\cos x} \cdot \frac{1 + \sin x}{1 + \sin x} = \frac{\cos x}{1 + \sin x}.$$

Then $$\lim_{x \to \frac{1}{2}\pi} f(x) = \lim_{x \to \frac{1}{2}\pi} \frac{\cos x}{1 + \sin x} = 0.$$

General methods for evaluating the indeterminate forms of Art. 191 depend upon the calculus.

193. Evaluation of the indeterminate form 0/0. Given a function of the form $f(x)/F(x)$ such that $f(a) = 0$ and $F(a) = 0$. The function is indeterminate when $x = a$. It is then required to find

$$\lim_{x \to a} \frac{f(x)}{F(x)}.$$

We shall prove the equation

(D) $$\lim_{x \to a} \frac{f(x)}{F(x)} = \lim_{x \to a} \frac{f'(x)}{F'(x)}.$$

Proof. Referring to **(A)**, Art. 190, and setting $b = x$, remembering that $f(a) = F(a) = 0$, we have

$$\frac{f(x)}{F(x)} = \frac{f'(x_1)}{F'(x_1)}. \qquad (a < x_1 < x) \qquad (1)$$

If $x \longrightarrow a$, so also $x_1 \longrightarrow a$. Hence, if the right-hand member of (1) approaches a limit when $x_1 \longrightarrow a$, then the left-hand member will approach the same limit. Thus **(D)** is proved.

From **(D)**, if $f'(a)$ and $F'(a)$ are not both zero, we shall have

$$\lim_{x \to a} \frac{f(x)}{F(x)} = \frac{f'(a)}{F'(a)}. \qquad (2)$$

Rule for evaluating the indeterminate form 0/0. Differentiate the numerator for a new numerator and the denominator for a new denominator. The value of this new fraction for the assigned value of the variable will be the limiting value of the original fraction.

This is known as L'Hôpital's rule.

In case it happens that $f'(a) = 0$ and $F'(a) = 0$, that is, the first derivatives also vanish for $x = a$, then **(D)** can be applied to the ratio

$$\frac{f'(x)}{F'(x)},$$

and the rule will give us $\lim_{x \to a} \dfrac{f(x)}{F(x)} = \dfrac{f''(a)}{F''(a)}.$

It may be necessary to repeat the process several times.

The student is warned against the very careless but common mistake of differentiating the whole expression as a fraction.

If $a = \infty$, the substitution $x = 1/z$ reduces the problem to the evaluation of the limit for $z = 0$.

Thus $\lim_{x \to \infty} \dfrac{f(x)}{F(x)} = \lim_{z \to 0} \dfrac{-f'\left(\dfrac{1}{z}\right)\dfrac{1}{z^2}}{-F'\left(\dfrac{1}{z}\right)\dfrac{1}{z^2}} = \lim_{z \to 0} \dfrac{f'\left(\dfrac{1}{z}\right)}{F'\left(\dfrac{1}{z}\right)} = \lim_{x \to \infty} \dfrac{f'(x)}{F'(x)}.$

Therefore the rule holds in this case also.

Example 1. Prove $\lim\limits_{x \to 0} \dfrac{\sin nx}{x} = n$.

Solution. Let $f(x) = \sin nx$, $F(x) = x$. Then $f(0) = 0$, $F(0) = 0$. Therefore, by (D),

$$\lim_{x \to 0} \frac{f(x)}{F(x)} = \lim_{x \to 0} \frac{f'(x)}{F'(x)} = \lim_{x \to 0} \frac{n \cos nx}{1} = n.$$

Example 2. Prove $\lim\limits_{x \to 1} \dfrac{x^3 - 3x + 2}{x^3 - x^2 - x + 1} = \dfrac{3}{2}$.

Solution. Let $f(x) = x^3 - 3x + 2$, $F(x) = x^3 - x^2 - x + 1$. Then $f(1) = 0$, $F(1) = 0$. Therefore, by (D),

$$\lim_{x \to 1} \frac{f(x)}{F(x)} = \lim_{x \to 1} \frac{f'(x)}{F'(x)} = \lim_{x \to 1} \frac{3x^2 - 3}{3x^2 - 2x - 1} = \frac{0}{0}. \quad \therefore \text{ indeterminate.}$$

$$= \lim_{x \to 1} \frac{f''(x)}{F''(x)} = \lim_{x \to 1} \frac{6x}{6x - 2} = \frac{3}{2}.$$

Example 3. Prove $\lim\limits_{x \to 0} \dfrac{e^x - e^{-x} - 2x}{x - \sin x} = 2$.

Solution. Let $f(x) = e^x - e^{-x} - 2x$, $F(x) = x - \sin x$. Then $f(0) = 0$, $F(0) = 0$. Therefore, by (D),

$$\lim_{x \to 0} \frac{f(x)}{F(x)} = \lim_{x \to 0} \frac{f'(x)}{F'(x)} = \lim_{x \to 0} \frac{e^x + e^{-x} - 2}{1 - \cos x} = \frac{0}{0}. \quad \therefore \text{ indeterminate.}$$

$$= \lim_{x \to 0} \frac{f''(x)}{F''(x)} = \lim_{x \to 0} \frac{e^x - e^{-x}}{\sin x} = \frac{0}{0}. \quad \therefore \text{ indeterminate.}$$

$$= \lim_{x \to 0} \frac{f'''(x)}{F'''(x)} = \lim_{x \to 0} \frac{e^x + e^{-x}}{\cos x} = 2.$$

PROBLEMS

Evaluate the following indeterminate forms by differentiation:*

1. $\lim\limits_{x \to 3} \dfrac{x^3 + x^2 - 7x - 15}{x^3 - 5x^2 + 8x - 6}$.

2. $\lim\limits_{x \to a} \dfrac{x^m - a^m}{x^n - a^n}$.

3. $\lim\limits_{x \to 0} \dfrac{\sqrt{a + x} - \sqrt{a - x}}{x}$.

4. $\lim\limits_{x \to 0} \dfrac{e^x - e^{-x}}{\sin x}$.

5. $\lim\limits_{x \to \phi} \dfrac{\sin x - \sin \phi}{x - \phi}$.

6. $\lim\limits_{x \to 0} \dfrac{\tan x - x}{x - \sin x}$.

7. $\lim\limits_{x \to 0} \dfrac{a^x - b^x}{x}$.

8. $\lim\limits_{\theta \to \frac{\pi}{2}} \dfrac{\cos \theta}{\pi - 2\theta}$.

*After differentiating, the student should in every case reduce the resulting expression to its simplest possible form before substituting the value of the variable.

9. $\lim\limits_{y \to 0} \dfrac{e^y + \sin y - 1}{\ln(1+y)}$.

10. $\lim\limits_{x \to \frac{\pi}{2}} \dfrac{\ln \sin x}{(\pi - 2x)^2}$.

11. $\lim\limits_{\theta \to 0} \dfrac{\theta - \arcsin \theta}{\sin^3 \theta}$.

12. $\lim\limits_{x \to 0} \dfrac{x - \sin x}{x^3}$.

13. $\lim\limits_{\theta \to 0} \dfrac{\sin \theta - \tan \theta}{\theta^2}$.

14. $\lim\limits_{x \to 0} \dfrac{e^{3x} - e^x - 2x}{e^{4x} - e^{2x} - 2x}$.

15. $\lim\limits_{\theta \to \frac{\pi}{2}} \dfrac{1 - \sin \theta}{1 + \cos 2\theta}$.

16. $\lim\limits_{x \to 0} \dfrac{xe^{nx} - x}{1 - \cos nx}$.

17. $\lim\limits_{x \to 0} \dfrac{e^x - e^{-x} - 2\sin x}{\sin^3 x}$.

18. Given a circle with center at O, radius r, and a tangent line AT. In the figure, AM equals arc AP, and B is the intersection of the line through M and P and the line through A and O. Find the limiting position of B as P approaches A as a limiting position.

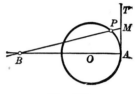

Fig. 193

194. Evaluation of the indeterminate form ∞/∞. In order to find

$$\lim_{x \to a} \frac{f(x)}{F(x)}$$

when both $f(x)$ and $F(x)$ become infinite when $x \longrightarrow a$, we follow the same rule as that given in Art. 193 for evaluating the indeterminate form $0/0$. Hence

Rule for evaluating the indeterminate form ∞/∞. Differentiate the numerator for a new numerator and the denominator for a new denominator. The value of this new fraction for the assigned value of the variable will be the limiting value of the original fraction.

A rigorous proof of this rule is beyond the scope of this book.

Example. Prove $\lim\limits_{x \to 0} \dfrac{\ln x}{\csc x} = 0$.

Solution. Let $f(x) = \ln x$, $F(x) = \csc x$. Then $f(0) = -\infty$, $F(0) = \infty$. Hence, by the rule,

$$\lim_{x \to 0} \frac{f(x)}{F(x)} = \lim_{x \to 0} \frac{f'(x)}{F'(x)} = \lim_{x \to 0} \frac{\dfrac{1}{x}}{-\csc x \operatorname{ctn} x} = \lim_{x \to 0} \frac{-\sin^2 x}{x \cos x}.$$

Substitution of $x = 0$ in the last fraction gives the indeterminate form $0/0$. Hence by the rule of Art. 193,

$$\lim_{x \to 0} \frac{-\sin^2 x}{x \cos x} = \lim_{x \to 0} \frac{-2\sin x \cos x}{\cos x - x \sin x} = 0.$$

195. Evaluation of the indeterminate form $0 \cdot \infty$. If a function $f(x) \cdot \phi(x)$ takes on the indeterminate form $0 \cdot \infty$ for $x = a$, we write the given function

$$f(x) \cdot \phi(x) = \frac{f(x)}{\dfrac{1}{\phi(x)}} \left(\text{or} = \frac{\phi(x)}{\dfrac{1}{f(x)}} \right)$$

so as to cause it to take on one of the forms $\dfrac{0}{0}$ or $\dfrac{\infty}{\infty}$, thus bringing it under Art. 193 or Art. 194.

As shown, the product $f(x) \cdot \phi(x)$ may be rewritten in either of the two forms given. As a rule, one of these forms is better than the other, and the choice will depend upon the example.

Example. Prove $\lim\limits_{x \to \frac{1}{2}\pi} (\sec 3x \cos 5x) = -\frac{5}{3}$.

Solution. Since $\sec \frac{3}{2}\pi = \infty$, $\cos \frac{5}{2}\pi = 0$, we write

$$\sec 3x \cos 5x = \frac{1}{\cos 3x} \cdot \cos 5x = \frac{\cos 5x}{\cos 3x}.$$

Let $f(x) = \cos 5x$, $F(x) = \cos 3x$. Then $f(\frac{1}{2}\pi) = 0$, $F(\frac{1}{2}\pi) = 0$. Hence, by **(D)**,

$$\lim_{x \to \frac{1}{2}\pi} \frac{f(x)}{F(x)} = \lim_{x \to \frac{1}{2}\pi} \frac{f'(x)}{F'(x)} = \lim_{x \to \frac{1}{2}\pi} \frac{-5 \sin 5x}{-3 \sin 3x} = -\frac{5}{3}.$$

196. Evaluation of the indeterminate form $\infty - \infty$. It is possible in general to transform the expression into a fraction which will assume either the form $0/0$ or ∞/∞.

Example. Prove $\lim\limits_{x \to 0} (\csc x - \operatorname{ctn} x) = 0$.

Solution. Direct substitution of $x = 0$ gives the indeterminate result $\infty - \infty$. We transform the expression as follows.

$$\csc x - \operatorname{ctn} x = \frac{1}{\sin x} - \frac{\cos x}{\sin x} = \frac{1 - \cos x}{\sin x}.$$

Let $f(x) = 1 - \cos x$, $F(x) = \sin x$. Then $f(0) = 0$, $F(0) = 0$. Hence, by **(D)**,

$$\lim_{x \to 0} \frac{f(x)}{F(x)} = \lim_{x \to 0} \frac{f'(x)}{F'(x)} = \lim_{x \to 0} \frac{\sin x}{\cos x} = 0.$$

PROBLEMS

Evaluate the following indeterminate forms.

1. $\lim\limits_{x \to 0} x^n \ln x. \quad (n > 0)$

2. $\lim\limits_{x \to \infty} \dfrac{\ln x}{e^x}$.

3. $\lim\limits_{x \to 0} \dfrac{\operatorname{ctn} x}{\operatorname{ctn} 2x}$.

4. $\lim\limits_{x \to \infty} \dfrac{x^3}{e^x}$.

5. $\lim\limits_{\theta \to \frac{\pi}{2}} \dfrac{\tan 3\,\theta}{\tan \theta}.$

6. $\lim\limits_{x \to 0} \dfrac{\ln x}{\text{ctn } x}.$

7. $\lim\limits_{x \to 0} x \ln \sin x.$

8. $\lim\limits_{\phi \to 0} \dfrac{\pi}{\phi} \tan \dfrac{\pi\phi}{2}.$

9. $\lim\limits_{x \to \frac{\pi}{2}} (\pi - 2\,x) \tan x.$

10. $\lim\limits_{\theta \to \frac{\pi}{4}} (1 - \tan \theta) \sec 2\,\theta.$

11. $\lim\limits_{\theta \to 0} \theta \csc 2\,\theta.$

12. $\lim\limits_{\phi \to 0} \left[\dfrac{2}{\sin^2 \phi} - \dfrac{1}{1 - \cos \phi} \right].$

13. $\lim\limits_{y \to 1} \left[\dfrac{y}{y - 1} - \dfrac{1}{\ln y} \right].$

14. $\lim\limits_{x \to 0} \left[\dfrac{1}{\sin^2 x} - \dfrac{1}{x^2} \right].$

15. $\lim\limits_{x \to 0} \left[\dfrac{1}{\sin^3 x} - \dfrac{1}{x^3} \right].$

197. Evaluation of the indeterminate forms 0^0, 1^∞, ∞^0. Given a func·tion of the form

$$f(x)^{\phi(x)}.$$

In order that the function shall take on one of the above three forms, we must have, for a certain value of x,

$$f(x) = 0, \quad \phi(x) = 0, \quad \text{giving } 0^0;$$

or $\qquad\qquad f(x) = 1, \quad \phi(x) = \infty, \quad \text{giving } 1^\infty;$

or $\qquad\qquad f(x) = \infty, \quad \phi(x) = 0, \quad \text{giving } \infty^0.$

Let $\qquad\qquad\qquad y = f(x)^{\phi(x)}.$

Taking the natural logarithm of both sides,

$$\ln y = \phi(x) \ln f(x).$$

In any of the above cases the natural logarithm of y (the function) will take on the indeterminate form

$$0 \cdot \infty.$$

Evaluating this by the process illustrated in Art. 195 gives the limit of the logarithm of the function. This being equal to the logarithm of the limit of the function, the limit of the function is known. For if limit $\ln y = a$, then $\lim y = e^a$.

Example. Prove $\lim\limits_{x \to 1} (2 - x)^{\tan \frac{1}{2} \pi x} = e^{\frac{2}{\pi}}.$

Solution. The function assumes the indeterminate form 1^∞ when $x = 1$.

Let $\qquad\qquad y = (2 - x)^{\tan \frac{1}{2} \pi x};$

then $\qquad \ln y = \tan \tfrac{1}{2} \pi x \ln (2 - x) = \infty \cdot 0, \qquad\qquad$ when $x = 1.$

By Art. 195, $\qquad \ln y = \dfrac{\ln (2 - x)}{\text{ctn } \frac{1}{2} \pi x} = \dfrac{0}{0}, \qquad\qquad$ when $x = 1$

By Art. 193, $\quad \lim\limits_{x \to 1} \dfrac{\ln (2 - x)}{\operatorname{ctn} \frac{1}{2} \pi x} = \lim\limits_{x \to 1} \dfrac{-\dfrac{1}{2 - x}}{-\frac{1}{2} \pi \csc^2 \frac{1}{2} \pi x} = \dfrac{2}{\pi}.$

Therefore $\quad \lim\limits_{x \to 1} \ln y = \dfrac{2}{\pi},\quad$ and $\quad \lim\limits_{x \to 1} y = \lim\limits_{x \to 1} (2 - x)^{\tan \frac{1}{2} \pi x} = e^{\frac{2}{\pi}}.$

PROBLEMS

Evaluate the following indeterminate forms.

1. $\lim\limits_{x \to \infty} \left(\dfrac{2}{x} + 1 \right)^{x}.$

2. $\lim\limits_{t \to 0} (1 + nt)^{\frac{1}{t}}.$

3. $\lim\limits_{x \to 0} x^{x}.$

4. $\lim\limits_{\theta \to \frac{\pi}{2}} (\sin \theta)^{\tan \theta}.$

5. $\lim\limits_{x \to 1} x^{\frac{1}{1-x}}.$

6. $\lim\limits_{\phi \to 0} (1 + \sin \phi)^{\operatorname{ctn} \phi}.$

7. $\lim\limits_{x \to 0} (e^x + x)^{\frac{1}{x}}.$

8. $\lim\limits_{x \to 0} (\operatorname{ctn} x)^{\sin x}.$

9. $\lim\limits_{x \to 0} \left(\dfrac{1}{x} \right)^{\sin x}.$

10. $\lim\limits_{x \to 0} (e^{2x} + 2 x)^{\frac{1}{4x}}.$

11. $\lim\limits_{x \to \infty} \left(\cos \dfrac{2}{x} \right)^{x}.$

12. $\lim\limits_{x \to \infty} \left(\cos \dfrac{2}{x} \right)^{x^2}.$

198. The extended theorem of mean value. Let $f(x)$ be continuous in the interval (a, b) and let n be some specified positive integer. Suppose that the nth derivative $f^{(n)}$ exists for every value such that $a < x < b$. Suppose first that $n = 2$ and let the constant R be defined by the equation

$$f(b) - f(a) - (b - a)f'(a) - \tfrac{1}{2}(b - a)^2 R = 0. \tag{1}$$

Let $F(x)$ be a function formed by replacing b by x in the left-hand member of (1); that is, in the interval (a, b),

$$F(x) = f(x) - f(a) - (x - a)f'(a) - \tfrac{1}{2}(x - a)^2 R. \tag{2}$$

From (1), $F(b) = 0$; and from (2), $F(a) = 0$; therefore, by Rolle's Theorem (Art. 189), at least one value of x between a and b, say x_1, will cause $F'(x)$ to vanish. Hence, since

$$F'(x) = f'(x) - f'(a) - (x - a)R,$$

we get $\quad F'(x_1) = f'(x_1) - f'(a) - (x_1 - a)R = 0.$

Since $F'(x_1) = 0$ and $F'(a) = 0$, it is evident that $F'(x)$ also satisfies the conditions of Rolle's Theorem, so that *its derivative*, namely $F''(x)$, must vanish for at least one value of x between a and x_1, say x_2, and therefore x_2 also lies between a and b. But

$$F''(x) = f''(x) - R; \text{ therefore } F''(x_2) = f''(x_2) - R = 0,$$

and
$$R = f''(x_2).$$

Substituting this result in (1), we get

(E) $\quad f(b) = f(a) + (b - a)f'(a) + \dfrac{1}{\underline{|2}}(b - a)^2 f''(x_2). \qquad (a < x_2 < b)$

By continuing this process we get the general result,

(F) $\quad f(b) = f(a) + \dfrac{(b - a)}{\underline{|1}} f'(a) + \dfrac{(b - a)^2}{\underline{|2}} f''(a)$

$$+ \dfrac{(b - a)^3}{\underline{|3}} f'''(a) + \cdots + \dfrac{(b - a)^{n-1}}{\underline{|n-1}} f^{(n-1)}(a)$$

$$+ \dfrac{(b - a)^n}{\underline{|n}} f^{(n)}(x_n). \qquad (a < x_n < b)$$

Equation **(F)** is called the *Extended Theorem of Mean Value*, or the *Extended Law of the Mean*, or *Taylor's Formula*.

199. Extreme values. The second derivative test (Art. 114) gives criteria for the determination of extreme values of a function $f(x)$. The theorem is applicable if $f'(c) = 0$ and $f''(c) \neq 0$. The criteria can be extended by use of the result in Art. 198.

Let $f(x)$ and its first n derivatives be continuous in the interval (a, b). Let $x = c$ be a point *within* the interval (a, b) and let I denote the interval $(c - h, c + h)$, where h is a conveniently chosen constant. Suppose that

$$f'(c) = f''(c) = \cdots = f^{(n-1)}(c) = 0 \quad \text{and} \quad f^{(n)}(c) \neq 0.$$

In **(F)** replace a by c, and b by x, and the interval (a, b) by I. The result is

$$f(x) - f(c) = \dfrac{(x - c)^n}{\underline{|n}} f^{(n)}(x_n). \qquad (c - h < x_n < c + h)$$

Suppose $f^{(n)}(c) > 0$. Then, by the continuity condition, h may be chosen small enough so that $f^{(n)}(x) > 0$ for all values of x in the interval I. There are two cases.

CASE 1. Suppose n is an even integer. Then $(x - c)^n > 0$ and $f(x) - f(c) > 0$ for all values of x except $x = c$, in the interval I. By definition (Art. 111), $f(c)$ is a minimum value.

CASE 2. Suppose n is an odd integer. Then $(x - c)^n > 0$ if $x > c$ and $(x - c)^n < 0$ if $x < c$. In this case $f(c)$ is neither a maximum nor a minimum value.

If this argument is repeated on the supposition that $f^{(n)}(c) < 0$, it follows readily that $f(c)$ is a maximum value if n is even and is neither a maximum nor a minimum value if n is odd. The results may be summarized as follows.

If the first of the derivatives of $f(x)$ which does not vanish for $x = c$ is of even order $(= n)$, then

$$f(c) \text{ is a maximum if } f^{(n)}(c) < 0;$$
$$f(c) \text{ is a minimum if } f^{(n)}(c) > 0.$$

If the first of the derivatives of $f(x)$ which does not vanish is of odd order, then $f(c)$ is neither a maximum nor a minimum value.

Example 1. Examine $f(x) = \frac{1}{5}x^5 - 2x^4 + 5x^3$ for extreme values.

Solution. Differentiating, we have

$$f'(x) = x^4 - 8x^3 + 15x^2,$$
$$f''(x) = 4x^3 - 24x^2 + 30x,$$
$$f'''(x) = 12x^2 - 48x + 30, \text{ etc.}$$

Setting $f'(x) = 0$, the critical values are found to be $x = 0$, $x = 3$, $x = 5$. For each of these values we have the following results.

$$f'(0) = 0, f''(0) = 0, f'''(0) = 30.$$

Hence $f(0) = 0$ is neither a maximum nor a minimum value.

$$f'(3) = 0, f''(3) = -18.$$

Hence $f(3) = 21.6$ is a maximum value.

$$f'(5) = 0, f''(5) = +50.$$

Hence $f(5) = 0$ is a minimum value.

Example 2. Examine $f(x) = e^x + 2\cos x + e^{-x}$ for the value $x = 0$.

Solution. Differentiating, we have

$$f'(x) = e^x - 2\sin x - e^{-x}, f'(0) = 0;$$
$$f''(x) = e^x - 2\cos x + e^{-x}, f''(0) = 0;$$
$$f'''(x) = e^x + 2\sin x - e^{-x}, f'''(0) = 0;$$
$$f^{IV}(x) = e^x + 2\cos x + e^{-x}, f^{IV}(0) = 4.$$

Hence $f(0) = 4$ is a minimum value.

PROBLEMS

Examine the following functions for extreme values.

1. $x^4 - 4x^3 + 15$.   2. $x^5 - 5x^4 + 128$.

3. $x^3(x - a)^2$.   4. $x^4(x - a)$.

5. Examine $4x^5 - 15x^4 + 20x^3 - 10x^2$ for the value $x = 1$.

6. Examine $e^{ax} + 2\cos ax + e^{-ax}$ for the value $x = 0$.

7. Examine $e^{2x} - 2\sin 2x - e^{-2x}$ for the value $x = 0$.

CHAPTER XV

INTEGRATION; STANDARD FORMS

200. Standard elementary forms. Certain formulas for integrating algebraic and exponential functions have already been given and, for convenience, they are repeated in the first six formulas of the following list. Others in the list are derived below.

STANDARD ELEMENTARY FORMS

(I) $\int du = u + C.$

(II) $\int [f(x) + g(x)]dx = \int f(x)dx + \int g(x)dx.$

(III) $\int a\,du = a \int du.$

(IV) $\int u^n\,du = \dfrac{u^{n+1}}{n+1} + C, \text{ if } n \neq -1.$

(V) $\int \dfrac{du}{u} = \ln u + C.$

(VI) $\int e^u\,du = e^u + C.$

(VI a) $\int a^u\,du = \dfrac{a^u}{\ln a} + C.$

(VII) $\int \sin u\,du = -\cos u + C.$

(VIII) $\int \cos u\,du = \sin u + C.$

(IX) $\int \sec^2 u\,du = \tan u + C.$

(X) $\int \csc^2 u\,du = -\operatorname{ctn} u + C.$

(XI) $\int \sec u \tan u\,du = \sec u + C.$

(XII) $\int \csc u \operatorname{ctn} u\,du = -\csc u + C.$

(XIII) $\int \tan u\,du = \ln \sec u + C.$

(XIV) $\int \operatorname{ctn} u\,du = \ln \sin u + C.$

(XV) $\int \sec u\,du = \ln (\sec u + \tan u) + C.$

323

(XVI) $\quad \displaystyle\int \csc u \, du = \ln (\csc u - \operatorname{ctn} u) + C.$

(XVII) $\quad \displaystyle\int \frac{du}{u^2 + a^2} = \frac{1}{a} \arctan \frac{u}{a} + C.$

(XVIII) $\quad \displaystyle\int \frac{du}{u^2 - a^2} = \frac{1}{2a} \ln \frac{u - a}{u + a} + C.$

(XVIII *a*) $\displaystyle\int \frac{du}{a^2 - u^2} = \frac{1}{2a} \ln \frac{a + u}{a - u} + C.$

(XIX) $\quad \displaystyle\int \frac{du}{\sqrt{a^2 - u^2}} = \arcsin \frac{u}{a} + C.$

(XX) $\quad \displaystyle\int \frac{du}{\sqrt{u^2 \pm a^2}} = \ln (u + \sqrt{u^2 \pm a^2}) + C.$

(XXI) $\quad \displaystyle\int \sqrt{a^2 - u^2} \, du = \frac{u}{2} \sqrt{a^2 - u^2} + \frac{a^2}{2} \arcsin \frac{u}{a} + C.$

(XXII) $\quad \displaystyle\int \sqrt{u^2 \pm a^2} \, du = \frac{u}{2} \sqrt{u^2 \pm a^2} \pm \frac{a^2}{2} \ln (u + \sqrt{u^2 \pm a^2}) + C.$

201. Proofs of formulas (VII)–(XVI). Formulas **(VII)–(XII)** follow directly from the corresponding formulas for differentiation.

Proof of (XIII). $\quad \displaystyle\int \tan u \, du = \int \frac{\sin u \, du}{\cos u}$

$$= - \int \frac{- \sin u \, du}{\cos u}$$

$$= - \int \frac{d(\cos u)}{\cos u}$$

$$= - \ln \cos u + C \qquad \text{by (V)}$$

$$= \ln \sec u + C.$$

$$\left[\text{Since} - \ln \cos u = - \ln \frac{1}{\sec u} = - \ln 1 + \ln \sec u = \ln \sec u.\right]$$

Proof of (XIV). $\quad \displaystyle\int \operatorname{ctn} u \, du = \int \frac{\cos u \, du}{\sin u} = \int \frac{d(\sin u)}{\sin u}$

$$= \ln \sin u + C. \qquad \text{By (V)}$$

Proof of (XV). Since

$$\sec u = \sec u \frac{\sec u + \tan u}{\sec u + \tan u}$$

$$= \frac{\sec u \tan u + \sec^2 u}{\sec u + \tan u},$$

$$\int \sec u \, du = \int \frac{\sec u \tan u + \sec^2 u}{\sec u + \tan u} \, du$$

$$= \int \frac{d(\sec u + \tan u)}{\sec u + \tan u}$$

$$= \ln (\sec u + \tan u) + C. \qquad \text{By (V)}$$

Proof of (XVI). Since

$$\csc u = \csc u \, \frac{\csc u - \operatorname{ctn} u}{\csc u - \operatorname{ctn} u}$$

$$= \frac{-\csc u \operatorname{ctn} u + \csc^2 u}{\csc u - \operatorname{ctn} u},$$

$$\int \csc u \, du = \int \frac{-\csc u \operatorname{ctn} u + \csc^2 u}{\csc u - \operatorname{ctn} u} \, du$$

$$= \int \frac{d(\csc u - \operatorname{ctn} u)}{\csc u - \operatorname{ctn} u}$$

$$= \ln (\csc u - \operatorname{ctn} u) + C. \qquad \text{By (V)}$$

Example 1. Find the value of $\int_0^1 \sin 2 x \, dx$.

Solution. Let $u = 2 x$; then $du = 2 \, dx$. Applying **(VII)**, we have, by substituting $2 x = u$, $dx = \frac{1}{2} du$ and changing the limits,

$$\int_0^1 \sin 2 x \, dx = -\tfrac{1}{2} \Big[\cos u\Big]_0^2$$

$$= -\tfrac{1}{2}[\cos 2 - \cos 0] = -\tfrac{1}{2}[-0.416 - 1]$$

$$= 0.708.$$

Example 2. Integrate $\int (\tan x + 2)^2 \, dx$.

Solution. $\int (\tan x + 2)^2 \, dx = \int (\tan^2 x + 4 \tan x + 4) dx$

$$= \int (\sec^2 x + 4 \tan x + 3) dx.$$

[Since $\tan^2 x = \sec^2 x - 1$.]

Applying **(IX)**, **(XIII)**, and **(I)**, we have

$$\int (\tan x + 2)^2 \, dx = \tan x + 4 \ln \sec x + 3 x + C.$$

PROBLEMS

1. Integrate the following expressions.

a. $\int \cos ax \, dx$.

b. $\int 2 \sin 3 x \, dx$.

c. $\int \tan 5 \phi \, d\phi$.

d. $\int \operatorname{ctn} \tfrac{1}{3} \theta \, d\theta$.

e. $\int \sec nx \, dx$.

f. $\int e^x \sin e^x \, dx$.

g. $\int \dfrac{dx}{\sin^2 x}$.

h. $\int \dfrac{dx}{\cos^2 x}$.

i. $\int \tan^2 2 x \, dx$.

j. $\int \operatorname{ctn}^2 x \, dx$.

k. $\int \dfrac{\cos x \, dx}{1 + \sin x}$.

l. $\int \dfrac{\sec^2 x \, dx}{2 + 3 \tan x}$.

m. $\int \dfrac{\csc^2 x\, dx}{\sqrt{3 - 2\operatorname{ctn} x}}.$

n. $\int (\tan \theta + \operatorname{ctn} \theta)^2\, d\theta.$

o. $\int (\sec \theta - \tan \theta)^2\, d\theta.$

p. $\int (\tan \theta + 2 \operatorname{ctn} \theta)^2\, d\theta.$

q. $\int (\sec \phi - \csc \phi)^2\, d\phi.$

r. $\int \dfrac{dx}{1 + \cos x}.$

s. $\int \dfrac{dx}{1 + \sin x}.$

HINT. In **r** multiply both numerator and denominator by $1 - \cos x$.

2. Find the values of the following definite integrals.

a. $\displaystyle\int_0^{\frac{\pi}{2}} \sin x\, dx.$

b. $\displaystyle\int_{\frac{\pi}{6}}^{\frac{\pi}{3}} \csc^2 x\, dx.$

c. $\displaystyle\int_0^1 \sec^2 \tfrac{1}{4}\, \pi x\, dx.$

d. $\displaystyle\int_{\frac{\pi}{6}}^{\frac{\pi}{4}} \operatorname{ctn} 2\, x\, dx.$

e. $\displaystyle\int_0^2 \tan \tfrac{1}{6}\, \pi x\, dx.$

f. $\displaystyle\int_0^{\pi} \tan \tfrac{1}{4}\, x\, dx.$

g. $\displaystyle\int_0^2 \sec^2 \tfrac{1}{2}\, x\, dx.$

h. $\displaystyle\int_0^1 \sin \pi x\, dx.$

i. $\displaystyle\int_0^1 \cos \tfrac{1}{2}\, \pi x\, dx.$

j. $\displaystyle\int_1^2 \sin \tfrac{1}{3}\, \pi x\, dx.$

k. $\displaystyle\int_0^1 \tan \tfrac{1}{4}\, \pi x\, dx.$

l. $\displaystyle\int_1^2 \operatorname{ctn} \tfrac{1}{4}\, \pi x\, dx.$

m. $\displaystyle\int_0^1 (x + \sin 2\, x)dx.$

n. $\displaystyle\int_0^2 (x + \cos x)dx.$

3. Sketch the following curves and find the area of one arch.

a. $y = 2 \cos x.$ **b.** $y = \cos 2\, x.$ **c.** $y = 2 \sin \tfrac{1}{2}\, \pi x.$

4. Find the volume generated when the area bounded by the x-axis and one arch of the curve $y = \sin x$ is revolved about the x-axis.

HINT. $2 \sin^2 x = 1 - \cos 2\, x.$

5. Find the area in the first quadrant bounded by the y-axis, the curve $y = \sin x$, and the curve $y = \cos x$.

6. Find the area between $x = 0$ and $x = 2\, \pi$ which is enclosed by the curves $y = \sin x$ and $y = \cos x$.

7. Find the area bounded by the curve $y = x + \cos \tfrac{1}{4}\, \pi x$, the coördinate axes, and the line $x = 2$.

8. The derivative of a certain function is $\sec^2 \theta + \tan \theta$, and its value is 5 when $\theta = 0$. Find the function.

9. Find the area bounded by the curve $y = \tan \tfrac{1}{4}\, \pi x$, the x-axis, and the line $x = 1$.

10. Find the volume generated if the area bounded by the curve $y = \sec \tfrac{1}{2}\, \pi x$, the x-axis, and the lines $x = \pm \tfrac{1}{2}$ is revolved about the x-axis.

11. The velocity of a point moving in a straight line is $v = 4 \cos \tfrac{1}{2}\, t$. Find the distance from the starting point when $t = \tfrac{1}{2}\, \pi$.

12. The acceleration of a point moving in a straight line is $a = -16 \cos 2t$. The point starts from rest. Find its distance from the starting point when $t = \frac{1}{2}\pi$.

13. Find the length of the arc of the curve $y = \ln \sec x$ from $x = 0$ to $x = \frac{1}{3}\pi$.

14. Find the volume generated if the area bounded by the curve $y = \operatorname{ctn} x$, the x-axis, and the line $x = \frac{1}{4}\pi$ is revolved about the x-axis.

15. Find the length of one arch of the curve $y = \sin x$. Use Simpson's rule, taking $n = 4$.

16. Find the length of the arc of the curve $y = \tan x$ from $x = 0$ to $x = \frac{1}{3}\pi$. Use Simpson's rule, taking $n = 6$.

202. Proofs of formulas (XVII)–(XXII). From the corresponding formula for differentiation, we have

Proof of (XVII). $\quad \displaystyle\int \frac{dv}{v^2 + 1} = \operatorname{arc\,tan} v + C.$

Setting $v = \dfrac{u}{a}$, $dv = \dfrac{du}{a}$, we get

$$\int \frac{\dfrac{du}{a}}{\dfrac{u^2}{a^2} + 1} = a \int \frac{du}{u^2 + a^2} = \operatorname{arc\,tan} \frac{u}{a} + C,$$

whence $\quad \displaystyle\int \frac{du}{u^2 + a^2} = \frac{1}{a} \operatorname{arc\,tan} \frac{u}{a} + C.$

Proof of (XVIII) and (XVIII a). To derive **(XVIII)** we proceed as follows. By algebra we have

$$\frac{1}{u - a} - \frac{1}{u + a} = \frac{2a}{u^2 - a^2},$$

whence $\quad \displaystyle\frac{1}{u^2 - a^2} = \frac{1}{2a}\left(\frac{1}{u - a} - \frac{1}{u + a}\right).$

Then $\quad \displaystyle\int \frac{du}{u^2 - a^2} = \frac{1}{2a}\int \left(\frac{1}{u - a} - \frac{1}{u + a}\right) du$

$$= \frac{1}{2a}\left[\ln (u - a) - \ln (u + a)\right] + C.$$

Hence $\quad \displaystyle\int \frac{du}{u^2 - a^2} = \frac{1}{2a}\ln \frac{u - a}{u + a} + C.$

Similarly **(XVIII a)** may be proved by starting from the relation

$$\frac{1}{a + u} + \frac{1}{a - u} = \frac{2a}{a^2 - u^2}.$$

Since
$$\int \frac{du}{u^2 - a^2} = -\int \frac{du}{a^2 - u^2},$$

we may use either **(XVIII)** or **(XVIII a)**. The choice depends upon the fact that a negative number has no real logarithm. If, for the values of a and u to be used, $\dfrac{u-a}{u+a}$ is positive, we use **(XVIII)**. If $\dfrac{a+u}{a-u}$ is positive, we use **(XVIII a)**.

The proof of **(XIX)** follows the same method as that used for **(XVII)**.

Proof of (XX). To derive **(XX)** we proceed as follows. When the $+$ sign occurs under the radical, we set $u = a \tan z$, where z is a new variable. Then $du = a \sec^2 z \, dz$, and

$$\int \frac{du}{\sqrt{u^2 + a^2}} = \int \frac{a \sec^2 z \, dz}{\sqrt{a^2 \tan^2 z + a^2}}$$

$$= \int \frac{\sec^2 z \, dz}{\sqrt{\tan^2 z + 1}}$$

$$= \int \sec z \, dz$$

[Since $\tan^2 z + 1 = \sec^2 z$.]

$$= \ln (\sec z + \tan z) + K.$$

To get the result in terms of u and a, we draw a right triangle with one acute angle z and call the opposite side u and the adjacent side a, so that $\tan z = u/a$. Then from the figure

$$\sec z = \frac{\sqrt{u^2 + a^2}}{a}.$$

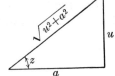

Fig. 202

The preceding result may now be written

$$\int \frac{du}{\sqrt{u^2 + a^2}} = \ln \left(\frac{u}{a} + \frac{\sqrt{u^2 + a^2}}{a} \right) + K$$

$$= \ln (u + \sqrt{u^2 + a^2}) - \ln a + K.$$

Hence, writing $C = -\ln a + K$,

$$\int \frac{du}{\sqrt{u^2 + a^2}} = \ln (u + \sqrt{u^2 + a^2}) + C.$$

When the $-$ sign occurs under the radical, we set $u = a \sec z$ and proceed as above.

Proof of (XXI). To derive **(XXI)** we set $u = a \sin z$. Then we have $du = a \cos z \, dz$ and $\sqrt{a^2 - u^2} = a \cos z$.

$$\int \sqrt{a^2 - u^2} \, du = \int a^2 \cos^2 z \, dz$$

$$= \int \tfrac{1}{2} a^2 (\cos 2 z + 1) dz$$

$$= \tfrac{1}{4} a^2 \sin 2 z + \tfrac{1}{2} a^2 z + C.$$

To get the result in terms of u and a, we write

$$z = \text{arc sin } \frac{u}{a}$$

and

$$\sin 2 z = 2 \sin z \cos z = \frac{2 u}{a} \sqrt{1 - \frac{u^2}{a^2}}$$

$$= \frac{2 u}{a^2} \sqrt{a^2 - u^2}.$$

The preceding result may be written

$$\int \sqrt{a^2 - u^2} \, du = \frac{u}{2} \sqrt{a^2 - u^2} + \frac{a^2}{2} \text{ arc sin } \frac{u}{a} + C.$$

To derive **(XXII)** there are two cases. When the $+$ sign occurs under the radical, we set $u = a \tan z$; when the $-$ sign occurs under the radical, we set $u = a \sec z$; and proceed as in the derivation of **(XXI)**.

Example 1. Integrate $\int \dfrac{dx}{4 \, x^2 + 9}$.

Solution. The integral resembles **(XVII)** if $4 \, x^2 = u^2$. Then $2 \, x = u$, and $x = \tfrac{1}{2} \, u$, and $dx = \tfrac{1}{2} \, du$. Hence, by substitution,

$$\int \frac{dx}{4 \, x^2 + 9} = \int \frac{\tfrac{1}{2} \, du}{u^2 + 9} = \tfrac{1}{2} \int \frac{du}{u^2 + 9} \qquad \text{by (III)}$$

$$= \tfrac{1}{2} \cdot \tfrac{1}{3} \text{ arc tan } \tfrac{1}{3} \, u + C \qquad \text{by (XVII)}$$

$$= \tfrac{1}{6} \text{ arc tan } \tfrac{2}{3} \, x + C.$$

Example 2. Evaluate $\int_0^3 \dfrac{dx}{x^2 - 25}$.

Solution.

$$\int_0^3 \frac{dx}{x^2 - 25} = - \int_0^3 \frac{dx}{25 - x^2}$$

$$= - \frac{1}{10} \left[\ln \frac{5 + x}{5 - x} \right]_0^3 \qquad \text{by (XVIII } a \text{)}$$

$$= - \tfrac{1}{10} \ln 4 = - 0.1386.$$

Example 3. Prove

$$\int \sqrt{16 - 3\,x^2}\; dx = \frac{x}{2}\sqrt{16 - 3\,x^2} + \frac{8\sqrt{3}}{3}\, \text{arc sin}\, \frac{\sqrt{3}\,x}{4} + C.$$

Solution. The integral resembles **(XXI)** with $3\,x^2 = u^2$. Then $\sqrt{3}\,x = u$, and

$$x = \frac{1}{\sqrt{3}}\,u, \text{ and } dx = \frac{1}{\sqrt{3}}\,du.$$

Hence, by substitution, we have

$$\int \sqrt{16 - 3\,x^2}\; dx = \int \sqrt{16 - u^2}\,\frac{du}{\sqrt{3}} = \frac{1}{\sqrt{3}}\int \sqrt{16 - u^2}\; du \quad \text{by (III)}$$

$$= \frac{1}{\sqrt{3}}\left[\frac{u}{2}\sqrt{16 - u^2} + \frac{16}{2}\, \text{arc sin}\, \frac{u}{4} + C\right]\cdot \text{ By (XXI)}$$

Substituting $u = \sqrt{3}\,x$, we get the answer.

PROBLEMS

1. Integrate the following expressions.

a. $\displaystyle\int \frac{dx}{x^2 - 25}\cdot$ **b.** $\displaystyle\int \frac{dy}{y^2 + 25}\cdot$ **c.** $\displaystyle\int \frac{dx}{\sqrt{25 - x^2}}\cdot$

d. $\displaystyle\int \frac{dx}{\sqrt{x^2 - 25}}\cdot$ **e.** $\displaystyle\int \frac{dx}{9\,x^2 + 4}\cdot$ **f.** $\displaystyle\int \frac{dx}{9\,x^2 - 4}\cdot$

g. $\displaystyle\int \frac{dt}{\sqrt{4\,t^2 - 1}}\cdot$ **h.** $\displaystyle\int \frac{dx}{\sqrt{16 - 9\,x^2}}\cdot$ **i.** $\displaystyle\int \frac{dx}{x\sqrt{4\,x^2 - 9}}\cdot$

j. $\displaystyle\int \frac{dx}{\sqrt{9\,x^2 + 25}}\cdot$ **k.** $\displaystyle\int \frac{d\theta}{5\,\theta^2 + 8}\cdot$ **l.** $\displaystyle\int \frac{dx}{3\,x^2 - 10}\cdot$

m. $\displaystyle\int \frac{dx}{\sqrt{7\,x^2 + 5}}\cdot$ **n.** $\displaystyle\int \frac{dx}{\sqrt{8 - 3\,x^2}}\cdot$ **o.** $\displaystyle\int \frac{ds}{\sqrt{16 - (s - 3)^2}}\cdot$

2. Evaluate the following definite integrals.

a. $\displaystyle\int_0^1 \frac{dx}{9\,x^2 + 25}\cdot$ **b.** $\displaystyle\int_0^1 \frac{dx}{\sqrt{9\,x^2 + 16}}\cdot$ **c.** $\displaystyle\int_3^5 \frac{dx}{\sqrt{x^2 - 9}}\cdot$

d. $\displaystyle\int_0^2 \frac{ds}{\sqrt{25 - 4\,s^2}}\cdot$ **e.** $\displaystyle\int_0^2 \sqrt{9 + 4\,t^2}\; dt.$ **f.** $\displaystyle\int_3^4 \sqrt{25 - x^2}\; dx.$

3. Find the area of the part of the circle $x^2 + y^2 = 25$ which lies between the lines $x = -2$ and $x = 3$.

4. Find the area of the part of the ellipse $\dfrac{x^2}{25} + \dfrac{y^2}{9} = 1$ which lies between the lines $x = -3$ and $x = 4$.

5. Find the area bounded by the hyperbola $x^2 - y^2 = a^2$ and the line $x = 2\,a$.

6. Find the area bounded by the hyperbola $x^2 - 4\,y^2 = 4$ and the line $x = 6$.

7. Find the length of the arc of the parabola $y^2 = 2\,px$ from the vertex to one extremity of the latus rectum.

8. Find the length of the arc of the parabola $6\,y = x^2$ from the vertex to the point $(4, \frac{8}{3})$.

9. Find the length of the arc of the parabola $2\,y = x^2$ from the vertex to the point $(4, 8)$.

10. The smaller segment of the circle whose equation is $x^2 + y^2 = 25$ cut off by the line whose equation is $x = 3$ is revolved about this line, generating a spindle-shaped solid. Find the volume.

11. The area cut from the hyperbola $x^2 - y^2 = 16$ by the line $x = 5$ is revolved about this line. Find the volume generated.

12. The smaller segment of the ellipse $4\,x^2 + y^2 = 4$ cut off by the line $y = 1$ is revolved about this line. Find the volume generated.

203. Integrals involving a trinomial quadratic expression. Formulas **(XVII–XXII)** involve quadratic expressions with two terms only. If an integral involves a quadratic expression containing three terms, it may be reduced to one containing only two terms by the process of completing the square, as shown in the following examples.

Example 1. Integrate $\displaystyle\int \frac{dx}{x^2 + 4\,x + 5}$.

Solution. Now, $x^2 + 4\,x + 5 = (x + 2)^2 + 1.$

Then $\displaystyle\int \frac{dx}{x^2 + 4\,x + 5} = \int \frac{dx}{(x + 2)^2 + 1}.$

This integral is in the form of **(XVII)** if $x + 2 = u$ and $a = 1$. Also $dx = du$.

Hence $\displaystyle\int \frac{dx}{x^2 + 4\,x + 5} = \text{arc tan } (x + 2) + C.$

Example 2. Integrate $\displaystyle\int \sqrt{x^2 - 3\,x + 2}\, dx$.

Solution. $x^2 - 3\,x + 2 = x^2 - 3\,x + \frac{9}{4} - \frac{1}{4} = (x - \frac{3}{2})^2 - \frac{1}{4}.$

Then $\displaystyle\int \sqrt{x^2 - 3\,x + 2}\, dx = \int \sqrt{(x - \frac{3}{2})^2 - \frac{1}{4}}\, dx.$

This integral is in the form of **(XXII)** if $x - \frac{3}{2} = u$ and $a = \frac{1}{2}$. Also $dx = du$. Hence

$\displaystyle\int \sqrt{x^2 - 3\,x + 2}\, dx = \frac{x - \frac{3}{2}}{2} \sqrt{(x - \frac{3}{2})^2 - \frac{1}{4}} - \frac{1}{8} \ln \left[x - \frac{3}{2} + \sqrt{(x - \frac{3}{2})^2 - \frac{1}{4}} \right] + C$

$\qquad = (\frac{1}{2}\,x - \frac{3}{4})\sqrt{x^2 - 3\,x + 2} - \frac{1}{8} \ln\, (x - \frac{3}{2} + \sqrt{x^2 - 3\,x + 2}) + C.$

Example 3. Prove $\int \dfrac{dx}{\sqrt{8 + 4x - 4x^2}} = \dfrac{1}{2} \text{ arc sin } \dfrac{2x - 1}{3} + C.$

Solution.

$$8 + 4x - 4x^2 = 8 - (4x^2 - 4x)$$
$$= 9 - (4x^2 - 4x + 1)$$
$$= 9 - (2x - 1)^2.$$

Then

$$\int \frac{dx}{\sqrt{8 + 4x - 4x^2}} = \int \frac{dx}{\sqrt{9 - (2x - 1)^2}}.$$

This integral is in the form of **(XIX)**, where $2x - 1 = u$ and $a = 3$. Hence $x = \frac{1}{2} + \frac{1}{2} u$ and $dx = \frac{1}{2} du$.

Therefore, by substitution,

$$\int \frac{dx}{\sqrt{9 - (2x - 1)^2}} = \int \frac{\frac{1}{2} du}{\sqrt{9 - u^2}} = \frac{1}{2} \text{ arc sin } \frac{u}{3} + C. \quad \text{By (III) and (XIX)}$$

Substituting in this result $u = 2x - 1$, we have the answer.

PROBLEMS

1. Integrate the following expressions.

a. $\int \dfrac{dx}{x^2 + 2x + 5}.$

b. $\int \dfrac{dt}{\sqrt{2 + t - t^2}}.$

c. $\int \dfrac{dx}{1 + x + x^2}.$

d. $\int \dfrac{dy}{\sqrt{3y - y^2 - 2}}.$

e. $\int \dfrac{dx}{x^2 - 6x + 5}.$

f. $\int \dfrac{ds}{\sqrt{s^2 + s + 1}}.$

g. $\int \dfrac{dx}{4x^2 + 4x + 5}.$

h. $\int \dfrac{dx}{\sqrt{4x^2 + 4x + 5}}.$

i. $\int \dfrac{dt}{\sqrt{15 + 6t - 9t^2}}.$

j. $\int \dfrac{dx}{9x^2 - 16x + 15}.$

k. $\int \sqrt{2x - x^2} \, dx.$

l. $\int \sqrt{x^2 + 2x + 2} \, dx.$

2. Evaluate the following definite integrals.

a. $\int_0^2 \dfrac{dx}{x^2 + 4x + 8}.$

b. $\int_0^2 \dfrac{dx}{\sqrt{x^2 + 4x + 8}}.$

c. $\int_0^{0.5} \dfrac{dt}{\sqrt{3 - 4t - 4t^2}}.$

d. $\int_0^1 \sqrt{2x + x^2} \, dx.$

e. $\int_1^2 \dfrac{ds}{4s^2 + 4s - 3}.$

f. $\int_0^1 \sqrt{3 - 2x - x^2} \, dx.$

3. Find the length of the arc of the parabola $y = 4x - x^2$ which lies above the x-axis.

204. Integrals involving certain fractional forms.

When the integrand is a fraction of which the numerator is an expression of the first degree and the denominator is an expression of the second degree or the square root of an expression of the second degree, the integral can be reduced to the standard forms by the process shown in the following examples. The method consists in completing the square in the expression of the second degree and using the substitution indicated by the result.

Example 1. Prove

$$\int \frac{x\,dx}{x^2+4\,x+5} = \tfrac{1}{2}\ln(x^2+4\,x+5) - 2\arctan(x+2) + C.$$

Solution. $\qquad\qquad x^2+4\,x+5 = (x+2)^2 + 1.$

Let $x+2 = u$. Then $x = u-2$ and $dx = du$.

Hence, by substitution,

$$\int \frac{x\,dx}{x^2+4\,x+5} = \int \frac{x\,dx}{(x+2)^2+1} = \int \frac{(u-2)du}{u^2+1}$$

$$= \int \frac{u\,du}{u^2+1} - 2\int \frac{du}{u^2+1} \qquad\qquad \text{by (II) and (III)}$$

$$= \tfrac{1}{2}\ln(u^2+1) - 2\arctan u + C. \quad \text{By (V) and (XVII)}$$

Substituting $u = x+2$, we obtain the answer.

Example 2. Prove

$$\int \frac{(x+2)dx}{\sqrt{3-6\,x-9\,x^2}} = -\frac{\sqrt{3-6\,x-9\,x^2}}{9} + \frac{5}{9}\arcsin\frac{3\,x+1}{2} + C.$$

Solution. $3-6\,x-9\,x^2 = 3-(9\,x^2+6\,x) = 3-9(x^2+\tfrac{2}{3}\,x+\tfrac{1}{9})+1$
$$= 4-9(x+\tfrac{1}{3})^2.$$

Hence

$$\int \frac{(x+2)dx}{\sqrt{3-6\,x-9\,x^2}} = \int \frac{(x+2)dx}{\sqrt{4-9(x+\tfrac{1}{3})^2}}.$$

Let $9(x+\tfrac{1}{3})^2 = u^2$. Hence $3(x+\tfrac{1}{3}) = u$; $x = \tfrac{1}{3}(u-1)$, and $dx = \tfrac{1}{3}\,du$.

Hence, by substitution,

$$\int \frac{(x+2)dx}{\sqrt{4-9(x+\tfrac{1}{3})^2}} = \int \frac{\tfrac{1}{3}u-\tfrac{1}{3}+2}{\sqrt{4-u^2}}\frac{du}{3} = \frac{1}{9}\int \frac{u+5}{\sqrt{4-u^2}}\,du \qquad \text{by (III)}$$

$$= \frac{1}{9}\int \frac{u\,du}{\sqrt{4-u^2}} + \frac{5}{9}\int \frac{du}{\sqrt{4-u^2}} \qquad \text{by (II) and (III)}$$

$$= -\frac{\sqrt{4-u^2}}{9} + \frac{5}{9}\arcsin\frac{u}{2} + C. \quad \text{By (IV) and (XIX)}$$

Substituting $u = 3\,x+1$, we get the answer.

PROBLEMS

Integrate the following expressions.

1. $\int \frac{(2+x)dx}{9+x^2}.$

2. $\int \frac{(1-3\,x)dx}{x^2+9}.$

3. $\int \frac{(3\,x-2)dx}{\sqrt{9-x^2}}.$

4. $\int \frac{(x+3)dx}{\sqrt{x^2+4}}.$

5. $\int \frac{(x-1)dx}{x^2+4\,x+8}.$

6. $\int \frac{(2\,x+3)dx}{5+4\,x-x^2}.$

7. $\int \frac{(1-x)dx}{\sqrt{x^2+2\,x+2}}.$

8. $\int \frac{(x+3)dx}{\sqrt{2\,x-x^2}}.$

9. $\int \frac{(x-3)dx}{x^2+4\,x}.$

10. $\int \frac{(3\,x-1)dx}{3-2\,x-x^2}.$

11. $\int \frac{(2\,x-5)dx}{\sqrt{9\,x^2-6\,x-15}}.$

12. $\int \frac{(6\,x+1)dx}{15+6\,x-9\,x^2}.$

205. Powers and products of trigonometric functions. Expressions of the form $\sin^m u \cos^n u \, du$, $\tan^m u \sec^n u \, du$, and $\mathrm{ctn}^m u \csc^n u \, du$ are frequently integrable by means of the Power Formula **(IV)**. The method will be illustrated by examples.

Example 1. Integrate $\int \cos^3 x \, dx$.

Solution.

$$\int \cos^3 x \, dx = \int (1 - \sin^2 x) \cos x \, dx$$

$$= \int \cos x \, dx - \int (\sin x)^2 \cos x \, dx$$

$$= \sin x - \tfrac{1}{3} \sin^3 x + C.$$

Example 2. Integrate $\int \sin^2 x \cos^5 x \, dx$.

Solution.

$$\int \sin^2 x \cos^5 x \, dx = \int \sin^2 x \cos^4 x \cos x \, dx$$

$$= \int \sin^2 x (1 - \sin^2 x)^2 \cos x \, dx$$

$$= \int (\sin^2 x - 2 \sin^4 x + \sin^6 x) \cos x \, dx$$

$$= \int (\sin x)^2 \cos x \, dx - 2 \int (\sin x)^4 \cos x \, dx + \int (\sin x)^6 \cos x \, dx$$

$$= \tfrac{1}{3} \sin^3 x - \tfrac{2}{5} \sin^5 x + \tfrac{1}{7} \sin^7 x + C.$$

The method illustrated in Examples 1 and 2 is effective for integrating $\sin^m u \cos^n u \, du$ when either m or n is a positive odd integer.

Example 3. Integrate $\int \tan^4 x \, dx$.

Solution.

$$\int \tan^4 x \, dx = \int \tan^2 x (\sec^2 x - 1) dx$$

$$= \int (\tan x)^2 \sec^2 x \, dx - \int \tan^2 x \, dx$$

$$= \tfrac{1}{3} \tan^3 x - \int (\sec^2 x - 1) dx$$

$$= \tfrac{1}{3} \tan^3 x - \tan x + x + C.$$

Example 4. Integrate $\int \sec^4 2 x \, dx$.

Solution.

$$\int \sec^4 2 x \, dx = \int (\tan^2 2 x + 1) \sec^2 2 x \, dx$$

$$= \int (\tan 2 x)^2 \sec^2 2 x \, dx + \int \sec^2 2 x \, dx$$

$$= \tfrac{1}{2} \int (\tan 2 x)^2 \sec^2 2 x \, 2 \, dx + \tfrac{1}{2} \int \sec^2 2 x \, 2 \, dx$$

$$= \tfrac{1}{6} \tan^3 2 x + \tfrac{1}{2} \tan 2 x + C.$$

Example 5. Integrate $\int \tan^4 x \sec^4 x \, dx$.

Solution. $\int \tan^4 x \sec^4 x \, dx = \int \tan^4 x \sec^2 x \sec^2 x \, dx$

$$= \int \tan^4 x (\tan^2 x + 1) \sec^2 x \, dx$$

$$= \int (\tan x)^6 \sec^2 x \, dx + \int (\tan x)^4 \sec^2 x \, dx$$

$$= \tfrac{1}{7} \tan^7 x + \tfrac{1}{5} \tan^5 x + C.$$

The essence of the method used above is to consider one of the trigonometric functions in the integrand as equal to v and factor out its differential. The other factor of the integrand must then be reducible to a polynomial involving only powers of v. Success in using the method depends upon familiarity with the differentials of the trigonometric functions and the three fundamental identities of trigonometry. These are given below, for convenience.

$$d \sin u = \cos u \, du, \qquad\qquad d \cos u = - \sin u \, du,$$
$$d \tan u = \sec^2 u \, du, \qquad\qquad d \operatorname{ctn} u = - \csc^2 u \, du,$$
$$d \sec u = \sec u \tan u \, du, \qquad\qquad d \csc u = - \csc u \operatorname{ctn} u \, du,$$

$$\sin^2 u + \cos^2 u = 1,$$
$$1 + \tan^2 u = \sec^2 u,$$
$$1 + \operatorname{ctn}^2 u = \csc^2 u.$$

The method of Example 2 fails when, in $\sin^m u \cos^n u \, du$, m and n are both positive even integers. In this case the integral may be transformed by the relations

$$\sin^2 u = \tfrac{1}{2} - \tfrac{1}{2} \cos 2 u,$$
$$\cos^2 u = \tfrac{1}{2} + \tfrac{1}{2} \cos 2 u.$$

Example 6. Integrate $\int \sin^2 x \cos^2 x \, dx$.

Solution. $\int \sin^2 x \cos^2 x \, dx = \int (\tfrac{1}{2} - \tfrac{1}{2} \cos 2 x)(\tfrac{1}{2} + \tfrac{1}{2} \cos 2 x) dx$

$$= \tfrac{1}{4} \int (1 - \cos^2 2 x) dx$$

$$= \tfrac{1}{4} \int (1 - \tfrac{1}{2} - \tfrac{1}{2} \cos 4 x) dx$$

$$= \tfrac{1}{8} \int dx - \tfrac{1}{32} \int \cos 4 x \, 4 \, dx$$

$$= \tfrac{1}{8} x - \tfrac{1}{32} \sin 4 x + C.$$

PROBLEMS

Integrate the following expressions.

1. $\int \sin^3 x \, dx$.

2. $\int \sin^5 x \, dx$.

3. $\int \cos^5 x \, dx$.

4. $\int \cos^2 x \, dx$.

5. $\int \sin^4 x \, dx$.

6. $\int \cos^4 x \, dx$.

7. $\int \sin^4 x \cos^2 x \, dx$.

8. $\int \sin^4 x \cos^4 x \, dx$.

9. $\int \operatorname{ctn}^2 \theta \, d\theta$.

10. $\int \tan^3 x \, dx$.

11. $\int \operatorname{ctn}^3 x \, dx$.

12. $\int \dfrac{\sin^3 \theta \, d\theta}{\sqrt{\cos \theta}}$.

206. The trigonometric substitutions. When an integral involves one of the three radicals $\sqrt{a^2 - u^2}$, $\sqrt{u^2 - a^2}$, or $\sqrt{a^2 + u^2}$, and is otherwise rational, the integrand can be rationalized by a proper trigonometric substitution. These substitutions are as follows.

To rationalize $\sqrt{a^2 - u^2}$, substitute $u = a \sin z$.

To rationalize $\sqrt{u^2 - a^2}$, substitute $u = a \sec z$.

To rationalize $\sqrt{a^2 + u^2}$, substitute $u = a \tan z$.

For in the first case

$$\sqrt{a^2 - u^2} = \sqrt{a^2 - a^2 \sin^2 z} = \sqrt{a^2(1 - \sin^2 z)} = \sqrt{a^2 \cos^2 z} = a \cos z.$$

The other statements can be verified in like manner.

Example 1. Integrate $\int \dfrac{\sqrt{4 x^2 - 9} \, dx}{x}$.

Solution. Here $2 x = u$ and $a = 3$. We therefore substitute $2 x = 3 \sec z$, whence $x = \frac{3}{2} \sec z$ and $dx = \frac{3}{2} \sec z \tan z \, dz$.

$$\int \frac{\sqrt{4 x^2 - 9} \, dx}{x} = \int \frac{\sqrt{9 \sec^2 z - 9} \cdot \frac{3}{2} \sec z \tan z \, dz}{\frac{3}{2} \sec z}$$

$$= 3 \int \tan^2 z \, dz = 3 \int (\sec^2 z - 1) dz$$

$$= 3 \tan z - 3 z + C.$$

In order to get the result in terms of x, we draw a right triangle with one acute angle z. Since $\sec z = \frac{2}{3} x$, the hypotenuse will be $2 x$ and the adjacent side will be 3. The opposite side is then $\sqrt{4 x^2 - 9}$ and $\tan z = \dfrac{\sqrt{4 x^2 - 9}}{3}$. Hence the result above gives

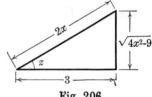

Fig. 206

$$\int \frac{\sqrt{4 x^2 - 9} \, dx}{x} = \sqrt{4 x^2 - 9} - 3 \text{ arc sec } \tfrac{2}{3} x + C.$$

When the integral is a definite integral, it is usually better to change the limits of integration when the substitution is made, as in the following example.

Example 2. Find the value of $\displaystyle\int_0^3 \frac{dx}{(25 - x^2)^{\frac{3}{2}}}$.

Solution. Let $x = 5 \sin z$; then $dx = 5 \cos z \, dz$. We next find the limits for z. When $x = 0$, $z = 0$; when $x = 3$, $\sin z = \frac{3}{5}$. Let $\alpha = \text{arc sin } \frac{3}{5}$. Then

$$\int_0^3 \frac{dx}{(25 - x^2)^{\frac{3}{2}}} = \int_0^\alpha \frac{5 \cos z \, dz}{(25 - 25 \sin^2 z)^{\frac{3}{2}}} = \int_0^\alpha \frac{5 \cos z \, dz}{125 \cos^3 z}$$

$$= \frac{1}{25} \int_0^\alpha \frac{dz}{\cos^2 z} = \frac{1}{25} \int_0^\alpha \sec^2 z \, dz$$

$$= \frac{1}{25} \left[\tan z \right]_0^\alpha = \frac{3}{100}.$$

[Since $\tan \alpha = \frac{3}{4}$.]

PROBLEMS

1. Integrate the following expressions.

a. $\displaystyle\int \frac{\sqrt{x^2 - a^2} \, dx}{x}$.

b. $\displaystyle\int \frac{\sqrt{t^2 + 4} \, dt}{t^2}$.

c. $\displaystyle\int \frac{dx}{x^2\sqrt{x^2 - 3}}$.

d. $\displaystyle\int \frac{\sqrt{4 - x^2} \, dx}{x^2}$.

e. $\displaystyle\int \frac{\sqrt{s^2 - 16} \, ds}{s}$.

f. $\displaystyle\int \frac{\sqrt{x^2 - 36} \, dx}{x^2}$.

g. $\displaystyle\int \frac{dx}{x^4\sqrt{x^2 + 1}}$.

h. $\displaystyle\int \frac{\sqrt{x^2 - a^2} \, dx}{x^4}$.

i. $\displaystyle\int \frac{t^2 \, dt}{\sqrt{4 \, t^2 + 1}}$.

2. Evaluate the following definite integrals.

a. $\displaystyle\int_0^3 x^3\sqrt{9 - x^2} \, dx$.

b. $\displaystyle\int_1^2 \frac{dx}{x^2\sqrt{5 - x^2}}$.

c. $\displaystyle\int_0^4 \frac{ds}{(s^2 + 9)^{\frac{3}{2}}}$.

207. Integration by parts. The formula for differentiating the product of two functions is

$$\frac{d}{dx}(uv) = u \frac{dv}{dx} + v \frac{du}{dx},$$

or, in differential notation,

$$d(uv) = u \, dv + v \, du;$$

whence

$$u \, dv = d(uv) - v \, du.$$

Integrating both sides of this equation, we have

(A) $$\int u \, dv = uv - \int v \, du.$$

This very important formula is known as the formula for *integration by parts.*

Among the most important applications of the method of integration by parts are the integration of (*a*) differentials involving products, (*b*) differentials involving logarithms, (*c*) differentials involving inverse trigonometric functions.

Example 1. Integrate $\int x \sin x \, dx$.

Solution. Comparing this integral with **(A)**, we see that $u \, dv = x \sin x \, dx$. This is the only relation which will guide us in the choice of the functions u and v. There are two possibilities.

1. If $u = \sin x$, then $dv = x \, dx$.
2. If $u = x$, then $dv = \sin x \, dx$.

Since the proper choice can be made only by experience, we shall try each of the possibilities.

1. If $u = \sin x$, $du = \cos x \, dx$, and if $dv = x \, dx$, $v = \frac{1}{2} x^2 + C$. Substitution in **(A)** gives

$$\int x \sin x \, dx = (\tfrac{1}{2} x^2 + C) \sin x - \int (\tfrac{1}{2} x^2 + C) \cos x \, dx.$$

Since the integral on the right-hand side is more difficult than the original problem, we conclude that the first choice of u and dv is not suitable.

2. If $u = x$, $du = dx$, and if $dv = \sin x \, dx$, $v = -\cos x + C_1$. Substitution in **(A)** gives

$$\int x \sin x \, dx = x(-\cos x + C_1) - \int (-\cos x + C_1) dx$$
$$= -x \cos x + C_1 x + \sin x - C_1 x + C$$
$$= -x \cos x + \sin x + C.$$

It will be observed that the constant C_1 cancels out of the final result. As this always happens, it is not necessary to add an arbitrary constant when integrating dv to find v.

Example 2. Integrate $\int x \ln x \, dx$.

Solution. Let $\qquad u = \ln x \quad$ and $\quad dv = x \, dx$;

then $\qquad\qquad\qquad du = \dfrac{dx}{x} \quad$ and $\quad v = \frac{1}{2} x^2.$

Substitution in **(A)** gives

$$\int x \ln x \, dx = \frac{1}{2} x^2 \ln x - \int \frac{1}{2} x^2 \cdot \frac{dx}{x}$$
$$= \frac{1}{2} x^2 \ln x - \frac{1}{4} x^2 + C.$$

Example 3. Integrate $\int x^2 e^x \, dx$.

Solution. Let $\qquad u = x^2 \qquad$ and $\quad dv = e^x \, dx$;
then $\qquad\qquad\qquad du = 2 \, x \, dx \quad$ and $\quad v = e^x$.

Substitution in **(A)** gives

$$\int x^2 e^x \, dx = x^2 e^x - 2 \int x e^x \, dx. \tag{1}$$

To find the integral on the right-hand side, we again use the method of integration by parts.

Let $\qquad\qquad\qquad u = x \qquad$ and $\quad dv = e^x \, dx$;
then $\qquad\qquad\qquad du = dx \quad$ and $\quad v = e^x$.

Substitution in **(A)** gives

$$\int x e^x \, dx = x e^x - \int e^x \, dx$$

$$= x e^x - e^x.$$

Substitution of this value in equation (1) gives the final result,

$$\int x^2 e^x \, dx = x^2 e^x - 2 \, x e^x + 2 \, e^x + C = e^x(x^2 - 2 \, x + 2) + C.$$

Example 4. Integrate $\int \arcsin x \, dx$.

Solution. Let $\qquad u = \arcsin x \quad$ and $\quad dv = dx$;
then $\qquad\qquad\qquad du = \dfrac{dx}{\sqrt{1 - x^2}} \quad$ and $\quad v = x$.

Substitution in **(A)** gives

$$\int \arcsin x \, dx = x \arcsin x - \int \frac{x \, dx}{\sqrt{1 - x^2}}.$$

To the last integral we apply the Power Formula **(IV)** with the final result

$$\int \arcsin x \, dx = x \arcsin x + \sqrt{1 - x^2} + C.$$

Example 5. Prove $\int e^{ax} \sin nx \, dx = \dfrac{e^{ax}(a \sin nx - n \cos nx)}{a^2 + n^2} + C.$

Solution. Let $\qquad u = e^{ax} \qquad$ and $\quad dv = \sin nx \, dx$;
then $\qquad\qquad\qquad du = a e^{ax} \, dx \quad$ and $\quad v = -\dfrac{\cos nx}{n}.$

Substituting in **(A)**, the result is

$$\int e^{ax} \sin nx \, dx = -\frac{e^{ax} \cos nx}{n} + \frac{a}{n} \int e^{ax} \cos nx \, dx.$$

Integrate the new integral by parts.

Let $\qquad\qquad\qquad u = e^{ax} \qquad$ and $\quad dv = \cos nx \, dx$;
then $\qquad\qquad\qquad du = a e^{ax} \, dx \quad$ and $\quad v = \dfrac{\sin nx}{n}.$

Hence, by **(4)**,

$$\int e^{ax} \cos nx \, dx = \frac{e^{ax} \sin nx}{n} - \frac{a}{n} \int e^{ax} \sin nx \, dx. \tag{3}$$

Substituting in (2), we obtain

$$\int e^{ax} \sin nx \, dx = \frac{e^{ax}}{n^2} (a \sin nx - n \cos nx) - \frac{a^2}{n^2} \int e^{ax} \sin nx \, dx. \tag{4}$$

The two integrals in (4) are the same. Transposing the one in the right-hand member and solving, the result is as above.

PROBLEMS

1. Integrate the following expressions.

a. $\int x \cos x \, dx.$ **b.** $\int t e^{-2t} \, dt.$ **c.** $\int x^n \ln x \, dx.$

d. $\int x^2 e^{-x} \, dx.$ **e.** $\int x \sin 2 x \, dx.$ **f.** $\int s \tan^2 s \, ds.$

g. $\int \text{arc} \cos x \, dx.$ **h.** $\int \text{arc} \tan x \, dx.$ **i.** $\int \text{arc} \operatorname{ctn} x \, dx.$

j. $\int \text{arc} \sec x \, dx.$ **k.** $\int \text{arc} \csc x \, dx.$ **l.** $\int x^2 \sin x \, dx.$

m. $\int x^2 \cos 2 x \, dx.$ **n.** $\int \text{arc} \tan \sqrt{s} \, ds.$ **o.** $\int x \, \text{arc} \tan x \, dx.$

p. $\int e^{2t} \cos 3 t \, dt.$ **q.** $\int e^{-x} \sin 3 x \, dx.$ **r.** $\int e^{-\frac{x}{2}} \sin \pi x \, dx.$

2. Find the area bounded by the curve $y = \ln x$, the x-axis, and the line $x = 10$.

3. Find the area bounded by the curve $y = x e^x$, the x-axis, and the line $x = 4$.

4. Find the volume generated by revolving the area of Problem 3 about the x-axis.

5. Find the coördinates of the centroid of the area bounded by the arch of the curve $y = \sin x$ between $x = 0$ and $x = \pi$, and by the x-axis.

6. Find the coördinates of the centroid of the area in the first quadrant bounded by the x-axis, the y-axis, and the curve $y = \cos x$.

7. Find the area under the curve $y = e^x \sin x$ from $x = 0$ to $x = \pi$.

8. Find the area in the first quadrant bounded by the coördinate axes and the curve $y = 4 e^{-\frac{x}{2}} \cos \frac{1}{2} \pi x$.

9. Find the volume generated by revolving the area of Problem 7 about the x-axis.

10. Find the volume generated by revolving the area of Problem 8 about the x-axis.

CHAPTER XVI

TABLE OF INTEGRALS; RATIONAL FRACTIONS

208. Table of integrals. From the definition of an integral (Art. 78) we see that from the derivative of any function can be found a corresponding formula for integration. Obviously the number of possible integration formulas is unlimited. A table, or list, of formulas which are useful for the purposes of this course is given in Art. 297. The formulas are arranged in groups, and this arrangement should be studied carefully in order to locate a particular formula when it is needed.

The formulas involve certain constants, a, b, c, n, etc. In general these constants may have any real values, but in some cases limitations are explicitly given in the statement of the formula (see formulas 4, 6, 18, 31, 32, 105, 106, etc.). When no explicit limitation is stated, it is understood that the formula is invalid in the two following cases: (1) When a zero value occurs in a denominator. For example, formula 166 is not valid when $m = n$. (2) When the formula involves the square root of a negative quantity. For example, formula 109 cannot be used if c is negative.

The use of the table is illustrated in the following examples.

Example 1. Find $\int \dfrac{dx}{x^2\sqrt{x-1}}$.

Solution. We apply first formula 33, where $m = 2$, $a = -1$, $b = 1$. Then

$$\int \frac{dx}{x^2\sqrt{x-1}} = \frac{-\sqrt{x-1}}{-x} - \frac{1}{-2}\int \frac{dx}{x\sqrt{x-1}}. \tag{1}$$

To work out the integral on the right side of (1) we apply formula 32. Observe that we should use formula 31 if a were positive.

$$\int \frac{dx}{x\sqrt{x-1}} = \frac{2}{\sqrt{1}} \text{ arc tan } \sqrt{x-1} + C.$$

Substituting this value in (1), the final result is

$$\int \frac{dx}{x^2\sqrt{x-1}} = \frac{\sqrt{x-1}}{x} + \text{ arc tan } \sqrt{x-1} + C.$$

Example 2. Evaluate $\displaystyle\int_0^1 \frac{x^2\,dx}{\sqrt{9-4\,x^2}}$.

Solution. This integral is similar to formula 66, but the formula cannot be applied immediately because the coefficient of u^2 in the denominator is -1.

while the corresponding coefficient of x^2 in the problem is -4. It is necessary to make a transformation so that the problem shall correspond exactly to the formula. In the integral to be evaluated we set $2x = u$. Then $4x^2 = u^2$, $dx = \frac{1}{2} du$ and the integral becomes, after changing the limits,

$$\int_0^1 \frac{x^2 \, dx}{\sqrt{9 - 4x^2}} = \frac{1}{8} \int_0^2 \frac{u^2 \, du}{\sqrt{9 - u^2}}.$$

To this integral we may apply formula 66, where $a = 3$. The result is

$$\frac{1}{8} \int_0^2 \frac{u^2 \, du}{\sqrt{9 - u^2}} = \frac{1}{8} \left[-\frac{u}{2}\sqrt{9 - u^2} + \frac{9}{2} \text{arc sin} \frac{u}{3} \right]_0^2$$

$$= \frac{1}{8}[- \sqrt{5} + \frac{9}{2} \text{arc sin } \frac{2}{3}] = 0.131.$$

Example 3. Integrate $\displaystyle\int \frac{dx}{1 + x + x^2}.$

Solution. For this integral we must choose between formulas 105 and 106. Since $a = 1$, $b = 1$, $c = 1$, we have $b^2 < 4\,ac$, and must use formula 105.

$$\int \frac{dx}{1 + x + x^2} = \frac{2}{\sqrt{3}} \text{arc tan} \frac{2x + 1}{\sqrt{3}} + C.$$

Example 4. Evaluate $\displaystyle\int_0^2 e^{-x} \cos^2 \left(\frac{x}{2}\right) dx.$

Solution. This integral may be transformed to the form of formula 171. In the integral we set $x = 2u$, $dx = 2\,du$, giving

$$\int_0^2 e^{-x} \cos^2 \left(\frac{x}{2}\right) dx = 2 \int_0^1 e^{-2u} \cos^2 u \, du.$$

To this integral we apply formula 171, where $a = -2$, $n = 2$. The result is

$$2 \int_0^1 e^{-2u} \cos^2 u \, du = 2 \left[\frac{e^{-2u} \cos u(-2\cos u + 2 \sin u)}{4 + 4} \right]_0^1 + \frac{2 \cdot 2 \cdot 1}{4 + 4} \int_0^1 e^{-2u} du.$$

Working out the last integral (formula 120), we have

$$2 \int_0^1 e^{-2u} \cos^2 u \, du = \left[\frac{e^{-2u} \cos u(\sin u - \cos u)}{2} - \frac{e^{-2u}}{4} \right]_0^1$$

$$= \frac{e^{-2}}{4} [2 \cos 1(\sin 1 - \cos 1) - 1] - [-\frac{1}{2} - \frac{1}{4}]$$

$$= \frac{0.135}{4}[1.08(0.841 - 0.540) - 1] + 0.75$$

$$= 0.727.$$

PROBLEMS

1. Verify the following formulas in the table of integrals, Art. 297, by differentiation: 14, 28, 41, 49, 107, 153.

2. Derive formula 25 by making the substitution $v^2 = a + bu$.

Work out the following integrals.

3. $\int \dfrac{dx}{x(3-2x)^2}.$

4. $\int \dfrac{x^2\,dx}{\sqrt{9x^2+1}}.$

5. $\int \dfrac{\sqrt{4x^2-9}\,dx}{x}.$

6. $\int \dfrac{dx}{(6x-4x^2)^{\frac{3}{2}}}.$

7. $\int \dfrac{\sqrt{3x^2+5}\,dx}{x^2}.$

8. $\int \dfrac{\sqrt{25-9x^2}\,dx}{x^2}.$

9. $\int \sqrt{\dfrac{7-2x}{9+2x}}\,dx.$

10. $\int \dfrac{dx}{5+4\sin x}.$

11. $\int \sin 2x \sin 3x\,dx.$

12. $\int \sqrt{\dfrac{1+3x}{5-3x}}\,dx.$

13. $\int \dfrac{x^2\,dx}{\sqrt{9x^2+4}}.$

14. $\int \dfrac{dx}{x\sqrt{12x-9x^2}}.$

15. $\int \dfrac{d\theta}{3\cos\theta+1}.$

16. $\int \dfrac{dx}{x^2\sqrt{9-4x^2}}.$

17. $\int \dfrac{\sqrt{2-x}\,dx}{x^3}.$

18. $\int_0^2 \dfrac{x^2\,dx}{(1+2x)^2}.$

19. $\int_0^4 \dfrac{x^2\,dx}{\sqrt{9+4x}}.$

20. $\int_0^1 \dfrac{dx}{(25-9x^2)^{\frac{3}{2}}}.$

21. $\int_1^2 \dfrac{dx}{x^2\sqrt{5+4x^2}}.$

22. $\int_1^2 \dfrac{dx}{x^2\sqrt{49-10x^2}}.$

23. $\int_1^3 \dfrac{dx}{x^3\sqrt{5x^2-2}}.$

24. $\int_0^{\frac{\pi}{2}} \dfrac{d\theta}{\cos^2\frac{1}{2}\theta+4\sin^2\frac{1}{2}\theta}.$

25. $\int_0^1 \sin\tfrac{1}{4}\pi t \cos\tfrac{1}{2}\pi t\,dt.$

209. Reduction formulas. By means of *reduction formulas* it is possible to evaluate certain integrals by successive steps. The standard reduction formulas for algebraic expressions involving binomials are formulas 96–104. (Special cases occur elsewhere in the list, as, for example, formula 30.) In these formulas it is understood that the exponents m, p, and q are positive numbers, and in the applications m and q are usually integers.

The standard reduction formulas for trigonometric functions are formulas 157–174. The exponents m and n are understood to be positive, and in the applications they are usually integers.

Example 1. Integrate $\int x^3\sqrt{x^2+1}\,dx.$

Solution. To this integral we may apply either formula 96 or formula 97, where $m=3$, $a=1$, $b=1$, $p=\frac{1}{2}$, $q=2$. (See also formula 39.) The choice is determined by the integral on the right side of the formula.

Observe that the effect of formula 96 is to decrease the exponent of u by q; the effect of formula 97 is to decrease the exponent of the binomial $(a+bu^q)$ by 1.

If we apply formula 96, we are led to the integral $\int x\sqrt{x^2+1}\,dx$, which can readily be found.

The result of using formula 96 is

$$\int x^3\sqrt{x^2+1}\,dx = \frac{x^2(x^2+1)^{\frac{3}{2}}}{5} - \frac{2}{5}\int x\sqrt{x^2+1}\,dx. \tag{1}$$

Using the Power Formula, **(IV)**, we find

$$\int x\sqrt{x^2+1}\,dx = \tfrac{1}{3}(x^2+1)^{\frac{3}{2}} + C.$$

Substituting this result in equation (1), we have

$$\int x^3\sqrt{x^2+1}\,dx = \frac{x^2(x^2+1)^{\frac{3}{2}}}{5} - \frac{2}{15}(x^2+1)^{\frac{3}{2}} + C$$

$$= \frac{(3\,x^2-2)(x^2+1)^{\frac{3}{2}}}{15} + C.$$

Example 2. Integrate $\displaystyle\int \frac{(1-x^2)^{\frac{3}{2}}\,dx}{x^2}$.

Solution. To this integral we may apply either formula 101 or formula 102, where $a=1$, $b=-1$, $q=2$, $m=2$, $p=\frac{3}{2}$. (See also formula 78.) Observe that the effect of formula 101 is to decrease the exponent of u by q; the effect of formula 102 is to decrease the exponent of the binomial $(a+bu^q)$ by 1.

We first apply formula 102, with the following result.

$$\int \frac{(1-x^2)^{\frac{3}{2}}}{x^2}\,dx = \frac{(1-x^2)^{\frac{3}{2}}}{2\,x} + \frac{3}{2}\int \frac{(1-x^2)^{\frac{1}{2}}}{x^2}\,dx.$$

The integral on the right side is now in the form of formula 76. The final result is

$$\int \frac{(1-x^2)^{\frac{3}{2}}\,dx}{x^2} = \frac{(1-x^2)^{\frac{3}{2}}}{2\,x} - \frac{3\sqrt{1-x^2}}{2\,x} - \frac{3}{2}\,\text{arc sin } x + C.$$

Example 3. Evaluate $\displaystyle\int_0^\pi \sec^4\left(\frac{x}{4}\right)dx$.

Solution. In order to apply formula 160, we first make the substitution $x = 4\,u$, $dx = 4\,du$.

Then

$$\int_0^\pi \sec^4\left(\frac{x}{4}\right)dx = 4\int_0^{\frac{\pi}{4}} \frac{du}{\cos^4 u}.$$

Applying formula 160 ($n=4$), we have

$$4\int_0^{\frac{\pi}{4}} \frac{du}{\cos^4 u} = \left[\frac{4\,\sin u}{3\,\cos^3 u}\right]_0^{\frac{\pi}{4}} + \frac{8}{3}\int_0^{\frac{\pi}{4}} \frac{du}{\cos^2 u}.$$

The last integral is found by formula 137. Hence

$$4\int_0^{\frac{\pi}{4}} \frac{du}{\cos^4 u} = \left[\frac{4\,\sin u}{3\,\cos^3 u} + \frac{8}{3}\,\tan u\right]_0^{\frac{\pi}{4}} = \frac{16}{3}.$$

Example 4. Integrate $\displaystyle\int x^3 \cos x\,dx$.

Solution. To this integral we apply formula 173, with $m=3$, $a=1$. The result is

$$\int x^3 \cos x\,dx = x^2(x\,\sin x + 3\,\cos x) - 6\int x\,\cos x\,dx. \qquad (2)$$

To find the last integral we again use formula 173, with $m = 1$, $a = 1$, and obtain

$$\int x \cos x \, dx = x \sin x + \cos x + C.$$

Substituting this result in equation (2), we have, finally,

$$\int x^3 \cos x \, dx = x^3 \sin x + 3 x^2 \cos x - 6 x \sin x + 6 \cos x + C.$$

PROBLEMS

Work out the following integrals.

1. $\displaystyle\int \frac{\sqrt{1 - x^2} \, dx}{x^4}$.

2. $\displaystyle\int \frac{x^5 \, dx}{\sqrt{2 + x^3}}$.

3. $\displaystyle\int \frac{x^3 \, dx}{\sqrt{x^2 - 4}}$.

4. $\displaystyle\int \frac{x^4 \, dx}{\sqrt{a^2 - x^2}}$.

5. $\displaystyle\int \frac{dx}{x^4 \sqrt{9 - 4 x^2}}$.

6. $\displaystyle\int \theta^2 \sin 2 \, \theta \, d\theta$.

7. $\displaystyle\int \theta^3 \cos 3 \, \theta \, d\theta$.

8. $\displaystyle\int \frac{\sin^2 2 \, x \, dx}{\cos^3 2 \, x}$.

9. $\displaystyle\int x^5 \sqrt{2 - 5 \, x^3} \, dx$.

10. $\displaystyle\int x^3 (1 + 4 \, x^2)^{\frac{3}{2}} \, dx$.

11. $\displaystyle\int x^3 \sqrt[3]{a^2 + x^2} \, dx$.

12. $\displaystyle\int \frac{x^3 \, dx}{(9 - 4 \, x^2)^{\frac{3}{2}}}$.

13. $\displaystyle\int e^{-t} \sin^2 \pi t \, dt$.

14. $\displaystyle\int \sec^2 \phi \csc^3 \phi \, d\phi$.

15. $\displaystyle\int \sin^2 \theta \tan^3 \theta \, d\theta$.

16. $\displaystyle\int_0^2 \frac{x^5 \, dx}{\sqrt{1 + x^3}}$.

17. $\displaystyle\int_0^1 x^3 (9 - 8 \, x^2)^{\frac{3}{2}} \, dx$.

18. $\displaystyle\int_1^2 \frac{dx}{x(2 \, x + 5)^2}$.

19. $\displaystyle\int_0^1 \frac{x^5 \, dx}{\sqrt{1 - x^2}}$.

20. $\displaystyle\int_3^5 \frac{dx}{x^3 \sqrt{25 - x^2}}$.

21. $\displaystyle\int_0^4 \frac{x^3 \, dx}{(9 + x^2)^{\frac{3}{2}}}$.

22. $\displaystyle\int_0^1 \frac{x^8 \, dx}{\sqrt{1 - x^3}}$.

23. $\displaystyle\int_0^\pi \cos^8 \theta \, d\theta$.

24. $\displaystyle\int_0^{\frac{\pi}{2}} t^2 \sin 2 \, t \, dt$.

210. Integration of rational fractions. A rational fraction is one in which the numerator and denominator are integral, rational functions of the same variable, that is, polynomials in which the variable is not affected by fractional or negative exponents. Certain rational fractions may be integrated directly by the tables (see formulas 7–19, 105–108); but the variety of rational fractions is so great that it would be impossible to include all types in a table, and we must consider some general methods for integrating them.

If the degree of the numerator is equal to or greater than the degree of the denominator, the fraction may be reduced to a mixed expression by dividing the numerator by the denominator. Thus

$$\frac{x^4 + 3 \, x^3}{x^2 + 2 \, x + 5} = x^2 + x - 7 + \frac{9 \, x + 35}{x^2 + 2 \, x + 5}.$$

The last term is a fraction reduced to its lowest terms, having the degree of the numerator less than that of the denominator. It appears that the other terms are at once integrable, and hence we need consider only the fraction. The general method consists in breaking up the given fraction into a sum of simpler fractions.

CASE I. *When the factors of the denominator are all of the first degree and none is repeated.*

Let us observe the result of adding the following fractions as indicated.

$$\frac{1}{x+1} - \frac{2}{x+2} + \frac{3}{x+3} = \frac{2\,x^2+6\,x+6}{(x+1)(x+2)(x+3)}.$$

The denominator of the sum is the product of the denominators of the fractions added, and the numerator is of lower degree than the denominator.

Suppose now that we start with a fraction whose numerator is of lower degree than the denominator and whose denominator is the product of factors of the first degree, all of which are different. It is shown in algebra that it is always possible to find a sum of simpler fractions, called *partial fractions*, which is equal to the given fraction. The method is illustrated in the following example.

Example. Resolve $\dfrac{8\,x+2}{x^3-x}$ into partial fractions.

Solution. The factors of the denominator are x, $x+1$, $x-1$. Assume

$$\frac{8\,x+2}{x^3-x} = \frac{A}{x} + \frac{B}{x+1} + \frac{C}{x-1}.$$

We must determine the values of the constants A, B, and C so that the sum of the fractions on the right will be equal to the given fraction. Clearing of fractions, we have

$$8\,x+2 \equiv A(x+1)(x-1) + Bx(x-1) + Cx(x+1). \tag{1}$$

The notation $\equiv$ indicates that the expression on the right side is *identically* equal to the expression on the left. There are two methods of determining the values of the constants.

First method. Since the relation (1) is an identity, it must be true for every value of x.

Setting $x = 0$ gives $2 = -A$, whence $A = -2$.
Setting $x = -1$ gives $-6 = 2B$, whence $B = -3$.
Setting $x = 1$ gives $10 = 2C$, whence $C = 5$.

Hence $\dfrac{8\,x+2}{x^3-x} = \dfrac{5}{x-1} - \dfrac{2}{x} - \dfrac{3}{x+1}.$

The correctness of the work may be tested by adding the fractions on the right.

Second method. Multiplying out the parentheses in (1), we have

$$8\,x + 2 \equiv A(x^2 - 1) + B(x^2 - x) + C(x^2 + x),$$
$$8\,x + 2 \equiv (A + B + C)x^2 + (-B + C)x - A.$$

Since this relation is an identity, we equate the coefficients of like powers of x in the two members according to the method of Undetermined Coefficients, and obtain three simultaneous equations.

$$A + B + C = 0,$$
$$-B + C = 8,$$
$$-A = 2.$$

Solving this system of equations, we find $A = -2$, $B = -3$, $C = 5$, as before.

In order to integrate a fraction of the type considered in Case I, we replace it by a sum of partial fractions and then integrate these separately.

Thus, in the above example,

$$\int \frac{(8\,x + 2)dx}{x^3 - x} = \int \left[\frac{5}{x-1} - \frac{2}{x} - \frac{3}{x+1} \right] dx$$
$$= 5 \ln (x - 1) - 2 \ln x - 3 \ln (x + 1) + C.$$

If more convenient, this result may be written also in the form

$$\ln \frac{c(x-1)^5}{x^2(x+1)^3}.$$

In every case the number of constants A, B, C, $\cdots$ to be found is equal to the degree of the denominator of the given fraction.

CASE II. *When the factors of the denominator are all of the first degree and some are repeated.*

The method in this case is the same as in Case I with this exception: if the denominator of the given fraction contains a factor $(x - a)^n$, there will correspond to this factor the sum of n partial fractions having the form

$$\frac{A}{(x-a)^n} + \frac{B}{(x-a)^{n-1}} + \cdots + \frac{N}{x-a}.$$

Example. Integrate $\int \frac{(2\,x^3 - 1)dx}{x^2(x+1)^2}$.

Solution. Assume

$$\frac{2\,x^3 - 1}{x^2(x+1)^2} \equiv \frac{A}{x^2} + \frac{B}{x} + \frac{C}{(x+1)^2} + \frac{D}{x+1}.$$

Clearing of fractions, we have

$$2\,x^3 - 1 \equiv (B + D)x^3 + (A + 2\,B + C + D)x^2 + (2\,A + B)x + A.$$

Equating the coefficients of the different powers of x on each side of the equation, we get

$$B + D = 2,$$
$$A + 2\,B + C + D = 0,$$
$$2\,A + B = 0,$$
$$A = -1.$$

The solution of this system of equations is

$$A = -1, \quad B = 2, \quad C = -3, \quad D = 0.$$

Therefore
$$\int \frac{(2\,x^3 - 1)dx}{x^2(x + 1)^2} = \int \left[-\frac{1}{x^2} + \frac{2}{x} - \frac{3}{(x + 1)^2} \right] dx$$

$$= \frac{1}{x} + 2\ln x + \frac{3}{x + 1} + C.$$

Note again here that the number of constants, A, B, C, D, to be found is equal to the degree of the denominator of the original fraction.

PROBLEMS

Work out the following integrals.

1. $\displaystyle\int \frac{(5\,x^2 - 15\,x + 12)dx}{x^3 - 5\,x^2 + 6\,x}.$

2. $\displaystyle\int \frac{(x^2 + 4)dx}{x^3 - 25\,x}.$

3. $\displaystyle\int \frac{(x^3 + 4\,x + 4)dx}{x^4 + x^3}.$

4. $\displaystyle\int \frac{(x + 1)dx}{x^2(x - 1)^2}.$

5. $\displaystyle\int \frac{(2\,t^2 - 5\,t)dt}{(t - 2)(t - 3)(t + 1)}.$

6. $\displaystyle\int \frac{dx}{(4\,x^2 - 1)(x + 2)}.$

7. $\displaystyle\int \frac{x^3\,dx}{(x + 1)^3}.$

8. $\displaystyle\int \frac{s^3\,ds}{(s^2 - 1)(s + 2)}.$

9. $\displaystyle\int \frac{(x^4 - 1)dx}{x^3 - 9\,x}.$

10. $\displaystyle\int \frac{(x^5 + x^4 - 8)dx}{x^3 - 4\,x}.$

11. $\displaystyle\int \frac{x^4\,dx}{(x - 1)^2(x - 2)}.$

12. $\displaystyle\int \frac{x^4\,dx}{(x - 2)(x^2 - 1)}.$

13. $\displaystyle\int_3^4 \frac{(x^2 + 6\,x - 8)dx}{x^3 - 4\,x}.$

14. $\displaystyle\int_3^6 \frac{(x^2 + x - 1)dx}{x^3 + x^2 - 6\,x}.$

15. $\displaystyle\int_3^5 \frac{dx}{(x - 1)^2(x - 2)}.$

16. $\displaystyle\int_0^4 \frac{x^2\,dx}{(x + 2)^2(x + 1)}.$

17. $\displaystyle\int_3^7 \frac{(y^3 + 4)dy}{y^3 - 4\,y}.$

18. $\displaystyle\int_2^4 \frac{(x^5 - x^3 - 1)dx}{x^4 - x^3}.$

CASE III. *When the denominator contains factors of the second degree but none repeated.*

The method in this case is similar to that in Case I except that to a factor of the form $x^2 + px + q$ in the denominator there corresponds a partial fraction of the form

$$\frac{Ax + B}{x^2 + px + q}.$$

The integration problem for a fraction of this form is discussed in Art. 203 (see also formula 108).

Example. Integrate $\int \frac{(4\,x^2 + x + 2)dx}{x^3 + x}$.

Solution. The factors of the denominator are x and $x^2 + 1$. Assume

$$\frac{4\,x^2 + x + 2}{x^3 + x} = \frac{Ax + B}{x^2 + 1} + \frac{C}{x}.$$

Clearing of fractions, we have

$$4\,x^2 + x + 2 \equiv (A + C)x^2 + Bx + C.$$

Equating the coefficients of the different powers of x on each side of the equation, we get

$$A + C = 4,$$
$$B = 1,$$
$$C = 2,$$

whence $\qquad A = 2, \quad B = 1, \quad C = 2.$

Therefore

$$\int \frac{(4\,x^2 + x + 2)dx}{x^3 + x} = \int \left[\frac{2\,x + 1}{x^2 + 1} + \frac{2}{x} \right] dx$$
$$= \ln (x^2 + 1) + \text{arc tan } x + 2 \ln x + C.$$

CASE IV. *When the denominator contains factors of the second degree, some of which are repeated.*

To every n-fold quadratic factor, such as $(x^2 + px + q)^n$, there will correspond the sum of n partial fractions,

$$\frac{Ax + B}{(x^2 + px + q)^n} + \frac{Cx + D}{(x^2 + px + q)^{n-1}} + \cdots + \frac{Lx + M}{x^2 + px + q}.$$

To carry out the integration, the reduction formula (see **(21)**, Art. 297)

$$\int \frac{du}{(u^2 + a^2)^n} = \frac{1}{2(n - 1)a^2} \left[\frac{u}{(u^2 + a^2)^{n-1}} + (2\,n - 3) \int \frac{du}{(u^2 + a^2)^{n-1}} \right] \qquad (2)$$

is necessary. If $n > 2$, repeated applications of (2) are necessary.

If p is not zero, we complete the square,

$$x^2 + px + q = (x + \tfrac{1}{2}p)^2 + \tfrac{1}{4}(4q - p^2) = u^2 + a^2.$$

Example. Prove

$$\int \frac{2x^3 + x + 3}{(x^2 + 1)^2}\, dx = \ln(x^2 + 1) + \frac{1 + 3x}{2(x^2 + 1)} + \frac{3}{2}\,\text{arc tan } x + C.$$

Solution. Since $x^2 + 1$ occurs twice as a factor, we assume

$$\frac{2x^3 + x + 3}{(x^2 + 1)^2} = \frac{Ax + B}{(x^2 + 1)^2} + \frac{Cx + D}{x^2 + 1}.$$

Clearing of fractions,

$$2x^3 + x + 3 = Ax + B + (Cx + D)(x^2 + 1).$$

Equating the coefficients of like powers of x and solving, we get

$$A = -1, \quad B = 3, \quad C = 2, \quad D = 0.$$

Hence $\displaystyle \int \frac{2x^3 + x + 3}{(x^2 + 1)^2}\, dx = \int \frac{-x + 3}{(x^2 + 1)^2}\, dx + \int \frac{2x\, dx}{x^2 + 1}$

$$= \ln(x^2 + 1) - \int \frac{x\, dx}{(x^2 + 1)^2} + 3\int \frac{dx}{(x^2 + 1)^2}.$$

The first of these two integrals is worked out by the Power Formula, **(IV)**, the second by (2) above, with $u = x$, $a = 1$, $n = 2$. Thus we obtain

$$\int \frac{2x^3 + x + 3}{(x^2 + 1)^2}\, dx = \ln(x^2 + 1) + \frac{1}{2(x^2 + 1)} + \frac{3}{2}\left[\frac{x}{x^2 + 1} + \text{arc tan } x\right] + C.$$

Reducing, we have the answer.

The result of the preceding discussion leads to the following statement.

The integral of every rational fraction whose denominator can be broken up into real linear and quadratic factors can be found, and is expressible in terms of algebraic, logarithmic, and inverse trigonometric functions, that is, in terms of the elementary functions.

PROBLEMS

Work out the following integrals.

1. $\displaystyle \int \frac{dx}{(x^2 + x)(x^2 + 1)}.$ 2. $\displaystyle \int \frac{x\, dx}{x^3 - 1}.$ 3. $\displaystyle \int \frac{dx}{x^3 + 1}.$

4. $\displaystyle \int \frac{dx}{x^3(x^2 + 4)}.$ 5. $\displaystyle \int \frac{(x^3 + x - 1)dx}{(x^2 + 2)^2}.$ 6. $\displaystyle \int \frac{(4x + 3)dx}{(4x^2 + 3)^3}.$

7. $\displaystyle \int \frac{dx}{(x^2 + 1)(x^2 - 1)}.$ 8. $\displaystyle \int \frac{x\, dx}{(x^2 + 1)(x^2 - 1)}.$ 9. $\displaystyle \int \frac{2x\, dx}{(1 + x)(1 + x^2)^2}.$

10. $\displaystyle \int_0^2 \frac{x\, dx}{(x + 1)(x^2 + 4)}.$ 11. $\displaystyle \int_0^{\frac{1}{2}} \frac{x^2\, dx}{1 - x^4}.$ 12. $\displaystyle \int_0^1 \frac{(2x^2 + x + 3)dx}{(x + 1)(x^2 + 1)}.$

CHAPTER XVII

POLAR COÖRDINATES AND APPLICATIONS

211. Polar coördinates. In this chapter we shall consider a second method of determining points of the plane by pairs of real numbers. We suppose given a fixed point O, called the *pole*, and a fixed line OA, passing through O, called the *polar axis*. Then any point P determines a length $OP = \rho$ (Greek letter "rho") and an angle $AOP = \theta$. The numbers ρ and θ are called the *polar coördinates* of P; ρ is called the *radius vector* and θ the *vectorial angle*. The vectorial angle θ may be positive or negative as in trigonometry. That is, when θ is positive (as in the figure), the angle formed by OA and OP is laid off from OA around to OP in a counterclockwise direction, and when negative, in a clockwise direction. The radius vector is positive if P lies on the terminal side of θ, and

negative if P lies on that line produced through the pole O. Thus in Fig. 211.1 the radius vector of P is positive, and that of P' is negative.

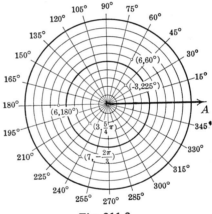

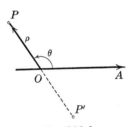

Fig. 211.1   Fig. 211.2

It is evident that every pair of real numbers (ρ, θ) determines a single point, which may be plotted by the

Rule for plotting a point whose polar coördinates (ρ, θ) are given. *Construct the terminal side of the vectorial angle θ, as in trigonometry. If the radius vector is positive, lay off a length $OP = \rho$ on the terminal side of θ; if negative, produce the terminal side through the pole and lay off OP equal to the numerical value of ρ. Then P is the required point.*

In Fig. 211.2 the points $(6, 60°)$, $(3, \frac{5}{4}\pi)$, $(-3, 225°)$, $(6, 180°)$, and $(7, -\frac{2}{3}\pi)$ are plotted.

351

A given point may be represented by an indefinite number of pairs of coördinates (ρ, θ). Thus, if $OB = \rho$, the coördinates of B (Fig. 211.3) may be written in any one of the forms (ρ, θ), $(-\rho, \theta + 180°)$, $(\rho, \theta + 360°)$, $(-\rho, \theta - 180°)$, etc. When possible, we keep θ within the limits $\pm \pi$ or 0 and 2π.

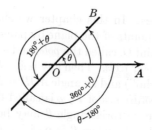

Fig. 211.3

212. Plotting polar equations. The *locus of an equation* in polar coördinates ρ and θ is a figure which contains all points whose coördinates satisfy the equation and no other points. Examples will now be given.

Example 1. Plot the locus of

$$\rho = 10 \cos \theta. \tag{1}$$

Solution. Assume values for θ, calculate the corresponding values of ρ, arrange the results in a table, plot the points, and draw a smooth curve through them. Since $\cos(-\theta) = \cos \theta$, the curve is symmetric to the polar axis OA. We note that the last point in the table is the same as the first point. Since $\cos(\theta + 180°) = -\cos \theta$, it is not necessary to take θ greater than 180°; that is, the complete locus is obtained by taking all values of θ such that $0 \leqq \theta < 180°$.

The maximum value of ρ is 10, and the curve is therefore closed. The curve is a circle (Art. 214).

$\rho = 10 \cos \theta$			
θ	ρ	θ	ρ
0	10	105°	-2.6
15°	9.7	120°	-5
30°	8.7	135°	-7.1
45°	7.1	150°	-8.7
60°	5	165°	-9.7
75°	2.6	180°	-10
90°	0		

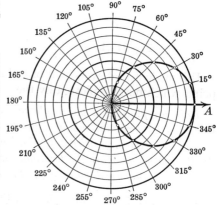

Fig. 212.1

Example 2. Plot the locus of

$$\rho = \frac{2}{1 + \cos \theta}.\qquad(2)$$

Solution. The calculation is shown in the table below.

θ	$\cos \theta$	$1 + \cos \theta$	ρ	θ	$\cos \theta$	$1 + \cos \theta$	ρ
			$\rho = 2 \div (1 + \cos \theta)$				
0	1	2	1	105°	− 0.259	0.741	2.7
15°	0.966	1.966	1.02	120°	− 0.500	0.500	4
30°	0.866	1.866	1.07	135°	− 0.707	0.293	6.7
45°	0.707	1.707	1.2	150°	− 0.866	0.134	15
60°	0.500	1.500	1.3	165°	− 0.966	0.034	50
75°	0.259	1.259	1.6	180°	− 1	0	∞
90°	0	1	2				

Here again the locus is symmetric with respect to OA, since $\cos(-\theta) = \cos\theta$. When $\theta \longrightarrow 180°$, $1 + \cos\theta \longrightarrow 0$, and ρ becomes infinite. Hence the curve is not closed but extends to infinity. The curve is a parabola.

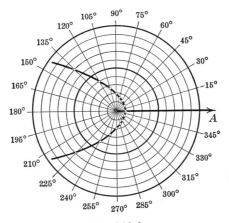

Fig. 212.2

Discussion of a polar equation. It is usually easy (see the above examples) to determine

1. *The symmetry of a curve with respect to the polar axis, the pole, or the line $\theta = 90°$.* (See Problem 4 below.)

2. *The extent of the curve (closed or not closed).*

Before plotting polar equations, the student should establish such simple facts as result from a discussion, as illustrated above.

PROBLEMS

Discuss, and plot the locus of, each of the following equations.

1. $\rho = 10$.   **2.** $\theta = 45°$.   **3.** $\rho = 10 \sin \theta$.

4. Show that the following pairs of points are symmetric in the way stated.

a. (ρ, θ) and $(-\rho, \theta)$, as well as (ρ, θ) and $(\rho, \pi + \theta)$, with respect to the pole;

b. (ρ, θ) and $(-\rho, \pi - \theta)$ with respect to the polar axis;

c. (ρ, θ) and $(\rho, \pi - \theta)$, and also (ρ, θ) and $(-\rho, -\theta)$, with respect to the line $\theta = \frac{1}{2} \pi$.

Discuss, and plot the locus of, each of the following equations.

5. $\rho = 5 \sin \theta + 4 \cos \theta$.   **6.** $\rho \cos \theta = 6$.   **7.** $\rho \sin \theta = 4$.

8. The cardioid $\rho = a(1 - \cos \theta)$.

9. The lemniscate $\rho^2 = a^2 \cos 2 \theta$.

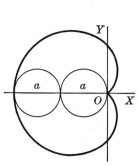

Fig. 212.3. Cardioid   Fig. 212.4. Lemniscate

10. $\rho^2 = 16 \sin 2 \theta$.   **11.** $\rho = a(1 + \sin \theta)$.

12. $\rho^2 \sin 2 \theta = 16$.   **13.** $\rho^2 \cos 2 \theta = 9$.

14. The three-leaved rose $\rho = a \cos 3 \theta$.

15. The four-leaved rose $\rho = a \sin 2 \theta$.

16. $\rho = \dfrac{8}{1 + \sin \theta}$.

17. $\rho = \dfrac{4}{2 - \cos \theta}$.

18. $\rho = \dfrac{10}{1 + \tan \theta}$.

19. The hyperbola $\rho(1 - 2 \cos \theta) = 4$.

20. The limaçon $\rho = a(2 + \cos \theta)$.

21. The limaçon $\rho = a(1 - 2 \cos \theta)$. (See Fig. 212.5.)

Fig. 212.5. Limaçon

$\rho = b - a \cos \theta; \ b < a$

213. Relations between rectangular and polar coördinates. In the figure the point P has the rectangular coördinates $OM = x$, $MP = y$, and polar coördinates $OP = \rho$, $\angle MOP = \theta$. Now $OM = OP \cos \theta$, $MP = OP \sin \theta$. Hence the relations

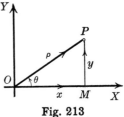

$(A) \qquad x = \rho \cos \theta, \quad y = \rho \sin \theta,$

expressing the rectangular coördinates of P in terms of its polar coördinates. When the rectangular equation of a curve is given, we find its polar equation by using formulas (A).

Fig. 213

Example. Find the polar equation of the equilateral hyperbola whose rectangular equation is $x^2 - y^2 = a^2$.

Solution. Substituting from (A), we have

$$\rho^2 \cos^2 \theta - \rho^2 \sin^2 \theta = a^2,$$

or

$$\rho^2 (\cos^2 \theta - \sin^2 \theta) = a^2,$$

whence

$$\rho^2 \cos 2\theta = a^2,$$

since $\cos^2 \theta - \sin^2 \theta = \cos 2\theta$.

Also, from the figure, we have easily

$(B)\ \rho = \sqrt{x^2 + y^2}, \ \tan \theta = \dfrac{y}{x}, \ \sin \theta = \dfrac{y}{\sqrt{x^2 + y^2}}, \ \cos \theta = \dfrac{x}{\sqrt{x^2 + y^2}}.$

Using (B), the rectangular equation of a curve is readily found from its polar equation.

Example 1. Find the rectangular equation of the circle

$$\rho = 3 \sin \theta + 5 \cos \theta.$$

Solution. Substituting from (B), we have

$$\sqrt{x^2 + y^2} = \frac{3y}{\sqrt{x^2 + y^2}} + \frac{5x}{\sqrt{x^2 + y^2}}, \quad \text{or} \quad x^2 + y^2 - 5x - 3y = 0.$$

Example 2. Show that the locus of

$$\rho = \frac{p}{1 - \cos \theta}$$

is a parabola.

Solution. Clearing of fractions, we get

$$\rho - \rho \cos \theta = p.$$

Substituting from (B) and (A), we have

$$\sqrt{x^2 + y^2} - x = p.$$

Transposing and squaring,

$$x^2 + y^2 = (p + x)^2.$$

Reducing, the result is $\quad\quad y^2 = 2\,px + p^2,$
a parabola by Art. 132.

214. Applications. Straight line and circle.

Theorem. *The general equation of a straight line in polar coördinates is*

$$\rho(A \cos \theta + B \sin \theta) + C = 0, \tag{1}$$

where A, B, and C are arbitrary constants.

Proof. The general equation of the straight line in rectangular coördinates is

$$Ax + By + C = 0.$$

By substitution, using **(A)**, we obtain (1).

Special cases of (1) are $\rho \cos \theta = a$, $\rho \sin \theta = b$, which result, respectively, when $B = 0$ or $A = 0$, that is, when the line is parallel to OY or OX.

By substitution, using **(A)**, in the general equation of the circle,

$$x^2 + y^2 + Dx + Ey + F = 0,$$

we may prove the

Theorem. *The general equation of a circle in polar coördinates is*

$$\rho^2 + \rho(D \cos \theta + E \sin \theta) + F = 0, \tag{2}$$

where D, E, and F are arbitrary constants.

If the pole is on the circumference and the polar axis is a diameter, the equation is

$$\rho = 2\,r \cos \theta, \tag{3}$$

where r is the radius of the circle.

Similarly, if the circle touches the polar axis at the pole, the equation is $\rho = 2\,r \sin \theta$. These results may easily be derived directly from Fig. 214.1 and Fig. 214.2.

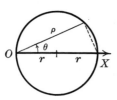

Fig. 214.1

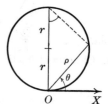

Fig. 214.2

215. Points of intersection of polar curves. The points of intersection of two curves whose polar equations are given are determined by solving the equations for ρ and θ.

Example. Find the points of intersection of

$$\rho = 1 + \cos \theta. \quad \text{(Cardioid)} \qquad (1)$$

$$\rho = \frac{1}{2(1 - \cos \theta)}. \quad \text{(Parabola)} \qquad (2)$$

Solution. Eliminating ρ,

$$1 + \cos \theta = \frac{1}{2(1 - \cos \theta)},$$

or $1 - \cos^2 \theta = \frac{1}{2}.$

$$\cos \theta = \pm \tfrac{1}{2} \sqrt{2}.$$

Therefore $\theta = \pm 45°, \pm 135°.$

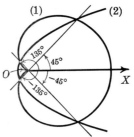

Substituting these values in either equation, we obtain the following four points.

$$\left(1 + \tfrac{1}{2}\sqrt{2}, \pm 45°\right), \qquad \left(1 - \tfrac{1}{2}\sqrt{2}, \pm 135°\right).$$

Fig. 215

The results check in the figure. The locus of (1) is a cardioid; of (2) a parabola. (See Fig. 215.)

PROBLEMS

Find polar equations for the following. Draw the locus.

1. Circle $x^2 + y^2 + 4x = 0.$

2. Circle $x^2 + y^2 + 4x + 6y = 0.$

3. Straight line $3x + 4y = 5.$

4. Parabola $y^2 - 4x - 4 = 0.$

5. Ellipse $3x^2 + 4y^2 - 6x - 9 = 0.$

6. Hyperbola $2xy = a^2.$

7. Hyperbola $2x^2 - y^2 = a^2.$

Find rectangular equations for the following. Draw the locus.

8. Circle $\rho = a \sin \theta + b \cos \theta.$

9. Lemniscate $\rho^2 = a^2 \cos 2\theta.$

10. Parabola $\rho(1 - \cos \theta) = 4.$

11. Ellipse $\rho(2 - \cos \theta) = 3.$

12. Hyperbola $\rho(1 - 2\cos \theta) = 4.$

13. Cardioid $\rho = a(1 - \cos \theta).$

14. $\rho(1 + \tan \theta) = a.$

15. Straight line $\rho \sin \left(\theta + \tfrac{1}{4}\pi\right) = 2.$

16. Straight line $\rho \cos(\theta + \alpha) = p$ (α and p constants).

17. Circle $\rho^2 + 2\rho \cos \theta = 3$.

18. Chords are drawn from a point on a circle. Find the mean value of their lengths. (Use (3), Art. 214.)

Find the points of intersection of the following pairs of curves, and check by drawing the figure.

19. $4 \rho \cos \theta = 3$,
  $2 \rho = 3$.

20. $4 \rho \cos \theta = 3$,
  $\rho = 3 \cos \theta$.

21. $2 \rho = 3$,
  $\rho = 3 \sin \theta$.

22. $\rho = \sqrt{3}$,
  $\rho = 2 \sin \theta$.

23. $\rho = \cos \theta$,
  $4 \rho = 3 \sec \theta$.

24. $\rho = 1 + \cos \theta$,
  $2 \rho = 3$.

25. $2 \rho = \sec^2 \frac{1}{2} \theta$,
  $\rho = 2$.

26. $3 \rho = 4 \cos \theta$,
  $2 \rho \cos^2 \frac{1}{2} \theta = 1$.

27. Find the points of intersection of

$$\rho = \tfrac{4}{3} \cos \theta. \quad \text{(Circle)}$$
$$\rho(1 + \cos \theta) = 1. \quad \text{(Parabola)}$$

28. Transform the equations in the preceding problem into rectangular coördinates, and from these find the rectangular coördinates of the points of intersection.

216. Loci using polar coördinates. When the required locus is traced by the end point of a line of variable length which rotates about its other extremity, assumed fixed, polar coördinates may be employed to advantage.

Example. The equation of a circle is $\rho = a \cos \theta$, where a is a positive constant, and B is any point on the circle. On the radius vector OB a point P is taken so that $BP = a$. Find the locus of P.

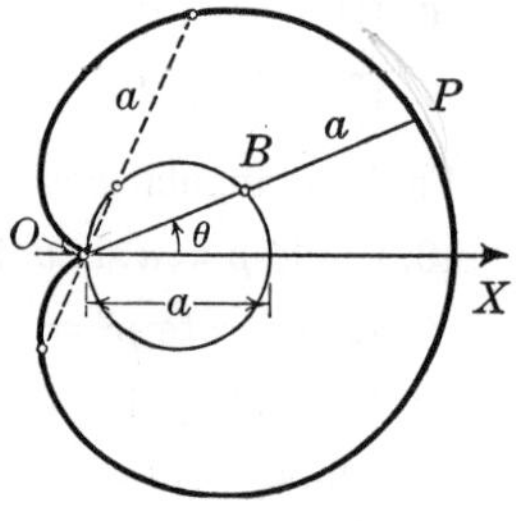

Solution. Denote the coördinates of B by (ρ_1, θ_1) and of P by (ρ, θ). From the statement of the problem, $\theta = \theta_1$, $OB = \rho_1 = a \cos \theta$ and $OP = OB + a$. Hence

$$\rho = a \cos \theta + a.$$

Fig. 216.1

This is the equation of a cardioid. (Compare Problem 8, page 354.) Note that θ must vary from $0°$ to $360°$ to plot the curve. The point B traverses the circle twice. One position on the first circuit is $(\frac{1}{2} a, 60°)$ and the corresponding coördinates of P are $(\frac{3}{2} a, 60°)$. The same position of B on the second circuit is $(-\frac{1}{2} a, 240°)$ and the corresponding coördinates of P are $(\frac{1}{2} a, 240°)$.

PROBLEMS

1. In Fig. 216.2 the line OP of variable length turns about the fixed extremity O. Find the polar equation of the curve traced by P and plot, under the given condition.

a. $AM = MP$.

b. $MP = OA = 4$.

c. $MP = OA + AM$.

d. $OP = 2\,AM$.

e. $\dfrac{OP}{OA} = \dfrac{AM}{OM}$.

f. $OP \cdot OM = 32$.

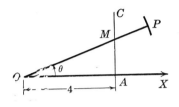

Fig. 216.2   Fig. 216.3

2. In Fig. 216.3, the line FP, of variable length, turns about the fixed extremity F so that the ratio of the distances of P from F and the fixed line CD remains constant and equal to e. If F is the pole and FA, the perpendicular drawn through F to CD, is the polar axis, show that the polar equation of the locus of P is $\rho(1 - e \cos \theta) = ep$, if $HF = p$.

3. In the example above (Fig. 216.1) suppose that $BP = b$, where b has no relation to the diameter a. The locus of P is a limaçon (compare Fig. 212.5). Find its equation and show that the curve has an inside loop if $b < a$.

4. *Spirals.* Three curves bearing this designation are defined as follows. Values of θ must be in radians. The figures are drawn for $\theta > 0$ on the assumption that the constants of proportionality are positive. Plot the curves for $\theta \leqq 0$.

a. *Spiral of Archimedes* (Fig. 216.4). The radius vector at any point is directly proportional to the vectorial angle, giving $\rho = a\theta$.

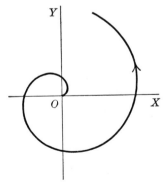

Fig. 216.4. Spiral of Archimedes

b. *Hyperbolic* or *Reciprocal Spiral* (Fig. 216.5). The radius vector at any point is inversely proportional to the vectorial angle, giving $\rho\theta = a$.

c. *Logarithmic* or *Equiangular Spiral* (Fig. 216.6). The natural logarithm of the radius vector at any point is directly proportional to the vectorial angle, giving $\ln \rho = a\theta$ or $\rho = e^{a\theta}$. Compare Problem 10, page 364.

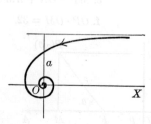

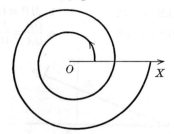

Fig. 216.5. Reciprocal spiral   **Fig. 216.6. Logarithmic or equiangular spiral**

217. Angle between the radius vector and the tangent line. Let the equation of a curve in polar coördinates ρ, θ be

$$\rho = f(\theta). \tag{1}$$

We proceed to prove the

Theorem. *If β is the angle between the radius vector OP and the tangent line at P, then*

(C) $$\tan \beta = \frac{\rho}{\rho'},$$

where $$\rho' = \frac{d\rho}{d\theta}.$$

Proof. The proof is based on Fig. 217.1. It is supposed that θ and $\Delta\theta$ are positive, that the positive direction along the radius vector is from O to P, the positive direction along the secant is from P to Q, and the positive direction along the tangent is from T to P. It is also supposed that $f'(\theta) > 0$, which implies $OQ > OR$.

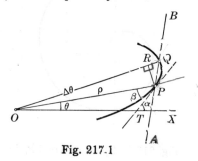

Fig. 217.1

Through P and a point $Q(\rho + \Delta\rho, \theta + \Delta\theta)$ on the curve near P draw the secant line AB. Draw PR perpendicular to OQ.

Then we have $OQ = \rho + \Delta\rho$, angle $POQ = \Delta\theta$, $PR = \rho \sin \Delta\theta$, and $OR = \rho \cos \Delta\theta$. Also,

$$\tan PQR = \frac{PR}{RQ} = \frac{PR}{OQ - OR} = \frac{\rho \sin \Delta\theta}{\rho + \Delta\rho - \rho \cos \Delta\theta}. \tag{2}$$

Denote by β the angle between the radius vector OP and the tangent line PT. If we now let $\Delta\theta$ approach zero as a limit, then

a. the point Q will approach P;

b. the secant AB will turn about P and approach the tangent line PT as a limiting position; and

c. the angle PQR will approach β as a limit.

Hence

$$\tan \beta = \lim_{\Delta\theta \to 0} \frac{\rho \sin \Delta\theta}{\rho + \Delta\rho - \rho \cos \Delta\theta}. \tag{3}$$

To get this fraction in a form so that the theorems on limits will apply, we transform it as shown in the following equations.

$$\frac{\rho \sin \Delta\theta}{\rho(1 - \cos \Delta\theta) + \Delta\rho} = \frac{\rho \sin \Delta\theta}{2\,\rho \sin^2 \dfrac{\Delta\theta}{2} + \Delta\rho}$$

[Since from (5), Art 296, $\rho - \rho \cos \Delta\theta = \rho(1 - \cos \Delta\theta) = 2\,\rho \sin^2 \tfrac{1}{2} \Delta\theta.$]

$$= \frac{\rho \cdot \dfrac{\sin \Delta\theta}{\Delta\theta}}{\rho \sin \dfrac{\Delta\theta}{2} \cdot \dfrac{\sin \dfrac{\Delta\theta}{2}}{\dfrac{\Delta\theta}{2}} + \dfrac{\Delta\rho}{\Delta\theta}}.$$

[Dividing both numerator and denominator by $\Delta\theta$ and factoring.]

When $\Delta\theta \longrightarrow 0$, then, by Art. 177,

$$\lim \frac{\sin \Delta\theta}{\Delta\theta} = 1, \quad \text{and} \quad \lim \frac{\sin \dfrac{\Delta\theta}{2}}{\dfrac{\Delta\theta}{2}} = 1.$$

Also, $\lim \sin \dfrac{\Delta\theta}{2} = 0, \quad \lim \dfrac{\Delta\rho}{\Delta\theta} = \dfrac{d\rho}{d\theta} = \rho'.$

Hence the limits of numerator and denominator are, respectively, ρ and ρ'. Thus **(C)** is proved.

The proof of **(C)** fails if $\rho = 0$. But if $\rho = 0$ when $\theta = \theta_0$, it is easy to see that the polar equation of the tangent line at $P(0, \theta_0)$ is $\theta = \theta_0$.

From the triangle OPT we get
$$\alpha = \theta + \beta. \tag{4}$$

Having found α, we may then find $\tan \alpha$, the slope of the tangent to the curve at P. Or since, from (4),
$$\tan \alpha = \tan (\theta + \beta) = \frac{\tan \theta + \tan \beta}{1 - \tan \theta \tan \beta}, \tag{5}$$

we may calculate $\tan \beta$ from (C), substitute, and thus find $\tan \alpha$.

NOTE. In Fig. 217.1 the angle α is the inclination of the tangent TP, as defined in Art. 13. Here, however, we do not limit α to values such that $0 \le \alpha \le 180°$. As θ increases and the point P traces the curve, the tangent line revolves continuously. See the example below. It is not feasible to give definitions and rigorous statements to cover all cases. Formula (C) gives a value for $\tan \beta$ which corresponds to the acute (obtuse) angle between the radius vector and the tangent line if $\tan \beta > 0$ ($\tan \beta < 0$).

In each problem the relations between the angles β, α, and θ should be determined by examining the signs of their trigonometric functions and drawing a figure.

Example 1. Find β and α in the cardioid $\rho = a(1 - \cos \theta)$. Also find the slope at $\theta = \pi/6$.

Solution. $\dfrac{d\rho}{d\theta} = \rho' = a \sin \theta$. Substituting in (C) gives
$$\tan \beta = \frac{\rho}{\rho'} = \frac{a(1 - \cos \theta)}{a \sin \theta}$$
$$= \frac{2 a \sin^2 \tfrac{1}{2} \theta}{2 a \sin \tfrac{1}{2} \theta \cos \tfrac{1}{2} \theta}$$
$$= \tan \tfrac{1}{2} \theta. \qquad \text{By (2), Art. 296}$$

Therefore $\beta = \tfrac{1}{2} \theta$. In Fig. 217.2,
$$\angle OPT = \tfrac{1}{2} \angle XOP.$$

Substituting in (4), $\alpha = \theta + \tfrac{1}{2} \theta = \tfrac{3}{2} \theta$.

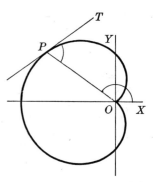

Fig. 217.2

The curve has a cusp at the origin, where the limiting position of the tangent line, as $\theta \longrightarrow 0$, is OX. As θ increases, the tangent line revolves in a counterclockwise direction. Thus, when $\theta = 90°$, $\alpha = 135°$, and the positive direction along the tangent line is upward; when $\theta = 180°$, $\alpha = 270°$, and the positive direction along the tangent line is downward.

To find the angle of intersection ϕ of two curves, C and C' (Fig. 217.3), whose equations are given in polar coördinates, we may proceed as follows.
$$\angle TPT' = \angle OPT' - \angle OPT,$$
or
$$\phi = \beta' - \beta. \tag{6}$$

Or, also, from (6),

$$\tan \phi = \frac{\tan \beta' - \tan \beta}{1 + \tan \beta' \tan \beta}, \tag{7}$$

where $\tan \beta'$ and $\tan \beta$ are calculated by (C) from the two curves and evaluated for the point of intersection.

The matter of directions may be complicated. In any case, $|\tan \phi|$ from (7) gives the acute angle between the tangents to the curves.

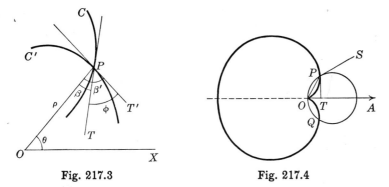

Fig. 217.3   Fig. 217.4

Example 2. Find the angles of intersection of the curves (Fig. 217.4)

$$\rho = 5(1 - \cos \theta), \quad \text{(Cardioid)} \tag{8}$$
$$\rho = 5 \cos \theta. \quad \text{(Circle)} \tag{9}$$

Solution. From the figure, $\beta' = \angle OPS$ and $\beta = \angle OPT$.

To find the points of intersection we have, by eliminating ρ from equations (8) and (9),

$$\cos \theta = 1 - \cos \theta,$$

whence $\cos \theta = \tfrac{1}{2},$

and $\theta = \pm 60° + n(360°).$

Equations (8) and (9) and Fig. 217.4 show that the points $P(\tfrac{5}{2}, 60°)$ and $Q(\tfrac{5}{2}, -60°)$ lie on both curves. For the cardioid we have, from Example 1,

$$\tan \beta = \tan \tfrac{1}{2} \theta.$$

For the circle we have, from (C),

$$\tan \beta' = \frac{5 \cos \theta}{-5 \sin \theta} = - \operatorname{ctn} \theta.$$

At P, $\theta = 60°$ and $\tan \beta = \tan 30° = \dfrac{1}{\sqrt{3}},$

$$\tan \beta' = - \operatorname{ctn} 60° = - \frac{1}{\sqrt{3}}.$$

Substitution in (7) gives

$$\tan \phi = \frac{-\dfrac{1}{\sqrt{3}} - \dfrac{1}{\sqrt{3}}}{1 - \frac{1}{3}} = -\sqrt{3}.$$

Hence $|\tan \phi| = \sqrt{3}$ and the acute angle of intersection at P is $60°$.

By symmetry, the acute angle of intersection at Q is also $60°$.

Fig. 217.4 shows that the origin lies on both curves. This fact does not appear from the preceding analysis because, when $\rho = 0$, (8) and (9) are not satisfied by the same values of θ. Thus, when $\rho = 0$, (8) is satisfied by $\theta = 0$ and (9) is satisfied by $\theta = 90°$. Formula (C) fails, but, as remarked above, the polar equation of the tangent to the cardioid at the origin is $\theta = 0$, and of the tangent to the circle is $\theta = 90°$. The angle of intersection is $90°$.

PROBLEMS

Find the angle β for the following curves for the given values of θ. In each case plot the curve and construct β.

1. $\rho = 4(1 + \sin \theta)$; $\theta = 0$, $90°$, $180°$.

2. $\rho(1 + \sin \theta) = 4$; $\theta = 0$, $90°$, $180°$.

3. $\rho = a \sin 2\,\theta$; $\theta = 30°$, $90°$, $150°$.

4. $\rho = a \sin \theta$; $\theta = 30°$, $90°$, $120°$.

5. $\rho = a \cos 2\,\theta$; $\theta = 30°$, $60°$, $120°$.

6. Find the angle of intersection for the following curves.

a. $4 \rho \cos \theta = 3$, $\rho = 3 \cos \theta$.
b. $\rho = 1 + \cos \theta$, $2 \rho = 3$.
c. $3 \rho = 10$, $\rho(2 - \sin \theta) = 5$.
d. $\rho(1 - \cos \theta) = 4$, $\rho(2 + \cos \theta) = 20$.

7. Show that the parabolas $\rho(1 + \cos \theta) = a$, $\rho(1 - \cos \theta) = b$ intersect at right angles.

8. In the cardioid $\rho = a(1 + \cos \theta)$ show that $\beta = \frac{1}{2}\,\pi + \frac{1}{2}\,\theta$.

9. Find the rectangular equation of the tangent line to $\rho = a(1 + \cos \theta)$ at the point where $\theta = 60°$.

10. Show that β is constant for the logarithmic spiral $\rho = be^{a\theta}$ (see Problem 4, c, page 360). This curve is also called the equiangular spiral.

11. Find the mean value of the radius vector of the cardioid $\rho = a(1 - \cos \theta)$ with respect to θ.

218. Areas using polar coördinates. Let it be required to find the area bounded by a polar curve and two of its radii vectors. Assume the equation of the curve to be

$$\rho = f(\theta).$$

and in the figure let OP_1 and OD be the two radii. Denote by α and β the angles which these radii make with the polar axis. Suppose that $f(\theta)$ is continuous and nowhere negative in the interval

$$\alpha \leqq \theta \leqq \beta.$$

Apply the fundamental theorem, Art. 92.

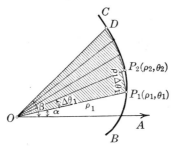

Fig. 218.1

The required area is clearly the limit of the sum of circular sectors constructed as in Fig. 218.1. Let the central angles of the successive sectors be $\Delta\theta_1$, $\Delta\theta_2$, etc., and their radii ρ_1, ρ_2, etc. Then the sum of the areas of the sectors is

$$\tfrac{1}{2}\,\rho_1{}^2\,\Delta\theta_1 + \tfrac{1}{2}\,\rho_2{}^2\,\Delta\theta_2 + \cdots + \tfrac{1}{2}\,\rho_n{}^2\,\Delta\theta_n = \sum_{i=1}^{n} \tfrac{1}{2}\,\rho_i{}^2\,\Delta\theta_i,$$

for the area of a circular sector is half its radius times its arc. Hence the area of the first sector is $\tfrac{1}{2}\,\rho_1 \cdot \rho_1\,\Delta\theta_1 = \tfrac{1}{2}\,\rho_1{}^2\,\Delta\theta_1$; etc.

Applying the fundamental theorem,

$$\lim_{n \to \infty} \sum_{i=1}^{n} \tfrac{1}{2}\,\rho_i{}^2\,\Delta\theta_i = \int_{\alpha}^{\beta} \tfrac{1}{2}\,\rho^2\,d\theta.$$

Hence the area swept over by the radius vector of the curve in moving from the position OP_1 to the position OD is given by the formula

(D) $$\text{Area} = \tfrac{1}{2}\int_{\alpha}^{\beta} \rho^2\,d\theta,$$

the value of ρ in terms of θ being substituted from the equation of the curve.

Compare this solution with the corresponding problem in rectangular coördinates, Art. 91. The element of area is now a circular sector of radius ρ and central angle $\Delta\theta$ (or $d\theta$) and is, in magnitude, equal to $\tfrac{1}{2}\,\rho^2\,d\theta$.

The area given by **(D)** is bounded by the given curve and the radial lines $\theta = \alpha$, $\theta = \beta$. The limits should be chosen so that the radius vector OP of the tracing point P of the curve rotates from $\theta = \alpha$ to $\theta = \beta$ *counterclockwise.*

REMARK. By using Art. 213, the polar equation of a curve can be found from the rectangular equation, and vice versa. If the polar equation is the one given, it may be easier to solve a problem in areas by means of the rectangular equation (see Problem 14, below). Conversely, if the rectangular equation is given, the solution of the given problem may be simpler by the use of polar coördinates. The convenience of selecting the kind of coördinates to be used should be kept in mind.

Example. Find the entire area of the lemniscate $\rho^2 = a^2 \cos 2\,\theta$.

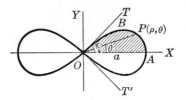

Fig. 218.2

Solution. Since the figure is symmetric with respect to both OX and OY, the whole area $= 4 \times$ area of OAB (the shaded area in Fig. 218.2). Since $\rho = 0$ when $\theta = \frac{1}{4}\pi$, we see that if θ varies from 0 at A to $\frac{1}{4}\pi$ at O, the radius vector OP sweeps over the area OAB. Hence, substituting in **(D)**, $\rho^2 = a^2 \cos 2\,\theta$ and $\alpha = 0$, $\beta = \frac{1}{4}\pi$,

$$\text{Entire area} = 4 \times \text{area } OAB = 2\,a^2 \int_0^{\frac{\pi}{4}} \cos 2\,\theta \, d\theta = a^2.$$

That is, the total area of the two loops equals the area of a square constructed on OA as one side.

PROBLEMS

Work out the following areas. Draw the figure and construct an element of the area in each case. The equation of the boundary curve and the values of θ for the bounding radii are given.

1. Circle $\rho = a \cos \theta$, between $\theta = 0$ and $\theta = 60°$.

2. Cardioid $\rho = a(1 - \cos \theta)$, entire area.

3. One loop of the three-leaved rose $\rho = a \cos 3\,\theta$.

4. Parabola $\rho = \dfrac{a}{1 + \cos \theta}$, between $\theta = 0$ and $\theta = 120°$.

5. Straight line $\rho = a \sec \theta$, between $\theta = 0$ and $\theta = 60°$.

6. $\rho = 2 + \tan \theta$ between $\theta = 0$ and $\theta = 45°$.

7. Limaçon $\rho = a(2 + \cos \theta)$, entire area.

8. Hyperbola $\rho^2 \cos 2\,\theta = a^2$, between $\theta = 0$ and $\theta = 30°$.

9. Three-leaved rose $\rho = a \sin 3\,\theta$, one loop.

10. Limaçon $\rho = a(1 + 2 \sin \theta)$, between $\theta = -\frac{1}{6}\,\pi$ and $\theta = \frac{7}{6}\,\pi$.

11. In Problem 10, find the area of the inside loop.

12. Find the area between the curves $2\,\rho(1 + \cos \theta) = 15$ and $\rho = 10 \cos \theta$.

13. Find the area swept over by the radius vector of the logarithmic spiral $\rho = be^{a\theta}$ in turning counterclockwise through a right angle from $\theta = 0$.

14. Find the area bounded by the ellipse $\rho(2 - \cos \theta) = 4$ and the lines $\theta = 0$, $\theta = \frac{1}{2}\,\pi$. (Use the rectangular equation.)

219. Differential and length of the arc in polar coördinates. From the relations

$$x = \rho \cos \theta, \quad y = \rho \sin \theta, \tag{1}$$

between the rectangular and polar coördinates of a point, we obtain, by differentiating,

$$dx = \cos \theta\, d\rho - \rho \sin \theta\, d\theta, \quad dy = \sin \theta\, d\rho + \rho \cos \theta\, d\theta. \tag{2}$$

In Art. 95, we had, in rectangular coördinates,

$$(ds)^2 = (dx)^2 + (dy)^2.$$

Squaring each of equations (2), adding them, reducing, and extracting the square root, the result is

(E) $$ds = \sqrt{(d\rho)^2 + \rho^2(d\theta)^2}.$$

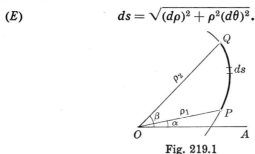

Fig. 219.1

This may be written

(F) $$ds = \left[\rho^2 + \left(\frac{d\rho}{d\theta}\right)^2\right]^{\frac{1}{2}} d\theta.$$

From **(F)** we obtain, by integration, the *length of arc* for a polar curve, namely,

(G) $$s = \int_a^\beta \left[\rho^2 + \left(\frac{d\rho}{d\theta}\right)^2\right]^{\frac{1}{2}} d\theta.$$

To use **(G)**, ρ and $d\rho/d\theta$ must be expressed in terms of θ from the polar equation of the curve.

Example. Find the perimeter of the cardioid $\rho = a(1 + \cos \theta)$.

Solution. Here $\dfrac{d\rho}{d\theta} = -a \sin \theta$. If we let θ vary from 0 to π, the point P will generate half of the curve. Substituting in **(G)**,

$$\frac{s}{2} = \int_0^\pi [a^2(1 + \cos \theta)^2 + a^2 \sin^2 \theta]^{\frac{1}{2}} \, d\theta$$

$$= a \int_0^\pi (2 + 2 \cos \theta)^{\frac{1}{2}} \, d\theta = 2 \, a \int_0^\pi \cos \tfrac{1}{2} \theta \, d\theta = 4 \, a.$$

Therefore $s = 8 \, a$.

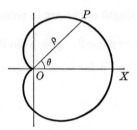

Fig. 219.2

PROBLEMS

Find the length of arc between the points indicated on each of the following curves.

1. Parabola $\rho = \frac{1}{2} \, a \sec^2 \frac{1}{2} \theta$; $\theta = 90°$, $\theta = 150°$.

2. Cissoid $\rho = 2 \, a \tan \theta \sin \theta$; $\theta = 0$, $\theta = 45°$.

3. Find the integral giving the length of arc of $\rho = 2 + \tan \theta$ from $\theta = 0$ to $\theta = 60°$, and evaluate it by the trapezoidal rule and Simpson's rule.

4. Derive the integral for the length of arc of the limaçon $\rho = 5(1 + 2 \sin \theta)$ from $\theta = 0$ to $\theta = 150°$, and evaluate it by Simpson's rule, taking $n = 6$.

5. Find the length of the arc of the spiral of Archimedes, $\rho = a\theta$, from $\theta = 0$ to $\theta = 2 \pi$.

6. Find the length of the arc of the reciprocal spiral, $\rho\theta = a$, from $\theta = \frac{1}{2} \pi$ to $\theta = 2 \pi$.

7. Given the equation $\rho = a \sin^3 \frac{1}{3} \theta$.

a. Show that the locus is symmetric with respect to the line $\theta = \frac{1}{2} \pi$.
b. Show that $\theta = \theta_0$ and $\theta = 3 \pi + \theta_0$ give the same point on the locus.
c. Show that the element of arc is given by $ds = a \sin^2 \frac{1}{3} \theta \, d\theta$.
d. Plot the locus and show that the length of the entire curve is $\frac{3}{2} \pi a$.

8. In Problem 7, show that $\tan \beta = \tan \frac{1}{3} \theta$. Under what angle does the curve in that problem intersect itself?

CHAPTER XVIII

PARAMETRIC EQUATIONS

220. Parametric equations. Suppose

$$x = f(t), \quad y = \phi(t), \tag{1}$$

where f and ϕ are differentiable in some interval. The variable t is called a *parameter* and each value of t gives a pair of values (x, y), which may be plotted as the rectangular coördinates of a point. The collection of all such points forms, in general, a curve. If values are assigned to t in some special order (usually increasing numerically), the points on the locus are arranged in a particular order.

Example 1. Plot the curve whose parametric equations are

$$x = \tfrac{1}{2} t^2, \quad y = \tfrac{1}{4} t^3. \tag{2}$$

t	x	y
-3	4.5	-6.75
-2	2	-2
-1	0.5	-0.25
0	0	0
1	0.5	0.25
2	2	2
3	4.5	6.75
etc.	etc.	etc.

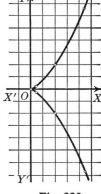

Fig. 220

Solution. The table of values is calculated from the given equations and the points whose coördinates are x and y are plotted (Fig. 220). A smooth curve drawn through these points represents the locus of (2).

Equations (2) imply a relation between x and y which may be found by eliminating the parameter. From the first equation, $t^2 = 2x$ and $t^6 = 8x^3$. From the second equation, $t^3 = 4y$ and $t^6 = 16 y^2$. Hence

$$16 y^2 = 8 x^3$$

and the result is

$$2 y^2 = x^3.$$

The curve is called a *semicubical parabola*.

If the functions in (1) satisfy the conditions for which **(VI)** and **(VIII)**, Art. 62, are valid,

$$(A) \qquad \frac{dy}{dx} = \frac{\dfrac{dy}{dt}}{\dfrac{dx}{dt}} = \frac{\phi'(t)}{f'(t)} = \textit{slope at } P(x, y).$$

369

Example 2. Find the slope of the curve in Example I at the point where $t = 1$.

Solution. From (2)

$$\frac{dx}{dt} = t,$$

$$\frac{dy}{dt} = \tfrac{3}{4} t^2,$$

and hence

$$\frac{dy}{dx} = \tfrac{3}{4} t.$$

When $t = 1$, $\dfrac{dy}{dx} = \dfrac{3}{4}$. Hence the slope at $(0.5, 0.25)$ is $\tfrac{3}{4}$.

Example 3. Find the rectangular equation of the curve whose parametric equations are

$$x = 3 + 4 \cos \theta, \quad y = 3 \sin \theta.$$

Solution. Remembering that $\sin^2 \theta + \cos^2 \theta = 1$, we solve the first equation for $\cos \theta$, the second for $\sin \theta$. This gives

$$\cos \theta = \tfrac{1}{4}(x - 3), \quad \sin \theta = \tfrac{1}{3} y.$$

Squaring and adding, the rectangular equation is

$$\frac{(x - 3)^2}{16} + \frac{y^2}{9} = 1,$$

an ellipse (Art. 141).

221. Various parametric equations for the same curve. Consider the equations

$$x = a \cos \pi t, \quad y = a \sin \pi t, \tag{1}$$

in which the parameter t represents time, measured in seconds. Since $x^2 + y^2 = a^2$ for every value of t, the locus is a circle and we may think of (1) as giving the motion of a point P on the rim of a flywheel. Allowing t to take positive values, beginning with 0, Table 1 indicates that P starts at $A(a, 0)$ and moves around the circle in a counterclockwise direction, completing a circuit in 2 seconds.

Consider the equations

$$x = a \sin 2 \pi t, \quad y = a \cos 2 \pi t. \tag{2}$$

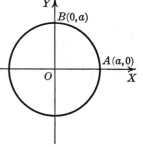

t	x	y
0	a	0
$\tfrac{1}{2}$	0	a
1	$-a$	0
$\tfrac{3}{2}$	0	$-a$
2	a	0

Table 1

t	x	y
0	0	a
$\tfrac{1}{4}$	a	0
$\tfrac{1}{2}$	0	$-a$
$\tfrac{3}{4}$	$-a$	0
1	0	a

Table 2

Fig. 221

The locus of (2) is the same as that of (1). However, Table 2 indicates that the point starts at $B(0, a)$ and moves around the circle in a clockwise direction, completing a circuit in 1 second.

It is now obvious that a given curve can be represented by an unlimited number of different parametric equations. In the rotation problem above each different starting point and each different speed will give different parametric equations of the circle.

In geometric problems it is sometimes advantageous to get parametric equations by assuming an expression for one of the coördinates. Suppose that the given curve is the circle

$$x^2 + y^2 = a^2 \tag{3}$$

and assume the relation

$$y = tx + a. \tag{4}$$

The result of solving (3) and (4) for x and y in terms of t is

$$x = \frac{-2at}{1+t^2}, \quad y = \frac{a(1-t^2)}{1+t^2},$$

which are parametric equations of the given circle.

222. The rectangular equation. A pair of equations of the form

$$x = f(t), \quad y = \phi(t) \tag{1}$$

implies a relation between x and y. If t can be eliminated from (1), there results an equation of the general form

$$F(x, y) = 0. \tag{2}$$

Every point with coördinates x and y obtained from (1) will lie on the locus of (2). However, the converse is not always true. As a simple example consider the equations

$$x = \sin t, \quad y = 2 \sin t. \tag{3}$$

Elimination of t gives

$$y = 2x. \tag{4}$$

The locus of (4) is an unlimited straight line.

Equations (3) show that, for all values of t,

$$|x| \leqq 1, \quad |y| \leqq 2.$$

The locus of (3) is the segment joining the points $A(-1, -2)$ and $B(1, 2)$. Interpreted as straight-line motion, (3) represents a simple harmonic vibration between A and B.

223. Horizontal and vertical tangents. With proper restrictions the slope of a curve given by parametric equations is determined by **(A)** in which we now assume that $f'(t)$ and $\phi'(t)$ are continuous for $t = t_1$.

If $\phi'(t_1) = 0$ and $f'(t_1) \neq 0$, then $dy/dx = 0$ and the curve has a horizontal tangent (parallel to the x-axis) at the point corresponding to $t = t_1$.

If $\phi'(t_1) \neq 0$ and $f'(t_1) = 0$, then $dx/dy = 0$ and the curve has a vertical tangent at the point corresponding to $t = t_1$.

If $\phi'(t_1) = 0$ and $f'(t_1) = 0$, the slope is indeterminate and a special investigation is required.

Example 1. Find the horizontal and vertical tangents for the curve
$$x = t^2 - 4\,t, \quad y = 2\,t + t^2.$$

Solution. $\dfrac{dx}{dt} = 2\,t - 4, \dfrac{dy}{dt} = 2 + 2\,t.$

Horizontal tangent. Setting $dy/dt = 0$, we find $t = -1$, whence $x = 5$, $y = -1$, $dx/dt = -6$. Hence the curve has a horizontal tangent at $M(5, -1)$.

Vertical tangent. Setting $dx/dt = 0$, we find $t = 2$, whence $x = -4$, $y = 8$, $dy/dt = 6$. Hence the curve has a vertical tangent at $N(-4, 8)$.

These tangents are drawn in Fig. 223.1. The locus is a parabola.

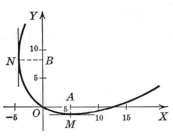
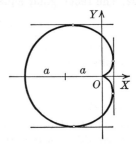

Fig. 223.1   Fig. 223.2

Example 2. Find the points of contact of the horizontal and vertical tangents to the cardioid (see Fig. 223.2)
$$\left. \begin{aligned} x &= a \cos \theta - \tfrac{1}{2} a \cos 2\,\theta - \tfrac{1}{2} a, \\ y &= a \sin \theta - \tfrac{1}{2} a \sin 2\,\theta. \end{aligned} \right\} \tag{1}$$

Solution. $\dfrac{dx}{d\theta} = a(-\sin \theta + \sin 2\,\theta);$

$\dfrac{dy}{d\theta} = a(\cos \theta - \cos 2\,\theta).$

Horizontal tangents. Then $\cos \theta - \cos 2\,\theta = 0.$

Substituting ((5), Art. 296)
$$\cos 2\,\theta = 2\cos^2\theta - 1,$$
and solving, we get $\theta = 0$, 120°, 240°.

Vertical tangents. Then $-\sin\theta + \sin 2\,\theta = 0$. Substituting ((5), Art. 296) $\sin 2\,\theta = 2\sin\theta\cos\theta$, and solving, $\theta = 0$, 60°, 300°.

The common root $\theta = 0$ should be rejected. For both numerator and denominator in (A) become zero, and the slope is indeterminate. From (1), $x = y = 0$ when $\theta = 0$. The point O is called a cusp.

Substituting the other values in (1), the results are:

Horizontal tangents: points of contact $\left(-\frac{3}{4}\,a,\ \pm\frac{3}{4}\,a\sqrt{3}\right)$.

Vertical tangents: points of contact $\left(\frac{1}{4}\,a,\ \pm\frac{1}{4}\,a\sqrt{3}\right)$.

The vertical tangents coincide, forming a "double tangent" line. These results agree with the figure.

PROBLEMS

1. Plot the loci of the following parametric equations, t and θ being variable parameters. Find the rectangular equation in each case. When a value is given for the parameter, find the slope and check in the figure.

a. $x = t - 1,\ y = 4 - t^2,\ t = 3.$ **b.** $x = 2\,t^2 - 2,\ y = t - 3,\ t = -1.$

c. $x = 3\cos\theta,\ y = \sin\theta,\ \theta = 60°.$ **d.** $x = 3\tan\theta,\ y = \sec\theta.$

e. $x = 2\,t,\ y = \dfrac{4}{t},\ t = 3.$ **f.** $x = 2 + \sin\theta,\ y = 2\cos\theta.$

g. $x = \frac{1}{2}\,t^3,\ y = \frac{1}{4}\,t.$ **h.** $x = t^2 - 2\,t,\ y = 1 - t^2,\ t = 2.$

i. $x = \cos\theta,\ y = \cos 2\,\theta.$ **j.** $x = \frac{1}{2}\sin\theta,\ y = \sin 2\,\theta.$

k. $x = 1 - \cos\theta,\ y = \frac{1}{2}\sin\frac{1}{2}\,\theta.$ **l.** $x = 3\,t^2,\ y = 3\,t - t^3,\ t = \frac{1}{2}.$

2. Derive parametric equations for the parabola $x^2 = 2\,py$ when the parameter t is the slope of the line joining $P(x,\,y)$ on the curve and the vertex.

3. Find the rectangular equation for each of the following. Plot, find the slope and its value when there is a given value of the parameter, and construct the tangent line.

a. $x = 2\sin\theta + 3\cos\theta,\ y = \sin\theta.$

b. $x = 2\cos\theta + 1,\ y = \sin\theta + 4\cos\theta.$

c. $x = t - t^2,\ y = t + t^2,\ t = 2.$

d. $x = 3 - 2\,t,\ y = 1 + \dfrac{2}{t},\ t = 4.$

4. Plot, determine the horizontal and vertical tangents, and draw them.

a. $x = t^2 - 1,\ y = 4\,t - t^2.$

b. $x = t^2 - 2\,t,\ y = 1 - t^2.$

c. $x = 2\cos\theta + 1,\ y = 3\sin\theta.$

d. $x = 12\,t - t^3,\ y = t^3 - 3\,t.$

e. $x = 2\sin\theta + \cos\theta,\ y = 2\sin\theta - 1.$

f. $x = 5(\theta - \sin\theta),\ y = 5(1 - \cos\theta).$

5. Show that $x = h + a \cos \theta$, $y = k + b \sin \theta$ are parametric equations for an ellipse with center (h, k). (See Art. 141.)

6. Show that $x = h + a \sec \theta$, $y = k + b \tan \theta$ are parametric equations for a hyperbola with center (h, k). (See Art. 151.)

7. Show that $x = h + a \tan \theta$, $y = k + a \operatorname{ctn} \theta$ are parametric equations for an equilateral hyperbola with center (h, k) and asymptotes respectively parallel to the axes of coördinates. (See Art. 150.)

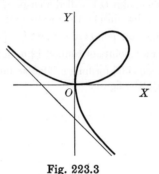

Fig. 223.3

8. Parametric equations for the folium of Descartes (Fig. 223.3) are

$$x = \frac{3\,at}{1 + t^3}, \quad y = \frac{3\,at^2}{1 + t^3}.$$

Find the points of contact of the horizontal and vertical tangents.

9. Find parametric equations by making the substitution given and solving for x and y in terms of the parameter t. Plot the locus from the parametric equations.

a. $y^2 = 4\,x^2 - x^3$; $y = tx$.

b. $x^2y^2 = 4\,x^2 + 9\,y^2$; $x = 3 \sec t$.

c. $x^2y^2 = 4\,y^2 - 9\,x^2$; $x = 2 \sin t$.

d. $y^3 = 10\,x^2 - x^3$; $y = tx$.

e. $x^{\frac{2}{3}} + y^{\frac{2}{3}} = a^{\frac{2}{3}}$; $x = a \sin^3 t$.

f. $x^{\frac{1}{2}} + y^{\frac{1}{2}} = a^{\frac{1}{2}}$; $x = a \cos^4 t$.

224. Curvilinear motion; velocity. When a point describes a plane curve, its coördinates, x and y, are functions of the time t, and the equations

$$x = f(t), \quad y = \phi(t), \tag{1}$$

which are called the *equations of motion* (curvilinear), are also parametric equations of the path of the point.

In Fig. 224, $P(x, y)$ is any position of the point on the path. Also, M is the projection of P on OX, and N its projection on OY. As P describes the path, M moves along OX and N moves along OY. Let

$$v_x = \text{velocity of } M \text{ along } OX,$$
$$v_y = \text{velocity of } N \text{ along } OY.$$

Then, using x and y as defined in (1),

(B) $$v_x = \frac{dx}{dt}, \quad v_y = \frac{dy}{dt},$$

since M and N have rectilinear motion. The velocity of P is compounded from the horizontal velocity v_x and the vertical velocity v_y as in the figure. That is, we lay off lengths to represent v_x and v_y from P, complete

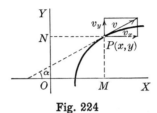

Fig. 224

the "velocity rectangle," and draw the diagonal from P. The length of this diagonal is the speed ($= v$) at P, and its direction from P is the direction of motion. If α is the inclination of the diagonal, obviously

$$\tan \alpha = \frac{v_y}{v_x}. \tag{2}$$

Using (B), above, and (A), Art. 220, we see that

$$\tan \alpha = \frac{dy}{dx}. \tag{3}$$

Hence *at any instant the direction of motion of P is along the tangent line at P.*

If $v_x > 0$ ($v_x < 0$), the point is moving toward the right (left); if $v_y > 0$ ($v_y < 0$), the point is moving upward (downward). Also, for the speed v, we have

(C) $$v = + \sqrt{v_x^2 + v_y^2}.$$

By using (B), we have also

$$v = + \sqrt{\frac{(dx)^2 + (dy)^2}{(dt)^2}} = \frac{ds}{dt}, \tag{4}$$

since $ds = + \sqrt{(dx)^2 + (dy)^2}$ (see Art. 95). The result in (4) should be compared with the formula for velocity in rectilinear motion (Art. 71).

Given the equations of motion for a material point, we may find by (2) and (C) the speed and direction of motion at any instant.

225. Curvilinear motion; acceleration. Referring to Fig. 225.1 and considering now component accelerations of P, let

$$a_x = \text{acceleration of } M \text{ along } OX,$$
$$a_y = \text{acceleration of } N \text{ along } OY.$$

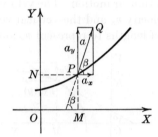

Fig. 225.1

Then, since M and N have rectilinear motion, we have

$$(D) \qquad \begin{cases} a_x = \dfrac{dv_x}{dt} = \dfrac{d^2x}{dt^2}, \\[2mm] a_y = \dfrac{dv_y}{dt} = \dfrac{d^2y}{dt^2}. \end{cases}$$

The acceleration vector PQ is constructed as shown in Fig. 225.1, a_x and a_y being, respectively, its horizontal and vertical components. If β is the inclination of the vector a, then

$$\tan \beta = \frac{a_y}{a_x}, \quad \text{and} \quad a = +\sqrt{a_x{}^2 + a_y{}^2}. \tag{1}$$

The direction of the acceleration vector may be determined from its components as in the case of the velocity.

The figure shows the "acceleration rectangle."

Given the equations of motion of a point, we may find by (D) and (1) its acceleration at any instant, in terms of both direction and magnitude.

From mechanics we know that the acceleration vector drawn from P cannot lie on the convex side of the path. In fact the motion of the point is maintained by a force whose direction is the same as the direction of the acceleration vector. But this force cannot be directed toward the convex side of the path. Thus the statement above is explained.

Example 1. Given the equations of a curvilinear motion

$$x = 3 \cos t, \quad y = 2 \sin t. \tag{2}$$

a. Find the position of the point, its speed, and the direction of motion when $t = 1$ sec.

b. Find the acceleration when $t = 1$ sec.

c. Find the rectangular equation of the path.

Solution. a. Differentiating (2),

$$v_x = -3 \sin t, \tag{3}$$
$$v_y = 2 \cos t.$$

*Substituting $t = 1$ in (2) and (3), the results are

$$x = 1.62, \qquad y = 1.68,$$
$$v_x = -2.52, \quad v_y = 1.08.$$

Then by (2), Art. 224, $\quad \tan \alpha = -\dfrac{1.08}{2.52} = -0.43.$

Hence $\qquad\qquad\qquad \alpha = 157°$ (approximately).

And, by **(C)**, $\qquad\qquad v = \sqrt{7.516} = 2.74.$

If the linear unit is 1 ft. and the unit time is one second, then the point is in the position (1.62, 1.68) when $t = 1$ sec. Its speed is 2.74 ft. per second, and it is moving upward in a direction making an angle of 157° with OX. (See Fig. 225.2.)

b. Differentiating (3), we get

$$a_x = -3 \cos t, \quad a_y = -2 \sin t. \tag{4}$$

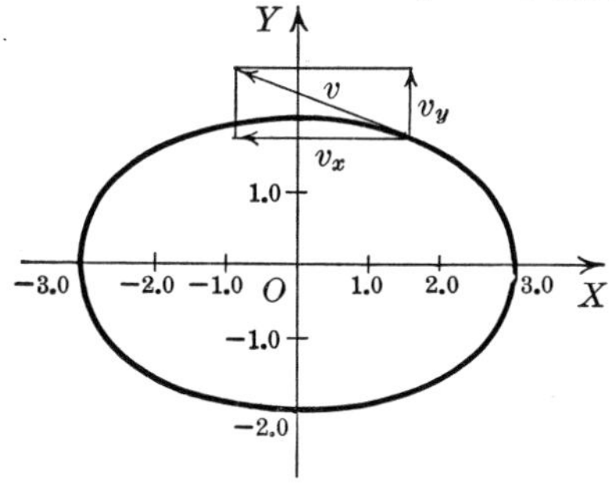

Fig. 225.2

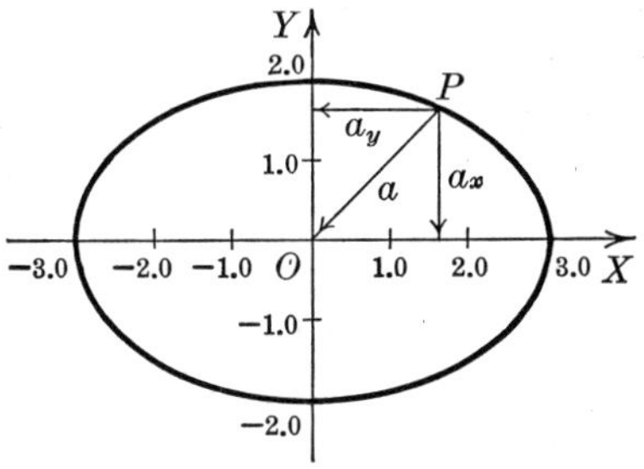

Fig. 225.3

Substituting $t = 1$, we find $a_x = -1.62$, $a_y = -1.68$. Then, by (1),

$$\tan \beta = \tfrac{1.68}{1.62} = 1.05,$$
$$a = +\sqrt{5.447} = 2.33.$$

We notice, by comparing (4) and (2), that $a_x = -x$, $a_y = -y$. Therefore

$$a = \sqrt{x^2 + y^2} = OP.$$
$$\tan \beta = \frac{y}{x}.$$

Hence the acceleration is equal to OP (numerically). It is directed along OP toward O. (Fig. 225.3.)

*In evaluating $x = 3 \cos 1$, it must be remembered that $\cos 1 = \cos$ (1 radian) $= \cos 57° 18'$ (see Table 6, Art. 298).

c. From (2), solving for $\cos t$ and for $\sin t$, and eliminating t by the relation $\cos^2 t + \sin^2 t = 1$, we obtain

$$\frac{x^2}{9} + \frac{y^2}{4} = 1. \quad \text{(An ellipse)}$$

The example illustrates *elliptic harmonic motion*. The elapsed time for a complete description of the ellipse is 2π seconds.

Example 2. The equations of a curvilinear motion are

$$x = \cos t, \quad y = \cos 2t.$$

Find the rectangular equation of the path and discuss the motion.

Solution. Since $\cos 2t = 2\cos^2 t - 1$, the equation of the path is $y = 2x^2 - 1$. Since $|x| \le 1$, $|y| \le 1$, the motion takes place on the arc of the parabola which joins the points $(1, 1)$ and $(-1, 1)$. From the given equations we have, by differentiation,

$$v_x = -\sin t, \quad v_y = -2\sin 2t,$$
$$a_x = -\cos t, \quad a_y = -4\cos 2t.$$

Referring to Art. 185, it is seen that the component of the motion in the x-direction is a simple harmonic vibration of amplitude 1 unit and period 2π units; the component of the motion in the y-direction is a simple harmonic vibration of amplitude 1 unit and period π units. The given equations represent a *composition* of these two vibrations, the result of which is a vibration along the arc of the parabola. The table enables us to follow the motion through one period.

t	x	y	v_x	v_y	a_x	a_y	Pt. on path
0	1	1	0	0	-1	-4	A
$\frac{1}{4}\pi$	0.71	0	-0.71	-2	-0.71	0	B
$\frac{1}{2}\pi$	0	-1	-1	0	0	4	C
$\frac{3}{4}\pi$	-0.71	0	-0.71	2	0.71	0	D
π	-1	1	0	0	1	-4	E
$\frac{5}{4}\pi$	-0.71	0	0.71	-2	0.71	0	D
$\frac{3}{2}\pi$	0	-1	1	0	0	4	C
$\frac{7}{4}\pi$	0.71	0	0.71	2	-0.71	0	B
2π	1	1	0	0	-1	-4	A

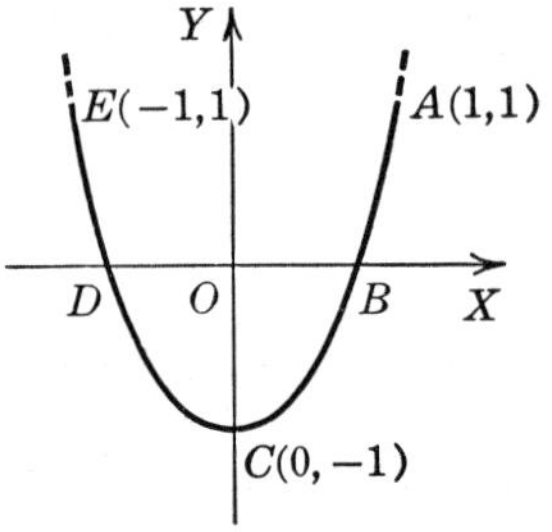

Fig. 225.4

When $t = 0$, the moving point is at $A(1, 1)$, its velocity is zero, and its acceleration vector is directed downwards along the tangent to the parabolic path. Starting from rest, the point begins to move in the direction of the acceleration. When $t = \frac{1}{2}\pi$ (quarter period), P is at $C(0, -1)$, is moving towards the left, and its acceleration vector is directed vertically upwards. When $t = \pi$ (half period), P is at $E(-1, 1)$, where it comes to rest. When $t = \frac{3}{2}\pi$, P is again at C and moving towards the right. When $t = 2\pi$, P has returned to rest at A and one of the periodic vibrations has been completed.

Example 3. *Rotation.* When the path of a moving point is a circle, the motion is called *rotation.* The time-rate of change of the central angle $AOP\ (= \theta)$, where OA is fixed, is called the *angular velocity.* That is,

$$\frac{d\theta}{dt} = \text{angular velocity of } P. \qquad (5)$$

If the arc $AP = s$, we have $s = r\theta$. Hence

$$v = \frac{ds}{dt} = r\frac{d\theta}{dt}.$$

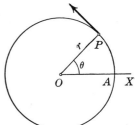

Fig. 225.5

Hence, in rotation, the speed (always positive) is numerically equal to the radius times the angular velocity.

Suppose the angular velocity is equal to a constant k. Then

$$\frac{d\theta}{dt} = k.$$

Integrating, assuming $\theta = \theta_0$ when $t = 0$, the result is

$$\theta = kt + \theta_0.$$

Let us find the equations of motion, taking the origin at O. Then since $x = r \cos \theta$, $y = r \sin \theta$, we have

$$\begin{cases} x = r \cos (kt + \theta_0) \\ y = r \sin (kt + \theta_0) \end{cases}$$

as the equations of motion. The point makes one revolution in $\frac{2\pi}{k}$ seconds. The speed is constant and equal to rk. The motion is called *uniform motion in a circle.*

Note that the motion of the projection of the moving point on any diameter is a simple harmonic vibration (see Art. 185).

Example 4. A material point moves in a plane with the component accelerations

$$a_x = 3 \sin t, \quad a_y = 3 \cos t. \qquad (6)$$

Furthermore, the point starts from rest at the origin. That is,

$$\text{when } t = 0, \ x = 0, \ y = 0, \ v_x = 0, \ v_y = 0. \qquad (7)$$

Find the component velocities and the equations of motion (parametric equations of the path), and draw the path.

Solution. This is an integration problem, and from (7) we shall be able to find the constants of integration. Substituting in (6) for a_x and a_y from (*D*), the result is

$$\frac{dv_x}{dt} = 3 \sin t, \quad \frac{dv_y}{dt} = 3 \cos t.$$

Multiply each equation by dt, integrate, and use the initial conditions (7). The steps are as follows.

$$v_x = -3 \cos t + c_1, \quad v_y = 3 \sin t + c_2,$$
$$0 = -3 + c_1, \qquad 0 = 0 + c_2.$$

Hence $c_1 = 3$, $c_2 = 0$, and the component velocities are

$$v_x = 3(1 - \cos t), \quad v_y = 3 \sin t. \tag{8}$$

Substituting for v_x and v_y from (B), Art. 224, the result is

$$\frac{dx}{dt} = 3(1 - \cos t), \quad \frac{dy}{dt} = 3 \sin t.$$

Multiply each equation by dt, integrate, and use the initial conditions (7). The steps are as follows.

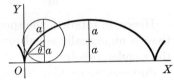

$$x = 3t - 3 \sin t + c_3,$$
$$y = -3 \cos t + c_4,$$
$$0 = 0 - 0 + c_3,$$
$$0 = -3 + c_4.$$

Fig. 225.6

Hence $c_3 = 0$, $c_4 = 3$, and the equations of motion are

$$x = 3(t - \sin t), \quad y = 3(1 - \cos t). \tag{9}$$

The path is a cycloid (see Art. 227 below). The motion is precisely that of a point on the circumference of a circle of diameter 6 linear units which rolls on OX so that its center moves parallel to OX with a constant speed equal to 3 linear units per second. In the figure $a = 3$. The value of t is the measure in radians of the central angle θ of the figure.

PROBLEMS

1. Given the equations of motion of a particle $x = t^2$, $y = 4t - t^3$.

a. Obtain the rectangular equation of the path and draw the curve.
b. Draw the velocity and acceleration vectors for $t = 0$, 1, and $\sqrt{3}$ sec. Is the speed increasing or decreasing at these times?
c. For what values of the time is the speed a minimum?
d. Where is the particle when the speed is $4\sqrt{5}$ units per second?
e. At what point on the curve is the acceleration the least?

2. The equations of a curvilinear motion are $x = 2 \cos 2t$, $y = 3 \cos t$.

a. Show that the moving point oscillates on an arc of the parabola whose equation is $4y^2 - 9x - 18 = 0$. Draw the path.
b. Draw the acceleration vectors at the points where $v = 0$.
c. Draw the velocity vector at the point where the speed is a maximum.

3. The equations of a curvilinear motion are

$$x = \sin 2t, \quad y = \cos t.$$

Show that the path is the figure-of-eight curve whose equation is $x^2 = 4y^2 - 4y^4$ and discuss the motion.

4. A particle moves on the ellipse $x^2 + 4\,y^2 = 16$. When $t = 0$, the particle is at the point $(-4, 0)$, and $v_y = 2 \cos t$. Find the equations of motion.

5. In Example 4 above, show that the path is the circle

$$(x - 1)^2 + (y - 2)^2 = 9$$

if $x = 1$, $y = -1$, $v_x = -3$, $v_y = 0$, when $t = 0$.

6. The component accelerations of a particle moving in a plane are given below, and also the initial position and velocity. Find the equations of motion and draw the path. In each case find the position and speed after 2 sec.

    a. $a_x = 0$, $a_y = -32$; $x = 0$, $y = 0$, $v_x = 4$, $v_y = 3$, when $t = 0$.
    b. $a_x = 6\,t$, $a_y = 2$; $x = 0$, $y = 0$, $v_x = 1$, $v_y = 2$, when $t = 0$.
    c. $a_x = 2 \cos t$, $a_y = \sin t$; $x = 0$, $y = 0$, $v_x = 1$, $v_y = 2$, when $t = 0$.
    d. $a_x = -3 \cos t$, $a_y = -8 \sin 2\,t$; $x = 3$, $y = 0$, $v_x = 0$, $v_y = 4$, when $t = 0$.

7. For a projectile, neglecting the resistance of the air and assuming the y-axis drawn vertically upward, we have for the components of the acceleration (in feet per second per second) $a_x = 0$, $a_y = -32$. If the projectile is hurled with the initial velocity v_0 in a direction making an angle α (called the "elevation") with the x-axis, show by integration that $x = v_0 t \cos \alpha$, $y = v_0 t \sin \alpha - 16\,t^2$ are the equations of motion if $x = y = 0$ when $t = 0$ (see Fig. 225.7).

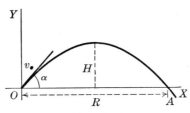

Fig. 225.7

8. Show that the rectangular equation of the path of the projectile in Problem 7 is

$$y = x \tan \alpha - \frac{16}{v_0^2}\,(1 + \tan^2 \alpha)x^2.$$

This curve is a parabola (see Art. 132).

9. The intercept OA of the parabola in Problem 8 (see Fig. 225.7) on the x-axis is called the *range*. Show that its mean value with respect to α is 0.6366 times the maximum range.

10. In Problem 8, $v_0 = 160$ ft. per second. The projectile is hurled at a wall 480 ft. away. Show that the highest point on this wall that can be hit is at a height above the x-axis of 256 ft. What is α for this height?

11. In uniform motion (speed constant) in a circle, show that the acceleration at any point P is constant in magnitude and directed along the radius from P toward O.

226. Integration problems using parametric equations. Some physical problems associated with a curve which is given by parametric equations may be solved by substitution. For example, the formulas in rectangular coördinates in Arts. 88 and 95,

Area under a curve

(I) $$A = \int_a^b y \, dx,$$

Length of a curve

(II) $$s = \int_a^b \sqrt{1 + \left(\frac{dy}{dx}\right)^2} \, dx,$$

are applied immediately to parametric equations by substituting the values of

$$y, \quad dx = \frac{dx}{dt} \, dt, \quad \frac{dy}{dx} = \frac{\dfrac{dy}{dt}}{\dfrac{dx}{dt}}$$

in terms of t, and at the same time replacing the limits $x = a$, $x = b$ by the corresponding values of t.

In each case care must be exercised to see that the range of values for t corresponds to the physical problem under consideration.

Example 1. Given the semicubical parabola whose parametric equations are (Art. 220)

$$x = \tfrac{1}{2} t^2, \quad y = \tfrac{1}{4} t^3. \tag{1}$$

a. Find the area between the curve, the x-axis, and ordinates at $t = 1$, $t = 3$.

b. Find the length of the curve between the points where $t = 1$, $t = 3$.

Solution.

a. From (1), $y = \tfrac{1}{4} t^3, \quad dx = \dfrac{dx}{dt} \, dt = t \, dt.$

Hence, by (I), $A = \displaystyle\int_1^3 \tfrac{1}{4} t^3 \cdot t \, dt = \tfrac{1}{4} \int_1^3 t^4 \, dt = \left[\dfrac{1}{4} \dfrac{t^5}{5}\right]_1^3$

$\qquad\qquad = \tfrac{1}{20}(243 - 1) = 12.1.$

b. From (1), $dx = t \, dt, \quad \dfrac{dy}{dt} = \tfrac{3}{4} t^2, \quad \dfrac{dy}{dx} = \tfrac{3}{4} t.$

Hence $s = \displaystyle\int_1^3 \sqrt{1 + \tfrac{9}{16} t^2} \, t \, dt = \tfrac{1}{4} \int_1^3 \sqrt{16 + 9 t^2} \, t \, dt = \tfrac{1}{108}\left[(16 + 9 t^2)^{\frac{3}{2}}\right]_1^3$

$\qquad = \dfrac{(97)^{\frac{3}{2}} - 125}{108} = 7.69.$

Formula (II) may also be written in the form

(III) $$s = \int_{t_1}^{2} \sqrt{\left(\frac{dx}{dt}\right)^2 + \left(\frac{dy}{dt}\right)^2}\, dt,$$

which is usually more convenient.

The formula for the element of volume of a solid of revolution in Art. 96,

(IV) $$dV = \pi r^2\, dx \quad (\text{or } \pi r^2\, dy),$$

or the formula for the element of area of a surface of revolution (Art. 100),

(V) $$dS = 2\,\pi y\, ds \quad (\text{or } 2\,\pi x\, ds),$$

is readily applied to parametric equations by expressing r, dx, dy, x, y, ds in terms of the parameter t and integrating between the proper limits for the parameter.

Example 2. The semicubical parabola of Example 1, Art. 220, is revolved about OY, forming a solid of revolution. Find dV and dS.

Solution. Using (IV), $r = x = \frac{1}{2}\,t^2$. Also $dy = \frac{3}{4}\,t^2\, dt$.

Hence $$dV = \pi x^2\, dy = \frac{3}{16}\,\pi t^6\, dt.$$

Using (V), $dS = 2\,\pi x\, ds$. By (III), since $\frac{dx}{dt} = t$, $\frac{dy}{dt} = \frac{3}{4}\,t^2$, we have

$$ds = \sqrt{t^2 + \frac{9}{16}\,t^4}\, dt.$$

Hence $$dS = \pi t^3 \sqrt{1 + \frac{9}{16}\,t^2}\, dt.$$

PROBLEMS

1. For the parabola $x = t - 1$, $y = 4 - t^2$, find

a. the area bounded by the curve and OY and OX in the second quadrant;

b. the length of the curve lying in the first quadrant;

c. the area of the curved surface obtained when the area of (**a**) is revolved about OX; about OY.

2. For the ellipse $x = 3\cos\theta$, $y = \sin\theta$, find

a. the area under the curve bounded by ordinates at $\theta = 0$, $\theta = 60°$;

b. the volume of the solid obtained when this area is revolved about OX;

c. the centroid of the area in (**a**).

3. For the equilateral hyperbola $x = 3 - 2\,t$, $y = 1 + \frac{2}{t}$, find

a. the area under the curve bounded by ordinates at $t = \frac{3}{2}$, $t = \frac{1}{2}$;

b. the length of arc of this part of the curve;

c. the centroid of the area.

4. Find the area between the parabola $x = t^2 - 2\,t$, $y = 1 - t^2$ and the x-axis.

5. Find the centroid of the area in Problem 4.

6. The area under the hyperbola $x = a \sec \theta$, $y = b \tan \theta$ from $x = a$ to $x = a\sqrt{2}$ is revolved about the x-axis. Find the volume of the solid of revolution generated.

7. Find the center of gravity of the solid in Problem 6.

8. Find the centroid of the area bounded by the x-axis and the parabola

$$x = \cos \theta, \quad y = -\cos 2\,\theta.$$

227. Locus problems solved by parametric equations; cycloid and involute of a circle. Parametric equations are important because it is sometimes easy in locus problems to express the coördinates of a point on the locus in terms of a parameter, when it is otherwise difficult to obtain the equation of the locus. The following examples illustrate this statement.

Example 1. The *cycloid* is the curve traced by a fixed point on the circumference of a circle which rolls without slipping on a straight line. Find parametric equations for the cycloid.

Solution. Let $P(x, y)$ be a point which is fixed on the circumference of the circle of radius a, which rolls without slipping on the x-axis. Take for origin a point O at which P touched the axis and take for the variable parameter θ the variable angle CBP (Fig. 227.1). The figure is drawn for $0 < \theta < \frac{1}{2}\pi$.

By the condition that the circle rolls without slipping,

$$OA = \text{arc } AP = a\theta.$$

From the figure, $PC = a \sin \theta$, $CB = a \cos \theta$.
Hence if (x, y) are the coördinates of P,

$$x = OD = OA - PC = a\theta - a \sin \theta,$$
$$y = DP = AB - CB = a - a \cos \theta.$$

Therefore
$$\begin{cases} x = a(\theta - \sin \theta), \\ y = a(1 - \cos \theta). \end{cases} \tag{1}$$

These are the parametric equations of the cycloid.

They have been derived on the assumption that $0 < \theta < \frac{1}{2}\pi$, but it can be shown that they hold for all values of θ.

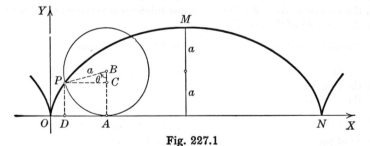

Fig. 227.1

The cycloid extends indefinitely to the right and left and consists of arcs congruent to OMN.

The length ON, called the base of the cycloid, is $2\pi a$. The coördinates of the maximum point M are $(\pi a, 2a)$. The points at which the cycloid touches the base line are cusps and the tangent at a cusp is perpendicular to the base line.

Hence to sketch one arch of a cycloid, lay off ON equal to the circumference of the rolling circle; at the middle point of ON erect a perpendicular equal to the diameter with extremity M, and draw the curve OMN. At O and N the curve is perpendicular to ON. When P moves from O to N along the cycloid, the parameter θ increases from 0° to 360°.

Example 2. If a string is wrapped around a circle and unwound, the free end traces a curve called an *involute of the circle*. Find parametric equations for this curve.

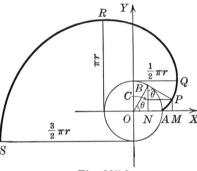

Fig. 227.2

Solution. The curve is shown in Fig. 227.2. Take the axes of coördinates as shown. Let $P(x, y)$ be the tracing point in any position. Then PB, the length unwound, is tangent to the circle at B. Take for parameter $\theta = \angle AOB$. Then $\angle PBC = \angle AOB = \theta$.

The figure is drawn for $0 < \theta < \frac{1}{2}\pi$ but it can be shown that the results hold for all positive values of θ.

By definition,
$$PB = \text{arc } AB = r\theta.$$

From the figure,
$$x = OM = ON + CP = OB \cos \theta + PB \sin \theta = r \cos \theta + r\theta \sin \theta,$$
$$y = MP = NC = NB - CB = OB \sin \theta - PB \cos \theta = r \sin \theta - r\theta \cos \theta.$$
Hence the equations of the involute of a circle are
$$\begin{cases} x = r \cos \theta + r\theta \sin \theta, \\ y = r \sin \theta - r\theta \cos \theta. \end{cases} \tag{2}$$

To sketch the curve, construct Q, R, S as shown in the figure The normal line at any point extends along the string.

PROBLEMS

1. Sketch the cycloid for the given value of a, compute x and y for the given value of the parameter, find the slope of the tangent, and draw the rolling circle and the tangent and normal. Check your figures by the theorem that the normal at P in the cycloid of Art. 227 will pass through A, the lowest point of the rolling circle.

    a. $a = 2$, $\theta = 45°$.   **b.** $a = 2$, $\theta = 270°$.   **c.** $a = 1$, $\theta = 90°$.

2. Prove the following properties for the cycloid of Art. 227.

    a. The normal at P passes through A.

    b. The area under one arch $OMN = 3\,\pi a^2$.

    c. The volume of the solid of revolution formed when the area OMN is revolved about OX is $5\,\pi^2 a^3$, and the area of its surface is $\frac{64}{3}\,\pi a^2$.

    d. The length of the arc $OMN = 8\,a$.

3. Sketch the involute for the given value of a, compute x and y for the given value of θ, find the slope of the tangent, and construct tangent and normal, checking by the fact that the latter is tangent to the fixed circle.

    a. $a = 2$, $\theta = 60°$.   **b.** $a = 2$, $\theta = 180°$.   **c.** $a = 1$, $\theta = 90°$.

4. In the involute of Art. 227,

    a. show that BP is the normal at P;   **b.** find the length of the arc $APQR$.

5. Find the centroid of the area under the cycloidal arc OMN, Art. 227.

6. In the cycloid of Art. 227 find the volume and area of the curved surface formed

    a. when the arc OM is revolved about OY;

    b. when the arc OM is revolved about the ordinate of M.

7. The equations of motion of a material point are

$$x = a(\pi t - \sin \pi t), \quad y = a(1 - \cos \pi t).$$

    a. Show that the path is a cycloid.

    b. Find the time elapsing when the point describes a complete arch.

    c. Show that the acceleration is constant in magnitude and that it is always directed toward the center of the rolling circle.

228. Hypocycloid and epicycloid. The *hypocycloid* is the curve traced by a point on the circumference of a circle which rolls without slipping on the *inside* of a fixed circle.

Example 1. Find parametric equations for the hypocycloid.

Solution. In Fig. 228.1, the tracing point P moves from A on the fixed circle along the arc AP. Draw OX along the radius OA. Take for the parameter

$$\angle AOB = \theta.$$

By definition,   Arc $AB = $ arc BP.   (1)

Let $R = OA$ = radius of the fixed circle, and $r = BC$ = radius of the rolling circle. Then, by (1),

$$R\theta = r \angle PCB.$$

Hence,
$$\angle PCB = \frac{R\theta}{r}. \tag{2}$$

From the figure,

$$\begin{aligned} x &= OF = OE + EF = OE + DP = OC \cos\theta + CP \cos\angle CPD. \\ y &= FP = ED = EC - DC = OC \sin\theta - CP \sin\angle CPD. \end{aligned} \Biggr\} \tag{3}$$

To find angle CPD in (3), extend CP to meet OX. Then angle PCB is seen to be the exterior angle of the triangle formed, whose opposite interior angles are θ and an angle equal to angle CPD. Hence

$$\angle PCB = \theta + \angle CPD.$$

Therefore, using (2), $\angle CPD = \angle PCB - \theta = \dfrac{R\theta}{r} - \theta = \dfrac{R - r}{r}\,\theta.$

In (3), $OC = R - r, \quad \text{and} \quad CP = r.$

Hence
$$x = (R - r)\cos\theta + r\cos\frac{R - r}{r}\,\theta,$$
$$y = (R - r)\sin\theta - r\sin\frac{R - r}{r}\,\theta \Biggr\} \tag{4}$$

are the required equations.

When the tracing point P meets the fixed circle a cusp is formed, as at A.

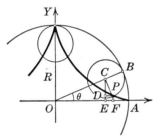

Fig. 228.1

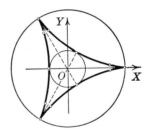

Fig. 228.2

Equations (4) are based on a figure in which $0 < \theta < \frac{1}{2}\pi$ and $0 < PCB < \frac{1}{2}\pi + \theta$ but they hold for all values of θ.

To sketch the hypocycloid, proceed as follows. Suppose, for example, $R = 3\,r$. There will be three cusps (see Fig. 228.2). Draw the fixed circle with radius R, and a concentric circle with radius $R - 2\,r = \frac{1}{3}\,R$. The cusps occur on the fixed circle when $\theta = 0°$, 120°, 240°. The hypocycloid touches the inner circle when $\theta = 60°$, 180°, 300°. Setting $R = 3\,r$ in (4), we obtain as the equations of the *hypocycloid of three cusps*

$$\begin{aligned} x &= 2\,r\cos\theta + r\cos 2\,\theta, \\ y &= 2\,r\sin\theta - r\sin 2\,\theta. \end{aligned} \Biggr\} \tag{5}$$

The *epicycloid* is the curve traced by a point on the circumference of a circle which rolls without slipping on the *outside* of a fixed circle.

Example 2. Find parametric equations for the epicycloid.

Using the same notation as in Example 1, $OA = R$, $BC = CP = r$, $\theta = \angle AOB$. Prove that

$$\begin{aligned}
x &= (R + r) \cos \theta - r \cos \frac{R + r}{r} \theta, \\
y &= (R + r) \sin \theta - r \sin \frac{R + r}{r} \theta.
\end{aligned} \tag{6}$$

The epicycloid is sketched by drawing the fixed circle with radius R, a concentric circle with radius $R + 2\,r$, marking the cusps on the fixed circle and the points of tangency with the larger circle, and sketching in the curve. In the figure, $R = 4\,r$, and there are 4 cusps. The cusps are on the fixed circle at the points $\theta = 0$, $\frac{1}{2}\,\pi$, π, $\frac{3}{2}\,\pi$. The curve touches the exterior circle with radius $6\,r$, or $\frac{3}{2}\,R$, and the values of θ at the points of contact are $\frac{1}{4}\,\pi$, $\frac{3}{4}\,\pi$, $\frac{5}{4}\,\pi$, $\frac{7}{4}\,\pi$. The figure shows part of the epicycloid, and the rolling circle in two special positions.

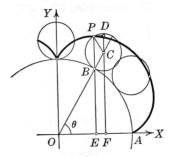

Fig. 228.3

PROBLEMS

1. Sketch the epicycloid (E), or hypocycloid (H), compute x and y, draw the rolling circle in the position determined by the given value of θ, find the slope, and draw the tangent and normal, checking the latter by the property that it passes through B in Figs. 228.1 and 228.3.

         a. H: $R = 2\,r$; locus a diameter, $y = 0$. (Why?)
         b. H: $R = 5\,r$; $\theta = 90°$.
         c. E: $R = r$; $\theta = 90°$, $210°$. (A cardioid)
         d. E: $R = 3\,r$: $\theta = 180°$.

2. In Fig. 228.3, $R = 4\,r$ and the tracing point P is in a position where the tangent is horizontal. Prove that $\theta = 60°$.

SUGGESTION. From (6), we find $y' = 0$ when $\cos 5\,\theta - \cos \theta = 0$. Use formula (6), Art. 296. Then $\sin 3\,\theta = 0$ or $\sin 2\,\theta = 0$. The latter equation gives the cusps.

3. In each of the following epicycloids find θ, x, y, for the first horizontal and the first vertical tangent. Check in Fig. 228.3.

         a. $R = 2\,r$.   **b.** $R = 3\,r$.   **c.** $R = 4\,r$.   **d.** $R = 5\,r$.

4. OB is the crank of an engine and AB the connecting rod. $AB > OB$. B moves on the crank circle whose center is O, and A moves on the fixed line OX.

With $\theta = \angle OAB$ as parameter, show that the parametric equations of the locus of a point P on AB are (Fig. 228.4)

$$x = b \cos \theta \pm \sqrt{[r^2 - (a+b)^2 \sin^2 \theta]}, \quad y = a \sin \theta.$$

Ellipse when $r = a + b$; otherwise an egg-shaped curve.

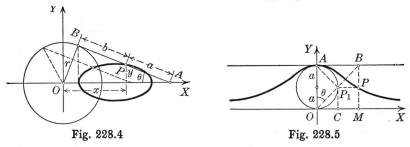

Fig. 228.4   Fig. 228.5

5. In Problem 4, take as parameter $\alpha = \angle XOB$, and prove that

$$x = r \cos \alpha + \frac{b}{a+b} \sqrt{[(a+b)^2 - r^2 \sin^2 \alpha]}, \quad y = \frac{ar \sin \alpha}{a+b}.$$

Using these parametric equations, plot when $a = b = 2$, $r = 1$, and locate the horizontal tangents.

6. Find the equation of the locus of a point P constructed as follows. Let OA be a diameter of the circle $x^2 + y^2 - 2\,ay = 0$, and let any line OB be drawn through O to meet the circle at P_1 and the tangent AB at B. Draw P_1P perpendicular to OA and BP parallel to OA. The locus of P is called the *witch of Agnesi*. Take as parameter $\theta = \angle AOB$ and show that the parametric equations are (Fig. 228.5)

$$x = 2\,a \tan \theta, \quad y = 2\,a \cos^2 \theta.$$

Show also that the rectangular equation is

$$x^2 y = 4\,a^2(2\,a - y).$$

229. Second derivative from parametric equations. Using y' as a symbol for the first derivative of y with respect to x, then **(A)**, Art. 220, will give y' as a function of t,

$$y' = h(t).$$

If the functions $h(t)$ and $f(t)$ (Art. 220) satisfy the conditions for which **(VI)** and **(VIII)**, Art. 62, are valid, then

(E)
$$y'' = \frac{d^2y}{dx^2} = \frac{dy'}{dx} = \frac{\dfrac{dy'}{dt}}{\dfrac{dx}{dt}} = \frac{h'(t)}{f'(t)}.$$

Formula **(E)** expresses the second derivative of y with respect to x as a function of t.

Example. Find y'' from the parametric equations (see Example 1, Art. 220)

$$x = \tfrac{1}{2} t^2, \quad y = \tfrac{1}{4} t^3.$$

Discuss the direction of bending.

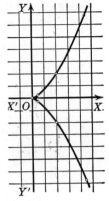

Solution. Differentiating with respect to t,

$$\frac{dx}{dt} = t, \frac{dy}{dt} = \tfrac{3}{4} t^2. \text{ Hence } y' = \tfrac{3}{4} t.$$

Therefore

$$\frac{dy'}{dt} = \frac{3}{4}, \text{ and by } (E), y'' = \frac{3}{4\,t}.$$

Hence y'' is positive when t is positive, that is, the arc in the first quadrant is concave upward. Also y'' is negative when t is negative; that is, the arc in the fourth quadrant is concave downward.

Fig. 229

PROBLEMS

1. Find y'' in terms of the parameter for each of the following curves.

    a. Ellipse, $x = a \cos \theta, y = b \sin \theta$.
    b. Parabola, $x = t + 2, y = 2\,t^2 - 3$.
    c. Cycloid, $x = a(\theta - \sin \theta), y = a(1 - \cos \theta)$.
    d. Hyperbola, $x = a \sec \theta, y = b \tan \theta$.

2. Investigate the direction of bending of the witch (Problem 6, page 389) whose equations are $x = 2\,a \tan \theta, y = 2\,a \cos^2 \theta$, and show that the coördinates of the points of inflection are $\left(\pm \tfrac{2}{3} a\sqrt{3}, \tfrac{3}{2} a\right)$.

3. Show in a curvilinear motion that

$$\frac{d^2y}{dx^2} = \frac{v_x a_y - v_y a_x}{v_x{}^3}.$$

230. Area bounded by a curve and two radii vectors. In certain problems in geometry and mechanics it is convenient to use parametric equations in connection with polar coördinates. If (ρ, θ) are the polar coördinates of a point and

$$\rho = g(t), \quad \theta = h(t),$$

where t is a variable parameter, these equations define, in general, a curve. If $g(t)$ and $h(t)$ satisfy proper conditions of continuity and differentiability, the formulas of Chapter XVII may be expressed in

terms of the parameter t by substitution. For example, the formula
of Art. 218,

$$\text{Area} = \tfrac{1}{2} \int_{\alpha}^{\beta} \rho^2 \, d\theta, \qquad (1)$$

which gives the area $OAPQB$ of Fig. 230.1
if $\angle XOA = \alpha$, $\angle XOB = \beta$, becomes

$$\text{Area} = \tfrac{1}{2} \int_{t_0}^{t_1} [g(t)]^2 h'(t) dt,$$

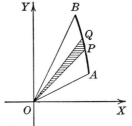

where t_0 and t_1 are determined by the equations

$$\alpha = h(t_0), \quad \beta = h(t_1).$$

Fig. 230.1

A formula of considerable convenience is obtained from (1) by transforming to rectangular coördinates. To do this, differentiate (Art. 213)

$$\theta = \text{arc tan } \frac{y}{x}, \qquad (2)$$

with respect to x, regarding y as a function of x which is determined by the curve. The result is

$$\frac{d\theta}{dx} = \frac{xy' - y}{x^2 + y^2}, \quad \text{where } y' = \frac{dy}{dx},$$

or, in differential form,

$$d\theta = \frac{x \, dy - y \, dx}{x^2 + y^2}. \qquad (3)$$

Hence, since $\rho^2 = x^2 + y^2$, (1) becomes

$$\text{Area} = \tfrac{1}{2} \int (x \, dy - y \, dx). \qquad (4)$$

The limits are determined by the coördinates of the extremities of the arc AB.

In deriving (1), the element of area was the circular sector with central angle POQ $(= d\theta)$ and radius OP $(= \rho)$. In (4) we may consider this elementary sector replaced by the triangle OPQ.

Example 1. Find the area between the equilateral hyperbola

$$x^2 - y^2 = a^2 \qquad (5)$$

and radii OA and OP drawn to $A(a, 0)$ and $P(x, y)$ (a hyperbolic sector). See Fig. 230.2.

Solution. From (5) $\qquad dy = \dfrac{x\,dx}{y}$

and $\qquad\qquad x\,dy - y\,dx = \dfrac{(x^2 - y^2)dx}{y} = \dfrac{a^2\,dx}{\sqrt{x^2 - a^2}}.$

The limits are $x = a$ and $x = x$. Then (4) gives

$$\text{Area } OAP = \frac{a^2}{2}\int_a^x \frac{dx}{\sqrt{x^2 - a^2}} \qquad (6)$$

$$= \frac{a^2}{2}\ln\frac{x + \sqrt{x^2 - a^2}}{a}.$$

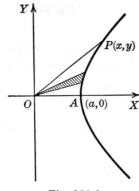

Fig. 230.2

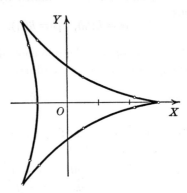

Fig. 230.3

Example 2. Find the entire area of the hypocycloid of three cusps (Art. 228 and Fig. 230.3).

$$\begin{array}{l} x = 2\,r\cos\theta + r\cos 2\,\theta, \\ y = 2\,r\sin\theta - r\sin 2\,\theta. \end{array} \Bigg\} \qquad (7)$$

Solution. Differentiating, we get

$$dx = 2\,r(-\sin\theta - \sin 2\,\theta)d\theta, \quad dy = 2\,r(\cos\theta - \cos 2\,\theta)d\theta. \qquad (8)$$

Substituting from (7) and (8), we find

$$x\,dy - y\,dx = 2\,r^2(1 - \cos\theta\cos 2\,\theta + \sin\theta\sin 2\,\theta)d\theta. \qquad (9)$$

The last two terms within the parenthesis reduce to a simple expression by noting that

$$\cos(2\,\theta + \theta) = \cos 2\,\theta\cos\theta - \sin 2\,\theta\sin\theta. \qquad (10)$$

Hence $x\,dy - y\,dx = 2\,r^2(1 - \cos 3\,\theta)d\theta.$
For the entire area the limits are $\theta = 0$, $\theta = 2\,\pi$. Hence, by (4),

$$\text{Entire area} = r^2\int_0^{2\pi}(1 - \cos 3\,\theta)d\theta = 2\,\pi r^2.$$

PROBLEMS

1. Find the area of a sector of the ellipse $x = a \cos \theta$, $y = b \sin \theta$ between radii drawn from the origin to $A(a, 0)$ and $B(\frac{1}{2} a, \frac{1}{2} b\sqrt{3})$.

2. Find the entire area of the hypocycloid for which $R = 4 r$.

3. Show that the length of the arc of the hypocycloid of three cusps (see Example 2) is $16 r$.

4. Find the length of arc for the hypocycloid of Problem 2.

5. Find the length of the arc of the epicycloid of three cusps, and the entire area.

6. Find the area bounded by the parabola $y^2 = 2 px + p^2$ and radii from the origin to $A(-\frac{1}{2} p, 0)$ and $B(4 p, 3 p)$.

7. Using the triangle OPQ (Fig. 230.1) as the element of area, show that the moments of area are

$$M_x = \tfrac{1}{3} \int y(x \, dy - y \, dx), \quad M_y = \tfrac{1}{3} \int x(x \, dy - y \, dx).$$

8. Find the centroid of the hyperbolic sector of Example 1 above by the formulas of Problem 7.

CHAPTER XIX

CURVATURE

231. Curvature. The shape of a curve at a point (its flatness or sharpness) depends upon the *rate of change of direction* and, when properly defined, this rate is called the *curvature at the point* and is denoted by K. Consider the curve of Fig. 231 where length of arc is measured from the fixed point A. Let P be a point on the curve determined by a given value of s and let P' be a second point near P.

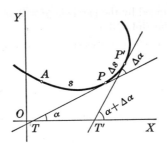

Fig. 231

When the point of contact of the tangent line describes the arc PP' ($= \Delta s$), the tangent line turns through the angle $\Delta \alpha$. That is, $\Delta \alpha$ is the change in the inclination of the tangent line. We now set down the following definitions.

$$\frac{\Delta \alpha}{\Delta s} = \text{average curvature of the arc } PP'.$$

The curvature at P ($= K$) *is the limiting value of the average curvature when* P' *approaches* P *as a limiting position; that is,*

$$(A) \qquad K = \lim_{\Delta s \to 0} \frac{\Delta \alpha}{\Delta s} = \frac{d\alpha}{ds} = \text{curvature at } P.$$

In formal terms the curvature is the *rate of change of the inclination with respect to the arc.*

Since the angle $\Delta \alpha$ is measured in radians and the length of arc Δs in units of length, it follows that *the unit of curvature at a point is one radian per unit of length.*

232. Curvature of a circle. Theorem. *The curvature of a circle at any point equals the reciprocal of the radius, and is therefore the same at all points.*

Proof. In Fig. 232 the angle $\Delta\alpha$ between the tangent lines at P and P' equals the central angle PCP' between the radii CP and CP'. Hence

$$\frac{\Delta\alpha}{\Delta s} = \frac{\text{angle } PCP'}{\Delta s} = \frac{\dfrac{\Delta s}{R}}{\Delta s} = \frac{1}{R},$$

since the angle PCP' is *measured in radians*. That is, the average curvature of the arc PP' is equal to a constant. Letting $\Delta s \longrightarrow 0$, we have the result stated in the theorem.

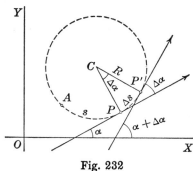

Fig. 232

From the standpoint of curvature, the circle is the simplest curve, since a circle bends at a uniform rate. Obviously the curvature of a straight line is everywhere zero.

233. Formulas for curvature; rectangular coördinates. *Theorem. When the equation of a curve is given in rectangular coördinates, then*

(B) $$K = \frac{y''}{(1 + y'^2)^{\frac{3}{2}}},$$

where y' and y'' are, respectively, the first and second derivatives of y with respect to x.

Proof. Since $\qquad \alpha = \text{arc tan } y', \qquad\qquad \left(y' = \dfrac{dy}{dx}\right)$

differentiating, we have

(1) $$\frac{d\alpha}{dx} = \frac{y''}{1 + y'^2}. \qquad\qquad \textbf{By (IX), Art. 188}$$

But

(2) $$\frac{ds}{dx} = (1 + y'^2)^{\frac{1}{2}}. \qquad\qquad \text{Art. 95}$$

Dividing (1) by (2) gives **(B)**. $\qquad\qquad$ By **(VIII)** and **(VI)**, Art. 62

Exercise. If y is the independent variable, show that

(C)
$$K = \frac{-x''}{(1 + x'^2)^{\frac{3}{2}}},$$

where x' and x'' are, respectively, the first and second derivatives of x with respect to y.

Formula (C) can be used as an alternative formula in cases where differentiation with respect to y is simpler. Also, (B) fails when y' becomes infinite, that is, when the tangent at P is vertical. Then in (C)

$$x' = 0 \quad \text{and} \quad K = -x''.$$

Sign of K. The formulas for K above and in the following articles contain a radical the sign of which depends upon the choice of the positive direction in measuring s along the curve and of the variable in which K is expressed. No simple rule covering all cases is practicable. For (B) the following statement may be made.

Choosing the positive sign in the denominator of (B), we see that K and y'' have like signs. That is, K is positive or negative according as the curve is concave upward or downward. See Arts. 236 and 237.

Example 1. Find the curvature of the parabola $y^2 = 4x$ (a) at the point $(1, 2)$; (b) at the vertex.

Solution.
$$y' = \frac{2}{y}, \quad y'' = \frac{d}{dx}\left(\frac{2}{y}\right) = -\frac{2y'}{y^2}.$$

a. When $x = 1$ and $y = 2$, then $y' = 1$, $y'' = -\frac{1}{2}$. Substituting in (B), $K = -\frac{1}{8}\sqrt{2} = -0.177$. Hence at $(1, 2)$ the curve is concave downward and the inclination of the tangent is changing at the rate of 0.177 radian per unit arc. Since 0.177 radian $= 10°\ 7'$, the angle between the tangent lines at $P(1, 2)$ and at a point Q such that arc $PQ = 1$ unit is approximately $10°$.

b. At the vertex $(0, 0)$, y' becomes infinite. Hence use (C).

$$x' = \frac{1}{2}y, \quad x'' = \frac{1}{2}\frac{dy}{dy} = \frac{1}{2}. \quad K = -\frac{1}{2}.$$

Example 2. Find K for the cycloid.
By (1), Art. 227, $x = a(\theta - \sin\theta)$, $y = a(1 - \cos\theta)$.

Solution. By (A), Art. 220, we find

$$y' = \frac{\sin\theta}{1 - \cos\theta}.$$

Hence
$$1 + y'^2 = \frac{2}{1 - \cos\theta}.$$

Also, by (**E**), Art. 229, we have, after simplification,

$$y'' = \frac{\frac{dy'}{d\theta}}{\frac{dx}{d\theta}} = \frac{-1}{a(1 - \cos\theta)^2}.$$

Substituting in (**B**),

$$K = -\frac{1}{2\,a\sqrt{2 - 2\cos\theta}} = -\frac{1}{4\,a\sin\frac{1}{2}\theta}.$$

As a practical application of the idea of curvature we mention *railroad* or *transition curves*. In laying out the curves on a railroad it will not do, on account of the high speed of trains, to pass abruptly from a straight stretch of track to a circular curve. In order to make the change of curvature gradual, engineers make use of transition curves to connect the straight part of a track with a circular track. This curve should have zero curvature at its point of junction with the straight track and the curvature of the circular track where it joins the latter. Arcs of cubical parabolas, or practicable approximations, are generally employed as transition curves.

Example. The transition curve on a railway track has the shape of an arc of the cubical parabola $y = \frac{1}{3}x^3$. At what rate is a car on this track changing its direction (1 mi. = unit of length) when it is passing through (**a**) the point $(3, 9)$? (**b**) the point $(2, \frac{8}{3})$? (**c**) the point $(1, \frac{1}{3})$?

Solution. $\qquad \dfrac{dy}{dx} = x^2, \quad \dfrac{d^2y}{dx^2} = 2\,x.$

Substituting in (**B**), $\qquad K = \dfrac{2\,x}{(1 + x^4)^{\frac{3}{2}}}.$

a. At $(3, 9)$, $K = \dfrac{6}{(82)^{\frac{3}{2}}}$ radian per mile $= 28'$ per mile.

b. At $(2, \frac{8}{3})$, $K = \dfrac{4}{(17)^{\frac{3}{2}}}$ radian per mile $= 3°\ 16'$ per mile.

c. At $(1, \frac{1}{3})$, $K = \dfrac{2}{(2)^{\frac{3}{2}}} = \dfrac{1}{\sqrt{2}}$ radian per mile $= 40°\ 30'$ per mile.

234. Special formula for parametric equations. From (**A**), Art. 220, we have, by differentiation,

$$\frac{dy'}{dt} = \frac{\dfrac{dx}{dt}\dfrac{d^2y}{dt^2} - \dfrac{dy}{dt}\dfrac{d^2x}{dt^2}}{\left(\dfrac{dx}{dt}\right)^2}. \tag{1}$$

Whence, using **(E)**, Art. 229, and substituting in **(B)** and reducing, we obtain

$$(D) \qquad K = \frac{x'y'' - y'x''}{(x'^2 + y'^2)^{\frac{3}{2}}},$$

where the accents indicate derivatives with respect to t, that is,

$$x' = \frac{dx}{dt}, \ x'' = \frac{d^2x}{dt^2}, \ y' = \frac{dy}{dt}, \ y'' = \frac{d^2y}{dt^2}.$$

Formula **(D)** is convenient, but it is often better to proceed as in Example 2, Art. 233, finding y' as in Art. 220, y'' as in Art. 229, and substituting directly in **(B)**.

235. Formula for curvature; polar coördinates. *Theorem. When the equation of a curve is given in polar coördinates,*

$$(E) \qquad K = \frac{\rho^2 + 2\,\rho'^2 - \rho\rho''}{(\rho^2 + \rho'^2)^{\frac{3}{2}}},$$

where ρ' and ρ'' are, respectively, the first and second derivatives of ρ with respect to θ.

Proof. By **(4)**, Art. 217, $\alpha = \theta + \beta$.

Hence
$$\frac{d\alpha}{d\theta} = 1 + \frac{d\beta}{d\theta}. \qquad (1)$$

Also, by **(C)**, Art. 217, $\beta = \arctan \dfrac{\rho}{\rho'}.$

Hence
$$\frac{d\beta}{d\theta} = \frac{\rho'^2 - \rho\rho''}{\rho'^2 + \rho^2}.$$

Then, by (1),
$$\frac{d\alpha}{d\theta} = \frac{\rho^2 + 2\,\rho'^2 - \rho\rho''}{\rho'^2 + \rho^2}. \qquad (2)$$

From **(F)**, Art. 219,
$$\frac{ds}{d\theta} = (\rho^2 + \rho'^2)^{\frac{1}{2}}. \qquad (3)$$

Dividing (2) by (3) gives **(E)**. By **(VIII)** and **(VI)**, Art. 62

Example. Find the curvature of the logarithmic spiral $\rho = e^{a\theta}$ at any point (Fig. 216.6).

Solution. $\dfrac{d\rho}{d\theta} = \rho' = ae^{a\theta} = a\rho;\ \dfrac{d^2\rho}{d\theta^2} = \rho'' = a^2 e^{a\theta} = a^2\rho.$

Substituting in **(E)**,
$$K = \frac{1}{\rho\sqrt{1 + a^2}}.$$

236. Radius of curvature. The radius of curvature R at a point on a curve equals the reciprocal of the curvature at that point. Hence, from (B),

(F)
$$R = \frac{1}{K} = \frac{(1 + y'^2)^{\frac{3}{2}}}{y''}.$$

The sign of R is the same as that of K (Art. 233) and the choice in (F) may be made as in (B). It is usually better, however, to consider only the numerical value of R and to construct its geometric representation as explained in Art. 237.

Example. Find the radius of curvature at any point of the catenary $y = \dfrac{a}{2}\left(e^{\frac{x}{a}} + e^{-\frac{x}{a}}\right)$ (Fig. 167.3).

Solution.
$$y' = \frac{1}{2}\left(e^{\frac{x}{a}} - e^{-\frac{x}{a}}\right); \quad y'' = \frac{1}{2a}\left(e^{\frac{x}{a}} + e^{-\frac{x}{a}}\right) = \frac{y}{a^2}.$$
$$1 + y'^2 = 1 + \frac{1}{4}\left(e^{\frac{x}{a}} - e^{-\frac{x}{a}}\right)^2 = \frac{1}{4}\left(e^{\frac{x}{a}} + e^{-\frac{x}{a}}\right)^2 = \frac{y^2}{a^2}.$$

Therefore
$$R = \frac{y^2}{a}.$$

237. Circle of curvature. Consider any point P on the curve L. The tangent line drawn to the curve at P has the same slope as the curve

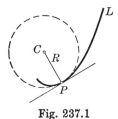

Fig. 237.1

itself at P. In an analogous manner we may construct for each point of the curve a tangent circle whose curvature is the same as the curvature of the curve itself at that point. To do this, proceed as follows Draw the normal to the curve at P on the concave side of the curve. Lay off on this normal the distance PC = radius of curvature ($= R$) at P. With C as a center draw the circle passing through P. The curvature of this circle is then

$$K = \frac{1}{R},$$

which also equals the curvature of the curve itself at P. The circle so constructed is called the *circle of curvature* for the point P on the curve.

In general, the circle of curvature of a curve at a point will cross the curve at that point. This is illustrated in Fig. 237.1 and follows from the fact that the direction of the circle changes at a *constant* rate, while the direction of the curve changes more rapidly on one side of P than on the other. (Compare with the tangent line at a point of inflection.)

Just as the tangent line at P shows the direction of the curve at P, so the circle of curvature at P aids us very materially in forming a geometric concept of the curvature of the curve at P, the rate of change of direction of the curve and of the circle being the same at P.

Example 1. Find the radius of curvature at the point $(3, 4)$ on the equilateral hyperbola $xy = 12$, and draw the corresponding circle of curvature.

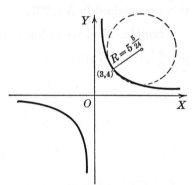

Fig. 237.2

Solution. $\dfrac{dy}{dx} = -\dfrac{y}{x}, \quad \dfrac{d^2y}{dx^2} = \dfrac{2\,y}{x^2}.$

For $(3, 4)$, $\dfrac{dy}{dx} = -\dfrac{4}{3}, \quad \dfrac{d^2y}{dx^2} = \dfrac{8}{9}.$

Therefore $R = \dfrac{(1 + \frac{16}{9})^{\frac{3}{2}}}{\frac{8}{9}} = \dfrac{125}{24} = 5\frac{5}{24}.$

The circle of curvature crosses the curve at two points.

Example 2. Find R at $(2, 1)$ for the curve $x^2 + 4\,xy - 2\,y^2 = 10$.

Solution. Differentiating, regarding y as an implicit function of x, we get
$$x + 2\,y + 2\,xy' - 2\,yy' = 0.$$

Differentiating this equation, regarding y and y' as implicit functions of x, we get
$$1 + 4\,y' - 2\,y'^2 + 2(x - y)y'' = 0.$$

Substituting the given values $x = 2$, $y = 1$, we find $y' = -2$, $y'' = \frac{15}{2}$.

Hence, by **(F)**, $\qquad\qquad R = \frac{2}{3}\sqrt{5}.$

The method of this example (namely, regarding y and y' as implicit functions of x) can often be used to advantage.

PROBLEMS

1. Find the radius of curvature for each of the following curves at the point indicated. Draw the curve and the corresponding circle of curvature.

a. $y = x^2$; $(0, 0)$.

b. $y = x^3$; $(1, 1)$.

c. $y^2 = x^3$; $(4, 8)$.

d. $y^2 = 8\,x$; $(\frac{9}{8}, 3)$.

e. $y = e^x$; $(0, 1)$.

f. $y = \ln x$; $(e, 1)$.

g. $y = \sin x$; $(\frac{1}{2}\,\pi, 1)$.

h. $y = 2\cos x$; $(\frac{1}{4}\,\pi, \sqrt{2})$.

i. $y = 2\sin 2\,x$; $(\frac{1}{4}\,\pi, 2)$.

j. $y = \tan x$; $(\frac{1}{4}\,\pi, 1)$.

2. Calculate the radius of curvature at any point on each of the following curves.

a. $y^2 = 2\,px$.   **b.** $b^2x^2 + a^2y^2 = a^2b^2$.

c. $b^2x^2 - a^2y^2 = a^2b^2$.   **d.** $x^{\frac{1}{2}} + y^{\frac{1}{2}} = a^{\frac{1}{2}}$.

e. $x^{\frac{2}{3}} + y^{\frac{2}{3}} = a^{\frac{2}{3}}$.   **f.** $y = \ln \sec x$.

3. If the point of contact of the tangent line at $(1, 4)$ to the parabola $y^2 = 16\,x$ moves along the curve a distance $\Delta s = 0.1$, through what angle, approximately, will the tangent line turn? (Use differentials.)

4. The inclination of the curve $2\,y = x^2$ at the point $A(1, \frac{1}{2})$ is $45°$. Use differentials to find approximately the inclination of the curve at the point B on the curve such that the distance along the curve from A to B is $\Delta s = 0.2$ unit.

5. Calculate the radius of curvature at any point (ρ_1, θ_1) on each of the following curves.

a. The circle $\rho = a \sin \theta$.   **b.** The spiral $\rho = a\theta$.

c. The cardioid $\rho = a(1 - \cos \theta)$.   **d.** The parabola $\rho = a \sec^2 \frac{1}{2}\,\theta$.

e. The lemniscate $\rho^2 = a^2 \cos 2\,\theta$.   **f.** The curve $\rho = a \sin^3 \frac{1}{3}\,\theta$.

    g. The equilateral hyperbola $\rho^2 \cos 2\,\theta = a^2$.

6. Find the radius of curvature for each of the following curves at the point indicated. Draw the curve and the corresponding circle of curvature.

    a. $x = 3\,t,\ y = 2\,t^2 - 1;\ t = 1.$

    b. $x = 4\,t,\ y = \dfrac{2}{t};\ t = 1.$

    c. $x = 3\,t^2,\ y = 3\,t - t^3;\ t = 1.$

    d. $x = 2\,e^t,\ y = e^{-t};\ t = 0.$

    e. $x = 4 \sin t,\ y = 2 \cos t;\ \text{where } x = 2.$

    f. $x = \sin t,\ y = \cos 2\,t;\ t = \frac{1}{6}\,\pi.$

    g. $x = 2 \cos t,\ y = \cos 2\,t;\ t = \frac{1}{2}\,\pi.$

7. Show that the radius of curvature at any point $(t = t_1)$ on the hypocycloid $x = a \cos^3 t,\ y = a \sin^3 t$ is $3\,a \sin t_1 \cos t_1$.

8. Show that the radius of curvature at any point $(t = t_1)$ on the involute of the circle $x = a(\cos t + t \sin t),\ y = a(\sin t - t \cos t)$ is at_1.

9. Find the point on the curve $y = e^x$ where the curvature is a maximum.

10. Find the points on the curve $3\,y = x^3 - 2\,x$ where the curvature is a maximum.

11. Show that the radius of curvature becomes infinite at a point of inflection

12. Given the curve $y = 3\,x - x^3$.

a. Find the radius of curvature at the maximum point of the curve and draw the corresponding circle of curvature.

b. Prove that the maximum point of the curve is not the point of maximum curvature.

c. Find to the nearest hundredth of a unit the abscissa of the point of maximum curvature.

13. Find the radius of curvature at each maximum and minimum point on the curve $y = x^4 - 2\,x^2$. Draw the curve and the circles of curvature. Find the points on the curve where the radius of curvature is a minimum.

14. Show that the curvature of the cubical parabola $3\,a^2y = x^3$ increases from zero to a maximum value when x increases from zero to $\frac{1}{5}\,a\sqrt[4]{125}$. Find the minimum value of the radius of curvature.

238. Center of curvature. The tangent line at $P(x, y)$ has the property that x, y, and y' have the same values at P for the tangent line and the curve. The circle of curvature at P has a similar property, namely, x, y, y', and y'' have the same values at P for the circle of curvature and the curve.

DEFINITION. **The center of curvature** (α, β) for a point $P(x, y)$ on a curve is the center of the circle of curvature.

Theorem. The coördinates (α, β) of the center of curvature for $P(x, y)$ are

(G) $\alpha = x - \dfrac{y'(1 + y'^2)}{y''}, \quad \beta = y + \dfrac{1 + y'^2}{y''}.$

Proof. Let $P(x, y)$ be a point on the curve L which has the equation $y = f(x)$. (See Fig. 237.1.) Let (ξ, η) be any point on the circle of curvature for P. Then

$$(\xi - \alpha)^2 + (\eta - \beta)^2 = R^2, \tag{1}$$

where R is given by **(F)**. Differentiating (1), regarding η and η' as functions (implicit) of ξ, the results are

$$\xi - \alpha + (\eta - \beta)\eta' = 0, \qquad \eta' = \frac{d\eta}{d\xi}, \tag{2}$$

$$1 + \eta'^2 + (\eta - \beta)\eta'' = 0. \quad \eta'' = \frac{d^2\eta}{d\xi^2}. \tag{3}$$

From (3) we have (if $\eta'' \neq 0$)

$$\eta - \beta = -\frac{1 + \eta'^2}{\eta''}. \tag{4}$$

Using this value in (2), we have

$$\xi - \alpha = \frac{\eta'(1 + \eta'^2)}{\eta''}. \tag{5}$$

Now at P

$$\xi = x, \; \eta = y, \; \eta' = y', \; \eta'' = y''.$$

Making these substitutions in (4) and (5) and solving for α and β, we have **(G)**.

Exercise. If x' and x'' are, respectively, the first and second derivatives of x with respect to y, derive **(G)** in the form

(H) $\alpha = x + \dfrac{1 + x'^2}{x''}, \quad \beta = y - \dfrac{x'(1 + x'^2)}{x''}.$

Formulas **(H)** may be used when y' becomes infinite, or if differentiation of the given equation with respect to y is simpler.

Example. Find the coördinates of the center of curvature of the parabola $y^2 = 2\,px$ corresponding **(a)** to any point on the curve; **(b)** to the vertex.

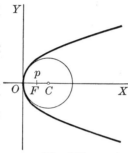

Fig. 238

Solution. Use **(H)**. Then $x' = \dfrac{y}{p}, \; x'' = \dfrac{1}{p}.$

Hence $\alpha = x + \dfrac{y^2 + p^2}{p} = 3\,x + p.$

$$\beta = y - \frac{y(y^2 + p^2)}{p^2} = -\frac{y^3}{p^2}.$$

Therefore **(a)** $\left(3\,x + p, \; -\dfrac{y^3}{p^2}\right)$ is the center of curvature corresponding to any point (x, y) on the curve.

(b) $(p, 0)$ is the center of curvature corresponding to the vertex $(0, 0)$.

COMMENTS. In deriving **(G)** it is assumed that y'' does not vanish. At a point of inflection, therefore, a special discussion is necessary. The radius of curvature becomes infinite at a point of inflection, since $K = 0$. The circle of curvature is now replaced by the tangent line at the point of inflection.

239. Evolutes. The locus of the centers of curvature of a given curve is called the *evolute* of that curve. Consider the circle of curvature at a point P on a curve. If P moves along the curve, we may suppose the corresponding circle of curvature to roll along the curve with it, its radius varying so as to be always equal to the radius of curvature of the curve at the point P. The curve CC_7 described by the center of the circle is the evolute of PP_7. (Fig. 239.1)

Formulas (*G*) or (*H*), Art. 238, give the coördinates of any point (α, β) on the evolute expressed in terms of the coördinates of the corresponding point (x, y) of the given curve. But y is a function of x; therefore these formulas give us at once the parametric equations of the evolute in terms of the parameter x.

To find the rectangular equation of the evolute we eliminate x and y between the expressions for α and β and the equation of the given curve. No general rule for elimination can be given that will apply in all cases. The method to be adopted must depend on the problem.

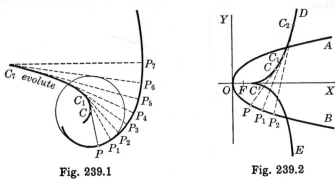

Fig. 239.1   Fig. 239.2

Example 1. Find the equation of the evolute of the parabola $y^2 = 2\,px$.

Solution. From the example of Art. 238,

$$\alpha = 3\,x + p, \quad \beta = -\frac{y^3}{p^2}.$$

Solving, $$x = \frac{\alpha - p}{3}, \quad y = -(p^2\beta)^{\frac{1}{3}}.$$

Substituting in $y^2 = 2\,px$,

$$(p^2\beta)^{\frac{2}{3}} = 2\,p\left(\frac{\alpha - p}{3}\right),$$

or $$p\beta^2 = \tfrac{8}{27}(\alpha - p)^3.$$

Remembering that α denotes the abscissa and β the ordinate of a rectangular system of coördinates, we see that the evolute of the parabola AOB is the semicubical parabola $DC'E$, the centers of curvature for O, P, P_1, P_2 being at C', C, C_1, C_2 respectively. (Fig. 239.2)

When the curve is defined by parametric equations,

$$x = f(t), \quad y = \phi(t),$$

formulas (*G*) or (*H*) will give parametric equations of the evolute. The rectangular equation of the evolute may be found by elimination of the parameter, as usual.

Example 2. Find parametric equations for the evolute of an ellipse.

Solution. From the parametric equations of the ellipse (Fig. 239.3)

$$x = a \cos \theta, \quad y = b \sin \theta,$$

we obtain
$$y' = -\frac{b \cos \theta}{a \sin \theta}, \quad y'' = -\frac{b}{a^2 \sin^3 \theta}.$$

Substituting in **(G)**, we get

$$\alpha = a \cos \theta - \frac{\cos \theta (a^2 \sin^2 \theta + b^2 \cos^2 \theta)}{a},$$

$$\beta = b \sin \theta - \frac{\sin \theta (a^2 \sin^2 \theta + b^2 \cos^2 \theta)}{b}.$$

Reducing, we obtain as the required parametric equations

$$\alpha = \frac{a^2 - b^2}{a} \cos^3 \theta, \quad \beta = \frac{b^2 - a^2}{b} \sin^3 \theta.$$

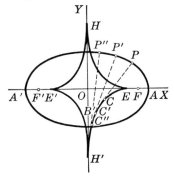

Fig. 239.3

Elimination of θ between these equations gives $(a\alpha)^{\frac{2}{3}} + (b\beta)^{\frac{2}{3}} = (a^2 - b^2)^{\frac{2}{3}}$, the rectangular equation of the evolute $EHE'H'$ of the ellipse $ABA'B'$. In the figure the points E, E', H', H are the centers of curvature corresponding to the points A, A', B, B' on the curve, and C, C', C'' correspond to the points P, P', P''.

Example 3. The parametric equations of a curve are

$$x = \frac{t^2 + 1}{4}, \quad y = \frac{t^3}{6}. \tag{1}$$

(a) Find the equations of the evolute in parametric form, and plot the curve and the evolute. (b) Find the radius of curvature at the point where $t = 1$, and draw the corresponding circle of curvature.

Solution. a.　　$\dfrac{dx}{dt} = \tfrac{1}{2} t, \quad \dfrac{dy}{dt} = \tfrac{1}{2} t^2. \quad \therefore y' = t.$　　By **(A)**, Art. 220

$\dfrac{dy'}{dt} = 1. \quad \therefore y'' = \dfrac{2}{t}.$　　　　　　By **(E)**, Art. 229

Substituting in **(G)** and reducing gives

$$\alpha = \frac{1 - t^2 - 2\,t^4}{4}, \quad \beta = \frac{4\,t^3 + 3\,t}{6}, \tag{2}$$

the parametric equations of the evolute. Assuming values of the parameter t, we calculate x, y from (1), and α, β from (2), and tabulate the results.

Now plot the curve and its evolute.

The point $(\frac{1}{4}, 0)$ is common to the given curve and its evolute. The given curve (semicubical parabola) lies entirely to the right and the evolute entirely to the left of $x = \frac{1}{4}$.

From **(F)**, Art. 236, we get

b. $R = \dfrac{t(1 + t^2)^{\frac{3}{2}}}{2} = \sqrt{2}$ when $t = 1$.

The circle of curvature at $A(\frac{1}{2}, \frac{1}{6})$, where $t = 1$, will have its center at $A'(-\frac{1}{2}, \frac{7}{6})$ on the evolute and radius $\sqrt{2}$.

This radius should equal the distance (Art. 4).

$$AA' = \sqrt{(\tfrac{1}{2} + \tfrac{1}{2})^2 + (\tfrac{1}{6} - \tfrac{7}{6})^2} = \sqrt{2}.$$

t	x	y	α	β
-3	$\frac{5}{2}$	$-\frac{9}{2}$		
-2	$\frac{5}{4}$	$-\frac{4}{3}$	$-\frac{35}{4}$	$-\frac{19}{3}$
$-\frac{3}{2}$	$\frac{13}{16}$	$-\frac{9}{16}$	$-\frac{91}{32}$	-3
-1	$\frac{1}{2}$	$-\frac{1}{6}$	$-\frac{1}{2}$	$-\frac{7}{6}$
0	$\frac{1}{4}$	0	$\frac{1}{4}$	0
1	$\frac{1}{2}$	$\frac{1}{6}$	$-\frac{1}{2}$	$\frac{7}{6}$
$\frac{3}{2}$	$\frac{13}{16}$	$\frac{9}{16}$	$-\frac{91}{32}$	3
2	$\frac{5}{4}$	$\frac{4}{3}$	$-\frac{35}{4}$	$\frac{19}{3}$
3	$\frac{5}{2}$	$\frac{9}{2}$		

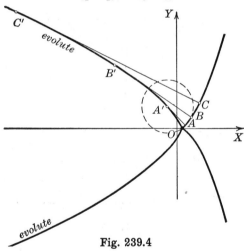

Fig. 239.4

Example 4. Find the parametric equations of the evolute of the cycloid

$$\begin{cases} x = a(t - \sin t), \\ y = a(1 - \cos t). \end{cases} \tag{3}$$

Solution. As in Example 2, Art. 233, we get

$$\frac{dy}{dx} = \frac{\sin t}{1 - \cos t}, \quad \frac{d^2y}{dx^2} = -\frac{1}{a(1 - \cos t)^2}.$$

Substituting these results in formulas **(G)**, we get

$$\begin{cases} \alpha = a(t + \sin t), \\ \beta = -a(1 - \cos t). \end{cases} \tag{4}$$

NOTE. If we eliminate t between equations (4), there results the rectangular equation of the evolute $OO'Q^V$ referred to the axes $O'\alpha$ and $O'\beta$ (Fig. 239.5). The coördinates of O with respect to these axes are $(-\pi a, -2\,a)$. Let us transform equations (4) to the new set of parallel axes OX and OY. Then, if the new coördinates of any point are (x, y),

$$\alpha = x - \pi a, \quad \beta = y - 2\,a.$$

Also, let $t = t' - \pi$.

Substituting in (4) and reducing, the equations of the evolute become

$$\begin{cases} x = a(t' - \sin t'), \\ y = a(1 - \cos t'). \end{cases} \tag{5}$$

Since (5) and (3) are identical in form, we have

The evolute of a cycloid is itself a cycloid whose generating circle equals that of the given cycloid.

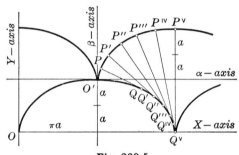

Fig. 239.5

240. Properties of the evolute. The evolute has two interesting properties.

Theorem I. *The normal at $P(x, y)$ to the given curve is tangent to the evolute at the center of curvature $C(\alpha, \beta)$ for P.* (See figures in the preceding article.)

Proof. From Fig. 240,*

$$\begin{cases} \alpha = x - R \sin \tau, \\ \beta = y + R \cos \tau. \end{cases} \tag{1}$$

The line PC lies along the normal at P, and

$$\text{Slope of } PC = \frac{y - \beta}{x - \alpha} = -\frac{1}{\tan \tau}. \tag{2}$$

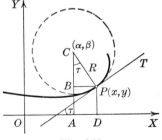

Fig. 240

*In the figure the inclination α of Art. 231 has been changed to τ to avoid confusion.

We show next that the slope of the evolute equals the slope of PC. Now the

$$\text{Slope of evolute} = \frac{d\beta}{d\alpha},$$

since α and β are the rectangular coördinates of any point on the evolute.

Let us choose as independent variable the length of arc on the given curve; then x, y, R, τ, α, β are functions of s. Differentiating (1) with respect to s gives

$$\frac{d\alpha}{ds} = \frac{dx}{ds} - R \cos \tau \frac{d\tau}{ds} - \sin \tau \frac{dR}{ds}, \tag{3}$$

$$\frac{d\beta}{ds} = \frac{dy}{ds} - R \sin \tau \frac{d\tau}{ds} + \cos \tau \frac{dR}{ds}. \tag{4}$$

Now $\dfrac{d\tau}{ds} = \dfrac{1}{R}$. We may also show that

$$\frac{dx}{ds} = \cos \tau, \quad \frac{dy}{ds} = \sin \tau.$$

For $\dfrac{dy}{dx} = y' = \tan \tau$. And, by Art. 95,

$$\frac{ds}{dx} = \sqrt{1 + y'^2} = \sqrt{1 + \tan^2 \tau} = \sec \tau.$$

Therefore, by **(VI)** and **(VIII)**, Art. 62,

$$\frac{dx}{ds} = \frac{1}{\sec \tau} = \cos \tau,$$

$$\frac{dy}{ds} = \frac{dy}{dx}\frac{dx}{ds} = \tan \tau \cos \tau = \sin \tau.$$

Substituting in (3) and (4), and reducing, we obtain

$$\frac{d\alpha}{ds} = - \sin \tau \frac{dR}{ds}, \quad \frac{d\beta}{ds} = \cos \tau \frac{dR}{ds}. \tag{5}$$

Dividing the second equation in (5) by the first gives

$$\frac{d\beta}{d\alpha} = - \operatorname{ctn} \tau = - \frac{1}{\tan \tau} = \text{slope of } PC. \tag{6}$$

Theorem II. *The length of an arc of the evolute is equal to the difference between the radii of curvature of the given curve which are tangent to this arc at its extremities, provided that along the arc of the given curve R increases or decreases.*

Proof. Squaring each of equations (5) and adding, we get

$$\left(\frac{d\alpha}{ds}\right)^2 + \left(\frac{d\beta}{ds}\right)^2 = \left(\frac{dR}{ds}\right)^2. \tag{7}$$

But if $s' =$ length of arc of the evolute,
$$ds'^2 = d\alpha^2 + d\beta^2,$$
by (2), Art. 95, if $s = s'$, $x = \alpha$, $y = \beta$. Hence (7) asserts that
$$\left(\frac{ds'}{ds}\right)^2 = \left(\frac{dR}{ds}\right)^2, \quad \text{or} \quad \frac{ds'}{ds} = \pm\frac{dR}{ds}. \tag{8}$$

Confining ourselves to an arc on the given curve for which the right-hand member does not change sign, we may write
$$ds' = + dR \quad \text{or} \quad ds' = - dR. \tag{9}$$
Integrating, we get
$$s' - s'_0 = \pm (R - R_0), \tag{10}$$
or (Figs. 239.1 and 239.2) Arc $CC_1 = \pm (P_1C_1 - PC)$.

Thus the theorem is proved.

In Example 4, Art. 239, we observe that at O', $R = 0$; at P^v, $R = 4\,a$. Hence arc $O'QQ^v = 4\,a$.

The length of one arch of the cycloid (as $OO'Q^v$) is eight times the length of the radius of the generating circle.

241. Involutes and their mechanical construction. Let a flexible ruler be bent in the form of the curve C_1C_9, the evolute of the curve P_1P_9, and suppose a string of length R_9, with one end fastened at C_9, to be stretched along the ruler (or curve). It is clear from the results of the last article that when the string is unwound and kept taut, the free end will describe the curve P_1P_9. Hence the name *evolute*.

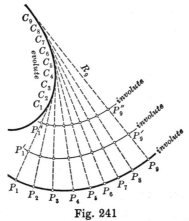

Fig. 241

The curve P_1P_9 is said to be an *involute* of C_1C_9. Obviously any point on the string will describe an involute, so that a given curve has an infinite number of involutes but only one evolute.

The involutes P_1P_9, $P'_1P'_9$, $P''_1P''_9$ are called *parallel curves* since the distance between any two of them measured along their common normals is constant.

PROBLEMS

1. Find the radius and center of curvature for each of the following curves at the given point. Check your results by proving (*a*) that the center of curvature lies on the normal to the curve at the given point, and (*b*) that the distance from the given point to the center of curvature is equal to the radius of curvature.

a. $4\,y = x^2 - 4$; $(0, -1)$.
c. $xy = 30$; $(3, 10)$.
e. $4\,y = x^4 - 8\,x^2$; $(2, -4)$.
g. $x^2 + 4\,y^2 = 25$; $(3, 2)$.
i. $y = e^x$; $(0, 1)$.
k. $y = \cos x$; $(0, 1)$.

b. $3\,y = x^3 - 3\,x^2 - 9\,x$; $(3, -9)$.
d. $xy = x^2 + 4$; $(2, 4)$.
f. $xy = x^3 + 8$; $(2, 8)$.
h. $x^3 + xy^2 - 6\,y^2 = 0$; $(3, 3)$.
j. $y = \ln x$; $(1, 0)$.
l. $y = 2 \sin 2\,x$; $(\tfrac{1}{4}\,\pi, 2)$.

2. Find the coördinates of the center of curvature at any point (x, y) of each of the following curves.

a. $y^2 = 2\,px$.
c. $y = x^3$.

b. $b^2x^2 - a^2y^2 = a^2b^2$.
d. $x^{\frac{2}{3}} + y^{\frac{2}{3}} = a^{\frac{2}{3}}$.

3. Find the radii and centers of curvature for the curve $xy = 4$ at the points $(1, 4)$ and $(2, 2)$. Draw the arc of the evolute between these centers. What is its length?

4. Find the parametric equations of the evolute of each of the following curves in terms of the parameter t. Draw the curve and its evolute, and draw at least one circle of curvature.

a. $x = 2\,t$, $y = 2\,t^2 - 1$.
c. $x = 3 - 2\,t$, $y = t^3 - 3$.
e. $x = 4\,t$, $y = 2/t$.
g. $x = 13 \sin t$, $y = 5 \cos t$.

i. $x = a \cos^3 t$,
 $y = a \sin^3 t$.

b. $x = 2\,t$, $y = t^3/3$.
d. $x = 2\,t + 1$, $y = t^3/3$.
f. $x = \sin t$, $y = t$.
h. $x = 2 \cos t + \cos 2\,t$,
 $y = 2 \sin t + \sin 2\,t$.
j. $x = a(\cos t + t \sin t)$,
 $y = a(\sin t - t \cos t)$.

5. Show that in the parabola $x^{\frac{1}{2}} + y^{\frac{1}{2}} = a^{\frac{1}{2}}$ we have the relation $\alpha + \beta = 3\,(x + y)$.

6. Given the equation of the equilateral hyperbola $2\,xy = a^2$, show that

$$\alpha + \beta = \frac{(y + x)^3}{a^2}, \quad \alpha - \beta = \frac{(y - x)^3}{a^2}.$$

From this derive the equation of the evolute

$$(\alpha + \beta)^{\frac{2}{3}} - (\alpha - \beta)^{\frac{2}{3}} = 2\,a^{\frac{2}{3}}.$$

CHAPTER XX

SERIES

242. Definitions. A *sequence* is a succession of terms formed according to some fixed rule or law.

Examples. $\quad$ 1, 4, 9, 16, 25.
$$1, -x, \tfrac{1}{2}x^2, -\tfrac{1}{3}x^3, \tfrac{1}{4}x^4, -\tfrac{1}{5}x^5.$$

A *series* is the indicated sum of the terms of a sequence.

Examples. $\quad$ $1 + 4 + 9 + 16 + 25.$
$$1 - x + \tfrac{1}{2}x^2 - \tfrac{1}{3}x^3 + \tfrac{1}{4}x^4 - \tfrac{1}{5}x^5.$$

When the number of terms is limited, the sequence or series is said to be *finite*. When the number of terms is unlimited, the sequence or series is said to be *infinite*.

The *general term*, or *n*th term, is an expression which indicates the law of formation of the terms.

Example 1. In the first example given above, the general term, or *n*th term, is n^2. The first term is obtained by setting $n = 1$, the tenth term by setting $n = 10$, etc. If the sequence is infinite, this fact is indicated by the following notation, using dots:

$$1, 4, 9, \cdots, n^2, \cdots$$

Example 2. In the second example given above, the *n*th term, except for $n = 1$, is $\dfrac{(-x)^{n-1}}{n-1}$.

Factorial numbers. An expression which occurs frequently in connection with series is the product of successive integers, beginning with 1. Thus, $1 \times 2 \times 3 \times 4 \times 5$ is called 5 *factorial* and is indicated by $\lfloor 5$ or $5!$.

In general
$$\lfloor n = 1 \times 2 \times 3 \times \cdots \times (n-1) \times n$$

is called *n factorial*. It is understood that n is a positive integer. An exception is sometimes made to include $n = 0$ by the arbitrary definition $\lfloor 0 = 1$. With this convention the *n*th term of the series

$$1 + x + \frac{x^2}{\lfloor 2} + \frac{x^3}{\lfloor 3} + \cdots$$

may be written as

$$\frac{x^{n-1}}{\lfloor n-1}.$$

411

PROBLEMS

Write out five terms of the sequence having the given nth term.

1. $\dfrac{1}{\sqrt{n}}$.   2. $\dfrac{1}{n^2}$.   3. $\dfrac{n}{n+1}$.   4. $\dfrac{(-x)^n}{n}$.   5. $\dfrac{2^n}{\lfloor n}$.

6. $\dfrac{x^n}{\lfloor n}$.   7. $\dfrac{(-1)^{n-1}}{2^n}$.   8. $\dfrac{(-1)^{n-1}x^n}{\lfloor n}$.   9. $\sqrt[n]{x}$.   10. $\dfrac{(2x)^n}{n+1}$.

Find the nth term of each of the following series.

11. $\frac{1}{3}+\frac{1}{9}+\frac{1}{27}+\frac{1}{81}+\cdots$.

12. $\frac{1}{2}-\frac{2}{3}+\frac{3}{4}-\frac{4}{5}+\cdots$.

13. $\frac{1}{2}+\frac{2}{4}+\frac{3}{8}+\frac{4}{16}+\cdots$.

14. $x-\frac{1}{2}x^2+\frac{1}{3}x^3-\frac{1}{4}x^4+\cdots$.

15. $1-\dfrac{x^2}{\lfloor 2}+\dfrac{x^4}{\lfloor 4}-\dfrac{x^6}{\lfloor 6}+\cdots$.

16. $x-\dfrac{x^3}{\lfloor 3}+\dfrac{x^5}{\lfloor 5}-\dfrac{x^7}{\lfloor 7}+\cdots$.

243. The geometric series. The following is a geometric series of n terms:

$$a+ar+ar^2+\cdots+ar^{n-1}.$$

In elementary algebra it is shown that the sum of the first n terms of a geometric series is given by the formula

$$S_n=\frac{a(1-r^n)}{1-r}, \quad\text{or}\quad S_n=\frac{a(r^n-1)}{r-1},$$

the first form being generally used if $|r|<1$, and the second form if $|r|>1$.

If $|r|<1$, then r^n decreases as n increases and

$$\lim_{n\to\infty} r^n=0.$$

From the formula above we see that if $|r|<1$,

$$\lim_{n\to\infty} S_n=\frac{a}{1-r}.$$

Hence if $|r|<1$ the sum of a geometric series approaches a definite fixed limit as the number of terms is increased indefinitely. In this case the series is said to be *convergent*.

If $|r|>1$ we see from the second formula above that as n increases the sum S_n increases and does not approach any limit, but becomes greater than any fixed number which may be assigned. In this case the series is said to be *divergent*.

A peculiar situation presents itself if $r=-1$. The series then becomes

$$a-a+a-a+a-a+\cdots.$$

If n is even, the sum is zero. If n is odd, the sum is a. As n increases indefinitely the sum does not increase indefinitely and it does not approach a fixed limit. Such a series is called an *oscillating* series.

244. Convergent and divergent series. Let u_k denote the kth term of an infinite series and let S_n denote the sum of the first n terms.

$$S_n = u_1 + u_2 + \cdots + u_n.$$

The variable S_n depends upon n and if we let n increase without limit one of two things may happen.

CASE I. S_n approaches a limit, say u, indicated by

$$\lim_{n \to \infty} S_n = u.$$

The infinite series is now said to be *convergent* and to converge to the value u, or to have the value u. An example is the geometric series with $|r| < 1$.

CASE II. S_n approaches no limit. The infinite series is now said to be *divergent*. Examples of divergent series are

$$1 + 2 + 3 + 4 + 5 + \cdots.$$
$$1 - 1 + 1 - 1 + \cdots.$$

No value is assigned to a divergent series.

For most practical applications a series must be convergent. The detailed consideration of the conditions under which a series is convergent is beyond the scope of this book. We give only the comparison tests below and the ratio test (Art. 245).

In considering the convergence of a series it should be noted that the sum of any given number of terms can always be found and does not affect the question of convergence. The convergence of a series depends upon the nature of the terms as the number of terms increases indefinitely. The following tests are stated without proof.

Test for convergence. Let

$$u_1 + u_2 + u_3 + \cdots \tag{1}$$

be a series of positive terms which it is desired to test for convergence. If a series of positive terms already known to be convergent, namely,

$$a_1 + a_2 + a_3 + \cdots, \tag{2}$$

can be found whose terms are never less than the corresponding terms in the series (1) to be tested, then (1) is a convergent series and its value does not exceed that of (2).

Example 1. Test the series

$$1 + \frac{1}{2^2} + \frac{1}{3^3} + \frac{1}{4^4} + \frac{1}{5^5} + \cdots. \tag{3}$$

Solution. Compare with the geometric series

$$1 + \frac{1}{2} + \frac{1}{2^2} + \frac{1}{2^3} + \frac{1}{2^4} + \cdots, \tag{4}$$

which is known to be convergent. The terms of (4) are never less than the corresponding terms of (3). Hence (3) also is convergent.

Following a line of reasoning similar to that applied to (1) and (2), we may prove the

Test for divergence. *Let*

$$u_1 + u_2 + u_3 + \cdots \tag{5}$$

be a series of positive terms to be tested, which are never less than the corresponding terms of a series of positive terms, namely,

$$b_1 + b_2 + b_3 + \cdots, \tag{6}$$

known to be divergent. Then (5) *is a divergent series*

Example 2. Show that the harmonic series

$$1 + \tfrac{1}{2} + \tfrac{1}{3} + \tfrac{1}{4} + \cdots \tag{7}$$

is divergent.

Solution. Rewrite (7) as below and compare with the series written under it. The square brackets are introduced to aid in the comparison.

$$1 + \tfrac{1}{2} + [\tfrac{1}{3} + \tfrac{1}{4}] + [\tfrac{1}{5} + \tfrac{1}{6} + \tfrac{1}{7} + \tfrac{1}{8}] + [\tfrac{1}{9} + \cdots + \tfrac{1}{16}] + \cdots. \tag{8}$$

$$\tfrac{1}{2} + \tfrac{1}{2} + [\tfrac{1}{4} + \tfrac{1}{4}] + [\tfrac{1}{8} + \tfrac{1}{8} + \tfrac{1}{8} + \tfrac{1}{8}] + [\tfrac{1}{16} + \cdots + \tfrac{1}{16}] + \cdots. \tag{9}$$

We observe the following facts. The terms in (8) are never less than the corresponding terms in (9).

But (9) is divergent. For the sum of the terms in each square bracket is $\tfrac{1}{2}$, and S_n will increase indefinitely as n becomes infinite.

Hence (8) is divergent.

A convenient series for use in connection with the tests above is the "*p* series,"

$$1 + \frac{1}{2^p} + \frac{1}{3^p} + \frac{1}{4^p} + \cdots + \frac{1}{n^p} + \cdots.$$

It can be shown that the p series is convergent when $p > 1$ and divergent for other values of p.

An *alternating* series is one whose terms are alternately positive and negative. It can be shown that *if*

$$u_1 - u_2 + u_3 - u_4 + \cdots,$$

is an alternating series in which each term is numerically less than or equal to the one which precedes it, and if $\lim\limits_{n \to \infty} | u_n | = 0$, *then the series is convergent.* See Art. 256.

Thus the series

$$1 - \tfrac{1}{2} + \tfrac{1}{3} - \tfrac{1}{4} + \cdots \tag{10}$$

is convergent since $\lim\limits_{n \to \infty} \dfrac{1}{n} = 0$.

A series is said to be *absolutely* or *unconditionally* convergent when the series formed from it by making all its terms positive is convergent. Other series are said to be *conditionally* convergent. Thus (10) is conditionally convergent since (7) is divergent.

A series with some positive and some negative terms is convergent if the series deduced from it by making all the signs positive is convergent.

245. Cauchy's test-ratio test. In the infinite geometric series

$$a + ar + ar^2 + \cdots + ar^n + ar^{n+1} + \cdots,$$

the ratio of the consecutive general terms ar^n and ar^{n+1} is the common ratio r. Moreover we know that the series is convergent when $| r | < 1$ and divergent for other values. We now explain a ratio test which may be applied to any series.

Theorem. *Let*

$$u_1 + u_2 + u_3 + \cdots + u_n + u_{n+1} + \cdots \tag{1}$$

be an infinite series of positive terms. Consider consecutive general terms u_n *and* u_{n+1}, *and form the **test ratio**.*

$$\textbf{Test ratio} = \frac{u_{n+1}}{u_n}.$$

Find the limit of this test ratio when n becomes infinite. Let this be

$$\rho = \lim\limits_{n \to \infty} \frac{u_{n+1}}{u_n}.$$

I. *When* $\rho < 1$, *the series is convergent.*
II. *When* $\rho > 1$, *the series is divergent.*
III. *When* $\rho = 1$, *the test fails.*

Proof. I. *When* $\rho < 1$. By the definition of a limit (Art. 40) we can choose n so large, say $n = m$, that when $n \geqq m$ the ratio $\dfrac{u_{n+1}}{u_n}$ will differ from ρ by as little as we please, and therefore be less than a proper fraction r. Hence

$$u_{m+1} < u_m r; \quad u_{m+2} < u_{m+1} r < u_m r^2; \quad u_{m+3} < u_m r^3;$$

and so on. Therefore, after the term u_m, each term of the series (1) is less than the corresponding term of the geometric series

$$u_m r + u_m r^2 + u_m r^3 + \cdots. \tag{2}$$

But since $r < 1$, the series (2), and therefore also the series (1), is convergent (Art. 244).

II. *When* $\rho > 1$ (*or* $\rho = \infty$). Following the same line of reasoning as in I, the series (1) may be shown to be divergent.

III. *When* $\rho = 1$, the series may be either convergent or divergent; that is, the test fails. For, consider the p series, namely,

$$1 + \frac{1}{2^p} + \frac{1}{3^p} + \frac{1}{4^p} + \cdots + \frac{1}{n^p} + \frac{1}{(n+1)^p} + \cdots.$$

The test ratio is $\dfrac{u_{n+1}}{u_n} = \left(\dfrac{n}{n+1}\right)^p = \left(1 - \dfrac{1}{n+1}\right)^p;$

and $\displaystyle\lim_{n \to \infty}\left(\frac{u_{n+1}}{u_n}\right) = \lim_{n \to \infty}\left(1 - \frac{1}{n+1}\right)^p = (1)^p = 1 \ (= \rho).$

Hence $\rho = 1$, no matter what value p may have. But

when $p > 1$, the series converges, and

when $p \leqq 1$, the series diverges.

Thus it appears that ρ can equal unity both for convergent and for divergent series. There are other tests to apply in cases like this, but the scope of our book does not admit of their consideration.

For convergence it is not enough that the test ratio is less than unity for all values of n. This test requires that the *limit* of the test ratio shall be less than unity. For instance, in the harmonic series the test ratio is always less than unity. The *limit*, however, equals unity.

The rejection of a group of terms at the beginning of a series will affect the *value* but not the *existence* of the limit.

PROBLEMS

Test the following series for convergence.

1. $\dfrac{1}{2} + \dfrac{2}{2^2} + \dfrac{3}{2^3} + \dfrac{4}{2^4} + \cdots.$

2. $\dfrac{2}{3} + 2(\tfrac{2}{3})^2 + 3(\tfrac{2}{3})^3 + 4(\tfrac{2}{3})^4 + \cdots$

3. $1 + \dfrac{1}{\lfloor 2} + \dfrac{1}{\lfloor 3} + \dfrac{1}{\lfloor 4} + \cdots.$

4. $\dfrac{\lfloor 2}{10} + \dfrac{\lfloor 3}{10^2} + \dfrac{\lfloor 4}{10^3} + \cdots.$

5. $\dfrac{5}{1} + \dfrac{5^2}{\lfloor 2} + \dfrac{5^3}{\lfloor 3} + \dfrac{5^4}{\lfloor 4} + \cdots$

6. $\dfrac{3}{2} + \dfrac{3^2}{2 \cdot 2^2} + \dfrac{3^3}{3 \cdot 2^3} + \dfrac{3^4}{4 \cdot 2^4} + \cdots.$

7. $\dfrac{1}{3} + \dfrac{1 \cdot 2}{3 \cdot 5} + \dfrac{1 \cdot 2 \cdot 3}{3 \cdot 5 \cdot 7} + \cdots + \dfrac{\lfloor n}{3 \cdot 5 \cdot 7 \cdots (2n+1)} + \cdots.$

8. $\dfrac{1}{1} + \dfrac{1 \cdot 3}{1 \cdot 4} + \dfrac{1 \cdot 3 \cdot 5}{1 \cdot 4 \cdot 7} + \cdots + \dfrac{1 \cdot 3 \cdot 5 \cdots (2n-1)}{1 \cdot 4 \cdot 7 \cdots (3n-2)} + \cdots.$

246. The binomial series. The binomial theorem ((3), Art. 295) furnishes another example of a series. This series is

$$(a+b)^m = a^m + \frac{m}{1} a^{m-1}b + \frac{m(m-1)}{1 \cdot 2} a^{m-2}b^2$$
$$+ \frac{m(m-1)(m-2)}{1 \cdot 2 \cdot 3} a^{m-3}b^3 + \cdots.$$

In elementary algebra the exponent m is always a positive integer and the series is finite because of the factor $(m-m)$ which enters the coefficients after $m+1$ terms. If m is a fraction or a negative number, the series is infinite. In this case the writing of the series is simplified as follows.

$$(a+b)^m = a^m \left(1 + \frac{b}{a}\right)^m,$$

or $$(a+b)^m = a^m(1+z)^m, \text{ where } z = \frac{b}{a}.$$

We need to consider then only the following series.

(A) $(1+z)^m = 1 + \dfrac{m}{1} z + \dfrac{m(m-1)}{1 \cdot 2} z^2 + \dfrac{m(m-1)(m-2)}{1 \cdot 2 \cdot 3} z^3 + \cdots.$

The infinite geometric series is a special case of the binomial series, as can be seen by setting $z = -r$ and $m = -1$. The series becomes

$$(1-r)^{-1} = 1 + \frac{(-1)}{1} (-r) + \frac{(-1)(-1-1)}{1 \cdot 2} (-r)^2$$
$$+ \frac{(-1)(-1-1)(-1-2)}{1 \cdot 2 \cdot 3} (-r)^3 + \cdots,$$

or, after simplification,

$$\frac{1}{1-r} = 1 + r + r^2 + r^3 + \cdots.$$

The nth term of the binomial series **(A)** is

$$u_n = \frac{m(m-1)(m-2) \cdots (m-n+2)}{1 \cdot 2 \cdot 3 \cdots (n-1)} z^{n-1}.$$

The next term is

$$u_{n+1} = \frac{m(m-1) \cdots (m-n+2)(m-n+1)}{1 \cdot 2 \cdots (n-1) \cdot n} z^n.$$

The test ratio is

$$\frac{u_{n+1}}{u_n} = \frac{m(m-1)\cdots(m-n+2)(m-n+1)}{1\cdot2\cdots(n-1)\cdot n}$$

$$\times \frac{1\cdot2\cdots(n-1)}{m(m-1)\cdots(m-n+2)} \times \frac{z^n}{z^{n-1}}$$

$$= \frac{m-n+1}{n}z = \left(\frac{m+1}{n}-1\right)z.$$

Hence
$$\rho = \lim_{n\to\infty}\left(\frac{m+1}{n}-1\right)z = -z.$$

The series is convergent if $\rho < 1$, that is, if $|z| < 1$. Hence the

Theorem. *The binomial series* **(A)**

$$(1+z)^m = 1 + \frac{m}{1}z + \frac{m(m-1)}{1\cdot2}z^2 + \cdots$$

is convergent if z has a fixed numerical value less than 1.

NOTE. If $|z| < 1$, the convergence does not depend on m and if $|z| > 1$, the series is divergent. If $|z| = 1$, the convergence depends on the value of m.

247. Numerical approximation by the binomial series. The binomial series may be used for numerical approximation, as illustrated in the following examples.

Example 1. Find $\sqrt{1.1}$, using the binomial series.

Solution. $\sqrt{1.1} = (1+0.1)^{\frac{1}{2}}$. Expanding by the binomial series, in which $z = 0.1$ and $n = \frac{1}{2}$, we have

$$\sqrt{1.1} = (1+0.1)^{\frac{1}{2}} = 1 + \frac{\frac{1}{2}}{1}(0.1) + \frac{\frac{1}{2}(-\frac{1}{2})}{1\cdot2}(0.1)^2 + \frac{(\frac{1}{2})(-\frac{1}{2})(-\frac{3}{2})}{1\cdot2\cdot3}(0.1)^3 + \cdots$$

$$= 1 + 0.05 - 0.00125 + 0.00006 + \cdots$$
$$= 1.04881.$$

The result is correct to five decimals.

Example 2. Find $\sqrt{101}$, using the binomial series.

Solution. $\sqrt{101} = \sqrt{100+1} = 10\sqrt{1+\frac{1}{100}}$. Expanding by the binomial series, in which $z = 0.01$ and $n = \frac{1}{2}$, we have

$$\sqrt{101} = 10(1+0.01)^{\frac{1}{2}} = 10\left[1 + \frac{1}{2}(0.01) + \frac{\frac{1}{2}(-\frac{1}{2})}{1\cdot2}(0.01)^2\right.$$
$$\left. + \frac{\frac{1}{2}(-\frac{1}{2})(-\frac{3}{2})}{1\cdot2\cdot3}(0.01)^3 + \cdots\right]$$

$$= 10[1 + 0.005000 - 0.000013 + 0.000001 + \cdots]$$
$$= 10.04988.$$

PROBLEMS

Using the binomial series, find approximately the values of the following numbers.

1. $\sqrt{630}$.   2. $\sqrt[3]{990}$.   3. $\sqrt[3]{130}$.   4. $\sqrt[5]{30}$.   5. $\sqrt[4]{80}$.

6. $\dfrac{1}{98}$.   7. $\dfrac{1}{10.3}$.   8. $\dfrac{1}{\sqrt{102}}$.   9. $\dfrac{1}{\sqrt[3]{65}}$.   10. $\dfrac{1}{\sqrt[4]{80}}$.

248. Approximate formulas from the binomial series. The binomial series can be used to obtain approximate formulas which are useful in many applications of mathematics. The following standard approximations are obtained immediately from the general series by using two or three terms. In these formulas it is understood that a is small and that *both a and n are positive.*

First Approximation

I. $(1+a)^n = 1 + na.$

II. $(1-a)^n = 1 - na.$

III. $\dfrac{1}{(1+a)^n} = 1 - na.$

IV. $\dfrac{1}{(1-a)^n} = 1 + na.$

Second Approximation

V. $(1+a)^n = 1 + na + \frac{1}{2}n(n-1)a^2.$

VI. $(1-a)^n = 1 - na + \frac{1}{2}n(n-1)a^2.$

VII. $\dfrac{1}{(1+a)^n} = 1 - na + \frac{1}{2}n(n+1)a^2.$

VIII. $\dfrac{1}{(1-a)^n} = 1 + na + \frac{1}{2}n(n+1)a^2.$

Example. Find approximately the value of $\displaystyle\int_0^1 \sqrt[3]{8+x^2}\,dx.$

Solution. A first approximation to the value of the integrand is

$$\sqrt[3]{8+x^2} = 2\left(1 + \frac{x^2}{8}\right)^{\frac{1}{3}} = 2\left(1 + \frac{1}{3}\frac{x^2}{8}\right). \qquad \text{By I}$$

Hence $\displaystyle\int_0^1 \sqrt[3]{8+x^2}\,dx = 2\int_0^1\left(1 + \frac{x^2}{24}\right)dx$, approximately,

$$= 2\left[x + \frac{x^3}{72}\right]_0^1 = 2.02778.$$

A second approximation is obtained as follows (using V).

$$\int_0^1 \sqrt[3]{8+x^2}\,dx = 2\int_0^1\left[1 + \frac{1}{3}\left(\frac{x^2}{8}\right) - \frac{1}{9}\left(\frac{x^2}{8}\right)^2\right]dx,\ \text{approximately,}$$

$$= 2\left[x + \frac{x^3}{72} - \frac{x^5}{2880}\right]_0^1 = 2.02708.$$

PROBLEMS

1. Derive formula VI from the general binomial series by setting $z = -a$.

2. Derive formula VIII from the general binomial series by replacing n by $-n$ and setting $z = -a$.

3. Work out the following developments.

a. $\dfrac{1}{\sqrt{1-x^2}} = 1 + \tfrac{1}{2}x^2 + \dfrac{1\cdot 3}{2\cdot 4}x^4 + \dfrac{1\cdot 3\cdot 5}{2\cdot 4\cdot 6}x^6 + \cdots.$

b. $\sqrt{1-x^2} = 1 - \tfrac{1}{2}x^2 - \dfrac{1}{2\cdot 4}x^4 - \dfrac{1\cdot 3}{2\cdot 4\cdot 6}x^6 - \cdots.$

4. Give to five figures the first and second approximations of the following numbers.

a. $\sqrt{404}.$ **b.** $\sqrt[3]{990}.$ **c.** $\sqrt[5]{30}.$ **d.** $\dfrac{1}{\sqrt{620}}.$

e. $\dfrac{1}{\sqrt{65}}.$ **f.** $\dfrac{1}{\sqrt[4]{630}}.$ **g.** $\dfrac{1}{98}.$ **h.** $\sqrt[4]{80}.$

5. Find approximately the values of the following integrals. In each case compute the value by using the exact formula and find the percentage error in the first approximation and the second approximation.

a. $\displaystyle\int_0^{\frac{1}{2}} \sqrt{1+x^2}\, dx.$ **b.** $\displaystyle\int_0^1 \dfrac{dx}{\sqrt{25+x^2}}.$ **c.** $\displaystyle\int_0^5 \sqrt{100-x^2}\, dx.$

6. Find approximately the values of the following integrals.

a. $\displaystyle\int_0^{\frac{1}{2}} \sqrt{1+x^3}\, dx.$ **b.** $\displaystyle\int_0^1 \dfrac{dx}{\sqrt{25-x^3}}.$

c. $\displaystyle\int_0^{\frac{1}{5}} \dfrac{dx}{\sqrt[3]{1-x^2}}.$ **d.** $\displaystyle\int_0^1 \sqrt[3]{27-x^2}\, dx.$

7. Given $f(x) = \displaystyle\int \sqrt{1-x^4}\, dx$ and $f(0) = 1$. Find first and second approximate formulas for $f(x)$.

8. Given $f(x) = \displaystyle\int \dfrac{dx}{\sqrt[3]{8-x^2}}$ and $f(0) = 3$. Find first and second approximate formulas for $f(x)$.

249. Application to an engineering problem. As an example of the use of approximate formulas in applied mathematics, let us consider the motion of the crosshead of the piston of an engine.

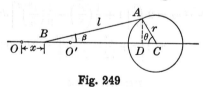

Fig. 249

In Fig. 249, AB represents the connecting rod and B the crosshead of the piston. As the crank AC revolves, the crosshead moves back and forth in a straight line from O when $\theta = 0$ to O' when $\theta = 180°$.

1. To find the position of the crosshead for any given position of the crank.

Let $\qquad OB = x, \quad AB = l, \quad AC = r.$

Now

$$x = OC - BC = OC - (BD + DC). \tag{1}$$
$$OC = r + l; \quad DC = r \cos \theta; \quad BD = l \cos \beta.$$

To express BD in terms of θ, we have

$$AD = l \sin \beta = r \sin \theta.$$

Hence $\qquad\qquad \sin \beta = \dfrac{r}{l} \sin \theta$

and $\qquad \cos \beta = \sqrt{1 - \sin^2 \beta} = \sqrt{1 - \left(\dfrac{r}{l} \sin \theta\right)^2}.$

Hence $\qquad\qquad BD = l \sqrt{1 - \left(\dfrac{r}{l} \sin \theta\right)^2}.$

Substituting the values of OC, BD, and DC in (1), the result is

$$x = r + l - r \cos \theta - l \sqrt{1 - \left(\dfrac{r}{l} \sin \theta\right)^2}. \tag{2}$$

This is an exact formula which permits us to calculate the value of x for any given value of θ. It has the disadvantage of making further calculations undesirably complicated.

2. To find an approximate equation of motion for the crosshead. Note that the motion of the crosshead is a straight-line motion. We write (II, Art. 248)

$$\sqrt{1 - \left(\dfrac{r}{l} \sin \theta\right)^2} = 1 - \dfrac{1}{2} \dfrac{r^2}{l^2} \sin^2 \theta, \text{ approximately.}$$

This approximation is justified because in standard-type engines the values of $\dfrac{r}{l}$ are about $\frac{1}{4}$, $\frac{1}{5}$, or $\frac{1}{6}$, and $\left(\dfrac{r}{l} \sin \theta\right)^2$ is necessarily small.

With this approximation (2) becomes

$$x = r - r \cos \theta + \dfrac{r^2}{2 l} \sin^2 \theta.$$

Setting $\sin^2 \theta = \frac{1}{2} - \frac{1}{2} \cos 2 \theta$, we obtain the common approximate formula used in engineering practice. This is

$$x = r(1 - \cos \theta) + \dfrac{r^2}{4 l} (1 - \cos 2 \theta). \tag{3}$$

If the angular velocity of the crank is constant and the crank makes n revolutions per second, then $\theta = 2 \pi n t$ and (3) becomes

$$x = r(1 - \cos 2 \pi n t) + \dfrac{r^2}{4 l} (1 - \cos 4 \pi n t). \tag{4}$$

Equation (4) is therefore an approximate equation of the motion.

3. In engineering practice the designer wishes to know the acceleration of the crosshead. This is found by differentiating (4). We have

$$\frac{dx}{dt} = 2\,\pi n r \sin 2\,\pi n t + \frac{\pi n r^2}{l} \sin 4\,\pi n t,$$

$$\frac{d^2x}{dt^2} = 4\,\pi^2 n^2 r \cos 2\,\pi n t + \frac{4\,\pi^2 n^2 r^2}{l} \cos 4\,\pi n t.$$

Hence if a is the acceleration of the crosshead, we may write

$$a = 4\,\pi^2 n^2 r \left[\cos\theta + \frac{r}{l}\cos 2\,\theta \right]. \tag{5}$$

It is of particular importance to the designer to know the value of θ for which the acceleration a is zero. The speed is then a maximum. This is given by the equation

$$\cos\theta + \frac{r}{l}\cos 2\,\theta = 0.$$

To solve this equation we set $\cos 2\,\theta = 2\cos^2\theta - 1$ and $\frac{r}{l} = \epsilon$, which gives

$$2\,\epsilon\cos^2\theta + \cos\theta - \epsilon = 0.$$

Solving this quadratic equation, we have

$$\cos\theta = \frac{-1 + \sqrt{1 + 8\,\epsilon^2}}{4\,\epsilon}, \tag{6}$$

where, to find the smallest value of θ, the positive sign is taken with the radical.

To obtain a simpler approximate formula for (6), we write, by V, Art. 248,

$$\sqrt{1 + 8\,\epsilon^2} = 1 + \tfrac{1}{2}(8\,\epsilon^2) + \tfrac{1}{2}(\tfrac{1}{2})(-\tfrac{1}{2})(8\,\epsilon^2)^2, \text{ approximately,}$$

and equation (6) becomes

$$\cos\theta = \epsilon - 2\,\epsilon^3. \tag{7}$$

PROBLEMS

1. Show that the crank is at right angles to the connecting rod when $\operatorname{ctn}\theta = \epsilon$, and hence $\cos\theta = \epsilon - \tfrac{1}{2}\,\epsilon^3$, approximately. Compare this result with (7).

2. Calculate the value of θ for which the connecting rod and the crank are perpendicular when (a) $\epsilon = \tfrac{1}{4}$; (b) $\epsilon = \tfrac{1}{5}$; (c) $\epsilon = \tfrac{1}{6}$. In each case find by (7) the value of θ for which the acceleration of the crosshead is zero, and compare results.

250. Power series. A series whose terms are monomials in ascending positive integral powers of a variable, say x, of the form

$$a_0 + a_1x + a_2x^2 + a_3x^3 + \cdots, \tag{1}$$

where the coefficients $a_0, a_1, a_2, \cdots$ are independent of x, is called a *power series in x*. Such series are of prime importance in the study of calculus.

A power series in x may converge for all values of x, or for no value except $x = 0$; or it may converge for some values of x different from 0 and be divergent for other values.

To test (1) for convergence we omit the first term, set $u_n = |a_nx^n|$, and form the test ratio (Art. 245).

Then we have

$$\frac{u_{n+1}}{u_n} = \left|\frac{a_{n+1}x^{n+1}}{a_nx^n}\right| = \left|\frac{a_{n+1}}{a_n}\right| \cdot |x|.$$

Hence, for any fixed value of x,

$$\rho = \lim_{n \to \infty} \frac{u_{n+1}}{u_n} = |x|L,$$

where

$$L = \lim_{n \to \infty} \left|\frac{a_{n+1}}{a_n}\right|.$$

We have two cases.

CASE I. If $L = 0$, the series (1) will converge for all finite values of x, since $\rho = 0$.

CASE II. If L is not zero, the series will converge when $|x|L$ is less than 1, that is, when x lies in the interval

$$-\frac{1}{L} < x < \frac{1}{L},$$

and will diverge for values of x outside this interval.

The end points of the interval, called the *interval of convergence*, must be examined separately. In any given series the test ratio should be formed and the interval of convergence determined by Art. 245.

Example 1. Find the interval of convergence for the series

$$x - \frac{x^2}{2^2} + \frac{x^3}{3^2} - \frac{x^4}{4^2} + \cdots. \tag{2}$$

Solution. The test ratio here is

$$\frac{u_{n+1}}{u_n} = -\frac{n^2}{(n+1)^2} x. \text{ Also, } \lim_{n \to \infty} \frac{n^2}{(n+1)^2} = 1.$$

Hence $\rho = -x$, and the series converges when x is numerically less than 1 and diverges when x is numerically greater than 1.

Now examine the end points. Substituting $x = 1$ in (2), we get

$$1 - \frac{1}{2^2} + \frac{1}{3^2} - \frac{1}{4^2} + \cdots,$$

which is an alternating series that converges.

Substituting $x = -1$ in (2), we get

$$-1 - \frac{1}{2^2} - \frac{1}{3^2} - \frac{1}{4^2} - \cdots,$$

which is convergent by comparison with the p series $(p > 1)$.

The series in the above example has $[-1, 1]$ as the *interval of convergence*. This may be written $-1 \le x \le 1$, or indicated graphically as in Fig. 250.

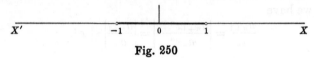

Fig. 250

Example 2. Determine the interval of convergence for the series

$$1 + \frac{x^2}{\underline{|2}} + \frac{x^4}{\underline{|4}} + \cdots + \frac{x^{2n}}{\underline{|2n}} + \frac{x^{2n+2}}{\underline{|2n+2}} + \cdots.$$

Solution. Omitting the first term, the test ratio is

$$\frac{u_{n+1}}{u_n} = \frac{\underline{|2n}}{\underline{|2n+2}} x^2 = \frac{1}{(2n+1)(2n+2)} x^2.$$

Also $\lim\limits_{n \to \infty} \dfrac{1}{(2n+1)(2n+2)} = 0$. Hence the series converges for all values of x.

Another type of power series has the form

$$b_0 + b_1(x - a) + b_2(x - a)^2 + \cdots + b_n(x - a)^n + \cdots \qquad (3)$$

in which a and the coefficients $b_0, b_1, \cdots, b_n, \cdots$ are constants. Such a series is called a *power series in* $(x - a)$.

Let us apply the test-ratio test to (1). Then, if

$$\lim_{n \to \infty} \left| \frac{b_{n+1}}{b_n} \right| = M,$$

we shall have, for any fixed value of x,

$$\rho = \lim_{n \to \infty} \frac{u_{n+1}}{u_n} = |x - a| M.$$

We have two cases.

Case I. If $M = 0$, series (3) is convergent for all values of x.

Case II. If M is not zero, series (1) will converge for the interval

$$a - \frac{1}{M} < x < a + \frac{1}{M}.$$

A convergent power series in x is adapted for computation when x is near zero. Series (3), if convergent, is useful when x is near the fixed value a, given in advance.

Example 3. Test the infinite series

$$1 - (x-1) + \frac{(x-1)^2}{2} - \frac{(x-1)^3}{3} + \cdots$$

for convergence.

Solution. Neglecting the first term, we have

$$\frac{u_{n+1}}{u_n} = \left| \frac{n}{n+1}(x-1) \right|$$

Also, $$\lim_{n \to \infty} \left(\frac{n}{n+1} \right) = 1.$$

Hence $\rho = |x - 1|$, and the series will converge when x lies between 0 and 2. The end point $x = 2$ may be included.

PROBLEMS

For what values of the variable are the following series convergent?

Graphical representations of intervals of convergence*

1. $1 + x + x^2 + x^3 + \cdots$. *Ans.* $-1 < x < 1$.

2. $x - \frac{x^2}{2} + \frac{x^3}{3} - \frac{x^4}{4} + \cdots$. *Ans.* $-1 < x \leqq 1$.

3. $x + x^4 + x^9 + x^{16} + \cdots$. *Ans.* $-1 < x < 1$.

4. $x + \frac{x^2}{\sqrt{2}} + \frac{x^3}{\sqrt{3}} + \cdots$. *Ans.* $-1 \leqq x < 1$.

5. $1 + x + \frac{x^2}{\lfloor 2} + \frac{x^3}{\lfloor 3} + \cdots$. *Ans.* All values of x.

6. $1 - \frac{\theta^2}{\lfloor 2} + \frac{\theta^4}{\lfloor 4} - \frac{\theta^6}{\lfloor 6} + \cdots$. *Ans.* All values of θ.

7. $\phi - \frac{\phi^3}{\lfloor 3} + \frac{\phi^5}{\lfloor 5} - \frac{\phi^7}{\lfloor 7} + \cdots$. *Ans.* All values of ϕ.

8. $x + \frac{1}{2} \cdot \frac{x^3}{3} + \frac{1 \cdot 3}{2 \cdot 4} \cdot \frac{x^5}{5} + \frac{1 \cdot 3 \cdot 5}{2 \cdot 4 \cdot 6} \cdot \frac{x^7}{7} + \cdots$. *Ans.* $-1 \leqq x \leqq 1$.

9. $1 - 2x + 3x^2 - 4x^3 + \cdots$.

10. $1 + x + \frac{x^2}{2^2} + \frac{x^3}{3^2} + \frac{x^4}{4^2} + \cdots$.

*End points that are not included in the interval of convergence have circles drawn about them.

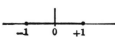

11. $2x + \dfrac{3x^2}{\lfloor 2} + \dfrac{4x^3}{\lfloor 3} + \dfrac{5x^4}{\lfloor 4} + \cdots$.

12. $\dfrac{x}{1\cdot 3} + \dfrac{x^2}{2\cdot 3^2} + \dfrac{x^3}{3\cdot 3^3} + \dfrac{x^4}{4\cdot 3^4} + \cdots$.

13. $1 + \dfrac{x^2}{2} + \dfrac{1\cdot 3\,x^4}{2\cdot 4} + \dfrac{1\cdot 3\cdot 5\,x^6}{2\cdot 4\cdot 6} + \cdots$.

14. $\dfrac{2x}{2} + \dfrac{2^2 x^2}{5} + \dfrac{2^3 x^3}{10} + \cdots + \dfrac{2^n x^n}{n^2 + 1} + \cdots$.

15. $1 + \dfrac{x}{2\cdot 1} + \dfrac{x^2}{2^2\cdot 2} + \dfrac{x^3}{2^3\cdot 3} + \dfrac{x^4}{2^4\cdot 4} + \cdots$.

16. $\dfrac{1}{3} + \dfrac{2x}{2\cdot 3^2} + \dfrac{3x^2}{2^2\cdot 3^3} + \dfrac{4x^3}{2^3\cdot 3^4} + \cdots$.

17. $\dfrac{x}{1\cdot 2} + \dfrac{2x^2}{2\cdot 2\cdot 3} + \dfrac{3x^3}{2^2\cdot 3\cdot 4} + \dfrac{4x^4}{2^3\cdot 4\cdot 5} + \cdots$.

18. $(x+1) - \dfrac{(x+1)^2}{2} + \dfrac{(x+1)^3}{3} - \dfrac{(x+1)^4}{4} + \cdots$.

19. $(x-1) + \dfrac{(x-1)^2}{\sqrt{2}} + \dfrac{(x-1)^3}{\sqrt{3}} + \dfrac{(x-1)^4}{\sqrt{4}} + \cdots$.

20. $2(2x+1) + \dfrac{3(2x+1)^2}{\lfloor 2} + \dfrac{4(2x+1)^3}{\lfloor 3} + \cdots$.

21. $1 + (x-2) + \dfrac{(x-2)^2}{2^2} + \dfrac{(x-2)^3}{3^2} + \dfrac{(x-2)^4}{4^2} + \cdots$.

251. Derivation of power series by differentiation. A convergent power series in x is obviously a function of x for all values in the interval of convergence. Thus we may write

$$f(x) = a_0 + a_1 x + a_2 x^2 + \cdots + a_n x^n + \cdots. \qquad (1)$$

If, then, a function is represented by a power series, what must be the form of the coefficients a_0, a_1, $\cdots$, a_n, etc.? To answer this question we proceed as follows.

Set $x = 0$ in (1). Then we must have

$$f(0) = a_0. \qquad (2)$$

Hence the first coefficient a_0 in (1) is determined. Now *assume* that the series in (1) may be differentiated term by term, and that this differentiation may be continued. Then we shall have

$$\left. \begin{aligned} f'(x) &= a_1 + 2\,a_2 x + 3\,a_3 x^2 + \cdots + n a_n x^{n-1} + \cdots \\ f''(x) &= 2\,a_2 + 6\,a_3 x + \cdots + n(n-1)a_n x^{n-2} + \cdots \\ f'''(x) &= 6\,a_3 + \cdots + n(n-1)(n-2)a_n x^{n-3} + \cdots \end{aligned} \right\} \qquad (3)$$

etc.

Letting $x = 0$, the results are

$$f'(0) = a_1, \quad f''(0) = \lfloor 2 \, a_2, \quad f'''(0) = \lfloor 3 \, a_3, \cdots, \quad f^{(n)}(0) = \lfloor n a_n. \quad (4)$$

Solving (4) for $a_1, a_2, \cdots, a_n$, etc., and substituting in (1), we obtain

$$\textbf{(B)} \quad f(x) = f(0) + f'(0) \frac{x}{\lfloor 1} + f''(0) \frac{x^2}{\lfloor 2} + \cdots + f^{(n)}(0) \frac{x^n}{\lfloor n} + \cdots.$$

This formula expresses $f(x)$ as a power series. We say, "the function $f(x)$ is developed (or expanded) in a power series in x." This is Maclaurin's series (or formula).*

It is now necessary to examine **(B)** critically. For this purpose refer to **(F)**, Art. 198, and rewrite it, letting $a = 0$, $b = x$. The result is

$$f(x) = f(0) + f'(0) \frac{x}{\lfloor 1} + f''(0) \frac{x^2}{\lfloor 2} + \cdots + f^{(n-1)}(0) \frac{x^{n-1}}{\lfloor n-1} + R, \quad (5)$$

where
$$R = f^{(n)}(x_n) \frac{x^n}{\lfloor n}. \qquad (0 < x_n < x)$$

The term R is called the *remainder after n terms*. The right-hand member of (5) agrees with Maclaurin's series **(B)** up to n terms. If we denote this sum by S_n, then (5) is

$$f(x) = S_n + R, \quad \text{or} \quad f(x) - S_n = R.$$

Now assume that, for a fixed value $x = x_0$, R approaches zero as a limit when n becomes infinite. Then S_n will approach $f(x_0)$ as a limit. That is, Maclaurin's series **(B)** will converge for $x = x_0$ and its value is $f(x_0)$. Thus we have the following result.

Theorem. *In order that the series **(B)** should converge and represent the function $f(x)$ it is necessary and sufficient that*

$$\lim_{n \to \infty} R = 0. \qquad (6)$$

It is frequently easier to determine the interval of convergence by the ratio test than that for which (6) holds. But in simple cases the two are identical.

To represent a function $f(x)$ by the power series **(B)**, it is obviously *necessary* that the function and its derivatives of all orders should be finite. This is, however, not *sufficient*.

Examples of functions that cannot be represented by a Maclaurin's series are

$$\ln x \quad \text{and} \quad \text{ctn } x,$$

since both become infinite when x is zero.

_____

*Named after Colin Maclaurin (1698–1746), and first published in his "Treatise of Fluxions" (Edinburgh, 1742). The series is really due to Stirling (1692–1770).

The student should not fail to note the importance of such an expansion as **(B)**. In all practical computations results correct to a certain number of decimal places are sought, and since the process in question replaces a function perhaps difficult to calculate by *an ordinary polynomial with constant coefficients*, it is very useful in simplifying such computations. Of course we must use terms enough to give the desired degree of accuracy.

In the case of an alternating series (Art. 256) the error made by stopping at any term is numerically less than that term.

Example 1. Expand cos x into an infinite power series and determine for what values of x it converges.

Solution. Differentiating first and then placing $x = 0$, we get

$$f(x) = \cos x, \qquad\qquad f(0) = 1,$$
$$f'(x) = -\sin x, \qquad\qquad f'(0) = 0,$$
$$f''(x) = -\cos x, \qquad\qquad f''(0) = -1,$$
$$f'''(x) = \sin x, \qquad\qquad f'''(0) = 0,$$
$$f^{iv}(x) = \cos x, \qquad\qquad f^{iv}(0) = 1,$$
$$f^{v}(x) = -\sin x, \qquad\qquad f^{v}_{\cdot}(0) = 0,$$
$$f^{vi}(x) = -\cos x, \qquad\qquad f^{vi}(0) = -1,$$
$$\text{etc.} \qquad\qquad\qquad \text{etc.}$$

Substituting in **(B)**, $\quad \cos x = 1 - \dfrac{x^2}{\underline{|2}} + \dfrac{x^4}{\underline{|4}} - \dfrac{x^6}{\underline{|6}} + \cdots.$ $\qquad\qquad$ (7)

Comparing with Problem 6, page 425, we see that the series converges for all values of x.

In the same way for sin x,

$$\sin x = x - \frac{x^3}{\underline{|3}} + \frac{x^5}{\underline{|5}} - \frac{x^7}{\underline{|7}} + \cdots,$$ $\qquad\qquad$ (8)

which converges for all values of x (Problem 7, page 425).

252. Series derived by integration. In the preceding article, series were derived by *differentiation*. A convergent power series may be *integrated* term by term, and the resulting series will converge also. Thus we may derive a new series by integrating a given convergent series.

Example. Derive the power series for arc sin x by integration.

Solution. Since $\dfrac{d}{dx}$ arc sin $x = \dfrac{1}{\sqrt{1-x^2}}$, we have

$$\text{arc sin } x + C = \int \frac{dx}{\sqrt{1-x^2}}.$$ $\qquad\qquad$ (1)

By Problem 3 **a**, page 420, we have

$$\frac{1}{\sqrt{1-x^2}} = 1 + \frac{x^2}{2,} + \frac{1\cdot 3}{2\cdot 4}x^4 + \frac{1\cdot 3\cdot 5}{2\cdot 4\cdot 6}x^6 + \cdots. \tag{2}$$

Substituting this series in the right-hand member of (1), and integrating term by term, the result is

$$\text{arc }\sin x + C = x + \frac{1}{2}\frac{x^3}{3} + \frac{1\cdot 3}{2\cdot 4}\frac{x^5}{5} + \frac{1\cdot 3\cdot 5}{2\cdot 4\cdot 6}\frac{x^7}{7} + \cdots.$$

To find C, let $x = 0$. Then $C = 0$, and we have the series required.

The series in (2) is convergent when x is numerically less than 1. The series for arc $\sin x$ is convergent also for the same values of x.

PROBLEMS

1. Verify the following expansions of functions by Maclaurin's series and show that they are convergent for all values of the variable.

a. $e^x = 1 + x + \frac{x^2}{\lfloor 2} + \frac{x^3}{\lfloor 3} + \cdots + \frac{x^{n-1}}{\lfloor n-1} + \cdots.$

b. $\sin\left(\tfrac{1}{4}\pi + x\right) = \frac{1}{\sqrt{2}}\left(1 + x - \frac{x^2}{\lfloor 2} - \frac{x^3}{\lfloor 3} + \frac{x^4}{\lfloor 4} + \frac{x^5}{\lfloor 5} - \frac{x^6}{\lfloor 6} - \frac{x^7}{\lfloor 7} + \cdots\right).$

2. Verify the following expansion and show that the interval of convergence is $-a < x \le a$.

$$\ln(a+x) = \ln a + \frac{x}{a} - \frac{x^2}{2\,a^2} + \cdots + \frac{(-1)^n x^{n-1}}{(n-1)a^{n-1}} + \cdots.$$

3. Verify the following expansions.

a. $\tan x = x + \frac{x^3}{3_{\rfloor}} + \frac{2\,x^5}{15} + \frac{17\,x^7}{315} + \cdots.$

b. $\sec x = 1 + \frac{x^2}{2} + \frac{5\,x^4}{24} + \frac{61\,x^6}{720} + \cdots.$

c. $\sin\left(\tfrac{1}{3}\pi + x\right) = \frac{1}{2}\left(\sqrt{3} + x - \frac{\sqrt{3}\,x^2}{\lfloor 2} - \frac{x^3}{\lfloor 3} + \frac{\sqrt{3}\,x^4}{\lfloor 4} + \frac{x^5}{\lfloor 5} - \cdots\right).$

d. $\tan\left(\tfrac{1}{4}\pi + x\right) = 1 + 2\,x + 2\,x^2 + \frac{8\,x^3}{3} + \cdots.$

e. $\tfrac{1}{2}(e^x + e^{-x}) = 1 + \frac{x^2}{\lfloor 2} + \frac{x^4}{\lfloor 4} + \frac{x^6}{\lfloor 6} + \cdots.$

4. Verify the following expansions by integration and show that in each case the interval of convergence is $-1 \le x \le 1$.

a. $\text{arc }\cos x = \frac{\pi}{2} - x - \frac{1}{2}\frac{x^3}{3} - \frac{1\cdot 3}{2\cdot 4}\frac{x^5}{5} - \cdots - \frac{1\cdot 3\cdots(2n-3)}{2\cdot 4\cdots(2n-2)}\frac{x^{2n-1}}{2\,n-1} - \cdots.$

b. $\text{arc }\tan x = x - \frac{x^3}{3} + \frac{x^5}{5} - \cdots + \frac{(-1)^{n-1}x^{2n-1}}{2\,n-1} + \cdots.$

Compute the values of the following functions by substituting directly in the corresponding power series, taking terms enough to make the results agree with those given below.

5. $e = 2.7182 \cdots$.

Solution. Let $x = 1$ in the series of Problem 1; then

$$e = 1 + 1 + \frac{1}{\underline{2}} + \frac{1}{\underline{3}} + \frac{1}{\underline{4}} + \frac{1}{\underline{5}} + \cdots.$$

First term	$= 1.00000$	
Second term	$= 1.00000$	
Third term	$= 0.50000$	
Fourth term	$= 0.16667 \cdots$	(Dividing third term by 3.)
Fifth term	$= 0.04167 \cdots$	(Dividing fourth term by 4.)
Sixth term	$= 0.00833 \cdots$	(Dividing fifth term by 5.)
Seventh term	$= 0.00139 \cdots$	(Dividing sixth term by 6.)
Eighth term	$= \underline{0.00020} \cdots$, etc.	(Dividing seventh term by 7.)
Adding,	$e = 2.71826 \cdots$	

6. arc tan $\frac{1}{5} = 0.1973 \cdots$; use series in Problem 4 **b**.

7. $\cos 1 = 0.5403 \cdots$; use series in Example 1, Art. 251.

8. $\cos 10° = 0.9848 \cdots$.

9. $\sin 0.1 = 0.0998 \cdots$.

10. arc sin $1 = 1.5708 \cdots$; use series in Example, Art. 252.

11. $\sqrt{e} = 1.6487 \cdots$; use series in Problem 1 **a**.

253. Operations with power series. Many of the operations of algebra and the calculus can be carried out with power series just as with polynomials. The following statements are given without proof.

Let
$$\text{1. } f(x) = a_0 + a_1 x + a_2 x^2 + \cdots,$$
$$\text{2. } g(x) = b_0 + b_1 x + b_2 x^2 + \cdots.$$

Then

I. $f(x) + g(x) = (a_0 + b_0) + (a_1 + b_1)x + (a_2 + b_2)x^2 + \cdots$.

II. $f(x)g(x) = a_0 b_0 + (a_0 b_1 + a_1 b_0)x + (a_0 b_2 + a_1 b_1 + a_2 b_0)x^2 + \cdots$.

Series (I) and (II) will be convergent at least for all values of x lying within the smaller of the two intervals of convergence of (1) and (2).

III. One power series may be divided by another as if they were polynomials.

IV. A power series may be differentiated term by term. Thus, from (1),

$$f'(x) = a_1 + 2 a_2 x + 3 a_3 x^2 + \cdots.$$

The result is convergent for all values of x lying within the interval of convergence of (1).

V. A power series may be integrated term by term if the limits lie within the interval of convergence, and the resulting series will be convergent.

Example 1. Obtain an approximate formula for $e^x \sin x$.

Solution. $\quad \sin x = x - \dfrac{x^3}{\lfloor 3} + \dfrac{x^5}{\lfloor 5} - \cdots.$

$$e^x = 1 + x + \dfrac{x^2}{\lfloor 2} + \dfrac{x^3}{\lfloor 3} + \dfrac{x^4}{\lfloor 4} + \cdots.$$

By multiplication, $\quad e^x \sin x = x + x^2 + \dfrac{x^3}{3} - \dfrac{x^5}{30} \cdots.$

The terms which have been omitted in this series involve x to at least the sixth power.

Example 2. Using series find approximately the value of

$$\int_0^1 \sin x^2 \, dx.$$

Solution. $\sin z = z - \dfrac{z^3}{\lfloor 3} + \dfrac{z^5}{\lfloor 5} - \cdots.$

Hence $\quad \sin x^2 = x^2 - \dfrac{x^6}{\lfloor 3} + \dfrac{x^{10}}{\lfloor 5} - \cdots$

and $\quad \displaystyle\int_0^1 \sin x^2 \, dx = \int_0^1 \left(x^2 - \dfrac{x^6}{\lfloor 3} + \dfrac{x^{10}}{\lfloor 5} \right) dx,$ approximately

$$= \left[\dfrac{x^3}{3} - \dfrac{x^7}{42} + \dfrac{x^{11}}{1320} \right]_0^1 = 0.3333 - 0.0238 + 0.0008$$

$$= 0.3103.$$

Example 3. Obtain an approximate formula for

$$\dfrac{\ln(1+x)}{\cos x}.$$

Solution. $\dfrac{\ln(1+x)}{\cos x} = \ln(1+x)[\cos x]^{-1},$

$$[\cos x]^{-1} = \left[1 - \dfrac{x^2}{\lfloor 2} + \dfrac{x^4}{\lfloor 4} \right]^{-1}, \text{ approximately}$$

$$= 1 - \left(-\dfrac{x^2}{\lfloor 2} + \dfrac{x^4}{\lfloor 4} \right) + \dfrac{(-1)(-2)}{1 \cdot 2} \left(-\dfrac{x^2}{\lfloor 2} + \dfrac{x^4}{\lfloor 4} \right)^2 + \cdots$$

$$= 1 + \dfrac{x^2}{2} + \dfrac{5 x^4}{24},$$

if all powers of x higher than the fifth are omitted. Now

$$\ln(1+x)[\cos x]^{-1} = \left(x - \dfrac{x^2}{2} + \dfrac{x^3}{3} - \dfrac{x^4}{4} \right) \left(1 + \dfrac{x^2}{2} + \dfrac{5 x^4}{24} \right), \text{ approximately}$$

and $\quad \dfrac{\ln(1+x)}{\cos x} = x - \dfrac{x^2}{2} + \dfrac{5 x^3}{6} - \dfrac{x^4}{2}.$

All terms which have been omitted will contain x to at least the fifth power.

254. Application to the period of a pendulum. When a simple pendulum of length l swings with an amplitude α, the period, or time for a complete oscillation, is given by the following integral.

$$P = 4 \sqrt{\frac{l}{g}} \int_0^{\frac{\pi}{2}} \frac{d\phi}{\sqrt{1 - k^2 \sin^2 \phi}},$$

where $k = \sin \frac{1}{2} \alpha$ and $\sin \phi = \dfrac{\sin \frac{1}{2} \theta}{k}$.

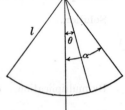

The integral cannot be found in terms of elementary functions. To obtain an approximate expression for P, we proceed as follows.

Fig. 254

$$P = 4 \sqrt{\frac{l}{g}} \int_0^{\frac{\pi}{2}} (1 - k^2 \sin^2 \phi)^{-\frac{1}{2}} d\phi$$

$$= 4 \sqrt{\frac{l}{g}} \int_0^{\frac{\pi}{2}} [1 + \tfrac{1}{2} k^2 \sin^2 \phi] d\phi, \text{ approximately}$$

$$= 4 \sqrt{\frac{l}{g}} \left[\phi + \frac{1}{2} k^2 \left(\frac{\phi}{2} - \frac{1}{4} \sin 2 \phi \right) \right]_0^{\frac{\pi}{2}}$$

$$= 4 \sqrt{\frac{l}{g}} \left[\frac{\pi}{2} + \frac{\pi}{8} k^2 \right].$$

Now $\qquad\qquad\qquad k = \sin \frac{1}{2} \alpha = \frac{1}{2} \alpha$, approximately.

Hence

$$P = 4 \sqrt{\frac{l}{g}} \left[\frac{\pi}{2} + \frac{\pi}{32} \alpha^2 \right] = 2 \pi \sqrt{\frac{l}{g}} \left[1 + \frac{\alpha^2}{16} \right].$$

The first approximation is the common formula $2 \pi \sqrt{\dfrac{l}{g}}$.

The second approximation gives an increase of

$$2 \pi \sqrt{\frac{l}{g}} \left(\frac{\alpha^2}{16} \right)$$

or a percentage increase of

$$\frac{100 \, \alpha^2}{16}.$$

A comparison of the results of the first and second approximations for different values of α is shown in the table at the right.

α	α (rad.)	α^2	Per Cent Increase
10°	0.175	0.031	0.2
20°	0.349	0.122	0.7
30°	0.524	0.275	1.7
40°	0.698	0.487	3.0

PROBLEMS

Verify the following series.

1. $e^{-\theta} \sin \theta = \theta - \theta^2 + \frac{1}{3} \theta^3 - \frac{1}{30} \theta^5 + \cdots$.

2. $e^{-x} \cos x = 1 - x + \frac{1}{3} x^3 + \cdots$.

3. $\dfrac{\sin x}{1-x} = x + x^2 + \frac{5}{6} x^3 + \frac{5}{6} x^4 + \frac{101}{120} x^5 + \cdots$.

4. $\dfrac{\cos x}{1+x} = 1 - x + \frac{1}{2} x^2 - \frac{1}{2} x^3 + \frac{13}{24} x^4 + \cdots$.

5. $\dfrac{\cos x}{1-x^2} = 1 + \frac{1}{2} x^2 + \frac{13}{24} x^4 + \cdots$.

6. $(1+x) \ln (1+x) = x + \frac{1}{2} x^2 - \frac{1}{6} x^3 + \frac{1}{12} x^4 + \cdots$.

7. $\dfrac{\ln (1+x)}{1+x} = x - \frac{3}{2} x^2 + \frac{11}{6} x^3 + \cdots$.

8. $\dfrac{e^x}{1+x} = 1 + \frac{1}{2} x^2 - \frac{1}{3} x^3 + \frac{3}{8} x^4 + \cdots$.

9. $e^x \ln (1+x) = x + \frac{1}{2} x^2 + \frac{1}{3} x^3 + \frac{3}{40} x^5 + \cdots$.

10. $\dfrac{e^{-x^2}}{\sqrt{1-x^2}} = 1 - \frac{1}{2} x^2 + \frac{3}{8} x^4 + \cdots$.

11. $\sin x \cos \sqrt{x} = x - \frac{1}{2} x^2 - \frac{1}{8} x^3 + \cdots$.

12. $\dfrac{\ln (1+x)}{\sec x} = x - \frac{1}{2} x^2 - \frac{1}{6} x^3 + \frac{3}{40} x^5 + \cdots$.

13. $(1-x) \arcsin x = x - x^2 + \frac{1}{6} x^3 - \frac{1}{6} x^4 + \frac{3}{40} x^5 - \frac{3}{40} x^6 + \cdots$.

14. $(1+x) \arctan x = x + x^2 - \frac{1}{3} x^3 - \frac{1}{3} x^4 + \frac{1}{5} x^5 + \frac{1}{5} x^6 + \cdots$.

15. $e^{-x} \cos \sqrt{x} = 1 - \frac{3}{2} x + \frac{25}{24} x^2 - \frac{331}{720} x^3 + \cdots$.

16. $e^{\frac{1}{2}x} \sin 2 x = 2 x + x^2 - \frac{13}{12} x^3 - \frac{5}{8} x^4 + \cdots$.

17. $\sqrt{1 + \sin x} = 1 + \frac{1}{2} x - \frac{1}{8} x^2 - \frac{1}{48} x^3 + \cdots$.

Verify the following approximate formulas.

18. $\displaystyle\int \cos \sqrt{x}\, dx = C + x - \frac{1}{4} x^2 + \frac{1}{72} x^3$.

19. $\displaystyle\int e^{-x^2}\, dx = C + x - \frac{1}{3} x^3 + \frac{1}{10} x^5$.

20. $\displaystyle\int \ln (1-x) dx = C - \frac{1}{2} x^2 - \frac{1}{6} x^3$.

21. $\displaystyle\int \arcsin x\, dx = C + \frac{1}{2} x^2 + \frac{1}{24} x^4$.

22. $\displaystyle\int e^{\theta} \sin \theta\, d\theta = C + \frac{1}{2} \theta^2 + \frac{1}{3} \theta^3$.

23. Derive the series for $\tan x$ and $\sec x$ (Problems 3 **a**, 3 **b**, page 429) from the series for $\sin x$ and $\cos x$.

Using series, find approximately the values of the following integrals.

24. $\displaystyle\int_0^1 \cos \sqrt{x}\, dx.$ **25.** $\displaystyle\int_0^1 e^{-x^2}\, dx.$ **26.** $\displaystyle\int_0^{\frac{1}{4}} \ln\left(1+\sqrt{x}\right) dx.$

27. $\displaystyle\int_0^1 e^{-\frac{1}{4}x^2}\, dx.$ **28.** $\displaystyle\int_0^1 \frac{\cos x\, dx}{\sqrt{4-x}}.$ **29.** $\displaystyle\int_0^{\frac{1}{5}} \frac{\sin x\, dx}{\sqrt{1-x}}.$

30. $\displaystyle\int_0^1 e^{\sqrt{x}}\, dx.$ **31.** $\displaystyle\int_0^1 e^x \cos \sqrt{x}\, dx.$ **32.** $\displaystyle\int_0^{\frac{1}{2}} \ln\left(1+x^2\right) dx.$

33. $\displaystyle\int_0^1 \sqrt{2 - \cos x}\, dx.$ **34.** $\displaystyle\int_0^{\frac{1}{2}} \frac{\ln\left(1-x\right)dx}{\cos x}.$

255. Taylor's series. A convergent power series in x is well adapted for calculating the value of the function which it represents for *small* values of x (near zero). We now derive an expansion in powers of $x - a$, where a is a fixed number. The series thus obtained is adapted for calculation of the function represented by it for values of x near a.

Assume that

$$f(x) = b_0 + b_1(x - a) + b_2(x - a)^2 + \cdots + b_n(x - a)^n + \cdots, \quad (1)$$

and that the series represents the function. The *necessary form* of the coefficients b_0, b_1, etc. is obtained as in Art. 251. That is, we differentiate (1) with respect to x, assuming that this is possible, and continue the process. Thus we have

$$f'(x) = b_1 + 2\,b_2(x - a) + \cdots + nb_n(x - a)^{n-1} + \cdots,$$
$$f''(x) = \qquad 2\,b_2 + \cdots + n(n - 1)b_n(x - a)^{n-2} + \cdots,$$

etc.

Substituting $x = a$ in these equations and in (1), and solving for b_0, b_1, b_2, $\cdots$, we obtain

$$b_0 = f(a), \quad b_1 = f'(a), \quad b_2 = \frac{f''(a)}{\underline{|2}}, \quad \cdots, \quad b_n = \frac{f^{(n)}(a)}{\underline{|n}}, \quad \cdots.$$

Replacing these values in (1), the result is

(C) $\qquad f(x) = f(a) + f'(a)\,\dfrac{(x - a)}{\underline{|1}} + f''(a)\,\dfrac{(x - a)^2}{\underline{|2}} + \cdots$

$$+ f^{(n)}(a)\,\frac{(x - a)^n}{\underline{|n}} + \cdots.$$

The series is called *Taylor's series* (or *formula*).*

*Published by Dr. Brook Taylor (1685–1731) in his "Methodus Incrementorum" (London, 1715).

Let us now examine (C) critically. Referring to (F), Art. 198, and letting $b = x$, the result may be written thus:

$$f(x) = f(a) + f'(a) \frac{(x-a)}{\lfloor 1} + \cdots + f^{(n-1)}(a) \frac{(x-a)^{n-1}}{\lfloor n-1} + R, \quad (2)$$

where
$$R = f^{(n)}(x_n) \frac{(x-a)^n}{\lfloor n}. \qquad (a < x_n < x)$$

The term R is called the *remainder after n terms.*

Now the series in the right-hand member of (2) agrees with Taylor's series (C) up to n terms. Denote the sum of these terms by S_n. Then, from (2), we have

$$f(x) = S_n + R, \quad \text{or} \quad f(x) - S_n = R.$$

Now assume that, for a fixed value $x = x_0$, the remainder R approaches zero as a limit when n becomes infinite. Then

$$\lim_{n \to \infty} S_n = f(x_0), \qquad (3)$$

and (C) converges for $x = x_0$ and its value is $f(x_0)$.

Theorem. *The infinite series (C) represents the function for those values of x, and those only, for which the remainder approaches zero as the number of terms increases without limit.*

It is usually easier to determine the interval of convergence of the series than that for which the remainder approaches zero; but in simple cases the two intervals are identical.

When the values of a function and its successive derivatives are known and finite for some fixed value of the variable, as $x = a$, then (C) is used for finding the value of the function for values of x near a, and (C) is also called *the expansion of $f(x)$ in the neighborhood of $x = a$.*

Another form of Taylor's series is obtained by setting $x - a = h$. Then $x = a + h$ and the series takes the form

$$(D) \quad f(a + h) = f(a) + f'(a)h + f''(a)\frac{h^2}{\lfloor 2} + f'''(a)\frac{h^3}{\lfloor 3} + \cdots.$$

Example 1. Expand $\ln x$ in powers of $(x - 1)$.

Solution.
$$f(x) = \ln x, \qquad f(1) = 0,$$
$$f'(x) = \frac{1}{x}, \qquad f'(1) = 1,$$
$$f''(x) = -\frac{1}{x^2}, \quad f''(1) = -1,$$
$$f'''(x) = \frac{2}{x^3}, \qquad f'''(1) = 2,$$
$$\text{etc.,} \qquad \text{etc.}$$

Substituting in **(C)**, $\ln x = x - 1 - \frac{1}{2}(x-1)^2 + \frac{1}{3}(x-1)^3 - \cdots$.

This converges for values of x between 0 and 2 and is the *expansion of ln x in the vicinity of* $x = 1$. See Example 3, Art. 250.

Example 2. Expand $\cos x$ in powers of $\left(x - \frac{\pi}{4}\right)$ to four terms.

Solution. Here $f(x) = \cos x$ and $a = \frac{\pi}{4}$. Then we have

$$f(x) = \cos x, \qquad f\left(\frac{\pi}{4}\right) = \frac{1}{\sqrt{2}},$$

$$f'(x) = -\sin x, \qquad f'\left(\frac{\pi}{4}\right) = -\frac{1}{\sqrt{2}},$$

$$f''(x) = -\cos x, \qquad f''\left(\frac{\pi}{4}\right) = -\frac{1}{\sqrt{2}},$$

$$f'''(x) = \sin x, \qquad f'''\left(\frac{\pi}{4}\right) = \frac{1}{\sqrt{2}},$$

$$\text{etc.,} \qquad\qquad \text{etc.}$$

The series is, therefore,

$$\cos x = \frac{1}{\sqrt{2}} - \frac{1}{\sqrt{2}}\left(x - \frac{\pi}{4}\right) - \frac{1}{\sqrt{2}}\frac{\left(x - \frac{\pi}{4}\right)^2}{\lfloor 2} + \frac{1}{\sqrt{2}}\frac{\left(x - \frac{\pi}{4}\right)^3}{\lfloor 3} + \cdots.$$

The result may be written in the form

$$\cos x = 0.70711\left[1 - \left(x - \frac{\pi}{4}\right) - \frac{1}{2}\left(x - \frac{\pi}{4}\right)^2 + \frac{1}{6}\left(x - \frac{\pi}{4}\right)^3 \cdots\right].$$

To check this result let us calculate $\cos 50°$. Then $x - \frac{\pi}{4} = 5°$ expressed in radians, or $x - \frac{\pi}{4} = 0.08727$, $\left(x - \frac{\pi}{4}\right)^2 = 0.00762$, $\left(x - \frac{\pi}{4}\right)^3 = 0.00066$. With these values the series above gives $\cos 50° = 0.64278$. Five-place tables give $\cos 50° = 0.64279$.

PROBLEMS

Verify the following series by Taylor's formula.

1. $e^x = e^a\left[1 + (x-a) + \frac{(x-a)^2}{\lfloor 2} + \frac{(x-a)^3}{\lfloor 3} + \cdots\right].$

2. $\sin x = \sin a + (x-a)\cos a - \frac{(x-a)^2}{\lfloor 2}\sin a - \frac{(x-a)^3}{\lfloor 3}\cos a + \cdots.$

3. $\cos x = \cos a - (x-a)\sin a - \frac{(x-a)^2}{\lfloor 2}\cos a + \frac{(x-a)^3}{\lfloor 3}\sin a + \cdots.$

4. $\ln(a+x) = \ln a + \frac{x}{a} - \frac{x^2}{2\,a^2} + \frac{x^3}{3\,a^3} + \cdots.$

5. $\cos (a + x) = \cos a - x \sin a - \dfrac{x^2}{\underline{|2}} \cos a + \dfrac{x^3}{\underline{|3}} \sin a + \cdots$.

6. $\tan (x + h) = \tan x + h \sec^2 x + h^2 \sec^2 x \tan x + \cdots$.

7. $(x + h)^n = x^n + nx^{n-1}h + \dfrac{n(n - 1)}{\underline{|2}} x^{n-2}h^2$

$$+ \dfrac{n(n - 1)(n - 2)}{\underline{|3}} x^{n-3}h^3 + \cdots.$$

8. $\tan x = 1 + 2\left(x - \dfrac{\pi}{4} \right) + 2\left(x - \dfrac{\pi}{4} \right)^2 + \cdots$.

9. Expand $\ln x$ in powers of $(x - 2)$ to four terms.

10. Expand e^x in powers of $(x - 1)$ to five terms.

11. Expand $\sin \left(\dfrac{\pi}{6} + x \right)$ in powers of x to four terms.

12. Expand $\operatorname{ctn} \left(\dfrac{\pi}{4} + x \right)$ in powers of x to three terms.

256. Alternating series. The usefulness of series as approximate formulas depends on the relative magnitude of the error in any particular application. When the signs of the terms of a convergent series are alternately positive and negative, there is a simple criterion for examining the error when a fixed number of terms is taken as an approximate formula.

The series for $\sin x$ and $\cos x$ derived in Art. 251 are examples of alternating series.

Suppose a function is represented by a convergent alternating series,

$$f(x) = a_1 - a_2 + a_3 - \cdots,$$

and that a certain fixed number m of terms is to be taken as an approximate formula. We may then write

$$f(x) = a_1 - a_2 + a_3 - a_4 + \cdots \pm a_m \mp E,$$

where E is equal to the sum of all the terms after a_m. Hence, if the first m terms are taken as an approximate formula, the error involved will be numerically equal to E.

If any series is convergent, the general term u_n must decrease and approach zero as a limit when $n \longrightarrow \infty$. If the terms of a series increase at first and then decrease, a good approximate formula should include at least all the terms which increase. We shall therefore assume that the discarded terms do not increase, and write

$$E = u_1 - u_2 + u_3 - u_4 + u_5 - u_6 + \cdots,$$

where $u_1 > u_2$, $u_2 > u_3$, etc. We assume that the value of x is some fixed value for which the series is convergent.

We now group the terms of E in two ways.

$$E = (u_1 - u_2) + (u_3 - u_4) + (u_5 - u_6) + \cdots,$$
$$E = u_1 - (u_2 - u_3) - (u_4 - u_5) - \cdots.$$

The quantity in every parenthesis is positive. The first grouping shows that E is positive because it is the sum of positive numbers. The second grouping shows that E is less than u_1. Hence

If an approximate formula is obtained by discarding from an alternating series all terms after some fixed term, the error will be numerically less than the first term discarded.

Example 1. How accurate is the approximate formula

$$\sin x = x - \tfrac{1}{6} x^3,$$

when $x = 30°$ (in radians)?

Solution. The error will be less than the first term of the sine series which has been discarded, that is, $\dfrac{x^5}{\underline{5}}$. When $x = 30°$, or 0.524 radian,

$$\frac{x^5}{\underline{5}} = 0.0003.$$

Hence the approximate formula will give the value of sin 30° with an error less than 0.0003.

Example 2. Compute the value of $\displaystyle\int_0^{\frac{1}{2}} \ln (1 + x^2)dx$ with an error less than 0.001.

Solution. $\displaystyle\int_0^{\frac{1}{2}} \ln (1 + x^2)dx = \int_0^{\frac{1}{2}} \left(x^2 - \frac{x^4}{2} + \frac{x^6}{3} - \frac{x^8}{4} + \frac{x^{10}}{5} \right) dx$, approximately

$$= \left[\frac{x^3}{3} - \frac{x^5}{10} + \frac{x^7}{21} - \frac{x^9}{36} + \frac{x^{11}}{55} \right]_0^{\frac{1}{2}}.$$

This is an alternating series, and when $x = \frac{1}{2}$ the value of the third term, $\dfrac{x^7}{21}$, is $\dfrac{1}{2688}$. Hence if all terms after the second are discarded, the error will be less than $\frac{1}{2688}$, or 0.0004. Taking the first two terms, we find

$$\left[\frac{x^3}{3} - \frac{x^5}{10} \right]_0^{\frac{1}{2}} = 0.0386.$$

The correct value must lie between

$$0.0386 - 0.0004 = 0.0382$$

and $\qquad 0.0386 + 0.0004 = 0.0390.$

The maximum percentage error is

$$\frac{0.0004}{0.0382} \times 100 = 1, \text{ approximately.}$$

PROBLEMS

1. How accurate is the approximate formula $\sin x = x - \frac{1}{6}x^3$ when **(a)** $x = 30°$? **(b)** $x = 60°$? **(c)** $x = 90°$?

2. How accurate is the approximate formula $\cos x = 1 - \frac{1}{2}x^2$ when **(a)** $x = 30°$? **(b)** $x = 60°$? **(c)** $x = 90°$?

3. How accurate is the approximate formula $e^{-x} = 1 - x$ when **(a)** $x = 0.1$? **(b)** $x = 0.5$?

4. How accurate is the approximate formula arc $\tan x = x - \frac{1}{3}x^3$ when **(a)** $x = 0.1$? **(b)** $x = 0.5$? **(c)** $x = 1$?

5. How many terms of the series $\sin x = x - \dfrac{x^3}{\lfloor 3} + \dfrac{x^5}{\lfloor 5} - \cdots$ must be taken to give $\sin 45°$ correct to five decimals?

6. How many terms of the series $\cos x = 1 - \dfrac{x^2}{\lfloor 2} + \dfrac{x^4}{\lfloor 4} - \cdots$ must be taken to give $\cos 60°$ correct to five decimals?

7. In the series $\ln (1 + x) = x - \frac{1}{2}x^2 + \frac{1}{3}x^3 - \cdots$ how many terms must be taken to give $\ln 1.2$ correct to five decimals?

8. Compute $\displaystyle\int_0^1 \cos \sqrt{x}\, dx$ with an error less than 0.001. What is the percentage error in your result?

9. Compute $\displaystyle\int_0^1 \sin x^2\, dx$ with an error less than 0.001. What is the percentage error in your result?

10. Compute $\displaystyle\int_0^{\frac{1}{4}} e^{-x^2}\, dx$ with an error less than 0.001. What is the percentage error in your result?

11. Compute $\displaystyle\int_0^{\frac{1}{4}} \ln (1 + \sqrt{x})\, dx$ with an error less than 0.001. What is the percentage error in your result?

12. Show that the approximate formula $\sin x = x$ is correct to three decimal places (error < 0.0005) if x is numerically less than 0.1442 radian.

CHAPTER XXI

DIFFERENTIAL EQUATIONS*

257. Differential equations; order and degree. A differential equation is an equation involving derivatives or differentials. Differential equations have appeared earlier in this book. See Art. 83. Thus, from the differential equation

$$\frac{dy}{dx} = 2\,x, \tag{1}$$

we find, by integrating,

$$y = x^2 + C. \tag{2}$$

Again, integration of the differential equation

$$\frac{dy}{dx} = -\frac{x}{y} \tag{3}$$

leads to the solution

$$x^2 + y^2 = 2\,C. \tag{4}$$

Equations (1) and (3) are examples of ordinary differential equations of the *first order*, and equations (2) and (4) are, respectively, the *complete solutions*.

Another example is

$$\frac{d^2y}{dx^2} + y = 0. \tag{5}$$

This is a differential equation of the *second order*, so named from the order of the derivative.

The *order of a differential equation* is the same as that of the derivative of highest order appearing in it.

If a differential equation has the form

$$f(y^{(k)},\ y^{(k-1)},\ \cdots y',\ y,\ x) = 0, \qquad \left(y^{(k)} = \frac{d^k y}{dx^k}\right)$$

where f is a polynomial in the derivatives, with coefficients which are functions of x and y, the *degree* of the equation is the highest exponent of $y^{(k)}$.

Thus the differential equation

$$y''^2 = (1 + y'^2)^3, \tag{6}$$

where y' and y'' are, respectively, the first and second derivatives of y with respect to x, is of the second degree and second order.

---

*A few types only of differential equations are treated in this chapter, namely, such types as the student is likely to encounter in elementary work in mechanics and physics.

258. Solutions of differential equations; constants of integration. A *solution* or *integral* of a differential equation is a relation between the variables involved by which the equation is satisfied. Thus

$$y = a \sin x \tag{1}$$

is a solution of the differential equation

$$\frac{d^2y}{dx^2} + y = 0. \tag{2}$$

For, differentiating (1),

$$\frac{d^2y}{dx^2} = - a \sin x. \tag{3}$$

Now, if we substitute from (1) and (3) in (2), we get

$$- a \sin x + a \sin x = 0,$$

and (2) is satisfied. Here a is an arbitrary constant. In the same manner

$$y = b \cos x \tag{4}$$

may be shown to be a solution of (2) for any value of b. The relation

$$y = c_1 \sin x + c_2 \cos x \tag{5}$$

is a still more general solution of (2). In fact, by giving particular values to c_1 and c_2 it is seen that the solution (5) includes the solutions (1) and (4). If $c_1 = c_2 = 0$, then $y = 0$. This is regarded as a solution of (2) although it does not involve the independent variable x.

The statement above defines a *formal* solution. The range of values of the variables and constants for which a formal solution is valid depends upon circumstances. For example, we see by differentiation that

$$x^2 + y^2 = c \tag{6}$$

is a formal solution of

$$y \frac{dy}{dx} + x = 0. \tag{7}$$

If we are dealing with real functions of real variables, then c can take no negative values. If a positive value has been assigned to c, then (6) shows that $c - x^2$ can take no negative values. Since it is naturally required that dy/dx shall exist for every valid value of x, (7) shows that $c - x^2$ must be positive. If we require that an explicit solution shall be single valued, then (6) furnishes two such solutions, valid for the ranges indicated.

$$y = \sqrt{c - x^2} \quad \text{and} \quad y = - \sqrt{c - x^2}; \ c > 0, \ x^2 < c.$$

We will be concerned with finding formal solutions, after which a separate discussion is required to determine the range of values for which the operations involved are legitimate.

The arbitrary constants c_1 and c_2 appearing in (5) are called *constants of integration.* A solution such as (5), which contains a number of essential arbitrary constants equal to the order of the equation (in this case two), is called the *complete* or *general solution.** Solutions obtained therefrom by giving particular values to the constants are called *particular solutions.* In practice a particular solution is obtained from the complete solution by using given conditions to be satisfied by the particular solution.

Example. The complete solution of the differential equation

$$y'' + y = 0$$

is
$$y = c_1 \cos x + c_2 \sin x. \quad \text{(See above.)}$$

Find a particular solution such that

$$y = 2, \quad y' = -1, \quad \text{when } x = 0. \tag{8}$$

Solution. From the complete solution

$$y = c_1 \cos x + c_2 \sin x, \tag{9}$$

by differentiation, we obtain

$$y' = -c_1 \sin x + c_2 \cos x. \tag{10}$$

Substituting in (9) and (10) from (8) we find $c_1 = 2$, $c_2 = -1$. Putting these values in (9) gives the particular solution required,

$$y = 2 \cos x - \sin x.$$

The solution of a differential equation is considered as having been effected when it has been reduced to an expression involving integrals, whether the actual integrations can be effected or not.

259. Verification of the solution of differential equations. The following examples show how to verify a solution of a differential equation.

Example 1. Is

$$y = a \cos 2x + b \sin 2x + c \tag{1}$$

a solution of

$$\frac{d^2y}{dx^2} + 4y = 8? \tag{2}$$

---

*It is shown in works on differential equations that the general solution has n arbitrary constants when the differential equation is of the nth order.

Solution. Differentiating (1), we get

$$\frac{dy}{dx} = -2\,a\sin 2\,x + 2\,b\cos 2\,x,$$

$$\frac{d^2y}{dx^2} = -4\,a\cos 2\,x - 4\,b\sin 2\,x. \tag{3}$$

Substituting in (2) from (1) and (3), we have

$$-4\,a\cos 2\,x - 4\,b\sin 2\,x + 4\,a\cos 2\,x + 4\,b\sin 2\,x + 4\,c = 8.$$

Hence (1) is a solution of (2) if, and only if, $c = 2$.

Example 2. **Show that**

$$y^2 - 4\,x = 0 \tag{4}$$

is a particular solution of the differential equation

$$xy'^2 - 1 = 0. \tag{5}$$

Solution. Differentiating (4), the result is

$$yy' - 2 = 0, \text{ whence } y' = \frac{2}{y}.$$

Substituting this value of y' in (5) and reducing, we obtain

$$4\,x - y^2 = 0,$$

which is true by (4).

PROBLEMS

Verify the following given solutions of the corresponding differential equations.

Differential Equations	Solutions
1. $(1 + x^2)\dfrac{dy}{dx} = xy.$	$cy^2 = 1 + x^2.$
2. $x\dfrac{dy}{dx} - 2\,y = 2\,x.$	$y = cx^2 - 2\,x.$
3. $\dfrac{du}{dv} = \dfrac{1 + u^2}{1 + v^2}.$	$u = \dfrac{c + v}{1 - cv}.$
4. $\left(\dfrac{dy}{dx}\right)^3 - 4\,xy\dfrac{dy}{dx} + 8\,y^2 = 0.$	$y = c(x - c)^2.$
5. $\dfrac{d^2y}{dx^2} - 4\dfrac{dy}{dx} + 3\,y = 0.$	$y = c_1 e^x + c_2 e^{3x}.$
6. $\dfrac{d^2x}{dt^2} - 2\dfrac{dx}{dt} + x = 0.$	$x = c_1 e^t + c_2 t e^t.$
7. $\dfrac{d^2s}{dt^2} + 2\dfrac{ds}{dt} + 2\,s = 0.$	$s = e^{-t}(c_1 \cos t + c_2 \sin t).$
8. $\dfrac{d^2y}{dx^2} - 2\dfrac{dy}{dx} - 3\,y = e^{2x}.$	$y = c_1 e^{3x} + c_2 e^{-x} - \frac{1}{3} e^{2x}.$

Which of the following equations are solutions of the corresponding differential equations?

Differential Equations	Equations
9. $x\dfrac{dy}{dx} - 2y + x = 0.$	$y = x^2 + x + 1.$
10. $\dfrac{ds}{dt} + s = \cos t - \sin t.$	$s = \cos t + ce^{-t}.$
11. $\dfrac{d^2s}{dt^2} + 9s = 6 \cos 3t.$	$s = (2 + t) \sin 3t.$
12. $\dfrac{d^2s}{dt^2} + 9s = 3t.$	$s = 2 \cos 3t + 3t.$
13. $\dfrac{d^2y}{dx^2} + 9y = e^x.$	$y = c_1 \cos 3x + c_2 \sin 3x + 8e^x.$
14. $\dfrac{d^2y}{dx^2} - 4\dfrac{dy}{dx} + 3y = 4e^x.$	$y = c_1 e^x + c_2 e^{3x} - 2xe^x.$

260. Differential equations of the first order and of the first degree. Such an equation, when solved for the derivative, has the form

$$\frac{dy}{dx} = f(x, y).$$

This equation may also be written in the form

(A) $M\,dx + N\,dy = 0,$

in which M and N are, in general, functions of x and y. The more common differential equations coming under this head may be divided into four types.

TYPE I. *Variables separable.* When the terms of a differential equation can be so arranged that it takes on the form

$$f(x)dx + F(y)dy = 0, \qquad (1)$$

where $f(x)$ is a function of x alone and $F(y)$ is a function of y alone, the process is called *separation of the variables*, and the solution is obtained by direct integration. Thus, integrating (1), we get the general solution

$$\int f(x)dx + \int F(y)dy = c, \qquad (2)$$

where c is an arbitrary constant.

When an equation can be arranged in the form (1), it is said that the *variables are separable*. Assuming that the values of the variables are such that the algebraic operations are legitimate, this arrangement may be accomplished by the following steps.

STEP I. *Clear of fractions, and if the equation involves derivatives, multiply through by the differential of the independent variable.*

STEP II. *Collect all the terms containing the same differential into a single term. If, then, the equation takes on the form*

$$XY\, dx + X'Y'dy = 0,$$

where X, X' are functions of x alone, and Y, Y' are functions of y alone, it may be brought to the form (1) by dividing through by $X'Y$.

STEP III. *Integrate each part separately, as in (2).*

Example 1. Solve the equation

$$\frac{dy}{dx} = \frac{1+y^2}{(1+x^2)xy}.$$

Solution. *Step I.* $(1+x^2)xy\, dy = (1+y^2)dx.$

Step II. $(1+y^2)dx - x(1+x^2)y\, dy = 0.$

To separate the variables we now divide by $x(1+x^2)(1+y^2)$, giving

$$\frac{dx}{x(1+x^2)} - \frac{y\, dy}{1+y^2} = 0.$$

Step III. $$\int \frac{dx}{x(1+x^2)} - \int \frac{y\, dy}{1+y^2} = C,$$

$$\int \frac{dx}{x} - \int \frac{x\, dx}{1+x^2} - \int \frac{y\, dy}{1+y^2} = C,$$

$$\ln x - \tfrac{1}{2}\ln(1+x^2) - \tfrac{1}{2}\ln(1+y^2) = C,$$

$$\ln[(1+x^2)(1+y^2)] = 2\ln x - 2C.$$

This result may be written in more compact form if we replace $-2C$ by $\ln c$, that is, give a new form to the arbitrary constant. Our solution then becomes

$$\ln[(1+x^2)(1+y^2)] = \ln x^2 + \ln c,$$

$$\ln[(1+x^2)(1+y^2)] = \ln cx^2,$$

$$(1+x^2)(1+y^2) = cx^2.$$

Example 2. Solve the equation

$$a\left(x\frac{dy}{dx} + 2\,y\right) = xy\frac{dy}{dx}.$$

Solution. *Step I.* $ax\, dy + 2\,ay\, dx = xy\, dy.$

Step II. $2\,ay\, dx + x(a-y)dy = 0.$

To separate the variables we divide by xy,

$$\frac{2\,a\, dx}{x} + \frac{(a-y)dy}{y} = 0.$$

Step III.
$$2a \int \frac{dx}{x} + a \int \frac{dy}{y} - \int dy = C,$$
$$2a \ln x + a \ln y - y = C,$$
$$a \ln x^2 y = C + y,$$
$$\ln x^2 y = \frac{C}{a} + \frac{y}{a}.$$

By passing from logarithms to exponentials this result may be written in the form
$$x^2 y = e^{\frac{C}{a} + \frac{y}{a}},$$

or
$$x^2 y = e^{\frac{C}{a}} \cdot e^{\frac{y}{a}}.$$

Denoting the constant $e^{\frac{C}{a}}$ by c, we get our solution in the form
$$x^2 y = c e^{\frac{y}{a}}.$$

Example 3. Solve the equation
$$\frac{dy}{dx} = \frac{1 + y^2}{1 + x^2}.$$

Solution. Separating the variables, we may write the equation in the form
$$\frac{dx}{1 + x^2} - \frac{dy}{1 + y^2} = 0,$$

and, by integration,   arc tan x − arc tan $y = C$.
This solution may be put in a different form as follows.
Let $x = \tan A, \quad y = \tan B.$
Then $A - B = C,$
and $\tan (A - B) = \tan C,$
$$\frac{\tan A - \tan B}{1 + \tan A \tan B} = \tan C,$$

or
$$\frac{x - y}{1 + xy} = k,$$

where $k = \tan C$ is an arbitrary constant.

The two solutions above are in the form of implicit relations between x and y. A solution giving y as an explicit function of x may be obtained from the last equation in the form
$$y = \frac{x - k}{1 + kx}.$$

Type II. *Homogeneous equations.* A function $f(x, y)$ is said to be homogeneous of degree n in x and y if
$$f(tx, ty) = t^n f(x, y), \tag{3}$$

where it is assumed that $t \neq 0$. For example, if
$$f(x, y) = ax^2 + bxy + cy^2,$$

then $\qquad f(tx,\ ty) = at^2x^2 + bt^2xy + ct^2y^2 = t^2f(x,\ y)$

and the function is homogeneous of degree 2.

If, in (3), $t = \dfrac{1}{x}$, the equation becomes

$$f\left(1, \frac{y}{x}\right) = \frac{1}{x^n}f(x,\ y)$$

or $\qquad\qquad f(x,\ y) = x^n f\left(1, \frac{y}{x}\right) = x^n\phi\left(\frac{y}{x}\right).$

This result may be stated as follows.

If $f(x,\ y)$ is homogeneous of degree n in x and y, then $f(x,\ y)$ can be expressed as the product of x^n and a function of v, where $v = y/x$.

A differential equation of the form **(A)** is said to be *homogeneous* if M and N are homogeneous functions of x and y of the same degree.

This definition does not imply that both x and y must occur in M and N. Thus the following differential equations are homogeneous.

$$x^2\ dx + y^2\ dy = 0.$$
$$y^2\ dx + (x^2 + xy)dy = 0.$$

Suppose

$$M(x,\ y)dx + N(x,\ y)dy = 0 \qquad\qquad (4)$$

is homogeneous and that the degree of M and N is n. Then

$$\frac{dy}{dx} = -\frac{M(x,\ y)}{N(x,\ y)} = -\frac{x^n\phi\left(\dfrac{y}{x}\right)}{x^n\psi\left(\dfrac{y}{x}\right)} = F\left(\frac{y}{x}\right). \qquad\qquad (5)$$

If we set

$$y = vx, \qquad \frac{dy}{dx} = x\frac{dv}{dx} + v,$$

(5) becomes

$$x\frac{dv}{dx} = F(v) - v, \qquad\qquad (6)$$

an equation in which the variables are separable.

The solution of (6) gives a relation involving x and v. If v is replaced by y/x, there results a solution of (4).

Example. Solve the equation

$$y^2 + x^2\frac{dy}{dx} = xy\frac{dy}{dx}.$$

Solution. $\qquad\qquad y^2\ dx + (x^2 - xy)dy = 0.$

Here $M = y^2$, $N = x^2 - xy$, and both are homogeneous and of the second degree in x and y. Also we have

$$\frac{dy}{dx} = \frac{y^2}{xy - x^2}.$$

Substitute $y = vx$. The result is

$$x\frac{dv}{dx} + v = \frac{v^2}{v-1},$$

which is in the form (6).

Hence

$$x\frac{dv}{dx} = \frac{v}{v-1}.$$

Changing to differentials, we get

$$v\,dx + x(1-v)dv = 0.$$

To separate the variables divide by vx. This gives

$$\frac{dx}{x} + \frac{(1-v)dv}{v} = 0,$$

$$\int\frac{dx}{x} + \int\frac{dv}{v} - \int dv = C,$$

$$\ln x + \ln v - v = C,$$

$$\ln vx = C + v,$$

$$vx = e^{C+v} = e^C \cdot e^v,$$

$$vx = ce^v.$$

But $v = \frac{y}{x}$. Hence the complete solution is $y = ce^{\frac{y}{x}}$.

PROBLEMS

Find the complete solution of each of the following differential equations.

1. $(a + y)dx - (b - x)dy = 0.$ **2.** $xy\,dx - (1 + x^2)dy = 0.$

3. $(a + 2y)x\,dx + (x^2 + b)dy = 0.$ **4.** $xy\,dx + \sqrt{1 - x^2}\,dy = 0.$

5. $d\rho + \rho\tan\theta\,d\theta = 0.$ **6.** $\sin n\theta\,d\rho = \rho\cos n\theta\,d\theta.$

7. $(a + x)dy + y^2\,dx = 0.$ **8.** $(x^2 + x)dy = (y - 1)dx.$

9. $(lx + my)dx + (mx + ny)dy = 0.$ **10.** $(3x + 5y)dx + (4x + 6y)dy = 0.$

11. $2(x + y)dx + y\,dy = 0.$ **12.** $(8y + 10x)dx + (5y + 7x)dy = 0.$

13. $(10x - 4y)dx = (x + 5y)dy.$ **14.** $(x^3 + y^3)dx = xy^2\,dy.$

15. $(m + n)(x\,dx - y\,dy) + (m - n)(x\,dy - y\,dx) = 0.$

16. $(x - 2y)dx + (2x + y)dy = 0.$

17. $2z(3z + 1)dw + (1 - 2w)dz = 0.$

18. $x\,dy - y\,dx = \sqrt{x^2 + y^2}\,dx.$ **19.** $(x^2 - 2y^2)dx + 2xy\,dy = 0.$

20. $(x^2 + y^2)dx = 2xy\,dy.$ **21.** $x\,dy - y\,dx = \sqrt{y^2 - x^2}\,dx.$

22. $(4x^3 + y^3)(x\,dy - y\,dx) - 3x^2y^2\,dx = 0.$

23. Show that the complete solution of

$$\sqrt{1 - y^2}\,dx + \sqrt{1 - x^2}\,dy = 0$$

can be put in the form $y\sqrt{1 - x^2} + x\sqrt{1 - y^2} = C.$

24. Show that the complete solution of

$$(1 + x^2)dy = \sqrt{1 - y^2}\, dx$$

can be put in the form $\dfrac{y}{\sqrt{1 - y^2}} = \dfrac{x + c}{1 - cx}.$

In the following problems find the particular solution which is determined by the given values of x and y.

    25. $xy\, dy = (1 + x^2)dx; \; x = 1, \, y = 1.$

    26. $xy\, dy = (1 + y^2)dx; \; x = 1, \, y = 0.$

    27. $(x^2 + y^2)dx = 2\, xy\, dy; \; x = 1, \, y = 0.$

    28. $x\, dy - y\, dx = \sqrt{x^2 + y^2}\, dx; \; x = \tfrac{1}{2}, \, y = 0.$

29. Find the equation of the curve whose slope at any point is equal to $-\dfrac{y}{x + y}$ and which passes through the point $(1, 1)$.

30. Find the equation of the curve whose slope at any point is equal to $\dfrac{\sqrt{1 - y^2}}{1 + x^2}$ and which passes through the origin.

Type III. *Linear equations.* The linear differential equation of the first order in y is of the form

(B) $$\frac{dy}{dx} + Py = Q,$$

where P, Q are functions of x alone or constants.

    Similarly, the equation

(C) $$\frac{dx}{dy} + Hx = J,$$

where H and J are functions of y or constants, is a linear differential equation in x.

    To integrate **(B)**, let

$$y = uz, \tag{7}$$

where z and u are functions of x to be determined.

    To find these two functions it will be necessary to have two conditions. The first condition is obtained by substitution in **(B)**.

    Differentiating (7),

$$\frac{dy}{dx} = u\frac{dz}{dx} + z\frac{du}{dx}. \tag{8}$$

Substituting from (8) and (7) in **(B)**, the result is

$$u\frac{dz}{dx} + z\frac{du}{dx} + Puz = Q,$$

or $$u\frac{dz}{dx} + \left(\frac{du}{dx} + Pu\right)z = Q. \tag{9}$$

As a second condition we now require that u satisfy the equation

$$\frac{du}{dx} + Pu = 0, \tag{10}$$

in which the variables x and u are separable. Using the value of u thus obtained, we find z by solving

$$u \frac{dz}{dx} = Q, \tag{11}$$

in which x and z can be separated. Obviously, the values of u and z thus found will satisfy (9), and the solution of **(B)** is then given by (7). The following examples show the details.

Example 1. Solve the equation

$$\frac{dy}{dx} - \frac{2\,y}{x+1} = (x+1)^{\frac{5}{2}}. \tag{12}$$

Solution. This is evidently in the linear form **(B)**, where

$$P = -\frac{2}{x+1} \quad \text{and} \quad Q = (x+1)^{\frac{5}{2}}.$$

Let $y = uz$; then

$$\frac{dy}{dx} = u \frac{dz}{dx} + z \frac{du}{dx}.$$

Substituting in the given equation (12), we get

$$u \frac{dz}{dx} + z \frac{du}{dx} - \frac{2\,uz}{1+x} = (x+1)^{\frac{5}{2}},$$

or

$$u \frac{dz}{dx} + \left(\frac{du}{dx} - \frac{2\,u}{1+x} \right) z = (x+1)^{\frac{5}{2}}. \tag{13}$$

To determine u we place the coefficient of z equal to zero. This gives

$$\frac{du}{dx} - \frac{2\,u}{1+x} = 0,$$

$$\frac{du}{u} = \frac{2\,dx}{1+x}.$$

Therefore

$$\ln u = 2 \ln (1 + x) = \ln (1 + x)^2.$$
$$u = (1 + x)^2.* \tag{14}$$

Equation (13) now becomes, since the term in z drops out,

$$u \frac{dz}{dx} = (x+1)^{\frac{5}{2}}.$$

Replacing u by its value from (14),

$$\frac{dz}{dx} = (x+1)^{\frac{1}{2}}, \quad dz = (x+1)^{\frac{1}{2}}\,dx.$$

*For the sake of simplicity we have assumed the particular value zero for the constant of integration (see Example 2).

Integrating, we get $\qquad z = \dfrac{2(x+1)^{\frac{3}{2}}}{3} + C.$ $\qquad\qquad$ (15)

Substituting from (15) and (14) in $y = uz$, we get the complete solution

$$y = \frac{2(x+1)^{\frac{7}{2}}}{3} + C(x+1)^2.$$

Example 2. Derive a formula for the complete solution of **(B)**.

Solution. Solving (10), we have

$$\ln u + \int P\, dx = \ln k,$$

where $\ln k$ is the constant of integration; whence

$$u = ke^{-\int P\, dx}.$$

Substituting this value of u in (11) and separating the variables z and x, the result is

$$dz = \left[\frac{Q}{k} e^{\int P\, dx} \right] dx.$$

Integrating, and substituting in (7), we obtain

$$y = e^{-\int P\, dx} \left(\int Q e^{\int P\, dx}\, dx + C \right).$$

It should be observed that the constant k cancels out of the final result. For this reason it is customary to omit the constant of integration in solving (10).

TYPE IV. *Equations reducible to the linear form.* Some equations that are not linear can be reduced to the linear form by means of a suitable transformation. One type of such equations is the equation of Bernoulli

(D) $\qquad\qquad \dfrac{dy}{dx} + Py = Qy^n,$

where P, Q are functions of x alone, or constants. Equation **(D)** may be reduced to the linear form **(B)**, Type III, by means of the substitution $z = y^{-n+1}$. Such a reduction, however, is not necessary if we employ the same method for finding the solution as that given under Type III. Let us illustrate this by means of an example.

Example. Solve the equation

$$\frac{dy}{dx} + \frac{y}{x} = (a \ln x)y^2. \qquad\qquad (16)$$

Solution. This is evidently in the form **(D)**, where

$$P = \frac{1}{x}, \quad Q = a \ln x, \quad n = 2.$$

Let $y = uz$; then $\quad\quad\quad \dfrac{dy}{dx} = u\dfrac{dz}{dx} + z\dfrac{du}{dx}.$

Substituting in (16), we get

$$u\frac{dz}{dx} + z\frac{du}{dx} + \frac{uz}{x} = (a\ln x)u^2z^2,$$

$$u\frac{dz}{dx} + \left(\frac{du}{dx} + \frac{u}{x}\right)z = (a\ln x)u^2z^2. \tag{17}$$

To determine u we place the coefficient of z equal to zero. This gives

$$\frac{du}{dx} + \frac{u}{x} = 0,$$

$$\frac{du}{u} = -\frac{dx}{x},$$

$$\ln u = -\ln x = \ln\frac{1}{x},$$

$$u = \frac{1}{x}. \tag{18}$$

Since the term in z drops out, (17) now becomes

$$u\frac{dz}{dx} = (a\ln x)u^2z^2,$$

$$\frac{dz}{dx} = (a\ln x)uz^2.$$

Replacing u by its value from (18),

$$\frac{dz}{dx} = (a\ln x)\frac{z^2}{x},$$

$$\frac{dz}{z^2} = (a\ln x)\frac{dx}{x}.$$

Integrating, $\quad\quad\quad\quad -\dfrac{1}{z} = \dfrac{a(\ln x)^2}{2} + C,$

$$z = -\frac{2}{a(\ln x)^2 + 2\,C}. \tag{19}$$

Substituting from (19) and (18) in $y = uz$, we get the complete solution

$$y = -\frac{1}{x}\cdot\frac{2}{a(\ln x)^2 + 2\,C},$$

or $\quad\quad\quad\quad xy[a(\ln x)^2 + 2\,C] + 2 = 0.$

PROBLEMS

Find the complete solution of each of the following differential equations.

1. $\dfrac{dy}{dx} + ay = b.$ $\quad\quad\quad\quad\quad\quad$ 2. $\dfrac{dy}{dx} + y = e^{-x}.$

3. $\dfrac{dy}{dx} + ay = f(x).$ $\quad\quad\quad\quad\quad$ 4. $\dfrac{dy}{dx} - y = -\,2\,e^{-x}.$

5. $\dfrac{ds}{dt} + s = \cos t - \sin t.$ **6.** $x\dfrac{dy}{dx} - y = (x-1)e^x.$

7. $t\dfrac{ds}{dt} + s = \sin t + t \cos t.$ **8.** $2x\dfrac{dy}{dx} + 2y = xy^3.$

9. $\dfrac{dy}{dx} + y \operatorname{ctn} x = \operatorname{ctn} x.$ **10.** $\dfrac{ds}{dt} - s \operatorname{ctn} t = e^t(1 - \operatorname{ctn} t).$

11. $\dfrac{ds}{dt} + s \tan t = \cos t.$ **12.** $x\dfrac{dy}{dx} - 2y = x^3 e^x.$

13. $2xy\dfrac{dy}{dx} - 2y^2 + x = 0.$ **14.** $\dfrac{dy}{dx} - \dfrac{2y}{x+1} = (x+1)^3.$

In each of the following problems find the particular solution which is determined by the given values of x and y.

15. $x\dfrac{dy}{dx} - 3y + 4x = 0;\ x=1,\ y=1.$

16. $x\dfrac{dy}{dx} + y = xy^3;\ x=2,\ y=-1.$

17. $\dfrac{dy}{dx} + y \tan x = e^{-x}(\tan x - 1);\ x=0,\ y=0.$

18. $\dfrac{dy}{dx} + y \tan x = 2x + x^2 \tan x;\ x=0,\ y=1.$

19. Find the equation of the curve whose slope at any point is equal to $2y - 2x + 1$ and which passes through the point $(0, 2)$.

20. Find the equation of the curve whose slope at any point is equal to $\dfrac{y^2 \ln x - y}{x}$ and which passes through the point $(1, 1)$.

261. Two special types of differential equations of higher order. The differential equations discussed in this article occur frequently.

(E) $$\frac{d^n y}{dx^n} = X,$$

where X is a function of x alone, or a constant.

To integrate, multiply both members by dx and integrate. Then we have

$$\frac{d^{n-1}y}{dx^{n-1}} = \int \frac{d^n y}{dx^n}\, dx = \int X\, dx + c_1.$$

Repeat the process $(n-1)$ times. Then the complete solution containing n arbitrary constants will be obtained.

Example. Solve $\dfrac{d^3y}{dx^3} = xe^x.$

Solution. Multiplying both members by dx and integrating,

$$\frac{d^2y}{dx^2} = \int xe^x \, dx + C_1,$$

or

$$\frac{d^2y}{dx^2} = xe^x - e^x + C_1.$$

Repeating the process,

$$\frac{dy}{dx} = \int xe^x \, dx - \int e^x \, dx + \int C_1 \, dx + C_2.$$

$$\frac{dy}{dx} = xe^x - 2\,e^x + C_1x + C_2.$$

$$y = \int xe^x \, dx - \int 2\,e^x \, dx + \int C_1x \, dx + \int C_2 \, dx + C_3$$

$$= xe^x - 3\,e^x + \frac{C_1x^2}{2} + C_2x + C_3.$$

$$y = xe^x - 3\,e^x + c_1x^2 + c_2x + c_3.$$

A second type of much importance is

(F)
$$\frac{d^2y}{dx^2} = Y,$$

where Y is a function of y alone.

To integrate, proceed thus. We have

$$\frac{d^2y}{dx^2} = \frac{dy'}{dx} = \frac{dy'}{dy}\frac{dy}{dx} = \frac{dy'}{dy}\,y'.$$

Hence **(F)** may be written, after multiplying through by dy,

$$y' \, dy' = Y \, dy.$$

The variables y and y' are now separated. Integrating, the result is

$$\tfrac{1}{2}\,y'^2 = \int Y \, dy + C_1.$$

The right-hand member is a function of y. Extract the square root of both members, separate the variables x and y, and integrate again. The following example illustrates the method.

Example. Solve $\dfrac{d^2y}{dx^2} + a^2y = 0.$

Solution. Here $\dfrac{d^2y}{dx^2} = -\,a^2y$, and hence the equation belongs to type **(F)**

Proceeding as above, the equation may be written

$$y' \, dy' = - a^2 y \, dy.$$
$$\tfrac{1}{2} y'^2 = - \tfrac{1}{2} a^2 y^2 + C.$$
$$y' = \sqrt{2 \, C - a^2 y^2},$$
$$\frac{dy}{dx} = \sqrt{C_1 - a^2 y^2},$$

setting $2 \, C = C_1$ and taking the positive sign of the radical. Separating the variables, we get

$$\frac{dy}{\sqrt{C_1 - a^2 y^2}} = dx.$$

Integrating,

$$\frac{1}{a} \arc \sin \frac{ay}{\sqrt{C_1}} = x + C_2,$$

or

$$\arc \sin \frac{ay}{\sqrt{C_1}} = ax + aC_2.$$

This is the same as $\dfrac{ay}{\sqrt{C_1}} = \sin (ax + aC_2)$

$$= \sin ax \cos aC_2 + \cos ax \sin aC_2,$$

or

$$y = \frac{\sqrt{C_1}}{a} \cos aC_2 \cdot \sin ax + \frac{\sqrt{C_1}}{a} \sin aC_2 \cdot \cos ax.$$

$$y = c_1 \sin ax + c_2 \cos ax.$$

PROBLEMS

Find the complete solution of each of the following differential equations.

1. $\dfrac{d^2 x}{dt^2} = t^2.$

2. $\dfrac{d^2 y}{dx^2} - a^2 y = 0.$

3. $\dfrac{d^2 x}{dt^2} = 4 \sin 2 \, t.$

4. $\dfrac{d^2 y}{dx^2} = e^{2x}.$

5. $\dfrac{d^2 s}{dt^2} = a \cos nt.$

6. $\dfrac{d^2 y}{dt^2} = \dfrac{a}{y^3}.$

7. $\dfrac{d^3 y}{dx^3} = x + \sin x.$

8. $\dfrac{d^2 y}{dx^2} = \dfrac{a + bx}{x^2}.$

9. Show that the complete solution of $\dfrac{d^2 s}{dt^2} = \dfrac{1}{\sqrt{s}}$ can be put in the form

$$3 \, t = 2(\sqrt{s} - 2 \, C_1)\sqrt{\sqrt{s} + C_1} + C_2.$$

10. Given $\dfrac{d^2 s}{dt^2} + \dfrac{1}{s^2} = 0.$

a. If $s = 1$, $s' = 0$, when $t = 0$, show that the solution is

$$\pm \sqrt{2} \, t = \sqrt{s(1 - s)} + \arc \sin \sqrt{1 - s}.$$

b. If $s = 1$, $s' = 2$, when $t = 0$, show that the solution is

$$\sqrt{2}(t + 1) = \sqrt{s(1 + s)} - \ln \left(\frac{\sqrt{s} + \sqrt{1 + s}}{1 + \sqrt{2}} \right).$$

262. Linear differential equations of the second order with constant coefficients. The *linear* equation of the second order is of the form

$$\frac{d^2y}{dx^2} + P\frac{dy}{dx} + Qy = X,$$

where P, Q, and X are functions of x alone. We consider only cases in which P and Q are constants and suppose first that $X = 0$.

Equations of the form

(G) $$\frac{d^2y}{dx^2} + p\frac{dy}{dx} + qy = 0,$$

where p and q are constants (independent of x and y), are important in applied mathematics.

To obtain a particular solution of (G), let us try to determine the value of the constant r so that (G) will be satisfied by

$$y = e^{rx}. \tag{1}$$

Differentiating (1), we obtain

$$\frac{dy}{dx} = re^{rx}, \quad \frac{d^2y}{dx^2} = r^2 e^{rx}. \tag{2}$$

Substituting from (1) and (2) in (G) and dividing out the factor e^{rx}, which cannot vanish, the result is

$$r^2 + pr + q = 0, \tag{3}$$

a quadratic equation whose roots are the values of r required. Equation (3) is called the *auxiliary equation* for (G). If (3) has distinct roots r_1 and r_2, then

$$y = e^{r_1 x} \quad \text{and} \quad y = e^{r_2 x} \tag{4}$$

are distinct particular solutions of (G), and the complete solution is

$$y = c_1 e^{r_1 x} + c_2 e^{r_2 x}. \tag{5}$$

In fact, (5) contains two essential arbitrary constants, and (G) is satisfied by this relation.

Example 1. Solve

$$\frac{d^2y}{dx^2} - 2\frac{dy}{dx} - 3y = 0. \tag{6}$$

Solution. The auxiliary equation is

$$r^2 - 2r - 3 = 0. \tag{7}$$

Solving (7), the roots are 3 and -1, and by (5) the complete solution is

$$y = c_1 e^{3x} + c_2 e^{-x}.$$

Check. Substituting this value of y in (6), the equation is satisfied.

Roots of (3) imaginary. If the roots of the auxiliary equation (3) are imaginary, the exponents in (5) will also be imaginary. A real complete solution may be found, however, by choosing imaginary values of c_1 and c_2 in (5).

In fact, let $\qquad r_1 = a + b\sqrt{-1}, \quad r_2 = a - b\sqrt{-1}$ $\qquad$ (8)

be the pair of conjugate imaginary roots of (3). Then

$$e^{r_1 x} = e^{(a+b\sqrt{-1})x} = e^{ax}e^{bx\sqrt{-1}},$$
$$e^{r_2 x} = e^{(a-b\sqrt{-1})x} = e^{ax}e^{-bx\sqrt{-1}}.$$

(9)

Substituting these values in (5), we obtain

$$y = e^{ax}\left(c_1 e^{bx\sqrt{-1}} + c_2 e^{-bx\sqrt{-1}}\right).$$

(10)

In the algebra of imaginary numbers it is shown that

$$e^{bx\sqrt{-1}} = \cos bx + \sqrt{-1}\,\sin bx,$$
$$e^{-bx\sqrt{-1}} = \cos bx - \sqrt{-1}\,\sin bx.$$

This can be shown formally as follows. Take the standard series (Problem 1, page 429).

$$e^z = 1 + z + \frac{z^2}{\underline{|2}} + \frac{z^3}{\underline{|3}} + \frac{z^4}{\underline{|4}} + \cdots.$$

Let $z = bx\sqrt{-1}$. Then the powers of z become

$$z^2 = -b^2x^2, \quad z^3 = -b^3x^3\sqrt{-1}, \quad z^4 = b^4x^4, \text{ etc.}$$

Hence $\qquad e^{bx\sqrt{-1}} = 1 + bx\sqrt{-1} - \frac{b^2x^2}{\underline{|2}} - \frac{b^3x^3}{\underline{|3}}\sqrt{-1} + \frac{b^4x^4}{\underline{|4}} + \cdots.$

Separating the real terms and imaginary terms,

$$e^{bx\sqrt{-1}} = 1 - \frac{b^2x^2}{\underline{|2}} + \frac{b^4x^4}{\underline{|4}} - \cdots + \sqrt{-1}\left(bx - \frac{b^3x^3}{\underline{|3}} + \cdots\right).$$

(10 a)

From the standard series for cos z and sin z, we obtain, by substituting $z = bx$,

$$\cos bx = 1 - \frac{b^2x^2}{\underline{|2}} + \frac{b^4x^4}{\underline{|4}} - \cdots,$$

$$\sin bx = bx - \frac{b^3x^3}{\underline{|3}} + \cdots.$$

Comparing these series with the right-hand member in (10 a), we see that

$$e^{bx\sqrt{-1}} = \cos bx + \sqrt{-1}\,\sin bx.$$

Similarly, by setting $z = -bx\sqrt{-1}$, we may show that

$$e^{-bx\sqrt{-1}} = \cos bx - \sqrt{-1}\,\sin bx.$$

When these values are substituted in (10), the complete solution may be written

$$y = e^{ax}(A \cos bx + B \sin bx),$$

(11)

if the new arbitrary constants A and B are determined from c_1 and c_2 by $A = c_1 + c_2$, $B = (c_1 - c_2)\sqrt{-1}$. That is, we now take for c_1 and c_2 in (5) the imaginary values $c_1 = \frac{1}{2}(A - B\sqrt{-1})$, $c_2 = \frac{1}{2}(A + B\sqrt{-1})$.

By giving to A and B in (11) the values 1 and 0, and 0, 1, in turn, we see that

$$y = e^{ax} \cos bx \quad \text{and} \quad y = e^{ax} \sin bx \tag{12}$$

are real particular solutions of **(G)**.

Example 2. Solve

$$\frac{d^2y}{dx^2} + k^2y = 0. \tag{13}$$

Solution. The auxiliary equation (3) is now

$$r^2 + k^2 = 0. \text{ Therefore } r = \pm k\sqrt{-1}.$$

Comparing with (8), we see that $a = 0$, $b = k$. Hence, by (11), the complete solution is

$$y = A \cos kx + B \sin kx.$$

Check. When this value of y is substituted in (13), the equation is satisfied. (Compare this method with that used for the same example in Art. 261.)

Roots of (3) real and equal. The roots of the auxiliary equation (3) will be equal if $p^2 = 4q$. Then (3) may be written, by substituting $q = \frac{1}{4}p^2$,

$$r^2 + pr + \frac{1}{4}p^2 = (r + \frac{1}{2}p)^2 = 0, \tag{14}$$

and $r_1 = r_2 = -\frac{1}{2}p$. In this case

$$y = e^{r_1 x}, \quad y = xe^{r_1 x}, \tag{15}$$

are distinct particular solutions, and

$$y = e^{r_1 x}(c_1 + c_2 x) \tag{16}$$

is the complete solution.

To corroborate this statement, it is only necessary to prove that the second equation in (15) gives a solution. But we have, by differentiating,

$$y = xe^{r_1 x}, \quad \frac{dy}{dx} = e^{r_1 x}(1 + r_1 x), \quad \frac{d^2y}{dx^2} = e^{r_1 x}(2r_1 + r_1^2 x). \tag{17}$$

Substituting from (17) into the left-hand member of **(G)**, the result is, after canceling $e^{r_1 x}$,

$$(r_1^2 + pr_1 + q)x + 2r_1 + p. \tag{18}$$

This vanishes, since r_1 satisfies (3) and equals $-\frac{1}{2}p$.

Example 3. Solve

$$\frac{d^2s}{dt^2} + 2\frac{ds}{dt} + s = 0. \tag{19}$$

Find the particular solution such that

$$s = 4, \quad \frac{ds}{dt} = -2, \quad \text{when } t = 0.$$

Solution. The auxiliary equation is

$$r^2 + 2r + 1 = 0, \quad \text{or} \quad (r+1)^2 = 0.$$

Hence the roots are equal, $r_1 = -1$, and, by (16),

$$s = e^{-t}(c_1 + c_2 t). \tag{20}$$

This is the complete solution.

To find the required particular solution, we must find values for the constants c_1 and c_2 such that the given conditions,

$$s = 4, \quad \frac{ds}{dt} = -2, \quad \text{when } t = 0,$$

are satisfied.

Substituting in the complete solution (20) the given values

$s = 4$, $t = 0$, we have $4 = c_1$ and hence

$$s = e^{-t}(4 + c_2 t). \tag{21}$$

Now differentiate (21) with respect to t. We get

$$\frac{ds}{dt} = e^{-t}(c_2 - 4 - c_2 t).$$

By the conditions given, $ds/dt = -2$ when $t = 0$. Substituting, the result is $-2 = c_2 - 4$, and hence $c_2 = 2$. Then the particular solution required is

$$s = e^{-t}(4 + 2t).$$

PROBLEMS

Find the complete solution of each of the following differential equations.

1. $\dfrac{d^2s}{dt^2} - 4\dfrac{ds}{dt} + 3s = 0.$

2. $\dfrac{d^2y}{dx^2} - \dfrac{dy}{dx} - 6y = 0.$

3. $\dfrac{d^2x}{dt^2} - 4\dfrac{dx}{dt} + 4x = 0.$

4. $\dfrac{d^2s}{dt^2} - 16s = 0.$

5. $\dfrac{d^2s}{dt^2} + 16s = 0.$

6. $\dfrac{d^2s}{dt^2} + 16\dfrac{ds}{dt} = 0.$

7. $\dfrac{d^2x}{dt^2} - 4\dfrac{dx}{dt} + 13x = 0.$

8. $\dfrac{d^2x}{dt^2} + 4\dfrac{dx}{dt} + 8x = 0.$

9. $\dfrac{d^2\rho}{d\theta^2} - 2\dfrac{d\rho}{d\theta} + \rho = 0.$

10. $\dfrac{d^2x}{dt^2} - 8\dfrac{dx}{dt} + 25x = 0.$

11. $\dfrac{d^2s}{dt^2} + 6\dfrac{ds}{dt} + 10s = 0.$

12. $2\dfrac{d^2s}{dt^2} + 3\dfrac{ds}{dt} + s = 0.$

13. $9\dfrac{d^2y}{dx^2} - 6\dfrac{dy}{dx} + y = 0.$

14. $2\dfrac{d^2y}{dx^2} + 3\dfrac{dy}{dx} - 2y = 0.$

15. $2\dfrac{d^2s}{dt^2} + 2\dfrac{ds}{dt} + s = 0.$

16. $3\dfrac{d^2s}{dt^2} - 4\dfrac{ds}{dt} + 2s = 0.$

In the following problems find the particular solutions which satisfy the given conditions.

17. $\dfrac{d^2x}{dt^2} - 5\dfrac{dx}{dt} + 6x = 0;\ x = \tfrac{1}{2},\ x' = 1,$ when $t = 0.$

18. $\dfrac{d^2y}{dx^2} + \dfrac{dy}{dx} - 2y = 0;\ y = 3,\ y' = 0,$ when $x = 0.$

19. $\dfrac{d^2s}{dt^2} - 4s = 0;\ s = 6,\ s' = 0,$ when $t = 0.$

20. $\dfrac{d^2s}{dt^2} + 4s = 0;\ s = 0,\ s' = 10,$ when $t = 0.$

21. $\dfrac{d^2y}{dx^2} + y = 0;\ y = 4$ when $x = 0,$ and $y = 0$ when $x = \tfrac{1}{2}\pi.$

22. $\dfrac{d^2y}{dx^2} - 6\dfrac{dy}{dx} + 9y = 0;\ y = 0,\ \dfrac{dy}{dx} = 2,$ when $x = 0.$

23. $\dfrac{d^2x}{dt^2} + 2\dfrac{dx}{dt} + 2x = 0;\ x = 3,\ \dfrac{dx}{dt} = -3,$ when $t = 0.$

24. $\dfrac{d^2s}{dt^2} + 2\dfrac{ds}{dt} + 5s = 0;\ s = 1,\ \dfrac{ds}{dt} = 1,$ when $t = 0.$

25. $\dfrac{d^2s}{dt^2} + 3\dfrac{ds}{dt} + 2s = 0;\ s = -1,\ \dfrac{ds}{dt} = 3,$ when $t = 0.$

26. $\dfrac{d^2y}{dx^2} + 4\dfrac{dy}{dx} + 4y = 0;\ y = 1,\ \dfrac{dy}{dx} = -1,$ when $x = 0.$

27. $\dfrac{d^2s}{dt^2} + n^2s = 0;\ s = 0,\ \dfrac{ds}{dt} = v_0,$ when $t = 0.$

28. $\dfrac{d^2s}{dt^2} - n\dfrac{ds}{dt} = 0;\ s = 0,\ \dfrac{ds}{dt} = n,$ when $t = 0.$

29. $\dfrac{d^2x}{dt^2} + 2\dfrac{dx}{dt} + 2x = 0;\ x = 0,\ \dfrac{dx}{dt} = 10,$ when $t = 0.$

30. $\dfrac{d^2y}{dx^2} - 2\dfrac{dy}{dx} + 2y = 0;\ y = 1,\ \dfrac{dy}{dx} = 3,$ when $x = 0.$

To solve the differential equation

(H) $$\frac{d^2y}{dx^2} + p\frac{dy}{dx} + qy = X,$$

where p and q are constants, as in **(G)**, and X is a function of the independent variable x, or a constant, various methods are available. In many applications X has one of the forms given below and the method of undetermined coefficients is most convenient. The following steps are employed.

STEP I. *Solve the equation* **(G)**. *Let the complete solution be*

$$y = u. \tag{22}$$

Then u is called the complementary function for **(H)**.

STEP II. *Obtain a particular solution*

$$y = v \tag{23}$$

of **(H)** *by trial.*

STEP III. *The complete solution of* **(H)** *is now*

$$y = u + v. \tag{24}$$

In fact, when the value of y from (24) is substituted in **(H)**, it is seen that the equation is satisfied, and (24) contains two essential arbitrary constants.

To determine the particular solution (23), the following directions are useful. In the formulas all letters except x, the independent variable, are constants.

GENERAL CASE. When $y = X$ is not a particular solution of **(G)**.

Form of X	*Form of v*
$X = a + bx$;	assume $y = v = A + Bx$.
$X = ae^{bx}$;	assume $y = v = Ae^{bx}$.
$X = a_1 \cos bx + a_2 \sin bx$;	assume $y = v = A_1 \cos bx + A_2 \sin bx$.

SPECIAL CASE. When $y = X$ is a particular solution of **(G)**, assume for v the corresponding above form multiplied by x (the independent variable). Note in this case that X may be obtained from the complementary function u by assigning to the constants of integration in u special values.

The method consists in substituting $y = v$, as given above, in **(H)**, and determining the constants A, B, A_1, A_2 so that **(H)** is satisfied.

Exceptional cases are noted below.

Example 4. Solve

$$\frac{d^2y}{dx^2} - 2\frac{dy}{dx} - 3y = 2x. \tag{25}$$

Solution. *Step I.* The complementary function u is found from the complete solution of

$$\frac{d^2y}{dx^2} - 2\frac{dy}{dx} - 3y = 0. \tag{26}$$

By Example 1, therefore,

$$y = u = c_1 e^{3x} + c_2 e^{-x}. \tag{27}$$

Step II. Since $y = X = 2x$ is not a particular solution of (26) (that is, cannot be obtained from the value of u in (27) by giving special values to c_1 and c_2) assume for a particular solution of (25),

$$y = v = A + Bx. \tag{28}$$

Substituting this value in (25) and collecting terms, the result is

$$-2B - 3A - 3Bx = 2x. \tag{29}$$

Equating coefficients of like powers of x in the two members, we get

$$-2B - 3A = 0, \quad -3B = 2.$$

Solving, $A = \frac{4}{9}$, $B = -\frac{2}{3}$, and, substituting in (28), we obtain the particular solution

$$y = v = \tfrac{4}{9} - \tfrac{2}{3}x. \tag{30}$$

Step III. Then, from (27) and (29), the complete solution is

$$y = u + v = c_1 e^{3x} + c_2 e^{-x} + \tfrac{4}{9} - \tfrac{2}{3}x.$$

Example 5. Solve

$$\frac{d^2y}{dx^2} - 2\frac{dy}{dx} - 3y = 2e^{-x}. \tag{31}$$

Solution. *Step I.* The complementary function is (27), or

$$y = u = c_1 e^{3x} + c_2 e^{-x}. \tag{32}$$

Step II. Here $y = X = 2e^{-x}$ is a particular solution of (26), for it is obtained from the complementary function in (32) by letting $c_1 = 0$, $c_2 = 2$. Hence assume for a particular solution v of (31),

$$y = v = Axe^{-x}. \tag{33}$$

Differentiating (33), we obtain

$$\frac{dy}{dx} = Ae^{-x}(1 - x), \quad \frac{d^2y}{dx^2} = Ae^{-x}(x - 2). \tag{34}$$

Substituting from (33) and (34) in (31), the result is

$$Ae^{-x}(x - 2) - 2Ae^{-x}(1 - x) - 3Axe^{-x} = 2e^{-x}.$$

Simplifying, we get $-4Ae^{-x} = 2e^{-x}$, and hence $A = -\frac{1}{2}$. Substituting in (33), we obtain

$$y = v = -\tfrac{1}{2}xe^{-x}.$$

Step III. The complete solution of (31) is, therefore,

$$y = u + v = c_1 e^{3x} + c_2 e^{-x} - \tfrac{1}{2}xe^{-x}.$$

Example 6. Determine the particular solution of

$$\frac{d^2s}{dt^2} + 4s = 2\cos 2t \tag{35}$$

such that $s = 0$, $\dfrac{ds}{dt} = 2$, when $t = 0$.

Solution. Find the complementary function first.

Step I. Solving

$$\frac{d^2s}{dt^2} + 4\,s = 0, \tag{36}$$

we find the complementary function

$$s = u = c_1 \cos 2\,t + c_2 \sin 2\,t. \tag{37}$$

Step II. Considering the right-hand member in (35), we observe that $s = 2 \cos 2\,t$ is a particular solution of (36) resulting from (37) when $c_1 = 2$, $c_2 = 0$. Hence for a particular solution $s = v$ of (35), assume

$$s = v = t(A_1 \cos 2\,t + A_2 \sin 2\,t). \tag{38}$$

Differentiating (38), we obtain

$$\frac{ds}{dt} = A_1 \cos 2\,t + A_2 \sin 2\,t - 2\,t(A_1 \sin 2\,t - A_2 \cos 2\,t).$$

$$\frac{d^2s}{dt^2} = -\,4\,A_1 \sin 2\,t + 4\,A_2 \cos 2\,t - 4\,t(A_1 \cos 2\,t + A_2 \sin 2\,t). \tag{39}$$

Substituting from (38) and (39) in (35), and simplifying, the result is

$$-\,4\,A_1 \sin 2\,t + 4\,A_2 \cos 2\,t = 2 \cos 2\,t. \tag{40}$$

This equation becomes an identity when $A_1 = 0$, $A_2 = \frac{1}{2}$. Substituting in (38), we get

$$s = v = \tfrac{1}{2}\,t \sin 2\,t. \tag{41}$$

Step III. By (37) and (41), the complete solution of (35) is

$$s = c_1 \cos 2\,t + c_2 \sin 2\,t + \tfrac{1}{2}\,t \sin 2\,t. \tag{42}$$

We must now determine c_1 and c_2 so that

$$s = 0, \quad \frac{ds}{dt} = 2, \quad \text{when } t = 0. \tag{43}$$

Differentiating (42),

$$\frac{ds}{dt} = -\,2\,c_1 \sin 2\,t + 2\,c_2 \cos 2\,t + \tfrac{1}{2} \sin 2\,t + t \cos 2\,t. \tag{44}$$

Substituting the given conditions (43) in (42) and (44), the results are

$$0 = c_1, \quad 2 = 2\,c_2. \quad \text{Hence } c_1 = 0, \quad c_2 = 1.$$

Putting these values back in (42), the particular solution required is

$$s = \sin 2\,t + \tfrac{1}{2}\,t \sin 2\,t. \tag{45}$$

Note, however, the following exceptional case not covered by the above. For example, to solve

$$\frac{d^2y}{dx^2} - 4\,\frac{dy}{dx} + 4\,y = e^{2x}, \tag{46}$$

we find, by (16),

$$y = u = c_1 e^{2x} + c_2 x e^{2x}.$$

for the complementary function. Then $X = e^{2x}$ is the value of u if $c_1 = 1$, $c_2 = 0$. But xe^{2x} is also a special case of u, namely, when $c_1 = 0$, $c_2 = 1$. Hence the above rule for the Special Case will fail. We now assume for a particular solution

$$y = v = Ax^2e^{2x},$$

and we find $A = \frac{1}{2}$. Hence the complete solution of (46) is

$$y = e^{2x}(c_1 + c_2x + \tfrac{1}{2}x^2).$$

An example of a second exceptional case is furnished by the equation

$$\frac{d^2y}{dx^2} - \frac{dy}{dx} = x. \tag{47}$$

The complementary function is found to be

$$y = u = c_1 + c_2e^x. \tag{48}$$

If we follow the directions above and assume $v = A + Bx$, it is apparent that A cannot be determined since (48) contains an arbitrary additive constant. In this case we try

$$y = v = Bx + Cx^2.$$

Substitution in (47) gives

$$2\,C - B - 2\,Cx = x.$$

Therefore, equating coefficients,

$$2\,C - B = 0, \quad -2\,C = 1, \quad \text{whence } B = -1, \quad C = -\tfrac{1}{2}.$$

Hence the complete solution of (47) is

$$y = c_1 + c_2e^x - x - \tfrac{1}{2}x^2.$$

PROBLEMS

Find the complete solutions of the following differential equations.

1. $\dfrac{d^2x}{dt^2} + 9\,x = t + \tfrac{1}{2}.$

2. $\dfrac{d^2x}{dt^2} + 9\,x = 9\,e^{3t}.$

3. $\dfrac{d^2x}{dt^2} + 9\,x = 5\cos 2\,t.$

4. $\dfrac{d^2x}{dt^2} + 9\,x = 3\cos 3\,t.$

5. $\dfrac{d^2y}{dx^2} - 9\,y = e^{3x}.$

6. $\dfrac{d^2y}{dx^2} - 9\,y = 6\cos 3\,x.$

7. $\dfrac{d^2s}{dt^2} + 4\,s = 10\sin 3\,t.$

8. $\dfrac{d^2s}{dt^2} + 4\,s = 8\sin 2\,t.$

9. $\dfrac{d^2y}{dx^2} + 2\dfrac{dy}{dx} = x + 3.$

10. $\dfrac{d^2y}{dx^2} - 2\dfrac{dy}{dx} = 4\,e^{2x}.$

11. $\dfrac{d^2x}{dt^2} - 2\dfrac{dx}{dt} - 3\,x = 2\,t + 1.$

12. $\dfrac{d^2x}{dt^2} - 6\dfrac{dx}{dt} + 13\,x = 39.$

13. $\dfrac{d^2x}{dt^2} - 4\dfrac{dx}{dt} + 7\,x = 14.$

14. $\dfrac{d^2y}{dx^2} - 2\dfrac{dy}{dx} - 3\,y = e^{2x}.$

15. $\dfrac{d^2y}{dx^2} - \dfrac{dy}{dx} - 2\,y = e^{2x}.$

16. $\dfrac{d^2y}{dt^2} - 2\dfrac{dy}{dt} - 3\,y = 8\cos 2\,t.$

17. $\dfrac{d^2x}{dt^2} + 6\dfrac{dx}{dt} + 13\,x = 30\sin t.$

18. $\dfrac{d^2x}{dt^2} - 2\dfrac{dx}{dt} + 5\,x = 17\sin 2\,t.$

In the following problems find the particular solutions which satisfy the given conditions.

19. $\dfrac{d^2s}{dt^2} - 4\,s = 4; \ s = 1, \ s' = 0,$ when $t = 0.$

20. $\dfrac{d^2s}{dt^2} + 4\,s = 8\,t; \ s = 0, \ s' = 4,$ when $t = 0.$

21. $\dfrac{d^2s}{dt^2} - 3\dfrac{ds}{dt} = 6; \ s = 1, \ s' = 1,$ when $t = 0.$

22. $\dfrac{d^2y}{dx^2} - 5\dfrac{dy}{dx} + 6\,y = 2\,e^x; \ y = 1, \ y' = 1,$ when $x = 0.$

23. $\dfrac{d^2x}{dt^2} + x = \sin 2\,t; \ x = 0, \ x' = 0,$ when $t = 0.$

24. $\dfrac{d^2x}{dt^2} + x = 2\cos t; \ x = 2, \ x' = 0,$ when $t = 0.$

25. $\dfrac{d^2y}{dx^2} - 9\,y = 2 - x; \ y = 0, \ y' = 1,$ when $x = 0.$

26. $\dfrac{d^2x}{dt^2} - 2\dfrac{dx}{dt} + x = 4; \ x = 4, \ x' = 2,$ when $t = 0.$

27. $\dfrac{d^2x}{dt^2} + 2\dfrac{dx}{dt} + x = e^t; \ x = 0, \ x' = -\,2,$ when $t = 0.$

28. $\dfrac{d^2s}{dt^2} + 4\,s = 4\sin t; \ s = 4, \ s' = 0,$ when $t = 0.$

29. $\dfrac{d^2s}{dt^2} + 4\,s = 2\cos 2\,t; \ s = 0, \ s' = 4,$ when $t = 0.$

30. $\dfrac{d^2y}{dx^2} - 4\,y = 4\,e^{2x}; \ y = 0, \ y' = 0,$ when $x = 0.$

31. $\dfrac{d^2s}{dt^2} + s = e^{-t} + 2; \ s = 0, \ s' = 0,$ when $t = 0.$

32. $\dfrac{d^2s}{dt^2} + s = \sin t + \cos 2\,t; \ s = 0, \ s' = 0,$ when $t = 0.$

263. Applications to problems in mechanics. Many important problems in mechanics and physics are solved by the methods explained in this chapter. For example, problems in rectilinear motion often lead to differential equations of the first or second order, and the solution of the problems depends upon solving these equations.

Before giving illustrative examples, it is to be recalled that

$$v = \frac{ds}{dt}, \quad a = \frac{d^2s}{dt^2} = \frac{dv}{dt} = v\frac{dv}{ds}, \tag{1}$$

where v and a are, respectively, the velocity and acceleration at any instant of time ($= t$), and s equals the distance of the moving point at this time from a fixed origin on the linear path.

Example 1. When a body falls to the earth from a great height, its acceleration due to the attraction of the earth is gR^2/s^2, where g is the acceleration at the surface of the earth, R is the radius of the earth and s ($s > R$) is the distance of the body from the center of the earth. Taking the origin at the center of the earth and the positive direction as upwards, the differential equation of motion is

$$v\frac{dv}{ds} = -\frac{gR^2}{s^2}. \tag{2}$$

Suppose a body should drop from rest at a height R above the surface of the earth. Find (a) the speed with which the body would strike the earth and (b) the time required for the fall.

Solution. The problem is to solve (2) with the initial conditions $s = 2R$, $v = 0$, when $t = 0$, and then to find v and t when $s = R$.

a. Separating the variables, (2) may be written

$$v \, dv = -\frac{gR^2 \, ds}{s^2},$$

and, by integration,

$$\tfrac{1}{2}v^2 = \frac{gR^2}{s} + c_1. \tag{3}$$

Setting $v = 0$, $s = 2R$, we find $c_1 = -\tfrac{1}{2}gR$ and (3) may be written

$$v^2 = gR\left(\frac{2R}{s} - 1\right). \tag{4}$$

Hence when $s = R$,

$$\text{Speed} = |v| = \sqrt{gR}.$$

For a numerical value of this result we take $R = 4000$ mi. and $g = 32$ ft. per sec. per sec. Then

$$gR = \frac{32(4000)}{5280} = \frac{800}{33},$$

and speed $= 4.92$ mi. per sec.

b. From (4)

$$\left(\frac{ds}{dt}\right)^2 = gR\left(\frac{2R - s}{s}\right).$$

Extracting the square root, we take the negative sign, since the body is falling, and

$$\frac{ds}{dt} = -\sqrt{gR}\sqrt{\frac{2R-s}{s}}. \tag{5}$$

Separating the variables, (5) may be written

$$dt = -\frac{1}{\sqrt{gR}}\sqrt{\frac{s}{2R-s}}\,ds.$$

The time of fall is then

$$T = -\frac{1}{\sqrt{gR}}\int_{2R}^{R}\sqrt{\frac{s}{2R-s}}\,ds \tag{6}$$

$$= \frac{1}{\sqrt{gR}}\left[\sqrt{s(2R-s)} + 2R \text{ arc sin }\sqrt{\frac{2R-s}{2R}}\right]_{2R}^{R}.$$

[By 117, Art. 297, with $a = 0$, $b = 2R$]

Substituting the limits, we get

$$T = \frac{1}{\sqrt{gR}}[R + 2R \text{ arc sin }\sqrt{\tfrac{1}{2}}] \tag{7}$$

$$= \sqrt{\frac{R}{g}}\left(1 + \frac{\pi}{2}\right) \text{ sec.}$$

$$= 35 \text{ min., approximately.}$$

An important type of rectilinear motion is that in which the acceleration is directly proportional to the distance and differs from it in sign. Then we may write

$$a = -k^2 s, \tag{8}$$

where $k^2 =$ magnitude of a at unit distance.

Remembering that a force and the acceleration caused by it differ in magnitude only, we see in the above case that the effective force is directed always toward the point $s = 0$ and is, in magnitude, directly proportional to the distance s. The motion is called *simple harmonic vibration* (see Art. 185).

From (8), we have, using (1),

$$\frac{d^2s}{dt^2} + k^2 s = 0, \tag{9}$$

a linear equation in s and t of the second order with constant coefficients. Integrating, we obtain the complete solution,

$$s = c_1 \cos kt + c_2 \sin kt. \tag{10}$$

From (10), by differentiation,

$$v = k(-c_1 \sin kt + c_2 \cos kt). \tag{11}$$

It is easy to see that the motion defined by (10) is a periodic oscillation between the extreme positions $s = c$, $s = -c$, determined by

$$c = \sqrt{c_1{}^2 + c_2{}^2}, \quad \text{period} = \frac{2\,\pi}{k}. \tag{12}$$

In fact, we may replace the constants c_1 and c_2 in (10) by other constants c and A, such that

$$c_1 = c \sin A, \quad c_2 = c \cos A. \tag{13}$$

Substituting these values in (10), it reduces to

$$s = c \sin (kt + A) \tag{14}$$

and now the truth of the above statement is manifest. (See Art. 182.)

In the following examples we give cases when the simple harmonic
motion is disturbed by other forces. In all cases the problem depends
upon the solution of an equation of the form **(G)** or **(H)**, discussed above.

Example 2. In a rectilinear motion

$$a = -\tfrac{5}{4} s - v. \tag{15}$$

Also, $v = 2$, $s = 0$, when $t = 0$.

a. Find the equation of motion (s in terms of t).

b. For what values of t will $v = 0$?

Solution. a. Using (1), we have, from (15),

$$\frac{d^2s}{dt^2} + \frac{ds}{dt} + \tfrac{5}{4} s = 0, \tag{16}$$

an equation of the form **(G)**. The roots of the auxiliary equation

$$r^2 + r + \tfrac{5}{4} = 0$$

are $$r_1 = -\tfrac{1}{2} + \sqrt{-1}, \quad r_2 = -\tfrac{1}{2} - \sqrt{-1}.$$

Hence the complete solution of (16) is

$$s = e^{-\frac{1}{2}t}(c_1 \cos t + c_2 \sin t). \tag{17}$$

By the given conditions, $s = 0$ when $t = 0$. Substituting these values in (17),
we find $c_1 = 0$, and hence

$$s = c_2 e^{-\frac{1}{2}t} \sin t. \tag{18}$$

Differentiating to find v, we get

$$= c_2 e^{-\frac{1}{2}t}(-\tfrac{1}{2} \sin t + \cos t). \tag{19}$$

Substituting the given value $v = 2$ when $t = 0$, we have $2 = c_2$.
With this value of c_2, (18) becomes

$$s = 2\,e^{-\frac{1}{2}t} \sin t. \tag{20}$$

b. When $v = 0$, the expression in the parenthesis of the right-hand member of (19) must vanish. Setting this equal to zero, we readily obtain

$$\tan t = 2. \tag{21}$$

For any value of t satisfying (21) v will vanish. These values are

$$t = 1.10 + n\pi. \quad (n = \text{an integer}) \tag{22}$$

Successive values of t in (22) differ by the constant interval of time π.

Discussion. This example illustrates *damped harmonic vibration* (see Art. 185). In fact, in (15) the acceleration is the sum of two components

$$a_1 = -\tfrac{5}{4}\, s, \quad a_2 = -v. \tag{23}$$

The simple harmonic vibration corresponding to the component a_1 is now disturbed by a *damping force* with the acceleration a_2, that is, by a force proportional to the velocity and opposite to the direction of motion. The effects of this damping force are twofold.

First, the interval of time between successive positions of the point where $v = 0$ is *lengthened*. In fact, for the simple harmonic vibration

$$a_1 = -\tfrac{5}{4}\, s, \tag{24}$$

we have, by comparison with (8), $k = \tfrac{1}{2}\sqrt{5} = 1.12$, and the half-period, by (12), is $0.89\,\pi$. As we have seen above, for the damped harmonic vibration the corresponding interval is π.

Second, the values of s for the successive extreme positions where $v = 0$, instead of being equal, now form a decreasing geometric progression. Proof is omitted.

Example 3. In a rectilinear motion

$$a = -4\, s + 2 \cos 2\, t. \tag{25}$$

Also, $s = 0$, $v = 2$, when $t = 0$.

a. Find the equation of motion.

b. For what values of t does $v = 0$?

Solution. a. By (1), we have, from (25),

$$\frac{d^2 s}{dt^2} + 4\, s = 2 \cos 2\, t.$$

The particular solution required was found in Ex. 6, Art. 262. Hence

$$s = \sin 2\, t + \tfrac{1}{2}\, t \sin 2\, t. \tag{27}$$

b. Differentiating (27) to find v, and setting the result equal to zero, we get

$$(2 + t) \cos 2\, t + \tfrac{1}{2} \sin 2\, t = 0; \tag{28}$$

or, also, dividing through by $\cos 2\, t$,

$$\tfrac{1}{2} \tan 2\, t + 2 + t = 0. \tag{29}$$

The roots of this equation may be found graphically. Draw the curves (see Fig. 263)

$$y = \tfrac{1}{2} \tan 2t, \quad y = -2 - t. \tag{30}$$

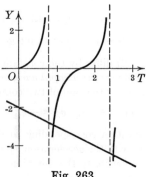

Fig. 263

The abscissas of the points of intersection are, approximately, $t = 0.9$, 2.4, etc.

Discussion. This example illustrates *forced harmonic vibration* (see Art. 185). In fact, in (25) the acceleration is the sum of two components

$$a_1 = -4s. \quad a_2 = 2 \cos 2t. \tag{31}$$

The simple harmonic vibration corresponding to the component a_1 with the period π is now disturbed by a force with the acceleration a_2, that is, by a periodic force whose period $(= \pi)$ is *the same* as the period of the undisturbed simple harmonic vibration. The effects of this disturbing force are twofold.

First, the interval of time between successive positions of the point where $v = 0$ is no longer constant, but increases and approaches $\tfrac{1}{2} \pi$. This is clear from the above figure.

Second, the values of s for the successive extreme positions where $v = 0$ now increase and eventually become, in numerical value, indefinitely great.

PROBLEMS

In each of the following problems the acceleration and initial conditions are given. Find the equation of motion.

1. $a = -4s$; $s = 0$, $v = 10$, when $t = 0$.

2. $a = -4s$; $s = 8$, $v = 0$, when $t = 0$.

3. $a = -4s$; $s = 2$, $v = 10$, when $t = 0$.

4. $a = -s + k$; $s = 0$, $v = 0$, when $t = 0$.

5. $a = -2v - 5s$; $s = 5$, $v = -5$, when $t = 0$.

6. $a = -2v - 5s$; $s = 0$, $v = 12$, when $t = 0$.

7. $a = \sin 2t - s$; $s = 0$, $v = 1$, when $t = 0$.

8. $a = \sin 2\,t - 4\,s$; $s = 0$, $v = 0$, when $t = 0$.

9. $a = -\dfrac{s}{4}$; $s = 0$, $v = 4$, when $t = 0$.

10. $a = 9(1 - s)$; $s = 0$, $v = 0$, when $t = 0$.

11. $a = -4\,v - 5\,s$; $s = 0$, $v = 5$, when $t = 0$.

12. $a = \cos t - 4\,s$; $s = 0$, $v = 0$, when $t = 0$.

13. $a = \cos 2\,t - 4\,s$; $s = 0$, $v = 0$, when $t = 0$.

14. $a = -4\,v - 13\,s$; $s = 0$, $v = 6$, when $t = 0$.

15. The acceleration of a particle is given by

$$a = -4\,s + 3\sin t.$$

a. If the particle starts from rest at the origin, show that its equation of motion is $s = \sin t - \tfrac{1}{2}\sin 2\,t$.

What is the greatest distance from the origin reached by the particle?

b. If the particle starts from the origin with velocity $v = 4$, show that its equation of motion is $s = \sin t + \tfrac{3}{2}\sin 2\,t$.

What is the greatest distance from the origin reached by the particle?

16. The acceleration of a particle is given by

$$a = -4\,s - 8\sin 2\,t.$$

a. If the particle starts from rest at the origin, show that its equation of motion is $s = 2\,t\cos 2\,t - \sin 2\,t$.

b. If the particle starts from the origin with velocity $v = 4$, show that its equation of motion is $s = 2\,t\cos 2\,t + \sin 2\,t$.

17. A body falls from rest under the action of its weight and a small resistance which varies as the velocity. Prove the following relations.

$$a = g - kv.$$

$$v = \frac{g}{k}\left(1 - e^{-kt}\right).$$

$$s = \frac{g}{k^2}\left(kt + e^{-kt} - 1\right).$$

$$ks + v + \frac{g}{k}\ln\left(1 - \frac{kv}{g}\right) = 0.$$

18. A body falls from rest a distance of 80 ft. Assuming $a = 32 - v$, show that the time required is approximately 3.47 sec.

19. In Example 1, Art. 263, suppose the body falls from a very great distance s_0 with a very small velocity v_0 given by $v_0 = -R\sqrt{2\,g/s_0}$. This condition is sometimes expressed loosely by saying that the body falls from rest at an infinite distance. Show that the speed with which the body would strike the earth is slightly less than 7 mi. per sec.

If a body could be projected upward from the surface of the earth with a speed equal to or greater than 7 mi. per sec., show that it would never stop.

CHAPTER XXII

SOLID ANALYTIC GEOMETRY

264. Function of two variables. If the values of a function z are obtained by assigning values to each of two independent variables x and y, then z is said to be a function of the two variables x and y. The notation used is

$$z = f(x, y). \tag{1}$$

For example, the volume of a right circular cylinder is a function of two independent variables, the altitude and the diameter.

To represent (1) graphically, we must have a system of three coordinates. The system most commonly used is the cartesian system of rectangular coördinates in space.

265. Cartesian coördinates in space. Let $X'X$, $Y'Y$, $Z'Z$ be three mutually perpendicular axes in space intersecting at the origin O. These three **coördinate axes** determine three mutually perpendicular **coördinate planes** XOY, YOZ, ZOX. In Fig. 265 the xy-plane is supposed to

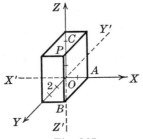

Fig. 265

be horizontal and the z-axis vertical. The position of a point in space is specified by giving its distances from each of the coördinate planes, the positive directions being indicated by the arrowheads on the axes. The point whose coördinates are (a, b, c) is a units from the yz-plane, b units from the zx-plane, and c units from the xy-plane.

The point $P(2, 6, 4)$ is shown in the figure. This point is found by starting at O and measuring 2 units to the right along the x-axis to the point A; then measuring 6 units from A to B in a direction parallel to the y-axis; then measuring 4 units from B to P in a direction parallel to the z-axis.

The axis OY in the figure, which represents the axis in space perpendicular to the plane XOZ, is drawn so that $\angle XOY = 135°$. Unit length on OY is equal to half of a unit of length on OX or OZ.

472

The origin O and the point P are opposite vertices of a rectangular parallelepiped whose edges are equal to the coördinates of P. This fact enables us to see at once that the distance of P

$$\text{from the } x\text{-axis} = \sqrt{6^2 + 4^2} = \sqrt{52};$$
$$\text{from the } y\text{-axis} = \sqrt{2^2 + 4^2} = \sqrt{20};$$
$$\text{from the } z\text{-axis} = \sqrt{2^2 + 6^2} = \sqrt{40};$$
$$\text{from the origin} = \sqrt{2^2 + 6^2 + 4^2} = \sqrt{56}.$$

266. Distances and directions. If P is a point whose coördinates are (x, y, z), then the line OP is the diagonal of a rectangular parallelepiped whose edges are $OA = x$, $OB = y$, $OC = z$ (see Fig. 266). The following formulas are evidently true for any position of the point P.

$$x = \text{perpendicular distance from the } yz\text{-plane.}$$
$$y = \text{perpendicular distance from the } zx\text{-plane.}$$
$$z = \text{perpendicular distance from the } xy\text{-plane.}$$
$$\sqrt{y^2 + z^2} = \text{perpendicular distance from the } x\text{-axis.}$$
$$\sqrt{z^2 + x^2} = \text{perpendicular distance from the } y\text{-axis.}$$
$$\sqrt{x^2 + y^2} = \text{perpendicular distance from the } z\text{-axis.}$$
$$\sqrt{x^2 + y^2 + z^2} = \text{distance from the origin.}$$

In solid geometry, as in plane geometry, the angle between two directed lines is the angle between their positive directions. The angle between two lines which do not intersect is defined as the angle between two intersecting lines which are parallel, respectively, to the given lines.

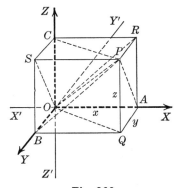

Fig. 266

The **direction angles** of a directed line are the angles between the line and the three coördinate axes. The direction angles are usually denoted by α, β, γ. That is, α is the angle between a line and the x-axis, β the angle between the line and the y-axis, and γ the angle between the line and the z-axis.

In Fig. 266, $\angle AOP = \alpha$, $\angle BOP = \beta$, $\angle COP = \gamma$.

The **direction cosines** of a line are the cosines of its direction angles. Since the angle between two directed lines cannot exceed $180°$, the angle is acute if its cosine is positive and obtuse if its cosine is negative.

267. Important relations. In Fig. 266 we have $\overline{OP}^2 = \overline{OQ}^2 + \overline{QP}^2 = \overline{OA}^2 + \overline{OB}^2 + \overline{OC}^2$. Hence, if $OP = \rho$,

(A) $$\rho^2 = x^2 + y^2 + z^2.$$

Let the angles between OP and OX, OY, and OZ be, respectively, α, β, and γ. The triangle OAP is a right triangle, for OA is perpendicular to AP. Also, $\angle XOP = \alpha$. Hence $OA = OP \cos \alpha$, or $x = \rho \cos \alpha$. Similar values for y and z are found from the right triangles OBP and OCP. Hence

(B) $$x = \rho \cos \alpha, \quad y = \rho \cos \beta, \quad z = \rho \cos \gamma.$$

Squaring these equations, adding them, using (A), and dividing by ρ^2, we obtain the important relation

(C) $$\cos^2 \alpha + \cos^2 \beta + \cos^2 \gamma = 1.$$

Hence *the sum of the squares of the direction cosines of a line is unity.*

Equations (B) may be written as a set of equal ratios, namely,

(D) $$\frac{\cos \alpha}{x} = \frac{\cos \beta}{y} = \frac{\cos \gamma}{z} = \frac{1}{\rho}.$$

Solving (B) for $\cos \alpha$, $\cos \beta$, $\cos \gamma$, and using the value of ρ from (A), we obtain

(E) $$\cos \alpha = \frac{x}{\sqrt{x^2 + y^2 + z^2}}, \quad \cos \beta = \frac{y}{\sqrt{x^2 + y^2 + z^2}},$$

$$\cos \gamma = \frac{z}{\sqrt{x^2 + y^2 + z^2}}.$$

We next find the length l of the line P_1P_2 joining two points $P_1(x_1, y_1, z_1)$ and $P_2(x_2, y_2, z_2)$, and the direction cosines of the line P_1P_2.

Theorem. *The length l of the line joining any two points $P_1(x_1, y_1, z_1)$ and $P_2(x_2, y_2, z_2)$ is given by*

(F) $$l = \sqrt{(x_1 - x_2)^2 + (y_1 - y_2)^2 + (z_1 - z_2)^2}.$$

Proof. Construct a rectangular parallelepiped by passing planes through P_1 and P_2 parallel to the coördinate planes. Its edges will be parallel to the axes and equal, respectively, to $x_2 - x_1$, $y_2 - y_1$, $z_2 - z_1$. P_1P_2 will be a diagonal of this parallelepiped, and hence l^2 will equal the sum of the squares of its three edges.

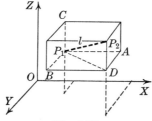

Fig. 267

Let α, β, γ be the direction angles of the line from P_1 to P_2. Then, from the figure, $\angle AP_1P_2 = \alpha$, and $P_1A = l \cos \alpha$. Similarly, $P_1B = l \cos \beta$, $P_1C = l \cos \gamma$.

Hence we have the equations

(G) $x_2 - x_1 = l \cos \alpha$, $y_2 - y_1 = l \cos \beta$, $z_2 - z_1 = l \cos \gamma$.

From these equations the values of the direction cosines may be found.

PROBLEMS

1. Find the direction angles of the line from the origin to the point $(3, 4, 5)$. What is meant by the angle which a line makes with a plane? Find the angle which OP makes with the xy-plane; the yz-plane; the zx-plane.

2. Describe the direction of a line if

a. $\cos \alpha = 0$.   b. $\cos \beta = 0$.   c. $\cos \alpha = \cos \beta = 0$.
d. $\cos \gamma = 0$.   e. $\cos \alpha = \cos \gamma = 0$.   f. $\cos \beta = \cos \gamma = 0$.

3. The direction angles of the line from the origin to the point P are α, β, γ. In what octant ($O\text{-}XYZ$, $O\text{-}X'Y'Z$, etc.) will P lie if

a. $\cos \alpha > 0$, $\cos \beta > 0$, $\cos \gamma > 0$?
b. $\cos \alpha > 0$, $\cos \beta > 0$, $\cos \gamma < 0$?
c. $\cos \alpha > 0$, $\cos \beta < 0$, $\cos \gamma < 0$?
d. $\cos \alpha > 0$, $\cos \beta < 0$, $\cos \gamma > 0$?
e. $\cos \alpha < 0$, $\cos \beta > 0$, $\cos \gamma > 0$?
f. $\cos \alpha < 0$, $\cos \beta < 0$, $\cos \gamma > 0$?
g. $\cos \alpha < 0$, $\cos \beta < 0$, $\cos \gamma < 0$?
h. $\cos \alpha < 0$, $\cos \beta > 0$, $\cos \gamma < 0$?

4. Find $\cos \gamma$ if $\cos \alpha = \frac{2}{3}$, $\cos \beta = -\frac{1}{3}$, and the line is directed downwards.

5. What angle does a diagonal make with an edge of a cube?

6. Find the volume of the tetrahedron whose vertices are $O(0, 0, 0)$, $A(4, 0, 0)$, $B(0, 6, 0)$, $C(0, 0, 5)$.

7. Find the volume of the pyramid whose vertices are $O(0, 0, 0)$, $A(0, 4, 0)$, $B(0, 0, 6)$, $C(0, 4, 6)$, $D(8, 2, 3)$. Find the length of each lateral edge of the pyramid.

8. Find the volume of the tetrahedron whose vertices are $A(0, 3, 0)$, $B(0, 0, 4)$ $C(0, 6, 4)$, $D(5, 4, 1)$. Find the length of each edge.

9. Show by the length formula that the triangle with vertices $(8, 4, 5)$, $(2, 1, 7)$, and $(5, 6, -1)$ is an isosceles right triangle.

10. Find the direction angles of the lines joining the following pairs of points.

a. $(1, 2, 3)$ to $(4, 5, 6)$.   **b.** $(1, 0, 4)$ to $(4, 1, 2)$.
c. $(0, 0, 5)$ to $(0, 4, 0)$.   **d.** $(2, -1, 3)$ to $(-1, 3, 2)$.
e. $(4, 0, -2)$ to $(0, 3, 4)$.   **f.** $(5, 1, 4)$ to $(-1, -2, -1)$.

11. Show that the following points lie on a sphere whose center is the origin: $(2, 2, 1)$, $(2, -1, 2)$, $(0, 3, 0)$.

12. Show that the following points lie on a circular cylinder whose axis is the z-axis: $(0, 5, 0)$, $(3, 4, 6)$, $(4, -3, -1)$.

13. By comparing directions and lengths determine the type of figure with the following vertices (given in order around the perimeter).

a. $(7, 3, -4)$, $(1, 0, -6)$, $(4, 5, 2)$.
b. $(2, -1, 5)$, $(3, 4, -2)$, $(6, 2, 2)$, $(5, -3, 9)$.
c. $(6, 7, 3)$, $(3, 11, 1)$, $(-3, 7, 2)$, $(0, 3, 4)$.
d. $(-6, 3, 2)$, $(3, -2, 4)$, $(5, 7, 3)$, $(-13, 17, -1)$.
e. $(2, 3, 0)$, $(4, 5, -1)$, $(3, 7, 1)$, $(1, 5, 2)$.
f. $(6, -6, 0)$, $(3, -4, 4)$, $(2, -9, 2)$, $(-1, -7, 6)$.
g. $(3, 2, 2)$, $(1, 2, 1)$, $(2, 4, -1)$, $(4, 4, 0)$.
h. $(2, 1, 4)$, $(0, 0, 0)$, $(3, -1, 2)$, $(5, 0, 6)$.

14. Examine each group of points given to determine if they lie on a straight line. Draw the figure.

a. $(3, 2, 7)$, $(1, 4, 6)$, $(7, -2, 9)$.
b. $(13, 4, 9)$, $(1, 7, 13)$, $(7, 5.5, 11)$, $(5, 6, 11\frac{2}{3})$.
c. $(3, 6, -2)$, $(7, -4, 3)$, $(-1, 16, -7)$, $(-5, 25, -12)$.
d. $(2, -15, -4)$, $(-3, -5, -9)$, $(3, -17, -3)$, $(4, -19, -2)$.

15. Given the line $P_1(x_1, y_1, z_1)$ $P_2(x_2, y_2, z_2)$. Show that the coördinates of the mid-point are

$$\left(\frac{x_1 + x_2}{2}, \quad \frac{y_1 + y_2}{2}, \quad \frac{z_1 + z_2}{2} \right).$$

268. Locus of an equation. *In space of three dimensions the locus of an equation is a surface containing all points whose coördinates satisfy the equation, and no other points.*

Let us consider some examples.

Example 1. The locus of the equation $x = 0$ is the yz-plane. The coördinates of every point in the yz-plane satisfy the equation $x = 0$, and the coördinates of no point outside of the yz-plane satisfy the equation.

Example 2. The locus of the equation $z = 2$ is the plane parallel to the xy-plane and 2 units above it. (See Fig. 268.1.)

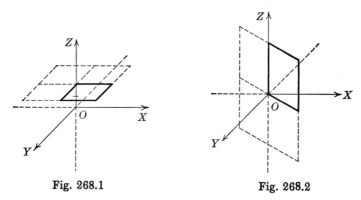

Fig. 268.1   Fig. 268.2

Example 3. The locus of the equation $x = y$ is the plane containing the z-axis and bisecting the dihedral angle between the planes XOZ and YOZ. (See Fig. 268.2.) From the equation $x = y$ and Art. 266, we see that the surface is the locus of a point e₁uidistant from the planes ZOX and YOZ.

Example 4. The locus of the equation $x^2 + y^2 = 25$ is the cylindrical surface of radius 5 units and having the z-axis for its axis. (See Fig. 268.3, in which the portion of the surface between the planes $z = 0$ and $z = 3$ is shown.) From the equation and Art. 266, we see that the surface is the locus of a point whose perpendicular distance from the z-axis is 5.

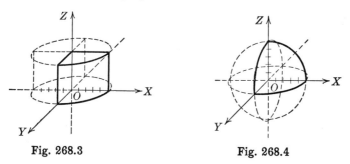

Fig. 268.3   Fig. 268.4

Example 5. The locus of the equation $x^2 + y^2 + z^2 = 25$ is the spherical surface of radius 5 units, having its center at the origin. (See Fig. 268.4.)

PROBLEMS

1. Find the equation of the locus of a point
   a. 3 units below the xy-plane;
   b. 8 units to the left of the yz-plane;
   c. 5 units in front of the xz-plane;
   d. 7 units above the xy-plane;
   e. 4 units back of the xz-plane;
   f. 2 units to the right of the yz-plane.

2. The locus of each of the following equations is a plane. Construct the locus.

a. $x = -3$.   **b.** $y = 5$.   **c.** $z = -1$.
d. $x + y = 0$.   **e.** $x + y = 6$.   **f.** $y + z = 5$.
g. $x + z = 8$.   **h.** $x - z = 4$.   **i.** $x - y + 3 = 0$.

3. Find the equation of the locus of a point
   a. at a distance of 4 units from $(4, 0, 0)$;
   b. at a distance of 5 units from $(1, -1, 3)$;
   c. equidistant from $(2, 0, 4)$ and $(4, -2, 8)$;
   d. equidistant from $(3, -1, 0)$ and $(5, 3, -2)$.

4. Write the equation of the circular cylinder
   a. of radius 5 with the z-axis for its axis;
   b. of radius 4 with the y-axis for its axis;
   c. of radius 3 with the x-axis for its axis.

5. Describe the locus of each of the following equations.

a. $y^2 + z^2 = 4$.   **b.** $x^2 + z^2 = 9$.   **c.** $x^2 + y^2 = 25$.
d. $x^2 + y^2 - 2x = 0$.   **e.** $y^2 + z^2 + 2z = 0$.   **f.** $x^2 + z^2 - 2z = 0$.

6. Show that the equation of the sphere with center at (a, b, c) and radius r is

$$(x - a)^2 + (y - b)^2 + (z - c)^2 = r^2.$$

7. Find the equation of the sphere

a. of radius 4 and center $(3, -4, -5)$;
b. having the line joining $(-3, 4, 2)$ and $(7, -2, 6)$ as a diameter;
c. with center $(2, 1, 4)$ and tangent to the yz-plane;
d. with center $(3, 2, 7)$ and passing through $(5, -3, 8)$;
e. of radius 3 and tangent to all three coördinate planes (eight cases).

8. Find the equation of the locus of a point

a. twice as far from $(5, 4, 0)$ as from $(-4, 3, 4)$;
b. the sum of whose distances from $(5, 0, 0)$ and $(-5, 0, 0)$ is 20;
c. the sum of the squares of whose distances from $(7, -5, 9)$ and $(5, -3, 8)$ is 6;
d. whose distance from $(-4, 3, 4)$ equals its distance from the xy-plane;
e. whose distance from the x-axis is 4 times its distance from $(4, -2, 4)$;
   f. the sum of whose distances from the coördinate planes equals its distance from the origin.

269. Equations of a curve. *In space of three dimensions the locus of two simultaneous equations is the curve, or curves, containing all points whose coördinates satisfy both equations, and no other points.*

Since the locus of each equation alone is a surface, the locus of two simultaneous equations is the curve, or curves, in which the two surfaces intersect.

Let us consider some examples.

Example 1. The locus of the simultaneous equations $x = 0$, $y = 0$ is the z-axis. The coördinates of every point on the z-axis satisfy both equations, and the coördinates of no point outside of the z-axis satisfy both equations. Also, the locus of the equation $x = 0$ is the yz-plane; the locus of the equation $y = 0$ is the zx-plane; these two planes intersect in the z-axis.

Example 2. The locus of the simultaneous equations $y = 4$, $z = 3$ is the straight line parallel to the x-axis in which the planes $y = 4$ and $z = 3$ intersect. (See Fig. 269.1.)

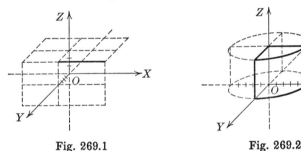

Fig. 269.1   Fig. 269.2

Example 3. The locus of the simultaneous equations

$$\begin{cases} x^2 + y^2 = 25, \\ z = 3 \end{cases}$$

is the circle cut from the cylindrical surface $x^2 + y^2 = 25$ by the plane $z = 3$. The circle lies in a plane parallel to the xy-plane and 3 units above; its center is on the z-axis, and its radius is 5 units. (See Fig. 269.2.)

Example 4. The locus of the simultaneous equations

$$\begin{cases} x^2 + y^2 + z^2 = 25, \\ x = 3 \end{cases}$$

is the circle cut from the spherical surface $x^2 + y^2 + z^2 = 25$ by the plane $x = 3$. The circle lies in a plane parallel to the yz-plane and 3 units to the right; its center is on the x-axis and its radius is 4 units. (See Fig. 269.3.) Substituting $x = 3$ in the equation of the spherical surface, we get $y^2 + z^2 = 16$. The result shows that this cylindrical surface passes through the locus.

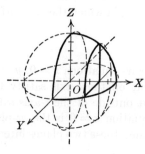

Fig. 269.3

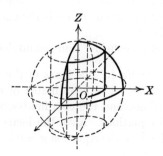

Fig. 269.4

Example 5. The locus of the simultaneous equations

$$\begin{cases} x^2 + y^2 + z^2 = 25, \\ x^2 + y^2 = 9 \end{cases}$$

consists of the two circles cut from the spherical surface $x^2 + y^2 + z^2 = 25$ by the cylindrical surface $x^2 + y^2 = 9$. These two circles lie in planes parallel to the xy-plane, one plane being 4 units above, the other 4 units below. In fact, if the equation of the cylindrical surface is subtracted from the equation of the spherical surface, the result is $z = \pm 4$, and the curves lie in these planes. The center of each circle is on the z-axis and its radius is 3 units. (See Fig. 269.4.)

PROBLEMS

Describe the locus of each of the following pairs of equations.

1. $x = 3$,
   $y = 2$.

2. $x = 4$,
   $z = 2$.

3. $x = 3$,
   $x = y$.

4. $x^2 + y^2 = 25$,
   $z = 0$.

5. $x^2 + y^2 = 25$,
   $x = 4$.

6. $y^2 + z^2 = 4$,
   $x = 4$.

7. $x^2 + z^2 = 9$,
   $y + 1 = 0$.

8. $x^2 + y^2 + z^2 = 9$,
   $z = 0$.

9. $x^2 + y^2 + z^2 = 16$,
   $y = 0$.

10. $x^2 + y^2 + z^2 = 100$,
    $z = 6$.

11. $x^2 + y^2 + z^2 = 100$,
    $y^2 + z^2 = 64$.

12. $z^2 = 4x$,
    $y = 2$.

13. $y^2 = 4x$,
    $x = 4$.

14. $x^2 = 4 - z$,
    $y = 0$.

15. $y^2 = 4 - z$,
    $z = 0$.

16. Find the equations of the locus of a point if it is

a. 5 units from the x-axis and also 5 units from the yz-plane;
b. 4 units from the z-axis and also 5 units from the origin;
c. 5 units from $(0, 0, 1)$ and also 3 units from the xy-plane;
d. equidistant from $(1, 3, 2)$ and $(0, 0, 1)$ and also from $(3, 0, 3)$ and $(0, -2, 0)$;
e. 3 units from $(1, 2, 1)$ and 2 units from $(2, 0, 1)$;
f. 2 units from the x-axis and from the y-axis.

270. Graphical representation of $z = f(x, y)$. To obtain a graphical representation of a function of two variables, we must construct a picture of the surface whose equation is $z = f(x, y)$. This is done by drawing sections of the surface made by conveniently chosen planes. The process is illustrated in the examples below.

Example 1. Construct the surface whose equation is $z = x^2 + y^2$.

Solution. 1. The section of the surface by the yz-plane is the curve whose equations are

$$\begin{cases} x = 0, \\ z = x^2 + y^2, \end{cases} \quad \text{or also} \quad \begin{cases} x = 0, \\ z = y^2. \end{cases}$$

This is a parabola in the yz-plane, and is the curve AOB in the figure.

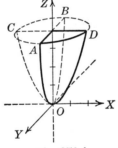

Fig. 270.1

2. The section of the surface by the xz-plane is the curve whose equations are

$$\begin{cases} y = 0, \\ z = x^2 + y^2, \end{cases} \quad \text{or also} \quad \begin{cases} y = 0, \\ z = x^2. \end{cases}$$

This is a parabola in the xz-plane, and is the curve COD in the figure.

3. The section of the surface by the plane $z = 0$ is a single point, the origin. Its equations are

$$\begin{cases} x^2 + y^2 = 0, \\ z = 0. \end{cases}$$

4. The section of the surface by a plane parallel to the yz-plane is the curve $x = K$, $z = x^2 + y^2$. This is a parabola for any positive or negative value of K. Its equations may be written

$$\begin{cases} x = K, \\ z = y^2 + K^2. \end{cases}$$

5. The section of the surface by a plane parallel to the xz-plane is the curve $y = K$, $z = x^2 + y^2$. This is a parabola for any positive or negative value of K. Its equations may be written

$$\begin{cases} y = K, \\ z = x^2 + K^2. \end{cases}$$

6. Finally, consider sections by planes parallel to the xy-plane. For any positive value of K the curve $z = K$, $z = x^2 + y^2$ is a circle of radius $\sqrt{K}$. Its equations may be written

$$\begin{cases} z = K, \\ x^2 + y^2 = K. \end{cases}$$

For any negative value of K the section is imaginary. The circle $ADBC$ in the figure is the section of the surface by the plane $z = 4$.

The three curves drawn are sufficient to give a general idea of the surface, which is a *paraboloid of revolution* about the z-axis.

GENERAL DIRECTIONS FOR DRAWING A SURFACE

STEP I. *Find the* **traces** *of the surface on the coördinate planes, that is, the sections of the surface by the planes* $x = 0$, $y = 0$, $z = 0$.

STEP II. *Examine the nature of the curves cut from the surface by planes parallel to the coördinate planes, that is, by the planes* $x = K$, $y = K$, *and* $z = K$, *for all values of* K. *This should give a general idea of the form of the surface.*

STEP III. *Select certain sections which will show the form of the surface and draw them.*

Symmetry. The following statement is obviously true: A surface is symmetric with respect to the xy-plane if its equation remains unchanged when z is replaced everywhere by $- z$. This statement affords a test for symmetry with respect to the xy-plane.

Similar statements apply to symmetry with respect to the planes YOZ and ZOX. These tests should be used in every example.

Example 2. Construct the surface whose equation is

$$z^2 = 4(x^2 + y^2).$$

Solution. *Step I.* Setting $x = 0$, the trace in the yz-plane is found to be two straight lines, $z = \pm 2 y$. Setting $y = 0$, the trace in the xz-plane is found to be two straight lines, $z = \pm 2 x$. Setting $z = 0$, the trace in the xy-plane is found to be a single point, the origin.

Step II. When $x = K$, the section of the surface is a hyperbola for every value of K. Its equations are

$$\begin{cases} x = K, \\ 4 y^2 - z^2 + 4 K^2 = 0. \end{cases}$$

When $y = K$, the section is a hyperbola for every value of K. Its equations are

$$\begin{cases} y = K, \\ 4\,x^2 - z^2 + 4\,K^2 = 0. \end{cases}$$

When $z = K$, the section is a circle of radius $\frac{1}{2}\,K$ for every value of K. Its equations are

$$\begin{cases} z = K, \\ x^2 + y^2 = \frac{1}{4}\,K^2. \end{cases}$$

Step III. The following sections are used to represent the surface.

$x = 0$, the straight lines AA', BB'
$y = 0$, the straight lines CC', DD'.
$z = 4$, the circle $ACBD$.
$z = -4$, the circle $A'C'B'D'$.

The surface is a cone of revolution about the z-axis. It is symmetric with respect to each of the coördinate planes. (Fig. 270.2)

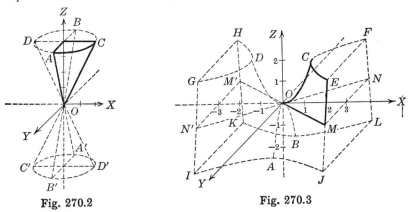

Fig. 270.2   Fig. 270.3

Example 3. Construct the surface whose equation is $z = x^2 - y^2$.

Solution. *Step I.* Setting $x = 0$, the trace in the yz-plane is found to be the parabola $z = -y^2$.

Setting $y = 0$, the trace in the xz-plane is found to be the parabola $z = x^2$.

Setting $z = 0$, the trace in the xy-plane is found to be two straight lines $x = \pm y$.

Step II. When $x = K$, the section of the surface is a parabola directed downward for every value of K. Its equations are

$$\begin{cases} x = K, \\ z = K^2 - y^2. \end{cases}$$

When $y = K$, the section is a parabola directed upward for every value of K. Its equations are

$$\begin{cases} y = K, \\ z = x^2 - K^2. \end{cases}$$

When $z = K$, the section is a hyperbola for every value of K. Its equations are

$$\begin{cases} z = K, \\ x^2 - y^2 = K. \end{cases}$$

When $K > 0$, that is, when the hyperbolic section is above the xy-plane, the transverse axis is parallel to the x-axis. When $K < 0$, that is, when the hyperbolic section is below the xy-plane, the transverse axis is parallel to the y-axis.

Step III. The following sections are used to show a portion of the surface.

$x = 0$, the parabola AOB.

$y = 0$, the parabola COD.

$z = 0$, the straight lines MM', NN'.

$z = 2$, the hyperbola ECF, GDH. Equations, $z = 2$, $x^2 - y^2 = 2$.

$z = -2$, the hyperbola IAJ, KBL. Equations, $z = -2$, $x^2 - y^2 + 2 = 0$.

$x = 3$, the parabola part of which is shown by the curves EJ and FL.

Equations, $x = 3$, $z = 9 - y^2$.

$x = -3$, the parabola part of which is shown by the curves GI and HK.

Equations, $x = -3$, $z = 9 - y^2$.

The surface is called a *hyperbolic paraboloid*. (Fig. 270.3)

PROBLEMS

Construct the surfaces represented by the following equations.

1. $4y^2 + 4z^2 = x$. (Paraboloid of revolution)

2. $4y^2 + 4z^2 = x^2$. (Cone)

3. $x^2 + 4y^2 = 4 - z$. (Elliptic paraboloid)

4. $6x^2 + 6y^2 + z^2 = 36$. (Ellipsoid)

5. $x^2 + y^2 - z^2 = 4$. (Hyperboloid of one sheet)

6. $x^2 - y^2 - z^2 = 4$. (Hyperboloid of two sheets)

7. $4x^2 + 4z^2 = y$.

8. $4x^2 + 4z^2 = y^2$.

9. $y^2 + z^2 = 9 - x$.

10. $x^2 + z^2 = 4 + y$.

11. $(z - 1)^2 = x^2 + y^2$.

12. $4y^2 + 4z^2 - x^2 = 16$.

13. $x^2 - 4y^2 - 4z^2 = 16$.

14. $x^2 + 4y^2 + 4z^2 = 16x$.

271. Equation of a plane. Let ON be the line perpendicular from the origin to the plane ABC intersecting the plane at D. Let the coördinates of D be (x_1, y_1, z_1) and denote the length of OD by p. Let the angles which ON makes with the x, y, and z axes, respectively, be denoted by α, β, γ. Then $x_1 = p \cos \alpha$, $y_1 = p \cos \beta$, $z_1 = p \cos \gamma$. (Fig. 271.1)

Let $P(x, y, z)$ be any point in the plane. Then ODP is a right angle and

$$\overline{OD}^2 + \overline{DP}^2 = \overline{OP}^2. \tag{1}$$

But $\qquad \overline{OD}^2 = p^2,$

$$\overline{DP}^2 = (x - x_1)^2 + (y - y_1)^2 + (z - z_1)^2, \qquad \text{By Art. 267}$$

$$\overline{OP}^2 = x^2 + y^2 + z^2.$$

Substituting in equation (1) and simplifying by **(A)** and **(B)**, Art. 267, the result is

(H) $\qquad x \cos \alpha + y \cos \beta + z \cos \gamma - p = 0.$

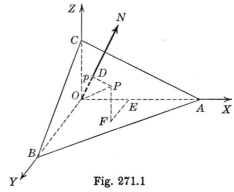

Fig. 271.1

Hence the equation of any plane is an equation of the first degree in x, y, z.

SPECIAL CASES. The equation of any plane perpendicular to the

x-axis is $x = K;$ (since β and γ are each 90°)
y-axis is $y = K;$
z-axis is $z = K.$

The equation of any plane parallel to the

x-axis is $ay + bz = c;$ (since $\alpha = 90°$)
y-axis is $az + bx = c;$ ⎫
z-axis is $ax + by = c.$

Conversely, the locus of any equation of the first degree,

$$Ax + By + Cz + D = 0, \tag{2}$$

is a plane.

Proof. Divide equation (2) by $\pm \sqrt{A^2 + B^2 + C^2}$, and choose the sign before the radical opposite to the sign of D.

Equation (2) now becomes

$$\frac{Ax}{\pm \sqrt{A^2 + B^2 + C^2}} + \frac{By}{\pm \sqrt{A^2 + B^2 + C^2}} + \frac{Cz}{\pm \sqrt{A^2 + B^2 + C^2}}$$

$$+ \frac{D}{\pm \sqrt{A^2 + B^2 + C^2}} = 0. \tag{3}$$

The sum of the squares of the coefficients of x, y, and z is unity and they are, therefore, the direction cosines of a line. Let

$$\frac{A}{\pm\sqrt{A^2+B^2+C^2}}=\cos\alpha;\quad\frac{B}{\pm\sqrt{A^2+B^2+C^2}}=\cos\beta;\quad(4)$$

$$\frac{C}{\pm\sqrt{A^2+B^2+C^2}}=\cos\gamma;\quad\frac{-D}{\pm\sqrt{A^2+B^2+C^2}}=p.$$

Substituting from (4) into (3), the latter becomes identical with **(H)**. Hence equation (2) represents a plane whose distance from the origin is p and whose normal makes angles α, β, γ with the coördinate axes, where α, β, γ, and p are given by (4).

By these formulas (4) we may find the direction cosines of the line drawn from the origin perpendicular to a given plane and the perpendicular distance from the origin to the plane. Note that the point (A, B, C) is on the line drawn through the origin perpendicular to the plane (2) (by **(E)**, Art. 267).

Example 1. Find the perpendicular distance from the origin to the plane

$$x-2y+2z-9=0,\quad(5)$$

and the direction cosines of this perpendicular line.

Solution. Comparing (5) with (2), we have

$$A=1,\quad B=-2,\quad C=2,\quad D=-9.$$

Hence $\pm\sqrt{A^2+B^2+C^2}=\pm 3$. Since D is negative, we must choose the positive sign in (4). Hence

$$p=3,\quad\cos\alpha=\tfrac{1}{3},\quad\cos\beta=-\tfrac{2}{3},\quad\cos\gamma=\tfrac{2}{3}.$$

Letting $y=z=0$ in the equation of a plane and solving for x, we obtain the *intercept* of the plane on the x-axis. Similarly for the other intercepts.

The straight lines in which a plane intersects the coördinate planes are called its *traces*. The equation of the trace on the xy-plane referred to OX and OY as axes is found by substituting $z=0$ in the equation of the plane; similarly for the other traces.

To construct a plane from its equation, find the intercepts, lay them off on the axes, and connect the points thus determined. These lines are the traces. If the intercepts are zero, plot one or more traces, and proceed as in the following examples.

Example 2. Find the intercepts and traces of the plane

$$2x+2y-z-6=0,\quad(6)$$

and draw the plane.

Solution. Following the rule and referring to Fig. 271.2, we find the intercepts to be

$$OA = 3, \quad OB = 3, \quad OC = -6.$$

The traces are $AB : x + y - 3 = 0; \quad BC : 2y - z - 6 = 0;$
$$CA : 2x - z - 6 = 0.$$

The line DE is drawn parallel to BA to set off part of the plane in the first octant.

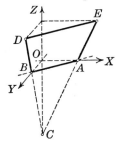

Fig. 271.2

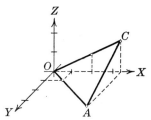

Fig. 271.3

Example 3. Find the traces, and draw the plane

$$x - y - 2z = 0. \tag{7}$$

Solution. The traces are (Fig. 271.3)
$$OA : x - y = 0; \quad OB : y + 2z = 0;$$
$$OC : x - 2z = 0.$$

Draw OA and OC, and the line CA to display part of the plane in the first octant.

PROBLEMS

1. Construct the planes having the following equations.

a. $x + y + z = 4$.   **b.** $x - y + z = 6$.
c. $x + y - z = 2$.   **d.** $2x - y + 2z - 12 = 0$.
e. $4x + 3y + 12z - 24 = 0$.   **f.** $3x - 12y - 4z + 36 = 0$.

2. Find the distance from the origin to each of the planes of Problem **1.**

3. Find the equations of the following planes and construct them.

        a. $p = 5, \alpha = 120°, \beta = 45°, \gamma = 120°$.
        b. $p = 7, \alpha = 45°, \beta = 60°, \gamma = 60°$.
        c. $p = 4, \alpha = 90°, \beta = 135°, \gamma = 45°$.
        d. $p = 2, \alpha = 60°, \beta = 45°, \gamma = 120°$.

        e. $p = 4, \dfrac{\cos \alpha}{6} = \dfrac{\cos \beta}{-2} = \dfrac{\cos \gamma}{3}$.

        f. $p = 6, \dfrac{\cos \alpha}{-2} = \dfrac{\cos \beta}{-1} = \dfrac{\cos \gamma}{-2}$.

        g. $p = 3, \dfrac{\cos \alpha}{6} = \dfrac{\cos \beta}{-3} = \dfrac{\cos \gamma}{2}$.

        h. $p = 0, \dfrac{\cos \alpha}{-3} = \dfrac{\cos \beta}{4} = \dfrac{\cos \gamma}{12}$.

4. Find the equation of the plane such that the foot of the perpendicular from the origin to the plane is

a. $(-2, -2, 1)$;   **b.** $(1, 4, 2)$;   **c.** $(-4, 3, -12)$;

d. $(2, -1, 3)$;   **e.** $(-1, 6, 3)$;   **f.** $(1, 0, 2)$.

5. Write the equation of the plane which is perpendicular to the line drawn from the origin

     a. to $(4, 5, 3)$ and passes through $(1, 3, 2)$;

     b. to $(2, -4, 3)$ and passes through $(3, -4, -5)$.

6. In each of the following show that the planes are parallel and find the perpendicular distance between them.

     a. $x - y - z + 5 = 0$, $2x - 2y - 2z - 7 = 0$.

     b. $3x - y + 2z + 10 = 0$, $3x - y + 2z - 7 = 0$.

     c. $6x + 2y - 3z - 63 = 0$, $6x + 2y - 3z + 49 = 0$.

     d. $x + 2y + 2z - 7 = 0$, $3x + 6y + 6z - 1 = 0$.

7. Find the point of intersection of the planes

     a. $x + 2y + z = 0$, $x - 2y - 8 = 0$, $x + y + z - 3 = 0$.

     b. $3x - 5y - 4z + 7 = 0$, $6x + 2y + 2z - 7 = 0$, $x + y = 5$.

8. Show that the four planes $4x + y + z + 4 = 0$, $y - 5z + 14 = 0$, $x + 2y - z + 3 = 0$, $x + y + z - 2 = 0$ have a common point.

9. a. Find the volume of the tetrahedron bounded by the coördinate planes and the plane $2x + y + 3z = 12$.

     b. Find the length of the altitude drawn from the origin.

     c. Find the area of each face.

10. Find the area of the triangle which the coördinate planes cut from each of the following planes.

HINT. Use intercepts and express the volume of the tetrahedron bounded by the given plane and the coördinate planes in two ways.

     a. $2x + 2y + z - 12 = 0$.   **b.** $7x - 7y + 2z - 6 = 0$.

     c. $2x - y - 3z + 12 = 0$.   **d.** $x + 5y + 7z - 3 = 0$.

272. Equations of a straight line. The locus of two simultaneous equations of the first degree,

$$\begin{cases} A_1x + B_1y + C_1z + D_1 = 0, \\ A_2x + B_2y + C_2z + D_2 = 0, \end{cases}$$

is a straight line. This follows immediately from the fact that the intersection of two planes is a straight line.

Example 1. Find the points where the line

$$\begin{cases} 2x - y - z - 1 = 0, \\ 3x - 2y - z = 0 \end{cases}$$

cuts the coördinate planes.

Solution. To find where the line cuts the xy-plane, we set $z = 0$ in each of the equations above, which gives

$$\begin{cases} 2x - y - 1 = 0, \\ 3x - 2y = 0. \end{cases}$$

Solving these equations, we find $x = 2$, $y = 3$. Hence the line cuts the xy-plane at the point $(2, 3, 0)$.

Similarly, if $y = 0$, we find $x = -1$, $z = -3$. Hence the line cuts the zx-plane at the point $(-1, 0, -3)$.

If $x = 0$, we find $y = 1$, $z = -2$. Hence the line cuts the yz-plane at the point $(0, 1, -2)$.

Example 2. Find the direction angles of the line

$$\begin{cases} x + 2y + z = 0, \\ x - y - z - 2 = 0, \end{cases}$$

assuming that the line is directed upward.

Solution. Any two points on the line will enable us to find its direction cosines. We will find the point P_1 where the line cuts the plane $z = 0$, and the point P_2 where the line cuts the plane $z = 1$.

Setting $z = 0$ in the given equations, we find $x = \frac{4}{3}$, $y = -\frac{2}{3}$.
Setting $z = 1$ in the given equations, we find $x = \frac{5}{3}$, $y = -\frac{4}{3}$.
Hence the line joins the point $(\frac{4}{3}, -\frac{2}{3}, 0)$ to the point $(\frac{5}{3}, -\frac{4}{3}, 1)$, and its direction cosines are as follows.

$$\cos \alpha = \frac{\frac{1}{3}}{\sqrt{\frac{14}{9}}} = \frac{1}{\sqrt{14}} = 0.2673.$$

$$\cos \beta = \frac{-\frac{2}{3}}{\sqrt{\frac{14}{9}}} = \frac{-2}{\sqrt{14}} = -0.5345.$$

$$\cos \gamma = \frac{1}{\sqrt{\frac{14}{9}}} = \frac{3}{\sqrt{14}} = 0.8018.$$

The positive sign was taken with the radical because it was assumed that the line is directed upward, and therefore $\cos \gamma$ is positive. The direction angles are found to be

$$\alpha = 74° \, 30', \quad \beta = 122° \, 19', \quad \gamma = 36° \, 42'.$$

273. System of planes passing through a straight line; the projecting planes. Consider the straight line which is the locus of the simultaneous equations

$$\begin{cases} 3x + 2y - z - 1 = 0, \\ 2x - y + 2z - 3 = 0 \end{cases} \tag{1}$$

The plane whose equation is

$$3x + 2y - z - 1 + k(2x - y + 2z - 3) = 0, \tag{2}$$

where k has any fixed arbitrary value, will pass through the line (1). For the coördinates of any point which satisfy both equations in (1) will certainly satisfy (2). That is, the plane (2) will pass through all points of the line of intersection of the two planes in (1). By giving to k in (2) a series of numerical values we obtain a variety of planes passing through the line (1).

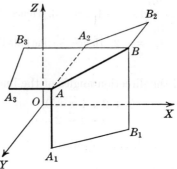

Fig. 273.1

The projecting planes of a line. A plane passing through a given line and perpendicular to one of the coördinate planes is called a *projecting plane.*

In Fig. 273.1 are shown the three projecting planes of a line AB. Multiplying out in (2) and collecting terms,

$$(3 + 2\,k)x + (2 - k)y + (-1 + 2\,k)z - 1 - 3\,k = 0. \qquad (3)$$

This plane will be perpendicular to the xy-plane when the coefficient of z equals zero, that is, if $k = \frac{1}{2}$ (Art. 271). Writing this value of k in (3) and reducing,

$$4\,x + \tfrac{3}{2}\,y - \tfrac{5}{2} = 0, \quad \text{or} \quad 8\,x + 3\,y - 5 = 0. \qquad (4)$$

This is therefore the equation of the projecting plane of the line (1) on XOY, that is, of the plane ABB_1A_1 (Fig. 273.1).

But equation (4) is simply the result obtained by *eliminating z from the equations* (1); namely, we multiply the first of equations (1) by 2 and add it to the second. Hence the result:

To find the equations of the projecting planes of a line, eliminate x, y, and z in turn from the given equations.

Thus, to finish the example begun, eliminating y from (1), we find $7\,x + 3\,z - 7 = 0$ for the projecting plane on the xz-plane. Eliminating x, we get $7\,y - 8\,z + 7 = 0$ for the equation of the projecting plane on the yz-plane.

Example. The equations of two projecting planes of a line are

$$2\,x + z = 6, \quad 3\,x + y = 6.$$

Draw the traces, find the point of intersection of the pair of traces on each coördinate plane, draw the line of intersection, and find the direction angle α.

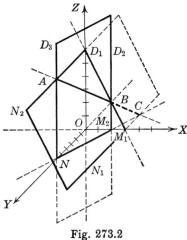

Fig. 273.2

Solution. The lines in the figure are as follows.
Plane $2\,x + z = 6$. Traces:

On ZX, $2\,x + z = 6$, line M_1D_1.
On XY, $\quad\quad x = 3$, line M_1N_1.
On YZ, $\quad\quad z = 6$, line D_1N_2.

Plane $3\,x + y = 6$. Traces:

On ZX, $\quad\quad\quad x = 2$, line M_2D_2.
On XY, $3\,x + y = 6$, line M_2N.
On YZ, $\quad\quad\quad y = 6$, line ND_3.

The points of intersection of the two traces on the respective coördinate planes are as follows.

M_2D_2 and M_1D_1 intersect at $B(2, 0, 2)$.
M_1N_1 and M_2N intersect at $C(3, -3, 0)$.
D_1N_2 and ND_3 intersect at $A(0, 6, 6)$.

To find $\cos \alpha$, take A and B as the points and the direction from B toward A (upward). Then (**(G)**, Art. 267),

$$\cos \alpha = \frac{-2}{\sqrt{56}} = -0.2673.$$

$$\alpha = 105° \, 30'.$$

PROBLEMS

1. Find the equations of the projecting planes and the direction angles of the following lines, assuming that the positive direction is upward. Construct the lines.

a. $\begin{cases} 2\,x - 2\,y + 2\,z = 0, \\ 2\,x + y - 2\,z = 0. \end{cases}$

b. $\begin{cases} x - 2\,z - 2 = 0, \\ y + z = 0. \end{cases}$

c. $\begin{cases} 2\,x - 4\,y - 3\,z - 2 = 0, \\ 4\,y - 3\,z = 0. \end{cases}$

d. $\begin{cases} x + y + z + 1 = 0, \\ x - y - z - 2 = 0. \end{cases}$

e. $\begin{cases} x + 2\,y = 6, \\ 2\,x + z = 5. \end{cases}$

f. $\begin{cases} y + 2\,z = 6, \\ 2\,x + y = 6. \end{cases}$

2. Show that the lines defined by the following pairs of equations are parallel, and construct the lines.

a. $2\,y + z = 0,\ 3\,y - 4\,z = 7,$ and $5\,y - 2\,z = 8,\ 4\,y + z = 44.$
b. $x + 2\,y - z = 7,\ y + z - 2\,x = 6,$ and $3\,x + 6\,y - 3\,z = 8,\ 2\,x - y - z = 0.$
c. $3\,x + z = 4,\ y + 2\,z = 9,$ and $6\,x - y = 7,\ 3\,y + 6\,z = 1.$
d. $6\,x + 3\,y - 4\,z + 6 = 0,\ 8\,x + 9\,y - 4\,z + 8 = 0,$ and $4\,x - 3\,y - 4\,z = 4,$
$7\,x + 6\,y - 4\,z = 9.$

3. Show that the lines of intersection of each of the following pairs of planes meet in a point.

a. $x + 2\,y = 1,\ 2\,y - z = 1,$ and $x - y = 1,\ x - 2\,z = 3.$
b. $4\,x + y - 3\,z + 24 = 0,\ z = 5,$ and $x + y + 3 = 0,\ x + 2 = 0.$
c. $3\,x + y - z = 1,\ 2\,x - z = 2,$ and $2\,x - y + 2\,z = 4,\ x - y + 2\,z = 3.$

4. Show that the lines of intersection of the planes $x + 2\,y + 3\,z = 3$ and $3\,x + 6\,y + 9\,z = 20$ with $4\,x - y + z = 0$ are parallel lines.

274. Examples of surfaces. A few standard equations of surfaces are given in this article. They are all *quadric surfaces*.

Ellipsoid. The locus of

$$\frac{x^2}{a^2} + \frac{y^2}{b^2} + \frac{z^2}{c^2} = 1$$

is called an ellipsoid. The surface is shown in Fig. 274.1. The traces (sections by the coördinate planes) are ellipses whose equations are

$$\frac{x^2}{a^2} + \frac{y^2}{b^2} = 1 \text{ in the } xy\text{-plane,}$$

$$\frac{y^2}{b^2} + \frac{z^2}{c^2} = 1 \text{ in the } yz\text{-plane,}$$

$$\frac{x^2}{a^2} + \frac{z^2}{c^2} = 1 \text{ in the } xz\text{-plane.}$$

Sections by planes parallel to a coördinate plane are ellipses diminishing in size as the plane of the section recedes, and finally reducing to a point ellipse. The surface is a closed surface.

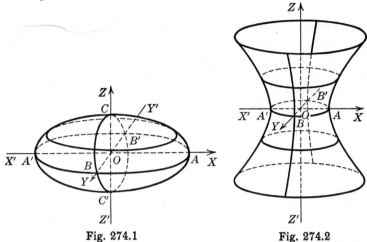

Fig. 274.1   Fig. 274.2

The intercepts $OA = a$, $OB = b$, $OC = c$ are called the *semi-axes* of the ellipsoid. If $a = b$, the surface is an *ellipsoid of revolution* generated by revolving the ellipse $ACA'C'$ (see Fig. 274.1) about the z-axis.

Hyperboloid of one sheet. The locus of

$$\frac{x^2}{a^2} + \frac{y^2}{b^2} - \frac{z^2}{c^2} = 1$$

is called a hyperboloid of one sheet (see Fig. 274.2). The traces on XOZ and YOZ are hyperbolas. Sections by planes parallel to the xy-plane are ellipses increasing in size as the plane of the section recedes. The surface is not closed. If $a = b$, the surface is a hyperboloid of revolution of one sheet about the z-axis.

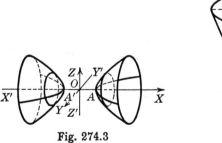

Fig. 274.3

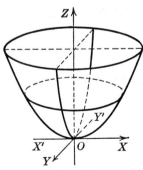

Fig. 274.4

Hyperboloid of two sheets. The locus of

$$\frac{x^2}{a^2} - \frac{y^2}{b^2} - \frac{z^2}{c^2} = 1$$

is called a hyperboloid of two sheets (see Fig. 274.3). The traces on XOZ and XOY are hyperbolas. There is no trace on YOZ. Sections by planes parallel to YOZ and at a distance numerically not less than a are ellipses increasing in size as the plane of the section recedes. If $b = c$, the surface is a hyperboloid of revolution of two sheets about the x-axis.

Elliptic paraboloid. The locus of

$$\frac{x^2}{a^2} + \frac{y^2}{b^2} = 2\,cz$$

is called an elliptic paraboloid (see Fig. 274.4, in which c is positive). The traces on XOZ and YOZ are parabolas. Sections by planes parallel to XOY and above it are ellipses increasing in size as the plane of the section recedes. Sections by planes parallel to XOY and below it are imaginary and the surface lies entirely above XOY. If $a = b$, the surface is a paraboloid of revolution about the z-axis.

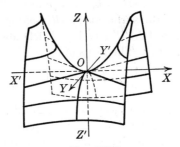

Fig. 274.5

Hyperbolic paraboloid. The locus of

$$\frac{x^2}{a^2} - \frac{y^2}{b^2} = 2\,cz$$

is called a hyperbolic paraboloid (see Fig. 274.5, in which c is positive). The traces on XOZ and YOZ are parabolas. The trace on XOY is the pair of lines

$$\frac{x}{a} + \frac{y}{b} = 0, \quad \frac{x}{a} - \frac{y}{b} = 0.$$

Sections by planes parallel to XOY are hyperbolas (see also Fig. 270.3).

Example. Calculate the volume of the ellipsoid

$$\frac{x^2}{a^2} + \frac{y^2}{b^2} + \frac{z^2}{c^2} = 1$$

by an integration by the method of Art. 98.

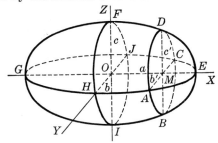

Fig. 274.6

Solution. Consider a plane section of the ellipsoid perpendicular to OX, as $ABCD$, with semiaxes b' and c'. The equation of the ellipse $HEJG$ in the xy-plane is

$$\frac{x^2}{a^2} + \frac{y^2}{b^2} = 1.$$

Solving this for $y(= b')$ in terms of $x(= OM)$ gives

$$b' = \frac{b}{a}\sqrt{a^2 - x^2}.$$

Similarly, from the equation of the ellipse $EFGI$ in the xz-plane we get

$$c' = \frac{c}{a}\sqrt{a^2 - x^2}.$$

Hence the area of the ellipse (section) $ABCD$ is

$$\pi b'c' = \frac{\pi bc}{a^2}(a^2 - x^2) = A(x).$$

Substituting in the formula of Art. 98,

$$V = \frac{\pi bc}{a^2} \int_{-a}^{+a} (a^2 - x^2)dx = \frac{4}{3}\pi abc.$$

PROBLEMS

1. Discuss, construct, and name the locus of each of the following equations.

a. $x^2 + 4y^2 + 9z^2 = 36.$
b. $x^2 + 4y^2 - 9z^2 = 36.$
c. $x^2 - 4y^2 - 9z^2 = 36.$
d. $x^2 + y^2 + z^2 = 36.$
e. $x^2 + y^2 = 4z.$
f. $y^2 - z^2 = 6x.$
g. $x^2 + y^2 = 4z^2.$
h. $x^2 - 4y^2 + 9z^2 = 36.$
i. $x^2 + y^2 + z = 4.$
j. $x^2 + 4y^2 - z = 4.$
k. $y^2 + z^2 = 1 - x.$
l. $4x^2 + 16y^2 - z^2 = 64.$
m. $y^2 - z^2 + 4x = 0.$
n. $y^2 - z^2 + 4x^2 = 0.$

2. Assuming that the equation of a central quadric surface is
$$Ax^2 + By^2 + Cz^2 + D = 0,$$
find its equation if it passes through the given points. Name the surface.

        a. $(2, -1, 1)$, $(-3, 0, 0)$, $(1, -1, -2)$.
        b. $(-1, 5, 4)$, $(-7, 1, -8)$, $(8, -2, 10)$.
        c. $(4, -2, -1)$, $(0, 1, -3)$, $(3, 5, 2)$.

3. Taking the equation of a noncentral quadric surface in the form
$$Ax^2 + By^2 + Cz = 0,$$
find its equation if the surface passes through the points given. Name the surface.

    a. $(1, 0, 1)$, $(0, 2, 1)$.   **b.** $(1, 0, 1)$, $(0, 2, -1)$.
    c. $(1, 2, 1)$, $(2, 1, 1)$.   **d.** $(3, 5, 8)$, $(4, -2, -6)$.

4. Find the equation of the locus of a point whose distance from the point $(1, 0, 0)$ is half its perpendicular distance from the plane $x = 4$. Name and draw the locus.

5. Find the equation of the locus of a point if its distance from $(0, -4, 0)$ is twice its perpendicular distance from the plane $y + 1 = 0$. Name and draw the locus.

6. Calculate the volume bounded by each of the following quadric surfaces and given planes.

        a. $z = x^2 + 4y^2$; $z = 1$.
        b. $4x^2 + 9z^2 + y = 0$; $y + 1 = 0$.
        c. $x^2 + 4y^2 = 1 + z^2$; $z + 1 = 0$; $z - 1 = 0$.
        d. $25y^2 + 9z^2 = 1 + x^2$; $x = 0$; $x - 2 = 0$.
        e. $x^2 + 9y^2 = z^2$; $z + 1 = 0$.

7. Show that the volume of the solid bounded by the elliptic paraboloid
$$2cz = \frac{x^2}{a^2} + \frac{y^2}{b^2}$$
and the plane $z = h$ is half the volume of the cylinder whose altitude is h and whose bases are equal to the section of the paraboloid by the plane.

8. A solid is bounded by the hyperboloid of one sheet
$$\frac{x^2}{a^2} + \frac{y^2}{b^2} - \frac{z^2}{c^2} = 1$$
and the planes $z = 0$, $z = h$. Show that its volume equals one third the volume of a cylinder with altitude h and base equal to the section by $z = h$ plus two thirds the volume of a cylinder with altitude h and base equal to the section by $z = 0$.

CHAPTER XXIII

PARTIAL DIFFERENTIATION

275. Partial derivatives. The preceding chapters have been concerned with applications of the calculus to functions of one variable. In this and the following chapter we consider briefly and informally some problems connected with functions of more than one independent variable. Given a function of two variables

$$z = f(x, y), \tag{1}$$

let the value of y remain constant and equal to y_0. Then in the relation

$$z = f(x, y_0) \tag{2}$$

z is a function of one variable x, and we may form its derivative in the usual manner. The notation is

$$\frac{\partial z}{\partial x} = partial\ derivative\ of\ z\ with\ respect\ to\ x\ (y\ remains\ constant).^*$$

Similarly,

$$\frac{\partial z}{\partial y} = partial\ derivative\ of\ z\ with\ respect\ to\ y\ (x\ remains\ constant).^*$$

Corresponding symbols are used for partial derivatives of functions of three or more variables.

In order to avoid confusion the round ∂† has been generally adopted to indicate partial differentiation.

Example 1. Find the partial derivatives of $z = x^3 + 3\,x^2y + y^2$.

Solution.
$$\frac{\partial z}{\partial x} = 3\,x^2 + 6\,xy.$$

$$\frac{\partial z}{\partial y} = 3\,x^2 + 2\,y.$$

Example 2. The slant height and diameter of a right circular cone are, respectively, 10 in. and 16 in. Find

a. the rate of change of the volume with respect to the slant height if the slant height varies and the diameter remains constant;

b. the rate of change of the volume with respect to the diameter if the diameter varies and the slant height remains constant.

*The constant values are substituted in the function before differentiating.
†Introduced by Jacobi (1804–1851).

Solution. Let slant height $= s$, diameter $= y$, volume $= v$. Then radius r of the cone $= \frac{1}{2} y$, altitude $h = \sqrt{s^2 - r^2} = \sqrt{s^2 - \frac{1}{4} y^2}$. But $v = \frac{1}{3} \pi r^2 h$. Hence, by substituting and reducing,

$$v = \tfrac{1}{24} \pi y^2 \sqrt{4 s^2 - y^2}.$$

a. Differentiating, holding y fast, we have

$$\frac{\partial v}{\partial s} = \frac{\pi y^2 s}{6 \sqrt{4 s^2 - y^2}}.$$

Substituting $s = 10$, $y = 16$, we get

$$\frac{\partial v}{\partial s} = \frac{320 \, \pi}{9}.$$

b. Differentiating, holding s fast, the result is

$$\frac{\partial v}{\partial y} = \frac{1}{12} \pi y \sqrt{4 s^2 - y^2} - \frac{\pi y^3}{24 \sqrt{4 s^2 - y^2}} = \frac{\pi y (8 s^2 - 3 y^2)}{24 \sqrt{4 s^2 - y^2}}.$$

Substituting $s = 10$, $y = 16$, we get

$$\frac{\partial v}{\partial y} = \frac{16 \, \pi}{9}.$$

Example 3. In the equation of the ellipsoid

$$\frac{x^2}{a^2} + \frac{y^2}{b^2} + \frac{z^2}{c^2} = 1$$

each coördinate is defined, implicitly, as a function of the other two. Find the partial derivatives of z with respect to x and with respect to y.

Solution. Holding y fast, and differentiating with respect to x term by term, we get

$$\frac{2 x}{a^2} + \frac{2 z}{c^2} \frac{\partial z}{\partial x} = 0. \quad \therefore \frac{\partial z}{\partial x} = - \frac{c^2 x}{a^2 z}.$$

Holding x fast, and differentiating with respect to y,

$$\frac{2 y}{b^2} + \frac{2 z}{c^2} \frac{\partial z}{\partial y} = 0. \quad \therefore \frac{\partial z}{\partial y} = - \frac{c^2 y}{b^2 z}.$$

Referring to (1), we have, in the notations commonly used,

$$\frac{\partial z}{\partial x} = \frac{\partial}{\partial x} f(x, y) = \frac{\partial f}{\partial x} = f_x(x, y) = f_x = z_x;$$

$$\frac{\partial z}{\partial y} = \frac{\partial}{\partial y} f(x, y) = \frac{\partial f}{\partial y} = f_y(x, y) = f_y = z_y.$$

Similar notations are used for functions of any number of variables.

PROBLEMS

Find the partial derivatives of the following functions.

1. $z = x^2 + 4\,xy + 7\,y^2.$

2. $u = x^2y + y^2z + z^2x.$

3. $z = \dfrac{x}{x-y}.$

4. $u = \dfrac{xyz}{x+2\,y+3\,z}.$

5. $z = x^2 \sin y.$

6. $z = \sin 2\,x \cos 2\,y.$

7. $\rho = \sin^2 \theta \cos \phi.$

8. $\rho = \cos \theta \tan \phi.$

9. $z = xye^{xy}.$

10. $\theta = \ln \sqrt{x^2 + y^2}.$

11. $z = x \ln \left(\dfrac{y}{x}\right).$

12. $\theta = \arc \tan \left(\dfrac{y}{x}\right).$

13. $u = \dfrac{x}{\sqrt{x^2 + y^2 + z^2}}.$

14. $\theta = \arc \sin \left(\dfrac{y}{\sqrt{x^2 + y^2}}\right)$

15. $z = \sqrt{x^2 + y^2 - 2\,xy \cos \theta}.$ (θ variable.)

16. $u = \frac{1}{2} \dfrac{h^2 \sin A \sin B}{\sin (A + B)}.$ (h, A, B variables.)

17. If $z = x^4 - 2\,x^3y + x^2y^2$, show that $x\,\dfrac{\partial z}{\partial x} + y\,\dfrac{\partial z}{\partial y} = 4\,z.$

18. If $z = \dfrac{xy}{x+y}$, show that $x\,\dfrac{\partial z}{\partial x} + y\,\dfrac{\partial z}{\partial y} = z.$

19. If $u = x^2y + y^2z + z^2x$, show that $\dfrac{\partial u}{\partial x} + \dfrac{\partial u}{\partial y} + \dfrac{\partial u}{\partial z} = (x + y + z)^2.$

20. If $u = \dfrac{x^3 - y^3}{xy}$, show that $x\,\dfrac{\partial u}{\partial x} + y\,\dfrac{\partial u}{\partial y} = u.$

21. The altitude and radius of a right circular cylinder are, respectively, 8 in. and 6 in. Find **(a)** the rate of change of the volume with respect to the altitude if the radius remains constant; **(b)** the rate of change of the volume with respect to the radius if the altitude remains constant.

22. The area of a triangle is given by the formula $K = \frac{1}{2}\,bc \sin A$. Given $b = 10$ in., $c = 20$ in., $A = 60°$.

a. Find the area.

b. Find the rate of change of the area with respect to the side b if c and A remain constant.

c. Find the rate of change of the area with respect to the angle A if b and c remain constant.

d. Using the rate found in **(c)** calculate approximately the change in area if the angle is increased by one degree.

e. Find the rate of change of c with respect to b if the area and the angle remain constant.

23. The law of cosines for any triangle is

$$a^2 = b^2 + c^2 - 2\,bc \cos A.$$

Given $b = 10$ in., $c = 15$ in., $A = 60°$.

a. Find a.

b. Find the rate of change of a with respect to b if c and A remain constant.

c. Using the rate found in (**b**), calculate approximately the change in a if b is decreased by 1 in.

d. Find the rate of change of a with respect to A if b and c remain constant.

e. Find the rate of change of c with respect to A if a and b remain constant.

276. Geometric interpretation of partial derivatives. Let the equation of the surface in Fig. 276 be

$$z = f(x, y). \tag{1}$$

Let the curve AP_1B be the section made by the plane $y = y_1$, where $y_1 = ON$. The curve AP_1B is the locus in space of the simultaneous equations

$$y = y_1, \quad z = f(x, y_1). \tag{2}$$

As P moves along the curve AP_1B from A to B, the coördinates x and z correspond to abscissa and ordinate referred to NX_1 and NZ_1 as axes. Then the meaning of

$$\frac{\partial z}{\partial x} = f_x(x, y), \text{ when } y = y_1,$$

is the same as the meaning of the derivative of the ordinate z **of** the curve AP_1B with respect to its abscissa x. That is, the value of this derivative at any point P_1 where $x = x_1$ is the "slope" of the tangent line P_1T. In the figure this slope is $\tan \angle X_1TP_1$. Hence

$$f_x(x_1, y_1) = \tan \angle X_1TP_1. \tag{3}$$

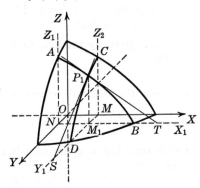

Fig. 276

Again, let the curve DP_1C be the section of the surface by the plane $x = x_1$, where $x_1 = OM$. As P moves along this curve from C to D, the coördinates y and z correspond to abscissa and ordinate referred to MY_1 and MZ_2 as axes. Hence it is clear that

$$f_y(x_1, y_1) = \tan \angle Y_1SP_1. \tag{4}$$

We can now write down equations for the tangent lines P_tT and P_1S as loci in space. Let $z_1 = M_1P_1$. Then the equation of P_1T referred to NX_1 and NZ_1 as axes is

$$z - z_1 = f_x(x_1, y_1)(x - x_1). \tag{5}$$

The locus of this equation (5) *in space* is the plane parallel to the y-axis passing through the line P_1T. Thus we see that the equations required are as follows.

$$\begin{cases} y = y_1, \\ z - z_1 = f_x(x_1, y_1)(x - x_1). \end{cases} \tag{6}$$

In the same manner, we find for the equations of the line P_1S

$$\begin{cases} x = x_1, \\ z - z_1 = f_y(x_1, y_1)(y - y_1). \end{cases} \tag{7}$$

277. Equation of a tangent plane to a surface. When a line is tangent to a curve on a surface, it is said to be tangent also to the surface. At any ordinary point of a surface all tangent lines lie in one plane* which is called the tangent plane to the surface at the given point. Any two lines which are tangent to a surface at the same point determine the tangent plane at that point.

Let $P_1(x_1, y_1, z_1)$ be a point on the surface

$$z = f(x, y).$$

The curve AP_1B of Fig. 276 is the section of the surface made by the plane $y = y_1$, and its equations are $z = f(x, y_1)$, $y = y_1$. The equations of the tangent line P_1T to this curve are

$$\begin{cases} y - y_1 = 0, \\ z - z_1 = f_x(x_1, y_1)(x - x_1). \end{cases} \tag{1}$$

The curve CP_1D is the section made by the plane $x = x_1$, and its equations are $z = f(x_1, y)$, $x = x_1$. The equations of the tangent line P_1S to this curve are

$$\begin{cases} x - x_1 = 0, \\ z - z_1 = f_y(x_1, y_1)(y - y_1). \end{cases} \tag{2}$$

---

*The proof of this statement is omitted. See Granville, Smith, and Longley's *Elements of Calculus* (Ginn and Company), p. 471.

Consider now the equation

$$z - z_1 = f_x(x_1, y_1)(x - x_1) + f_y(x_1, y_1)(y - y_1). \tag{3}$$

This equation represents a plane because it is of the first degree in x, y, and z. We shall prove that it passes through the tangent lines $P_1 T$ and $P_1 S$.

Let (ξ, η, ζ) be any point on the line $P_1 T$. Then ξ, η, ζ satisfy both of equations (1) when substituted for x, y, z and consequently will satisfy equation (3). Hence the plane (3) contains every point of the line $P_1 T$.

Similarly, the plane (3) contains every point of the line $P_1 S$.

Hence (3) is the equation of the tangent plane to the surface at the point P_1, since it contains two tangent lines to the surface at this point.

Example 1. Find the equation of the tangent plane to the surface $z = x^2 + y^2$ at the point where $x = 1$, $y = 2$.

Solution. When $x_1 = 1$, $y_1 = 2$, we find $z_1 = 5$.

$$\frac{\partial z}{\partial x} = 2x, \quad \text{and} \quad f_x(x_1, y_1) = 2.$$

$$\frac{\partial z}{\partial y} = 2y, \quad \text{and} \quad f_y(x_1, y_1) = 4.$$

Hence, substituting in (3) above, the equation of the tangent plane is found to be

$$z - 5 = 2(x - 1) + 4(y - 2),$$

or, after simplification, $2x + 4y - z - 5 = 0.$

Example 2. Find the equation of the tangent plane to the sphere $x^2 + y^2 + z^2 = 9$ at the point $(2, 1, 2)$.

Solution. The partial derivatives may be obtained as in Example 3, Art. 275. To find the partial derivative of z with respect to x, we regard y as a constant and differentiate the equation, as it stands, with respect to x.

$$2x + 2z \frac{\partial z}{\partial x} = 0, \quad \text{whence} \quad \frac{\partial z}{\partial x} = -\frac{x}{z}.$$

Similarly, $2y + 2z \dfrac{\partial z}{\partial y} = 0, \quad \text{whence} \quad \dfrac{\partial z}{\partial y} = -\dfrac{y}{z}.$

Hence, since $x_1 = 2$, $y_1 = 1$, we have

$$f_x(x_1, y_1) = -1, \qquad f_y(x_1, y_1) = -\tfrac{1}{2}.$$

Substituting in (3) above, the equation of the tangent plane is found to be

$$z - 2 = -1(x - 2) - \tfrac{1}{2}(y - 1),$$

or, after simplification, $2x + y + 2z - 9 = 0.$

PROBLEMS

Find the equation of the tangent plane to each of the following surfaces at the point indicated.

1. $z = x^2 + 2\,y^2$; $x = 2$, $y = -1$.
2. $z = 4 - x^2 - y^2$; $x = 2$, $y = 0$.
3. $z = x^2 + y^2 - 9$; $x = 2$, $y = -2$.
4. $x^2 + y^2 - z^2 = 0$; $x = 3$, $y = 4$, $z = 5$.
5. $z = (x - 2)^2 + 3(y - 1)^2 - 4$; $x = 2$, $y = 1$.
6. $z = x^2 + xy + y^2$; $x = 1$, $y = 2$.
7. $z = xy - y^2$; $x = 3$, $y = 1$.
8. $z = 2\,x - x^2 - 4\,y - y^2$; $x = 1$, $y = -2$.

278. Maximum and minimum values of functions of two variables.
A function of two variables $z = f(x, y)$ is represented geometrically by a surface. If the function has a minimum value (Fig. 278.1) or a maximum value (Fig. 278.2), the corresponding surface will have a horizontal tangent plane, that is, the tangent plane at a maximum or a minimum point will be parallel to the xy-plane. Hence, if the function has a maximum or a minimum value $z = z_1$ when $x = x_1$, $y = y_1$, the tangent plane to the surface at this point will have the equation $z = z_1$. The above argument is based upon geometric intuition.

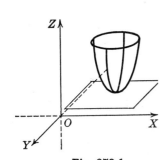

Fig. 278.1

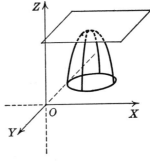

Fig. 278.2

The equation of the tangent plane to the surface $z = f(x, y)$ at any point (x_1, y_1, z_1) is, by (3), Art. 277,

$$z - z_1 = f_x(x_1, y_1)(x - x_1) + f_y(x_1, y_1)(y - y_1).$$

If this plane is parallel to the xy-plane, we must have

$$f_x(x_1, y_1) = 0, \quad f_y(x_1, y_1) = 0. \tag{1}$$

These two conditions are *necessary* if the function has either a maximum or a minimum value. The conditions are not sufficient. Equations (1) may be true and yet $z_1 = f(x_1, y_1)$ be neither a maximum nor a minimum value. Sufficient conditions are derived in more advanced textbooks.*

To examine a function of two variables

$$z = f(x, y) \tag{2}$$

for maximum and minimum values, we obtain the partial derivatives, and solve the simultaneous equations

$$f_x(x, y) = 0, \quad f_y(x, y) = 0 \tag{3}$$

for x and y. These common solutions will give all values of x and y for which z in (2) is a maximum or a minimum. From the nature of the given problem we must determine for each common solution of (3) whether z has a maximum or a minimum value.

Example 1. Examine

$$f(x, y) = x^2 + xy + y^2 + 4x + 5y + 10 \tag{4}$$

for maximum or minimum values.

Solution. Setting the partial derivatives equal to zero, we have

$$f_x(x, y) = 2x + y + 4 = 0,$$
$$f_y(x, y) = x + 2y + 5 = 0.$$

Solving, $x = -1$, $y = -2$. For these values, from (4),

$$f(-1, -2) = 1 + 2 + 4 - 4 - 10 + 10 = 3.$$

Is, then, this value 3 a maximum or a minimum value? To decide this, we calculate the value of (4) for values of x and y near $(-1, -2)$. Let

$$x = -1 + h, \qquad y = -2 + k.$$

Substituting in (4), we have

$$f(-1 + h, -2 + k) = (-1 + h)^2 + (-1 + h)(-2 + k) + (-2 + k)^2$$
$$+ 4(-1 + h) + 5(-2 + k) + 10$$
$$= 3 + h^2 + hk + k^2.$$

But $h^2 + hk + k^2 = (h + \frac{1}{2}k)^2 + \frac{3}{4}k^2$, and is therefore *positive* for *all* values of h and k, positive or negative (except $h = k = 0$). Hence the value of the function (4) is greater than 3 for all values of x and y near $x = -1$, $y = -2$. Therefore the function (4) has a minimum value equal to 3.

*See Granville, Smith, and Longley's *Elements of Calculus* (Ginn and Company), p. 481.

Example 2. A long piece of tin 24″ wide is to be made into a trough by bending up two sides. Find the width and inclination of each side if the carrying capacity is a maximum.

Solution. The area of the cross section shown in Fig. 278.3 must be a maximum. The cross section is a trapezoid of upper base $24 - 2x + 2x\cos\alpha$, lower base $24 - 2x$, and altitude $x\sin\alpha$. The area A is given by

$$A = 24x\sin\alpha - 2x^2\sin\alpha + x^2\sin\alpha\cos\alpha.$$

By differentiation we have

$$\frac{\partial A}{\partial x} = 24\sin\alpha - 4x\sin\alpha + 2x\sin\alpha\cos\alpha.$$

$$\frac{\partial A}{\partial \alpha} = 24x\cos\alpha - 2x^2\cos\alpha + x^2(\cos^2\alpha - \sin^2\alpha).$$

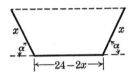

$$\longleftarrow 24 - 2x \longrightarrow$$

Fig. 278.3

Setting these partial derivatives equal to zero, we have the two equations

$$2\sin\alpha(12 - 2x + x\cos\alpha) = 0,$$
$$x[24\cos\alpha - 2x\cos\alpha + x(\cos^2\alpha - \sin^2\alpha)] = 0.$$

One solution of this system is $\alpha = 0$, $x = 0$, which has no meaning in the physical problem. Assuming $\alpha \neq 0$, $x \neq 0$, and solving the equations, we get $\cos\alpha = \frac{1}{2}$, $x = 8$.

A consideration of the physical problem shows that there must exist a maximum value of the area. Hence this maximum value occurs when $\alpha = 60°$ and $x = 8''$.

PROBLEMS

1. Discuss the following functions for maxima and minima.

a. $x^2 + xy + y^2 - 6x - 4y$.   b. $x^2 + xy + y^2 - x - 5y$.

c. $x^2 - 6xy + y^3 + 3x + 6y - 7$.   d. $4xy + \dfrac{1}{x} + \dfrac{1}{y}$.

e. $\sin x + \sin y + \sin(x + y)$.   f. $x^3 - 6xy + y^3$.

g. $x^3 + y^3 - x^2y^2 - \frac{1}{2}(x^2 + y^2)$.

2. Find the minimum value of $x^2 + xy + y^2 - ax - by$.

3. Show that the maximum value of $\dfrac{(ax + by + c)^2}{x^2 + y^2 + 1}$ is $a^2 + b^2 + c^2$.

4. Find the rectangular parallelepiped of maximum volume which has three faces in the coördinate planes and one vertex in the plane $\frac{x}{a} + \frac{y}{b} + \frac{z}{c} = 1$.

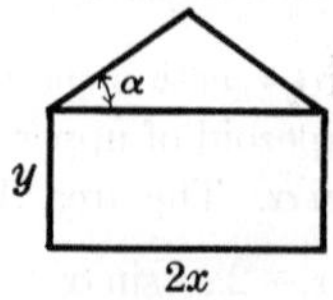

$2x$

Fig. 278.4

5. Find the volume of the largest rectangular parallelepiped that can be inscribed in the ellipsoid $\frac{x^2}{a^2} + \frac{y^2}{b^2} + \frac{z^2}{c^2} = 1$.

6. A pentagon is composed of a rectangle surmounted by an isosceles triangle. If the perimeter of the pentagon has a given value P, find the dimensions for maximum area. (Fig. 278.4)

$$Ans. \ \alpha = 30°, \ 2\,x = \frac{P}{2 + 2\sec\alpha - \tan\alpha},$$

$$y = \frac{P}{2} - x(1 + \sec\alpha).$$

7. If x, y, z are the lengths of the perpendiculars dropped from any point P to the three sides a, b, c, respectively, of a triangle of area K, show that the minimum value of $x^2 + y^2 + z^2$ is equal to $\dfrac{4\,K^2}{a^2 + b^2 + c^2}$.

8. A point P lies within a triangle. Show that the sum of the squares of the distances from P to the vertices is a minimum if P is the point of intersection of the medians.

9. A floating anchorage is to be made with a cylindrical body and equal conical ends. Find the dimensions that make the surface least for a given volume.

10. Find the shortest distance between the lines $x = \frac{1}{2}\,y = \frac{1}{3}\,z$ and $x = y - 3 = z$.

11. If $(x + y)^2 + (y + z)^2 + (z + x)^2 = 3$, show that the greatest and least values of z are $\frac{3}{2}$ and $-\frac{3}{2}$.

279. Total differential and total derivative. If $y = f(x)$ and x is changed to $x + \Delta x$, the increment of y is given by the formula

$$\Delta y = f(x + \Delta x) - f(x). \tag{1}$$

An approximate value of the right-hand member is

$$f'(x)\Delta x, \quad \text{or} \quad \left(\frac{dy}{dx}\right)\Delta x. \tag{2}$$

This quantity equals dy, the differential of y. Also, dx and Δx are equal (see Art. 73). Formula (2) may be described as follows.

If $y = f(x)$, and x is changed by the amount Δx, then y will change by the amount

$$\left(\frac{dy}{dx}\right)\Delta x, \text{ approximately.}$$

If z is a function of two independent variables x and y, and if x and y are given the increments Δx and Δy respectively, the corresponding increment of z is calculated by the formula

$$\Delta z = f(x + \Delta x, y + \Delta y) - f(x, y). \tag{3}$$

It is shown in more advanced texts* that an *approximate* value of the right-hand member is given by

$$\left(\frac{\partial z}{\partial x}\right)\Delta x + \left(\frac{\partial z}{\partial y}\right)\Delta y. \tag{4}$$

The quantity (4) is called the *total differential of the dependent variable* z, and is denoted by dz. Hence

$$dz = \left(\frac{\partial z}{\partial x}\right)\Delta x + \left(\frac{\partial z}{\partial y}\right)\Delta y. \tag{5}$$

If $z = x$, then (5) gives $dx = \Delta x$. If $z = y$, then (5) gives $dy = \Delta y$. Hence, finally,

(A) $$dz = \left(\frac{\partial z}{\partial x}\right)dx + \left(\frac{\partial z}{\partial y}\right)dy.$$

If y remains constant, then $\Delta y = 0$, and (4) becomes

$$\left(\frac{\partial z}{\partial x}\right)\Delta x.$$

Now z is a function of one variable x, and this term in (4) agrees with (2). Hence we see that (4) is the sum of two terms of which the first is, approximately, the change in z when y is held fast and x changes by the amount Δx while the second is the approximate change in z when x is held fast and y changes by the amount Δy.

Summing up: In **(A)**, dx and dy are equal to the increments Δx and Δy respectively. The formula gives an approximate value for Δz, the increment of z, when small increments Δx and Δy are given to x and y.

Example. Given

$$z = x^2 + xy + y^2, \tag{6}$$

and $x = 1$, $y = 2$, $\Delta x = 0.1$, $\Delta y = 0.2$. Find dz and Δz and compare results.

*Granville, Smith, and Longley's *Elements of Calculus*, p. 445.

Solution. Find dz by **(A)**. Differentiating (6),

$$\frac{\partial z}{\partial x} = 2\,x + y, \quad \frac{\partial z}{\partial y} = x + 2\,y.$$

Substituting these values in **(A)**, the result is

$$dz = (2\,x + y)dx + (x + 2\,y)dy. \tag{7}$$

Using the given values,

$$x = 1, \quad y = 2, \quad dx = \Delta x = 0.1, \quad dy = \Delta y = 0.2,$$

we get $dz = 4 \times 0.1 + 5 \times 0.2 = 1.4.$

Now find Δz. In (6), replace x, y, and z by $x + \Delta x$, $y + \Delta y$, $z + \Delta z$, respectively. Then

$$z + \Delta z = (x + \Delta x)^2 + (x + \Delta x)(y + \Delta y) + (y + \Delta y)^2$$
$$= x^2 + xy + y^2 + (2\,x + y)\Delta x + (x + 2\,y)\Delta y + (\Delta x)^2 + \Delta x\Delta y + (\Delta y)^2$$
$$= z + dz + (\Delta x)^2 + \Delta x\Delta y + (\Delta y^2),$$

using the values of z in (6) and dz in (7). Hence

$$\Delta z = dz + (\Delta x)^2 + \Delta x\Delta y + (\Delta y)^2. \tag{8}$$

Substituting $dz = 1.4$, $\Delta x = 0.1$, $\Delta y = 0.2$, we get

$$\Delta z = 1.4 + 0.01 + 0.02 + 0.04 = 1.47.$$

Then $\Delta z - dz = 1.47 - 1.4 = 0.07$. The percentage error made in taking the value 1.4 of dz as the value of Δz is, therefore, $7/1.47 = 4.76\%$.

This example shows by (8) that the difference between Δz and dz is of the second degree in Δx and Δy. The given values of Δx and Δy are assumed to be small compared with the given values of x and y. The quantity $(\Delta x)^2 + \Delta x \cdot \Delta y + (\Delta y)^2$ in (8) will then be relatively a very small number. Thus we see that Δz and dz are equal, approximately.

The total differential of a function u of three independent variables x, y, z is

$$\textbf{(B)} \qquad\qquad du = \left(\frac{\partial u}{\partial x}\right)dx + \left(\frac{\partial u}{\partial y}\right)dy + \left(\frac{\partial u}{\partial z}\right)dz.$$

A similar formula holds for a function of any number of independent variables.

280. Approximation of the total increment; small errors; rates. Formula **(B)** is used to calculate Δu approximately. Also, when the values of x, y, and z are determined by measurement or experiment and hence subject to small errors Δx, Δy, and Δz, a close approximation to the error in u can be found by **(B)** (compare Art. 77).

Example 1. Find the approximate volume of a thin cylindrical can without a top if the inside diameter and height are, respectively, 6 in. and 8 in., and the thickness is $\frac{1}{8}$ in.

Solution. The volume v of a solid right circular cylinder with diameter x and height y is
$$v = \tfrac{1}{4} \pi x^2 y. \tag{1}$$

Obviously, the exact volume of the can is the difference Δv between the volumes of two solid cylinders for which $x = 6\tfrac{1}{4}$, $y = 8\tfrac{1}{8}$, and $x = 6$, $y = 8$, respectively. Since only an approximate value is required, we calculate dv instead of Δv.

Differentiating (1), and using **(B)**, we get
$$dv = \tfrac{1}{2} \pi xy \, dx + \tfrac{1}{4} \pi x^2 \, dy. \tag{2}$$

Substituting in (2) $x = 6$, $y = 8$, $dx = \tfrac{1}{4}$, $dy = \tfrac{1}{8}$, the result is
$$dv = 7\tfrac{1}{8} \pi = 22.4 \text{ cu. in.}$$

The exact value, to the nearest tenth, is $\Delta v = 23.1$ cu. in.

Example 2. Two sides of an oblique plane triangle measured, respectively, 63 ft. and 78 ft., and the included angle measured 60°. These measurements are subject to errors whose maximum values are 0.1 ft. in each length and 1° in the angle. Find the approximate maximum error and the percentage error made in calculating the third side from these measurements.

Solution. Use the law of cosines (**(7)**, Art. 296)
$$u^2 = x^2 + y^2 - 2\,xy \cos \alpha, \tag{3}$$

where x, y are the given sides, α the included angle, and u the third side. The given data are
$$x = 63,\ y = 78,\ \alpha = 60° = \tfrac{1}{3} \pi,\ dx = dy = 0.1,\ d\alpha = 0.01745 \text{ (radian)}. \tag{4}$$

Differentiating (3), we get
$$\frac{\partial u}{\partial x} = \frac{x - y \cos \alpha}{u}, \quad \frac{\partial u}{\partial y} = \frac{y - x \cos \alpha}{u}, \quad \frac{\partial u}{\partial \alpha} = \frac{xy \sin \alpha}{u}.$$

Hence, using **(B)**,
$$du = \frac{(x - y \cos \alpha)dx + (y - x \cos \alpha)dy + xy \sin \alpha \, d\alpha}{u}.$$

Substituting the values from (4), we find
$$du = \frac{2.4 + 4.65 + 74.25}{71.7} = 1.13 \text{ ft.}$$

The percentage error $= 100 \dfrac{du}{u} = 1.6\%$.

Suppose x and y are both functions of a third variable t. Then dx/dt represents the rate of change of x with respect to t, and dy/dt the rate of change of y with respect to t. If z is a function of x and y, it is also a function of t, and the proper limiting process shows* that the rate of change of z with respect to t is given by the following expression:

$$(C) \qquad \frac{dz}{dt} = \left(\frac{\partial z}{\partial x}\right)\frac{dx}{dt} + \left(\frac{\partial z}{\partial y}\right)\frac{dy}{dt}.$$

This expression, dz/dt, is called the *total derivative of z with respect to t*.

Example 1. Given $z = x^2 + 3\,xy - 2\,y^2$, $x = e^t$, $y = \sin t$; find $\dfrac{dz}{dt}$.

Solution. $\qquad \dfrac{\partial z}{\partial x} = 2\,x + 3\,y,$

$$\frac{\partial z}{\partial y} = 3\,x - 4\,y,$$

$$\frac{dx}{dt} = e^t, \quad \frac{dy}{dt} = \cos t.$$

Hence $\qquad \dfrac{dz}{dt} = (2\,x + 3\,y)e^t + (3\,x - 4\,y)\cos t$

$$= (2\,e^t + 3\sin t)e^t + (3\,e^t - 4\sin t)\cos t$$
$$= e^t(3\sin t + 3\cos t + 2\,e^t) - 4\sin t\,\cos t.$$

Example 2. The characteristic equation of a perfect gas is

$$R\theta = pv,$$

where θ is the temperature, p the pressure, v the volume, and R a constant. At a certain instant a given amount of gas has a volume of 15 cu. ft. and is under a pressure of 25 lb. per square inch. Assuming $R = 96$, find the temperature and the rate at which the temperature is changing if the volume is increasing at the rate of $\frac{1}{2}$ cu. ft. per second and the pressure is decreasing at the rate of $\frac{1}{10}$ lb. per square inch per second.

Solution. $\qquad\qquad \theta = \dfrac{1}{R}\,pv,$

$$\frac{d\theta}{dt} = \frac{1}{R}\left(p\frac{dv}{dt} + v\frac{dp}{dt}\right).$$

From the data of the problem, we have

$$p = 25, \quad v = 15, \quad R = 96, \quad \frac{dp}{dt} = -\frac{1}{10}, \quad \frac{dv}{dt} = \frac{1}{2}.$$

$$\frac{d\theta}{dt} = \frac{1}{96}\left(\frac{25}{2} - \frac{15}{10}\right) = \frac{1}{96} \times 11 = \frac{11}{96}.$$

Hence the temperature is increasing at the rate of $\frac{11}{96}$ degrees per second.

---

*For proofs see Granville, Smith, and Longley's *Elements of Calculus* (Ginn and Company), p. 451.

PROBLEMS

Find the total differential of each of the following functions.

1. $z = x^3 - 2xy + 3y^2$.　　　　**2.** $u = \dfrac{x+y}{x-y}$.　　　　**3.** $\rho = \theta \sin \phi$.

4. If $x^2 + y^2 + z^2 = a^2$, show that $dz = -\dfrac{x\,dx + y\,dy}{z}$.

5. Find dz if $xy + yz + zx = a^2$.

6. Compute Δu and du for the function $u = x^2 - xy + y^2$ when $x = 4$, $y = -2$, $\Delta x = 0.2$, $\Delta y = -0.2$.

7. Compute du for the function $u = x + \sqrt{xy}$ when $x = 8$, $y = 2$, $dx = \frac{1}{2}$, $dy = \frac{1}{4}$.

8. Compute Δu and du for the function $u = x^2 - 4y^2$ when $x = 5$, $y = 3$, $\Delta x = -0.1$, $\Delta y = 0.3$.

9. Compute $d\rho$ for the function $\rho = \sin \frac{1}{2}\theta \cos \phi$ when $\theta = \frac{1}{2}\pi$, $\phi = \frac{1}{2}\pi$, $\Delta\theta = 0.2$, $\Delta\phi = 0.1$.

10. The legs of a right triangle measured 5.3 ft. and 12.6 ft. respectively, with maximum errors in each of 0.1 ft. Find the maximum error and percentage error in calculating (**a**) the area, (**b**) the hypotenuse, from these measurements.

11. One side of a triangle measures 2000 ft., and the adjacent angles measure 30° and 60° respectively, with a maximum error in each angle of 30′. The maximum error in the measurement of the side is ± 1 ft. Find the maximum error and percentage error in calculating from these measurements (**a**) the altitude on the given side; (**b**) the area of the triangle.

12. The diameter and altitude of a right circular cylinder are found by measurement to be 10 in. and 12 in. respectively. If there is a probable error of 0.2 in. in each measurement, what is the greatest possible error in the computed volume?

13. The dimensions of a box are found by measurement to be 3 ft., 4 ft., $5\frac{1}{2}$ ft. If there is a probable error of 0.01 ft., (**a**) what is the greatest possible error in the computed volume? (**b**) what is the percentage error?

14. The specific gravity of a solid is given by the formula $s = P/w$, where P is the weight in a vacuum and w is the weight of an equal volume of water. How is the computed specific gravity affected by an error of $\pm \frac{1}{10}$ in weighing P and $\pm \frac{1}{20}$ in weighing w, assuming $P = 8$ and $w = 1$ in the experiment, (**a**) if both errors are positive? (**b**) if one error is negative? (**c**) What is the largest percentage error?

15. The diameter and the slant height of a right circular cone are found by measurement to be 10 in. and 20 in. respectively. If there is a probable error of 0.2 in. in each measurement, what is the greatest possible error in the computed value of (**a**) the volume? (**b**) the curved surface?

16. Two sides of a triangle are found by measurement to be 63 ft. and 78 ft. and the included angle to be 60°. If there is a probable error of 0.5 ft. in measuring the sides and of 2° in measuring the angle, what is the greatest possible error in the computed value of the area? (See **(7)**, Art. 296.)

17. If specific gravity is determined by the formula $s = \dfrac{A}{A - W}$, where A is the weight in air and W the weight in water, what is **(a)** approximately the largest error in s if A can be read within 0.01 lb. and W within 0.02 lb., the actual readings being $A = 9$ lb., $W = 5$ lb.? **(b)** the largest relative error?

18. If the formula $\sin (x + y) = \sin x \cos y + \cos x \sin y$ were used to calculate $\sin (x + y)$, what approximate error would result if an error of 0.1° were made in measuring both x and y, the measurements of the two acute angles giving $\sin x = \frac{3}{5}$ and $\sin y = \frac{5}{13}$?

19. The acceleration of a particle down an inclined plane is given by $a = g \sin i$. If g varies by 0.1 ft. per second per second, and i, which is measured as 30°, may be in error 1°, what is the error in the computed value of a? Take the normal value of g to be 32 ft. per second per second.

20. The period of a pendulum is $P = 2 \pi \sqrt{l/g}$. **(a)** What is the greatest error in the period if there is an error of ± 0.1 ft. in measuring a 10-foot suspension and g, taken as 32 ft. per second per second, may be in error by 0.05 ft. per second per second? **(b)** What is the percentage error?

21. The dimensions of a cone are radius of base $= 4$ in., altitude $= 6$ in. What is the error in volume and in total surface if there is a shortage of 0.01 in. per inch in the measure used?

22. The length l and the period P of a simple pendulum are connected by the equation $4 \pi^2 l = P^2 g$. If l is calculated assuming $P = 1$ sec. and $g = 32$ ft. per second per second, what is approximately the error in l if the true values are $P = 1.02$ sec. and $g = 32.01$ ft. per second per second? What is the percentage error?

281. Differentiation of implicit functions. The equation

$$f(x, y) = 0 \tag{1}$$

defines either of the variables x or y as an implicit function of the other. It represents an equation containing x and y when all its terms have been transposed to the first member. Let

$$u = f(x, y). \tag{2}$$

In **(C)**, Art. 280, let $z = u$, $t = x$. Then we have

$$\frac{du}{dx} = \frac{\partial f}{\partial x} + \frac{\partial f}{\partial y} \frac{dy}{dx},$$

in which y is an arbitrary function of x. Now let y be the function of x satisfying (1). Then $u = 0$ and $du = 0$ for all values of x, and hence

$$\frac{\partial f}{\partial x} + \frac{\partial f}{\partial y}\frac{dy}{dx} = 0. \tag{3}$$

Solving, we get

(D) $$\frac{dy}{dx} = -\frac{\dfrac{\partial f}{\partial x}}{\dfrac{\partial f}{\partial y}}. \qquad \left(\frac{\partial f}{\partial y} \neq 0\right)$$

Thus we have a formula for differentiating implicit functions. This formula in the form (3) is equivalent to the process employed in Art. 63 for differentiating implicit functions.

When the equation of a curve is in the form (1), formula **(D)** a**ᶠᵕ**rds an easy way for getting the slope.

Example. Find $\dfrac{dy}{dx}$ if $x^3 - 3\,xy^2 + y^3 = 10$.

Solution. $f(x,\, y) = x^3 - 3\,xy^2 + y^3 - 10$.

$$\frac{\partial f}{\partial x} = 3\,x^2 - 3\,y^2, \quad \frac{\partial f}{\partial y} = -6\,xy + 3\,y^2.$$

$$\frac{dy}{dx} = -\frac{\dfrac{\partial f}{\partial x}}{\dfrac{\partial f}{\partial y}} = -\frac{3\,x^2 - 3\,y^2}{-6\,xy + 3\,y^2} = \frac{x^2 - y^2}{2\,xy - y^2}.$$

PROBLEMS

In Problems 1–6 find $\dfrac{dy}{dx}$ by **(D)**.

1. $x^3 - x^2y + xy^2 + y^3 = 0$.   2. $x + \sqrt{xy} - 4\,y = 7$.

3. $x^3 + y^3 - 3\,axy = 0$.   4. $y \sin x - x \cos y = 0$.

5. $y^2 - \dfrac{x+y}{x-y} = 0$.   6. $x^2 - \dfrac{x^2+y^2}{x^2-y^2} = 0$.

In Problems 7–11 verify that the given values of x and y satisfy the equation, and find the corresponding value of dy/dx.

7. $x^2 + 3\,xy - y^2 + 12 = 0$; $x = 2,\ y = -2$.

8. $x^3 + y^3 - 6\,xy - 19 = 0$; $x = 2,\ y = -1$.

9. $Ax^2 + 2\,Bxy + Cy^2 + Dx + Ey = 0$; $x = 0,\ y = 0$.

10. $x + 2\sqrt{xy} - 4\,y = 8$; $x = 8,\ y = 2$.

11. $2\,x + 4\,y + 3\,e^{xy} = 3$; $x = 0,\ y = 0$.

282. Successive partial differentiation. If $z = f(x, y)$ is a function of x and y, then $\dfrac{\partial z}{\partial x}$ and $\dfrac{\partial z}{\partial y}$ are also functions of x and y and the process of partial differentiation may be continued. The following notation is employed.

a. $\dfrac{\partial}{\partial x}\left(\dfrac{\partial z}{\partial x}\right) = \dfrac{\partial^2 z}{\partial x^2} = f_{xx}.$
$\qquad$
b. $\dfrac{\partial}{\partial y}\left(\dfrac{\partial z}{\partial x}\right) = \dfrac{\partial^2 z}{\partial y\,\partial x} = f_{yx}.$

c. $\dfrac{\partial}{\partial x}\left(\dfrac{\partial z}{\partial y}\right) = \dfrac{\partial^2 z}{\partial x\,\partial y} = f_{xy}.$
$\qquad$
d. $\dfrac{\partial}{\partial y}\left(\dfrac{\partial z}{\partial y}\right) = \dfrac{\partial^2 z}{\partial y^2} = f_{yy}.$

The derivatives above are called second partial derivatives of z, (a) with respect to x, (b) with respect to y and x, etc. A similar notation is applied to higher derivatives.

The order of differentiation is immaterial. That is,

$$\frac{\partial^2 z}{\partial y\,\partial x} = \frac{\partial^2 z}{\partial x\,\partial y}.$$

The proof of this statement is omitted.

Example. Find the first and second partial derivatives of

$$x^3 - 2\,x^2 y + 4\,xy^2 - 3\,y^3.$$

Solution. $\qquad \dfrac{\partial z}{\partial x} = 3\,x^2 - 4\,xy + 4\,y^2. \qquad\qquad (1)$

$$\frac{\partial z}{\partial y} = -\,2\,x^2 + 8\,xy - 9\,y^2. \qquad\qquad (2)$$

$$\frac{\partial^2 z}{\partial x^2} = 6\,x - 4\,y. \qquad\qquad (3)$$

$$\frac{\partial^2 z}{\partial x\,\partial y} = -\,4\,x + 8\,y. \qquad\qquad (4)$$

$$\frac{\partial^2 z}{\partial y^2} = 8\,x - 18\,y. \qquad\qquad (5)$$

Equation (3) is obtained by differentiating (1) with respect to x. Equation (4) is obtained by differentiating (2) with respect to x or by differentiating (1) with respect to y. Equation (5) is obtained by differentiating (2) with respect to y.

PROBLEMS

1. If $f(x, y) = x^3 + 3\,x^2 y + 6\,xy^2 - y^3$, show that
$f_{xx}(x, y) = 6\,x + 6\,y; \; f_{xy}(x, y) = 6\,x + 12\,y; \; f_{yy}(x, y) = 12\,x - 6\,y.$

2. If $z = \dfrac{x + y}{x - y}$, show that
$$\frac{\partial^2 z}{\partial v^2} = \frac{4\,y}{(x - y)^3}; \; \frac{\partial^2 z}{\partial x\,\partial y} = \frac{-\,2(x + y)}{(x - y)^3}; \; \frac{\partial^2 z}{\partial y^2} = \frac{4\,x}{(x - y)^3}.$$

Find the second partial derivatives of the following functions.

3. $f(x, y) = Ax^2 + 2Bxy + Cy^2$.

4. $f(x, y) = \dfrac{Ax + By}{Cx + Dy}$.

5. $f(x, y) = Ax^3 + Bx^2y + Cxy^2 + Dy^3$.

6. $f(x, y) = Ax + By + Ce^{xy}$.

7. If $f(x, y) = x^4 - 8x^2y^2 + 3y^4$, show that
$$f_{xx}(2, -1) = 32, \quad f_{xy}(2, -1) = 64, \quad f_{yy}(2, -1) = -28.$$

8. If $f(x, y) = \sin x \ln(y + 1) + \cos y \ln(1 - x)$, show that
$$f_x(0, 0) = -1, \quad f_y(0, 0) = 0, \quad f_{xx}(0, 0) = -1, \quad f_{xy}(0, 0) = 1, \quad f_{yy}(0, 0) = 0.$$

9. If $f(x, y) = x^3 - 3x^2y + 2y^2$, find the values of
$$f_{xx}(-1, 2), \quad f_{xy}(-1, 2), \quad f_{yy}(-1, 2).$$

10. If $u = 2x^3 - 4x^2y + 5xy^2 - 8xy + 7y^2$, verify the following results.
$$\frac{\partial^3 u}{\partial x^3} = 12, \quad \frac{\partial^3 u}{\partial x^2 \partial y} = -8, \quad \frac{\partial^3 u}{\partial x \partial y^2} = 10, \quad \frac{\partial^3 u}{\partial y^3} = 0.$$

11. If $u = (ax^2 + by^2 + cz^2)^3$, show that
$$\frac{\partial^3 u}{\partial x^2 \partial y} = \frac{\partial^3 u}{\partial x \partial y \partial x} = \frac{\partial^3 u}{\partial y \partial x^2}.$$

12. If $u = \dfrac{xy}{x + y}$, show that $x^2 \dfrac{\partial^2 u}{\partial x^2} + 2xy \dfrac{\partial^2 u}{\partial x \partial y} + y^2 \dfrac{\partial^2 u}{\partial y^2} = 0$.

13. If $u = \ln \sqrt{x^2 + y^2}$, show that $\dfrac{\partial^2 u}{\partial x^2} + \dfrac{\partial^2 u}{\partial y^2} = 0$.

14. If $u = \dfrac{1}{\sqrt{x^2 + y^2 + z^2}}$, show that $\dfrac{\partial^2 u}{\partial x^2} + \dfrac{\partial^2 u}{\partial y^2} + \dfrac{\partial^2 u}{\partial z^2} = 0$.

CHAPTER XXIV

DOUBLE INTEGRALS

283. Successive integration. Corresponding to partial differentiation in the differential calculus we have the inverse process of partial integration in the integral calculus. If u is a function of two independent variables x and y, then $\partial u/\partial x$ is calculated by differentiating with respect to x, regarding y as constant. If $\partial u/\partial x$ is known, then u is found by integrating with respect to x, regarding y as constant. Thus, suppose

$$\frac{\partial u}{\partial x} = 4\,x + 3\,y + 5.$$

Integrating with respect to x, considering y as a constant, we have

$$u = 2\,x^2 + 3\,xy + 5\,x + \alpha,$$

where α is the constant of integration.

But since y was regarded as constant during this integration, α may depend on y. Hence the most general form of u is

$$u = 2\,x^2 + 3\,xy + 5\,x + \alpha(y),$$

where $\alpha(y)$ *denotes an arbitrary function of* y.

If $\partial^2 u/\partial x\,\partial y$ is known, then u is found by *successive integration*, as illustrated by the following example. Suppose

$$\frac{\partial^2 u}{\partial x\,\partial y} = Ax^2 + Bxy + Cy^2.$$

Integrating first with respect to y, regarding x as constant, we get

$$\frac{\partial u}{\partial x} = Ax^2 y + \tfrac{1}{2}\,Bxy^2 + \tfrac{1}{3}\,Cy^3 + \alpha(x),$$

where $\alpha(x)$ is an arbitrary function of x.

Now integrating this result with respect to x, regarding y as constant, we have

$$u = \tfrac{1}{3}\,Ax^3 y + \tfrac{1}{4}\,Bx^2 y^2 + \tfrac{1}{3}\,Cxy^3 + \gamma(x) + \beta(y),$$

where $\beta(y)$ is an arbitrary function of y and

$$\gamma(x) = \int \alpha(x)dx \text{ is an arbitrary function of } x.$$

The result will be the same if the integration is performed first with respect to x and then with respect to y.

NOTATION. If
$$\frac{\partial^2 u}{\partial x\,\partial y} = f(x,\,y),$$

then
$$u = \int \int f(x,\,y)dy\,dx = \int \int f(x,\,y)dx\,dy.$$

The order in which the integrations are to be performed is shown by the order in which the differentials are written. Thus, in the first form above, the order $dy\,dx$ means that the integration is to be performed first with respect to y, regarding x as constant.

284. Definite double integral. Geometric interpretation. In the xy-plane consider a region S bounded by a closed curve. To avoid complications we suppose that this curve has a tangent at every point and that any secant cuts the curve in two points only. The tangent at the extreme left point of the curve (Fig. 284.1) is parallel to the y-axis and cuts the x-axis at the point A. The tangent at the extreme right point of the curve is parallel to the y-axis and cuts the x-axis at B. The equation of the upper part of the curve between the two points of tangency is $y = u_1(x)$ and the equation of the lower part is $y = u_2(x)$.

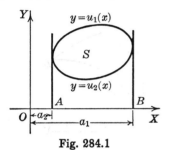

Fig. 284.1   Fig. 284.2

Similarly, the tangent at the highest point of the curve (Fig. 284.2) is parallel to the x-axis and cuts the y-axis at V. The tangent at the lowest point of the curve is parallel to the x-axis and cuts the y-axis at C. The equation of the right-hand portion of the curve between the points of tangency is $x = v_1(y)$ and the equation of the left-hand portion is $x = v_2(y)$.

Let $f(x, y)$ be a function which, for every pair of values (x, y) in S (including the boundary) is continuous, single-valued, and nowhere negative. Geometrically,

$$z = f(x, y) \tag{1}$$

is the equation of a surface as KL (Fig. 284.3). Construct upon the boundary of S as a base the cylindrical surface whose elements are parallel to the z-axis and let this cylinder enclose the area S' on KL. We wish to find the volume V of the solid bounded by S, S', and the cylindrical surface. The result is made plausible by the following informal discussion.

At equal distances apart ($= \Delta x$) in the area S draw a set of lines parallel to OY, and then a second set parallel to OX at equal distances apart ($= \Delta y$). Through these lines pass planes parallel to YOZ and XOZ respectively. Then within the areas S and S' we have a network of lines, as in the figure, that in S being composed of rectangles, each of area $\Delta x\, \Delta y$. This construction divides the cylinder into a number of vertical columns, such as $MNPQ$, whose upper and lower bases are corresponding portions of the networks in S' and S respectively. As the upper bases of these columns are curvilinear, we of course cannot calculate the volume of the columns directly. Let us replace these columns by prisms whose upper bases are found thus: each column is cut through by a plane parallel to XOY passed through that vertex of the upper base for which x and y have the least numerical values. Thus the column $MNPQ$ is replaced by the right prism $MNPR$, the upper base being in a plane through P parallel to the XOY-plane.

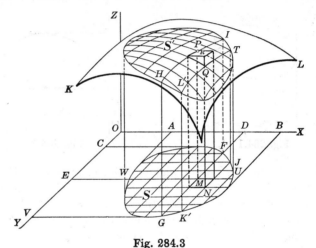

Fig. 284.3

If the coördinates of P are (x, y, z), then $MP = z = f(x, y)$ and therefore

$$\text{Volume of } MNPR = f(x, y)\Delta y\, \Delta x. \tag{2}$$

Calculating the volume of each of the other prisms formed in the same way by replacing x and y in (2) by corresponding values, and adding the results, we obtain a volume V' approximately equal to V; that is,

$$V' = \sum \sum f(x, y)\Delta y\, \Delta x; \tag{3}$$

where the double summation sign $\sum\sum$ indicates that values of *two variables* x, y must be taken account of in the quantity to be summed up.

If now in the figure we increase the number of divisions of the network in S indefinitely by letting Δx and Δy diminish indefinitely, and calculate in each case the double sum (3), then obviously V' will approach V as a limit, and hence we have the fundamental result

$$V = \lim_{\substack{\Delta x \to 0 \\ \Delta y \to 0}} \sum \sum f(x, y)\Delta y \, \Delta x. \tag{4}$$

We show now that this limit can be found by successive integration.

The required volume may be found as follows. Consider any one of the slices into which the solid is divided by two successive planes parallel to YOZ; for example, the slice whose faces are $FIHG$ and $JTL'K'$. The thickness of this slice is Δx. Now the values of z along the curve HI are found by writing $x = OD$ in the equation $z = f(x, y)$; that is, along HI

$$z = f(OD, y).$$

Hence the area $\qquad FIHG = \int_{DF}^{DG} f(OD, y)dy.$

The volume of the slice under discussion is approximately equal to that of a prism with base $FIHG$ and altitude Δx; that is, equal to

$$\Delta x \cdot \text{area } FIHG = \Delta x \int_{DF}^{DG} f(OD, y)dy.$$

The required volume of the whole solid is evidently the limit of the sum of all prisms constructed in like manner, as $x \ (= OD)$ varies from OA to OB; that is,

$$V = \int_{OA}^{OB} dx \int_{DF}^{DG} f(x, y)dy. \tag{5}$$

Similarly, it may be shown that

$$V = \int_{OC}^{OV} dy \int_{EW}^{EU} f(x, y)dx. \tag{6}$$

The integrals (5) and (6) are also written in the more compact form

$$\int_{OA}^{OB} \int_{DF}^{DG} f(x, y)dy \, dx \quad \text{and} \quad \int_{OC}^{OV} \int_{EW}^{EU} f(x, y)dx \, dy.$$

In (5) the limits DF and DG are functions of x, since they are found by solving the equation of the boundary curve of the base of the solid for y.

Similarly, in (6) the limits EW and EU are functions of y. Now comparison of (4), (5), and (6) gives the result

$$(A) \qquad V = \lim_{\substack{\Delta x \to 0 \\ \Delta y \to 0}} \sum \sum f(x, y)\Delta y \cdot \Delta x = \int_{a_2}^{a_1} \int_{u_2}^{u_1} f(x, y)dy \, dx$$

$$= \int_{b_2}^{b_1} \int_{v_2}^{v_1} f(x, y)dx \, dy,$$

where v_1 and v_2 are, in general, functions of y, and u_1 and u_2 functions of x. The second integral sign in each case applies to the first differential.

Equation (A) is an extension of the Fundamental Theorem of Art. 92 to double sums.

Our result may be stated in the following form.

The definite double integral

$$\int_{a_2}^{a_1}\int_{u_2}^{u_1} f(x,\,y)dy\;dx$$

may be interpreted as that portion of the volume of a right cylinder which is included between the plane XOY and the surface

$$z = f(x,\,y),$$

the base of the cylinder being the area in the XOY-plane bounded by the curves

$$y = u_1, \quad y = u_2, \quad x = a_1, \quad x = a_2.$$

A similar statement holds for the second integral.

It is instructive to look upon the above process of finding the volume of the solid as follows.

Consider a column with rectangular base $dy\;dx$ and of altitude z as an element of the volume. Summing up all such elements from $y = DF$ to $y = DG$, x in the meanwhile being constant (say $= OD$), gives the volume of a thin slice having $FGHI$ as one face. The volume of the whole solid is then found by summing up all such slices from $x = OA$ to $x = OB$.

In successive integration involving two variables the order of integration denotes that the limits on the second integral sign correspond to the variable whose differential is written first, the differentials of the variables and their corresponding limits being written in the reverse order. Before attempting to apply successive integration to practical problems it is best that the student should acquire by practice some facility in evaluating definite multiple integrals.

Example 1. Find the value of the definite double integral

$$\int_0^a \int_0^{\sqrt{a^2-x^2}} (x+y)dy\;dx.$$

Solution. $\displaystyle\int_0^a \int_0^{\sqrt{a^2-x^2}} (x+y)dy\;dx$

$$=\int_0^a \left[\int_0^{\sqrt{a^2-x^2}} (x+y)dy\right]dx$$

$$=\int_0^a \left[xy + \frac{y^2}{2}\right]_0^{\sqrt{a^2-x^2}} dx$$

$$=\int_0^a \left(x\sqrt{a^2-x^2} + \frac{a^2-x^2}{2}\right)dx$$

$$=\tfrac{2}{3}\,a^3$$

Interpreting this result geometrically, we have found the volume of the solid of cylindrical shape standing on OAB as base and bounded at the top by the surface (plane) $z = x + y$.

The solid here stands on a base in the xy-plane bounded by

$$\left. \begin{array}{l} y = 0 \text{ (line } OB) \\ y = \sqrt{a^2 - x^2} \text{ (quadrant of circle } AB) \end{array} \right\} \text{ from } y \text{ limits;}$$

$$\left. \begin{array}{l} x = 0 \text{ (line } OA) \\ x = a \text{ (line } BE) \end{array} \right\} \text{ from } x \text{ limits.}$$

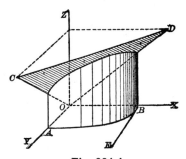

Fig. 284.4

Example 2. Verify $\displaystyle\int_b^{2b} \int_0^a (a - y)x^2 \, dy \, dx = \frac{7 \, a^2 b^5}{6}$.

Solution. $\displaystyle\int_b^{2b} \int_0^a (a - y)x^2 \, dy \, dx = \int_b^{2b} \left[ay - \frac{y^2}{2} \right]_0^a x^2 \, dx = \int_b^{2b} \frac{a^2}{2} x^2 \, dx = \frac{7 \, a^2 b^5}{6}$

Example 3. Verify $\displaystyle\int_0^a \int_{-\sqrt{a^2 - x^2}}^{\sqrt{a^2 - x^2}} x \, dy \, dx = \frac{2 \, a^3}{3}$.

Solution. $\displaystyle\int_0^a \int_{-\sqrt{a^2 - x^2}}^{\sqrt{a^2 - x^2}} x \, dy \, dx = \int_0^a \left[xy \right]_{-\sqrt{a^2 - x^2}}^{\sqrt{a^2 - x^2}} dx$

$$= \int_0^a 2 \, x \sqrt{a^2 - x^2} \, dx = \left[-\frac{2}{3} \left(a^2 - x^2 \right)^{\frac{3}{2}} \right]_0^a = \frac{2}{3} a^3.$$

PROBLEMS

Work out the following integrals.

1. $\displaystyle\int_0^2 \int_0^1 (x + y) dy \, dx$.

2. $\displaystyle\int_0^2 \int_0^3 xy(x - y) dy \, dx$.

3. $\displaystyle\int_{\frac{1}{2}}^1 \int_0^y y \, dx \, dy$.

4. $\displaystyle\int_0^a \int_0^{x^2} y \, dy \, dx$.

5. $\displaystyle\int_1^2 \int_0^1 x^2 y^2 \, dy \, dx$.

6. $\displaystyle\int_0^2 \int_{\sqrt{2x}}^{x - x^2} dy \, dx$.

7. $\displaystyle\int_0^{\sqrt{2}} \int_{x^3 - x}^{3x - x^3} dy \, dx$.

8. $\displaystyle\int_0^\pi \int_0^{a \cos \theta} \rho \sin \theta \, d\rho \, d\theta$.

9. $\int_0^b \int_t^{10\,t} \sqrt{st - t^2}\ ds\ dt.$

10. $\int_{-1}^2 \int_{2\,x^2 - 2}^{x^2 + x} x\ dy\ dx.$

11. $\int_{-1}^2 \int_{2\,x^2 - 2}^{x^2 + x} y\ dy\ dx.$

12. $\int_0^1 \int_{\sqrt{y}}^{2 - y} y^2\ dx\ dy.$

13. $\int_0^1 \int_{x^2}^{2 - x} x^2\ dy\ dx.$

14. $\int_0^1 \int_{x^2}^{2 - x} y^2\ dy\ dx.$

15. $\int_0^1 \int_{\sqrt{y}}^{2 - y} x^2\ dx\ dy.$

16. $\int_0^{\frac{\pi}{2}} \int_{a \cos \theta}^a \rho^4\ d\rho\ d\theta.$

17. $\int_0^\pi \int_0^{a(1 + \cos \theta)} \rho^2 \sin \theta\ d\rho\ d\theta.$

In the following problems a few triple definite integrals are given.* In Problem 18, integrate with respect to z, holding y and x fast, between the limits $z = 0$, $z = c$. Next integrate with respect to y, holding x fast; etc.

18. $\int_0^a \int_0^b \int_0^c (x^2 + y^2 + z^2)\,dz\ dy\ dx.$

19. $\int_0^c \int_0^z \int_0^y z\ dx\ dy\ dz.$

20. $\int_0^1 \int_0^{\sqrt{2\,x - x^2}} \int_0^{2 - x} dz\ dy\ dx.$

21. $\int_0^1 \int_{y^2}^1 \int_0^{1 - x} x\ dz\ dx\ dy.$

22. $\int_0^1 \int_0^{1 - x} \int_0^{1 - y^2} z\ dz\ dy\ dx.$

23. $\int_1^2 \int_0^z \int_0^{x\sqrt{3}} \frac{x\ dy\ dx\ dz}{x^2 + y^2}.$

285. Value of a definite double integral taken over a plane region S. In the last article the definite double integral appeared as a volume. This does not necessarily mean that every definite double integral is a volume, for the physical interpretation of the result depends on the nature

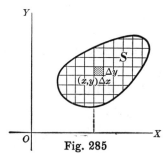

Fig. 285

of the quantities represented by x, y, z. If x, y, z are the coördinates of a point in space, then the result is indeed a volume. In order to give the definite double integral in question an interpretation not necessarily involving the geometric concept of volume, we observe that the variable z does not occur explicitly in the integral, and therefore we may confine

---

* See Appendix, page 579, for a discussion of Triple Integration

ourselves to the xy-plane. In fact, let us consider simply a region S in the xy-plane, and a given function $f(x, y)$. Within this region construct rectangular elements of area by drawing a network of lines, as in Art. 284. Choose a point (x, y) of the rectangular element of area $\Delta x \, \Delta y$, *either within the rectangle or on its perimeter.* Form the product

$$f(x, y)\Delta x \, \Delta y,$$

and similar products for all other rectangular elements. Sum up these products. The result is

$$\sum\sum f(x, y)\Delta x \, \Delta y.$$

Finally let $\Delta x \longrightarrow 0$, and $\Delta y \longrightarrow 0$.

We write the result

$$\lim_{\substack{\Delta x \to 0 \\ \Delta y \to 0}} \sum\sum f(x, y)\Delta x \, \Delta y = \iint\limits_{S} f(x, y)dx \, dy, \tag{1}$$

and call it *the double integral of the function $f(x, y)$ taken over the region S.*

By **(A)** the value of the left-hand member in (1) was found by successive integration when $f(x, y)$ had no negative values for the region S. The reasoning of Art. 284 will hold, however, if the portion S' of the surface $z = f(x, y)$ lies below the xy-plane. The limit of the double sum will then be the volume with a negative sign. The integrals in **(A)** will give the same negative number. Finally, if $f(x, y)$ is sometimes positive, sometimes negative for points of S, we may divide S into subregions in which $f(x, y)$ will be either always positive or always negative. The reasoning will hold for each subregion and therefore for the combined region S. Hence the conclusion; *the double integral in (1) may be evaluated in all cases by successive integration.* It remains to explain the method of determining the *limits* of integration. This is done in the next article.

286. Plane area as a definite double integral. Rectangular coördinates. The problem of plane areas has been solved by single integration in Art. 94. The discussion using double integration is useful chiefly because the determination of limits for the general problem of Art. 285 is made clear. To set up the desired double integral, proceed as follows.

Draw a network of rectangles as before. Then, in Fig. 285,

$$\text{Element of area} = \Delta x \, \Delta y. \tag{1}$$

If A is the entire area of the region S, obviously, by (1), Art. 285,

$$\textbf{(B)} \qquad A = \lim_{\substack{\Delta x \to 0 \\ \Delta y \to 0}} \sum\sum \Delta x \, \Delta y = \iint\limits_{S} dx \, dy.$$

Referring to the result stated in Art. 285, we may say:

The area of any region is the value of the double integral of the function $f(x, y) = 1$ *taken over that region.*

Or, also, from Art. 284,

The area equals numerically the volume of a right cylinder of unit height erected on the base S.

The examples show how the limits of integration are found.

Example 1. Calculate that portion of the area above OX which is bounded by the semicubical parabola $y^2 = x^3$ and the straight line $y = x$.

Solution. The order of integration is indicated in Fig. 286.1. Integrate first with respect to x. That is, sum up first the elements $dx\,dy$ in a horizontal strip. Then we have

$$\int_{AB}^{AC} dx\,dy = dy\int_{AB}^{AC} dx = \text{area of a horizontal strip of altitude } dy.$$

Next, integrate this result with respect to y. This corresponds to summing up all horizontal strips. In this way we obtain

$$A = \int_0^{OD}\int_{AB}^{AC} dx\,dy.$$

The limits AB and AC are found by solving each of the equations of the bounding curves for x. Thus from the equation of the line, $x = AB = y$, and from the equation of the curve, $x = AC = y^{2/3}$. To determine OD, solve the two equations *simultaneously* to find the point of intersection E. This gives the point $(1, 1)$; hence $OD = 1$. Therefore

$$A = \int_0^1\int_y^{y^{\frac{2}{3}}} dx\,dy = \int_0^1 (y^{\frac{2}{3}} - y)dy = \left[\tfrac{3}{5} y^{\frac{5}{3}} - \tfrac{1}{2} y^2\right]_0^1$$

$$= \tfrac{3}{5} - \tfrac{1}{2} = \tfrac{1}{10}.$$

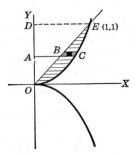

Fig. 286.1

Or we may begin by summing up the elements $dx\,dy$ in a vertical strip, and then sum up these strips. We shall then have

$$A = \int_0^1 \int_{x^{\frac{3}{2}}}^x dy\,dx = \int_0^1 \left(x - x^{\frac{3}{2}}\right)dx = \tfrac{1}{2} - \tfrac{2}{5} = \tfrac{1}{10}.$$

In this example either order of integration may be chosen. This is not always the case, as the following example shows.

Example 2. Find the area in the first quadrant bounded by the x-axis and the curves $\qquad x^2 + y^2 = 10, \quad y^2 = 9\,x.$

Solution. Here we first integrate with respect to x to cover a horizontal strip, that is, from the parabola to the circle. We then have, for the entire area,

$$A = \int_0^3 \int_{HG}^{HI} dx\,dy,$$

since the point of intersection S is $(1, 3)$. To find HG, solve $y^2 = 9\,x$ for x. Then

$$x = HG = \tfrac{1}{9}\,y^2.$$

To find HI, solve $x^2 + y^2 = 10$ for x. We get

$$x = HI = +\sqrt{10 - y^2}.$$

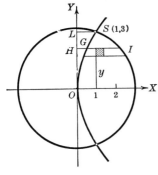

Fig. 286.2

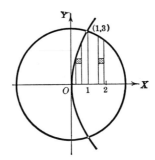

Fig. 286.3

Hence

$$A = \int_0^3 \int_{\frac{1}{9}y^2}^{\sqrt{10-y^2}} dx\,dy = \left[\frac{y}{2}\sqrt{10 - y^2} + 5 \arcsin\frac{y}{\sqrt{10}} - \frac{1}{27}y^3\right]_0^3 = 6.75.$$

If we integrate first with respect to y, using vertical strips, two integrals are necessary. Then

$$A = \int_0^1 \int_0^{3\sqrt{x}} dy\,dx + \int_1^{\sqrt{10}} \int_0^{\sqrt{10-x^2}} dy\,dx = 6.75.$$

The order of integration should be such that the area is given by one integral, if this is possible.

The examples above show that we set

$$A = \iint dx\, dy \quad \text{or} \quad A = \iint dy\, dx$$

according to the nature of the curves bounding the area. Figures 286.4 illustrate, in a general way, the difference in the summation processes indicated by the two integrals.

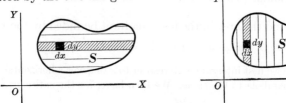

Fig. 286.4

PROBLEMS

1. Show that the area between the two parabolas $3\, y^2 = 25\, x$ and $5\, x^2 = 9\, y$ is given by either of the following double integrals.

$$\textbf{a.}\ \int_0^3 \int_{\frac{5\,x^2}{9}}^{\sqrt{\frac{25\,x}{3}}} dy\, dx = 5 \qquad\qquad \textbf{b.}\ \int_0^5 \int_{\frac{3\,y^2}{25}}^{\sqrt{\frac{9\,y}{5}}} dx\, dy = 5$$

Calculate by double integration the finite area bounded by the following pairs of curves.

2. $y^2 = x + 1,\ x + y = 1.$
3. $y = 9 - x^2,\ y = x + 7.$
4. $xy = 4,\ x + y = 5.$
5. $y^2 = 5 - x,\ y^2 = 4\, x.$
6. $y = 2\, x - x^2,\ y = 3\, x^2 - 6\, x.$
7. $y = x^3 - 3\, x,\ 4\, y = x^3.$
8. $y = x^3 - 2\, x,\ y = 6\, x - x^3.$
9. $4\, y = x^3,\ x = y - y^2 + 4.$
10. $xy = 2\, y - 6,\ y - x = 5.$
11. $x^2 + y^2 = 10\, x,\ 4\, x + y^2 = 24.$
12. $y^2 = 4 - x,\ y^2 + 2\, y = x.$
13. $y = x^2 + x,\ y = 2\, x^2 - 2$
14. $9\, y = (x + 3)^2,\ y = (x - 1)^2.$
15. $y^2 = 2\, x,\ y = x - x^2.$
16. $x^{\frac{1}{2}} + y^{\frac{1}{2}} = a^{\frac{1}{2}},\ x + y = a.$
17. $x^{\frac{2}{3}} + y^{\frac{2}{3}} = a^{\frac{2}{3}},\ x + y = a.$

287. Volume under a surface. In Art. 284 we discussed the volume of a solid bounded by a surface

$$z = f(x, y), \tag{1}$$

the xy-plane, and a cylinder. The elements of the cylinder were parallel to OZ, and its base was a region S in the xy-plane. The volume of this solid is, by (A),

$$V = \iint_S z\, dx\, dy = \iint_S f(x, y)\, dx\, dy. \tag{2}$$

The order of integration and the limits are the same as for the area of the region S. The volume of a solid of this type is the "volume under the surface (1)." The analogous problem for the plane, "area under a curve," has been treated in Chapter VII. As a special case the volume may be bounded by the surface and the xy-plane itself.

Note that the element of volume in (2) is a right prism with base $dx\, dy$ and altitude z.

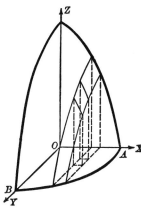

Fig. 287.1

Example 1. Find the volume bounded by the elliptic paraboloid

$$4\,z = 16 - 4\,x^2 - y^2 \tag{3}$$

and the xy-plane.

Solution. Solving (3) for z, we get

$$z = 4 - x^2 - \tfrac{1}{4}\, y^2. \tag{4}$$

Letting $z = 0$, we obtain $\qquad 4\,x^2 + y^2 = 16, \tag{5}$

which is the equation of the perimeter of the base of the solid in the xy-plane. Hence by (2), using the value of z in (4),

$$V = 4 \int_0^2 \int_0^{2\sqrt{4-x^2}} (4 - x^2 - \tfrac{1}{4}\, y^2)\,dy\, dx = 16\,\pi. \tag{6}$$

The limits are taken for the area OAB of the ellipse (5) lying in the first quadrant.

Example 2. Find the volume of the solid bounded by the paraboloid of revolution

$$x^2 + y^2 = az, \tag{7}$$

the xy-plane, and the cylinder

$$x^2 + y^2 = 2\,ax. \tag{8}$$

Solution. Solving (7) for z, and finding the limits for the area of the base of the cylinder (8) in the xy-plane, we get, using (2),

$$V = 2 \int_0^{2a} \int_0^{\sqrt{2\,ax\,-\,x^2}} \frac{x^2 + y^2}{a}\, dy\, dx = \frac{3}{2}\,\pi a^3.$$

For the area ONA (Fig. 287.2), $MN = \sqrt{2\,ax - x^2}$, (solving (8) for y), and $OA = 2\,a$. These are the limits.

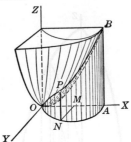

Fig. 287.2

PROBLEMS

1. Find by double integration the volume of one of the wedges cut from the cylinder $x^2 + y^2 = r^2$ by the planes $z = 0$ and $z = mx$.

2. Find the volume bounded by the cylindrical surface $y^2 = 1 - x$, the plane $z = x$, and the plane $z = 0$.

3. Find by double integration the volume of the tetrahedron bounded by the coördinate planes and the plane $\dfrac{x}{a} + \dfrac{y}{b} + \dfrac{z}{c} = 1$.

4. Find the volume in the first octant bounded by the cylinder $x^2 + y^2 = 9$ and the planes $y = 0$, $z = 0$, $z = x$.

5. Find the volume in the first octant bounded by the surfaces $y^2 = x$, $x + z = 1$, $y = 0$, $z = 0$.

6. Find the volume below the cylindrical surface whose equation is $y^2 = 16 - 4z$, above the plane $z = 0$, and within the cylindrical surface whose equation is $x^2 + y^2 = 4\,x$.

7. Find the volume in the first octant bounded by the surfaces $y^2 = x$, $x + y + z = 2$, $y = 0$, $z = 0$.

8. Find the volume in the first octant bounded by the surfaces $y^2 + z = 1$, $x + y = 1$, $x = 0$, $z = 0$.

9. Find the volume common to the two cylinders $x^2 + y^2 = r^2$ and $x^2 + z^2 = r^2$.

10. Find the volume in the first octant bounded by the surfaces $x^2 + y^2 - 2\,x = 0$, $2\,x + z - 2 = 0$, $y = 0$, $z = 0$.

11. Compute the volume of a cylindrical column standing on the area common to the two parabolas $x = y^2$, $y = x^2$ as base and cut off by the surface $z = 12 + y - x^2$.

12. Find the volume bounded by the following surfaces: $y^2 = 2x + 4$, $x + z = 1$, $z = 0$.

13. Find the volume bounded by the following surfaces: $x^2 + y^2 = 4$, $z + y = 3$, $z = 0$.

14. Find the volume bounded by the coördinate planes and the surface

$$\left(\frac{x}{a}\right)^{\frac{1}{2}} + \left(\frac{y}{b}\right)^{\frac{1}{2}} + \left(\frac{z}{c}\right)^{\frac{1}{2}} = 1.$$

15. Find the entire volume of the solid bounded by the surface $x^{\frac{2}{3}} + y^{\frac{2}{3}} + z^{\frac{2}{3}} = a^{\frac{2}{3}}$.

288. Directions for setting up a double integral. We shall now state a rule for forming the double integral which will give a required property for a given area. Applications are made in the following articles. For single integration the corresponding method is explained in Chapter VIII.

STEP I. *Draw the curves which bound the area concerned.*

STEP II. *At any point $P(x, y)$ **within the area** construct the rectangular element of area $\Delta x \, \Delta y$.*

STEP III. *Work out the function $f(x, y)$, which, when multiplied by $\Delta x \, \Delta y$, gives the required property for the rectangular element of area.*

STEP IV. *The required integral is*

$$\iint f(x, y) dx \, dy$$

taken over the given region, or area. The order of integration and limits are determined in the same manner as in finding the area itself.

289. Moment of area and centroids. This problem is treated in Art. 101 by single integration. Double integration is often more convenient.

We follow the rule of the preceding article. The moments of area for the rectangular element of area are, respectively,

$$x \, \Delta x \, \Delta y, \text{ with respect to } OY,$$
$$y \, \Delta x \, \Delta y, \text{ with respect to } OX.$$

Hence for the entire area, using the notation of Art. 101, we have

(C) $\qquad M_x = \iint y \, dx \, dy, \quad M_y = \iint x \, dx \, dy.$

The centroid of the area is given by

(D) $\qquad \bar{x} = \dfrac{M_y}{\text{area}}, \quad \bar{y} = \dfrac{M_x}{\text{area}}.$

In **(C)**, the integrals give the values of the functions

$$f(x, y) = y \quad \text{and} \quad f(x, y) = x,$$

respectively, taken over the area.

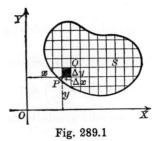

Fig. 289.1

For an area bounded by a curve, the x-axis, and two ordinates (the "area under a curve"), we derive from **(C)**,

$$M_x = \int_a^b \int_0^y y \, dy \, dx = \tfrac{1}{2} \int_a^b y^2 \, dx, \quad M_y = \int_a^b \int_0^y x \, dy \, dx = \int_a^b xy \, dx.$$

These agree with Art. 101. Note that y is the ordinate of a point on the curve; and its value in terms of x must be found from the equation of the curve and substituted in the integrand before integration.

Example 1. Find the centroid of the area in the first quadrant bounded by the semicubical parabola $y^2 = x^3$ and the straight line $y = x$.

Solution. The order and the limits of integration were found in Example 1, Art. 286. Hence, using **(C)**,

$$M_x = \int_0^1 \int_y^{y^{\frac{2}{3}}} y \, dx \, dy = \int_0^1 (y^{\frac{5}{3}} - y^2) dy = \tfrac{1}{24}.$$

$$M_y = \int_0^1 \int_y^{y^{\frac{2}{3}}} x \, dx \, dy = \tfrac{1}{2} \int_0^1 (y^{\frac{4}{3}} - y^2) dy = \tfrac{1}{21}.$$

Since $A = \text{area} = \tfrac{1}{10}$, we have, from **(D)**, $\bar{x} = \tfrac{10}{21} = 0.48$, $\bar{y} = \tfrac{5}{12} = 0.42$.

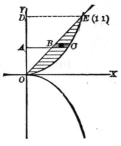

Fig. 289.2

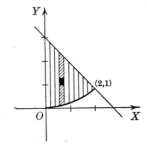

Fig. 289.3

Example 2. Find the coördinates of the centroid of the area bounded by the y-axis and the curves $4\,y = x^2$, $x + y = 3$. (Fig. 289.3)

Solution. We have

$$\text{Area} = \int_0^2 \int_{\frac{1}{4}x^2}^{3-x} dy\, dx$$

$$= \int_0^2 (3 - x - \tfrac{1}{4}\,x^2)dx = 3\tfrac{1}{3}.$$

$$M_x = \int_0^2 \int_{\frac{1}{4}x^2}^{3-x} y\, dy\, dx$$

$$= \tfrac{1}{2}\int_0^2 [(3-x)^2 - \tfrac{1}{16}\,x^4]dx = \tfrac{62}{15}.$$

$$M_y = \int_0^2 \int_{\frac{1}{4}x^2}^{3-x} x\, dy\, dx$$

$$= \int_0^2 (3\,x - x^2 - \tfrac{1}{4}\,x^3)dx = \tfrac{7}{3}.$$

Hence, by **(D)**, $\bar{x} = \tfrac{7}{10}$, $\bar{y} = \tfrac{62}{50} = 1.24$.

PROBLEMS

Find the centroid of the area bounded by the following pairs of curves.

    1. $y = x^3$, $y = 4\,x$. (Area in first quadrant.)

    2. $y^2 = x$, $x + y = 2$, $y = 0$. (Area in first quadrant.)

3. $y = 6\,x - x^2$, $y = x$.

4. $y = 4\,x - x^2$, $y = 2\,x - 3$.

5. $x^2 = 4\,y$, $x - 2\,y + 4 = 0$.

6. $y = x^2$, $2\,x - y + 3 = 0$.

7. $y = x^2 - 2\,x - 3$, $y = 2\,x - 3$.

8. $y^2 = x$, $x + y = 2$, $x = 0$.

9. $y^3 = x^2$, $2\,y = x$.

10. $4\,y = 3\,x^2$, $2\,y^2 = 9\,x$.

11. $y^2 = 2\,x$, $y = x - x^2$.

12. $y^2 = 8\,x$, $x + y = 6$.

13. $y^2 = 4\,x$, $y^2 = 5 - x$.

14. $y = 6\,x - x^2$, $x + y = 6$.

15. $x = 4\,y - y^2$, $y = x$.

16. $y = 4\,x - x^2$, $y = 5 - 2\,x$.

17. $y^2 = 4\,x$, $2\,x - y = 4$.

18. $y = x^2 - 2\,x - 3$, $y = 6\,x - x^2 - 3$.

19. $x^2 + y^2 = 1$, $x + y = 1$.

20. $x^2 + y^2 = 32$, $y^2 = 4\,x$.

21. $x^2 + y^2 = 10\,x$, $x^2 = y$.

22. $x^{\frac{1}{2}} + y^{\frac{1}{2}} = a^{\frac{1}{2}}$, $x = 0$, $y = 0$.

    23. $x^{\frac{2}{3}} + y^{\frac{2}{3}} = a^{\frac{2}{3}}$. (Area in first quadrant.)

    24. Find the centroid of the area under one arch of the cycloid $x = a(\theta - \sin \theta)$, $y = a(1 - \cos \theta)$.

290. Center of fluid pressure. Moment of inertia of an area. The problem of calculating the pressure of a fluid on a vertical wall was discussed in Art. 103.

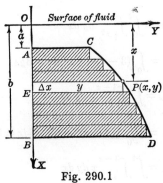

Fig. 290.1

The pressures on the rectangular elements of the figure constitute a system of parallel forces, since they are perpendicular to the plane of the area XOY. The resultant of this system of forces is the total fluid pressure P, given by

$$P = w \int_a^b yx \, dx. \tag{1}$$

The point of application of P is called the *center of fluid pressure*. We wish to find the x-coördinate ($= x_0$) of this point.

To this end we use the *principle of force moments*. This may be stated as follows.

The sum of the turning moments of a system of parallel forces about an axis is equal to the turning moment of their resultant about this axis.

Now the force of pressure dP on the rectangular element EP is, by Art. 103,

$$dP = wxy \, \Delta x. \tag{2}$$

The turning moment of this force about the axis OY is the product of dP by its lever arm OE ($= x$), or, using (2),

$$\text{Turning moment of } dP = x \, dP = wx^2y \, \Delta x. \tag{3}$$

Hence we have, for the entire turning moment for the distributed fluid pressure,

$$\text{Total turning moment} = \int_a^b wx^2y \, dx. \tag{4}$$

But the turning moment of the resultant fluid pressure P is x_0P. Hence

$$x_0P = w \int_a^b x^2y \, dx. \tag{5}$$

Solving for x_0 and using (1), we get the formula for the *depth of the center of pressure*

$$x_0 = \frac{\displaystyle\int_a^b x^2 dA}{\displaystyle\int_a^b x\, dA}, \tag{6}$$

where dA = element of area = $y\, dx$.

The denominator in (6) is the moment of area of $ABCD$ with respect to OY (see Art. 289). The numerator is an integral not met with hitherto. It is called the *moment of inertia of the area $ABCD$ about OY.*

The letter I is commonly used for moment of inertia about an axis, and a subscript is attached to designate the axis. Thus (6) becomes

$$x_0 = \frac{I_y}{M_y}. \tag{7}$$

The usual notation for moment of inertia about an axis l is

$$I_l = \int r^2\, dA, \tag{8}$$

in which

$$r = \text{distance of the element } dA \text{ from the axis } l. \tag{9}$$

The preceding problem is one of many which lead to moments of inertia. In mechanics the moment of inertia of an area about an axis is an important concept. The calculation of moments of inertia will now be explained. We follow the rule of Art. 288.

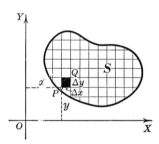

Fig. 290.2

For the elementary rectangle PQ at $P(x, y)$ the moment of inertia about OX is defined as

$$y^2\, \Delta x\, \Delta y,$$

and about the y-axis it is

$$x^2\, \Delta x\, \Delta y.$$

Then, if I_x and I_y are the corresponding moments of inertia for the entire area, we have

(E) $$I_x = \int\int y^2 \, dx \, dy, \quad I_y = \int\int x^2 \, dx \, dy.$$

The **radii of gyration** r_x and r_y are given by

(F) $$r_x{}^2 = \frac{I_x}{\text{area}}, \quad r_y{}^2 = \frac{I_y}{\text{area}}.$$

In (E) the functions whose integrals are taken over the area are, respectively, $f(x, y) = y^2$, and $f(x, y) = x^2$.

Formulas (E) become simple for an area "under a curve," that is, an area bounded by a curve, the x-axis and two ordinates. Thus we obtain

$$I_x = \int_a^b \int_0^y y^2 \, dy \, dx = \tfrac{1}{3} \int_a^b y^3 \, dx,$$

$$I_y = \int_a^b \int_0^y x^2 \, dy \, dx = \int_a^b x^2 y \, dx.$$

(10)

In these equations y is the ordinate of a point on the curve, and its value in terms of x must be found from the equation of this curve and substituted in the integrand.

Formulas for moments of inertia I are written in the form

(G) $$I = Ar^2,$$

where A = area and r = radius of gyration. Solving (F) for I_x and I_y will give this form.

Dimensions. If the linear unit is 1 in., the moment of inertia has the dimensions in.[4]. By (F), r_x and r_y are lengths, in inches.

Example 1. Find I_x, I_y, and the corresponding radii of gyration for the area of Illustrative Example 1, Art. 286.

Solution. Using the same order of integration and the same limits as before, we have, by (E),

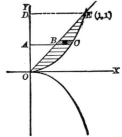

$$I_x = \int_0^1 \int_y^{y^{\frac{2}{3}}} y^2 \, dx \, dy = \int_0^1 (y^{\frac{8}{3}} - y^3) \, dy = \tfrac{1}{44}.$$

$$I_y = \int_0^1 \int_y^{y^{\frac{2}{3}}} x^2 \, dx \, dy = \tfrac{1}{3} \int_0^1 (y^2 - y^3) \, dy = \tfrac{1}{36}.$$

Since A = area = $\tfrac{1}{10}$, we find, by (F),

$$r_x = 0.48, \quad r_y = 0.53.$$

Fig. 290.3

Example 2. Find I_x and I_y for the parabolic segment BOC shown in Fig. 290.4.

Solution. With the axes of coördinates as drawn, the equation of the bounding parabola is

$$y^2 = 2\,px. \tag{11}$$

Since $B(a, b)$ is a point on the curve, we get, by substituting $x = a$, $y = b$ in (11), $b^2 = 2\,pa$. Solving this equation for $2\,p$ and substituting its value in (11), we obtain

$$y^2 = \frac{b^2 x}{a},$$

or

$$y = \frac{bx^{\frac{1}{2}}}{a^{\frac{1}{2}}}. \tag{12}$$

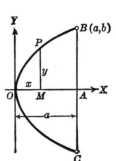

Fig. 290.4

The moments of inertia for the area under the parabola OPB in the first quadrant will be half the required moments. Hence, using (10), and substituting the value of y from (12), we get

$$\frac{1}{2}\,I_x = \frac{1}{3}\int_0^a \frac{b^3}{a^{\frac{3}{2}}}\,x^{\frac{3}{2}}\,dx = \frac{2}{15}\,ab^3. \quad \therefore\ I_x = \frac{4}{15}\,ab^3.$$

$$\frac{1}{2}\,I_y = \int_0^a x^2 \frac{b}{a^{\frac{1}{2}}}\,x^{\frac{1}{2}}\,dx = \frac{2}{7}\,a^3b. \quad \therefore\ I_y = \frac{4}{7}\,a^3b.$$

For the area of the segment, we find

$$\frac{1}{2}\,A = \int_0^a y\,dx = \int_0^a \frac{b}{a^{\frac{1}{2}}}\,x^{\frac{1}{2}}\,dx = \frac{2}{3}\,ab. \quad \therefore\ A = \frac{4}{3}\,ab.$$

Hence, by ***(F)***,

$$r_x{}^2 = \frac{I_x}{A} = \frac{1}{5}\,b^2, \quad \text{and} \quad I_x = \frac{1}{5}\,Ab^2,$$

$$r_y{}^2 = \frac{I_y}{A} = \frac{3}{7}\,a^2, \quad \text{and} \quad I_y = \frac{3}{7}\,Aa^2.$$

The results are in the form ***(G)***.

In Fig. 290.1 the axis OY lies in the surface of the fluid. If we denote this axis in any figure by s, then the depth of the center of pressure is, by (7),

$$x_0 = \frac{I_s}{M_s} = \frac{r_s{}^2}{h_s}, \tag{13}$$

if $r_s = $ radius of gyration about the axis s,

and $h_s = $ depth of centroid below the axis s.

Example 3. Find the depth of the center of pressure on the trapezoidal water gate of Fig. 290.5.

Solution. Choose axes OX and OY as shown, and draw an elementary horizontal strip. Let the distance of this strip from the axis s at the water level be r. Then

$$r = 8 - y, \quad dA = 2\,x\,dy.$$

Hence, by (8), and by the definition of moment of area, we have

$$I_s = \int r^2\,dA = \int (8 - y)^2\,2\,x\,dy, \tag{14}$$

$$M_s = \int r\,dA = \int (8 - y)2\,x\,dy. \tag{15}$$

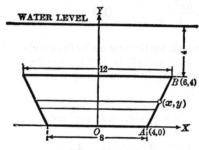

Fig. 290.5

The equation of AB is $y = 2\,x - 8$. Solving this for x, substituting in (14) and (15), and integrating with limits $y = 0$, $y = 4$, we obtain

$$I_s = \int_0^4 (8 - y)^2(8 + y)dy = 1429\tfrac{1}{3}, \quad M_s = \int_0^4 (64 - y^2)dy = 234\tfrac{2}{3}.$$

Hence, by (13), $x_0 = 6.09$.

291. Polar moment of inertia. The moment of inertia of the elementary rectangle PQ about the origin O is the product of the area and $\overline{OP}^2$, that is,

$$(x^2 + y^2)\Delta x\,\Delta y. \tag{1}$$

Hence, by Art. 288, for the entire area

$$I_0 = \iint (x^2 + y^2)dx\,dy. \tag{2}$$

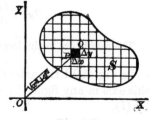

Fig. 291

We may, however, write the right-hand member as the sum of two integrals, for (2) is clearly the same as

$$I_0 = \iint x^2\,dx\,dy + \iint y^2\,dx\,dy = I_x + I_y. \tag{3}$$

Hence we have the following theorem.

The moment of inertia of an area about the origin equals the sum of its moments of inertia about the x-axis and the y-axis.

PROBLEMS

Find I_x, I_y, and I_0 for each of the areas described below.

1. The semicircle which is to the right of the y-axis and which is bounded by $x^2 + y^2 = r^2$.

2. The isosceles triangle of height h and base a whose vertices are $(0, 0)$, $(h, \frac{1}{2} a)$, $(h, -\frac{1}{2} a)$.

3. The right triangle whose vertices are $(0, 0)$, (b, a), $(b, 0)$.

4. The ellipse $\dfrac{x^2}{a^2} + \dfrac{y^2}{b^2} = 1$.

5. The area in the first quadrant bounded by $y^2 = 4 x$, $x = 4$, $y = 0$.

6. The area included between the ellipse $4 x^2 + 9 y^2 = 36$ and the circle $x^2 + y^2 = 2 y$.

7. The area included between the ellipses $9 x^2 + 16 y^2 = 144$ and $4 x^2 + y^2 = 4$.

8. The area included between the circle $x^2 + y^2 = 16$ and the circle $x^2 + (y + 2)^2 = 1$.

9. The area included between the circle $x^2 + y^2 = 36$ and the circle $x^2 + (y + 3)^2 = 4$.

10. The area between the circle $x^2 + y^2 = 4$ and the ellipse $4 x^2 + 9 y^2 = 144$.

11. The entire area bounded by $x^{\frac{2}{3}} + y^{\frac{2}{3}} = a^{\frac{2}{3}}$.

12. Find the depth of the center of pressure on a triangular water gate having its vertex below the base, which is horizontal and on a level with the surface of the water.

13. Find the depth of the center of pressure on a rectangular water gate 8 ft. wide and 4 ft. deep when the level of the water is 5 ft. above the top of the gate.

14. Find the depth of the center of pressure on the end of a horizontal cylindrical oil tank of diameter 5 ft. when the depth of oil is (**a**) 2.5 ft.; (**b**) 4 ft.

292. Polar coördinates. Plane area. When the equations of the curves bounding an area are given in polar coördinates, certain modifications in the preceding integrals are necessary.

The area is now divided into elementary portions, as follows.

Draw arcs of circles about the common center O with successive radii differing by $\Delta\rho$. Thus, in Fig. 292.1, $OP = \rho$, $OS = \rho + \Delta\rho$. Then draw radial lines from O such that the angle between any two consecutive lines is the same and equal to $\Delta\theta$. Thus, in Fig. 292.1, angle $POR = \Delta\theta$.

The area will now contain a large number of rectangular portions, such as $PSQR$ in Fig. 292.1.

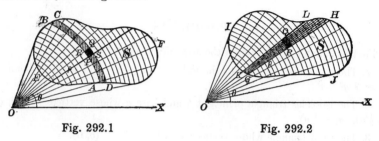

Fig. 292.1 Fig. 292.2

Let $PSQR = \Delta A$. Now ΔA is the difference of the areas of the circular sectors POR and SOQ. Hence

$$\Delta A = \tfrac{1}{2}(\rho + \Delta\rho)^2\,\Delta\theta - \tfrac{1}{2}\,\rho^2\,\Delta\theta = \rho\,\Delta\rho\,\Delta\theta + \tfrac{1}{2}\,(\Delta\rho)^2\,\Delta\theta. \quad (1)$$

The function $f(x, y)$ of Art. 285 is to be replaced by a function using polar coördinates. Let this be $F(\rho, \theta)$. Then, proceeding as in Art. 285, we choose a point (ρ, θ) of ΔA, form the product

$$F(\rho, \theta)\Delta A$$

for each ΔA within the region S, add these products, and finally let $\Delta\rho \longrightarrow 0$ and $\Delta\theta \longrightarrow 0$. It is shown in more advanced treatises that ΔA may be replaced by the first term of the right-hand member of (1) when this limit is taken. We now write (compare (1), Art. 285)

$$\iint\limits_{S} F(\rho, \theta)\rho\,d\rho\,d\theta, \quad (2)$$

and call it *the double integral of the function* $F(\rho, \theta)$ *taken over the region* S.

A discussion of (2) analogous to that in Art. 284 proves that this double integral is computed by successive integration.

The simplest case of (2) is that of finding the area of the region S. We then have

(H) $\qquad\qquad A = \int\int\rho\,d\rho\,d\theta = \int\int\rho\,d\theta\,d\rho.$

These are easily remembered if we think of the elements (checks) as being rectangles with dimensions $\rho\,d\theta$ and $d\rho$, and hence of area $\rho\,d\theta\,d\rho$. The figures illustrate, in a general way, the difference in the processes indicated by the two integrals.

In the first, we integrate first with respect to ρ, since $d\rho$ precedes $d\theta$, keeping θ constant. This process will cover the radial strip $KGHL$ in

Fig. 292.2. The limits for ρ are $\rho = OG$ and $\rho = OH$, found by solving the equation (or equations) of the bounding curve (or curves) for ρ in terms of θ. Then integrate varying θ, the limits being $\theta = \angle JOX$ and $\theta = \angle IOX$.

The second integral in (H) is worked out by integrating with respect to θ, ρ remaining constant. This step covers the circular strip $ABCD$ in Fig. 292.1, between two consecutive circular arcs. Then integrate varying ρ.

When the area is bounded by a curve and two of its radii vectors (area swept over by the radius vector), we obtain from the first form in (H)

$$A = \int_\alpha^\beta \int_0^\rho \rho \, d\rho \, d\theta = \tfrac{1}{2} \int_\alpha^\beta \rho^2 \, d\theta,$$

which agrees with Art. 218.

Double integrals in polar coördinates have one of the forms

$$\int\!\!\int F(\rho, \theta) \rho \, d\rho \, d\theta \quad \text{or} \quad \int\!\!\int F(\rho, \theta) \rho \, d\theta \, d\rho. \tag{3}$$

Example 1. Find the limits for the double integral giving some required property related to the area inside the circle $\rho = 2\,r \cos\theta$ and outside the circle $\rho = r$.

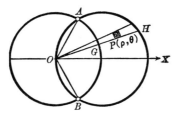

Fig. 292.3

Solution. The points of intersection are $A(r, \tfrac{1}{3}\pi)$, and $B(r, -\tfrac{1}{3}\pi)$. Use the first form in (3).

The limits for ρ are
$$\rho = OG = r,$$
$$\rho = OH = 2\,r \cos\theta;$$

for θ they are $\tfrac{1}{3}\pi$ and $-\tfrac{1}{3}\pi$.

Example 2. Find the area inside the circle $\rho = 2\,r \cos\theta$ and outside the circle $\rho = r$.

Solution. From Example 1 above, we have

$$A = \int_{-\frac{1}{3}\pi}^{\frac{1}{3}\pi} \int_r^{2\,r\cos\theta} \rho \, d\rho \, d\theta = \int_{-\frac{1}{3}\pi}^{\frac{1}{3}\pi} \tfrac{1}{2}(4\,r^2\cos^2\theta - r^2)d\theta = r^2(\tfrac{1}{3}\pi + \tfrac{1}{2}\sqrt{3})$$
$$= 1.91\,r^2.$$

293. Problems using polar coördinates. There should now be no difficulty in establishing the following formulas (use **(A)**, Art. 213).

$$M_x = \iint \rho^2 \sin\theta \, d\rho \, d\theta. \tag{1}$$

$$M_y = \iint \rho^2 \cos\theta \, d\rho \, d\theta. \tag{2}$$

$$I_x = \iint \rho^3 \sin^2\theta \, d\rho \, d\theta. \tag{3}$$

$$I_y = \iint \rho^3 \cos^2\theta \, d\rho \, d\theta. \tag{4}$$

$$I_0 = \iint \rho^3 \, d\rho \, d\theta. \tag{5}$$

The order of the differentials will have to be changed if integration with respect to θ is performed first.

Example 1. On account of important applications the moments of inertia of a circle are now worked out.

Solution. Let a = radius. Then, by (5), the polar moment of inertia with respect to the center is

$$I_0 = \int_0^a \left[\int_0^{2\pi} d\theta \right] \rho^3 \, d\rho \tag{6}$$
$$= \tfrac{1}{2}\pi a^4 = \tfrac{1}{2} A a^2,$$

where A = area of the circle.

Also, since $I_x = I_y$, by symmetry, we have, by (3), Art. 291,

$$I_x = \tfrac{1}{2} I_0 = \tfrac{1}{4} A a^2. \tag{7}$$

In words: *The polar moment of inertia of a circle with respect to its center equals the product of half the area and the square of the radius; the moment of inertia with respect to any diameter equals the product of one fourth the area and the square of the radius.*

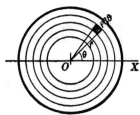

Fig. 293.1

Example 2. Find the centroid of a loop of the lemniscate

$$\rho^2 = a^2 \cos 2\theta.$$

Solution. Since OX is an axis of symmetry, we have $\bar{y} = 0$.

$$\tfrac{1}{2} A = \int_0^{\frac{1}{4}\pi} \int_0^{a\sqrt{\cos 2\theta}} \rho \, d\rho \, d\theta$$

$$= \tfrac{1}{2} a^2 \int_0^{\frac{1}{4}\pi} \cos 2\theta \, d\theta = \tfrac{1}{4} a^2.$$

$$\tfrac{1}{2} M_y = \int_0^{\frac{1}{4}\pi} \int_0^{a\sqrt{\cos 2\theta}} \rho^2 \cos\theta \, d\rho \, d\theta = \tfrac{1}{3} a^3 \int_0^{\frac{1}{4}\pi} (\cos 2\theta)^{\frac{3}{2}} \cos\theta \, d\theta$$

$$= \tfrac{1}{3} a^3 \int_0^{\frac{1}{4}\pi} (1 - 2\sin^2\theta)^{\frac{3}{2}} \cos\theta \, d\theta \qquad \text{(by (5), Art. 296)}$$

$$= \frac{\sqrt{2}}{6} a^3 \int_0^1 (1 - z^2)^{\frac{3}{2}} dz \left(\text{if } \sin\theta = \tfrac{1}{2} z\sqrt{2} \right) = \frac{\pi}{32} a^3\sqrt{2}.$$

Hence $x = \dfrac{M_y}{A} = \dfrac{\pi}{8} a\sqrt{2} = 0.55\, a.$

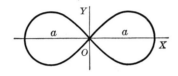

Fig. 293.2

Example 3. Find I_0 over the region bounded by the circle $\rho = 2\, r \cos\theta.$

Solution. Summing up for the elements in the triangular-shaped strip OP, the ρ limits are zero and $2\, r \cos\theta$ (found from the equation of the circle). Summing up for all such strips, the θ limits are $-\tfrac{1}{2}\pi$ and $\tfrac{1}{2}\pi$. Hence, by (5),

$$I_0 = \int_{-\frac{1}{2}\pi}^{\frac{1}{2}\pi} \int_0^{2\, r \cos\theta} \rho^3 \, d\rho \, d\theta = \tfrac{3}{2}\, \pi r^4.$$

Or, summing up first for the elements in a circular strip (as QR), we have

$$I_0 = \int_0^{2\, r} \int_{-\arccos\frac{\rho}{2\, r}}^{\arccos\frac{\rho}{2\, r}} \rho^3 \, d\theta \, d\rho = \tfrac{3}{2}\, \pi r^4.$$

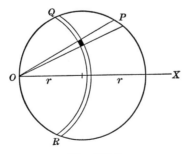

Fig. 293.3

PROBLEMS

1. Find the area inside the circle $\rho = \frac{3}{2}$ and to the right of the line $4\,\rho\cos\theta = 3$.

2. Find the area which is inside the circle $\rho = 3\cos\theta$ and outside the circle $\rho = \frac{3}{2}$.

3. Find the area which is inside the circle $\rho = 3\cos\theta$ and outside the circle $\rho = \cos\theta$.

4. Find the area inside the cardioid $\rho = 1 + \cos\theta$ and to the right of the line $4\,\rho\cos\theta = 3$.

5. Find the area which is inside the cardioid $\rho = 1 + \cos\theta$ and outside the circle $\rho = 1$.

6. Find the area which is inside the circle $\rho = 1$ and outside the cardioid $\rho = 1 + \cos\theta$.

7. Find the area which is inside the circle $\rho = 3\cos\theta$ and outside the cardioid $\rho = 1 + \cos\theta$.

8. Find the area which is inside the circle $\rho = 1$ and outside the parabola $\rho(1 + \cos\theta) = 1$.

9. Find the area which is inside the cardioid $\rho = 1 + \cos\theta$ and outside the parabola $\rho(1 + \cos\theta) = 1$.

10. Find the area which is inside the circle $\rho = \cos\theta + \sin\theta$ and outside the circle $\rho = 1$.

11. Find the area which is inside the circle $\rho = \sin\theta$ and outside the cardioid $\rho = 1 - \cos\theta$.

12. Find the area which is inside the lemniscate $\rho^2 = 2\,a^2\cos 2\,\theta$ and outside the circle $\rho = a$.

13. Find the area which is inside the cardioid $\rho = 4(1 + \cos\theta)$ and outside the parabola $\rho(1 - \cos\theta) = 3$.

14. Find the area which is inside the circle $\rho = 2\,a\cos\theta$ and outside the circle $\rho = a$. Find the centroid of the area and I_x and I_y.

15. Find the centroid of the area bounded by the cardioid

$$\rho = a(1 + \cos\theta).$$

16. Find the centroid of the area bounded by a loop of the curve $\rho = a\cos 2\,\theta$.

17. Find the centroid of the area bounded by a loop of the curve $\rho = a\cos 3\,\theta$.

18. Find I_y for the lemniscate $\rho^2 = a^2\cos 2\,\theta$.

19. Find I_x for the cardioid $\rho = a(1 + \cos\theta)$.

20. Find I_x and I_y for one loop of the curve $\rho = a\cos 2\,\theta$.

21. Show that the volume of the solid bounded by the ellipsoid of revolution $b^2(x^2 + y^2) + a^2 z^2 = a^2 b^2$ and the cylindrical surface $x^2 + y^2 - ax = 0$ is given by

$$V = 4\frac{b}{a}\int_0^{\frac{1}{2}\pi}\int_0^{a\cos\theta}\sqrt{a^2 - \rho^2}\;\rho\,d\rho\,d\theta.$$

Evaluate this integral.

294. Volumes using cylindrical coördinates. In many problems involving integration the work is much simplified
by employing cylindrical coördinates (ρ, θ, z) as
indicated in Fig. 294.1.

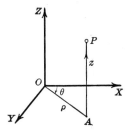

If the rectangular coördinates of P are x, y, z,
then, from the definitions and the figure, we have

$$x = \rho \cos \theta, \quad y = \rho \sin \theta, \quad z = z;$$

$$\rho^2 = x^2 + y^2, \quad \theta = \text{arc tan } \frac{y}{x}.$$

The cylindrical equation of any one of the
bounding surfaces may often be written down

Fig. 294.1

directly from its definition. In any case it may be found from its rectangular equation by the substitution

$$x = \rho \cos \theta, \quad y = \rho \sin \theta. \tag{1}$$

Cylindrical coördinates are especially useful when a bounding surface is

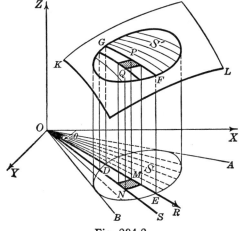

Fig. 294.2

a surface of revolution. For the equation of such a surface, when the axis
is OZ, will have the form $z = f(\rho)$; that is, the coördinate θ will be absent.

Volume under a surface. Let

$$z = F(\rho, \theta) \tag{2}$$

be the cylindrical equation of a surface, as KL in the figure. We wish
to find the volume of the solid bounded above by this surface, below by
the plane XOY, and laterally by the cylindrical surface whose right section by the plane XOY is the region S. This cylindrical surface intercepts
on the surface (2) the region S'.

Divide the solid into elements of volume as follows. Divide S into elements of area ΔA by drawing radial lines from O and arcs of circles about O, as in Art. 292. Pass planes through the radial lines and OZ. Pass cylindrical surfaces of revolution about OZ standing on the circular arcs within S. Then the solid is divided into columns such as $MNPQ$, where area $MN = \Delta A$, and $MP = z$. The element of volume is then a right prism with base ΔA and altitude z. Hence

$$\Delta V = z\,\Delta A. \tag{3}$$

The volume V is found by summing up the prisms (3) whose bases lie within S and finding the limit of this sum when the radial lines and circular arcs within S increase in number so that $\Delta\rho \longrightarrow 0$ and $\Delta\theta \longrightarrow 0$. That is,

$$V = \lim_{\substack{\Delta\rho \to 0 \\ \Delta\theta \to 0}} \sum\sum z\,\Delta A. \tag{4}$$

It can be shown* that the double limit in (4) may be found by successive integration, giving the formulas

(I) $$V = \int_S\int z\rho\,d\rho\,d\theta = \int_S\int z\rho\,d\theta\,d\rho.$$

The limits are found as in Art. 292 for the area S.

Example. Show that the volume of the solid bounded by the ellipsoid of revolution $b^2(x^2 + y^2) + a^2z^2 = a^2b^2$ and the cylindrical surface $x^2 + y^2 - ax = 0$ is given by

$$V = 4\frac{b}{a}\int_0^{\frac{1}{2}\pi}\int_0^{a\cos\theta}\sqrt{a^2 - \rho^2}\,\rho\,d\rho\,d\theta. \tag{5}$$

Evaluate this integral.

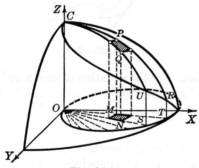

Fig. 294.3

---

*See Granville, Smith, and Longley's *Elements of Calculus* (Ginn and Company), p. 521.

Solution. By (1) the cylindrical equation of the ellipsoid is $b^2\rho^2 + a^2z^2 = a^2b^2$. Hence

$$z = \frac{b}{a}\sqrt{a^2 - \rho^2}. \tag{6}$$

The polar equation of the circle $x^2 + y^2 - ax = 0$ in the XY-plane bounding S is, by (1),

$$\rho = a\cos\theta. \tag{7}$$

For the semicircle the limits for ρ are zero and $a\cos\theta$, when θ is held fast, and for θ, zero and $\frac{1}{2}\pi$. Substituting in (I) the value of z from (6) and the above limits, we get (5). Integrating,

$$V = \tfrac{2}{9} a^2b(3\pi - 4) = 1.206 \, a^2b.$$

PROBLEMS

1. Show that the volume described in Example 2, Art. 287, is given in cylindrical coördinates by

$$V = \frac{2}{a}\int_0^{\frac{1}{2}\pi}\int_0^{2a\cos\theta} \rho^3 \, d\rho \, d\theta.$$

2. Two planes forming an angle α radians with each other meet along a diameter of a sphere of radius a. Find the volume of the *spherical wedge* included between the planes and the spherical surface, using cylindrical coördinates.

3. Find the volume bounded by the sphere $\rho^2 + z^2 = a^2$ within the cylinder $\rho = a\cos\theta$.

4. Find the volume above $z = 0$, below the cone $z^2 = x^2 + y^2$, and within the cylinder $x^2 + y^2 = 2\,ax$, using cylindrical coördinates.

5. Find the volume below the plane $2\,z = 4 + \rho\cos\theta$, above $z = 0$, and within the cylinder $\rho = 2\cos\theta$.

6. A solid is bounded by the paraboloid of revolution $az = \rho^2$ and the plane $z = c$. Find the centroid.

7. A solid is bounded by the hyperboloid $z^2 = a^2 + \rho^2$ and the upper nappe of the cone $z^2 = 2\,\rho^2$. Find the volume.

8. Find the centroid of the solid in Problem 7.

9. Find the centroid of the solid in Problem 1.

10. Find the centroid of the solid in Problem 4.

11. Find the volume of the solid bounded below by $z = 0$, above by the cone $z = a - \rho$, and laterally by $\rho = a\cos\theta$.

12. Find the centroid of the solid in the preceding problem.

13. Show that the volume below the plane $z = x$ and above the paraboloid $z = x^2 + y^2$ is given by

$$V = 2\int_0^{\frac{1}{2}\pi}\int_0^{\cos\theta} (\cos\theta - \rho)\rho^2 \, d\rho \, d\theta = \frac{\pi}{32}.$$

14. Find the volume of the solid below the spherical surface $\rho^2 + z^2 = 25$ and above the upper nappe of the conical surface $z = \rho + 1$.

CHAPTER XXV

FORMULAS AND TABLES

295. Formulas from elementary algebra and geometry.

(1) Quadratic $\qquad Ax^2 + Bx + C = 0.$

Solution. 1. By factoring: Factor $Ax^2 + Bx + C$, set each factor equal to zero, and solve for x.

2. By completing the square: Transpose C, divide by the coefficient of x^2, add to both members the square of half the coefficient of x, and extract the square root.

3. By the formula

$$x = \frac{-B \pm \sqrt{B^2 - 4\,AC}}{2\,A}.$$

Nature of the roots. The expression $B^2 - 4\,AC$ beneath the radical in the formula is called the *discriminant*. The two roots are real and unequal, real and equal, or imaginary, according as the discriminant is positive, zero, or negative.

(2) Logarithms

$$\log ab = \log a + \log b.$$

$$\log \frac{a}{b} = \log a - \log b.$$

$$\log a^n = n \log a.$$

$$\log \sqrt[n]{a} = \frac{1}{n} \log a.$$

$$\log 1 = 0.$$

$$\log_a a = 1.$$

(3) Binomial theorem (n being a positive integer)

$$(a+b)^n = a^n + na^{n-1}b + \frac{n(n-1)}{\lfloor 2} a^{n-2}b^2 + \frac{n(n-1)(n-2)}{\lfloor 3} a^{n-3}b^3 + \cdots$$

$$+ \frac{n(n-1)(n-2)\cdots(n-r+2)}{\lfloor r-1} a^{n-r+1}b^{r-1} + \cdots + b^n.$$

(4) Factorial numbers. $n! = \lfloor n = 1 \cdot 2 \cdot 3 \cdot 4 \cdots (n-1)n$

In the following formulas from elementary geometry, r or R denotes radius, a altitude, B area of base, and s slant height.

(5) Circle. Circumference $= 2\pi r$. Area $= \pi r^2$.

(6) Circular sector. Area $= \frac{1}{2} r^2 \alpha$, where $\alpha =$ central angle of the sector measured in radians.

(7) Prism. Volume $= Ba$.

(8) Pyramid. Volume $= \frac{1}{3} Ba$.

(9) Right circular cylinder. Volume $= \pi r^2 a$. Lateral surface $= 2\pi r a$. Total surface $= 2\pi r(r + a)$.

(10) Right circular cone. Volume $= \frac{1}{3}\pi r^2 a$. Lateral surface $= \pi r s$. Total surface $= \pi r(r + s)$.

(11) Sphere. Volume $= \frac{4}{3}\pi r^3$. Surface $= 4\pi r^2$.

(12) Frustum of a right circular cone. Volume $= \frac{1}{3}\pi a(R^2 + r^2 + Rr)$. Lateral surface $= \pi s(R + r)$.

296. Formulas from plane trigonometry. Many of the following formulas will be found useful.

(1) Measurement of angles. There are two common methods of measuring angular magnitude; that is, there are two unit angles.

Degree measure. The unit angle is $\frac{1}{360}$ of a complete revolution and is called a *degree*.

Radian measure. The unit angle is an angle whose intercepted arc is equal to the radius of that arc, and is called a *radian*.

The relation between the unit angles is given by the equation

$$180 \; degrees = \pi \; radians \; (\pi = 3.14159 \cdots),$$

the solution of which gives

$$1 \text{ degree} = \frac{\pi}{180} = 0.0174 \cdots \text{radian}; \; 1 \text{ radian} = \frac{180}{\pi} = 57.29 \cdots \text{degrees}.$$

From the above definition we have

Number of radians in an angle $=$ intercepted arc $\div$ radius.

These equations enable us to change from one measurement to another

(2) Relations

$$\operatorname{ctn} x = \frac{1}{\tan x}; \; \sec x = \frac{1}{\cos x}; \; \csc x = \frac{1}{\sin x}.$$

$$\tan x = \frac{\sin x}{\cos x}; \; \operatorname{ctn} x = \frac{\cos x}{\sin x}.$$

$$\sin^2 x + \cos^2 x = 1; \; 1 + \tan^2 x = \sec^2 x; \; 1 + \operatorname{ctn}^2 x = \csc^2 x.$$

(3) Formulas for reducing angles

Angle'	Sine	Cosine	Tangent	Cotangent	Secant	Cosecant
$-x$	$-\sin x$	$\cos x$	$-\tan x$	$-\operatorname{ctn} x$	$\sec x$	$-\csc x$
$90° - x$	$\cos x$	$\sin x$	$\operatorname{ctn} x$	$\tan x$	$\csc x$	$\sec x$
$90° + x$	$\cos x$	$-\sin x$	$-\operatorname{ctn} x$	$-\tan x$	$-\csc x$	$\sec x$
$180° - x$	$\sin x$	$-\cos x$	$-\tan x$	$-\operatorname{ctn} x$	$-\sec x$	$\csc x$
$180° + x$	$-\sin x$	$-\cos x$	$\tan x$	$\operatorname{ctn} x$	$-\sec x$	$-\csc x$
$270° - x$	$-\cos x$	$-\sin x$	$\operatorname{ctn} x$	$\tan x$	$-\csc x$	$-\sec x$
$270° + x$	$-\cos x$	$\sin x$	$-\operatorname{ctn} x$	$-\tan x$	$\csc x$	$-\sec x$
$360° - x$	$-\sin x$	$\cos x$	$-\tan x$	$-\operatorname{ctn} x$	$\sec x$	$-\csc x$

(4) Functions of $(x + y)$ and $(x - y)$

$$\sin (x + y) = \sin x \cos y + \cos x \sin y$$
$$\sin (x - y) = \sin x \cos y - \cos x \sin y.$$
$$\cos (x + y) = \cos x \cos y - \sin x \sin y.$$
$$\cos (x - y) = \cos x \cos y + \sin x \sin y.$$

$$\tan (x + y) = \frac{\tan x + \tan y}{1 - \tan x \tan y}. \qquad \tan (x - y) = \frac{\tan x - \tan y}{1 + \tan x \tan y}.$$

(5) Functions of $2x$ and $\frac{1}{2}x$

$$\sin 2x = 2 \sin x \cos x; \quad \cos 2x = \cos^2 x - \sin^2 x; \quad \tan 2x = \frac{2 \tan x}{1 - \tan^2 x}.$$

$$\sin \frac{x}{2} = \pm \sqrt{\frac{1 - \cos x}{2}}; \quad \cos \frac{x}{2} = \pm \sqrt{\frac{1 + \cos x}{2}}; \quad \tan \frac{x}{2} = \pm \sqrt{\frac{1 - \cos x}{1 + \cos x}}.$$

$$\sin^2 x = \tfrac{1}{2} - \tfrac{1}{2} \cos 2x; \quad \cos^2 x = \tfrac{1}{2} + \tfrac{1}{2} \cos 2x.$$

(6) Addition theorems

$$\sin x + \sin y = 2 \sin \tfrac{1}{2}(x + y) \cos \tfrac{1}{2}(x - y).$$
$$\sin x - \sin y = 2 \cos \tfrac{1}{2}(x + y) \sin \tfrac{1}{2}(x - y).$$
$$\cos x + \cos y = 2 \cos \tfrac{1}{2}(x + y) \cos \tfrac{1}{2}(x - y).$$
$$\cos x - \cos y = - 2 \sin \tfrac{1}{2}(x + y) \sin \tfrac{1}{2}(x - y).$$

(7) Relations for any triangle

Law of sines.
$$\frac{a}{\sin A} = \frac{b}{\sin B} = \frac{c}{\sin C}.$$

Law of cosines. $\qquad a^2 = b^2 + c^2 - 2\,bc \cos A.$

Formulas for area. $K = \frac{1}{2} bc \sin A.$

$$K = \frac{\frac{1}{2} a^2 \sin B \sin C}{\sin (B + C)}.$$

$$K = \sqrt{s(s - a)(s - b)(s - c)}, \quad \text{where} \quad s = \frac{1}{2}(a + b + c).$$

297. Table of integrals. In this article is given a short table of integrals. Explanation of its use is made in Art. 208.

Some Elementary Forms

1. $\int df(x) = \int f'(x)dx = f(x) + C.$

2. $\int a\, du = a \int du.$

3. $\int (du \pm dv \pm dw \pm \cdots) = \int du \pm \int dv \pm \int dw \pm \cdots.$

4. $\int u^n\, du = \dfrac{u^{n+1}}{n+1} + C.$ $(n \neq -1)$

5. $\int \dfrac{du}{u} = \ln u + C.$

Rational Forms containing $a + bu$

See also the Binomial Reduction Formulas 96–104.

6. $\int (a + bu)^n\, du = \dfrac{(a + bu)^{n+1}}{b(n + 1)} + C.$ $(n \neq -1)$

7. $\int \dfrac{du}{a + bu} = \dfrac{1}{b} \ln (a + bu) + C.$

8. $\int \dfrac{u\, du}{a + bu} = \dfrac{1}{b^2} [a + bu - a \ln (a + bu)] + C.$

9. $\int \dfrac{u^2\, du}{a + bu} = \dfrac{1}{b^3} [\frac{1}{2}(a + bu)^2 - 2\, a(a + bu) + a^2 \ln (a + bu)] + C.$

10. $\int \dfrac{u\, du}{(a + bu)^2} = \dfrac{1}{b^2}\left[\dfrac{a}{a + bu} + \ln (a + bu)\right] + C.$

11. $\int \dfrac{u^2\, du}{(a + bu)^2} = \dfrac{1}{b^3}\left[a + bu - \dfrac{a^2}{a + bu} - 2\, a \ln (a + bu)\right] + C.$

12. $\int \dfrac{u\, du}{(a + bu)^3} = \dfrac{1}{b^2}\left[-\dfrac{1}{a + bu} + \dfrac{a}{2(a + bu)^2}\right] + C.$

13. $\int \dfrac{du}{u(a + bu)} = -\dfrac{1}{a} \ln \left(\dfrac{a + bu}{u}\right) + C.$

14. $\int \dfrac{du}{u^2(a + bu)} = -\dfrac{1}{au} + \dfrac{b}{a^2} \ln \left(\dfrac{a + bu}{u}\right) + C.$

15. $\int \dfrac{du}{u(a + bu)^2} = \dfrac{1}{a(a + bu)} - \dfrac{1}{a^2} \ln \left(\dfrac{a + bu}{u}\right) + C.$

Rational Forms containing $a^2 \pm b^2u^2$

16. $\int \dfrac{du}{a^2 + b^2u^2} = \dfrac{1}{ab} \text{ arc tan } \dfrac{bu}{a} + C.$

17. $\int \dfrac{du}{a^2 - b^2u^2} = \dfrac{1}{2\,ab} \ln \left(\dfrac{a + bu}{a - bu} \right) + C.$ $\hspace{2em}(a^2 > b^2u^2)$

$\int \dfrac{du}{b^2u^2 - a^2} = \dfrac{1}{2\,ab} \ln \left(\dfrac{bu - a}{bu + a} \right) + C.$ $\hspace{2em}(a^2 < b^2u^2)$

18. $\int u(a^2 \pm b^2u^2)^n du = \dfrac{(a^2 \pm b^2u^2)^{n+1}}{\pm 2\,b^2(n + 1)} + C.$ $\hspace{2em}(n \neq -1)$

19. $\int \dfrac{u\,du}{a^2 \pm b^2u^2} = \dfrac{1}{\pm 2\,b^2} \ln (a^2 \pm b^2u^2) + C.$

20. $\int \dfrac{u^m\,du}{(a^2 \pm b^2u^2)^p} = \dfrac{u^{m-1}}{\pm b^2(m - 2\,p + 1)(a^2 \pm b^2u^2)^{p-1}}$
$\hspace{6em} - \dfrac{a^2(m - 1)}{\pm b^2(m - 2\,p + 1)} \int \dfrac{u^{m-2}\,du}{(a^2 \pm b^2u^2)^p}.$

21. $\int \dfrac{u^m\,du}{(a^2 \pm b^2u^2)^p} = \dfrac{u^{m+1}}{2\,a^2(p - 1)(a^2 \pm b^2u^2)^{p-1}}$
$\hspace{6em} - \dfrac{m - 2\,p + 3}{2\,a^2(p - 1)} \int \dfrac{u^m\,du}{(a^2 \pm b^2u^2)^{p-1}}.$

22. $\int \dfrac{du}{u(a^2 \pm b^2u^2)} = \dfrac{1}{2\,a^2} \ln \left(\dfrac{u^2}{a^2 \pm b^2u^2} \right) + C.$

23. $\int \dfrac{du}{u^m(a^2 \pm b^2u^2)^p} = -\dfrac{1}{a^2(m - 1)u^{m-1}(a^2 \pm b^2u^2)^{p-1}}$
$\hspace{6em} - \dfrac{\pm b^2(m + 2\,p - 3)}{a^2(m - 1)} \int \dfrac{du}{u^{m-2}(a^2 \pm b^2u^2)^p}.$

24. $\int \dfrac{du}{u^m(a^2 \pm b^2u^2)^p} = \dfrac{1}{2\,a^2(p - 1)u^{m-1}(a^2 \pm b^2u^2)^{p-1}}$
$\hspace{6em} + \dfrac{m + 2\,p - 3}{2\,a^2(p - 1)} \int \dfrac{du}{u^m(a^2 \pm b^2u^2)^{p-1}}.$

Forms containing $\sqrt{a + bu}$

The integrand may be rationalized by setting $a + bu = v^2$. See also the Binomial Reduction Formulas 96–104.

25. $\int u\sqrt{a + bu}\,du = -\dfrac{2(2\,a - 3\,bu)(a + bu)^{\frac{3}{2}}}{15\,b^2} + C.$

26. $\int u^2\sqrt{a + bu}\,du = \dfrac{2(8\,a^2 - 12\,abu + 15\,b^2u^2)(a + bu)^{\frac{3}{2}}}{105\,b^3} + C.$

27. $\int u^m\sqrt{a + bu}\,du = \dfrac{2\,u^m(a + bu)^{\frac{3}{2}}}{b(2\,m + 3)} - \dfrac{2\,am}{b(2\,m + 3)} \int u^{m-1}\sqrt{a + bu}\,du.$

28. $\int \dfrac{u\,du}{\sqrt{a + bu}} = -\dfrac{2(2\,a - bu)\sqrt{a + bu}}{3\,b^2} + C.$

29. $\int \dfrac{u^2\,du}{\sqrt{a+bu}} = \dfrac{2(8\,a^2 - 4\,abu + 3\,b^2u^2)\sqrt{a+bu}}{15\,b^3} + C.$

30. $\int \dfrac{u^m\,du}{\sqrt{a+bu}} = \dfrac{2\,u^m\sqrt{a+bu}}{b(2\,m+1)} - \dfrac{2\,am}{b(2\,m+1)}\int \dfrac{u^{m-1}\,du}{\sqrt{a+bu}}.$

31. $\int \dfrac{du}{u\sqrt{a+bu}} = \dfrac{1}{\sqrt{a}}\ln\left(\dfrac{\sqrt{a+bu}-\sqrt{a}}{\sqrt{a+bu}+\sqrt{a}}\right) + C,$ for $a > 0.$

32. $\int \dfrac{du}{u\sqrt{a+bu}} = \dfrac{2}{\sqrt{-a}}\ \text{arc tan}\sqrt{\dfrac{a+bu}{-a}} + C,$ for $a < 0.$

33. $\int \dfrac{du}{u^m\sqrt{a+bu}} = -\dfrac{\sqrt{a+bu}}{a(m-1)u^{m-1}} - \dfrac{b(2\,m-3)}{2\,a(m-1)}\int \dfrac{du}{u^{m-1}\sqrt{a+bu}}.$

34. $\int \dfrac{\sqrt{a+bu}\,du}{u} = 2\sqrt{a+bu} + a\int \dfrac{du}{u\sqrt{a+bu}}.$

35. $\int \dfrac{\sqrt{a+bu}\,du}{u^m} = -\dfrac{(a+bu)^{\frac{3}{2}}}{a(m-1)u^{m-1}} - \dfrac{b(2\,m-5)}{2\,a(m-1)}\int \dfrac{\sqrt{a+bu}\,du}{u^{m-1}}.$

Forms containing $\sqrt{u^2 \pm a^2}$

In this group of formulas we may replace

$$\ln\left(u + \sqrt{u^2 + a^2}\right)\quad \text{by}\quad \sinh^{-1}\dfrac{u}{a},$$

$$\ln\left(u + \sqrt{u^2 - a^2}\right)\quad \text{by}\quad \cosh^{-1}\dfrac{u}{a},$$

$$\ln\left(\dfrac{a + \sqrt{u^2 + a^2}}{u}\right)\quad \text{by}\quad \sinh^{-1}\dfrac{a}{u}.$$

36. $\int (u^2 \pm a^2)^{\frac{1}{2}}\,du = \dfrac{u}{2}\sqrt{u^2 \pm a^2} \pm \dfrac{a^2}{2}\ln\left(u + \sqrt{u^2 \pm a^2}\right) \mp C.$

37. $\int (u^2 \pm a^2)^{\frac{n}{2}}\,du = \dfrac{u(u^2 \pm a^2)^{\frac{n}{2}}}{n+1} \pm \dfrac{na^2}{n+1}\int (u^2 \pm a^2)^{\frac{n}{2}-1}\,du.\ (n \neq -1)$

38. $\int u(u^2 \pm a^2)^{\frac{n}{2}}\,du = \dfrac{(u^2 \pm a^2)^{\frac{n}{2}+1}}{n+2} + C.$ $(n \neq -2)$

39. $\int u^m(u^2 \pm a^2)^{\frac{n}{2}}\,du = \dfrac{u^{m-1}(u^2 \pm a^2)^{\frac{n}{2}+1}}{n+m+1}$
$$-\dfrac{\pm a^2(m-1)}{n+m+1}\int u^{m-2}(u^2 \pm a^2)^{\frac{n}{2}}\,du.$$

40. $\int \dfrac{du}{(u^2 \pm a^2)^{\frac{1}{2}}} = \ln\left(u + \sqrt{u^2 \pm a^2}\right) + C.$

41. $\int \dfrac{du}{(u^2 \pm a^2)^{\frac{3}{2}}} = \dfrac{u}{\pm a^2\sqrt{u^2 \pm a^2}} + C.$

42. $\displaystyle\int \frac{u\,du}{(u^2 \pm a^2)^{\frac{n}{2}}} = \frac{(u^2 \pm a^2)^{1-\frac{n}{2}}}{2-n} + C.$

43. $\displaystyle\int \frac{u^2\,du}{(u^2 \pm a^2)^{\frac{1}{2}}} = \frac{u}{2}\sqrt{u^2 \pm a^2} - \frac{\pm a^2}{2}\ln\left(u + \sqrt{u^2 \pm a^2}\right) + C.$

44. $\displaystyle\int \frac{u^2\,du}{(u^2 \pm a^2)^{\frac{3}{2}}} = -\frac{u}{\sqrt{u^2 \pm a^2}} + \ln\left(u + \sqrt{u^2 \pm a^2}\right) + C.$

45. $\displaystyle\int \frac{u^m\,du}{(u^2 \pm a^2)^{\frac{n}{2}}} = \frac{u^{m-1}}{(m-n+1)(u^2 \pm a^2)^{\frac{n}{2}-1}} - \frac{\pm a^2(m-1)}{m-n+1}\int \frac{u^{m-2}\,du}{(u^2 \pm a^2)^{\frac{n}{2}}}.$

46. $\displaystyle\int \frac{u^m\,du}{(u^2 \pm a^2)^{\frac{n}{2}}} = \frac{u^{m+1}}{\pm a^2(n-2)(u^2 \pm a^2)^{\frac{n}{2}-1}}$
$\qquad\qquad - \frac{m-n+3}{\pm a^2(n-2)}\int \frac{u^m\,du}{(u^2 \pm a^2)^{\frac{n}{2}-1}}.$

47. $\displaystyle\int \frac{du}{u(u^2 + a^2)^{\frac{1}{2}}} = -\frac{1}{a}\ln\left(\frac{a + \sqrt{u^2 + a^2}}{u}\right) + C.$

48. $\displaystyle\int \frac{du}{u(u^2 - a^2)^{\frac{1}{2}}} = \frac{1}{a}\operatorname{arc\,sec}\frac{u}{a} + C.$

49. $\displaystyle\int \frac{du}{u^2(u^2 \pm a^2)^{\frac{1}{2}}} = -\frac{\sqrt{u^2 \pm a^2}}{\pm a^2 u} + C.$

50. $\displaystyle\int \frac{du}{u^3(u^2 + a^2)^{\frac{1}{2}}} = -\frac{\sqrt{u^2 + a^2}}{2\,a^2 u^2} + \frac{1}{2\,a^3}\ln\left(\frac{a + \sqrt{u^2 + a^2}}{u}\right) + C.$

51. $\displaystyle\int \frac{du}{u^3(u^2 - a^2)^{\frac{1}{2}}} = \frac{\sqrt{u^2 - a^2}}{2\,a^2 u^2} + \frac{1}{2\,a^3}\operatorname{arc\,sec}\frac{u}{a} + C.$

52. $\displaystyle\int \frac{du}{u^m(u^2 \pm a^2)^{\frac{n}{2}}} = -\frac{1}{\pm a^2(m-1)u^{m-1}(u^2 \pm a^2)^{\frac{n}{2}-1}}$
$\qquad\qquad - \frac{m+n-3}{\pm a^2(m-1)}\int \frac{du}{u^{m-2}(u^2 \pm a^2)^{\frac{n}{2}}}.$

53. $\displaystyle\int \frac{du}{u^m(u^2 \pm a^2)^{\frac{n}{2}}} = \frac{1}{\pm a^2(n-2)u^{m-1}(u^2 \pm a^2)^{\frac{n}{2}-1}}$
$\qquad\qquad + \frac{m+n-3}{\pm a^2(n-2)}\int \frac{du}{u^m(u^2 \pm a^2)^{\frac{n}{2}-1}}.$

54. $\displaystyle\int \frac{(u^2 + a^2)^{\frac{1}{2}}\,du}{u} = \sqrt{u^2 + a^2} - a\ln\left(\frac{a + \sqrt{u^2 + a^2}}{u}\right) + C.$

55. $\displaystyle\int \frac{(u^2 - a^2)^{\frac{1}{2}}\,du}{u} = \sqrt{u^2 - a^2} - a\operatorname{arc\,sec}\frac{u}{a} + C.$

56. $\int \dfrac{(u^2 \pm a^2)^{\frac{1}{2}} du}{u^2} = -\dfrac{\sqrt{u^2 \pm a^2}}{u} + \ln \left(u + \sqrt{u^2 \pm a^2}\right) + C.$

57. $\int \dfrac{(u^2 \pm a^2)^{\frac{n}{2}} du}{u^m} = -\dfrac{(u^2 \pm a^2)^{\frac{n}{2}+1}}{\pm a^2(m-1)u^{m-1}}$

$$-\dfrac{m-n-3}{\pm a^2(m-1)} \int \dfrac{(u^2 \pm a^2)^{\frac{n}{2}} du}{u^{m-2}}.$$

58. $\int \dfrac{(u^2 \pm a^2)^{\frac{n}{2}} du}{u^m} = \dfrac{(u^2 \pm a^2)^{\frac{n}{2}}}{(n-m+1)u^{m-1}} + \dfrac{\pm a^2 n}{n-m+1} \int \dfrac{(u^2 \pm a^2)^{\frac{n}{2}-1} du}{u^m}.$

Forms containing $\sqrt{a^2 - u^2}$

59. $\int (a^2 - u^2)^{\frac{1}{2}} du = \dfrac{u}{2} \sqrt{a^2 - u^2} + \dfrac{a^2}{2} \arcsin \dfrac{u}{a} + C.$

60. $\int (a^2 - u^2)^{\frac{n}{2}} du = \dfrac{u(a^2 - u^2)^{\frac{n}{2}}}{n+1} + \dfrac{a^2 n}{n+1} \int (a^2 - u^2)^{\frac{n}{2}-1} du.$ $(n \neq -1)$

61. $\int u(a^2 - u^2)^{\frac{n}{2}} du = -\dfrac{(a^2 - u^2)^{\frac{n}{2}+1}}{n+2} + C.$ $(n \neq -2)$

62. $\int u^m (a^2 - u^2)^{\frac{n}{2}} du = -\dfrac{u^{m-1}(a^2 - u^2)^{\frac{n}{2}+1}}{n+m+1}$

$$+\dfrac{a^2(m-1)}{n+m+1} \int u^{m-2}(a^2 - u^2)^{\frac{n}{2}} du.$$

63. $\int \dfrac{du}{(a^2 - u^2)^{\frac{1}{2}}} = \arcsin \dfrac{u}{a} + C.$

64. $\int \dfrac{du}{(a^2 - u^2)^{\frac{3}{2}}} = \dfrac{u}{a^2 \sqrt{a^2 - u^2}} + C.$

65. $\int \dfrac{u \, du}{(a^2 - u^2)^{\frac{n}{2}}} = \dfrac{(a^2 - u^2)^{1-\frac{n}{2}}}{n-2} + C.$

66. $\int \dfrac{u^2 \, du}{(a^2 - u^2)^{\frac{1}{2}}} = -\dfrac{u}{2} \sqrt{a^2 - u^2} + \dfrac{a^2}{2} \arcsin \dfrac{u}{a} + C.$

67. $\int \dfrac{u^2 \, du}{(a^2 - u^2)^{\frac{3}{2}}} = \dfrac{u}{\sqrt{a^2 - u^2}} - \arcsin \dfrac{u}{a} + C.$

68. $\int \dfrac{u^m \, du}{(a^2 - u^2)^{\frac{n}{2}}} = -\dfrac{u^{m-1}}{(m-n+1)(a^2 - u^2)^{\frac{n}{2}-1}} + \dfrac{a^2(m-1)}{m-n+1} \int \dfrac{u^{m-2} du}{(a^2 - u^2)^{\frac{n}{2}}}.$

69. $\int \dfrac{u^m \, du}{(a^2 - u^2)^{\frac{n}{2}}} = \dfrac{u^{m+1}}{a^2(n-2)(a^2 - u^2)^{\frac{n}{2}-1}} - \dfrac{m-n+3}{a^2(n-2)} \int \dfrac{u^m du}{(a^2 - u^2)^{\frac{n}{2}-1}}.$

70. $\displaystyle\int \frac{du}{u(a^2 - u^2)^{\frac{1}{2}}} = -\frac{1}{a} \ln\left(\frac{a + \sqrt{a^2 - u^2}}{u}\right) + C = -\frac{1}{a} \cosh^{-1}\frac{a}{u} + C.$

71. $\displaystyle\int \frac{du}{u^2(a^2 - u^2)^{\frac{1}{2}}} = -\frac{\sqrt{a^2 - u^2}}{a^2 u} + C.$

72. $\displaystyle\int \frac{du}{u^3(a^2 - u^2)^{\frac{1}{2}}} = -\frac{\sqrt{a^2 - u^2}}{2\,a^2 u^2} - \frac{1}{2\,a^3} \ln\left(\frac{a + \sqrt{a^2 - u^2}}{u}\right) + C$

$\displaystyle\qquad\qquad = -\frac{\sqrt{a^2 - u^2}}{2\,a^2 u^2} - \frac{1}{2\,a^3} \cosh^{-1}\frac{a}{u} + C.$

73. $\displaystyle\int \frac{du}{u^m(a^2 - u^2)^{\frac{n}{2}}} = -\frac{1}{a^2(m - 1)u^{m-1}(a^2 - u^2)^{\frac{n}{2}-1}}$

$\displaystyle\qquad\qquad + \frac{m + n - 3}{a^2(m - 1)} \int \frac{du}{u^{m-2}(a^2 - u^2)^{\frac{n}{2}}}.$

74. $\displaystyle\int \frac{du}{u^m(a^2 - u^2)^{\frac{n}{2}}} = \frac{1}{a^2(n - 2)u^{m-1}(a^2 - u^2)^{\frac{n}{2}-1}}$

$\displaystyle\qquad\qquad + \frac{m + n - 3}{a^2(n - 2)} \int \frac{du}{u^m(a^2 - u^2)^{\frac{n}{2}-1}}.$

75. $\displaystyle\int \frac{(a^2 - u^2)^{\frac{1}{2}}\,du}{u} = \sqrt{a^2 - u^2} - a \ln\left(\frac{a + \sqrt{a^2 - u^2}}{u}\right) + C$

$\displaystyle\qquad\qquad = \sqrt{a^2 - u^2} - a \cosh^{-1}\frac{a}{u} + C.$

76. $\displaystyle\int \frac{(a^2 - u^2)^{\frac{1}{2}}\,du}{u^2} = -\frac{\sqrt{a^2 - u^2}}{u} - \text{arc sin}\frac{u}{a} + C.$

77. $\displaystyle\int \frac{(a^2 - u^2)^{\frac{n}{2}}\,du}{u^m} = -\frac{(a^2 - u^2)^{\frac{n}{2}+1}}{a^2(m - 1)u^{m-1}} + \frac{m - n - 3}{a^2(m - 1)} \int \frac{(a^2 - u^2)^{\frac{n}{2}}\,du}{u^{m-2}}.$

78. $\displaystyle\int \frac{(a^2 - u^2)^{\frac{n}{2}}\,du}{u^m} = \frac{(a^2 - u^2)^{\frac{n}{2}}}{(n - m + 1)u^{m-1}} + \frac{a^2 n}{n - m + 1} \int \frac{(a^2 - u^2)^{\frac{n}{2}-1}\,du}{u^m}.$

Forms containing $\sqrt{2\,au \pm u^2}$

The Binomial Reduction Formulas 96–104 may be applied by writing $\sqrt{2\,au \pm u^2} = u^{\frac{1}{2}}(2\,a \pm u)^{\frac{1}{2}}.$

79. $\displaystyle\int \sqrt{2\,au - u^2}\,du = \frac{u - a}{2}\sqrt{2\,au - u^2} + \frac{a^2}{2} \text{ arc cos}\left(1 - \frac{u}{a}\right) + C.$

80. $\displaystyle\int u\sqrt{2\,au - u^2}\,du = -\frac{3\,a^2 + au - 2\,u^2}{6}\sqrt{2\,au - u^2}$

$\displaystyle\qquad\qquad + \frac{a^3}{2} \text{ arc cos}\left(1 - \frac{u}{a}\right) + C.$

$$81. \int u^m \sqrt{2\,au - u^2}\, du = -\frac{u^{m-1}(2\,au - u^2)^{\frac{3}{2}}}{m+2}$$
$$+ \frac{a(2\,m + 1)}{m + 2} \int u^{m-1} \sqrt{2\,au - u^2}\, du.$$

$$82. \int \frac{\sqrt{2\,au - u^2}\, du}{u} = \sqrt{2\,au - u^2} + a \arccos\left(1 - \frac{u}{a}\right) + C.$$

$$83. \int \frac{\sqrt{2\,au - u^2}\, du}{u^2} = -\frac{2\sqrt{2\,au - u^2}}{u} - \arccos\left(1 - \frac{u}{a}\right) + C.$$

$$84. \int \frac{\sqrt{2\,au - u^2}\, du}{u^3} = -\frac{(2\,au - u^2)^{\frac{3}{2}}}{3\,au^3} + C.$$

$$85. \int \frac{\sqrt{2\,au - u^2}\, du}{u^m} = -\frac{(2\,au - u^2)^{\frac{3}{2}}}{a(2\,m - 3)u^m} + \frac{m - 3}{a(2\,m - 3)} \int \frac{\sqrt{2\,au - u^2}\, du}{u^{m-1}}.$$

$$86. \int \frac{du}{\sqrt{2\,au - u^2}} = \arccos\left(1 - \frac{u}{a}\right) + C.$$

$$87. \int \frac{du}{\sqrt{2\,au + u^2}} = \ln\left(u + a + \sqrt{2\,au + u^2}\right) + C.$$

$$88. \int F\left(u, \sqrt{2\,au + u^2}\right)\, du = \int F\left(z - a, \sqrt{z^2 - a^2}\right)\, dz, \text{ where } z = u + a.$$

$$89. \int \frac{u\, du}{\sqrt{2\,au - u^2}} = -\sqrt{2\,au - u^2} + a \arccos\left(1 - \frac{u}{a}\right) + C.$$

$$90. \int \frac{u^2\, du}{\sqrt{2\,au - u^2}} = -\frac{(u + 3\,a)\sqrt{2\,au - u^2}}{2} + \frac{3\,a^2}{2} \arccos\left(1 - \frac{u}{a}\right) + C.$$

$$91. \int \frac{u^m\, du}{\sqrt{2\,au - u^2}} = -\frac{u^{m-1}\sqrt{2\,au - u^2}}{m} + \frac{a(2\,m - 1)}{m} \int \frac{u^{m-1}\, du}{\sqrt{2\,au - u^2}}.$$

$$92. \int \frac{du}{u\sqrt{2\,au - u^2}} = -\frac{\sqrt{2\,au - u^2}}{au} + C.$$

$$93. \int \frac{du}{u^m\sqrt{2\,au - u^2}} = -\frac{\sqrt{2\,au - u^2}}{a(2\,m - 1)u^m} + \frac{m - 1}{a(2\,m - 1)} \int \frac{du}{u^{m-1}\sqrt{2\,au - u^2}}.$$

$$94. \int \frac{du}{(2\,au - u^2)^{\frac{3}{2}}} = \frac{u - a}{a^2\sqrt{2\,au - u^2}} + C.$$

$$95. \int \frac{u\, du}{(2\,au - u^2)^{\frac{3}{2}}} = \frac{u}{a\sqrt{2\,au - u^2}} + C.$$

Binomial Reduction Formulas

$$96. \int u^m (a + bu^q)^p\, du = \frac{u^{m-q+1}(a + bu^q)^{p+1}}{b(pq + m + 1)}$$
$$- \frac{a(m - q + 1)}{b(pq + m + 1)} \int u^{m-q}(a + bu^q)^p\, du.$$

97. $\int u^m(a + bu^q)^p\, du = \dfrac{u^{m+1}(a + bu^q)^p}{pq + m + 1}$

$\qquad\qquad + \dfrac{apq}{pq + m + 1}\int u^m(a + bu^q)^{p-1}\, du.$

98. $\int \dfrac{du}{u^m(a + bu^q)^p} = -\dfrac{1}{a(m - 1)u^{m-1}(a + bu^q)^{p-1}}$

$\qquad\qquad - \dfrac{b(m - q + pq - 1)}{a(m - 1)}\int \dfrac{du}{u^{m-q}(a + bu^q)^p}.$

99. $\int \dfrac{du}{u^m(a + bu^q)^p} = \dfrac{1}{aq(p - 1)u^{m-1}(a + bu^q)^{p-1}}$

$\qquad\qquad + \dfrac{m - q + pq - 1}{aq(p - 1)}\int \dfrac{du}{u^m(a + bu^q)^{p-1}}.$

100. $\int \dfrac{du}{u(a + bu^q)} = \dfrac{1}{aq}\ln\left(\dfrac{u^q}{a + bu^q}\right) + C.$

101. $\int \dfrac{(a + bu^q)^p\, du}{u^m} = -\dfrac{(a + bu^q)^{p+1}}{a(m - 1)u^{m-1}}$

$\qquad\qquad - \dfrac{b(m - q - pq - 1)}{a(m - 1)}\int \dfrac{(a + bu^q)^p\, du}{u^{m-q}}.$

102. $\int \dfrac{(a + bu^q)^p\, du}{u^m} = \dfrac{(a + bu^q)^p}{(pq - m + 1)u^{m-1}}$

$\qquad\qquad + \dfrac{apq}{pq - m + 1}\int \dfrac{(a + bu^q)^{p-1}\, du}{u^m}.$

103. $\int \dfrac{u^m\, du}{(a + bu^q)^p} = \dfrac{u^{m-q+1}}{b(m - pq + 1)(a + bu^q)^{p-1}}$

$\qquad\qquad - \dfrac{a(m - q + 1)}{b(m - pq + 1)}\int \dfrac{u^{m-q}\, du}{(a + bu^q)^p}.$

104. $\int \dfrac{u^m\, du}{(a + bu^q)^p} = \dfrac{u^{m+1}}{aq(p - 1)(a + bu^q)^{p-1}}$

$\qquad\qquad - \dfrac{m + q - pq + 1}{aq(p - 1)}\int \dfrac{u^m\, du}{(a + bu^q)^{p-1}}.$

Forms containing $a + bu \pm cu^2$ $(c > 0)$

The expression $a + bu + cu^2$ may be reduced to a binomial by writing
$u = z - \dfrac{b}{2c}, \quad k = \dfrac{b^2 - 4ac}{4c^2}.$

Then $\qquad\qquad a + bu + cu^2 = c(z^2 - k).$

The expression $a + bu - cu^2$ may be reduced to a binomial by writing
$u = z + \dfrac{b}{2c}, \quad k = \dfrac{b^2 + 4ac}{4c^2}.$

Then $\qquad\qquad a + bu - cu^2 = c(k - z^2).$

105. $\int \dfrac{du}{a + bu + cu^2} = \dfrac{2}{\sqrt{4ac - b^2}} \arctan\left(\dfrac{2cu + b}{\sqrt{4ac - b^2}}\right) + C,$

$\qquad\qquad\qquad\qquad\qquad\qquad\qquad\qquad \text{when } b^2 < 4ac.$

106. $\int \dfrac{du}{a + bu + cu^2} = \dfrac{1}{\sqrt{b^2 - 4\,ac}} \ln\left(\dfrac{2\,cu + b - \sqrt{b^2 - 4\,ac}}{2\,cu + b + \sqrt{b^2 - 4\,ac}}\right) + C,$

$$\text{when } b^2 > 4\,ac$$

107. $\int \dfrac{du}{a + bu - cu^2} = \dfrac{1}{\sqrt{b^2 + 4\,ac}} \ln\left(\dfrac{\sqrt{b^2 + 4\,ac} + 2\,cu - b}{\sqrt{b^2 + 4\,ac} - 2\,cu + b}\right) + C.$

108. $\int \dfrac{(Mu + N)du}{a + bu \pm cu^2} = \dfrac{\pm M}{2\,c} \ln\left(a + bu \pm cu^2\right))$

$$+ \left(N \mp \dfrac{bM}{2\,c}\right) \int \dfrac{du}{a + bu \pm cu^2}.$$

109. $\int \sqrt{a + bu + cu^2}\, du = \dfrac{2\,cu + b}{4\,c} \sqrt{a + bu + cu^2}$

$$- \dfrac{b^2 - 4\,ac}{8\,c^{\frac{3}{2}}} \ln\left(2\,cu + b + 2\sqrt{c}\sqrt{a + bu + cu^2}\right) + C.$$

110. $\int \sqrt{a + bu - cu^2}\, du = \dfrac{2\,cu - b}{4\,c} \sqrt{a + bu - cu^2}$

$$+ \dfrac{b^2 + 4\,ac}{8\,c^{\frac{3}{2}}} \arcsin\left(\dfrac{2\,cu - b}{\sqrt{b^2 + 4\,ac}}\right) + C.$$

111. $\int \dfrac{du}{\sqrt{a + bu + cu^2}} = \dfrac{1}{\sqrt{c}} \ln\left(2\,cu + b + 2\sqrt{c}\sqrt{a + bu + cu^2}\right) + C.$

112. $\int \dfrac{du}{\sqrt{a + bu - cu^2}} = \dfrac{1}{\sqrt{c}} \arcsin\left(\dfrac{2\,cu - b}{\sqrt{b^2 + 4\,ac}}\right) + C.$

113. $\int \dfrac{u\, du}{\sqrt{a + bu + cu^2}} = \dfrac{\sqrt{a + bu + cu^2}}{c}$

$$- \dfrac{b}{2\,c^{\frac{3}{2}}} \ln\left(2\,cu + b + 2\sqrt{c}\sqrt{a + bu + cu^2}\right) + C.$$

114. $\int \dfrac{u\, du}{\sqrt{a + bu - cu^2}} = - \dfrac{\sqrt{a + bu - cu^2}}{c}$

$$+ \dfrac{b}{2\,c^{\frac{3}{2}}} \arcsin\left(\dfrac{2\,cu - b}{\sqrt{b^2 + 4\,ac}}\right) + C.$$

Other Algebraic Forms

115. $\int \sqrt{\dfrac{a + u}{b + u}}\, du = \sqrt{(a + u)(b + u)}$

$$+ (a - b) \ln\left(\sqrt{a + u} + \sqrt{b + u}\right) + C.$$

116. $\int \sqrt{\dfrac{a - u}{b + u}}\, du = \sqrt{(a - u)(b + u)}$

$$+ (a + b) \arcsin\sqrt{\dfrac{u + b}{a + b}} + C.$$

117. $\int \sqrt{\dfrac{a + u}{b - u}}\, du = - \sqrt{(a + u)(b - u)} - (a + b) \arcsin\sqrt{\dfrac{b - u}{a + b}} + C.$

118. $\int \sqrt{\dfrac{1 + u}{1 - u}}\, du = - \sqrt{1 - u^2} + \arcsin u + C.$

119. $\int \dfrac{du}{\sqrt{(u - a)(b - u)}} = 2 \arcsin\sqrt{\dfrac{u - a}{b - a}} + C.$

Exponential and Logarithmic Forms

120. $\int e^{au}\,du = \dfrac{e^{au}}{a} + C.$

121. $\int b^{au}\,du = \dfrac{b^{au}}{a\ln b} + C.$

122. $\int u e^{au}\,du = \dfrac{e^{au}}{a^2}(au-1) + C.$

123. $\int u^n e^{au}\,du = \dfrac{u^n e^{au}}{a} - \dfrac{n}{a}\int u^{n-1} e^{au}\,du.$

124. $\int u^n b^{au}\,du = \dfrac{u^n b^{au}}{a\ln b} - \dfrac{n}{a\ln b}\int u^{n-1} b^{au}\,du + C.$

125. $\int \dfrac{b^{au}\,du}{u^n} = -\dfrac{b^{au}}{(n-1)u^{n-1}} + \dfrac{a\ln b}{n-1}\int \dfrac{b^{au}\,du}{u^{n-1}}.$

126. $\int \ln u\,du = u\ln u - u + C.$

127. $\int u^n \ln u\,du = u^{n+1}\left[\dfrac{\ln u}{n+1} - \dfrac{1}{(n+1)^2}\right] + C.$

128. $\int u^m \ln^n u\,du = \dfrac{u^{m+1}}{m+1}\ln^n u - \dfrac{n}{m+1}\int u^m \ln^{n-1} u\,du.$

129. $\int e^{au}\ln u\,du = \dfrac{e^{au}\ln u}{a} - \dfrac{1}{a}\int \dfrac{e^{au}}{u}\,du.$

130. $\int \dfrac{du}{u\ln u} = \ln(\ln u) + C.$

Trigonometric Forms

In forms involving $\tan u$, $\operatorname{ctn} u$, $\sec u$, $\csc u$, which do not appear below, first use the relations

$$\tan u = \frac{\sin u}{\cos u}, \quad \operatorname{ctn} u = \frac{\cos u}{\sin u}, \quad \sec u = \frac{1}{\cos u}, \quad \csc u = \frac{1}{\sin u}.$$

131. $\int \sin u\,du = -\cos u + C.$

132. $\int \cos u\,du = \sin u + C.$

133. $\int \tan u\,du = -\ln \cos u + C = \ln \sec u + C.$

134. $\int \operatorname{ctn} u\,du = \ln \sin u + C.$

135. $\int \sec u\,du = \int \dfrac{du}{\cos u} = \ln(\sec u + \tan u) + C$
$$= \ln \tan\left(\frac{u}{2} + \frac{\pi}{4}\right) + C.$$

136. $\int \csc u\,du = \int \dfrac{du}{\sin u} = \ln(\csc u - \operatorname{ctn} u) + C$
$$= \ln \tan \frac{u}{2} + C.$$

137. $\int \sec^2 u \, du = \tan u + C.$

138. $\int \csc^2 u \, du = - \operatorname{ctn} u + C.$

139. $\int \sec u \tan u \, du = \sec u + C.$

140. $\int \csc u \operatorname{ctn} u \, du = - \csc u + C.$

141. $\int \sin^2 u \, du = \tfrac{1}{2} u - \tfrac{1}{4} \sin 2 u + C.$

142. $\int \cos^2 u \, du = \tfrac{1}{2} u + \tfrac{1}{4} \sin 2 u + C.$

143. $\int \cos^n u \sin u \, du = - \dfrac{\cos^{n+1} u}{n+1} + C.$

144. $\int \sin^n u \cos u \, du = \dfrac{\sin^{n+1} u}{n+1} + C.$

145. $\int \sin mu \sin nu \, du = - \dfrac{\sin (m+n)u}{2(m+n)} + \dfrac{\sin (m-n)u}{2(m-n)} + C.$

146. $\int \cos mu \cos nu \, du = \dfrac{\sin (m+n)u}{2(m+n)} + \dfrac{\sin (m-n)u}{2(m-n)} + C.$

147. $\int \sin mu \cos nu \, du = - \dfrac{\cos (m+n)u}{2(m+n)} - \dfrac{\cos (m-n)u}{2(m-n)} + C.$

148. $\int \dfrac{du}{1 + \cos a \cos u} = 2 \csc a \arctan (\tan \tfrac{1}{2} a \tan \tfrac{1}{2} u) + C.$

149. $\int \dfrac{du}{\cos a + \cos u} = \csc a \ln \left(\dfrac{1 + \tan \tfrac{1}{2} a \tan \tfrac{1}{2} u}{1 - \tan \tfrac{1}{2} a \tan \tfrac{1}{2} u} \right) + C \quad (\tan^2 \tfrac{1}{2} u < \operatorname{ctn}^2 \tfrac{1}{2} a)$

$$= 2 \csc a \tanh^{-1}(\tan \tfrac{1}{2} a \tan \tfrac{1}{2} u) + C \quad (\tan^2 \tfrac{1}{2} u < \operatorname{ctn}^2 \tfrac{1}{2} a)$$

150. $\int \dfrac{du}{1 + \cos a \sin u} = 2 \csc a \arctan (\csc a \tan \tfrac{1}{2} u + \operatorname{ctn} a) + C.$

151. $\int \dfrac{du}{\cos a + \sin u} = \csc a \ln \left(\dfrac{\tan a - \tan \tfrac{1}{2} u - \sec a}{\tan a + \tan \tfrac{1}{2} u + \sec a} \right) + C$

$$[(\operatorname{ctn} a \tan \tfrac{1}{2} u + \csc a)^2 < 1]$$

$$= - 2 \csc a \tanh^{-1}(\operatorname{ctn} a \tan \tfrac{1}{2} u + \csc a) + C$$

$$[(\operatorname{ctn} a \tan \tfrac{1}{2} u + \csc a)^2 < 1]$$

152. $\int \dfrac{du}{a^2 \cos^2 u + b^2 \sin^2 u} = \dfrac{1}{ab} \arctan \left(\dfrac{b \tan u}{a} \right) + C.$

153. $\int e^{au} \sin nu \, du = \dfrac{e^{au}(a \sin nu - n \cos nu)}{a^2 + n^2} + C.$

154. $\int e^{au} \cos nu \, du = \dfrac{e^{au}(n \sin nu + a \cos nu)}{a^2 + n^2} + C.$

155. $\int u \sin u \, du = \sin u - u \cos u + C.$

156. $\int u \cos u \, du = \cos u + u \sin u + C.$

Trigonometric Reduction Formulas

157. $\int \sin^n u \, du = -\dfrac{\sin^{n-1} u \cos u}{n} + \dfrac{n-1}{n} \int \sin^{n-2} u \, du.$

158. $\int \cos^n u \, du = \dfrac{\cos^{n-1} u \sin u}{n} + \dfrac{n-1}{n} \int \cos^{n-2} u \, du.$

159. $\int \dfrac{du}{\sin^n u} = -\dfrac{\cos u}{(n-1) \sin^{n-1} u} + \dfrac{n-2}{n-1} \int \dfrac{du}{\sin^{n-2} u}.$

160. $\int \dfrac{du}{\cos^n u} = \dfrac{\sin u}{(n-1) \cos^{n-1} u} + \dfrac{n-2}{n-1} \int \dfrac{du}{\cos^{n-2} u}.$

161. $\int \cos^m u \sin^n u \, du = \dfrac{\cos^{m-1} u \sin^{n+1} u}{m+n} + \dfrac{m-1}{m+n} \int \cos^{m-2} u \sin^n u \, du.$

162. $\int \cos^m u \sin^n u \, du = -\dfrac{\sin^{n-1} u \cos^{m+1} u}{m+n} + \dfrac{n-1}{m+n} \int \cos^m u \sin^{n-2} u \, du.$

163. $\int \dfrac{du}{\cos^m u \sin^n u} = \dfrac{1}{(m-1) \sin^{n-1} u \cos^{m-1} u}$
$\qquad\qquad + \dfrac{m+n-2}{m-1} \int \dfrac{du}{\cos^{m-2} u \sin^n u}.$

164. $\int \dfrac{du}{\cos^m u \sin^n u} = -\dfrac{1}{(n-1) \sin^{n-1} u \cos^{m-1} u}$
$\qquad\qquad + \dfrac{m+n-2}{n-1} \int \dfrac{du}{\cos^m u \sin^{n-2} u}.$

165. $\int \dfrac{\cos^m u \, du}{\sin^n u} = -\dfrac{\cos^{m+1} u}{(n-1) \sin^{n-1} u} - \dfrac{m-n+2}{n-1} \int \dfrac{\cos^m u \, du}{\sin^{n-2} u}.$

166. $\int \dfrac{\cos^m u \, du}{\sin^n u} = \dfrac{\cos^{m-1} u}{(m-n) \sin^{n-1} u} + \dfrac{m-1}{m-n} \int \dfrac{\cos^{m-2} u \, du}{\sin^n u}.$

167. $\int \dfrac{\sin^n u \, du}{\cos^m u} = \dfrac{\sin^{n+1} u}{(m-1) \cos^{m-1} u} - \dfrac{n-m+2}{m-1} \int \dfrac{\sin^n u \, du}{\cos^{m-2} u}.$

168. $\int \dfrac{\sin^n u \, du}{\cos^m u} = -\dfrac{\sin^{n-1} u}{(n-m) \cos^{m-1} u} + \dfrac{n-1}{n-m} \int \dfrac{\sin^{n-2} u \, du}{\cos^m u}.$

169. $\int \tan^n u \, du = \dfrac{\tan^{n-1} u}{n-1} - \int \tan^{n-2} u \, du.$

170. $\int \operatorname{ctn}^n u \, du = -\dfrac{\operatorname{ctn}^{n-1} u}{n-1} - \int \operatorname{ctn}^{n-2} u \, du.$

171. $\int e^{au} \cos^n u \, du = \dfrac{e^{au} \cos^{n-1} u (a \cos u + n \sin u)}{a^2 + n^2}$
$\qquad\qquad + \dfrac{n(n-1)}{a^2 + n^2} \int e^{au} \cos^{n-2} u \, du.$

172. $\int e^{au} \sin^n u \, du = \dfrac{e^{au} \sin^{n-1} u (a \sin u - n \cos u)}{a^2 + n^2}$
$\qquad\qquad + \dfrac{n(n-1)}{a^2 + n^2} \int e^{au} \sin^{n-2} u \, du.$

173. $\int u^m \cos au\, du = \dfrac{u^{m-1}}{a^2}\,(au \sin au + m \cos au)$

$$-\,\frac{m(m-1)}{a^2}\int u^{m-2} \cos au\, du.$$

174. $\int u^m \sin au\, du = \dfrac{u^{m-1}}{a^2}\,(m \sin au - au \cos au)$

$$-\,\frac{m(m-1)}{a^2}\int u^{m-2} \sin au\, du.$$

Inverse Trigonometric Functions

175. $\int \arcsin u\, du = u \arcsin u + \sqrt{1-u^2} + C.$

176. $\int \arccos u\, du = u \arccos u - \sqrt{1-u^2} + C.$

177. $\int \arctan u\, du = u \arctan u - \ln \sqrt{1+u^2} + C.$

178. $\int \operatorname{arc\,ctn} u\, du = u \operatorname{arc\,ctn} u + \ln \sqrt{1+u^2} + C.$

179. $\int \operatorname{arc\,sec} u\, du = u \operatorname{arc\,sec} u - \ln\left(u + \sqrt{u^2-1}\right) + C.$
$$= u \operatorname{arc\,sec} u - \cosh^{-1} u + C.$$

180. $\int \operatorname{arc\,csc} u\, du = u \operatorname{arc\,csc} u + \ln\left(u + \sqrt{u^2-1}\right) + C$
$$= u \operatorname{arc\,csc} u + \cosh^{-1} u + C.$$

Hyperbolic Functions

181. $\int \sinh u\, du = \cosh u + C.$

182. $\int \cosh u\, du = \sinh u + C.$

183. $\int \tanh u\, du = \ln \cosh u + C.$

184. $\int \operatorname{ctnh} u\, du = \ln \sinh u + C.$

185. $\int \operatorname{sech} u\, du = \arctan(\sinh u) + C = \operatorname{gd} u + C.$

186. $\int \operatorname{csch} u\, du = \ln \tanh \tfrac{1}{2} u + C.$

187. $\int \operatorname{sech}^2 u\, du = \tanh u + C.$

188. $\int \operatorname{csch}^2 u\, du = -\operatorname{ctnh} u + C.$

189. $\int \operatorname{sech} u \tanh u\, du = -\operatorname{sech} u + C.$

190. $\int \operatorname{csch} u \operatorname{ctnh} u\, du = - \operatorname{csch} u + C.$

191. $\int \sinh^2 u\, du = \frac{1}{4} \sinh 2u - \frac{1}{2} u + C.$

192. $\int \cosh^2 u\, du = \frac{1}{4} \sinh 2u + \frac{1}{2} u + C.$

193. $\int \tanh^2 u\, du = u - \tanh u + C.$

194. $\int \operatorname{ctnh}^2 u\, du = u - \operatorname{ctnh} u + C.$

195. $\int u \sinh u\, du = u \cosh u - \sinh u + C.$

196. $\int u \cosh u\, du = u \sinh u - \cosh u + C.$

197. $\int \sinh^{-1} u\, du = u \sinh^{-1} u - \sqrt{1 + u^2} + C.$

198. $\int \cosh^{-1} u\, du = u \cosh^{-1} u - \sqrt{u^2 - 1} + C.$

199. $\int \tanh^{-1} u\, du = u \tanh^{-1} u + \frac{1}{2} \ln (1 - u^2) + C.$

200. $\int \operatorname{ctnh}^{-1} u\, du = u \operatorname{ctnh}^{-1} u + \frac{1}{2} \ln (1 - u^2) + C.$

201. $\int \operatorname{sech}^{-1} u\, du = u \operatorname{sech}^{-1} u + \operatorname{gd} (\tanh^{-1} u) + C$
$$= u \operatorname{sech}^{-1} u + \operatorname{arc\,sin} u + C.$$

202. $\int \operatorname{csch}^{-1} u\, du = u \operatorname{csch}^{-1} u + \sinh^{-1} u + C.$

203. $\int \sinh mu \sinh nu\, du = \dfrac{\sinh (m+n)u}{2(m+n)} - \dfrac{\sinh (m-n)u}{2(m-n)} + C. \left(m \gtrless n\right)$

204. $\int \cosh mu \cosh nu\, du = \dfrac{\sinh (m+n)u}{2(m+n)} + \dfrac{\sinh (m-n)u}{2(m-n)} + C. \left(m \gtrless n\right)$

205. $\int \sinh mu \cosh nu\, du = \dfrac{\cosh (m+n)u}{2(m+n)} + \dfrac{\cosh (m-n)u}{2(m-n)} + C. \left(m \gtrless n\right)$

206. $\int \dfrac{du}{\cosh a + \cosh u} = 2 \operatorname{csch} a \tanh^{-1}(\tanh \frac{1}{2} u \tanh \frac{1}{2} a) + C.$

207. $\int \dfrac{du}{\cos a + \cosh u} = 2 \csc a \operatorname{arc\,tan} (\tanh \frac{1}{2} u \tan \frac{1}{2} a) + C.$

208. $\int \dfrac{du}{1 + \cos a \cosh u} = 2 \csc a \tanh^{-1}(\tanh \frac{1}{2} u \tan \frac{1}{2} a) + C$
$$(\tanh^2 \tfrac{1}{2} u < \operatorname{ctn}^2 \tfrac{1}{2} a)$$

209. $\int e^{au} \sinh nu\, du = \dfrac{e^{au}(a \sinh nu - n \cosh nu)}{a^2 - n^2} + C.$

210. $\int e^{au} \cosh nu\, du = \dfrac{e^{au}(a \cosh nu - n \sinh nu)}{a^2 - n^2} + C.$

298. Numerical tables.* As an aid to the student in computation, some selected numerical tables are given in the following pages. A few words of explanation are set down here.

TABLE 1. *Numerical constants and their common logarithms.*

An examination of this table will repay the reader. Many constants occurring frequently in numerical computation are given to five places of decimals. Others are given with a greater degree of precision.

TABLE 2. *Mantissas of common logarithms, base 10.*

The mantissas are given to four decimal places. On pages 567–568 are tabulated the mantissas of numbers with four significant figures and first significant figure 1. Interpolation is carried out here in the usual manner by proportion. But on pages 569–570 the correction for a fourth significant figure of the number may readily be made by using the tables for proportional parts at the right of the pages. The corrections to be added for a fourth significant figure 1, 2, 3, 4, 5 are tabulated.

Thus $\log 478.2 = 2.6794 + 0.0002 = 2.6796.$

If the fourth significant figure of N is greater than 5, take the next greater mantissa in the tables and *subtract* the number under "Prop. parts" corresponding to 1, 2, 3, 4 for a fourth figure 9, 8, 7, 6.

Thus $\log 478.6 = 2.6803 - 0.0004 = 2.6799.$
                   $\log 478.8 = 2.6803 - 0.0002 = 2.6801.$

TABLES 3, 4, 5. *Minutes to radians. Degrees to radians. Radians to degrees and minutes.*

By these tables the change from degree measure to radian measure or from radian measure to degree measure is quickly made (see Art. 173).

Thus $128° 31' = 2.234 + 0.009 = 2.243$ radians.
       2.18 radians $= 124° 54'.$
       4 radians $= (3 + 1)$ radians $= 171° 53' + 57° 18' = 229° 11'.$

TABLE 6. *Natural values of trigonometric functions for angles in radians.*

When an angle is given directly in radians, its functions (three significant figures) may be found by this table without converting the angle

*The tables in this article are reprinted by permission from *Mathematical Tables and Formulas,* by P. F. Smith and W. R. Longley, published by John Wiley and Sons, Inc.

into degrees. For angles greater than 3.20 radians, we first use formulas **(3)**, Art. 296, for reducing angles, and recall that 180 degrees $= \pi$ (3.14) radians.

Thus
$$4 \text{ radians} = 3.14 + 0.86 = \pi + 0.86.$$
$$\sin 4 = \sin (\pi + 0.86) = - \sin 0.86 = - 0.757.$$
$$\tan 4 = \tan (\pi + 0.86) = \tan 0.86 = 1.16.$$

A value of the angle directly in radians may be found for an inverse circular function (see Art. 186).

Thus
$$\text{arc cos } 0.796 = 0.65 \text{ radian.}$$
$$\text{arc ctn } 0.600 = 1.03 \text{ radians.}$$

TABLES 7 and 8. *Napierian (natural) logarithms. Base $e = 2.71828$.*

Logarithms to the base e (Art. 167) are found by Table 7. The correction for a fifth significant figure in N is made, using the table for proportional parts as explained above under Table 2.

For a value of N greater than 10, we express N as the product of a number between 1 and 10 by a power of 10; use the formula

$$\ln ab = \ln a + \ln b,$$

and find $\ln a$ and $\ln b$ from Table 7 and Table 8.

Thus
$$\ln 45.62 = \ln (4.562 \times 10) = \ln 4.562 + \ln 10$$
$$= 1.5177 + 2.3026 = 3.8203.$$

If N is less than 1, we express N as the quotient of a number between 1 and 10 by a power of 10 and use the formula

$$\ln \frac{a}{b} = \ln a - \ln b.$$

Thus
$$\ln 0.4562 = \ln \frac{4.562}{10} = \ln 4.562 - \ln 10$$
$$= 1.5177 - 2.3026 = - 0.7849.$$

TABLE 9. *Exponentials e^u and e^{-u}.*

By this table we find values of the exponential function (see Art. 165).

Thus
$$e^{0.88} = 2.411.$$
$$e^{-0.88} = 0.4148.$$

TABLE 10. *Hyperbolic functions.*

This table gives the values of certain important functions involving the exponential function. These are

$$\text{Hyperbolic sine of } u = \sinh u = \tfrac{1}{2}(e^u - e^{-u}).$$
$$\text{Hyperbolic cosine of } u = \cosh u = \tfrac{1}{2}(e^u + e^{-u}).$$
$$\text{Hyperbolic tangent of } u = \tanh u = \frac{e^u - e^{-u}}{e^u + e^{-u}}.$$

Thus

$$\sinh 0.80 = 0.8881,$$
$$\tanh 4.9 = 0.9999.$$

The relation
$$z = \sinh u$$
is also written
$$u = \sinh^{-1} z.$$

This is read, "u is the inverse hyperbolic sine of z." Similarly for $\cosh^{-1} z$ and $\tanh^{-1} z$. Reference is made to these functions on page 551. The relations between Napierian logarithms and the inverse hyperbolic functions are established as in the following illustration.

Thus, to prove
$$\ln (z + \sqrt{z^2 + 1}) = \sinh^{-1} z,$$

let
$$\sinh^{-1} z = u. \qquad (1)$$

Now
$$\ln (z + \sqrt{z^2 + 1}) = u. \qquad (2)$$
$$z + \sqrt{z^2 + 1} = e^u. \qquad \text{By Art. 166}$$

Solving,
$$z = \tfrac{1}{2}(e^u - e^{-u}) = \sinh u.$$

Then
$$u = \sinh^{-1} z.$$

Substituting this value of u in (2) gives (1).

Table 1. Numerical constants and their common logarithms

Constant		Logarithm
$\pi = 3.14159$	. . . .	0.49715
$\frac{1}{2}\pi = 1.57080$	. . . .	0.19612
$\frac{1}{3}\pi = 1.04720$	. . . .	0.02003
$\frac{1}{4}\pi = 0.78540$	. . . .	9.89509–10
$\frac{1}{6}\pi = 0.52360$	. . . .	9.71900–10
$\pi^2 = 9.86960$	. . . .	0.99430
$\pi^3 = 31.00628$	. . . .	1.49145
$\sqrt{\pi} = 1.77245$	. . . .	0.24857
$\sqrt[3]{\pi} = 1.46459$	. . . .	0.16572
$\dfrac{1}{\pi} = 0.31831$	. . . .	9.50285–10
$\dfrac{1}{\pi^2} = 0.10132$	. . . .	9.00570–10
$\dfrac{1}{\pi^3} = 0.03225$	. . . .	8.50855–10
$\dfrac{1}{\sqrt{\pi}} = 0.56419$	. . . .	9.75143–10
$\dfrac{1}{\sqrt[3]{\pi}} = 0.68277$	. . . .	9.83427–10
$e = 2.71828$	. . . .	0.43429

$$1 \text{ radian} = 57.295780 \text{ degrees}$$
$$= 3437.7468 \text{ minutes}$$
$$1 \text{ degree} = 0.017453 \text{ radians}$$
$$1 \text{ minute} = 0.0002909 \text{ radians}$$
$$\log_e \pi = 1.144730$$
$$\log_e 10 = \frac{1}{\log_{10} e} = 2.302585$$

Angle	sin	cos	tan	ctn	sec	csc
0	0	1	0	∞	1	∞
30°	$\frac{1}{2}$	$\frac{1}{2}\sqrt{3}$	$\frac{1}{3}\sqrt{3}$	$\sqrt{3}$	$\frac{2}{3}\sqrt{3}$	2
45°	$\frac{1}{2}\sqrt{2}$	$\frac{1}{2}\sqrt{2}$	1	1	$\sqrt{2}$	$\sqrt{2}$
60°	$\frac{1}{2}\sqrt{3}$	$\frac{1}{2}$	$\sqrt{3}$	$\frac{1}{3}\sqrt{3}$	2	$\frac{2}{3}\sqrt{3}$
90°	1	0	∞	0	∞	1

Table 2. Mantissas of common logarithms, base 10

N	0	1	2	3	4	5	6	7	8	9
100	.0000	.0004	.0009	.0013	.0017	.0022	.0026	.0030	.0035	.0039
101	.0043	.0048	.0052	.0056	.0060	.0065	.0069	.0073	.0077	.0082
102	.0086	.0090	.0095	.0099	.0103	.0107	.0111	.0116	.0120	.0124
103	.0128	.0133	.0137	.0141	.0145	.0149	.0154	.0158	.0162	.0166
104	.0170	.0175	.0179	.0183	.0187	.0191	.0195	.0199	.0204	.0208
105	.0212	.0216	.0220	.0224	.0228	.0233	.0237	.0241	.0245	.0249
106	.0253	.0257	.0261	.0265	.0269	.0273	.0278	.0282	.0286	.0290
107	.0294	.0298	.0302	.0306	.0310	.0314	.0318	.0322	.0326	.0330
108	.0334	.0338	.0342	.0346	.0350	.0354	.0358	.0362	.0366	.0370
109	.0374	.0378	.0382	.0386	.0390	.0394	.0398	.0402	.0406	.0410
110	.0414	.0418	.0422	.0426	.0430	.0434	.0438	.0441	.0445	.0449
111	.0453	.0457	.0461	.0465	.0469	.0473	.0477	.0481	.0484	.0488
112	.0492	.0496	.0500	.0504	.0508	.0512	.0515	.0519	.0523	.0527
113	.0531	.0535	.0538	.0542	.0546	.0550	.0554	.0558	.0561	.0565
114	.0569	.0573	.0577	.0580	.0584	.0588	.0592	.0596	.0599	.0603
115	.0607	.0611	.0615	.0618	.0622	.0626	.0630	.0633	.0637	.0641
116	.0645	.0648	.0652	.0656	.0660	.0663	.0667	.0671	.0674	.0678
117	.0682	.0686	.0689	.0693	.0697	.0700	.0704	.0708	.0711	.0715
118	.0719	.0722	.0726	.0730	.0734	.0737	.0741	.0745	.0748	.0752
119	.0755	.0759	.0763	.0766	.0770	.0774	.0777	.0781	.0785	.0788
120	.0792	.0795	.0799	.0803	.0806	.0810	.0813	.0817	.0821	.0824
121	.0828	.0831	.0835	.0839	.0842	.0846	.0849	.0853	.0856	.0860
122	.0864	.0867	.0871	.0874	.0878	.0881	.0885	.0888	.0892	.0896
123	.0899	.0903	.0906	.0910	.0913	.0917	.0920	.0924	.0927	.0931
124	.0934	.0938	.0941	.0945	.0948	.0952	.0955	.0959	.0962	.0966
125	.0969	.0973	.0976	.0980	.0983	.0986	.0990	.0993	.0997	.1000
126	.1004	.1007	.1011	.1014	.1017	.1021	.1024	.1028	.1031	.1035
127	.1038	.1041	.1045	.1048	.1052	.1055	.1059	.1062	.1065	.1069
128	.1072	.1075	.1079	.1082	.1086	.1089	.1092	.1096	.1099	.1103
129	.1106	.1109	.1113	.1116	.1119	.1123	.1126	.1129	.1133	.1136
130	1139	.1143	.1146	.1149	.1153	.1156	.1159	.1163	.1166	.1169
131	1173	.1176	.1179	.1183	.1186	.1189	.1193	.1196	.1199	.1202
132	.1206	.1209	.1212	.1216	.1219	.1222	.1225	.1229	.1232	.1235
133	.1239	.1242	.1245	.1248	.1252	.1255	.1258	.1261	.1265	.1268
134	.1271	.1274	.1278	.1281	.1284	.1287	.1290	.1294	.1297	.1300
135	.1303	.1307	.1310	.1313	.1316	.1319	.1323	.1326	.1329	.1332
136	.1335	.1339	.1342	.1345	.1348	.1351	.1355	.1358	.1361	.1364
137	.1367	.1370	.1374	.1377	.1380	.1383	.1386	.1389	.1392	.1396
138	.1399	.1402	.1405	.1408	.1411	.1414	.1418	.1421	.1424	.1427
139	.1430	.1433	.1436	.1440	.1443	.1446	.1449	.1452	.1455	.1458
140	.1461	.1464	.1467	.1471	.1474	.1477	.1480	.1483	.1486	.1489
141	.1492	.1495	.1498	.1501	.1504	.1508	.1511	.1514	.1517	.1520
142	.1523	.1526	.1529	.1532	.1535	.1538	.1541	.1544	.1547	.1550
143	.1553	.1556	.1559	.1562	.1565	.1569	.1572	.1575	.1578	.1581
144	.1584	.1587	.1590	.1593	.1596	.1599	.1602	.1605	.1608	.1611
145	.1614	.1617	.1620	.1623	.1626	.1629	.1632	.1635	.1638	.1641
146	.1644	.1647	.1649	.1652	.1655	.1658	.1661	.1664	.1667	.1670
147	.1673	.1676	.1679	.1682	.1685	.1688	.1691	.1694	.1697	.1700
148	.1703	.1706	.1708	.1711	.1714	.1717	.1720	.1723	.1726	.1729
149	.1732	.1735	.1738	.1741	.1744	.1746	.1749	.1752	.1755	.1758
150	.1761	.1764	.1767	.1770	.1772	.1775	.1778	.1781	.1784	.1787
N	0	1	2	3	4	5	6	7	8	9

Table 2. Mantissas of common logarithms, base 10 — *Continued*

N	0	1	2	3	4	5	6	7	8	9
150	.1761	.1764	.1767	.1770	.1772	.1775	.1778	.1781	.1784	.1787
151	.1790	.1793	.1796	.1798	.1801	.1804	.1807	.1810	.1813	.1816
152	.1818	.1821	.1824	.1827	.1830	.1833	.1836	.1838	.1841	.1844
153	.1847	.1850	.1853	.1855	.1858	.1861	.1864	.1867	.1870	.1872
154	.1875	.1878	.1881	.1884	.1886	.1889	.1892	.1895	.1898	.1901
155	.1903	.1906	.1909	.1912	.1915	.1917	.1920	.1923	.1926	.1928
156	.1931	.1934	.1937	.1940	.1942	.1945	.1948	.1951	.1953	.1956
157	.1959	.1962	.1965	.1967	.1970	.1973	.1976	.1978	.1981	.1984
158	.1987	.1989	.1992	.1995	.1998	.2000	.2003	.2006	.2009	.2011
159	.2014	.2017	.2019	.2022	.2025	.2028	.2030	.2033	.2036	.2038
160	.2041	.2044	.2047	.2049	.2052	.2055	.2057	.2060	.2063	.2066
161	.2068	.2071	.2074	.2076	.2079	.2082	.2084	.2087	.2090	.2092
162	.2095	.2098	.2101	.2103	.2106	.2109	.2111	.2114	.2117	.2119
163	.2122	.2125	.2127	.2130	.2133	.2135	.2138	.2140	.2143	.2146
164	.2148	.2151	.2154	.2156	.2159	.2162	.2164	.2167	.2170	.2172
165	.2175	.2177	.2180	.2183	.2185	.2188	.2191	.2193	.2196	.2198
166	.2201	.2204	.2206	.2209	.2212	.2214	.2217	.2219	.2222	.2225
167	.2227	.2230	.2232	.2235	.2238	.2240	.2243	.2245	.2248	.2251
168	.2253	.2256	.2258	.2261	.2263	.2266	.2269	.2271	.2274	.2276
169	.2279	.2281	.2284	.2287	.2289	.2292	.2294	.2297	.2299	.2302
170	.2304	.2307	.2310	.2312	.2315	.2317	.2320	.2322	.2325	.2327
171	.2330	.2333	.2335	.2338	.2340	.2343	.2345	.2348	.2350	.2353
172	.2355	.2358	.2360	.2363	.2365	.2368	.2370	.2373	.2375	.2378
173	.2380	.2383	.2385	.2388	.2390	.2393	.2395	.2398	.2400	.2403
174	.2405	.2408	.2410	.2413	.2415	.2418	.2420	.2423	.2425	.2428
175	.2430	.2433	.2435	.2438	.2440	.2443	.2445	.2448	.2450	.2453
176	.2455	.2458	.2460	.2463	.2465	.2467	.2470	.2472	.2475	.2477
177	.2480	.2482	.2485	.2487	.2490	.2492	.2494	.2497	.2499	.2502
178	.2504	.2507	.2509	.2512	.2514	.2516	.2519	.2521	.2524	.2526
179	.2529	.2531	.2533	.2536	.2538	.2541	.2543	.2545	.2548	.2550
180	.2553	.2555	.2558	.2560	.2562	.2565	.2567	.2570	.2572	.2574
181	.2577	.2579	.2582	.2584	.2586	.2589	.2591	.2594	.2596	.2598
182	.2601	.2603	.2605	.2608	.2610	.2613	.2615	.2617	.2620	.2622
183	.2625	.2627	.2629	.2632	.2634	.2636	.2639	.2641	.2643	.2646
184	.2648	.2651	.2653	.2655	.2658	.2660	.2662	.2665	.2667	.2669
185	.2672	.2674	.2676	.2679	.2681	.2683	.2686	.2688	.2690	.2693
186	.2695	.2697	.2700	.2702	.2704	.2707	.2709	.2711	.2714	.2716
187	.2718	.2721	.2723	.2725	.2728	.2730	.2732	.2735	.2737	.2739
188	.2742	.2744	.2746	.2749	.2751	.2753	.2755	.2758	.2760	.2762
189	.2765	.2767	.2769	.2772	.2774	.2776	.2778	.2781	.2783	.2785
190	.2788	.2790	.2792	.2794	.2797	.2799	.2801	.2804	.2806	.2808
191	.2810	.2813	.2815	.2817	.2819	.2822	.2824	.2826	.2828	.2831
192	.2833	.2835	.2838	.2840	.2842	.2844	.2847	.2849	.2851	.2853
193	.2856	.2858	.2860	.2862	.2865	.2867	.2869	.2871	.2874	.2876
194	.2878	.2880	.2882	.2885	.2887	.2889	.2891	.2894	.2896	.2898
195	.2900	.2903	.2905	.2907	.2909	.2911	.2914	.2916	.2918	.2920
196	.2923	.2925	.2927	.2929	.2931	.2934	.2936	.2938	.2940	.2942
197	.2945	.2947	.2949	.2951	.2953	.2956	.2958	.2960	.2962	.2964
198	.2967	.2969	.2971	.2973	.2975	.2978	.2980	.2982	.2984	.2986
199	.2989	.2991	.2993	.2995	.2997	.2999	.3002	.3004	.3006	.3008
200	.3010	.3012	.3015	.3017	.3019	.3021	.3023	.3025	.3028	.3030
N	0	1	2	3	4	5	6	7	8	9

Table 2. Mantissas of common logarithms, base 10 — *Continued*

N	0	1	2	3	4	5	6	7	8	9	Prop. parts 1	2	3	4	5
20	.3010	.3032	.3054	.3075	.3096	.3118	.3139	.3160	.3181	.3201	2	4	6	8	11
21	.3222	.3243	.3263	.3284	.3304	.3324	.3345	.3365	.3385	.3404	2	4	6	8	10
22	.3424	.3444	.3464	.3483	.3502	.3522	.3541	.3560	.3579	.3598	2	4	6	8	10
23	.3617	.3636	.3655	.3674	.3692	.3711	.3729	.3747	.3766	.3784	2	4	6	7	9
24	.3802	.3820	.3838	.3856	.3874	.3892	.3909	.3927	.3945	.3962	2	4	5	7	9
25	.3979	.3997	.4014	.4031	.4048	.4065	.4082	.4099	.4116	.4133	2	3	5	7	9
26	.4150	.4166	.4183	.4200	.4216	.4232	.4249	.4265	.4281	.4298	2	3	5	7	8
27	.4314	.4330	.4346	.4362	.4378	.4393	.4409	.4425	.4440	.4456	2	3	5	6	8
28	.4472	.4487	.4502	.4518	.4533	.4548	.4564	.4579	.4594	.4609	2	3	5	6	8
29	.4624	.4639	.4654	.4669	.4683	.4698	.4713	.4728	.4742	.4757	1	3	4	6	7
30	.4771	.4786	.4800	.4814	.4829	.4843	.4857	.4871	.4886	.4900	1	3	4	6	7
31	.4914	.4928	.4942	.4955	.4969	.4983	.4997	.5011	.5024	.5038	1	3	4	6	7
32	.5051	.5065	.5079	.5092	.5105	.5119	.5132	.5145	.5159	.5172	1	3	4	5	7
33	.5185	.5198	.5211	.5224	.5237	.5250	.5263	.5276	.5289	.5302	1	3	4	5	6
34	.5315	.5328	.5340	.5353	.5366	.5378	.5391	.5403	.5416	.5428	1	3	4	5	6
35	.5441	.5453	.5465	.5478	.5490	.5502	.5514	.5527	.5539	.5551	1	2	4	5	6
36	.5563	.5575	.5587	.5599	.5611	.5623	.5635	.5647	.5658	.5670	1	2	4	5	6
37	.5682	.5694	.5705	.5717	.5729	.5740	.5752	.5763	.5775	.5786	1	2	3	5	6
38	.5798	.5809	.5821	.5832	.5843	.5855	.5866	.5877	.5888	.5899	1	2	3	5	6
39	.5911	.5922	.5933	.5944	.5955	.5966	.5977	.5988	.5999	.6010	1	2	3	4	5
40	.6021	.6031	.6042	.6053	.6064	.6075	.6085	.6096	.6107	.6117	1	2	3	4	5
41	.6128	.6138	.6149	.6160	.6170	.6180	.6191	.6201	.6212	.6222	1	2	3	4	5
42	.6232	.6243	.6253	.6263	.6274	.6284	.6294	.6304	.6314	.6325	1	2	3	4	5
43	.6335	.6345	.6355	.6365	.6375	.6385	.6395	.6405	.6415	.6425	1	2	3	4	5
44	.6435	.6444	.6454	.6464	.6474	.6484	.6493	.6503	.6513	.6522	1	2	3	4	5
45	.6532	.6542	.6551	.6561	.6571	.6580	.6590	.6599	.6609	.6618	1	2	3	4	5
46	.6628	.6637	.6646	.6656	.6665	.6675	.6684	.6693	.6702	.6712	1	2	3	4	5
47	.6721	.6730	.6739	.6749	.6758	.6767	.6776	.6785	.6794	.6803	1	2	3	4	5
48	.6812	.6821	.6830	.6839	.6848	.6857	.6866	.6875	.6884	.6893	1	2	3	4	4
49	.6902	.6911	.6920	.6928	.6937	.6946	.6955	.6964	.6972	.6981	1	2	3	4	4
50	.6990	.6998	.7007	.7016	.7024	.7033	.7042	.7050	.7059	.7067	1	2	3	3	4
51	.7076	.7084	.7093	.7101	.7110	.7118	.7126	.7135	.7143	.7152	1	2	3	3	4
52	.7160	.7168	.7177	.7185	.7193	.7202	.7210	.7218	.7226	.7235	1	2	2	3	4
53	.7243	.7251	.7259	.7267	.7275	.7284	.7292	.7300	.7308	.7316	1	2	2	3	4
54	.7324	.7332	.7340	.7348	.7356	.7364	.7372	.7380	.7388	.7396	1	2	2	3	4
55	.7404	.7412	.7419	.7427	.7435	.7443	.7451	.7459	.7466	.7474	1	2	2	3	4
56	.7482	.7490	.7497	.7505	.7513	.7520	.7528	.7536	.7543	.7551	1	2	2	3	4
57	.7559	.7566	.7574	.7582	.7589	.7597	.7604	.7612	.7619	.7627	1	2	2	3	4
58	.7634	.7642	.7649	.7657	.7664	.7672	.7679	.7686	.7694	.7701	1	1	2	3	4
59	.7709	.7716	.7723	.7731	.7738	.7745	.7752	.7760	.7767	.7774	1	1	2	3	4
60	.7782	.7789	.7796	.7803	.7810	.7818	.7825	.7832	.7839	.7846	1	1	2	3	4
61	.7853	.7860	.7868	.7875	.7882	.7889	.7896	.7903	.7910	.7917	1	1	2	8	4
62	.7924	.7931	.7938	.7945	.7952	.7959	.7966	.7973	.7980	.7987	1	1	2	3	3
63	.7993	.8000	.8007	.8014	.8021	.8028	.8035	.8041	.8048	.8055	1	1	2	3	3
64	.8062	.8069	.8075	.8082	.8089	.8096	.8102	.8109	.8116	.8122	1	1	2	3	3
65	.8129	.8136	.8142	.8149	.8156	.8162	.8169	.8176	.8182	.8189	1	1	2	3	3
66	.8195	.8202	.8209	.8215	.8222	.8228	.8235	.8241	.8248	.8254	1	1	2	3	3
67	.8261	.8267	.8274	.8280	.8287	.8293	.8299	.8306	.8312	.8319	1	1	2	3	3
68	.8325	.8331	.8338	.8344	.8351	.8357	.8363	.8370	.8376	.8382	1	1	2	3	3
69	.8388	.8395	.8401	.8407	.8414	.8420	.8426	.8432	.8439	.8445	1	1	2	3	3
70	.8451	.8457	.8463	.8470	.8476	.8482	.8488	.8494	.8500	.8506	1	1	2	2	3
N	0	1	2	3	4	5	6	7	8	9					

Table 2. Mantissas of common logarithms, base 10 — *Concluded*

N	0	1	2	3	4	5	6	7	8	9	Prop. parts				
											1	2	3	4	5
70	.8451	.8457	.8463	.8470	.8476	.8482	.8488	.8494	.8500	.8506	1	1	2	2	3
71	.8513	.8519	.8525	.8531	.8537	.8543	.8549	.8555	.8561	.8567	1	1	2	2	3
72	.8573	.8579	.8585	.8591	.8597	.8603	.8609	.8615	.8621	.8627	1	1	2	2	3
73	.8633	.8639	.8645	.8651	.8657	.8663	.8669	.8675	.8681	.8686	1	1	2	2	3
74	.8692	.8698	.8704	.8710	.8716	.8722	.8727	.8733	.8739	.8745	1	1	2	2	3
75	.8751	.8756	.8762	.8768	.8774	.8779	.8785	.8791	.8797	.8802	1	1	2	2	3
76	.8808	.8814	.8820	.8825	.8831	.8837	.8842	.8848	.8854	.8859	1	1	2	2	3
77	.8865	.8871	.8876	.8882	.8887	.8893	.8899	.8904	.8910	.8915	1	1	2	2	3
78	.8921	.8927	.8932	.8938	.8943	.8949	.8954	.8960	.8965	.8971	1	1	2	2	3
79	.8976	.8982	.8987	.8993	.8998	.9004	.9009	.9015	.9020	.9025	1	1	2	2	3
80	.9031	.9036	.9042	.9047	.9053	.9058	.9063	.9069	.9074	.9079	1	1	2	2	3
81	.9085	.9090	.9096	.9101	.9106	.9112	.9117	.9122	.9128	.9133	1	1	2	2	3
82	.9138	.9143	.9149	.9154	.9159	.9165	.9170	.9175	.9180	.9186	1	1	2	2	3
83	.9191	.9196	.9201	.9206	.9212	.9217	.9222	.9227	.9232	.9238	1	1	2	2	3
84	.9243	.9248	.9253	.9258	.9263	.9269	.9274	.9279	.9284	.9289	1	1	2	2	3
85	.9294	.9299	.9304	.9309	.9315	.9320	.9325	.9330	.9335	.9340	1	1	2	2	3
86	.9345	.9350	.9355	.9360	.9365	.9370	.9375	.9380	.9385	.9390	1	1	2	2	3
87	.9395	.9400	.9405	.9410	.9415	.9420	.9425	.9430	.9435	.9440	0	1	1	2	2
88	.9445	.9450	.9455	.9460	.9465	.9469	.9474	.9479	.9484	.9489	0	1	1	2	2
89	.9494	.9499	.9504	.9509	.9513	.9518	.9523	.9528	.9533	.9538	0	1	1	2	2
90	.9542	.9547	.9552	.9557	.9562	.9566	.9571	.9576	.9581	.9586	0	1	1	2	2
91	.9590	.9595	.9600	.9605	.9609	.9614	.9619	.9624	.9628	.9633	0	1	1	2	2
92	.9638	.9643	.9647	.9652	.9657	.9661	.9666	.9671	.9675	.9680	0	1	1	2	2
93	.9685	.9689	.9694	.9699	.9703	.9708	.9713	.9717	.9722	.9727	0	1	1	2	2
94	.9731	.9736	.9741	.9745	.9750	.9754	.9759	.9763	.9768	.9773	0	1	1	2	2
95	.9777	.9782	.9786	.9791	.9795	.9800	.9805	.9809	.9814	.9818	0	1	1	2	2
96	.9823	.9827	.9832	.9836	.9841	.9845	.9850	.9854	.9859	.9863	0	1	1	2	2
97	.9868	.9872	.9877	.9881	.9886	.9890	.9894	.9899	.9903	.9908	0	1	1	2	2
98	.9912	.9917	.9921	.9926	.9930	.9934	.9939	.9943	.9948	.9952	0	1	1	2	2
99	.9956	.9961	.9965	.9969	.9974	.9978	.9983	.9987	.9991	.9996	0	1	1	2	2
100	.0000	.0004	.0009	.0013	.0017	.0022	.0026	.0030	.0035	.0039					
N	0	1	2	3	4	5	6	7	8	9					

Table 3. Minutes to radians

Min.	Rad.	Min.	Rad.	Min.	Rad.	Min.	Rad.	Min.	Rad.	Min.	Rad.
1	.0003	11	.0032	21	.0061	31	.0090	41	.0119	51	.0148
2	.0006	12	.0035	22	.0064	32	.0093	42	.0122	52	.0151
3	.0009	13	.0038	23	.0067	33	.0096	43	.0125	53	.0154
4	.0012	14	.0041	24	.0070	34	.0099	44	.0128	54	.0157
5	.0015	15	.0044	25	.0073	35	.0102	45	.0131	55	.0160
6	.0017	16	.0047	26	.0076	36	.0105	46	.0134	56	.0163
7	.0020	17	.0049	27	.0079	37	.0108	47	.0137	57	.0166
8	.0023	18	.0052	28	.0081	38	.0111	48	.0140	58	.0169
9	.0026	19	.0055	29	.0084	39	.0113	49	.0143	59	.0172
10	.0029	20	.0058	30	.0087	40	.0116	50	.0145	60	.0175

Table 4. Degrees to radians

Deg.	Rad.	Deg.	Rad.	Deg.	Rad.	Deg.	Rad.	Deg.	Rad.	Deg.	Rad.
1	0.017	61	1.065	121	2.112	181	3.159	241	4.206	301	5.253
2	0.035	62	1.082	122	2.129	182	3.177	242	4.223	302	5.271
3	0.052	63	1.100	123	2.147	183	3.194	243	4.241	303	5.288
4	0.070	64	1.117	124	2.164	184	3.211	244	4.258	304	5.306
5	0.087	65	1.134	125	2.182	185	3.229	245	4.276	305	5.323
6	0.105	66	1.152	126	2.199	186	3.246	246	4.294	306	5.341
7	0.122	67	1.169	127	2.217	187	3.264	247	4.311	307	5.358
8	0.140	68	1.187	128	2.234	188	3.281	248	4.328	308	5.376
9	0.157	69	1.204	129	2.251	189	3.299	249	4.346	309	5.393
10	0.175	70	1.222	130	2.269	190	3.316	250	4.363	310	5.411
11	0.192	71	1.239	131	2.286	191	3.334	251	4.381	311	5.428
12	0.209	72	1.257	132	2.304	192	3.351	252	4.398	312	5.445
13	0.227	73	1.274	133	2.321	193	3.368	253	4.416	313	5.463
14	0.244	74	1.292	134	2.339	194	3.386	254	4.433	314	5.480
15	0.262	75	1.309	135	2.356	195	3.403	255	4.451	315	5.498
16	0.279	76	1.326	136	2.374	196	3.421	256	4.468	316	5.515
17	0.297	77	1.344	137	2.391	197	3.438	257	4.485	317	5.533
18	0.314	78	1.361	138	2.409	198	3.456	258	4.503	318	5.550
19	0.332	79	1.379	139	2.426	199	3.473	259	4.520	319	5.568
20	0.349	80	1.396	140	2.443	200	3.491	260	4.538	320	5.585
21	0.367	81	1.414	141	2.461	201	3.508	261	4.555	321	5.603
22	0.384	82	1.431	142	2.478	202	3.526	262	4.573	322	5.620
23	0.401	83	1.449	143	2.496	203	3.543	263	4.590	323	5.637
24	0.419	84	1.466	144	2.513	204	3.560	264	4.608	324	5.655
25	0.436	85	1.484	145	2.531	205	3.578	265	4.625	325	5.672
26	0.454	86	1.501	146	2.548	206	3.595	266	4.643	326	5.690
27	0.471	87	1.518	147	2.566	207	3.613	267	4.660	327	5.707
28	0.489	88	1.536	148	2.583	208	3.630	268	4.677	328	5.725
29	0.506	89	1.553	149	2.601	209	3.648	269	4.695	329	5.742
30	0.524	90	1.571	150	2.618	210	3.665	270	4.712	330	5.760
31	0.541	91	1.588	151	2.635	211	3.683	271	4.730	331	5.777
32	0.559	92	1.606	152	2.653	212	3.700	272	4.747	332	5.794
33	0.576	93	1.623	153	2.670	213	3.718	273	4.765	333	5.812
34	0.593	94	1.641	154	2.688	214	3.735	274	4.782	334	5.829
35	0.611	95	1.658	155	2.705	215	3.752	275	4.800	335	5.847
36	0.628	96	1.676	156	2.723	216	3.770	276	4.817	336	5.864
37	0.646	97	1.693	157	2.740	217	3.787	277	4.835	337	5.882
38	0.663	98	1.710	158	2.758	218	3.805	278	4.852	338	5.899
39	0.681	99	1.728	159	2.775	219	3.822	279	4.869	339	5.917
40	0.698	100	1.745	160	2.793	220	3.840	280	4.887	340	5.934
41	0.716	101	1.763	161	2.810	221	3.857	281	4.904	341	5.952
42	0.733	102	1.780	162	2.827	222	3.875	282	4.922	342	5.969
43	0.750	103	1.798	163	2.845	223	3.892	283	4.939	343	5.986
44	0 768	104	1.815	164	2.862	224	3.910	284	4.957	344	6.004
45	0.785	105	1.833	165	2.880	225	3.927	285	4.974	345	6.021
46	0.803	106	1.850	166	2.897	226	3.944	286	4.992	346	6.039
47	0.820	107	1.868	167	2.915	227	3.962	287	5.009	347	6.056
48	0.838	108	1.885	168	2.932	228	3.979	288	5.027	348	6.074
49	0.855	109	1.902	169	2.950	229	3.997	289	5.044	349	6.091
50	0.873	110	1.920	170	2.967	230	4.014	290	5.061	350	6.109
51	0.890	111	1.937	171	2.985	231	4.032	291	5.079	351	6.126
52	0.908	112	1.955	172	3.002	232	4.049	292	5.096	352	6.144
53	0.925	113	1.972	173	3.019	233	4.067	293	5.114	353	6.161
54	0.942	114	1.990	174	3.037	234	4.084	294	5.131	354	6.178
55	0.960	115	2.007	175	3.054	235	4.102	295	5.149	355	6.196
56	0.977	116	2.025	176	3.072	236	4.119	296	5.166	356	6.213
57	0.995	117	2.042	177	3.089	237	4.136	297	5.184	357	6.231
58	1.012	118	2.059	178	3.107	238	4.154	298	5.201	358	6.248
59	1.030	119	2.077	179	3.124	239	4.171	299	5.219	359	6.266
60	1.047	120	2.094	180	3.142	240	4.189	300	5.236	360	6.283

Table 5. Radians to degrees and minutes

Rad.	Deg. and min.	Rad.	Deg. and min.	Rad.	Deg. and min.	Rad.	Deg. and min.	Rad.	Deg. and min.	Rad.	Deg. and min.
0.001	0° 3′	0.47	26° 56′	1.02	58° 27′	1.57	89° 57′	2.12	121° 28′	2.67	152° 59′
0.002	0 7	0.48	27 30	1.03	59 1	1.58	90 32	2.13	122 2	2.68	153 33
0.003	0 10	0.49	28 4	1.04	59 35	1.59	91 6	2.14	122 37	2.69	154 8
0.004	0 14	0.50	28 39	1.05	60 10	1.60	91 40	2.15	123 11	2.70	154 42
0.005	0 17	0.51	29 13	1.06	60 44	1.61	92 15	2.16	123 46	2.71	155 16
0.006	0 21	0.52	29 48	1.07	61 18	1.62	92 49	2.17	124 20	2.72	155 51
0.007	0 24	0.53	30 22	1.08	61 53	1.63	93 24	2.18	124 54	2.73	156 25
0.008	0 28	0.54	30 56	1.09	62 27	1.64	93 58	2.19	125 29	2.74	156 59
0.009	0 31	0.55	31 31	1.10	63 2	1.65	94 32	2.20	126 3	2.75	157 34
0.01	0 34	0.56	32 5	1.11	63 36	1.66	95 6	2.21	126 37	2.76	158 8
0.02	1 9	0.57	32 40	1.12	64 10	1.67	95 41	2.22	127 12	2.77	158 43
0.03	1 43	0.58	33 14	1.13	64 45	1.68	96 15	2.23	127 46	2.78	159 17
0.04	2 18	0.59	33 48	1.14	65 19	1.69	96 50	2.24	128 21	2.79	159 51
0.05	2 52	0.60	34 23	1.15	65 53	1.70	97 24	2.25	128 55	2.80	160 26
0.06	3 26	0.61	34 57	1.16	66 28	1.71	97 59	2.26	129 29	2.81	161 0
0.07	4 1	0.62	35 31	1.17	67 2	1.72	98 33	2.27	130 4	2.82	161 34
0.08	4 35	0.63	36 6	1.18	67 37	1.73	99 7	2.28	130 38	2.83	162 9
0.09	5 9	0.64	36 40	1.19	68 11	1.74	99 42	2.29	131 12	2.84	162 43
0.10	5 44	0.65	37 15	1.20	68 45	1.75	100 16	2.30	131 47	2.85	163 18
0.11	6 18	0.66	37 49	1.21	69 20	1.76	100 50	2.31	132 21	2.86	163 52
0.12	6 53	0.67	38 23	1.22	69 54	1.77	101 25	2.32	132 56	2.87	164 26
0.13	7 27	0.68	38 58	1.23	70 28	1.78	101 59	2.33	133 30	2.88	165 1
0.14	8 1	0.69	39 32	1.24	71 3	1.79	102 34	2.34	134 4	2.89	165 35
0.15	8 36	0.70	40 6	1.25	71 37	1.80	103 8	2.35	134 39	2.90	166 9
0.16	9 10	0.71	40 41	1.26	72 12	1.81	103 42	2.36	135 13	2.91	166 44
0.17	9 44	0.72	41 15	1.27	72 46	1.82	104 17	2.37	135 47	2.92	167 18
0.18	10 19	0.73	41 50	1.28	73 20	1.83	104 51	2.38	136 22	2.93	167 53
0.19	10 53	0.74	42 24	1.29	73 55	1.84	105 25	2.39	136 56	2.94	168 27
0.20	11 28	0.75	42 58	1.30	74 29	1.85	106 0	2.40	137 31	2.95	169 1
0.21	12 2	0.76	43 33	1.31	75 3	1.86	106 34	2.41	138 5	2.96	169 36
0.22	12 36	0.77	44 7	1.32	75 38	1.87	107 9	2.42	138 39	2.97	170 10
0.23	13 11	0.78	44 41	1.33	76 12	1.88	107 43	2.43	139 14	2.98	170 44
0.24	13 45	0.79	45 16	1.34	76 47	1.89	108 17	2.44	139 48	2.99	171 19
0.25	14 19	0.80	45 50	1.35	77 21	1.90	108 52	2.45	140 22	3.00	171 53
0.26	14 54	0.81	46 25	1.36	77 55	1.91	109 26	2.46	140 57	3.01	172 28
0.27	15 28	0.82	46 59	1.37	78 30	1.92	110 0	2.47	141 31	3.02	173 2
0.28	16 3	0.83	47 33	1.38	79 4	1.93	110 35	2.48	142 6	3.03	173 36
0.29	16 37	0.84	48 8	1.39	79 38	1.94	111 9	2.49	142 40	3.04	174 11
0.30	17 11	0.85	48 42	1.40	80 13	1.95	111 44	2.50	143 14	3.05	174 45
0.31	17 46	0.86	49 16	1.41	80 47	1.96	112 18	2.51	143 49	3.06	175 20
0.32	18 20	0.87	49 51	1.42	81 22	1.97	112 52	2.52	144 23	3.07	175 54
0.33	18 54	0.88	50 25	1.43	81 56	1.98	113 27	2.53	144 57	3.08	176 28
0.34	19 29	0.89	51 0	1.44	82 30	1.99	114 1	2.54	145 32	3.09	177 3
0.35	20 3	0.90	51 34	1.45	83 5	2.00	114 35	2.55	146 6	3.10	177 37
0.36	20 38	0.91	52 8	1.46	83 39	2.01	115 10	2.56	146 41	3.11	178 11
0.37	21 12	0.92	52 43	1.47	84 13	2.02	115 44	2.57	147 15	3.12	178 46
0.38	21 46	0.93	53 17	1.48	84 48	2.03	116 19	2.58	147 49	3.13	179 20
0.39	22 21	0.94	53 51	1.49	85 22	2.04	116 53	2.59	148 24	3.14	179 55
0.40	22 55	0.95	54 26	1.50	85 57	2.05	117 27	2.60	148 58	3.15	180 29
0.41	23 29	0.96	55 0	1.51	86 31	2.06	118 2	2.61	149 32	3.1416	180 π
0.42	24 4	0.97	55 35	1.52	87 5	2.07	118 36	2.62	150 7	6.2832	360 2π
0.43	24 38	0.98	56 9	1.53	87 40	2.08	119 11	2.63	150 41	9.4248	540 3π
0.44	25 13	0.99	56 43	1 54	88 14	2.09	119 45	2.64	151 16	12.5664	720 4π
0.45	25 47	1.00	57 18	1.55	88 49	2.10	120 19	2.65	151 50	15.7080	900 5π
0.46	26 21	1.01	57 52	1.56	89 23	2.11	120 54	2.66	152 24	18.8496	1080 6π

Table 6. Natural values of trigonometric functions for angles in radians (0.02–1.60 radians)

Rad.	Sine	Cosine	Tangent	Cotangent	Secant	Cosecant
0.02	0.020	1.000	0.020	50.0 *	1.00	50.0 *
0.04	0.040	0.999	0.040	25.0 *	1.00	25.0 *
0.06	0.060	0.998	0.060	16.6 *	1.00	16.7 *
0.08	0.080	0.997	0.080	12.5 *	1.00	12.5 *
0.10	0.100	0.995	0.100	9.97 *	1.00	10.0 *
0.12	0.120	0.993	0.121	8.29 *	1.01	8.35 *
0.14	0.140	0.990	0.141	7.10 *	1.01	7.17 *
0.16	0.159	0.987	0.161	6.20 *	1.01	6.28 *
0.18	0.179	0.984	0.182	5.50 *	1.02	5.59 *
0.20	0.199	0.980	0.203	4.93	1.02	5.03
0.22	0.218	0.976	0.224	4.47	1.02	4.58
0.24	0.238	0.971	0.245	4.09	1.03	4.21
0.26	0.257	0.966	0.266	3.76	1.03	3.89
0.28	0.276	0.961	0.288	3.48	1.04	3.62
0.30	0.296	0.955	0.309	3.23	1.05	3.38
0.35	0.343	0.939	0.365	2.74	1.06	2.92
0.40	0.389	0.921	0.423	2.37	1.09	2.57
0.45	0.435	0.900	0.483	2.07	1.11	2.30
0.50	0.479	0.878	0.546	1.83	1.14	2.09
0.55	0.523	0.853	0.613	1.63	1.17	1.91
0.60	0.565	0.825	0.684	1.46	1.21	1.77
0.65	0.605	0.796	0.760	1.32	1.26	1.65
0.70	0.644	0.765	0.842	1.19	1.31	1.55
0.75	0.682	0.732	0.932	1.073	1.37	1.47
0.80	0.717	0.697	1.030	0.971	1.44	1.39
0.85	0.751	0.660	1.14	0.878	1.52	1.33
0.90	0.783	0.622	1.26	0.794	1.61	1.28
0.95	0.813	0.582	1.40	0.715	1.72	1.23
1.00	0.841	0.540	1.56	0.642	1.85	1.19
1.05	0.867	0.498	1.74	0.574	2.01	1.15
1.10	0.891	0.454	1.96	0.509	2.20	1.12
1.15	0.913	0.408	2.23	0.448	2.45	1.10
1.20	0.932	0.362	2.57	0.389	2.76	1.07
1.22	0.939	0.344	2.73	0.366	2.91	1.06
1.24	0.946	0.325	2.91	0.343	3.07	1.06
1.26	0.952	0.306	3.11	0.321	3.27	1.05
1.28	0.958	0.287	3.34	0.299	3.49	1.04
1.30	0.963	0.267	3.60	0.278	3.74	1.04
1.32	0.969	0.248	3.90	0.256	4.03	1.03
1.34	0.973	0.229	4.26	0.235	4.37	1.03
1.36	0.978	0.209	4.67	0.214	4.78	1.02
1.38	0.982	0.190	5.18 *	0.193	5.28 *	1.02
1.40	0.985	0.170	5.80 *	0.172	5.88 *	1.01
1.42	0.989	0.150	6.58 *	0.152	6.66 *	1.01
1.44	0.991	0.130	7.60 *	0.132	7.67 *	1.01
1.46	0.994	0.111	8.99 *	0.111	9.04 *	1.01
1.48	0.996	0.091	10.98 *	0.091	11.0 *	1.00
1.50	0.997	0.071	14.1 *	0.071	14.1 *	1.00
1.52	0.999	0.051	19.7 *	0.051	19.7 *	1.00
1.54	1.000	0.031	32.5 *	0.031	32.5 *	1.00
1.56	1.000	0.011	92.6 *	0.011	92.6 *	1.00
1.58	1.000	− 0.009	− 109.0 *	− 0.009	− 109.0 *	1.00
1.60	1.000	− 0.029	− 34.2 *	− 0.029	− 34.2 *	1.00

* Do not interpolate here.

Table 6. Natural values of trigonometric functions for angles in radians (1.62–3.20 radians) — *Continued*

Rad.	Sine	Cosine	Tangent	Cotangent	Secant	Cosecant
1.62	0.999	− 0.049	− 20.3 *	− 0.049	− 20.4 *	1.00
1.64	0.998	− 0.069	− 14.4 *	− 0.069	− 14.5 *	1.00
1.66	0.996	− 0.089	− 11.2 *	− 0.089	− 11.3 *	1.00
1.68	0.994	− 0.109	− 9.13 *	− 0.110	− 9.19 *	1.01
1.70	0.992	− 0.129	− 7.70 *	− 0.130	− 7.76 *	1.01
1.72	0.989	− 0.149	− 6.65 *	− 0.150	− 6.73 *	1.01
1.74	0.986	− 0.168	− 5.85 *	− 0.171	− 5.94 *	1.01
1.76	0.982	− 0.188	− 5.23 *	− 0.191	− 5.32 *	1.02
1.78	0.978	− 0.208	− 4.71 *	− 0.212	− 4.81 *	1.02
1.80	0.974	− 0.227	− 4.29 *	− 0.233	− 4.40 *	1.03
1.82	0.969	− 0.247	− 3.93	− 0.255	− 4.05	1.03
1.84	0.964	− 0.266	− 3.63	− 0.276	− 3.76	1.04
1.86	0.959	− 0.285	− 3.36	− 0.297	− 3.51	1.04
1.88	0.953	− 0.304	− 3.13	− 0.319	− 3.29	1.05
1.90	0.946	− 0.323	− 2.93	− 0.342	− 3.09	1.06
1.95	0.929	− 0.370	− 2.51	− 0.398	− 2.70	1.08
2.00	0.909	− 0.416	− 2.19	− 0.458	− 2.40	1.10
2.05	0.887	− 0.461	− 1.92	− 0.520	− 2.17	1.13
2.10	0.863	− 0.505	− 1.71	− 0.585	− 1.98	1.16
2.15	0.837	− 0.547	− 1.53	− 0.654	− 1.83	1.20
2.20	0.808	− 0.588	− 1.37	− 0.728	− 1.70	1.24
2.25	0.778	− 0.628	− 1.24	− 0.807	− 1.59	1.28
2.30	0.746	− 0.666	− 1.12	− 0.894	− 1.50	1.34
2.35	0.711	− 0.703	− 1.01	− 0.988	− 1.43	1.41
2.40	0.675	− 0.737	− 0.916	− 1.09	− 1.36	1.48
2.45	0.638	− 0.770	− 0.828	− 1.21	− 1.30	1.57
2.50	0.598	− 0.801	− 0.747	− 1.34	− 1.25	1.67
2.55	0.558	− 0.830	− 0.672	− 1.49	− 1.20	1.79
2.60	0.515	− 0.857	− 0.602	− 1.66	− 1.17	1.94
2.65	0.472	− 0.881	− 0.535	− 1.87	− 1.13	2.12
2.70	0.427	− 0.904	− 0.473	− 2.12	− 1.11	2.34
2.75	0.382	− 0.924	− 0.413	− 2.42	− 1.08	2.62
2.80	0.335	− 0.942	− 0.356	− 2.81	− 1.06	2.99
2.82	0.316	− 0.949	− 0.333	− 3.00	− 1.05	3.16
2.84	0.297	− 0.955	− 0.311	− 3.21	− 1.05	3.37
2.86	0.278	− 0.961	− 0.289	− 3.46	− 1.04	3.60
2.88	0.259	− 0.966	− 0.268	− 3.73	− 1.04	3.87
2.90	0.239	− 0.971	− 0.246	− 4.06	− 1.03	4.18
2.92	0.220	− 0.976	− 0.225	− 4.44	− 1.03	4.55
2.94	0.200	− 0.980	− 0.204	− 4.89	− 1.02	4.88
2.96	0.181	− 0.984	− 0.184	− 5.45	− 1.02	5.54
2.98	0.161	− 0.987	− 0.163	− 6.13 *	− 1.01	6.21 *
3.00	0.141	− 0.990	− 0.143	− 7.01 *	− 1.01	7.08 *
3.02	0.121	− 0.993	− 0.122	− 8.18 *	− 1.01	8.24 *
3.04	0.101	− 0.995	− 0.102	− 9.82 *	− 1.01	9.87 *
3.06	0.081	− 0.997	− 0.082	− 12.3 *	− 1.00	12.3 *
3.08	0.062	− 0.998	− 0.062	− 16.2 *	− 1.00	16.2 *
3.10	0.042	− 0.999	− 0.042	− 24.0 *	− 1.00	24.0 *
3.12	0.022	− 1.000	− 0.022	− 46.4 *	− 1.00	46.5 *
3.14	0.001	− 1.000	− 0.001	− 688.0 *	− 1.00	688.0 *
3.16	− 0.018	− 0.999	0.018	54.6 *	− 1.00	− 54.6 *
3.18	− 0.038	− 0.999	0.038	26.0 *	− 1.00	− 26.0 *
3.20	− 0.058	− 0.998	0.059	17.0 *	− 1.00	− 17.1 *

* Do not interpolate here.

Table 7. Napierian (natural) logarithms. Base $e = 2.71828$

N	0	1	2	3	4	5	6	7	8	9	Prop. parts				
											1	2	3	4	5
1.0	0.0000	0.0100	0.0198	0.0296	0.0392	0.0488	0.0583	0.0677	0.0770	0.0862					
1.1	0.0953	0.1044	0.1133	0.1222	0.1310	0.1398	0.1484	0.1570	0.1655	0.1740					
1.2	0.1823	0.1906	0.1989	0.2070	0.2151	0.2231	0.2311	0.2390	0.2469	0.2546					
1.3	0.2624	0.2700	0.2776	0.2852	0.2927	0.3001	0.3075	0.3148	0.3221	0.3293					
1.4	0.3365	0.3436	0.3507	0.3577	0.3646	0.3716	0.3784	0.3853	0.3920	0.3988			Interpolate		
1.5	0.4055	0.4121	0.4187	0.4253	0.4318	0.4383	0.4447	0.4511	0.4574	0.4637					
1.6	0.4700	0.4762	0.4824	0.4886	0.4947	0.5008	0.5068	0.5128	0.5188	0.5247					
1.7	0.5306	0.5365	0.5423	0.5481	0.5539	0.5596	0.5653	0.5710	0.5766	0.5822					
1.8	0.5878	0.5933	0.5988	0.6043	0.6098	0.6152	0.6206	0.6259	0.6313	0.6366					
1.9	0.6419	0.6471	0.6523	0.6575	0.6627	0.6678	0.6729	0.6780	0.6831	0.6881					
2.0	0.6931	0.6981	0.7031	0.7080	0.7129	0.7178	0.7227	0.7275	0.7324	0.7372					
2.1	0.7419	0.7467	0.7514	0.7561	0.7608	0.7655	0.7701	0.7747	0.7793	0.7839	5	9	14	19	23
2.2	0.7885	0.7930	0.7975	0.8020	0.8065	0.8109	0.8154	0.8198	0.8242	0.8286	4	9	13	18	22
2.3	0.8329	0.8372	0.8416	0.8459	0.8502	0.8544	0.8587	0.8629	0.8671	0.8713	4	9	13	17	21
2.4	0.8755	0.8796	0.8838	0.8879	0.8920	0.8961	0.9002	0.9042	0.9083	0.9123	4	8	12	16	20
2.5	0.9163	0.9203	0.9243	0.9282	0.9322	0.9361	0.9400	0.9439	0.9478	0.9517	4	8	12	16	20
2.6	0.9555	0.9594	0.9632	0.9670	0.9708	0.9746	0.9783	0.9821	0.9858	0.9895	4	8	11	15	19
2.7	0.9933	0.9969	1.0006	1.0043	1.0080	1.0116	1.0152	1.0188	1.0225	1.0260	4	7	11	15	18
2.8	1.0296	1.0332	1.0367	1.0403	1.0438	1.0473	1.0508	1.0543	1.0578	1.0613	4	7	11	14	18
2.9	1.0647	1.0682	1.0716	1.0750	1.0784	1.0818	1.0852	1.0886	1.0919	1.0953	3	7	10	14	17
3.0	1.0986	1.1019	1.1053	1.1086	1.1119	1.1151	1.1184	1.1217	1.1249	1.1282	3	7	10	13	16
3.1	1.1314	1.1346	1.1378	1.1410	1.1442	1.1474	1.1506	1.1537	1.1569	1.1600	3	6	10	13	16
3.2	1.1632	1.1663	1.1694	1.1725	1.1756	1.1787	1.1817	1.1848	1.1878	1.1909	3	6	9	12	15
3.3	1.1939	1.1969	1.2000	1.2030	1.2060	1.2090	1.2119	1.2149	1.2179	1.2208	3	6	9	12	15
3.4	1.2238	1.2267	1.2296	1.2326	1.2355	1.2384	1.2413	1.2442	1.2470	1.2499	3	6	9	12	14
3.5	1.2528	1.2556	1.2585	1.2613	1.2641	1.2669	1.2698	1.2726	1.2754	1.2782	3	6	8	11	14
3.6	1.2809	1.2837	1.2865	1.2892	1.2920	1.2947	1.2975	1.3002	1.3029	1.3056	3	5	8	11	14
3.7	1.3083	1.3110	1.3137	1.3164	1.3191	1.3218	1.3244	1.3271	1.3297	1.3324	3	5	8	11	13
3.8	1.3350	1.3376	1.3403	1.3429	1.3455	1.3481	1.3507	1.3533	1.3558	1.3584	3	5	8	10	13
3.9	1.3610	1.3635	1.3661	1.3686	1.3712	1.3737	1.3762	1.3788	1.3813	1.3838	3	5	8	10	13
4.0	1.3863	1.3888	1.3913	1.3938	1.3962	1.3987	1.4012	1.4036	1.4061	1.4085	2	5	7	10	12
4.1	1.4110	1.4134	1.4159	1.4183	1.4207	1.4231	1.4255	1.4279	1.4303	1.4327	2	5	7	10	12
4.2	1.4351	1.4375	1.4398	1.4422	1.4446	1.4469	1.4493	1.4516	1.4540	1.4563	2	5	7	9	12
4.3	1.4586	1.4609	1.4633	1.4656	1.4679	1.4702	1.4725	1.4748	1.4770	1.4793	2	5	7	9	11
4.4	1.4816	1.4839	1.4861	1.4884	1.4907	1.4929	1.4951	1.4974	1.4996	1.5019	2	4	7	9	11
4.5	1.5041	1.5063	1.5085	1.5107	1.5129	1.5151	1.5173	1.5195	1.5217	1.5239	2	4	7	9	11
4.6	1.5261	1.5282	1.5304	1.5326	1.5347	1.5369	1.5390	1.5412	1.5433	1.5454	2	4	6	9	11
4.7	1.5476	1.5497	1.5518	1.5539	1.5560	1.5581	1.5602	1.5623	1.5644	1.5665	2	4	6	8	11
4.8	1.5686	1.5707	1.5728	1.5748	1.5769	1.5790	1.5810	1.5831	1.5851	1.5872	2	4	6	8	10
4.9	1.5892	1.5913	1.5933	1.5953	1.5974	1.5994	1.6014	1.6034	1.6054	1.6074	2	4	6	8	10
5.0	1.6094	1.6114	1.6134	1.6154	1.6174	1.6194	1.6214	1.6233	1.6253	1.6273	2	4	6	8	10
5.1	1.6292	1.6312	1.6332	1.6351	1.6371	1.6390	1.6409	1.6429	1.6448	1.6467	2	4	6	8	10
5.2	1.6487	1.6506	1.6525	1.6544	1.6563	1.6582	1.6601	1.6620	1.6639	1.6658	2	4	6	8	10
5.3	1.6677	1.6696	1.6715	1.6734	1.6752	1.6771	1.6790	1.6808	1.6827	1.6845	2	4	6	7	9
5.4	1.6864	1.6882	1.6901	1.6919	1.6938	1.6956	1.6974	1.6993	1.7011	1.7029	2	4	6	7	9
5.5	1.7047	1.7066	1.7084	1.7102	1.7120	1.7138	1.7156	1.7174	1.7192	1.7210	2	4	5	7	9
5.6	1.7228	1.7246	1.7263	1.7281	1.7299	1.7317	1.7334	1.7352	1.7370	1.7387	2	4	5	7	9
5.7	1.7405	1.7422	1.7440	1.7457	1.7475	1.7492	1.7509	1.7527	1.7544	1.7561	2	3	5	7	9
5.8	1.7579	1.7596	1.7613	1.7630	1.7647	1.7664	1.7681	1.7699	1.7716	1.7733	2	3	5	7	9
5.9	1.7750	1.7766	1.7783	1.7800	1.7817	1.7834	1.7851	1.7867	1.7884	1.7901	2	3	5	7	8
N	0	1	2	3	4	5	6	7	8	9	1	2	3	4	5
											Prop. parts				

Table 7. Napierian logarithms — *Continued*

N	0	1	2	3	4	5	6	7	8	9	1	2	3	4	5
6.0	1.7918	1.7934	1.7951	1.7967	1.7984	1.8001	1.8017	1.8034	1.8050	1.8066	2	3	5	7	8
6.1	1.8083	1.8099	1.8116	1.8132	1.8148	1.8165	1.8181	1.8197	1.8213	1.8229	2	3	5	7	8
6.2	1.8245	1.8262	1.8278	1.8294	1.8310	1.8326	1.8342	1.8358	1.8374	1.8390	2	3	5	6	8
6.3	1.8405	1.8421	1.8437	1.8453	1.8469	1.8485	1.8500	1.8516	1.8532	1.8547	2	3	5	6	8
6.4	1.8563	1.8579	1.8594	1.8610	1.8625	1.8641	1.8656	1.8672	1.8687	1.8703	2	3	5	6	8
6.5	1.8718	1.8733	1.8749	1.8764	1.8779	1.8795	1.8810	1.8825	1.8840	1 8856	2	3	5	6	8
6.6	1.8871	1.8886	1.8901	1.8916	1.8931	1.8946	1.8961	1.8976	1.8991	1 9006	2	3	5	6	8
6.7	1.9021	1.9036	1.9051	1.9066	1.9081	1.9095	1.9110	1.9125	1.9140	1 9155	1	3	4	6	7
6.8	1.9169	1.9184	1.9199	1.9213	1.9228	1.9242	1.9257	1.9272	1.9286	1 9301	1	3	4	6	7
6.9	1.9315	1.9330	1.9344	1.9359	1.9373	1.9387	1.9402	1.9416	1.9430	1.9445	1	3	4	6	7
7.0	1.9459	1.9473	1.9488	1.9502	1.9516	1.9530	1.9544	1.9559	1.9573	1.9587	1	3	4	6	7
7.1	1.9601	1.9615	1.9629	1.9643	1.9657	1.9671	1.9685	1.9699	1.9713	1.9727	1	3	4	6	7
7.2	1.9741	1.9755	1.9769	1.9782	1.9796	1.9810	1.9824	1.9838	1.9851	1.9865	1	3	4	6	7
7.3	1.9879	1.9892	1.9906	1.9920	1.9933	1.9947	1.9961	1.9974	1.9988	2.0001	1	3	4	5	7
7.4	2.0015	2.0028	2.0042	2.0055	2.0069	2.0082	2.0096	2.0109	2.0122	2.0136	1	3	4	5	7
7.5	2.0149	2.0162	2.0176	2.0189	2.0202	2.0215	2.0229	2.0242	2.0255	2.0268	1	3	4	5	7
7.6	2.0281	2.0295	2.0308	2.0321	2.0334	2.0347	2.0360	2.0373	2.0386	2.0399	1	3	4	5	7
7.7	2.0412	2.0425	2.0438	2.0451	2.0464	2.0477	2.0490	2.0503	2.0516	2.0528	1	3	4	5	6
7.8	2.0541	2.0554	2.0567	2.0580	2.0592	2.0605	2.0618	2.0631	2.0643	2.0656	1	3	4	5	6
7.9	2.0669	2.0681	2.0694	2.0707	2.0719	2.0732	2.0744	2.0757	2.0769	2.0782	1	3	4	5	6
8.0	2.0794	2.0807	2.0819	2.0832	2.0844	2.0857	2.0869	2.0882	2.0894	2.0906	1	2	4	5	6
8.1	2.0919	2.0931	2.0943	2.0956	2.0968	2.0980	2.0992	2.1005	2.1017	2.1029	1	2	4	5	6
8.2	2.1041	2.1054	2.1066	2.1078	2.1090	2.1102	2.1114	2.1126	2.1138	2.1150	1	2	4	5	6
8.3	2.1163	2.1175	2.1187	2.1199	2.1211	2.1223	2.1235	2.1247	2.1258	2.1270	1	2	4	5	6
8.4	2.1282	2.1294	2.1306	2.1318	2.1330	2.1342	2.1353	2.1365	2.1377	2.1389	1	2	4	5	6
8.5	2.1401	2.1412	2.1424	2.1436	2.1448	2.1459	2.1471	2.1483	2.1494	2.1506	1	2	4	5	6
8.6	2.1518	2.1529	2.1541	2.1552	2.1564	2.1576	2.1587	2.1599	2.1610	2.1622	1	2	3	5	6
8.7	2.1633	2.1645	2.1656	2.1668	2.1679	2.1691	2.1702	2.1713	2.1725	2.1736	1	2	3	5	6
8.8	2.1748	2.1759	2.1770	2.1782	2.1793	2.1804	2.1815	2.1827	2.1838	2.1849	1	2	3	5	6
8.9	2.1861	2.1872	2.1883	2.1894	2.1905	2.1917	2.1928	2.1939	2.1950	2.1961	1	2	3	4	6
9.0	2.1972	2.1983	2.1994	2.2006	2.2017	2.2028	2.2039	2.2050	2.2061	2.2072	1	2	3	4	6
9.1	2.2083	2.2094	2.2105	2.2116	2.2127	2.2138	2.2148	2.2159	2.2170	2.2181	1	2	3	4	5
9.2	2.2192	2.2203	2.2214	2.2225	2.2235	2.2246	2.2257	2.2268	2.2279	2.2289	1	2	3	4	5
9.3	2.2300	2.2311	2.2322	2.2332	2.2343	2.2354	2.2364	2.2375	2.2386	2.2396	1	2	3	4	5
9.4	2.2407	2.2418	2.2428	2.2439	2.2450	2.2460	2.2471	2.2481	2.2492	2.2502	1	2	3	4	5
9.5	2.2513	2.2523	2.2534	2.2544	2.2555	2.2565	2.2576	2.2586	2.2597	2.2607	1	2	3	4	5
9.6	2.2618	2.2628	2.2638	2.2649	2.2659	2.2670	2.2680	2.2690	2.2701	2.2711	1	2	3	4	5
9.7	2.2721	2.2732	2.2742	2.2752	2.2762	2.2773	2.2783	2.2793	2.2803	2.2814	1	2	3	4	5
9.8	2.2824	2.2834	2.2844	2.2854	2.2865	2.2875	2.2885	2.2895	2.2905	2.2915	1	2	3	4	5
9.9	2.2925	2.2935	2.2946	2.2956	2.2966	2.2976	2.2986	2.2996	2.3006	2.3016	1	2	3	4	5

Table 8. Napierian logarithms of powers of 10

u	Nap. log. 10^u	u	Nap. log. 10^u	u	Nap. log. 10^u	u	Nap. log. 10
0	0.000 000	2.5	5.756 463	5.0	11.512 925	7.5	17.269 388
0.5	1.151 293	3.0	6.907 755	5.5	12.664 218	8.0	18.420 681
1.0	2.302 585	3.5	8.059 048	6.0	13.815 511	8.5	19.571 973
1.5	3.453 878	4.0	9.210 340	6.5	14.966 803	9.0	20.723 266
2.0	4.605 170	4.5	10.361 633	7.0	16.118 096	9.5	21.874 558

Table 9. Exponentials e^u and e^{-u}

u	e^u	e^{-u}	u	e^u	e^{-u}	u	e^u	e^{-u}
0.00	1.000	1.0000	0.50	1.649	0.6065	1.0	2.718	0.3679
0.01	1.010	0.9900	0.51	1.665	0.6005	1.1	3.004	0.3329
0.02	1.020	0.9802	0.52	1.682	0.5945	1.2	3.320	0.3012
0.03	1.030	0.9704	0.53	1.699	0.5886	1.3	3.669	0.2725
0.04	1.041	0.9608	0.54	1.716	0.5827	1.4	4.055	0.2466
0.05	1.051	0.9512	0.55	1.733	0.5769	1.5	4.482	0.2231
0.06	1.062	0.9418	0.56	1.751	0.5712	1.6	4.953	0.2019
0.07	1.073	0.9324	0.57	1.768	0.5655	1.7	5.474	0.1827
0.08	1.083	0.9231	0.58	1.786	0.5599	1.8	6.050	0.1653
0.09	1.094	0.9139	0.59	1.804	0.5543	1.9	6.686	0.1496
0.10	1.105	0.9048	0.60	1.822	0.5488	2.0	7.389	0.1353
0.11	1.116	0.8958	0.61	1.840	0.5434	2.1	8.166	0.1225
0.12	1.127	0.8869	0.62	1.859	0.5379	2.2	9.025	0.1108
0.13	1.139	0.8781	0.63	1.878	0.5326	2.3	9.974	0.1003
0.14	1.150	0.8694	0.64	1.896	0.5273	2.4	11.02	0.09072
0.15	1.162	0.8607	0.65	1.916	0.5220	2.5	12.18	0.08209
0.16	1.174	0.8521	0.66	1.935	0.5169	2.6	13.46	0.07427
0.17	1.185	0.8437	0.67	1.954	0.5117	2.7	14.88	0.06721
0.18	1.197	0.8353	0.68	1.974	0.5066	2.8	16.44	0.06081
0.19	1.209	0.8270	0.69	1.994	0.5016	2.9	18.17	0.05502
0.20	1.221	0.8187	0.70	2.014	0.4966	3.0	20.09	0.04979
0.21	1.234	0.8106	0.71	2.034	0.4916	3.1	22.20	0.04505
0.22	1.246	0.8025	0.72	2.054	0.4868	3.2	24.53	0.04076
0.23	1.259	0.7945	0.73	2.075	0.4819	3.3	27.11	0.03688
0.24	1.271	0.7866	0.74	2.096	0.4771	3.4	29.96	0.03337
0.25	1.284	0.7788	0.75	2.117	0.4724	3.5	33.12	0.03020
0.26	1.297	0.7711	0.76	2.138	0.4677	3.6	36.60	0.02732
0.27	1.310	0.7634	0.77	2.160	0.4630	3.7	40.45	0.02472
0.28	1.323	0.7558	0.78	2.181	0.4584	3.8	44.70	0.02237
0.29	1.336	0.7483	0.79	2.203	0.4538	3.9	49.40	0.02024
0.30	1.350	0.7408	0.80	2.226	0.4493	4.0	54.60	0.01832
0.31	1.363	0.7334	0.81	2.248	0.4449	4.5	90.02	0.01111
0.32	1.377	0.7261	0.82	2.271	0.4404	5.0	148.4	0.00674
0.33	1.391	0.7189	0.83	2.293	0.4360	5.5	244.7	0.00409
0.34	1.405	0.7118	0.84	2.316	0.4317	6.0	403.4	0.00248
0.35	1.419	0.7047	0.85	2.340	0.4274	6.5	665.1	0.00150
0.36	1.433	0.6977	0.86	2.363	0.4232	7.0	1097	0.00091
0.37	1.448	0.6907	0.87	2.387	0.4190	8.0	2981	0.00034
0.38	1.462	0.6839	0.88	2.411	0.4148	9.0	8103	0.00012
0.39	1.477	0.6771	0.89	2.435	0.4107	10.0	22026	0.00005
0.40	1.492	0.6703	0.90	2.460	0.4066	$\pi/4$	2.193	0.45594
0.41	1.507	0.6637	0.91	2.484	0.4025	$2\pi/4$	4.811	0.20788
0.42	1.522	0.6570	0.92	2.509	0.3985	$3\pi/4$	10.55	0.09478
0.43	1.537	0.6505	0.93	2.535	0.3946	$4\pi/4$	23.14	0.04321
0.44	1.553	0.6440	0.94	2.560	0.3906	$5\pi/4$	50.75	0.01970
0.45	1.568	0.6376	0.95	2.586	0.3867	$6\pi/4$	111.3	0.00898
0.46	1.584	0.6313	0.96	2.612	0.3829	$7\pi/4$	244.2	0.00410
0.47	1.600	0.6250	0.97	2.638	0.3791	$8\pi/4$	535.5	0.00187
0.48	1.616	0.6188	0.98	2.664	0.3753	$9\pi/4$	1175	0.00085
0.49	1.632	0.6126	0.99	2.691	0.3716	$10\pi/4$	2576	0.00039

Table 10. Hyperbolic functions

u	Sinh u	Cosh u	Tanh u	u	Sinh u	Cosh u	Tanh u	u	Sinh u	Cosh u	Tanh u
.00	.0000	1.000	.0000	.50	.5211	1.128	.4621	1.0	1.175	1.543	.7616
.01	.0100	1.000	.0100	.51	.5324	1.133	.4700	1.1	1.336	1.669	.8005
.02	.0200	1.000	.0200	.52	.5438	1.138	.4777	1.2	1.509	1.811	.8337
.03	.0300	1.000	.0300	.53	.5552	1.144	.4854	1.3	1.698	1.971	.8617
.04	.0400	1.001	.0400	.54	.5666	1.149	.4930	1.4	1.904	2.151	.8854
.05	.0500	1.001	.0500	.55	.5782	1.155	.5005	1.5	2.129	2.352	.9052
.06	.0600	1.002	.0599	.56	.5897	1.161	.5080	1.6	2.376	2.577	.9217
.07	.0701	1.002	.0699	.57	.6014	1.167	.5154	1.7	2.646	2.828	.9354
.08	.0801	1.003	.0798	.58	.6131	1.173	.5227	1.8	2.942	3.107	.9468
.09	.0901	1.004	.0898	.59	.6248	1.179	.5299	1.9	3.268	3.418	.9562
.10	.1002	1.005	.0997	.60	.6367	1.185	.5370	2.0	3.627	3.762	.9640
.11	.1102	1.006	.1096	.61	.6485	1.192	.5441	2.1	4.022	4.144	.9705
.12	.1203	1.007	.1194	.62	.6605	1.198	.5511	2.2	4.457	4.568	.9757
.13	.1304	1.008	.1293	.63	.6725	1.205	.5581	2.3	4.937	5.037	.9801
.14	.1405	1.010	.1391	.64	.6846	1.212	.5649	2.4	5.466	5.557	.9837
.15	.1506	1.011	.1489	.65	.6967	1.219	.5717	2.5	6.050	6.132	.9866
.16	.1607	1.013	.1587	.66	.7090	1.226	.5784	2.6	6.695	6.769	.9890
.17	.1708	1.014	.1684	.67	.7213	1.233	.5850	2.7	7.406	7.473	.9910
.18	.1810	1.016	.1781	.68	.7336	1.240	.5915	2.8	8.192	8.253	.9926
.19	.1911	1.018	.1878	.69	.7461	1.248	.5980	2.9	9.060	9.115	.9940
.20	.2013	1.020	.1974	.70	.7586	1.255	.6044	3.0	10.02	10.07	.9951
.21	.2115	1.022	.2070	.71	.7712	1.263	.6107	3.1	11.08	11.12	.9960
.22	.2218	1.024	.2165	.72	.7838	1.271	.6169	3.2	12.25	12.29	.9967
.23	.2320	1.027	.2260	.73	.7966	1.278	.6231	3.3	13.54	13.57	.9973
.24	.2423	1.029	.2355	.74	.8094	1.287	.6291	3.4	14.97	15.00	.9978
.25	.2526	1.031	.2449	.75	.8223	1.295	.6352	3.5	16.54	16.57	.9982
.26	.2629	1.034	.2543	.76	.8353	1.303	.6411	3.6	18.29	18.31	.9985
.27	.2733	1.037	.2636	.77	.8484	1.311	.6469	3.7	20.21	20.24	.9988
.28	.2837	1.039	.2729	.78	.8615	1.320	.6527	3.8	22.34	22.36	.9990
.29	.2941	1.042	.2821	.79	.8748	1.329	.6584	3.9	24.69	24.71	.9992
.30	.3045	1.045	.2913	.80	.8881	1.337	.6640	4.0	27.29	27.31	.9993
.31	.3150	1.048	.3004	.81	.9015	1.346	.6696	4.1	30.16	30.18	.9995
.32	.3255	1.052	.3095	.82	.9150	1.355	.6751	4.2	33.34	33.35	.9996
.33	.3360	1.055	.3185	.83	.9286	1.365	.6805	4.3	36.84	36.86	.9996
.34	.3466	1.058	.3275	.84	.9423	1.374	.6858	4.4	40.72	40.73	.9997
.35	.3572	1.062	.3364	.85	.9561	1.384	.6911	4.5	45.00	45.01	.9998
.36	.3678	1.066	.3452	.86	.9700	1.393	.6963	4.6	49.74	49.75	.9998
.37	.3785	1.069	.3540	.87	.9840	1.403	.7014	4.7	54.97	54.98	.9998
.38	.3892	1.073	.3627	.88	.9981	1.413	.7064	4.8	60.75	60.76	.9999
.39	.4000	1.077	.3714	.89	1.012	1.423	.7114	4.9	67.14	67.15	.9999
.40	.4108	1.081	.3800	.90	1.027	1.433	.7163	5.0	74.20	74.21	.9999
.41	.4216	1.085	.3885	.91	1.041	1.443	.7211	5.1	82.01	82.01	.9999
.42	.4325	1.090	.3969	.92	1.055	1.454	.7259	5.2	90.63	90.64	.9999
.43	.4434	1.094	.4053	.93	1.070	1.465	.7306	5.3	100.17	100.17	1.0000
.44	.4543	1.098	.4136	.94	1.085	1.475	.7352	5.4	110.70	110.71	1.0000
.45	.4653	1.103	.4219	.95	1.099	1.486	.7398	5.5	122.34	122.35	1.0000
.46	.4764	1.108	.4301	.96	1.114	1.497	.7443	5.6	135.21	135.22	1.0000
.47	.4875	1.112	.4382	.97	1.129	1.509	.7487	5.7	149.43	149.44	1.0000
.48	.4986	1.117	.4462	.98	1.145	1.520	.7531	5.8	165.15	165.15	1.0000
.49	.5098	1.122	.4542	.99	1.160	1.531	.7574	5.9	182.52	182.52	1.0000

APPENDIX. TRIPLE INTEGRATION

1. Definite triple integral. An extension of the ideas discussed in Art. 285 leads to the concept of the triple integral of a function throughout a three-dimensional region R. Suppose that a closed region R is bounded by simple surfaces (that is, surfaces having a tangent plane at every point) and, as in Art. 284, that any secant cuts the bounding surfaces in two points only. Using rectangular coördinates, suppose that R is divided by planes parallel to the coördinate planes into rectangular parallelepipeds having the dimensions Δx, Δy, Δz. The element of volume is

$$\Delta V = \Delta z \, \Delta y \, \Delta x.$$

This form of ΔV indicates that the integration is to be performed first with respect to z, next with respect to y, and finally with respect to x. Other forms with obvious meanings are

$$\Delta z \, \Delta x \, \Delta y, \ \Delta y \, \Delta z \, \Delta x, \ \Delta y \, \Delta x \, \Delta z, \ \Delta x \, \Delta y \, \Delta z, \ \Delta x \, \Delta z \, \Delta y.$$

The choice depends on convenience for the problem in hand.

Let $f(x, y, z)$ be a continuous function of the variables for all points inside and on the boundary of R. Choose a point (x, y, z) of the element of volume $\Delta z \, \Delta y \, \Delta x$. Form the product

$$f(x, y, z)\Delta z \, \Delta y \, \Delta x,$$

and similar products for all other elements of the volume within R. Summing up these products, we have

$$\sum \sum \sum f(x, y, z)\Delta z \, \Delta y \, \Delta x.$$

It can be shown that this sum has a unique limit as $\Delta z \longrightarrow 0$, $\Delta y \longrightarrow 0$, $\Delta x \longrightarrow 0$. This is designated by writing

$$(A) \qquad \lim_{\substack{\Delta z \to 0 \\ \Delta y \to 0 \\ \Delta x \to 0}} \sum \sum \sum f(x, y, z)\Delta z \, \Delta y \, \Delta x$$

$$= \iiint\limits_{R} f(x, y, z)dz \, dy \, dx,$$

which is called the triple integral of the function $f(x, y, z)$ taken over the region R. The method of evaluating definite triple integrals is explained on page 522.

When $f(x, y, z) = 1$, (A) gives the volume of the region R. In rectangular coördinates

$$(B) \qquad V = \iiint\limits_{R} dz \, dy \, dx.$$

579

It is readily seen that if cylindrical coördinates are used the volume is given by

(C) $$V = \iiint_R \rho \, dz \, d\rho \, d\theta.$$

Extending (D) on page 529 to three dimensions the formulas for the centroid (center of gravity) $(\bar{x}, \bar{y}, \bar{z})$ of a homogeneous solid may be written in the form

(D) $$V\bar{x} = \iiint_R x \, dz \, dy \, dx, \quad V\bar{y} = \iiint_R y \, dz \, dy \, dx,$$

$$V\bar{z} = \iiint_R z \, dz \, dy \, dx.$$

If the density of a solid at a point (x, y, z) is a function $\partial(x, y, z)$, the mass of the solid is given by

(E) $$\textbf{Mass} = \iiint_R \partial(x, y, z) dz \, dy \, dx.$$

In space of three dimensions it is possible theoretically to consider the moment of inertia of a solid with respect to a plane, or a line, or a point. If the density is given by the function $\partial(x, y, z)$, the moment of inertia with respect to the xy-plane is given by

(F) $$I_{xy} = \iiint_R \partial(x, y, z) z^2 \, dz \, dy \, dx.$$

Similarly,

(G) $$I_{yz} = \iiint_R \partial(x, y, z) x^2 \, dz \, dy \, dx.$$

(H) $$I_{xz} = \iiint_R \partial(x, y, z) y^2 \, dz \, dy \, dx.$$

With respect to the coördinate axes the formulas are

(I) $$I_x = \iiint_R \partial(x, y, z)(y^2 + z^2) dz \, dy \, dx.$$

(J) $$I_y = \iiint_R \partial(x, y, z)(z^2 + x^2) dz \, dy \, dx.$$

(K) $$I_z = \iiint_R \partial(x, y, z)(x^2 + y^2) dz \, dy \, dx.$$

The moment of inertia with respect to the origin is given by

(L) $$I_o = \iiint_R \partial(x, y, z)(x^2 + y^2 + z^2)dz\, dy\, dx.$$

Example 1. Find the volume of the solid bounded by the surfaces

$$z = 4 - x^2 - \tfrac{1}{4} y^2 \qquad\qquad (1)$$
$$z = 3\,x^2 + \tfrac{1}{4} y^2 \qquad\qquad (2)$$

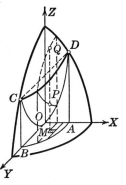

Fig. 1

Solution. Fig. 1 shows one fourth of the volume bounded by the two elliptic paraboloids. Eliminating z between (1) and (2), we have

$$4\,x^2 + \tfrac{1}{2} y^2 = 4, \qquad\qquad (3)$$

which is the equation of the cylinder $ABCD$ that passes through the intersection of (1) and (2) and has its elements parallel to OZ.

The volume is given by

$$V = 4\int_0^1 \int_0^{2\sqrt{2(1-x^2)}} \int_{3\,x^2+\frac{1}{4}y^2}^{4-x^2-\frac{1}{4}y^2} dz\, dy\, dx.$$

The notation indicates that the integration is to be performed first with respect to z, that is, by summing the elements of volume $dz\, dy\, dx$ in a column of base $dy\, dx$ from the surface (2) to the surface (1) (MP to MQ in the figure). The limits for z are, then, the right-hand members of these equations. Thus we find

$$V = 4\int_0^1 \int_0^{2\sqrt{2(1-x^2)}} (4 - 4\,x^2 - \tfrac{1}{2} y^2)dy\, dx. \qquad\qquad (4)$$

The limits on this double integral are those for the region OAB, the portion of the area of the base of the cylinder (3) which lies in the first quadrant. Working out the double integral (4), the result is

$$V = 4\sqrt{2}\,\pi = 17.77 \text{ cubic units.}$$

Example 2. Find the centroid of the volume bounded by the surface $az = a^2 - y^2$ and the planes $y + z = a$, $x = 0$, $x = 2\,a$.

Solution. Proceeding as in Example 1, the volume is given by

$$V = \int_0^{2a} \int_0^a \int_{a-y}^{a-\frac{y^2}{a}} dz\, dy\, dx.$$

Evaluating this integral, we find $V = \tfrac{1}{3} a^3$. Refer to Fig. 2, page 582.

It is clear from the symmetry of the figure that $\bar{x} = a$. The second of formulas (D), page 580, gives

$$\tfrac{1}{3} a^3 \bar{y} = \int_0^{2a} \int_0^a \int_{a-y}^{a-\frac{y^2}{a}} y\, dz\, dy\, dx = \tfrac{1}{6} a^4,$$

whence $\bar{y} = \tfrac{1}{2} a.$

Similarly, $\frac{1}{3} a^3 \bar{z} = \int_0^{2a} \int_0^a \int_{a-y}^{a-\frac{y^2}{a}} z \, dz \, dy \, dx = \frac{1}{5} a^4,$

whence $\bar{z} = \frac{3}{5} a.$

Hence the centroid is the point $(a, \frac{1}{2} a, \frac{3}{5} a).$

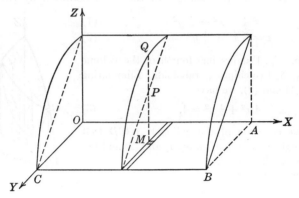

Fig. 2

2. Spherical coördinates. A point in space may be located by spherical coördinates (r, θ, ϕ) as follows. (See Fig. 3.)

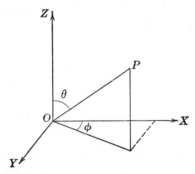

Fig. 3

$r = OP =$ distance from the origin = radius vector,

$\theta =$ angle between the z-axis and $OP =$ co-latitude,

$\phi =$ angle between the x-axis and the projection of OP on the xy-plane = longitude.

The variation of these coördinates is subject to the following restrictions.

$$r \geqq 0, \, 0° \leqq \theta \leqq 180°, \, 0° \leqq \phi < 360°.$$

If the rectangular coördinates of P are (x, y, z), the following relations appear from the figure:

$$x = r \sin \theta \cos \phi, \; y = r \sin \theta \sin \phi, \; z = r \cos \theta.$$

In spherical coördinates an element of volume is bounded as follows:
By two spheres of radii r and $r + \Delta r$;
by two cones with half vertex angles θ and $\theta + \Delta\theta$;
by two planes through the z-axis making with the xz-plane angles
     ϕ and $\phi + \Delta\phi$.

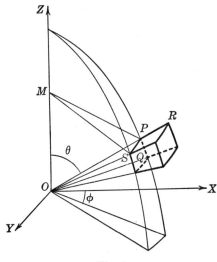

Fig. 4

It appears from Fig. 4 that three edges of the element of volume are

$$PR = \Delta r, \quad \text{(straight line)}$$
$$PS = r \sin \theta \, \Delta\phi, \quad \text{(arc of circle)}$$
$$PQ = r \, \Delta\theta. \quad \text{(arc of circle)}$$

The volume of a rectangular parallelepiped with these edges is

$$r^2 \sin \theta \, \Delta r \, \Delta\theta \, \Delta\phi$$

and this is an approximation to the value of ΔV.

     The proper limiting process shows that the volume of a region R is given by

(M) $$V = \iiint\limits_{R} r^2 \sin \theta \, dr \, d\theta \, d\phi.$$

Example. Find the mass of a spherical shell of inside radius a and outside radius b if the density at any point is proportional to the distance from the center.

Solution. Here $\partial = kr$, and by (E), p. 580,

$$\text{Mass} = \iiint\limits_{R} kr \, dV.$$

In spherical coördinates this becomes

$$\text{Mass} = \int_0^{2\pi} \int_0^{\pi} \int_a^b kr^3 \sin\theta \, dr \, d\theta \, d\phi.$$

Working out the integral, the result is

$$\text{Mass} = k\pi(b^4 - a^4) \text{ units.}$$

For a solid sphere, $a = 0$ and the result may be written

$$\text{Mass} = \tfrac{3}{4} kbV, \text{ where } V \text{ is the volume.}$$

PROBLEMS

1. Find the volume of the solid bounded above by the cylinder $z = 4 - x^2$ and below by the elliptic paraboloid $z = 3x^2 + y^2$.   *Ans.* 4π.

2. a. Find the volume of the solid below the cylindrical surface $x^2 + z = 4$, above the plane $x + z = 2$, and included between the planes $y = 0$, $y = 2$.

Ans. 9.

b. Find the centroid.

Ans. $(\tfrac{1}{2}, 1, \tfrac{12}{5})$.

3. Using cylindrical coördinates, find by triple integration the volume bounded by the cylinder $x^2 + y^2 = 2a^2$ and the hyperboloid $x^2 + y^2 - z^2 = a^2$.

Ans. $\tfrac{4}{3}\pi a^3$.

4. Find the mass of a spherical shell of inside radius a and outside radius b if the density at any point is inversely proportional to the distance from the center.   *Ans.* $2\pi k(b^2 - a^2)$.

5. Show that the volume inside a sphere of radius $2a$ and outside a cylinder of radius a whose axis is a diameter of the sphere is $4\pi\sqrt{3}\,a^3$. Use spherical coördinates.

6. The vertex angle of a cone is $90°$ and the altitude is h. If the density at a point is k times the distance from the vertex of the cone, show that the mass is $\tfrac{1}{6}\pi kh^4(2\sqrt{2} - 1)$.

7. Show that I_z for the sphere $x^2 + y^2 + z^2 = a^2$ is $\tfrac{2}{5} Va^2$.

8. In (G), page 580, let $\partial(x, y, z) = 1$ and R be the volume of the cylinder $x^2 + y^2 = a^2$ between the planes $z = 0$ and $z = h$. Use cylindrical coördinates to show that $I_{yz} = \tfrac{1}{4}\pi ha^4$.

9. A homogeneous tetrahedron is bounded by the coördinate planes and the plane $\frac{x}{a} + \frac{y}{b} + \frac{z}{c} = 1$. Show that the centroid is $(\frac{1}{4} a, \frac{1}{4} b, \frac{1}{4} c)$.

10. Assume that the density of the tetrahedron in Problem 9 is $\partial = kx$.

a. Show that the mass is $M = \frac{1}{24} ka^2bc$.
b. Show that $\bar{x} = \frac{2}{5} a$.
c. Show that $I_{yz} = \frac{1}{5} Ma^2$.

11. In a hemisphere of radius a the density varies as the distance from the center. $\partial = kr$.

a. Show that the mass is $\frac{1}{2} \pi ka^4$.
b. Show that the distance of the centroid from the base is $\frac{2}{5} a$.

12. A solid lies above $z = 0$, below the cone $z^2 = x^2 + y^2$, and within the cylinder $x^2 + y^2 = 2 ax$.

a. Find the volume.   *Ans.* $\frac{32}{9} a^3$.
b. Find the centroid.   *Ans.* $(\frac{6}{5} a, 0, \frac{27}{128} \pi a)$.

13. A sphere of radius a is cut by a cone whose axis is a diameter OZ of the sphere and whose vertex O lies on the surface. The vertex angle of the cone is 60°.

a. Show that the volume of that part of the sphere which lies within the cone is $\frac{7}{12} \pi a^3$.
b. Show that the distance of the centroid from the vertex of the cone is $\frac{37}{28} a$.
c. Show that the moment of inertia with respect to OZ is $\frac{67}{480} \pi a^5$.

14. The base of a homogeneous solid lies in the xy-plane. The solid lies inside the cylinder $x^2 + y^2 = 2 ax$ and under the sphere $x^2 + y^2 + z^2 = 4 a^2$. Show that the distance from the base to the center of gravity is approximately $0.8 a$.

15. In Problem 1 show that integration with respect to z gives (without further integration) $V = 4 A - 4 I_y - I_x$, where A is the area of the ellipse $4 x^2 + y^2 = 4$, and I_x and I_y are the moments of inertia for this ellipse as given by *(E)*, page 534.

16. Using rectangular coördinates, find the volume of the solid bounded by the paraboloid $x^2 + y^2 = az$, the cylinder $x^2 + y^2 = 2 ax$, and the plane $z = 0$.

Ans. $\dfrac{3 \pi a^3}{2}$.

17. Solve the preceding problem using cylindrical coördinates.

18. Find the centroid of the solid described in Problem 16.

Ans. $\left(\dfrac{4 a}{3}, 0, \dfrac{10 a}{9}\right)$.

19. A homogeneous tetrahedron is bounded by the coördinate planes and the plane $x + y + z = a$. Find I_z.   *Ans.* $\dfrac{Ma^2}{5}$.

20. Assume that the density of the tetrahedron in Problem 19 is $\partial = kx$.

$$Ans. \quad \frac{4\,Ma^2}{15}.$$

21. Find the volume below the plane $z = \rho \cos \theta$ and above the paraboloid $z = \rho^2$.

$$Ans. \quad \frac{\pi}{32}.$$

22. A homogeneous tetrahedron is bounded by the planes $x + y + z = 2\,a$, $x + y - z = 0$, $x = 0$, $y = 0$.

a. Find the volume.

$$Ans. \quad \frac{a^3}{3}.$$

b. Find the centroid.

$$Ans. \quad \left(\frac{a}{4},\ \frac{a}{4},\ a\right).$$

23. A solid is bounded by the hyperboloid $z^2 = a^2 + \rho^2$ and the upper nappe of the cone $z^2 = 2\,\rho^2$.

a. Find the volume.

$$Ans. \quad \frac{2\,\pi a^3(\sqrt{2} - 1)}{3}.$$

b. Find the centroid.

$$Ans. \quad \bar{z} = \frac{3\,a(\sqrt{2} + 1)}{8}.$$

24. The radius of the base of a right circular cone (of density $\partial = k$) is a and the altitude is h. If the slant height is a given constant, show that the maximum moment of inertia of the cone with respect to its axis is $\dfrac{\pi k a^5}{20}$.

ANSWERS TO SELECTED PROBLEMS

Page 8

1. c. $x > 4$, $x < -5$. **e.** $-2 < x < 3$. **f.** $x > 4$.
2. a. 2, 4. **b.** ± 6. **c.** None. **d.** $\pm 2, \pm 3$.
3. a. $2 < x < 4$. **b.** $-2.5 < x < -1.5$. **c.** $2 < x < 6$. **9. a.** 16. **b.** 30. **c.** 15. **d.** 20.

Pages 16–17

7. b. $\sqrt{10}$. **c.** 10. **10.** (1, 11). **11.** (12, 11). **13.** (0, 4).

Pages 24–25

4. $77° 37'$. **5.** $\frac{1}{2}$ or -2. **6. a.** $\frac{7}{3}$ or $-\frac{3}{7}$. **b.** 3 or $-\frac{1}{3}$. **c.** 2 or $-\frac{1}{2}$. **7.** -13 or $\frac{7}{11}$.
8. 2. **12. a.** $63° 26', 75° 58', 40° 36'$. **b.** $49° 46', 35° 50', 94° 24'$. **c.** $76° 12', 55° 37', 48° 11'$.
d. $22° 22', 119° 45', 37° 53'$. **15.** (2, 20). **17.** (5, 7).

Page 32

13. c. $15° 15', 81° 52', 82° 53'$.
15. $4x + y - 22 = 0$, $4x + y + 12 = 0$, $x - 4y + 20 = 0$, $x - 4y - 14 = 0$.
16. $3x + 2y = 24$ or $27x + 2y = 72$.

Page 34

5. a. $(\frac{8}{3}, \frac{8}{3})$, $r = \frac{1}{3}\sqrt{170}$. **b.** (2, 1), $r = 5$. **c.** $(-\frac{61}{9}, \frac{47}{3})$, $r = \frac{1}{9}\sqrt{37{,}570}$. **6. a.** (0, 4).
8. $5\sqrt{5}$. **10.** $(\frac{100}{37}, \frac{171}{37})$, $(\frac{260}{37}, -\frac{29}{37})$, $(-\frac{160}{37}, \frac{15}{37})$.

Page 43

4. a. 16. **b.** $\frac{23}{2}$. **c.** 12. **d.** 28. **5.** 134.5.
7. $77x + 99y - 254 = 0$, $7x - 56y - 3 = 0$, $7x - y - 20 = 0$.
8. $27x - 21y + 46 = 0$, $32x + 4y - 153 = 0$, $x + 7y + 4 = 0$.
12. $4x + 7y - 12 = 0$. **13.** (0, 3), $r = \sqrt{5}$.
14. a. $59° 29', 30° 31', 90°$.
b. $x + 8y - 16 = 0$, $7x - 9y + 18 = 0$, $7x + y - 2 = 0$.
c. (0, 2), $r = 2$. **e.** $(-\frac{8}{5}, \frac{16}{5})$, $(\frac{6}{5}, \frac{18}{5})$, $(\frac{10}{13}, \frac{2}{13})$.

Page 49

8. $x + 2$. **9.** $\dfrac{\pm 4y}{x(x - 4)}$.

Page 53

8. $\frac{1}{2}$. **9.** $a = \sqrt{28}$, $b = \sqrt{\frac{28}{3}}$. **10.** $a = 5$, $b = 3$, $c = -1$.
16. $7x^2 + 14xy - 7y^2 - 12x + 36y - 36 = 0$.

Pages 56–57

1. a. $26x - 16y = 59$. **b.** $x - y = 0$. **2. a.** $y^2 + 44x - 220 = 0$. **b.** $x^2 - 6y + 39 = 0$.
3. a. $y^2 + 48x - 288 = 0$. **b.** $x^2 - 10y + 55 = 0$. **4.** $xy = 16$.
5. a. $x^2 + y^2 - 8x + 10y - 59 = 0$. **e.** $x^2 + y^2 - 2hx - 2ky + h^2 + k^2 - r^2 = 0$.
6. $x^2 + y^2 - 8x = 0$. **7.** $x^2 + y^2 - 9x - 9y + 36 = 0$. **8.** $x^2 - 6y - 9 = 0$
9. $x^2 - y^2 - 2 = 0$. **10.** $x^2 + y^2 - 6x + 4 = 0$.

i

11. a. $20\,x = k.$ **b.** $x^2 + y^2 + 25 = k/2.$ **c.** $x(k-1) = 5(k+1).$ **12.** $x^2 + y^2 = 100.$
13. a. $3\,x^2 - y^2 \pm 8\,x = 16.$ **b.** $x^2 + y^2 \pm 8\,y = 16.$ **e.** $9\,x^2 + 25\,y^2 = 225.$
14. a. $x^2 - 3\,x - 4\,y + 2 = 0.$ **b.** $xy - x - 14\,y + 46 = 0.$ **c.** $xy - 4\,x - 2\,y + 4 = 0.$
$\qquad$ **15. a.** $x = 3.$ **b.** $x^2 - 6\,x + 18\,y = 72.$ **c.** $x^2 + y^2 = 36.$

Page 58

1. $(2, 4), (\tfrac{1}{2}, -2).$ $\qquad$ **2.** $(4, -5).$ $\qquad\qquad$ **4.** $(4, -9), (-4, -9)\dot{}$
$\qquad$ **7.** $(3.45, 2.03), (-3.45, -2.03), (2.03, 3.45), (-2.03, -3.45).$
11. $(2, 1), (-2, 1).$ $\quad$ **13.** $(0, 0), (1.73, 0.43), (-1.73, -0.43).$ $\quad$ **15.** $(6, -4), (-1, -5)$
16. $15.748.$ $\quad$ **17.** $2.121.$ $\quad$ **18.** $(1, 1), (-\tfrac{15}{17}, \tfrac{145}{17}).$ $\quad$ **19.** $m^2 < 20.$

Page 62

1. a. $y = 64\,x - 16\,x^2,\ x = 2 \pm \tfrac{1}{4}\sqrt{64 - y}.$ **d.** $y = \pm x\sqrt{x^2 - 2},\ x = \pm\sqrt{1 \pm \sqrt{y^2 + 1}}.$

Page 66

3. $\tfrac{3}{4}.$ $\qquad$ **6.** $2\,a^2.$ $\qquad$ **7.** $3.$ $\qquad$ **8.** $\tfrac{1}{3}.$ $\qquad$ **11. a.** $2\,x.$ **b.** $3\,x^2.$ **c.** $-1/x^2.$

Pages 72–73

1. b. $\Delta y = -0.158.$ **e.** $\Delta y = 0.312.$ $\qquad\qquad$ **3. b.** $\Delta y = 3\,x^2\,\Delta x + 3\,x(\Delta x)^2 + (\Delta x)^3.$
d. $\Delta y = \dfrac{-36\,\Delta x}{x(x + \Delta x)}.$ $\quad$ **4. b.** $2 - \Delta x.$ **d.** $\dfrac{\sqrt{9 + \Delta x} - 3}{\Delta x}.$ $\quad$ **5. a.** $-1.$ $\quad$ **6. a.** $8\,x.$
c. $3\,x^2 - 1.$ **d.** $-\dfrac{8}{x^2}.$ **f.** $\dfrac{-4}{(x-2)^2}.$ **g.** $1 - \dfrac{1}{x^2}.$ **h.** $-\dfrac{2}{x^3}.$ $\quad$ **8.** $\dfrac{-a^2}{(t-a)^2}.$ $\quad$ **9.** $-\dfrac{c}{t^2}.$
12. $\dfrac{2}{(z+1)^2}.$ $\quad$ **13.** $1 - \dfrac{1}{t^2}.$ $\quad$ **14.** $\dfrac{1}{(\theta+1)^2}.$ $\quad$ **15.** $\dfrac{-2\,az}{(z^2+a^2)^2}.$

Pages 75–76

2. a. $3.$ **b.** $3, -3.$ **d.** $-\tfrac{4}{3}.$ **g.** $-\tfrac{1}{2}, -\tfrac{8}{25}, -\tfrac{8}{9}.$ $\qquad$ **3. a.** $(6, 13).$ **b.** $(\pm 2, \mp \tfrac{8}{3}).$
c. $(2, -\tfrac{7}{3}), (-1, \tfrac{13}{6}).$ **d.** $(\pm 2, \pm 4).$ $\quad$ **4. a.** $y = 3\,x - 2,\ y = 3\,x + 2.$ **b.** $y = 3\,x - 12,$
$y = -3\,x + 3.$ $\quad$ **d.** $4\,x + 3\,y + 24 = 0.$ $\quad$ **g.** $x + 2\,y = 4,\ 8\,x + 25\,y = 48,\ 8\,x + 9\,y = 16.$
7. $\dfrac{dy}{dx} = \dfrac{1}{2\sqrt{x}}.$

Pages 80–81

2. a. $20\,x^4 - 6\,x^2.$ **c.** $6\,x - 5.$ **e.** $33\,x^{10} - 63\,x^8.$
3. a. $-\dfrac{6}{x^3} + \dfrac{10}{x^6}.$ **b.** $3\,x^2 + \dfrac{1}{x^2}.$ **c.** $40\,t^4 - \dfrac{21}{t^4} + \dfrac{12}{t^5}.$ **d.** $-\dfrac{3}{t^2} + \dfrac{12}{t^3}.$ **e.** $\dfrac{20}{x^6} + \dfrac{3}{\sqrt{x}}.$
4. a. $1 + \dfrac{2}{x^3}.$ **b.** $-\tfrac{1}{2}x^{-\frac{3}{2}} + \tfrac{1}{2}x^{-\frac{1}{2}}.$ **c.** $4\,t(t^2 - 1).$ **d.** $3(t - 1)^2.$ **g.** $-\tfrac{4}{3}.$
5. a. $28.$ **b.** $\tfrac{3}{16}.$ **c.** $\tfrac{364}{81}.$ **d.** $\tfrac{160}{27}.$ $\qquad$ **6. a.** $-\dfrac{3}{x^2} - \dfrac{1}{2}.$ **b.** $-\dfrac{18}{x^2}.$ **c.** $-\dfrac{5}{x^2}.$ **d.** $1 - \dfrac{1}{\sqrt{x}}.$
7. a. $-1.$ **b.** $\tfrac{5}{3}.$ **c.** $6.$ **d.** $-2.$ **e.** $-\tfrac{32}{27}.$ **f.** $\tfrac{21}{2}.$ **g.** $-3, -\tfrac{4}{3}, -\tfrac{3}{4}, -\tfrac{1}{3}.$ **h.** $\pm\tfrac{1}{2}.$

Pages 86–87

2. a. $14\,x^6 + 12\,x^3 + \dfrac{9}{x^4}.$ $\qquad$ **b.** $\dfrac{3}{(3 - 5\,x)^2}.$ $\qquad$ **c.** $\dfrac{-2\,x^2 + 2\,x - 2}{(x^2 - 2\,x)^2}.$

d. $5\,x^{-\frac{1}{6}} + x^{-\frac{1}{2}} + x^{-\frac{2}{3}}.$ $\qquad$ **e.** $2 - \dfrac{2}{x^2}.$ $\qquad$ **f.** $\dfrac{2\,t^2 - 2}{(1 + t + t^2)^2}.$

g. $3\,t^2 + 4\,t - 1.$ $\qquad$ **h.** $\dfrac{9\,z^2}{(3 - z^3)^2}.$ $\qquad$ **i.** $\dfrac{4 - 4\,x + 6\,x^2}{(1 - 3\,x)^2}.$

j. $\dfrac{1}{\sqrt{x}(\sqrt{x} + 1)^2}.$ $\qquad$ **k.** $\dfrac{-4}{3\,x^{\frac{2}{3}}(2 + x^{\frac{1}{3}})^2}.$ $\qquad$ **l.** $\dfrac{-4\,a^2t}{(a^2 + t^2)^2}.$

2. *m.* $\dfrac{2\,a^2 z}{(z^2+a^2)^2}$.

p. $\dfrac{-16\,a^3 x}{(x^2+4\,a^2)^2}$.

n. $\dfrac{4-3\,x^2}{(x^2+4)^3}$.

q. $\dfrac{-32\,x^3}{(1-x^4)^2}$.

o. $\dfrac{-7}{\sqrt{x}(7+\sqrt{x})^2}$.

r. $\dfrac{a^2-x^2}{(x^2+a^2)^2}$.

3. *a.* 3. *b.* $\frac{3}{8}$. *c.* $\frac{7}{50}$. *d.* $\frac{15}{8}$. *e.* 1. *f.* -0.139.

4. *a.* $\dfrac{-1}{(2\,x-1)^2}$. *b.* $\dfrac{-2}{(x-1)^2}$.

c. $\dfrac{2\,a^3 x}{(x^2+a^2)^2}$. *d.* $\dfrac{-6}{(x+2)^2}$. **5.** -1. **6.** $x+2\,y=4$. **7.** $(-1,-2)$. **8.** $45°$.

9. $0,\pm\frac{1}{4}$. **10.** $-2, 3, 6.$ **11.** $-\frac{1}{27}$. **12.** $(1, 3), (-1, -3).$ **13.** 1.

Pages 91–92

1. *e.* $\dfrac{x}{(a^2-x^2)^{\frac{3}{2}}}$.

h. $\dfrac{-(x+2)}{(x^2+4\,x+10)^{\frac{3}{2}}}$.

k. $-6\,u(4-u^2)^2$.

f. $-\frac{9}{2}\sqrt{2-3\,x}$.

i. $\dfrac{3(3\,x+1)}{2(3\,x)^{\frac{3}{2}}}$.

l. $\dfrac{2\,z}{(1-z)^3}$.

g. $\dfrac{3\,x^2}{2\sqrt{1+x^3}}$.

j. $\dfrac{2\,a-3\,t}{2\sqrt{a-t}}$.

3. *a.* $\dfrac{-x}{\sqrt{9-x^2}}$. *b.* $\dfrac{-bx}{a\sqrt{a^2-x^2}}$. *c.* $\sqrt{\dfrac{p}{2\,x}}$. *d.* $\dfrac{-\sqrt{a^{\frac{2}{3}}-x^{\frac{2}{3}}}}{x^{\frac{1}{3}}}$.

4. *a.* $\dfrac{4\,u^7}{\sqrt{x}}$. *b.* $\dfrac{(1-2\sqrt{u})(x+1)}{\sqrt{u}}$. *d.* $\dfrac{-24}{(2\,x+9)^2}$.

6. *a.* $(\pm 1, \pm 2)$. *b.* $(0, 0), (1, 1), (2, 0)$. *c.* $(0, -1), (2, -5), (-2, -5)$.

7. *a.* Hor., $x=\frac{4}{3}$; ver., $x=0$. *b.* Hor., $x=0, x=\frac{3}{2}$; ver., $x=2$.

9. $153° 26'$. **10.** $3\,x-4\,y=10$.

Pages 94–95

1. *a.* $\dfrac{(x+5)(x-1)^2}{(x+1)^3}$. *b.* $\dfrac{4\,at}{(a-t^2)^2}$. *c.* $\dfrac{x(2-5\,x)}{\sqrt{1-2\,x}}$. *d.* $\dfrac{2(1+u^2)}{\sqrt{2+u^2}}$.

e. $\dfrac{2(1+2\,x)}{(2+3\,x)^{\frac{2}{3}}}$. *f.* $\dfrac{4-t}{2(2-t)^{\frac{3}{2}}}$. *g.* $\dfrac{a}{(a-x)\sqrt{a^2-x^2}}$. *h.* $\dfrac{-2\,at}{(a+t^2)\sqrt{a^2-t^4}}$.

i. $\dfrac{x(2\,x^2-3\,a^2)}{(a^2-x^2)^{\frac{3}{2}}}$. *j.* $\dfrac{2\,x^2-3\,x-2\,x^4}{(2-x^2)^{\frac{1}{2}}(3+x^3)^{\frac{2}{3}}}$. *k.* $\dfrac{-(a+2\,x)\sqrt{a-x}}{\sqrt{a+x}}$.

2. *a.* 3. *b.* 3. *c.* 1.624. *d.* 0. *e.* 2. **3.** *a.* $\dfrac{6\sqrt{y}}{3+4\,y^{\frac{1}{6}}}$. *b.* $\dfrac{\sqrt{4+y^2-y^4}}{y-2\,y^3}$. *c.* $\dfrac{\sqrt{4-y^2}}{-3\,y}$.

Pages 98–99

1. *a.* $\dfrac{2\,x+y}{5\,y^4-x}$. *b.* $\dfrac{y^2-2\,xy}{x^2-2\,xy+3\,y^2}$. *c.* $\dfrac{-(y+1)}{3\,y^2+x+1}$. *d.* $\dfrac{-(2\,x+3\,y)}{3\,x+2\,y}$.

e. $\dfrac{-4\sqrt{x^3 y}+y}{4\sqrt{xy^3}+x}$. *f.* $-\dfrac{2\sqrt{xy}+y}{2\sqrt{xy}+x}$. *g.* $-\dfrac{2}{9}\sqrt[3]{\dfrac{9\,y}{2\,x}}$. *h.* $\dfrac{ay-x^2}{y^2-ax}$.

8. -1. **9.** *a.* -8. *b.* $100\,x^3-18\,x$. *c.* $-6\,x$. *d.* $12\,u-4$. *e.* $18\,x^{-4}-24\,x^{-5}$.

f. $2+6\,t^{-4}$. *g.* $\dfrac{4\,a}{(x-a)^3}$. *h.* $12\,t^2-4$. *i.* $\dfrac{3\,x-16}{4(4-x)^{\frac{3}{2}}}$. *j.* $\dfrac{64-48\,x^2}{(x^2+4)^3}$. *k.* $\dfrac{1-2\,x}{(x-1)^{\frac{3}{2}}(x+1)^{\frac{5}{2}}}$.

l. $\dfrac{2\,t^3-4\,a^2 t-a^3}{(a^2-t^2)^{\frac{3}{2}}}$. *m.* $\dfrac{-6\,s^4+4\,s^2-1}{(s^2-1)^{\frac{3}{2}}(s^2+1)^{\frac{1}{2}}}$.

10. a. $-\dfrac{p^2}{y^3}$. **b.** $\dfrac{-2\,y^3}{(y^2-1)^3}$. **c.** $\dfrac{12\,y^2-8}{(8\,y-4\,y^3)^2}$. **d.** $\dfrac{2\,xy+y^2}{(x+y)^3}$. **e.** $\dfrac{1}{2(\sqrt{y}+1)^3}$.

f. $\dfrac{-b^4}{a^2y^3}$. **g.** $\dfrac{\sqrt{a}}{2\,x\sqrt{x}}$. **h.** $\dfrac{2\,a^3x}{y^5}$. **i.** $\dfrac{a^{\frac{2}{3}}}{3\,x^{\frac{4}{3}}y^{\frac{1}{3}}}$. **j.** $\dfrac{-12\,ax^2y^2}{(y^2-ax)^3}$.

Pages 101–102

1. a. $(-\frac{3}{2}, -\frac{9}{4})$. **b.** $(-\frac{1}{2}, \frac{17}{4})$. **c.** $(\pm 1, \pm 2)$. **2.** $(4, 2)$. **3.** $(2, 1), (-2, 1), (2\sqrt{3}, 3)$.
4. $(-1, \frac{7}{10}), (5, \frac{5}{2})$. **5. a.** $2\,x+y=0$, $x-2\,y=0$. **b.** $2\,x-y=5$, $x+2\,y+5=0$.
c. $x-3\,y+10=0$, $3\,x+y=0$. **6. a.** $(2, 0)$. **b.** $(-1, -3)$. **c.** $(\pm 1, \pm 3)$. **7. a.** $(0, 0)$,
$(6, 24)$. **b.** $(1, -3), (7, 45)$. **c.** $\left(\dfrac{55+\sqrt{280}}{61}, \dfrac{5-11\sqrt{280}}{61}\right), \left(\dfrac{55-\sqrt{280}}{61}, \dfrac{5+11\sqrt{280}}{61}\right)$.
8. $\sqrt{80}$. **9.** $\sqrt{5}$. **10.** $\frac{5}{2}\sqrt{17}$. **11. a.** $30°\,58'$, $108°\,26'$. **b.** $36°\,52'$. **c.** $45°$, $135°$.
d. $126°\,52'$. **e.** $71°\,1'$. **f.** $108°\,26'$. **g.** $16°\,16'$, $163°\,44'$.

Page 106

3. $l=999.62+0.019\,T$. **4.** $l=4+0.125\,w$. **6.** $p=759.17-0.0833\,h$.
7. $188°$. **8.** $d=15-0.034\,t$. **9.** 0.64 in./min.

Pages 111–113

1. a. 5.625. **b.** 3. **c.** 1.89. **2.** Approx. change in $y=-1$. **6.** 80 ft./sec., 128 ft./sec.,
12.96 ft. **8. a.** $-\frac{26}{27}$ ft./sec.^{2} **b.** $-\frac{1}{16}\sqrt{2}$ ft./sec.^{2}. **10.** 2463 cu. in./in., 739 cu. in.
12. 18.8 mi./hr. **13. a.** 51.5 mi./hr. **b.** 3.45 P.M. **c.** 70.7 mi. **14.** 7.9 ft./sec.
16. 0.054 sec./in., 0.0108 sec. **17.** -0.035 lb. per sq. in. per cu. in., 0.35 lb./sq. in.

Pages 115–116

1. 1.06 ft./sec. **2.** When bottom is 16.97 ft. from wall. **3.** 0.52 in./sec.
4. a. 5 mi./hr. **b.** 2 mi./hr. **5.** 3 mi./hr. **6.** 6.66 ft./sec. **8.** 5.33 sq. in./sec.
9. $\dfrac{6}{25\,\pi}$ ft./min. **10. a.** $-156\,\pi$ cu. in./min. **b.** -118.6 sq. in./min. **11.** $10\,\pi$ cu. in./sec.
12. b. 1 unit/min. **14.** -0.1 ft. per sec. per min. **15.** $-\frac{1}{256}$ unit/min.

Pages 119–120

1. a. $(1-7\,x^2)(1-x^2)^2dx$. **b.** $\dfrac{-(2+x)dx}{8\,x^2\sqrt{1+x}}$. **d.** $\dfrac{2\,dy}{(y+1)^2}$. **e.** $\dfrac{(2+3\,t)dt}{2\sqrt{1+t}}$.
f. $\dfrac{(2\,a-3\,bx)dx}{2\sqrt{a-bx}}$. **2. a.** $\dfrac{-\sqrt{y}\,dx}{\sqrt{x}}$. **e.** $\dfrac{-2\,xy^3\,dx}{3\,x^2y^2+4}$. **f.** $\dfrac{-(4+y)dx}{x+2\,y}$. **3.** 0.031. **4.** 2.
6. a. -0.6. **b.** 0.2. **c.** 0.6. **d.** 0.4. **8.** 1060.

Pages 123–124

3. $2\,\pi rht$. **4. a.** $\dfrac{32.2(21,000,000)^2}{s^2}$. **b.** -0.064. **5.** Error $\leqq 0.0058$ in.
6. a. 3.26 ft. **b.** 0.00153 sec. **c.** -2 min. 12 sec. **7.** 1.2, 65.2, 62.8. **8.** 0.198.
11. 25.12 cu. in. **12.** 1.3 HP. **13.** $\dfrac{1}{18\,\pi}$ in. **14.** Error $\leqq 0.5\%$. **18.** 2%.
19. 32.20, 0.022, 0.07%. **20.** 493 ft.

Pages 129–130

16. $\frac{3}{2}x^{\frac{4}{3}} - \frac{3}{7}x^{\frac{7}{3}} + C$. **18.** $ax - \frac{4}{3}x\sqrt{ax} + \frac{1}{2}x^2 + C$.
19. $a^2x - \frac{9}{5}ax\sqrt[3]{ax^2} + \frac{9}{7}x^2\sqrt[3]{a^2x} - \frac{1}{3}x^3 + C$.

22. $-\frac{1}{2}\left(a^{\frac{2}{3}}-x^{\frac{2}{3}}\right)^3+C.$ **25.** $\frac{2}{3}(2\,ax+x^2)^{\frac{4}{3}}+C.$ **27.** $-\frac{9}{4}(1-2\,s^2)^{\frac{2}{3}}+C.$

28. $\frac{2}{3}\,x\sqrt{5\,x}+C.$ **31.** $-2\sqrt{5-x}+C.$ **34.** $-\frac{1}{9}(4-3\,x)^3+C.$

35. $\frac{1}{2}\,x^2+\frac{2}{5}\,x^{\frac{5}{2}}+C.$ **37.** $\frac{14}{3}\,x\sqrt{x}+14\sqrt{x}+C.$ **39.** $\dfrac{-5}{2\,x^2-6}+C.$

43. $-\frac{1}{3}(a^2-x^2)^{\frac{3}{2}}+C.$ **44.** $\frac{1}{6}(2\,s^2-3)^{\frac{3}{2}}+C.$

46. $\frac{1}{3}(y^4+a^4)^{\frac{3}{2}}+C.$ **47.** $\dfrac{3}{16(9-4\,v^2)^2}+C.$

Page 131

1. $\frac{2}{5}(a-x)^{\frac{5}{2}}-\frac{2}{3}\,a(a-x)^{\frac{3}{2}}+C.$ **2.** $\frac{2}{3}(x-2)\sqrt{1+x}+C.$

4. $\frac{2}{105}(15\,x^2-12\,x+8)(1+x)^{\frac{3}{2}}+C.$ **6.** $-\frac{2}{15}(3\,x^2+4\,x+8)\sqrt{1-x}+C.$

8. $\dfrac{2(bt-2\,a)\sqrt{a+bt}}{3\,b^2}+C.$ **9.** $\dfrac{2(3\,by-2\,a)(a+by)^{\frac{3}{2}}}{15\,b^2}+C.$

12. $\frac{2}{9}(2-t)^{\frac{9}{2}}-\frac{12}{7}(2-t)^{\frac{7}{2}}+\frac{24}{5}(2-t)^{\frac{5}{2}}-\frac{16}{3}(2-t)^{\frac{3}{2}}+C.$

13. $\frac{1}{140}(10\,x^2-6\,x+3)(4\,x+3)^{\frac{3}{2}}+C.$

14. $\frac{3}{10}(2\,x-3)(x+1)^{\frac{2}{3}}+C.$ **15.** $\frac{1}{7}(x-1)(3+4\,x)^{\frac{3}{4}}+C.$

16. $\frac{3}{28}(4\,x+3)(x-1)^{\frac{4}{3}}+C.$ **18.** $\frac{2}{9}(x-2)(5+2\,x)^{\frac{5}{4}}+C.$

Page 132

2. $y=\frac{2}{3}\left(x^{\frac{3}{2}}-5\right).$ **3.** $y=\frac{1}{3}(x^2+16)^{\frac{3}{2}}-\frac{110}{3}.$

5. $y=\frac{3}{4}\,x^{\frac{4}{3}}+\frac{3}{2}\,x^{\frac{2}{3}}-18.$ **6.** $s=\frac{2}{3}\,t\sqrt{7\,t}.$

8. $s=\frac{1}{3}\,t^3-\frac{4}{3}\,t^{\frac{3}{2}}+7.$ **10.** $y=\frac{2}{3}(x-10)\sqrt{5+x}+13.$

12. $y=\frac{125}{3}-\frac{1}{3}(25-x^2)^{\frac{3}{2}}.$ **13.** $s=4-2\sqrt{5-x}.$

15. $y=\frac{2}{3}(x-2)\sqrt{1+x}+2.$ **16.** $z=\frac{3}{2}\,x^{\frac{4}{3}}-\frac{3}{7}\,x^{\frac{7}{3}}+\frac{209}{7}.$

17. 4.   **18.** $-0.8856.$ **19.** $\frac{8}{3}\,p^2.$ **20.** $-\frac{66}{35}.$ **21.** 13.   **22.** 0.   **23.** $\frac{42}{6}$

24. $\frac{298}{15}.$

Pages 134–135

1. **b.** $2\sqrt{y}=x.$ **c.** $3\,y=16-2(4-x)^{\frac{3}{2}}.$ **e.** $3\sqrt{y}=x^{\frac{3}{2}}+2.$ **f.** $xy=-1.$

g. $4\,x^2+9\,y^2=324.$ **h.** $4\,y^{\frac{3}{2}}=3(x^2-1).$ **j.** $ay=b\sqrt{a^2-x^2}.$ **k.** $y=\sqrt{x^2+4}$

m. $\sqrt{x}+\sqrt{y}=\sqrt{a}.$ **n.** $x^{\frac{2}{3}}+y^{\frac{2}{3}}=a^{\frac{2}{3}}.$ **2.** **b.** $y^2-4\,x^2=C.$ **c.** $4\,x^2+y^2=C.$

e. $y^2=x+C.$ **f.** $3\,y=x^3-3\,x+C.$ **h.** $(x-4)^2+(y+2)^2=C.$ **i.** $\frac{1}{x}-\frac{1}{y}=C.$

3. $2\,y=2\,x-3\,x^2.$ **4.** $x^2y=4.$ **5.** $3\,y=3\,x^3+2\,x-6.$ **6.** $y=4\,x^{\frac{5}{2}}-79\,x+188.$

7. $y=2\,x-4+3/x.$

Pages 137–138

1. 87.64 ft./sec.   **2.** 89.89 ft./sec.   **3.** 400 ft.   **4.** 904 ft.   **5.** $7\frac{1}{16}$ sec.

6. 126.7 ft.   **7.** **a.** 12,500 ft.   **b.** 1924 ft./min.   **c.** 6944 ft.   **8.** **a.** $v=t-\frac{1}{3}\,t^3+2.$

b. $v=\frac{41}{6}-\frac{1}{2}\,t^2-1/t.$ **c.** $v=2+2\sqrt{3}-6/\sqrt{t}.$ **d.** $v=4\,t-\frac{1}{3}\,t^3-1.$ **9.** **a.** $s=\frac{2}{3}(t-1)^{\frac{3}{2}}+2.$

b. $s=\frac{1}{3}\,t^3+\frac{8}{3}-1/t.$ **12.** **a.** 12 ft.   **b.** 24 ft./sec.

Pages 140–141

1. 2.   **2.** $\frac{16}{3}.$ **4.** $\frac{38}{3}.$ **5.** 13.73.   **7.** $\frac{50}{3}.$ **8.** $\frac{2}{15}.$ **10.** $-42\frac{2}{3}.$ **11.** $-\frac{27}{4}\,a^{\frac{4}{3}}.$

13. $\frac{1}{3}\,a^3.$ **14.** 8.   **16.** 399.9.   **17.** 6.95.   **19.** $\frac{1}{6}\,a^2.$ **20.** $\frac{52}{9}.$ **22.** $\frac{15}{32}.$

23. $\frac{2}{3}\,a^{\frac{3}{2}}.$ **25.** 4.   **26.** $\dfrac{1}{4\,a^2}.$ **28.** $-\frac{2}{3}.$ **29.** $\frac{26}{3}.$ **31.** $\frac{4}{15}\,a^{\frac{5}{2}}.$ **32.** $\frac{76}{15}.$

33. $\frac{26}{3}.$ **34.** $\frac{3072}{35}.$ **35.** $-\frac{300}{7}.$ **36.** $\frac{2}{3}(2-\sqrt{2})a^{\frac{3}{2}}.$

Pages 146–147

1. $a.$ $\frac{16}{3}$. $c.$ $\frac{81}{4}$. $e.$ $\frac{14}{3}$. $g.$ 7. $i.$ 14.48. $k.$ $\frac{27}{4}$. 2. $a.$ 30. $c.$ $\frac{28}{3}$. 3. $a.$ $\frac{32}{3}$. $c.$ $\frac{1}{2}$.
$e.$ $-\frac{123}{15}\sqrt{2}$. 4. $\frac{8}{3}$. 5. $\frac{64}{3}$. 6. 1. 7. $\frac{1}{6}a^2$. 8. $\frac{32}{3}$. 9. $\frac{16}{3}\sqrt{2}$. 10. 226.8.
12. $\frac{128}{15}$. 13. $2\sqrt{3}$. 14. $\frac{648}{5}$. 15. $\frac{32}{15}$. 16. $\frac{32}{15}\sqrt{2}$. 17. $\frac{2048}{105}$.

Page 149

1. Correct value $= 0.8813+$. 2. $a.$ 16.48. $b.$ 1.791. $c.$ 5.502. $d.$ 45.25. $e.$ 17.08.
$f.$ 65.28.

Page 151

1. $a.$ 9.84. $b.$ 36.39. $c.$ 6.89. $d.$ 18.10. 2. $a.$ 9π. $c.$ $42\frac{2}{3}$. $d.$ 72. $e.$ $\frac{4}{5}$. $g.$ 4.184.
3. 9.2 acres.

Pages 158–159

1. $a.$ 8. $c.$ $\frac{9}{2}$. $e.$ 9. $g.$ $\frac{32}{3}$. $i.$ $\frac{4096}{75}$. 2. $\frac{4}{3}p^2$. 3. $\frac{1}{3}ab$. 4. 36. 5. $1\frac{1}{3}a^2$.
6. 36. 7. 36. 8. $\frac{27}{10}$. 9. $\frac{7}{6}$. 10. 16.64. 11. 44. 12. $\frac{128}{5}$. 13. $\frac{256}{15}$.
14. $\frac{3}{4}$. 15. 108.

Page 162

1. $\frac{335}{27}$. 2. $\frac{14}{3}$. 3. $1.44\,a$. 4. 9.07. 5. $6\,a$. 6. 9.07. 7. $\frac{14}{3}$. 9. 4.65.
10. 9.30. 11. 8.61. 12. 4.56. 13. 9.38.

Pages 167–168

1. $a.$ $\frac{32}{5}\pi$. $b.$ $\frac{8}{3}\pi$. $c.$ $\frac{224}{15}\pi$. $d.$ 8π. $e.$ 8π. $f.$ $\frac{256}{15}\pi$. $g.$ $\frac{40}{3}\pi$. $h.$ $\frac{128}{3}\pi$.
2. $a.$ $\frac{32}{7}\pi$. $b.$ $\frac{8}{5}\pi$. $c.$ $\frac{80}{7}\pi$. $d.$ $\frac{32}{5}\pi$. $e.$ $\frac{48}{5}\pi$. $f.$ $\frac{144}{5}\pi$. $g.$ $\frac{72}{5}\pi$. $h.$ $\frac{192}{7}\pi$.
8. $\frac{4}{3}\pi a^2 b$. 10. $\frac{4}{3}\pi a^3$. 11. $\frac{32}{105}\pi a^3$. 12. $\frac{1}{15}\pi a^3$. 14. $\frac{416}{3}\pi$ cu. in. 16. 18π.
17. $\frac{128}{15}\pi$. 18. 4π. 19. 57.44. 20. $\frac{1250}{3}\pi$.

Pages 173–174

1. $a.$ 2309 cu. in. $b.$ 1333 cu. in. $c.$ 2667 cu. in. $d.$ 1571 cu. in. $e.$ 2667 cu. in.
2. $a.$ 1333 cu. in. $b.$ 577.3 cu. in. $c.$ 392.7 cu. in. 3. $a.$ 1024 cu. in. $b.$ 443.4 cu. in.
$c.$ 426.7 cu. in. 4. $a.$ 228.7 cu. in. $b.$ 359.2 cu. in. 5. $\frac{16}{3}r^3$. 6. 1066.
7. $\frac{250}{3}$ cu. in. 8. $a.$ $\frac{625}{3}\pi$. $b.$ $\frac{128}{5}\pi$. $c.$ $\frac{128}{105}\pi$. $d.$ $\frac{64}{3}\pi$. 9. $a.$ $\frac{108}{5}\pi$. $b.$ $\frac{256}{15}\pi$.
$c.$ $\frac{12}{5}\pi p^3$. $d.$ $\frac{45}{2}\pi$.

Pages 177–178

1. $\frac{13}{3}\pi$. 2. $\frac{52}{3}\pi p^2$. 3. 203.0. 4. 36.18. 5. $\frac{12}{5}\pi a^2$. 6. 53. 7. 77.
8. 217. 10. 410.3. 11. 141.5. 13. 131.2. 14. $\frac{208}{9}\pi$. 15. 3π. 16. $\frac{56}{5}\sqrt{3}\pi$.

Page 182

2. $(\frac{3}{2}h, 0)$. 4. $(\frac{8}{5}, \frac{16}{7})$. 6. $(\frac{20}{7}, \frac{5}{2})$. 8. $(\frac{16}{15}, \frac{64}{21})$.
10. $(\frac{12}{5}, \frac{3}{2})$. 12. $(\frac{3}{5}, 1)$. 14. $(1, 1\frac{1}{5})$. 15. $(2, 1)$.
16. $(\frac{1}{3}a, \frac{1}{3}b)$. 17. $(\frac{1}{5}a, \frac{1}{5}a)$. 18. $(\frac{16}{7}, 0)$. 19. $\left(\dfrac{4\,a}{3\,\pi}, \dfrac{4\,b}{3\,\pi}\right)$.

Page 183

1. $\frac{3}{8}r$. 2. $2r/\pi$. 3. $\frac{1}{2}r$.

Pages 186–187

1. 2560 lb. 2. 1302 lb. 3. 3771 lb. 4. 7800 lb. 5. 533 lb. 6. 1667 lb.
7. 3375 lb. 8. 20,333 lb. 9. $a.$ 41,250 lb. $b.$ 11 ft. 11. 3682 T.

Pages 189–190

1. 800,000 π ft.-lb.
2. 84,375 π ft.-lb.
3. 432,000 π ft.-lb.
4. 2,500,000 $\pi/3$ ft.-lb.
5. 9375 π ft.-lb.
6. 78,000 ft.-lb.
7. 327,600 ft.-lb.
8. 13,333 ft.-lb.
9. 8000 ft.-lb.
10. 29,333 ft.-lb.

Pages 192–193

4. $\frac{8}{3}$. **5.** $\frac{4}{3}$. **6. a.** 80 ft./sec. **b.** $81\frac{2}{3}$ ft./sec. **7.** $\frac{1}{4}\pi r$. **8.** $\frac{4}{3}r^2$.

Pages 199–200

4. $C = \frac{1}{2} + 500/n$. **5.** $M = 2\pi r^2 + 116/r$; 91.47, 83.13, 85.67. **6.** $A = \frac{8}{3}x\sqrt{9-x^2}$.
7. $w = 1,600,000,000/x^2$; 95.18 lb. **8.** $t = 0.5547\sqrt{l}$; 0.8125 ft. **9.** $P = 2x + 640/x$.
11. $A = x^2 + 24/x$. **13.** $V = 4x^3 - 50x^2 + 150x$. **14.** $A = \frac{1}{2}x\sqrt{256 - x^2}$.
15. $P = 2x + \sqrt{256 - x^2}$.

Page 203

4. a. Increasing when $t < \frac{25}{8}$; decreasing when $t > \frac{25}{8}$. **c.** Increasing when $0 < r < 4$; decreasing when $r < 0$ and $r > 4$. **d.** Increasing when $x > 2$; decreasing when $x < 0$ and $0 < x < 2$. **f.** Increasing when $x < 1.962$ and $x > 6.371$; decreasing when $1.962 < x < 6.371$. **5. a.** All values except $x = 0$. **b.** $x < 1$ and $x > 7$.

Pages 214–217

1. a. Min. $(\frac{5}{2}, -\frac{13}{4})$. **b.** Max. $(\frac{7}{4}, \frac{57}{8})$.
c. Min. $(2, -\frac{4}{3})$, max. $(-1, \frac{19}{6})$. **d.** Min. $(-5, -2)$, max. $(-3, 2)$.
e. Min. $(3, -\frac{27}{2})$, max. $(-2, \frac{22}{3})$. **f.** No max. or min.
g. Min. $(-1, -1)$, $(1, -1)$, max. $(0, 0)$.
h. Min. $(-1, -\frac{38}{15})$, $(2, \frac{16}{15})$, max. $(-2, -\frac{16}{15})$, $(1, \frac{38}{15})$.
i. Min. $(0, -4)$, $(2, -4)$, max. $(1, -3)$. **j.** Min. $(0, 0)$, max. $(\pm\sqrt{3}, 9)$.
k. Min. $(2, 12)$. **l.** Min. $(\pm 1, 2)$.
m. Min. $(1, 4)$, max. $(-1, -4)$. **n.** Min. $(-1, -3)$, max. $(1, 3)$.
2. a. 10, 10. **b.** 10, 10. **c.** 10, 10. **d.** 1, 19. **3.** $40\sqrt{2}$ rd. $\times 80\sqrt{2}$ rd. **5.** $20\sqrt{2}$ in
6. 3.5 in. **7.** 24 ft. **8.** 2.94 in. **9.** 7.7 ft. $\times$ 7.7 ft. $\times$ 13.5 ft. **10.** 4.90 ft. $\times$ 3.27 ft.
11. 550.8 cu. ft. **12.** $\frac{1}{3}\sqrt{3}\,a \times \frac{1}{3}\sqrt{6}\,a$. **13.** $a \times \sqrt{3}\,a$. **14.** $\frac{1}{2}\sqrt{2}\,a \times \sqrt{2}\,a$.
15. 100 sq. rd. **16.** $\sqrt{2}\,a \times \sqrt{2}\,a$. **17. a.** **18.** 0.64 ft. **19. a.** $\sqrt[3]{464/\pi} = 5.29$ in.
b. $\sqrt[3]{232/\pi} = 4.20$ in. **20.** $10\sqrt[3]{16/\pi} = 17.2$ ft. **21.** $\frac{4}{27}\pi r^2 h$. **22.** $\frac{1}{3}\sqrt{3}\,a$.
23. $h = \frac{16}{3}\sqrt{3}$ in., $r = \frac{8}{3}\sqrt{6}$ in. **24.** $h = \frac{2}{3}\sqrt{3}\,r$. **25.** Altitude of cylinder = altitude
of cone. **26.** 13.6 mi./hr. **27.** 75. **28.** 16 sq. units. **29.** $\sqrt{2}\,a \times \sqrt{2}\,b$.
30. $4a^2$. **31.** $(6, 8)$. **32.** $(\pm\sqrt{8}, 2)$. **33.** $2r$. **34.** $AC = 76.14$ ft. **35.** 5.18 ft.
42. a. $2a, 2b$. **b.** $a + a^{\frac{1}{3}}b^{\frac{2}{3}}$, $b + a^{\frac{2}{3}}b^{\frac{1}{3}}$. **c.** $a + \sqrt{ab}$, $b + \sqrt{ab}$. **d.** $\dfrac{a^2 + b^2}{a}$, $\dfrac{a^2 + b^2}{b}$.

Pages 221–223

4. e. $x^2 + y^2 \pm 4x - 14y + 49 = 0$. **i.** $x^2 + y^2 = 18$. **6. a.** Length $= 5\sqrt{2}$.
b. Length $= 2\sqrt{5}$. **9.** $2x^2 + 2y^2 = k - 2c^2$. **10.** $(x + 3c)^2 + y^2 = 4k^2$ or
$(x - 3c)^2 + y^2 = 4k^2$. **11.** $x^2 + y^2 - 16x + 48 = 0$. **12.** $x^2 + y^2 - 5x + 5y + 8 = 0$.
13. $C\left(\dfrac{c(1+k^2)}{1-k^2}, 0\right)$, $r = \dfrac{2ck}{k^2 - 1}$. **16. a.** $4x - 3y = 8$, $3x + 4y = 6$. **b.** $y = -2$,
$y = -4$; $x = 4$. **c.** $5x \pm 12y = -52$, $12x \pm 5y = 78$. **17. a.** 45°. **b.** 90°. **c.** 45°.
d. 45°. **e.** 45°. **f.** 66° 15′. **g.** 71° 34′. **18. a.** 63° 26′. **b.** 90°.

Pages 227–228

2. e. $x^2 + y^2 - 20x - 8y + 16 = 0$ or $x^2 + y^2 - 52x + 24y + 144 = 0$.
f. $x^2 + y^2 - 4x - 4y = 0$. **g.** $x^2 + y^2 - 26x + 26y + 169 = 0$. **h.** $x^2 + y^2 + 4x - 8y = 52$.
4. a. $4x + 3y - 36 = 0$, $4x + 3y + 14 = 0$. **b.** $x + 2y + 17 = 0$, $x + 2y - 13 = 0$.

5. *a.* $(2, -3)$, $(3, 2)$. *b.* $(0, 0)$, $(8, 4)$. *c.* $(3, -1)$, $(4, -8)$. *d.* $(0, 0)$, $(1, 7)$.   **6.** Length $\sqrt{2}\,a$, width $\frac{1}{2}\sqrt{2}\,a$.   **7.** *a.*   **8.** 5.81×2.07.   **11.** $45°$.   **12.** *a.* $(x-2)^2 + (y+2)^2 = 25$ or $(x-2)^2 + (y+2)^2 = 225$. *b.* $x^2 + y^2 + 14x + 4y + 17 = 0$ or $x^2 + y^2 + 38x + 4y + 329 = 0$.

Pages 231–232

3. *a.* $x^2 + 16y = 0$. *e.* $y^2 + 18x = 0$. *f.* $y^2 = 9x$ or $3x^2 = 8y$.   **4.** $(\frac{1}{4}, 1)$.
5. $x^2 + y^2 - 10x = 0$.   **6.** $0°$, $8°8'$.   **8.** *a.* $x - 4y + 6 = 0$, $4x + y = 27$.
b. $x - y = 4$, $x + y = 12$.   **9.** *a.* $(\frac{2}{3}, 1)$. *b.* $(4, 1)$. *c.* $(-\frac{9}{5}, -3)$. *d.* $(\frac{1}{2}, \frac{3}{2})$.
11. *a.* $12°32'$, $63°26'$. *b.* $76°6'$. *c.* $108°26'$.   **12.** $\tan\theta = \frac{3}{4}$.   **13.** 13.86×4.

Pages 237–238

4. *a.* $x^2 - 4x + 16y - 44 = 0$. *b.* $x^2 - 8x - 12y + 28 = 0$. *c.* $y^2 - 6y + 16x - 23 = 0$.
d. $y^2 - 4y - 20x + 24 = 0$.   **5.** Vertex $(\frac{1}{4}a, \frac{1}{4}a)$.   **6.** $(2, 4)$.   **7.** $(5, -7)$.
9. $18°56'$.   **10.** $4x + 4y = 25$.   **11.** $71°34'$, $30°58'$.   **12.** $(4 \pm \frac{1}{2}\sqrt{62}, \frac{1}{2})$.
16. 60, 47.8, 37.8, 30, 24.4, 21.1, 20, 21.1, etc.

Pages 242–244

3. *a.* $16x^2 + 25y^2 = 400$. *b.* $x^2 + 2y^2 = 32$.   **4.** *a.* $x^2 + 4y^2 = 52$. *b.* $9x^2 + y^2 = 45$.
8. $x^2 + 4y^2 = 64$.   **9.** $b^2x^2 + a^2y^2 = a^2b^2$.   **11.** $x - 2y = 7$, $2x + y = 4$.
12. $x + 2y = 6$, $x + 14y = 30$.   **13.** $(0, 8)$.   **14.** *a.* $90°$. *b.* $42°5'$. *c.* $18°26'$, $21°22'$.
d. $71°34'$. *e.* $100°27'$.   **15.** $(3, \frac{16}{3})$, $(-3, -\frac{16}{3})$.   **17.** $2ab$.   **18.** $2y + b = 0$.
21. *a.* $100(x+3)^2 + 36(y-2)^2 = 3600$.   *b.* $16(x-6)^2 + 25(y+2)^2 = 400$.
       c. $x^2 + 2(y-6)^2 = 100$.   *d.* $9(x-5)^2 + 25y^2 = 225$.
22. *a.* $\dfrac{(x-4)^2}{16} + \dfrac{y^2}{4} = 1$.   *b.* $\dfrac{(x-1)^2}{4} + \dfrac{(y+2)^2}{64} = 1$.   *d.* $\dfrac{(x+2)^2}{36} + \dfrac{(y-3)^2}{81} = 1$.

Pages 249–250

3. *a.* $x^2 - 4y^2 = 20$. *b.* $7x^2 - 4y^2 = 12$.   **4.** $25x^2 - 144y^2 + 3600 = 0$.
7. *a.* $5x - 4y = 9$, $4x + 5y = 40$. *b.* $16x + 5y = 39$, $5x - 16y = 100$.   **8.** $2x - y = \pm 3$.
9. $3x - 5y = \pm 16$.   **10.** $11x - 24y + 7 = 0$, $3x - 8y = 1$.   **11.** $22°37'$.
14. Decreasing $\frac{9}{4}$ units/sec.   **15.** 2.54×7.44.   **16.** $(5, 3)$.

Pages 251–252

3. $4x^2 - y^2 = 3$.   **4.** $(8, 6)$, $(-8, -6)$.   **5.** *a.* $x + y = 4$, $x - y + 12 = 0$.
b. $3x - 2y = 24$, $2x + 3y + 10 = 0$.   *c.* $3x - 2y + 25 = 0$, $2x + 3y = 18$.
8. *a.* $49°24'$. *b.* $72°15'$. *c.* $30°58'$, $18°26'$. *d.* $16°16'$.   **11.** $k = \pm 8$.
   14. *a.* $9x^2 - 16y^2 - 54x + 657 = 0$.   *b.* $9x^2 - 16y^2 - 36x + 96y - 252 = 0$.
        c. $9x^2 - 4y^2 + 54x - 16y - 79 = 0$.

Page 262

2. Max. pt. $(\sqrt{5}, \frac{1}{3}(3 + 10\sqrt{5}))$; min. pt. $(-\sqrt{5}, \frac{1}{3}(3 - 10\sqrt{5}))$; infl. pt. $(0, 1)$, slope 5.
4. Max. pt. $(-1, \frac{1}{2})$; min. pt. $(0, 0)$; infl. pt. $(-\frac{1}{2}, \frac{1}{4})$, slope $-\frac{3}{4}$.
6. Max. pt. $(1 - \frac{1}{3}\sqrt{3}, \frac{2}{9}\sqrt{3})$; min. pt. $(1 + \frac{1}{3}\sqrt{3}, -\frac{2}{9}\sqrt{3})$; infl. pt. $(1, 0)$, slope -1.
8. Max. pt. $(0, -\frac{3}{2})$; min. pts. $(\pm 2\sqrt{3}, -\frac{39}{2})$; infl. pts. $(\pm 2, -\frac{23}{2})$, slopes $\mp \frac{16}{5}$.
10. Max. pt. $(0, 4)$; min. pts. $(\pm \frac{1}{2}\sqrt{10}, -\frac{9}{4})$; infl. pts. $(\pm \frac{1}{6}\sqrt{30}, \frac{19}{36})$; slopes $\mp \frac{19}{9}\sqrt{30}$.
12. Max. pts. $(1, \frac{38}{15})$, $(-2, -\frac{16}{15})$; min. pts. $(-1, -\frac{38}{15})$, $(2, \frac{16}{15})$; infl. pts. $(0, 0)$,
     $(\pm \frac{1}{2}\sqrt{10}, \pm \frac{13}{24}\sqrt{10})$, slopes 4, $-\frac{9}{4}$.
14. Min. pt. $(1, 3)$; infl. pt. $(-\sqrt[3]{2}, 0)$, slope $-3\sqrt[3]{2}$.
18. $(0, 0)$, $(\pm\sqrt{3}, \pm\frac{1}{4}\sqrt{3})$.

Page 266

2. Max. pts. $(\pm 2\sqrt{3}, 14)$; min. pt. $(0, 2)$; infl. pts. $(\pm 2, \frac{26}{3})$.   **4.** Min. pt. $(0, 0)$;
infl. pts. $(\pm \frac{1}{3}\sqrt{3}, \frac{1}{4})$; asympt. $y = 1$.   **6.** Max. pts. $(\pm \sqrt{2}, 4)$; min. pt. $(0, 0)$;
infl. pts. $(\pm \frac{1}{3}\sqrt{6}, \frac{20}{9})$ **8.** Min. pt. $(2, 3)$; asympt. $x = 0$.

9. Max. pt. $\left(-2\sqrt[5]{\frac{2}{3}}, -\frac{5}{3}\sqrt[5]{\frac{9}{4}}\right)$; infl. pt. $(2, 0)$; asympt. $x = 0$.

12. Max. pt. $(0, 2\,a)$; infl. pts. $\left(\pm\frac{2}{3}\sqrt{3}\,a, \frac{3}{2}\,a\right)$; asympt. $y = 0$.

14. Max. pt. $\left(-\frac{3}{5}, \frac{162}{3125}\right)$; min. pt. $\left(\frac{3}{5}, -\frac{162}{3125}\right)$; infl. pts. $(0, 0)$, $\left(\pm\frac{3}{10}\sqrt{2}, \mp\frac{567}{25,000}\sqrt{2}\right)$.

16. Max. pt. $(0, 0)$; min. pt. $\left(\frac{1}{5}\sqrt[3]{50}, -\frac{3}{25}\sqrt[3]{20}\right)$; infl. pt. $\left(\frac{1}{10}\sqrt[3]{100}, -\frac{9}{100}\sqrt[3]{10}\right)$.

Page 268

2. Max. pt. $\left(\frac{1}{3}\sqrt{3}, \frac{1}{3}\sqrt[4]{12}\right)$; min. pt. $\left(\frac{1}{3}\sqrt{3}, -\frac{1}{3}\sqrt[4]{12}\right)$.   **4.** No max. or min. pts.

6. Max. pts. $(\pm\sqrt{8}, 8)$; min. pts. $(\pm\sqrt{8}, -8)$.

8. Max. pt. $\left(-\frac{2}{3}\sqrt{3}, \frac{4}{9}\sqrt{3}\right)$; min. pt. $\left(\frac{2}{3}\sqrt{3}, -\frac{4}{9}\sqrt{3}\right)$.

10. No max. or min. pts.; asympt. $x = 2\,a$.   **12.** Max. pt. $x = \frac{16}{25}$; infl. pt. $x = \frac{64}{225}$

14. Asympt. $x = 1, x = 3$.   **16.** Max. pt. $(0, 4)$; min. pt. $(0, -4)$.

Pages 272–273

1. b. $x = \frac{1}{5}e^{2y}$.   **d.** $x = -2\ln y$.   **13. a.** $x = \ln\left(y \pm \sqrt{y^2 - 1}\right)$.   **b.** $x = \ln\left(y + \sqrt{y^2 + 1}\right)$.

Pages 275–276

1. $y' = \dfrac{a}{ax + b}$.   **2.** $y' = \dfrac{16\,x}{x^4 - 16}$.   **4.** $y' = \dfrac{0.434(2 - 4\,x^3)}{2\,x - x^4}$.   **6.** $f'(x) = \dfrac{3}{x}$.

7. $f'(x) = \dfrac{3\ln^2 x}{x}$.   **9.** $f'(x) = \dfrac{1}{\sqrt{1 + x^2}}$.   **11.** $\dfrac{ds}{dt} = \dfrac{1}{2}(1 + \ln t)$.   **13.** $\dfrac{du}{dy} = \dfrac{1}{1 - y^2}$.

17. $f'(x) = \dfrac{x + 3}{2\,x^2 - 2}$.   **18.** $\dfrac{dy}{dx} = \dfrac{1 - \ln x}{x^2}$.   **19.** $\dfrac{dy}{dx} = \dfrac{4 + 5\,x^2}{x + x^3}$.   **21. a.** $y = 1.9356$,

$y' = \frac{1}{12}$.   **b.** $y = 0.6931$, $y' = 3$.   **c.** $y = 4.9437$, $y' = 4.7958$.   **d.** $y = 0.1733$, $y' = -0.0483$.

e. $y = 2.1072$, $y' = 0.2545$.   **f.** $y = 0.6901$, $y' = -0.0362$.   **22. a.** $1, \frac{1}{2}$.   **b.** $\pm 4\sqrt{2}$, 0.

c. $\pm 2, \frac{2}{3}$.   **d.** 0.434.   **e.** $-1, -\frac{1}{4}$.   **f.** $\pm\sqrt{3}, 0$.   **g.** $-\frac{1}{2}, -\frac{1}{6}$.   **h.** $\pm 0.8686\sqrt{15}$, 0.

23. a. $(4, 2.7726)$.   **b.** $(2, 1.3863)$.   **24. a.** Min. pt. $(0, 0)$; infl. pts. $(\pm 1, \ln 2)$.

b. Min. pt. (e, e); infl. pt. $(e^2, \frac{1}{2}\,e^2)$.   **c.** Min. pt. $(1/e, -1/e)$.   **d.** Max. pt. $(4, \ln 16)$.

25. 2.323.

Page 277

1. $y' = 2(x + 2)(x + 3)^2(3\,x^2 + 11\,x + 9)$.   **3.** $y' = \dfrac{1 + 3\,x^2 - 2\,x^4}{(1 - x^2)^{\frac{3}{2}}}$.

5. $y = 6\sqrt{5}$, $y' = \frac{67}{20}\sqrt{5}$.   **6.** $y = 2\sqrt{6}$, $y' = \frac{23}{12}\sqrt{6}$.

7. $y = 3$, $y' = -\frac{1}{4}$.   **8.** $y = 49\sqrt{6}$, $y' = \frac{91}{24}\sqrt{6}$.

Page 279

14. $2\sqrt{ab}$.   **15.** Max. pt. $(0, 1)$; infl. pts. $\left(\pm\frac{1}{2}\sqrt{2}, 1/\sqrt{e}\right)$.

17. Min. pt. $(-1, -1/e)$; infl. pt. $(-2, -2/e^2)$.   **18.** 8.13.

Pages 280–281

1. $\dfrac{e^{ax}}{a} + C$.   **3.** $\frac{1}{2}e^{2s} + C$.   **5.** $\dfrac{a^{2x}}{2\ln a} + C$.   **8.** $\frac{1}{2}x^2 - x + \ln(x + 1) + C$.

9. $\ln\sqrt{x^2 - 2\,x - 5} + C$.   **11.** $2\,y^{-4} - 6\,y^{-2} + \frac{1}{2}\,y^2 - \ln y^6 + C$.   **13.** $\frac{1}{4}\ln^4 x + C$.

14. $2\ln(e^x + 1) + C$.   **16.** $\frac{1}{2}(e^{2y} - e^{-2y}) + 2\,y + C$.   **19.** 3.195.   **20.** 0.5596.

21. 1.   **22.** 25.94.   **23.** 1.1513.   **24.** $\ln\sqrt{2}$.   **25.** $\frac{8}{3} - \ln 3$.   **26.** 0.3167.

27. 8.318.   **28.** 8.318.   **30.** 0.3181.   **31.** $e^x - 1$.   **32.** $a^2(e - e^{-1})$.   **33.** 1.070.

34. 5.751.   **35.** $\frac{1}{2}a(e - e^{-1})$.   **36.** 4.443.   **37.** $2.594\,\pi$.   **38.** $0.2115\,\pi a^3$.

39. $\frac{1}{2}\pi(e^2 - 1)$.   **40.** $\left(\frac{15}{2} - 4\ln 4\right)\pi$.   **41.** $\frac{1}{4}(e^2 + 4 - e^{-2})\pi a^3$.   **42.** $\frac{1}{2}(1 - e^{-20})\pi$.

Pages 283–284

1. 6931 gal.   **2.** 61.25 deg.; 2.16 min.   **3. a.** 12.2 lb./sq. in. **b.** 8.2 lb./sq. in.
4. 586 lb.   **6.** 6931 gal.   **7.** 5.9 sec.   **8.** 92 sec.

Pages 294–296

1. a. $dy/dx = \frac{3}{2}\cos\frac{1}{2}x$.   **c.** $dr/d\theta = \frac{1}{2}\sec^2\frac{1}{2}\theta$.   **e.** $dy/dx = 6\sec 2x \tan 2x$.
g. $dy/dx = x\cos x + \sin x$.   **h.** $ds/dt = -e^{-2t}(\sin t + 2\cos t)$.   **j.** $\dfrac{dy}{dx} = -\dfrac{x\sin x + \cos x}{x^2}$.
k. $\dfrac{ds}{dt} = \dfrac{a}{t^2}\sin\dfrac{a}{t}$.   **n.** $dy/dx = \sec x$.   **p.** $\dfrac{dr}{d\theta} = \dfrac{-2\sin\theta}{(1-\cos\theta)^2}$.

2. a. 1.081. **b.** -3.637. **c.** $\frac{1}{2}\pi$. **d.** 1.382. **e.** -0.5708. **f.** 4.981. **g.** 1. **h.** -0.2580.
3. a. $\rho'' = -16\sin 4\theta$. **b.** $s'' = -\frac{1}{4}\pi^2\cos\frac{1}{2}\pi t$. **c.** $y'' = 2\sec^2 x\tan x$. **d.** $\rho'' = -\csc^2\theta$.
e. $\rho'' = -\theta\cos\theta - 2\sin\theta$. **f.** $s'' = e^t(4\cos 2t - 3\sin 2t)$.   **5.** 109° 28′.   **6.** 53° 8′.
7. 90°, 26° 34′.   **9.** $t=0$, $s=0$, $v=2\pi$, $a=0$; $t=1$, $s=4$, $v=0$, $a=-\pi^2$; $t=2$,
$s=0$, $v=-2\pi$, $a=0$; $t=3$, $s=-4$, $v=0$, $a=\pi^2$; $t=4$, $s=0$, $v=2\pi$, $a=0$.
10. $-1200\,\pi r$ ft./min.; 0.   **12.** 15.7 mi./min.   **13.** 12 ft.   **15.** 11.22 ft.
16. a. 0.8835. **b.** 0.4849. **c.** 0.8573. **d.** 0.5302.   **17.** $y = \sqrt{25 - 24\cos\theta}$, 3.606 ft.,
0.0026 ft.

Page 298

1. Max. pt. $(\frac{2}{3}\pi, 3.826)$; min. pt. $(\frac{4}{3}\pi, 2.457)$; infl. pts. $(0, 0)$, (π, π), $(2\pi, 2\pi)$.
2. Max. pt. $(\frac{1}{3}\pi, 1.913)$; min. pt. $(\frac{2}{3}\pi, 1.228)$; infl. pts. $(0, 0)$, $(\frac{1}{2}\pi, \frac{1}{2}\pi)$, (π, π).
3. Min. pt. $(\frac{1}{3}\pi, -0.342)$; max. pt. $(\frac{5}{3}\pi, 3.484)$; infl. pts. $(0, 0)$, $(\pi, \frac{1}{2}\pi)$, $(2\pi, \pi)$.
4. Max. pt. $(\frac{1}{4}\pi, 1.414)$; min. pt. $(\frac{5}{4}\pi, -1.414)$; infl. pts. $(\frac{3}{4}\pi, 0)$, $(\frac{7}{4}\pi, 0)$.
5. Max. pt. $(\frac{1}{4}\pi, 0.571)$; min. pt. $(\frac{3}{4}\pi, 5.712)$; infl. pts. $(0, 0)$, $(\pi, 2\pi)$.
6. Min. pt. $(\frac{1}{3}\pi, -2.457)$; max. pt. $(\frac{2}{3}\pi, -10.11)$; infl. pts. $(0, 0)$, $(\pi, -4\pi)$.
7. Max. pt. $(2.498, 5)$; min. pt. $(5.640, -5)$; infl. pts. $(0.927, 0)$, $(4.069, 0)$.

Page 299

1. $y = \sqrt{2}\sin\left(x + \frac{1}{4}\pi\right)$.   **2.** $y = 5\sin(x + 0.93)$.   **3.** $y = 2.24\sin(x - 0.46)$.

Page 301

6. Max. pt. $(0.66, 1.39)$; min. pt. $(2.23, -0.63)$; infl. pts. $(1.33, 0.48)$, $(2.90, -0.22)$.
7. Min. pt. $(2.90, -1.36)$; max. pt. $(6.04, 0.62)$; infl. pts. $(1.08, 1.08)$, $(4.22, -0.49)$.
8. Max. pt. $(0.64, 4.45)$; min. pt. $(2.64, -0.63)$; infl. pts. $(1.28, 2.53)$, $(3.28, -0.34)$.
9. Min. pt. $(2.82, -2.80)$; max. pt. $(5.82, 1.54)$; infl. pts. $(1.15, 1.42)$, $(4.15, 0.78)$.

Pages 310–311

2. a. $\dfrac{dy}{dx} = \dfrac{1}{\sqrt{a^2 - x^2}}$.   **b.** $\dfrac{dy}{dx} = \dfrac{-1}{\sqrt{4 - x^2}}$.   **c.** $\dfrac{dy}{dx} = \dfrac{a}{a^2 + x^2}$.   **d.** $\dfrac{dy}{dx} = \dfrac{-a}{x\sqrt{x^2 - a^2}}$.
e. $\dfrac{dy}{dx} = \dfrac{-2x}{x^4 - 6x^2 + 10}$.   **f.** $\dfrac{dy}{dx} = \dfrac{2}{1 + x^2}$.   **g.** $\dfrac{dy}{dx} = \dfrac{2}{\sqrt{9 - 4x^2}}$.   **h.** $\dfrac{dy}{dx} = 2\sqrt{a^2 - x^2}$.
i. $\dfrac{d\theta}{dr} = \dfrac{3}{\sqrt{6r - 9r^2}}$. **j.** $\dfrac{d\theta}{dr} = \dfrac{1}{1 + r^2}$. **k.** $\dfrac{d\theta}{dr} = \dfrac{1}{\sqrt{1 - r^2}}$. **l.** $\dfrac{dy}{dx} = \dfrac{2}{e^x + e^{-x}}$. **m.** $\dfrac{ds}{dt} = \dfrac{-2}{e^t + e^{-t}}$.
3. a. $y = 0.785$, $y' = 1$.   **b.** $y = 1.047$, $y' = -0.577$.   **c.** $y = 0.785$, $y' = 0.707$.
d. $y = 2.356$, $y' = -0.5$.   **e.** $y = -0.524$, $y' = 2.309$.   **f.** $y = -2.618$, $y' = 0.144$.
g. $y = 2.356$, $y' = -0.5$.   **h.** $y = 0.251$, $y' = 0.509$.

4. a. $x = 1$, $y = \frac{1}{6}\pi$, $y' = \frac{1}{3}\sqrt{3}$; $x = -0.5$, $y = -0.253$, $y' = 0.516$.
b. $x = 0.5$, $y = \frac{1}{3}$, $y' = \dfrac{-2\sqrt{3}}{3\pi}$; $x = -0.6$, $y = 0.705$, $y' = -0.398$.
c. $x = 1$, $y = 1.106$, $y' = 0.4$; $x = -\frac{1}{2}$, $y = -\frac{1}{4}\pi$, $y' = 1$.
d. $x = 1$, $y = 0.5$, $y' = -0.318$; $x = -2$, $y = 1.702$, $y' = -0.127$.
e. $x = 2$, $y = \frac{2}{3}\pi$, $y' = \frac{1}{3}\sqrt{3}$; $x = -2$, $y = -\frac{1}{3}\pi$, $y' = -\frac{1}{3}\sqrt{3}$.

Pages 316–317

3. $1/\sqrt{a}$.　　**4.** 2.　　**5.** $\cos\phi$.　　**6.** 2.　　**7.** $\ln(a/b)$.　　**8.** $\frac{1}{2}$.　　**9.** 2.　　**10.** $-\frac{1}{8}$.
11. $-\frac{1}{8}$.　　**18.** $OB = 2\,r$.

Pages 318–319

3. 2.　　**4.** 0.　　**5.** $\frac{1}{3}$.　　**6.** 0.　　**7.** 0.　　**8.** $\frac{1}{2}\pi^2$.　　**9.** 2.　　**10.** 1.　　**12.** $\frac{1}{2}$.
13. $\frac{1}{2}$.　　**14.** $\frac{1}{3}$.

Page 320

1. e^2.　　**2.** e^n.　　**4.** 1.　　**5.** e^{-1}.　　**6.** e.　　**7.** e^2.　　**9.** 1.　　**10.** e.　　**11.** 1.
12. e^{-2}.

Page 322

1. $x = 0$ gives neither, $x = 3$ gives min.　　　　**2.** $x = 0$ gives max., $x = 4$ gives min.
5. $x = 1$ gives min.

Pages 325–327

1. a. $\dfrac{1}{a}\sin ax + C$.　**d.** $3\ln\sin\frac{1}{3}\theta + C$.　**e.** $\dfrac{1}{n}\ln(\sec nx + \tan nx) + C$.　**g.** $-\operatorname{ctn} x + C$.
h. $\tan x + C$.　**n.** $\tan\theta - \operatorname{ctn}\theta + C$.　**o.** $2(\tan\theta - \sec\theta) - \theta + C$.　**r.** $\csc x - \operatorname{ctn} x + C$.
s. $\tan x - \sec x + C$.　　**2. a.** 1.　**b.** $\frac{2}{3}\sqrt{3}$.　**c.** $4/\pi$.　**d.** 0.0719.　**e.** 1.324.　**m.** 1.2081.
3. a. 4.　**b.** 1.　**c.** $8/\pi$.　　**4.** $\frac{1}{2}\pi^2$.　　**5.** 0.414.　　**6.** 2.828.　　**7.** 3.273.
8. $\tan\theta + \ln\sec\theta + 5$.　　**9.** 0.441.　　**10.** 4.　　**11.** 5.656.　　**12.** -8.　　**13.** 1.317.
14. $\pi - \frac{1}{4}\pi^2$.　　**15.** 3.83.　　**16.** 2.059.

Pages 330–331

1. f. $\dfrac{1}{12}\ln\dfrac{3x-2}{3x+2} + C$.　　**g.** $\frac{1}{2}\ln(2t + \sqrt{4t^2 - 1}) + C$.　　**h.** $\frac{1}{3}\arcsin\frac{3}{4}x + C$.
i. $\frac{1}{3}\operatorname{arc\,sec}\frac{2}{3}x + C$.　　**o.** $\arcsin\frac{1}{4}(s - 3) + C$.　　**2. a.** 0.036.　　**b.** 0.231.　　**c.** 1.099.
d. 0.463.　**e.** 7.472.　　**3.** 47.54.　　**5.** $[2\sqrt{3} - \ln(2 + \sqrt{3})]a^2$.　　**7.** $\frac{1}{2}[\sqrt{2} + \ln(1 + \sqrt{2})]p$.
8. 4 98.　　**10.** 57.28.　　**11.** $2(57 - 80\ln 2)\pi = 9.73$.

Page 332

1. a. $\frac{1}{2}\arctan\frac{1}{2}(x + 1) + C$.　　**b.** $\arcsin\frac{1}{3}(2t - 1) + C$.　　　　**2. a.** 0.161.

c. $\dfrac{2}{\sqrt{3}}\arctan\dfrac{2x+1}{\sqrt{3}} + C$.　　**d.** $\arcsin(2y - 3) + C$.　　　　**b.** 0.563.

　　　　　　　　　　　　　　　　　　　　　　　　　　　　c. $\frac{1}{6}\pi$.

e. $\frac{1}{4}\ln\dfrac{x-5}{x-1} + C$.　　**f.** $\ln(s + \frac{1}{2} + \sqrt{s^2 + s + 1}) + C$.　　**d.** 1.073.

3. 9.29.

Page 333

1. $\frac{2}{3}\arctan\frac{1}{3}x + \frac{1}{2}\ln(9 + x^2) + C$.　　**2.** $\frac{1}{3}\arctan\frac{1}{3}x - \frac{3}{2}\ln(9 + x^2) + C$.
3. $-3\sqrt{9 - x^2} - 2\arcsin\frac{1}{3}x + C$.　　**4.** $\sqrt{x^2 + 4} + 3\ln(x + \sqrt{x^2 + 4}) + C$.

Page 336

1. $\frac{1}{3}\cos^3 x - \cos x + C$.　　　　　　　　**2.** $-\frac{1}{5}\cos^5 x + \frac{2}{3}\cos^3 x - \cos x + C$
3. $\frac{1}{5}\sin^5 x - \frac{2}{3}\sin^3 x + \sin x + C$.　　**4.** $\frac{1}{2}x + \frac{1}{4}\sin 2x + C$
5. $\frac{3}{8}x - \frac{1}{4}\sin 2x + \frac{1}{32}\sin 4x + C$.　　**6.** $\frac{3}{8}x + \frac{1}{4}\sin 2x + \frac{1}{32}\sin 4x + C$.
8. $\frac{3}{128}x - \frac{1}{128}\sin 4x + \frac{1}{1024}\sin 8x + C$.　　**9.** $-\theta - \operatorname{ctn}\theta + C$.
10. $\frac{1}{2}\tan^2 x + \ln\cos x + C$.　　**11.** $-\frac{1}{2}\operatorname{ctn}^2 x - \ln\sin x + C$.

Page 337

1. a. $\sqrt{x^2 - a^2} - a \operatorname{arc sec} \dfrac{x}{a} + C.$ **b.** $\ln\left(t + \sqrt{t^2 + 4}\right) - \dfrac{\sqrt{t^2 + 4}}{t} + C.$ **2. a.** $\frac{162}{5}.$

c. $\dfrac{\sqrt{x^2 - 3}}{3\,x} + C.$ **d.** $-\dfrac{\sqrt{4 - x^2}}{x} - \operatorname{arc\,sin} \tfrac{1}{2}\,x + C.$ **b.** $\frac{3}{10}.$

e. $\sqrt{s^2 - 16} - 4 \operatorname{arc\,sec} \tfrac{1}{4}\,s + C.$ **f.** $\ln\left(x + \sqrt{x^2 - 36}\right) - \dfrac{\sqrt{x^2 - 36}}{x} + C.$ **c.** $\frac{4}{45}.$

g. $\dfrac{(2\,x^2 - 1)\sqrt{x^2 + 1}}{3\,x^3} + C.$ **h.** $\dfrac{(x^2 - a^2)^{\frac{3}{2}}}{3\,a^2 x^3} + C.$

Page 340

1. a. $x \sin x + \cos x + C.$ **b.** $-\tfrac{1}{4}\,e^{-2t}(2\,t + 1) + C.$

c. $\dfrac{x^{n+1}}{n+1}\left(\ln x - \dfrac{1}{n+1}\right) + C.$ **d.** $-e^{-x}(2 + 2\,x + x^2) + C.$

g. $x \operatorname{arc\,cos} x - \sqrt{1 - x^2} + C.$ **h.** $x \operatorname{arc\,tan} x - \tfrac{1}{2} \ln\left(1 + x^2\right) + C.$

i. $x \operatorname{arc\,ctn} x + \tfrac{1}{2} \ln\left(1 + x^2\right) + C.$ **j.** $x \operatorname{arc\,sec} x - \ln\left(x + \sqrt{x^2 - 1}\right) + C.$

k. $x \operatorname{arc\,csc} x + \ln\left(x + \sqrt{x^2 - 1}\right) + C.$ **p.** $\tfrac{1}{13}\,e^{2t}(3 \sin 3\,t + 2 \cos 3\,t) + C.$

q. $-\tfrac{1}{10}\,e^{-x}(\sin 3\,x + 3 \cos 3\,x) + C.$

2. 14.026.   **3.** 164.8.   **4.** $18{,}631\,\pi.$ **5.** $(\tfrac{1}{2}\,\pi,\ \tfrac{1}{8}\,\pi).$ **7.** $\tfrac{1}{2}(e^{\pi} + 1) = 12.07.$

Page 343

3. $\dfrac{1}{3(3 - 2\,x)} - \dfrac{1}{9} \ln \dfrac{3 - 2\,x}{x} + C.$ **4.** $\tfrac{1}{18}x\,\sqrt{9\,x^2 + 1} - \tfrac{1}{54} \ln\left(3\,x + \sqrt{9\,x^2 + 1}\right) + C.$

5. $\sqrt{4\,x^2 - 9} - 3 \operatorname{arc\,sec} \tfrac{2}{3}\,x + C.$ **6.** $\dfrac{4\,x - 3}{9\sqrt{6\,x - 4\,x^2}} + C.$

7. $-\dfrac{\sqrt{3\,x^2 + 5}}{x} + \sqrt{3} \ln\left(\sqrt{3}\,x + \sqrt{3\,x^2 + 5}\right) + C.$

8. $-\dfrac{\sqrt{25 - 9\,x^2}}{x} - 3 \operatorname{arc\,sin} \tfrac{3}{5}\,x + C.$

9. $\tfrac{1}{2}\sqrt{(7 - 2\,x)(9 + 2\,x)} + 8 \operatorname{arc\,sin} \tfrac{1}{4}\sqrt{9 + 2\,x} + C.$ **10.** $\tfrac{2}{3} \operatorname{arc\,tan} \tfrac{1}{3}(5 \tan \tfrac{1}{2}\,x + 4) + C.$

11. $\tfrac{1}{2} \sin x - \tfrac{1}{10} \sin 5\,x + C.$ **18.** 0.1977.   **19.** $\frac{47}{10}.$ **20.** $\frac{1}{100}.$ **21.** 0.142.

22. 0.097.   **23.** 0.167.   **24.** 1.11.   **25.** 0.176.

Page 345

1. $-\dfrac{(1 - x^2)^{\frac{3}{2}}}{3\,x^3} + C.$ **2.** $\tfrac{2}{9}(x^3 - 4)\sqrt{2 + x^3} + C.$ **3.** $\tfrac{1}{3}(x^2 + 8)\sqrt{x^2 - 4} + C.$

4. $\tfrac{3}{8}\,a^4 \operatorname{arc\,sin} \dfrac{x}{a} - \dfrac{1}{8}\,x(3\,a^2 + 2\,x^2)\sqrt{a^2 - x^2} + C.$ **5.** $-\dfrac{(8\,x^2 + 9)\sqrt{9 - 4\,x^2}}{243\,x^3} + C.$

6. $\tfrac{1}{4} \cos 2\,\theta + \tfrac{1}{2}\,\theta \sin 2\,\theta - \tfrac{1}{2}\,\theta^2 \cos 2\,\theta + C.$

7. $\tfrac{1}{3}\,\theta^3 \sin 3\,\theta + \tfrac{1}{3}\,\theta^2 \cos 3\,\theta - \tfrac{2}{9}\,\theta \sin 3\,\theta - \tfrac{2}{27} \cos 3\,\theta + C.$

8. $\tfrac{1}{4} \tan 2\,x \sec 2\,x - \tfrac{1}{4} \ln\left(\sec 2\,x + \tan 2\,x\right) + C.$

16. $\frac{40}{9}.$ **17.** $\frac{1079}{560}.$ **18.** 0.0113.   **19.** $\frac{8}{15}.$ **20.** 0.0133.   **21.** $\frac{4}{5}.$ **22.** $\frac{16}{45}.$

23. $\frac{35}{128}\,\pi.$ **24.** 0.734.

Page 348

1. $\ln \dfrac{x^2(x - 3)^4}{x - 2} + C.$ **2.** $\tfrac{23}{50} \ln\left(x^2 - 25\right) - \tfrac{4}{25} \ln x + C.$

3. $\ln(x + 1) - \dfrac{2}{x^2} + C.$ **4.** $3 \ln \dfrac{x}{x - 1} - \dfrac{3\,x - 1}{x(x - 1)} + C.$

5. $\tfrac{2}{3} \ln(t - 2) + \tfrac{3}{4} \ln(t - 3) + \tfrac{7}{12} \ln(t + 1) + C.$

6. $\tfrac{1}{15} \ln(x + 2) + \tfrac{1}{10} \ln(2\,x - 1) - \tfrac{1}{6} \ln(2\,x + 1) + C.$

7. $x + \dfrac{1}{2(x + 1)^2} - \dfrac{3}{x + 1} - 3 \ln(x + 1) + C.$

8. $s + \tfrac{1}{6} \ln(s - 1) + \tfrac{1}{2} \ln(s + 1) - \tfrac{2}{3} \ln(s + 2) + C.$

9. $\tfrac{1}{2}\,x^2 + \tfrac{1}{9} \ln x + \tfrac{40}{9} \ln\left(x^2 - 9\right) + C.$

13. 0.9038.   **14.** 0.9438.   **15.** 0.1555.   **16.** 0.2761.   **17.** 5.2730.   **18.** 7.9383.

Page 350

1. $\ln x - \frac{1}{2}\ln (x+1) - \frac{1}{4}\ln (x^2+1) - \frac{1}{2}\arctan x + C.$

2. $\frac{1}{3}\ln (x-1) - \frac{1}{6}\ln (x^2+x+1) + \frac{1}{\sqrt{3}}\arctan\frac{2x+1}{\sqrt{3}} + C.$

3. $\frac{1}{3}\ln (x+1) - \frac{1}{6}\ln (x^2-x+1) + \frac{1}{\sqrt{3}}\arctan\frac{2x-1}{\sqrt{3}} + C.$

4. $\frac{1}{32}\ln (x^2+4) - \frac{1}{16}\ln x - \frac{1}{8x^2} + C.$

5. $\frac{2-x}{4(x^2+2)} + \frac{1}{2}\ln (x^2+2) - \frac{\sqrt{2}}{8}\arctan\frac{x}{\sqrt{2}} + C.$

6. $\frac{4x^3+5x-2}{8(4x^3+3)^2} + \frac{\sqrt{3}}{48}\arctan\frac{2x}{\sqrt{3}} + C.$

9. $\frac{1}{4}\ln (x^2+1) - \frac{1}{2}\ln (x+1) + \frac{x-1}{2(x^2+1)} + C.$ **10.** 0.1638.   **11.** 0.047.   **12.** 2.171

Pages 357–358

4. $\rho(1-\cos\theta) = 2.$ **6.** $\rho^2\sin 2\theta = a^2.$

18. 0.637 times the diameter.   **27.** $(\frac{2}{3}, 60°), (\frac{2}{3}, -60°).$

Page 359

1. a. $\rho = 4(\tan\theta + \sec\theta).$

Page 364

1. 45°, 90°, 135°.   **6. a.** 120°. **b.** 30°. **c.** 30°.

Pages 366–367

1. $0.37\,a^2.$ **2.** $\frac{3}{2}\pi a^2.$ **3.** $\frac{1}{12}\pi a^2.$ **4.** $0.866\,a^2.$ **7.** $\frac{9}{2}\pi a^2.$ **8.** $0.33\,a^2.$

10. $\frac{1}{3}(4\pi + 3\sqrt{3})a^2.$ **11.** $\frac{1}{2}(2\pi - 3\sqrt{3})a^2.$ **14.** 7.8.

Page 368

1. $s = \frac{1}{2}a\int_{\frac{1}{2}\pi}^{\frac{5}{6}\pi}\frac{d\theta}{\cos^3\frac{1}{2}\theta} = 7.08\,a.$ **2.** $1.46\,a.$ **3.** 3.34.   **4.** 36.56.

Pages 373–374

1. a. $y = 3 - 2x - x^2.$ **d.** $x^2 - 9y^2 + 9 = 0.$ **i.** A segment of $y = 2x^2 - 1.$

2. $x = 2pt, y = 2pt^2.$ **3. b.** $17x^2 - 16xy + 4y^2 - 34x + 16y + 13 = 0.$

9. a. $x = 4 - t^2, y = 4t - t^3.$ **b.** $x = 3\sec\theta, y = 2\csc\theta.$

Pages 380–381

1. a. $y^2 = x(4-x)^2.$ **c.** $\frac{1}{3}\sqrt{10}.$ **e.** $(0, 0).$ **4.** $x = -4\cos t, y = 2\sin t$

6. a. (in part) $x = 4t, y = 3t - 16t^2, 4y = 3x - 4x^2.$

b. (in part) When $t = 2, x = 10, y = 8, v = \sqrt{205}.$

c. (in part) $x = t + 2(1 - \cos t), y = 3t - \sin t.$

d. Equation of path, $81y^2 = 16x^2(9 - x^2).$ **10.** $\alpha = 59°.$

Pages 383–384

1. a. 9. **b.** 3.16. **c.** $30.2\,\pi; 22.1\,\pi.$ **6.** $\frac{1}{3}(2 - \sqrt{2})\pi ab^2.$

Page 386

4. b. $\frac{1}{2}\pi^2 r.$ **6. a.** Vol. $= \frac{1}{6}(3\pi^2 - 16)\pi a^3,$ Sur. $= \frac{32}{3}\pi a^2.$

b. Vol. $= \frac{1}{6}(9\pi^2 - 16)\pi a^3,$ Sur. $= \frac{8}{3}(3\pi - 4)\pi a^2.$

Page 388

3. b. Hor., $\theta = 72°;$ ver., $\theta = 36°.$

1. a. $\dfrac{-b \csc^3 \theta}{a^2}$. **c.** $\dfrac{-1}{a(1 - \cos \theta)^2}$.

1. $\frac{1}{6} \pi a b$. **2.** $6 \pi r^2$.

1. a. $\frac{1}{2}$. **b.** $\frac{5}{3}\sqrt{10}$. **c.** $\frac{40}{3}\sqrt{40}$. **d.** $\frac{125}{16}$. **2. b.** $\dfrac{(b^4 x_1{}^2 + a^4 y_1{}^2)^{\frac{3}{2}}}{a^4 b^4}$. **d.** $\dfrac{2(x_1 + y_1)^{\frac{3}{2}}}{\sqrt{a}}$.

e. $3(a x_1 y_1)^{\frac{1}{3}}$. **f.** $\sec x_1$. **5. a.** $\frac{1}{2} a$. **b.** $\dfrac{(\rho_1{}^2 + a^2)^{\frac{3}{2}}}{\rho_1{}^2 + 2 a^2}$. **c.** $\frac{2}{3}\sqrt{2 a \rho_1}$. **d.** $2 a \sec^3 \frac{1}{2} \theta_1$.

z. $\dfrac{a^2}{3 \rho_1}$. **f.** $\frac{3}{4} a \sin^2 \frac{1}{3} \theta_1$. **g.** $\dfrac{\rho_1{}^3}{a^2}$. **6. a.** $\frac{125}{12}$. **b.** $\frac{5}{2}\sqrt{5}$. **c.** 6. **e.** $\frac{13}{8}\sqrt{13}$. **9.** $x = -0.347$.

10. $x = \pm 0.931$. **12.** 1.01. **14.** $0.983\,a$.

1. a. $(0, 1)$. **b.** $(3, -\frac{35}{4})$. **c.** $(\frac{127}{6}, \frac{309}{20})$. **d.** $(2, 5)$. **g.** $(\frac{81}{100}, -\frac{96}{25})$. **h.** $(-7, 8)$.
i. $(-2, 3)$. **l.** $(\frac{1}{4}\pi\ \frac{15}{8})$.

2. b. $\alpha = \dfrac{(a^2 + b^2) x^3}{a^4}$, $\beta = -\dfrac{(a^2 + b^2) y^3}{b^4}$.

c. $\alpha = \frac{1}{2}(x - 9 x^5)$, $\beta = \dfrac{1 + 15 x^4}{6 x}$.

d. $\alpha = x + 3 x^{\frac{1}{3}} y^{\frac{2}{3}}$, $\beta = y + 3 x^{\frac{2}{3}} y^{\frac{1}{3}}$.

3. At $(1, 4)$, $R_1 = \frac{17}{8}\sqrt{17}$, $\alpha = \frac{19}{2}$, $\beta = \frac{49}{8}$; at $(2, 2)$, $R_2 = 2\sqrt{2}$, $\alpha = 4$, $\beta = 4$; $R_1 - R_2 = 5.933$.

4. a. $\alpha = -8 t^3$, $\beta = 6 t^2$. **b.** $\alpha = t - \frac{1}{4} t^5$, $\beta = \frac{5}{6} t^3 + 2/t$.

c. $\alpha = 3 - t + \frac{9}{4} t^5$, $\beta = \dfrac{2}{3 t} + \dfrac{5}{2} t^3 - 3$. **d.** $\alpha = 1 + t - \frac{1}{4} t^5$, $\beta = \frac{5}{6} t^3 + 2/t$.

e. $\alpha = \dfrac{12 t^4 + 1}{2 t^3}$, $\beta = \dfrac{4 t^4 + 3}{t}$. **f.** $\alpha = -2 \cos t \operatorname{ctn} t$, $\beta = t + \operatorname{ctn} t(1 + \cos^2 t)$.

g. $\alpha = \frac{144}{13} \sin^3 t$, $\beta = -\frac{144}{5} \cos^3 t$.
h. $\alpha = \frac{1}{3}(2 \cos t - \cos 2 t)$, $\beta = \frac{1}{3}(2 \sin t - \sin 2 t)$.
i. $\alpha = a \cos t(\cos^2 t + 3 \sin^2 t)$, $\beta = a \sin t(\sin^2 t + 3 \cos^2 t)$.
j. $\alpha = a \cos t$, $\beta = a \sin t$.

5. Convergent. **6.** Divergent. **7.** Convergent. **8.** Convergent.

4. a. 20.1, 20.0998. **b.** 9.9667, 9.9666. **c.** 1.9750, 1.9744. **d.** 0.04016, 0.040162.
6. a. 0.508. **b.** 0.201. **c.** 0.201. **d.** 2.988.

9. $-1 < x < 1$. **10.** $-1 \leqq x \leqq 1$. **11.** All values of x. **12.** $-3 \leqq x < 3$.
13. $-1 < x < 1$. **14.** $-\frac{1}{2} \leqq x \leqq \frac{1}{2}$. **15.** $-2 \leqq x < 2$. **16.** $-6 < x < 6$.
17. $-2 \leqq x < 2$. **18.** $-2 < x \leqq 0$. **19.** $0 \leqq x < 2$. **20.** All values of x.

24. 0.764. **25.** 0.747. **26.** 0.071. **27.** 0.9226. **29.** 0.0214.

1. a. Error < 0.00033. **b.** Error < 0.01. **c.** Error < 0.08.
2. a. Error < 0.0032. **b.** Error < 0.05. **c.** Error < 0.25.
4. a. Error < 0.000002. **b.** Error < 0.006. **c.** Error < 0.2.
5. Four. **7.** Six.

9. No. **10.** Yes. **11.** Yes. **12.** No. **13.** No. **14.** Yes.

ANSWERS

Pages 448–449

1. $(b - x)(a + y) = c.$ **2.** $cy^2 = 1 + x^2.$ **3.** $(x^2 + b)(a + 2y) = c$

4. $y = ce^{\sqrt{1-x^2}}.$ **5.** $\rho = c \cos \theta.$ **6.** $\rho^n = c \sin n\theta.$

7. $y \ln c(a + x) = 1.$ **9.** $lx^2 + 2 mxy + ny^2 = c.$ **10.** $(x + y)^2(x + 2y) = c$

11. $\frac{1}{2} \ln (2x^2 + 2xy + y^2) - \arctan \dfrac{x + y}{x} = c.$

12. $(x + y)^2(2x + y)^3 = c.$ **13.** $(2x + y)^3(x - y)^2 = c.$

14. $y^3 = x^3 \ln cx^3.$ **15.** $(x + y)^m(x - y)^n = c.$

16. $\ln (x^2 + y^2) + 4 \arctan \dfrac{y}{x} = c.$ **17.** $(2w - 1)(3z + 1) = cz.$

19. $y^2 + x^2 \ln cx = 0.$

18. $1 + 2cy - c^2x^2 = 0.$ **25.** $y^2 = x^2 + \ln x^2.$

20. $y^2 = x^2 + cx.$ **27.** $y^2 = x^2 - x.$

26. $x^2 - y^2 = 1.$

28. $1 \pm 4y - 4x^2 = 0.$ **29.** $y^2 + 2xy = 3.$ **30.** $x\sqrt{1 - y^2} = y.$

Pages 452–453

1. $y = ce^{-ax} + b/a.$ **2.** $y = (x + c)e^{-x}.$ **3.** $y = e^{-ax}(\int e^{ax}f(x)dx + c).$

4. $y = ce^x + e^{-x}.$ **5.** $s = ce^{-t} + \cos t.$ **6.** $y = cx + e^x.$

7. $s = \sin t + c/t.$ **8.** $cx^2y^2 + xy^2 = 1.$ **9.** $y = c \csc x + 1.$

10. $s = c \sin t + e^t.$ **11.** $s = (t + c)\cos t.$ **12.** $y = x^2(e^x + c).$

13. $y^2 = cx^2 + x.$ **14.** $y = c(x + 1)^2 + \frac{1}{2}(x + 1)^4.$ **15.** $y = 2x - x^3.$

16. $y^2(8x - 3x^2) = 4.$ **17.** $y = e^{-x} - \cos x.$ **18.** $y = x^2 + \cos x.$

19. $y = x + 2e^{2x}.$ **20.** $y(1 + \ln x) = 1.$

Page 455

1. $y = \frac{1}{12}t^4 + c_1t + c_2.$ **2.** $y = c_1e^{ax} + c_2e^{-ax}.$

3. $x = c_1t + c_2 - \sin 2t.$ **4.** $y = \frac{1}{4}e^{2x} + c_1x + c_2.$

5. $s = c_1t + c_2 - \dfrac{a \cos nt}{n^2}.$ **6.** $c_1y^2 = a + (c_1t + c_2)^2.$

7. $y = \frac{1}{24}x^4 + \cos x + c_1x^2 + c_2x + c_3.$ **8.** $y = (bx - a)\ln x + c_1x + c_2.$

Pages 459–460

1. $s = c_1e^t + c_2e^{3t}.$ **2.** $y = c_1e^{-2x} + c_2e^{3x}.$

3. $x = c_1e^{2t} + c_2te^{2t}.$ **4.** $s = c_1e^{4t} + c_2e^{-4t}.$

5. $s = c_1 \cos 4t + c_2 \sin 4t.$ **6.** $s = c_1 + c_2e^{-16t}.$

7. $x = e^{2t}(c_1 \cos 3t + c_2 \sin 3t).$ **8.** $x = e^{-2t}(c_1 \cos 2t + c_2 \sin 2t).$

9. $\rho = c_1e^\theta + c_2\theta e^\theta.$ **10.** $x = e^{4t}(c_1 \cos 3t + c_2 \sin 3t).$

11. $s = e^{-3t}(c_1 \cos t + c_2 \sin t).$ **12.** $s = c_1e^{-\frac{1}{2}t} + c_2e^{-t}.$

13. $y = e^{\frac{1}{3}x}(c_1 + c_2x).$ **14.** $y = c_1e^{\frac{1}{2}x} + c_2e^{-2x}.$

15. $s = e^{-\frac{1}{2}t}(c_1 \cos \frac{1}{2}t + c_2 \sin \frac{1}{2}t).$ **16.** $s = e^{\frac{2}{3}t}(c_1 \cos \frac{1}{3}\sqrt{2}t + c_2 \sin \frac{1}{3}\sqrt{2}t).$

17. $x = \frac{1}{2}e^{2t}.$ **18.** $y = 2e^x + e^{-2x}.$ **19.** $s = 3e^{2t} + 3e^{-2t}.$

20. $s = 5 \sin 2t.$ **21.** $y = 4 \cos x.$ **22.** $y = 2xe^{3x}.$

23. $x = 3e^{-t} \cos t.$ **24.** $s = e^{-t}(\cos 2t + \sin 2t).$

Pages 464–465

1. $x = c_1 \cos 3t + c_2 \sin 3t + \frac{1}{9}t + \frac{1}{18}.$ **2.** $x = c_1 \cos 3t + c_2 \sin 3t + \frac{1}{2}e^{3t}.$

3. $x = c_1 \cos 3t + c_2 \sin 3t + \cos 2t.$ **4.** $x = c_1 \cos 3t + c_2 \sin 3t + \frac{1}{2}t \sin 3t.$

5. $y = c_1e^{3x} + c_2e^{-3x} + \frac{1}{6}xe^{3x}.$ **6.** $y = c_1e^{3x} + c_2e^{-3x} - \frac{1}{3}\cos 3x.$

7. $s = c_1 \cos 2t + c_2 \sin 2t - 2 \sin 3t.$ **8.** $s = c_1 \cos 2t + c_2 \sin 2t - 2t \cos 2t.$

9. $y = c_1 + c_2e^{-2x} + \frac{5}{4}x + \frac{1}{4}x^2.$ **10.** $y = c_1 + c_2e^{2x} + 2xe^{2x}.$

11. $x = c_1e^{3t} + c_2e^{-t} - \frac{2}{3}t + \frac{1}{9}.$ **12.** $x = e^{3t}(c_1 \cos 2t + c_2 \sin 2t) + 3.$

13. $x = e^{2t}(c_1 \cos \sqrt{3}t + c_2 \sin \sqrt{3}t) + 2.$ **14.** $y = c_1e^{3x} + c_2e^{-x} - \frac{1}{3}e^{2x}.$

15. $y = c_1e^{2x} + c_2e^{-x} + \frac{1}{3}xe^{2x}.$

16. $y = c_1e^{3t} + c_2e^{-t} - \frac{56}{65}\cos 2t - \frac{32}{65}\sin 2t.$

17. $x = e^{-3t}(c_1 \cos 2t + c_2 \sin 2t) - \cos t + 2 \sin t.$

18. $x = e^t(c_1 \cos 2t + c_2 \sin 2t) + 4 \cos 2t + \sin 2t.$

19. $s = e^{2t} + e^{-2t} - 1$.   **20.** $s = \sin 2t + 2t$.   **21.** $s = e^{3t} - 2t$.
22. $y = e^x$.   **23.** $x = \frac{2}{3}\sin t - \frac{1}{3}\sin 2t$.   **24.** $x = 2\cos t + t\sin t$.

Pages 470–471

1. $s = 5\sin 2t$.   **2.** $s = 8\cos 2t$.   **3.** $s = 2\cos 2t + 5\sin 2t$.
4. $s = k(1 - \cos t)$.   **5.** $s = 5e^{-t}\cos 2t$.   **6.** $s = 6e^{-t}\sin 2t$.
7. $s = \frac{5}{3}\sin t - \frac{1}{3}\sin 2t$.   **8.** $s = \frac{1}{8}\sin 2t - \frac{1}{4}t\cos 2t$.

Pages 475–476

4. $-\frac{2}{3}$.   **5.** $54° 44'$.   **10.** *b.* $\alpha = 36° 42'$, $\beta = 74° 30'$, $\gamma = 122° 19'$.   *e.* $\alpha = 120° 48'$, $\beta = 67° 25'$, $\gamma = 39° 48'$.   **13.** *b.* Parallelogram. *d.* Trapezoid. *e.* Square.

Page 478

3. *a.* Sphere, $x^2 + y^2 + z^2 = 8x$.   *c.* Plane, $x - y + 2z = 16$.
7. *b.* $x^2 + y^2 + z^2 - 4x - 2y - 8z - 17 = 0$.   *d.* $x^2 + y^2 + z^2 - 6x - 4y - 14z + 32 = 0$.
8. *b.* $3x^2 + 4y^2 + 4z^2 = 300$.   *f.* $xy + yz + zx = 0$.

Pages 487–488

3. *a.* $x - \sqrt{2}y + z + 10 = 0$.   *c.* $y - z + 4\sqrt{2} = 0$.
e. $6x - 2y + 3z + 28 = 0$.   *h.* $3x - 4y - 12z = 0$.
4. *a.* $2x + 2y - z + 9 = 0$.   *b.* $x + 4y + 2z - 21 = 0$.
6. *c.* 16.   **7.** *a.* $(2, -3, 4)$.   **10.** *a.* 54 sq. units. *c.* $12\sqrt{14}$ sq. units.

Page 492

1. *a.* $\alpha = 78° 41'$, $\beta = 38° 16'$, $\gamma = 53° 55'$.   *b.* $\alpha = 35° 16'$, $\beta = 114° 6'$, $\gamma = 65° 54'$.
c. $\alpha = 22° 37'$, $\beta = 76° 39'$, $\gamma = 72° 5'$.   *d.* $\alpha = 90°$, $\beta = 135°$, $\gamma = 45°$.
e. $\alpha = 115° 53'$, $\beta = 77° 24'$, $\gamma = 29° 12'$.   *f.* $\alpha = 65° 54'$, $\beta = 144° 44'$, $\gamma = 65° 54'$.

Page 496

2. *a.* $x^2 + 4y^2 + z^2 = 9$.   **3.** *a.* $4x^2 + y^2 - 4z = 0$.   **4.** $3x^2 + 4y^2 + 4z^2 = 12$.
6. *a.* $\frac{1}{4}\pi$. *b.* $\frac{1}{12}\pi$. *c.* $\frac{4}{3}\pi$. *d.* $\frac{14}{45}\pi$. *e.* $\frac{1}{9}\pi$.

Page 499

12 and 14. $\dfrac{\partial\theta}{\partial x} = \dfrac{-y}{x^2 + y^2}$, $\dfrac{\partial\theta}{\partial y} = \dfrac{x}{x^2 + y^2}$.   **22.** *a.* $50\sqrt{3}$ sq. in. *b.* $5\sqrt{3}$ sq. in. per inch.
c. 50 sq. in. per radian. *d.* 0.85 sq. in. *e.* -2 in. per inch.

Page 503

1. $4x - 4y - z = 6$.   **3.** $4x - 4y - z = 17$.   **5.** $z + 4 = 0$.   **7.** $x + y - z = 2$.

Pages 505–506

1. *a.* $x = \frac{8}{3}$, $y = \frac{2}{3}$ gives min.   *b.* $x = -1$, $y = 3$ gives min.
c. $x = \frac{27}{2}$, $y = 5$ gives min.   *d.* $x = y = \frac{1}{2}\sqrt{2}$ gives min.
e. $x = y = \frac{1}{4}\pi$ gives max.; $x = y = \frac{5}{4}\pi$ gives min.   *f.* $x = y = 2$ gives min.
2. $\frac{1}{3}(ab - a^2 - b^2)$.   **4.** Vol. $= \frac{1}{27}abc$.   **5.** $\frac{8}{9}\sqrt{3}\,abc$.

Pages 511–512

1. $dz = (3x^2 - 2y)dx + (6y - 2x)dy$.   **2.** $du = \dfrac{-2y\,dx + 2x\,dy}{(x - y)^2}$.
3. $d\rho = \sin\phi\,d\theta + \theta\cos\phi\,d\phi$.   **6.** $\Delta u = 3.72$, $du = 3.6$.   **7.** $\frac{7}{8}$.
10. *a.* 0.895 sq. ft., 2.7%. *b.* 0.13 ft., 0.96%.   **11.** *a.* 4.8 ft., 0.55%.
12. 17π cu. in.   **13.** *a.* 0.505 cu. ft. *b.* $\frac{101}{132}$%.
14. *a.* 0.3. *b.* 0.5. *c.* $\frac{25}{4}$%.   **15.** *a.* $\frac{37}{18}\sqrt{15}\,\pi = 25$ cu. in. *b.* 3π sq. in.
16. 74 sq. ft.   **17.** *a.* 0.0144. *b.* $\frac{23}{3600}$.   **18.** 0.0018.   **19.** 0.534 ft./sec.²
20. *a.* 0.0204 sec. *b.* $\frac{37}{64}$%.   **21.** $dV = 3.0159$ cu. in., $dS = 2.818$ sq. in.

Page 513

1. $\dfrac{dy}{dx} = \dfrac{3\,x^2 - 2\,xy + y^2}{x^2 - 2\,xy - 3\,y^2}$.

4. $\dfrac{dy}{dx} = \dfrac{\cos y - y \cos x}{\sin x + x \sin y}$.

3. $\dfrac{dy}{dx} = \dfrac{ay - x^2}{y^2 - ax}$.

5. $\dfrac{dy}{dx} = \dfrac{y^2 - 1}{3\,y^2 - 2\,xy + 1}$.

7. $\frac{1}{5}$.

8. 2.

9. $-D/E$.

Pages 514–515

3. $f_{xx} = 2\,A,\ f_{xy} = 2\,B,\ f_{yy} = 2\,C$.

5. $f_{xx} = 6\,Ax + 2\,By,\ f_{xy} = 2\,Bx + 2\,Cy,\ f_{yy} = 2\,Cx + 6\,Dy$.

6. $f_{xx} = Cy^2 e^{xy},\ f_{xy} = C(1 + xy)e^{xy},\ f_{yy} = Cx^2 e^{xy}$.

Pages 521–522

1. 3. 2. -6. 3. $\frac{7}{24}$. 4. $\frac{1}{10}a^5$. 5. $\frac{7}{9}$. 6. $-\frac{10}{3}$. 7. 2. 8. $\frac{1}{3}a^2$.
9. $6\,b^3$. 10. $\frac{9}{4}$. 11. $\frac{27}{20}$. 12. $\frac{11}{84}$. 13. $\frac{13}{60}$. 14. $\frac{101}{84}$. 15. $\frac{67}{60}$.
16. $\frac{1}{150}(15\,\pi - 16)a^5$. 17. $\frac{4}{3}a^3$. 18. $\frac{1}{3}abc(a^2 + b^2 + c^2)$. 19. $\frac{1}{8}c^4$. 20. $\frac{1}{3} + \frac{1}{4}\,\pi$.
21. $\frac{4}{35}$. 22. $\frac{11}{60}$. 23. $\frac{1}{2}\,\pi$.

Page 526

2. $\frac{9}{2}$. 3. $\frac{9}{2}$. 4. $7.5 - 4\ln 4 = 1.956$. 5. $\frac{40}{3}$. 6. $\frac{16}{3}$. 7. 6. 8. 16.
9. $\frac{32}{3}$. 16. $\frac{1}{3}a^2$. 17. $\frac{1}{32}(16 - 3\,\pi)a^2$.

Pages 528–529

1. $\frac{2}{3}mr^3$. 2. $\frac{8}{15}$. 3. $\frac{1}{6}abc$. 4. 9. 5. $\frac{4}{15}$. 6. $15\,\pi$. 7. $\frac{17}{20}$. 8. $\frac{5}{12}$.
9. $\frac{16}{3}r^3$. 10. $\frac{2}{3}$. 11. $\frac{569}{140}$. 12. $\frac{24}{5}\sqrt{6}$. 13. $12\,\pi$. 14. $\frac{1}{90}abc$. 15. $\frac{4}{35}\pi a^3$.

Page 531

1. $(\frac{16}{15}, \frac{64}{21})$. 2. $(\frac{32}{35}, \frac{5}{14})$. 3. $(\frac{5}{2}, 5)$. 4. $(1, \frac{3}{5})$. 5. $(1, \frac{8}{5})$. 6. $(1, \frac{17}{5})$.
7. $(2, -\frac{2}{5})$. 8. $(\frac{8}{25}, \frac{11}{10})$. 9. $(\frac{10}{9}, \frac{40}{21})$. 10. $(\frac{9}{10}, \frac{27}{20})$. 11. $(\frac{14}{15}, -\frac{11}{15})$.
12. $(\frac{34}{5}, -4)$. 13. $(\frac{11}{5}, 0)$. 14. $(\frac{7}{2}, 5)$. 15. $(\frac{12}{5}, \frac{3}{2})$. 16. $(3, \frac{3}{2})$. 17. $(\frac{8}{5}, 1)$.
18. $(2, 1)$. 19. $(0.585, 0.585)$. 22. $(\frac{1}{5}a, \frac{1}{5}a)$. 23. $\left(\dfrac{256\,a}{315\,\pi}, \dfrac{256\,a}{315\,\pi}\right)$. 24. $(\pi a, \frac{5}{6}a)$.

Page 537

1. $I_x = I_y = \frac{1}{4}Ar^2$. 2. $I_x = \frac{1}{24}Aa^2,\ I_y = \frac{1}{2}Ah^2$. 3. $I_x = \frac{1}{3}Aa^2,\ I_y = \frac{1}{3}Ab^2$.
4. $I_x = \frac{1}{4}Ab^2,\ I_y = \frac{1}{4}Aa^2$. 5. $I_x = \frac{16}{5}Ab^2,\ I_y = \frac{48}{7}A$. 6. $I_x = \frac{19}{20}A,\ I_y = \frac{53}{20}A$.
7. $I_x = \frac{5}{3}A,\ I_y = \frac{19}{4}A$. 8. $I_x = \frac{239}{60}A,\ I_y = \frac{17}{4}A$. 9. $I_x = \frac{71}{64}A,\ I_y = 10\,A$.
10. $I_x = \frac{23}{5}A,\ I_y = \frac{53}{5}A$. 11. $I_x = I_y = \frac{7}{4}A$.
13. 7.19 ft. below surface of water. 14. *a.* $\frac{15}{32}\,\pi = 1.47$ ft. *b.* 2.4 ft.

Page 542

1. $\frac{3}{16}(4\,\pi - 3\sqrt{3})$. 2. $\frac{3}{8}(2\,\pi + 3\sqrt{3})$. 3. $2\,\pi$. 4. $\frac{1}{2}\,\pi + \frac{9}{16}\sqrt{3}$. 5. $\frac{1}{4}\,\pi + 2$.
6. $2 - \frac{1}{4}\,\pi$. 7. π. 8. $\frac{1}{2}\,\pi - \frac{2}{3}$. 9. $\frac{3}{4}\,\pi + \frac{4}{3}$. 10. $\frac{1}{2}$. 11. $1 - \frac{1}{4}\,\pi$.
12. $0.684\,a^2$. 13. 5.504.

14. $A = \frac{1}{6}(2\,\pi + 3\sqrt{3})a^2$, $\bar{x} = \dfrac{(8\,\pi + 3\sqrt{3})a}{4\,\pi + 6\sqrt{3}}$,

$I_x = \frac{1}{48}(4\,\pi + 9\sqrt{3})a^4$, $I_y = \frac{1}{16}(12\,\pi + 11\sqrt{3})a^4$.

15. $\bar{x} = \frac{5}{6}a$. 16. $\bar{x} = \dfrac{128\sqrt{2}\,a}{105\,\pi}$. 17. $\bar{x} = \dfrac{81\sqrt{3}\,a}{80\,\pi}$. 18. $\frac{1}{48}(3\,\pi + 8)Aa^2$. 21. $1.206\,a^2 b$.

Page 545

2. $\frac{2}{3}\alpha a^3$. 3. $\frac{2}{3}(3\,\pi - 4)a^3$. 4. $\frac{32}{9}a^3$. 5. $\frac{5}{2}\,\pi$. 6. $(0, 0, \frac{2}{3}c)$.
7. $\frac{2}{3}(\sqrt{2} - 1)\pi a^3$. 8. $(0, 0, \frac{3}{8}(\sqrt{2} + 1)a)$. 9. $(\frac{4}{9}a, 0, \frac{10}{9}a)$. 11. $\frac{1}{36}(9\,\pi - 16)a^3$.

INDEX

D E F G H I J K   0 6 9 8 7 6 5
PRINTED IN THE UNITED STATES OF AMERICA